Fascinating Life Sciences

This interdisciplinary series brings together the most essential and captivating topics in the life sciences. They range from the plant sciences to zoology, from the microbiome to macrobiome, and from basic biology to biotechnology. The series not only highlights fascinating research; it also discusses major challenges associated with the life sciences and related disciplines and outlines future research directions. Individual volumes provide in-depth information, are richly illustrated with photographs, illustrations, and maps, and feature suggestions for further reading or glossaries where appropriate.

Interested researchers in all areas of the life sciences, as well as biology enthusiasts, will find the series' interdisciplinary focus and highly readable volumes especially appealing.

David G. Ainley • Rory P. Wilson

The Aquatic World of Penguins

Biology of Fish-Birds

Springer

David G. Ainley
HT Harvey and Associates Ecological Consultants
Los Gatos, CA, USA

Rory P. Wilson
Department of Biosciences
Swansea University
Swansea, UK

ISSN 2509-6745 ISSN 2509-6753 (electronic)
Fascinating Life Sciences
ISBN 978-3-031-33989-9 ISBN 978-3-031-33990-5 (eBook)
https://doi.org/10.1007/978-3-031-33990-5

This Springer imprint is published by the registered company Springer Nature Switzerland AG
The registered company address is: Gewerbestrasse 11, 6330 Cham, Switzerland

Paper in this product is recyclable.

To the smallest, loveliest, and most venerable penguins we know: Cai, the Elf, and the Bear RPW; and Susan and Ian DGA.

Penguin Transitions

Silver dimples, restless blue
Oceans heaving darkness through
Monstrous power, but fickle whim
Best sing praises, swells and hymn

Warm life came, in times long gone
A bird, a being with flighted song
Merger plea but ice-heart sea –
"You'll need to change to ride with me"

Gentle morphing, nuptial vow
Such sweeping change, ethereal now
Unfettered diver, vortex kiss
And sheen sublime, what spirit this?

Weaves soothing spells with measured beat
Speed spins, cavorts in shadows deep
Twirling, whirling, bubbles make
The incantations in its wake

What games are played, what fusions made?
In depths beyond, the darker glade
In pressured paths for would-be fish
Imagined waltz? abyssal wish?

But bubbles burst and eddies tire
Eroding paths, soon ash from fire
The void has called, snatched ocean song
The murmur passed, the mage now gone

(Anonymous)

Preface

The idea for this book was born of excitement, but also of frustration. We had been working intensively with penguins for many years (we think it germane not to mention how many!), both of us blown away by the simple fact that these birds – astonishing to think that they are indeed that – had departed so far from "bird norms" that earlier explorers wondered what, exactly, they really were. Indeed, centuries ago, when penguins were first encountered by European explorers, they were thought to be a fish-like thing, perhaps a relative of birds.

Experiencing wild penguins on land is the easiest thing, if you're in the right place at the right time. They can be observed so readily sitting on nests, or waddling slowly along, precariously balancing their weird torpedo shape on tiny terminal legs. And it is this seeming instability and gracelessness that incited people to write all sorts of things that continue to mold public perceptions of this group.

> "Comical" – "Adorable" – "The little fellow in the dress suit," et cetera. It is tempting to be anthropomorphic about penguins. Using human comparisons, it is easy to think of them as little clowns, the ridiculous dwarfs that enliven the circus, waddling with baggy pants across the area for our amusement. They are far from that; they are not little people dressed in feathers. (R.T. Peterson, 1979, p. 1)

This is the penguin the media love to sell. It's probably also self-indulgent – anything that looks like us has got to be interesting, especially if it trips and stumbles.

But that very penguin shape, and apparent physical awkwardness on land tell you, if you are a right-thinking biologist, all you need to know. For every penguin ineptitude manifest at their nesting sites, or in the local zoo, there is an acquired grace, poise, and competence in water – yin and yang. No other bird has transformed from the "standard bird model" so completely as penguins (we hear swift fanatics clamor – hmmm, perhaps…) and the reason is because they are not flighted or land birds at all. They have morphed to become as much a part of the sea as a bird can seemingly be (we'd love to see a more evolved version). That is the "real penguin" – and here all credit to Lloyd Spencer Davis who was instrumental in producing a film entitled "Meet the Real Penguin," in a bid to make people understand. Despite this, the message still seems to be lost.

Some native people obviously did view penguins in their aquatic element, and long before Europeans encountered them, but, given a lack of written language, we have little idea of what they said or thought. We do know that the Maori referred to Yellow-eyed and Little penguins as *hoiho* and *kororaa*, respectively (Simpson 1976), and they, a sea-faring folk, did appreciate their aquatic nature:

Ko te Moana-Tapolopoko-a-Tawhaki Ko te marae o Hoiho

is a proverb meaning, "The Southern Ocean is the marae [sacred place] of hoiho" (Webster 2018; T. Webster, pers. comm.).

In 1620, in a marvelously insightful moment, Admiral Beaulieu, passing Cape of Good Hope suggested (in French) "…for my part, I take them to be feathered fish." In those days, with the knowledge that he had, he was quite close to the truth, but for much of the following centuries, we have slipped down the slope of comparing all things animals to us. Our science has defined them to be birds, warm-blooded and all the rest, and so their relationship with the sea and abilities in water must somehow mirror ours, more or less, because we are "warm-blooded and all the rest too." One of us (Rory) chatted to a penguin researcher in the very early 1980s who had attempted to track Little Penguins at sea by tying a balloon with a length of string to their legs so that they could be followed in a boat. The idea is laudable, the execution perhaps less so. The cord, 8″ long (not even 3 m) was considered to be enough! We will show you in this book that 8″ is not enough, and why it is never a good idea to associate balloons with penguins until the purpose is theoretical.

Since that time, a huge amount of information on the aquatic aspects of penguins has been amassed, primarily thanks to dramatic advances in microelectronics – which allow smart tags, mini computers, to be attached to penguins when they go to sea – and maturation of some long-term studies of penguin life history. But, although these approaches have shown scientists what penguins can "do" in their real element, much of this has not filtered through to the general public, perhaps despite TV companies showing the odd tantalizing clip of wild penguins underwater (though actually only "a drop in the ocean" compared to the broad spectrum of their aquatic lives). It's fair enough. How can we expect a camera team to track a bird that "ambles" along at speeds in excess of 2 m/s, flip-flopping and slipping between distant, deep, light-poor ocean water layers with an ease akin to a magician shuffling cards?

But we also feel that, for scientists and the layperson alike, it is time to play with more overarching frameworks and examine how the various aspects of penguin morphology, physiology, behavior, etc. gel and make sense for penguins in their ocean home. They are birds after all, and birds shouldn't really be playing in cold water, some diving to depths in excess of half a kilometer where the pressure would kill us in an instant. But extreme specialization is a dangerous game because, if conditions change, if the very ocean to which penguins have become wedded, changes, the glorious monument to ocean living that is the penguin can come tumbling down. The modern public does not abide by the Maori proverb. Overfishing and climate change spring to mind, with penguins now being increasingly used as sentinels of what can happen if big systems, like oceans, are overly messed with.

And, within this context, it might be germane to reconsider penguins as manifestations of man – a crystal ball into our own future.

So, we are releasing our frustration about the true nature of penguins in this book – whew! It does feel good. But in attempting to do so, we have tried to produce something that appeals to scientists and laypeople alike (our inability to do this will show, at once, our incompetence in writing and the vast distance between the two camps). This means that we have many references to the literature (though scientists will want more and laypeople less), but we are in no way exhaustive, which we think is ok if we want to deal in principles rather than concentrating on specifics. We should also note that we have reached out to some wonderful penguin people to ask for snippets of their tantalizing data so that we can expand our own musings and datasets and present the broadest front possible. Bless them all. As a result, we have hidden, within this book, original data from no less than 12 species of penguin – that's well over half the penguin species on the planet. The bottom line, though, is that our frustration has driven our own dive into understanding the real penguin, and we have loved it. If we have managed to convey just a fraction of our awe and fascination with these incredible birds in our text, we can't see how you cannot agree.

Los Gatos, CA, USA David G. Ainley
Swansea, UK Rory P. Wilson

Acknowledgments

We are, of course, responsible for the content of this book (cough) but, by way of incrimination, though seriously, huge thanks and applause, we have to mention the people that helped us fabulously and happily along the way. You are the best. Were the world made up of folk like you, we'd be flying high or more properly venturing wide and deep. But where to start?

First of all, we do have to thank our families who for more than a year put up with our periods of being "distant" as ideas cascaded around in our heads. Otherwise, in no particular order, we have those who used matchsticks to hold their eyes open while they reviewed our chapters to provide feedback and criticisms (yes, those too – things like "you've got to be joking"). Those people included Belinda Cannell, Theresa Cole, Kyle Elliot (who has to study auks but who we believe is really a closet penguin fan), Steve Emslie, Andreas – "you've got to be joking" – Fahlman, Dan Ksepka, Amélie Lescroël, Guillermo (Memito) Luna, Eric Woehler, and also those remaining anonymous. Similarly, thanks are due to many others for discussions about penguin life history, including Sandy Bartle, Theresa Cole, Ursula Ellenberg, Dan Ksepka, Thomas Mattern, Scott Terrill, Sue Townsend, and Trudi Webster (Yellow-eyed Penguin Trust).

Then, there are those who provided photos, many stunning, of penguins doing their thing, mostly at sea (the hardest ones to get). Your photos show that you have an eye for the cool and the dramatic. You know who you are, but for the sake of the rest of the world, there's Peter Ryan (whose photos are so wicked that Rory uses them as screen savers), Tui De Roy, Jean Pennycook, Louise Chilvers, Eric Woehler, Virginia Morandini, Paul Ponganis, Robert Pitman, Holly Fearnbach, Viola Toniolo, David Roberts (SANCCOB), Valeria Ruopollo, Arvind Varsani, and Gerald Kooyman. As part of that, we would also like to thank Mirian Scadeng and Paul Ponganis for help with graphics and CT scans, respectively. Abra Kaiser and Joy DeBruyn, of H.T. Harvey & Associates, helped with graphics, too, and indispensably steered us through EndNote and the like.

Ashwin Sanghi said "never judge a book by its cover." The book cover is thanks to, and courtesy of, Neil Wilson (Rory's brother), who simply had the vision and skill to put it together beautifully. So the reader will have to judge whether what's inside the book is any good. We think that what's on the front, though, is terrific. Read on to find out if Ashwin Sanghi was right in our case!

Data, data, data. We have a lot of our unpublished data in this book, from quite a few species, and which species these are is obvious from reading the book, these being the species we've both most worked with. In this respect, and from a personal perspective, Rory has to thank Flavio Quintana, with whom he has worked on Magellanic Penguins for years now (shock horror – how did it turn into years? – seems like 3 weeks). Flavio is a model of efficiency and competence (as well as being fab company) and a huge part of the Magellanic Penguin data in this book could only be collected because of him. Gracias che Fla. And in that context, Rory has to extend personal thanks to Rolex who believed that helping support smart tag initiatives can make a difference (when so many scientific funding bodies laughed at it). But we also wanted to cast the net as wide as possible, to make the book as broadly penguiny as possible, and so asked some of the great and the good for data on other species. So we've got wicked (unpublished) data from some other penguin species, and for that we have to thank Antje Steinfurth (who spent many an over-hot month in Galápagos to get them), Thomas Mattern and Ursula Ellenberg (who work with really zany species – seen from our perspective at least), Cassondra Williams and Paul Ponganis (of penguin ranch fame), Belinda Cannell (again ☺), Charly Bost and Camille Lemonnier (for some king penguin stuff), and Onno Huyser. And leaving the penguins for a while, we also got data on other species for comparison; stuff on auks from Kyle Elliott (of course) and David Grémillet (merci Daviiid) as well as Emma Cole and Emily Shepard. Perhaps as part of that, we should also mention the people that actively went out to get data, or help us get data, that we felt we needed for the book too (to close a few knowledge holes). Phil Hopkins whisked the first featherometer in the world out of his hat (he does that sort of thing all the time) and Sue Townsend was magnificent in using it, harassing penguin skins in various museums (if you've not read the book you should be burning with curiosity by now). Thanks Sue. And actually, within that context, we have to thank Carla Cicero (Museum of Vertebrate Zoology, University of California), Maureen Flannery (California Academy of Sciences), and Paul Sweet (American Museum of Natural History), who gave us access to penguin specimens.

The analysis of penguin data, much of which was gathered from our very hi-tech tags, conceived by the ever-eclectic Mark Holton in his batcave, was also made possible by Mark's amazing software, also straight from the batcave. But in addition, many people helped by putting in time to analyze some of the data (tut – and we made it seem as if it was just us). Pivotal in this were Richard – if it moves I'll dead-reckon it – Gunner, and James Redcliffe but also Elliot Dee, Izzy Stuart, Olivia Shott, and Katie Bambridge, all of whom have sat through reams of penguin data magically visualized by DDMT or dealt with videos taken by penguins as they swam underwater (which is fun, but rather sick-making with all the motion).

The list goes on, though. There's Javi Ciancio for providing energy density of prey data as well as Siobhan Sheridan (for being the fastest, and most curious, needle in the West!).

Well ending the list for now, in our efforts, we are also grateful for some indispensable help that we have received, beginning with the editors at Springer, Kenneth Teng, Shina Harshavardhan, and team, all of whom provided much guidance in how to stitch this stuff together.

It's also fair to acknowledge that some of David Ainley's time in drafting this book was covered under grants from the National Science Foundation (grant OPP 1935870) and the National Aeronautics and Space Administration.

All this shows that, though it's fun to "do" research, and you can do much on your own, you absolutely cannot beat a team. Would that the planet could figure that out, too!

Contents

Part III The Hardware of a Fish-Bird

Part I
In the Beginning

Chapter 1
Wading In: Introduction to Fish-Birds

Admiral Beaulieu, in 1620, upon rounding the Cape of Good Hope and first encountering penguins at sea, said "*Ils sont un mélange de la bete, de l'oiseau et du poisson – mais c'est de l'oiseau qu'ils se rapprochent le plus!*" Translated, this reads "They are a mix of the beast, the bird and the fish – but it is the bird that they are most close to." However, he added (translated): "The Pinguins are Fowls without Wings, which have two fins and two broad Paws, upon which they walk upright…They have nothing of the taste of Flesh, and for my part, I take them to be feather'd Fish." (from Sparks and Soper [58], pp. 166 and 167).

When European explorers, during their first ventures in the late 1400s to waters and lands well south in the Southern Hemisphere, encountered penguins on land, they assumed they were some sort of duck or goose. For instance, a member of Vasco da Gama's crew, A. Vello, wrote in 1497 on a South African island: "There are birds as big as ducks, but they cannot fly because they have no feathers on their wings. These birds of whom we killed as many as we chose…bray like asses." (from R.T. Peterson [50], p. 19). Similar observations were made for the next 100 or more years, including off Patagonia. The mariners, thus, made no connection between what they saw on land and what they saw in the ocean. For the longest time, penguins seen at sea were not considered birds at all but rather a fish-like relative, as noted in the quotes above. One mariner described them as: "So perfect a fische." It was not until 1758 that they were first scientifically classified, by Linnaeus [58]. Subsequently, as another aspect of their initially confusing anatomy and behavior seemingly gave way to admiration, some fossilized penguins were given the generic name of *Delphinornis* or "dolphin birds," acknowledging their similarity with the Delphinidae, i.e., dolphins and porpoises [36]. The eventual accumulation of knowledge has shown why penguins, with a suite of radical adaptations, are unrivaled among birds and most mammals in terms of their aquatic attributes – thus exonerating the early impressions a bit. They have a remarkable bullet-like shape (though they also change shape with speed), phenomenally low drag, and fascinating drag-reducing surface "compliancy" due to their particular scale-like featheration

D. G. Ainley, R. P. Wilson, *The Aquatic World of Penguins*, Fascinating Life Sciences, https://doi.org/10.1007/978-3-031-33990-5_1

Fig. 1.1 Porpoising Adélie penguins catch a breath (as indicated by their open beaks) as they travel like bullets past ice floes along which predators may be lurking. Penguins traveling in this manner, i.e., "porpoising," take advantage of their bullet shape and breathe about once every 10–15 s or so, as the aerial component of their movement is less than 1 s. Therefore, only a small portion of the flock is above the water surface at any given time, and the true size of this flock is much larger (likely exceeds 20 birds). (Photo courtesy of Jean Pennycook)

(an attribute that is now being applied to the design of surfboards and ocean-racing sailboats) (Fig. 1.1). They also possess the ability to regulate their buoyancy, which helps them exploit variable depths with minimal effort (Figs. 1.2 and 1.3). Penguins are certainly the most extremely specialized diving group of birds on the planet, which alone makes them unique and sensational. Indeed, the translations of their Latin names emphasize their aquatic powers, e.g., *Eudyptes* = "good diver," *Megadyptes* = "large diver," *Eudyptula* = "good little diver," and *Aptenodytes* = "without wings diver" – these are names given by people who had no idea just how extraordinarily "good" penguins are at diving. However, it is specifically because penguins have given up aerial flight (which is hugely advantageous) that they have been able (forced by evolution?) to realize astonishing underwater prowess such as breathtaking speed, stunning maneuverability, and an ability to exploit an incredible range of depths. Of the ~10,500 presently existing species of birds, only 57 cannot fly, with more than a third of them being penguins. Like penguins exploiting ocean, other flightless birds such as ostriches, rheas, and the like gave up flight long ago to better exploit any available terrestrial niche [41], and the majority of those that have done so relatively recently are oceanic island forms. To be a bit anthropocentric, predecessors of non-penguin flightless birds found refuge on an island after having been blown off course during migration and, subsequently, through morphological

Fig. 1.2 Emperor penguins surfacing and breathing at the edge of the fast ice at Cape Washington, Ross Sea. It should be noted how the birds exhale, in part, underwater and that substantial amounts of air are also lost from the plumage, especially in the case of the two birds on the right side. This release of air happens as the penguins regulate their buoyancy in preparation for diving (see Chaps. 6, 7, and 14 for more on this). (Photo courtesy of Paul Ponganis)

changes, avoided becoming airborne to prevent being blown back to the sea again and getting out of the gene pool (e.g., rails, ducks [43, 44]).

Penguins' union with the sea is so complete, on the other hand, that they may spend months without returning to land, feeding, sleeping, and interacting with distant waters in wild oceanic regions. Adélie penguins, for example, remain entirely at sea during their first 2–3 years and spend about 90% of their existence at sea during their lifetimes [1] (see Chap. 2). Many authors describe penguins as being flightless, but George Simpson [56] pointed out the error in doing so, stating that penguins just changed their capabilities to fly in a denser medium than did most birds. This change, i.e., flying underwater, however, does come at a cost that would be substantial for most seabirds, such as an inability to cover large distances quickly (and therefore to react to the ephemeral concentrations of prey that characterize oceanic food webs). Moreover, given that much energy is needed to travel through a viscous aqueous environment, seabirds are so much more resistant to motion other than in air. These are the subjects that we will thoroughly cover in this book.

Most people identify penguins as not only being associated with snow and sea ice (which actually applies to just two species; see below) but also by the images they project from their time on land, where they are the most accessible to scientists,

Fig. 1.3 Penguins are positively buoyant but considerably less so than normal waterbirds such that most of their body is submerged when they are on the water's surface (cf. the bird on the bottom left). The apparent variance in buoyancy in these Snares penguins is due to some individuals (particularly the bird on the far right of the photograph) raising themselves out of the water, probably to get a better view of the photographer. (Photo courtesy of Louise Chilvers)

the public, and popular media alike. However, these periods, when they are on land, represent only a small fraction of their lives compared to their interactions with the sea and so these birds are depicted from their most uncomfortable side – "fishes out of water" (see Chap. 2). This is all the more important because penguins are viewed as a "sentinel species" or an "indicator species" as a means to measure the pulse of the marine ecosystem's well-being (see Chap. 13). This designation, in part, is not only because of their relatively easy access to researchers (they do not run away and return to their nests repeatedly) and, also, their relatively large size that allows attachment of sophisticated recording tags (aka biologgers; see Chaps. 3, 6, 7, 8, 9, 10, and 14) but also because of their acknowledged links to the sea, a playground for many of man's dubious exploitation practices (Chaps. 3 and 5). With little doubt, and just a few exceptions, penguins' well-being is largely determined by their marine ecology and their interaction with various food webs, competitors, and predators. Indeed, their underwater prowess and associated energetic costs (Chaps. 6, 7, 8, 9, and 10), as they exploit their oceanic environment (Chaps. 3, 4, and 5), make them especially sensitive to temporal and spatial changes in food availability (Chaps. 5 and 13). It comes as no surprise, therefore, given the state of the world oceans, that the majority of penguin species are considered to have a special status – "near threatened," "threatened," or "endangered" – and in several cases, other than three taxa, they have been hunted to extinction in New Zealand (NZ) or are facing difficulties with the introduction of alien predators (see below, Chap. 2); their plight is primarily due to depletion of prey species brought about by commercial fishing [26, 45, 46] (Chap. 5). Heightening that indirect effect is the direct mortality resulting from entanglement of Humboldt penguins in commercial and artisanal fishing nets along the coast of Peru and Chile, which has been a cause of major mortality in recent decades [55]. The same is true for the yellow-eyed penguins in New Zealand's coastal waters [66].

An Impressive Number of Penguin Species: Evolution of Unique Capabilities

Currently, there are 19–23 species of penguins swimming the oceans of the Southern Hemisphere, with the number depending on the degree to which a taxonomist is either a "lumper" or a "splitter" in consideration of what constitutes a species (Table 1.1). Over past eons, more than 50 species of penguins have come and gone,

Table 1.1 The population status among the present-day penguin avifauna

	Breeding pairs	Colonies/locations	IUCN (ESA)a	Source
King penguins	1,600,000	18 L	LC	[7]
Emperor penguins	265,000	62 C	NT(T)	[22]
Adélie penguins	3,790,000	251 C	LC	[34]
Chinstrap penguins	3,420,000	375 C	LC	[62]
Gentoo penguins	337,000	12 L	LC	[33]
Macaroni penguins	6, 310,000	>300 C	V	[15]
Royal penguins	500,000	1 L	NT	[15]
Southern rockhopper penguins	780,768	>15 L	V	[52]
Eastern rockhopper penguins	423,780	9 L	V	[52]
Northern rockhopper penguins	329,000	7 L	E	[16]
Fiordland penguins	2750	1 L	V(T)	[37]
Snares penguins	25,000	1 L	V	[38]
Erect-crested penguins	81,000	2 L	E(T)	[17]
Yellow-eyed penguins	1670	4 L	E(T)	[54]
Australian little penguins	202,600	18 L	LC	[39]
New Zealand little penguins	32,350	5 L	LC	[39]
Magellanic penguins	1,400,000	138 C	LC	[4]
Humboldt penguins	24,000	60 C	V(T)	[18]
Galápagos penguins	3000	5 L	E(E)	[5]
African penguins	25,000	13 C	E(E)	[13]

[a]Some information on status was provided by P.N. Trathan. The status of emperor penguins under the ESA (Endangered Species Act) was revised by the US Fish and Wildlife Service to "threatened" as of 25 October 2022

C colonies, *IUCN* International Union for Conservation of Nature, *L* locations, *LC* least concern, *V* vulnerable, *NT* near threatened, *T* threatened, *E* endangered all within the same climatic/latitudinal zone occupied by present-day penguins – strong agreement exists among taxonomists that penguins derived or split from an aerial seabird. There are, however, two somewhat competing hypotheses (the stuff of how science works) about why, in the first place, birds developed flight [28, 70]. It begins with many dinosaurs or non-dinosaur lizards having feathers, ostensibly as insulation to improve thermoregulation. The first line of thought, which is the most popular among paleontologists, is that through evolution (essentially a step-like series of genetic mutations over time, with only a few being passed on to progeny), one (or more) species' radiations occurred among groups of tiny reptiles that climbed trees (note that the well-known *Archaeopteryx*, the so-called first bird, had fingers and claws at the wrist" of its wings). As a result of morphological changes, these arboreal lizards developed long, overlapping arm feathers that broadened their forelimbs to facilitate better gliding, thus improving their movement on trees and the ground. The second, though less popular, competing hypothesis, theorizes that a group of small lizards may have improved arm feathers to increase surface area and, by flapping, to increase their speed on the ground to help them better escape from predators or catch prey such as insects. A domestic hen running from a dog while vigorously flapping comes to mind. Subsequently, one genetic change led to another, so to speak, and birds eventually morphed sufficiently to actually fly, that is, without needing the assistance of a tree or a rock for the initial altitude. Evolutionary radiation continued, and, today, we have many different kinds of birds, found in different habitats, with different types of flight depending on wing morphology, including wing shape (aspect ratio, i.e., the ratio of length to width) and wing loading (body mass relative to wing area). By being able to fly, birds have been able to exploit habitats that terrestrial creatures cannot without a lot of effort and they can respond better to the dramatic seasonal changes in habitat characteristics prevalent over much of Earth. That is, they can migrate between breeding and wintering grounds depending on the place and season seeking resources to make life easier.

Another fascinating evolutionary issue, and directly related to penguins, is why some birds of the air lost their capacity to fly, especially since flight has so many advantages. As noted above, perhaps for some, it was because as wings were not needed for their lifestyle (e.g., ostriches), rising into air might have resulted in being blown to an unsuitable habitat (the ocean, e.g., terrestrial/freshwater birds that initially got blown to islands during migration), thus endangering them, or, perhaps, it was for better use of a productive niche available in the ocean that other birds could not utilize. The latter reason would be apt for penguins, and it is pretty certain that they evolved or split from a procellariform origin (albatrosses, petrels, and shearwaters), based on genetic evidence. In addition, one should consider the fact that today some petrels and shearwaters belonging to the larger avian group do "fly" in water, in a manner that is vaguely similar to their flight in air (especially diving petrels [30, 48, 49]). In fact, the very earliest "penguin" may have had similar wings to those of the procellariforms and not the flipper-like wings of modern penguins [40]. Moreover, yes, as noted above, penguins fly in water rather than air.

Penguins are unique in the animal world with respect to passing through fluids (air and water). The difference in viscosity between air and water poses a problem for any creature trying to deal with both for travel, and there are not many groups of species that do that regularly. Wings that work well in the thin medium of air, with their large surface area and low wing loading that helps keep birds aloft, are of poor use underwater because these large wings have to be dragged up and down in a fluid that opposes their every movement. This costs a great deal of energy and makes aerial wings inefficient underwater. A partial solution is for air-flighted birds, among some petrels and Northern Hemisphere auks, to half open their wings for underwater propulsion – not ideal, but substantially better than trying to row with a full wing area. An extreme example of the latter would be the flightless steamer ducks of Patagonia, birds that are too heavy to fly but that vigorously row their wings in a spectacularly splashy manner as they progress over the water surface. They use their wings and not their feet, which other ducks use to swim. Another option is to evolve reduced wing areas, which helps underwater "flight" but then makes aerial flight energetically more onerous, if not impossible.

Today, among air-flighted seabirds, we see all manner of solutions for aerial and underwater flight in wing morphologies, all of which, however, make it clear that the easier a species can fly in air, the worse it can propel itself underwater with its wings, and vice versa. So, at one end of the spectrum, there are gulls, for example, with large, broad wings and low wing loading and no capacity to swim underwater. "Transition" species include some petrels and shearwaters, as noted, with narrow wings and higher wing loadings (design features requiring the use of wind as an energy source) but with the compromised ability to power themselves underwater with their wings (sometimes helped by their feet). This leads to the most specialized wing-propelled divers of the flighted bird taxa, the auks (and some petrels [69]). These birds have remarkable diving abilities; one species, the thick-billed murre, even reaches depths of 210 m [14] and executes dives as long as 4 min [20]. Auks manage this performance by having narrow or short wings, pretty well-suited for underwater swimming. This also means that they have the highest energy expenditure for aerial flight of any bird group [21], in the process of furiously beating their wings, almost like a bee, to grapple with air. Given the greater energy expended in aerial flight, their range is limited compared to that of other seabirds. Interestingly, these auks become flightless in air when they shed all their wing feathers at once during their annual molt. It is perhaps no coincidence, therefore, that one auk, the great auk, abandoned aerial flight altogether to be able to exploit water depths cleanly and efficiently, without having to paddle a compromised wing. This bird, with the Latin name of *Pinguinus impennis*, actually was the basis for the naming of penguins by the first Europeans who encountered them on Southern Hemisphere cruises. Sadly, today, we can only speculate about the diving capacities of this penguin-similar species, with the last ones being exterminated by humans in 1850 – they were too easy to catch and too close at hand to the ravenous Europeans [23].

In a manner similar to the great auk, penguins, in their evolutionary history, lost the power of aerial flight but gained a huge advantage with their aqueous flight. Today, this makes them unrivaled in the bird world for the ease with which they can exploit oceanic depths, both shallow and deep (exploiting shallow depths is more problematic than one might believe – see Chap. 6). The speeds that they can attain underwater and their maneuverability are breathtaking (our breath; Chaps. 6, 9, 10, and 11). Some penguins can reach depths of 600 m or more and stay submerged for more than 20 min! They can also "turn on a dime" as the saying goes. The gains achieved by penguins are substantial because they can exploit particular aspects and regions of the ocean's resources better than other birds (see Chaps. 10 and 11), but the ecological price is high. Aerial flight is much faster than swimming to cover great distances usually needed to find food patches at sea, especially while having to return to nests to provide for the young (see Chap. 2). Penguins are no longer privy to the easy solution of aerial birds, so searching for prey is an issue. This lack of wide searching affects their natural history in a significant way. Moreover, living in dangerous seas, they are not on the top of the food web (Chap. 12).

The evolutionary history of penguins is among the best, if not the best known of all birds, due to their large, solid bones that make better fossils than those of aerial birds, which have thin-walled and hollow bones that do not last long enough to be easily fossilized [56, 57]. The dense bones possessed by penguins facilitate some of their aquatic abilities. It is believed, based on available fossilized evidence, that the immediate precursors of penguins appeared about 62 million years ago or just after what is called the Cretaceous–Tertiary (or K–T boundary) extinction event. At that time, owing to dramatically altered climate caused by a large asteroid colliding with Earth (in the vicinity of what is now the Veracruz Peninsula, Mexico), a huge proportion of the animal species existing on Earth, including the sea, went extinct, involving many whole taxonomic orders and families. Among those that went extinct were most of the aquatic giant lizards, the mosasaurs, which would have been formidable predators of penguins had penguins existed with them. Moreover, from an adaptive, evolutionary perspective, it would seem that the survivors of the cataclysm "enjoyed" an Earth that offered many niches that were no longer occupied, waiting for them to be exploited. We will get much more into this below. Maybe it is rather like the crows, ravens, or gulls in one's neighborhood that were "preadapted" to find food in garbage dumpsters, having the flight morphology that allows them access by diving into bins without lids, once those bins became common, i.e., once the niche appeared; nowadays, as a result, gull, crow, and, especially, raven populations are larger than ever before around suburbia. More so, though, what happened to the asteroid collision survivors is more like the few lost flocks of finches that found their way to the isolated, oceanic Hawaiian and Galápagos islands. With no other forest birds present against which to compete, they radiated into spectacular arrays of new species by taking advantage of "unoccupied niches" [25, 31, 51]. Competition among and between existing penguins, and their mammalian and fish competitors, owing to penguin species radiation, still looms large in penguins' aquatic lives, a subject discussed below and more exhaustively elsewhere in this book.

Of course, there is more to the evolution of a water-flying specialist bird than just its bones and wing structure, i.e., those items that most commonly appear as fossils [12]. Penguins also had to change many other aspects of their morphology and physiology [28]. These included their feathers (to become rather "scale-like," especially on the flippers) and featheration – evolving complete cover, with dozens of feathers per square centimeter, rather than having them in tracts separated by unfeathered areas as is the case in aerial birds (thus to lighten the load so to speak). Penguins' molting process as well as their musculature and physiology had to have changed as well (Chaps. 2, 7, and 8). Their feathers, and featheration, relate to their buoyancy, which can have both positive and negative effects with which penguins have to deal (Figs. 1.2 and 1.3; Chap. 6). Even their blood flow system had to change, allowing their extremities to receive less blood when subjected to cold temperatures, especially that of the body heat-sapping ocean, thus to conserve that heat. Conversely, to expel heat (especially after vigorous exercise in the ocean), their extremities receive more blood (Chap. 7). Ironically, this blood flow system appears to have evolved in penguins well after the primordial penguins appeared and when Earth was undergoing a warm period. Moreover, it appeared in the sorts of penguins that these days frequent the cold, but not icy, waters of the upwelling currents in temperate latitudes [63]. It is worth noting here, though, as will become abundantly clear further on in this book, that not all penguins are alike in terms of their aquatic abilities.

In fact, temperate locations were where penguins first appeared and evolved, with the most fossils having been found thus far in the New Zealand region, seconded by Patagonia [56, 57]. Through much of their geological-scale evolutionary history, the appearance and disappearance of penguin species were closely linked to major shifts in plate tectonics and climate as well as the development of modern oceanic currents [9, 12, 28, 65]. When penguins first began to appear, the southern continents were still drifting apart (plate tectonics), and, although Antarctica had drifted to its present position, it was polar in name only and not in climate (it was temperate). New Zealand, then in the form of Zealandia, was far closer to Patagonia than it is now and was several times larger than it is now, with most of it having since sunk or submerged by rising sea levels. All of the several dozen paleo-penguins known to have existed then for a number of reasons have become extinct, i.e., more than half of all penguins known to have existed. Most of the 19–23 extant penguin species (Table 1.1; [10, 12, 47]) are fairly recent, geologically speaking (Fig. 1.4), and the vast majority of them are also found in locations having a temperate oceanic climate (cool to cold but not cold enough to have sea ice). The reason for the profusion of penguins in cooler water is that it holds more oxygen and nutrients, and thus cool-temperate to polar waters are far more productive than are subtropical to tropical waters. This is where the lack of the prey-searching capability among birds unable to fly in air becomes important. The energy-sapping lifestyle of penguins requires high availability and quality of food, a subject that will be extensively covered elsewhere in this book (Chaps. 5, 10, and 11). This is one reason why they do not exist in subtropical or warmer waters, which can be productive but in which food patches are too widely spread for penguins to search for them efficiently (Chaps. 3 and 5). Another reason might be that in warm water, penguins' ectothermic (cold-blooded) prey (fish) appear to have an advantage in terms of increased speed and maneuverability to escape compared to their cold-water piscine relatives [8]. In fact, in general, penguins living at the warm end of the temperature spectrum are less capable of dealing with temperature variations than are those at the cold end (Chaps. 7 and 13).

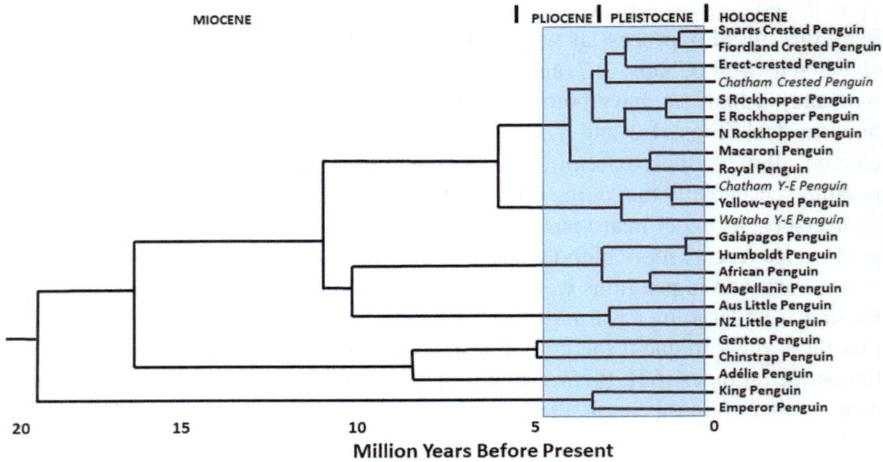

Fig. 1.4 An evolutionary tree outlining radiation of the current penguin avifauna; recently extirpated species are in italics, and the period of frequent subantarctic island emergence is shaded [10, 12]. A number of penguin species arose and became extinct during the period depicted but are not shown. The emperor–king penguin lineage may have been the sister to all penguins currently present and not just the *Pygoscelis* species (Adélie, gentoo, chinstrap) line [2, 3, 47, 65]. *Aptenodytes* persisted through several glacial and warm interlacing periods

The immediate precursors of modern penguins appeared "only" about 15–16 million years ago, with most of the present penguin fauna appearing only within the last 4 million years (Fig. 1.4 [10]). Among the existing penguin species, the first were *Aptenodytes*, then came *Pygoscelis* and *Spheniscus/Eudyptula,* followed by, or contemporaneous with, *Eudyptes/Megadyptes*. We will review the subsequent ecological groupings in later chapters as "ecotypes," on the basis of the foraging habitat and mode, as well as diet and interspecific compatibility (competitive or not; Chaps. 3, 4, and 5). For these reasons, they do not necessarily fit well together in the same aqueous space (Fig. 1.5). Moreover, we will also get into body size, below, as well as later in this book (Chaps. 7 and 8). Otherwise, there now are 19–23 species within 6 living genera, depending on whether one lumps or splits *Eudyptula* into two species (we do so in Table 1.1; some split into three) and lumps or splits various *Eudyptes*, particularly rockhopper penguins, into three (we do so in Table 1.1). Proposals exist [64] to split gentoo penguins into four species (but not by us). Not long ago, there were three subspecies of *Megadyptes*, but two were extirpated from New Zealand and Chatham Islands by the newly arrived Polynesians some 800 years ago, leaving just the present-day yellow-eyed penguins. When that extirpation happened, the current yellow-eyed penguins expanded from subantarctic islands to the South Island of New Zealand [6, 10]. Another NZ species of the *Eudyptes* on the Chatham Islands also went extinct at about that time, mostly due to hunting. These penguins that were extirpated in the modern age have been among the most recently appearing penguins, geologically speaking (more below).

Fig. 1.5 Penguins occasionally travel at sea in mixed assemblages. Here, an Adélie penguin is shown amongst gentoo penguins – although interspecific groups are seldom stable over time (see Chap. 14). The turbulence on the water surface as these birds flap their wings while breathing for a few seconds during their travel should be noted. Such turbulence represents an energetic cost paid for by the birds, which is a compelling reason to do most of their traveling underwater. (Photo courtesy of Jean Pennycook)

Penguin Species Radiation and the Ontogeny of their Watery World

When the first known penguin-like bird appeared, around the time of the K–T boundary, Gondwanaland, the super continent, had only partially broken up. It was not until Australia and South America began to separate from Antarctica around 40–30 million years ago that the Antarctic Circumpolar Current began to appear. Its flow, and cooler temperatures, enhanced productivity. There was further evolution of the "penguin wing" and a diversification of penguins, especially of species much larger in body size than those of today. Those large penguins disappeared with the appearance of dolphins during the middle Miocene, 30–10 million years ago. Appearing as well in the Southern Hemisphere during the Miocene were the modern seals and sea lions, also competitors (even predators) of penguins [28, 57] (see Chaps. 5 and 12). These complementary shifts among penguins and marine mammals therefore may not have been a mere coincidence. Since those early shifts in the penguin species assemblage, i.e., reduced diversity (unlike most vertebrates in the fossil record), only relatively small penguins have appeared and continue to exist, seemingly representing a predator size–prey size continuum in the air-breathing creature competition (with penguins at the lower end of the size trend; more on this later). A number of penguin species appeared but then went extinct within the last 20 million years [12, 28], but the factors involved, other than perhaps interspecific competition, are not known. Interspecific competition, however, is an important factor in the ecology of present-day penguins (see Chap. 5) as well as in their evolution (see below). Not to be forgotten is the large array of petrels and albatross (several dozens of species) that extract prey from the upper few meters of the water column, thus becoming somewhat of a penguin competitor. These aerial competitors, too, were undergoing species radiation during the same period as were the modern penguins [42]. Current-day penguins, in one sense though, do something that aerial seabirds cannot do: they forage deeper to exploit a much larger volume of the oceanic habitat than do the surface foragers.

About two-thirds of the living penguin species arose only during the last few million years (Fig. 1.4). This had likely to do with repeated glacial–interglacial periods characteristic of the Pleistocene [24], which introduced several major swings in climate, encouraging altered adaptation, possibly to different temperature regimes [9, 65]. The present-day oceanic current systems and water masses also developed, further facilitating penguin speciation [12, 29, 65]. In addition, confirming the importance of physiographic changes in the penguin world, it appears that during the Last Glacial Maximum, 12–110 thousand years ago, a number of penguin species exhibited increasing population sizes as judged from genetic analysis [9, 65]. This makes sense, as with the Ice Caps taking up a lot of water, a sea level that was lowered 200 m would provide more coastal nesting habitat, perhaps relaxing the interspecific competition. When the Ice Caps melted, sea levels rose and "reclaimed" a lot of the coastal habitats, and these elevated penguin populations then decreased. Sea levels fluctuated repeatedly with the various glacial periods that occurred during modern penguins' existence.

Penguin species radiation also had to do with an increased emergence of islands, owing to volcanism, in the subpolar/low-latitude polar Southern Hemisphere. On a geological timescale, not long after an island or island group emerged, penguins would colonize it and a "new" penguin species appearance would follow [10, 11]. As reviewed by the latter authors, below are a few examples (see Fig. 1.4):

- In the Southwest Pacific, Antipodes Island emerged ~5 Ma (million years ago) with the Antipodes-endemic erect-crested penguins separating from the extinct Chatham Island penguins (Richdale's). The crested penguins appeared soon after, 3.5–1.7 Ma.
- The Snares penguins separated from the Fiordland penguins 1.4–0.5 Ma, which is likely coincident with the emergence of the Snares Islands.
- The Chatham Islands emerged ~3 Ma with the recently extirpated *Megadyptes* and *Eudyptes* appearing 2.5–0.1 Ma.
- In the Atlantic, Gough Island appeared ~2.5 Ma, with divergence of the northern rockhoppers from the southern and eastern rockhoppers at about that time.
- Between New Zealand and Tasmania, Macquarie Island uplifted only about 0.7 Ma, with the royal penguins separating from the macaroni penguins less than 0.2 Ma.

In the eastern Pacific, certain islands of the Galápagos emerged ~3 Ma, with the Galápagos penguins diverging from the Humboldt penguins 1.7–0.6 Ma. Thus, the modern radiation of the island-endemic penguin species has been facilitated by having a lot of islands, newly emerged, in the cooler, more productive parts of the Southern Hemisphere ocean (Chap. 3). Colonization of new islands could have actually been rapid, as indicated by what happened not long ago at the Macdonald Islands (53°S, 73°E), Kerguelen Plateau [59], southwest of Australia. The volcano that formed this island had been quiescent from the time of its discovery by Europeans in 1874 until 1997, when volcanic activity doubled the size of the main island and also provided a gradual slope to the sea, i.e., a beach, which allowed the king penguins to come ashore (more on that below). Within a few years, a colony of "thousands" of king penguins was established. Where they came from is a mystery, but, they were seemingly present in the area, though not breeding because they were too clumsy to access the site previously (see below) As it is, the number and spread of oceanic islands in the subantarctic and low-latitude Antarctic appears to be quite unusual, in comparison to much of the remaining world ocean where isolated, oceanic islands are much fewer (except in the Bering Sea). These Southern Hemisphere islands are spread widely enough through the climatic zones of the Southern Ocean that even a little bit of philopatry (tendency to stay or return home) among the resident penguins, plus reduced traveling range (compared to aerial birds), would have produced enough genetic isolation to facilitate speciation. Along the continental margins of South America and Africa, in contrast, where cool, productive waters also facilitate penguin presence (but only for the last few million years [29]), there are very few such islands. This is because continental plates are being subducted (along with any islands) beneath the west coasts of those continents. Penguin species diversity is somewhat reduced (one genus) along thousands of kilometers of these continental margins compared to the subantarctic/low Antarctic with all the islands and climatic zones (five genera). That one genus, *Spheniscus*, spread eastward from South America to Africa, perhaps also leading to an eastward spread and the presence of what would become *Eudyptula* in Australia–New Zealand.

Penguin Evolution: Body Size and Climate

Having mentioned the subject of body size, above, we note that one of the longest-established maxims of ecology is Bergman's Rule, which states that body sizes of birds (or warm-blooded animals in general) increase as the temperature of their environment decreases. The reason is that as the volume of an animal increases, its relative surface area decreases, thus helping conserve heat (Chap. 7). Specifically, it is the body and its metabolic rate that generates heat, in direct proportion to that volume, while heat is lost over the surface area [32]. So, small birds have little capacity to generate heat and a great capacity to lose it – and they therefore need more energy to maintain body temperatures. The opposite is true for large birds.

Apparently, modern penguins demonstrate this better than do most avian taxonomic groups (explained in great detail in Chap. 7). Not only do penguin species living in colder temperatures tend to have a greater body size but also the length of their flippers relative to body size decreases [60]. Penguins with larger flippers lose heat more rapidly – again, a surface area issue. In cold temperatures, of course, heat loss becomes particularly critical, especially in water because water has an extremely high volumetric heat capacity, requiring 3200 times more energy to heat a given volume than air. However, this characteristic of water also sucks heat from warm bodies. In addition, the ocean is cooler than the atmosphere just about anywhere that penguins occur, except in Antarctica and some subantarctic islands during winter and spring. Owing to the penguin' characteristic prolate spheroid shape (shape of a rugby or gridiron football), they have a surface area, relative to mass, that is 15% smaller than theory would predict and that is great for conserving heat.

Nonetheless, Bergman's Rule, though generally true within species, is well-illustrated as one progresses in body size among the penguin species, from emperor penguins to little penguins, but it is also well-illustrated within penguin groups, most notably by the species of the genus *Eudyptes* (Table 1.2; Figs. 1.6 and 1.7). With the exception of the erect-crested penguins, the macaroni/royal penguins are the largest (frequenting polar and subpolar waters), and the remaining *Eudyptes* species are close to half their size (frequenting temperate to near subtropical waters). It should be noted that the recently evolved, but the more recently extirpated, *Eudyptes* at the warmish Chatham Islands, off the east coast of New Zealand (closely adjacent to the Subtropical Front), was among the smallest of the genus. The same goes for the two recently evolved but recently extirpated *Megadyptes*, which also occurred in proximity to the Subtropical Front and were derived from a larger, cool-water species from the subantarctic (now the yellow-eyed penguin [10]). In addition, according to Bernard Stonehouse [60], Humboldt, African, erect-crested, and both Australian and NZ little penguins have wings that are 10% larger in size than what would be expected from their body mass, and all these occur in the warmer portions of the oceans where penguins are found (Table 1.2). This would be an example of Allen's rule. In accordance, Magellanic, yellow-eyed, king, and emperor penguins have flippers 10% smaller than expected, and these tend to occur in colder oceans (see below). This is not the same as wing shape, i.e., the aspect ratio, with the warm-water penguins having broader wings relative to length (Table 1.2). Broader wings may lead to greater maneuverability, required to catch the quicker fish of warmer oceans (Chaps. 4 and 10).

Stonehouse [60] also found that feather length increases in penguins as climatic temperature decreases, with the feathers being important for providing a matrix in which air can reside for insulation, especially when in water (further explored in Chap. 7). Polar penguins also have increased fat beneath their skin, which provides added insulation, especially deep underwater because feathers become compressed, losing much of their insulative capacity. Subdermal fat also helps increase the overall body volume. Warmish-water (and particularly out of water) *Spheniscus*

Table 1.2 Measurements of penguins exhibited as averages, mostly from Williams [67] (girth from Sato [53]); rockhoppers from Pütz et al. [52] and Cuthbert [16])

	Body length	Weight		Girth	Flipper length		Flipper aspect ratio	Bill length		DI	Climate zone	SST
		M	F		M	F		M	F			
Emperor penguins	1150	23–38	22–30	930	358	347	3.9	82	81	1.2	P	−1.7 to 3
King penguins	950	13–16	11–14	660	350	340		130	124	4.7	SA–P	3–9
Gentoo penguins	780	5.6–6.8	5.1–7.5	560	243	233		58	53	9.1	SA–P	3–9
Yellow-eyed penguins	720	4.4–8.5	4.2–7.5		212	205	3.2	55	53	3.7	ST–T	9–13
Erect-crested penguins	670	4.4–6.4	2.9–5.6		212	204		59	52	12.7	SA	6–9
Adélie penguins	700	4.3–5.4	3.9–4.7	490	196	193	3.4	40	36	10.5	P	−1.7 to 3
Chinstrap penguins	680	3.6–5.0	3.4–4.8	500	186	187		49	46	10.6	P	2–7
Macaroni penguins	700	3.4–6.4	3.2–5.7	460	211	204		61	53	14.0	SA–P	3–9
Royal penguins	700	3.0–8.1	3.0–6.3		199	185		66	59	11.3	SA	6–9
Magellanic penguins	700	4.0–4.7	3.2–3.9		195	186		58	53	8.9	T	7–17
Humboldt penguins	650	4.7	4.0		174	164	2.9	65	59	9.7	T	10–14
African penguins	700	3.0–4.0	3.0–3.5		174	164		65	59	9.7	T	10–14
Fiordland penguins	550	2.7–4.9	2.5–4.8		185	177		51	45	12.5	ST–T	10–14
Snares penguins	560	2.6–3.4	2.5–2.8		185	178		59	52	12.5	ST–T	7–10
Southern rockhopper penguins	550	1.6–3.4	1.8–3.3		176	168	2.9	43	39	9.8	ST–T	7–10
Northern rockhopper penguins	650	3.0	3.1		185	184		49	44	10.9	ST–T	11–13
Eastern rockhopper penguins	580	2.0–3.2	2.0–3.2		165	161		46	41	11.6	ST–T	11–13
Galápagos penguins	530	2.1–2.3	1.7–1.9		150	142	3.2	58	54	7.1	ST–T	11–13
Little penguins	420	1.4	1.0	310	124	117	2.8	36	34	5.7	ST–T	15–17

Size is length, taken from Lynch [35], supplemented by Harrison et al. [27] for additional splits in taxonomy; caution is required, though, as measuring a specimen is affected by the preparator's technique and it is best to compare also using mass. Body weight is expressed in kilograms and linear measurements in millimeters. Climate zones: *P* polar, *SA* subantarctic, *T* temperate, *ST* subtropical (see Chap. 2), *DI* dimorphism index of bill length [61]; mean aspect ratio (wing length divided by width at wrist = widest portion) for species with sufficient sample size (>10), and measured by us; SST, °C from Earle and Gover [19]

Fig. 1.6 The markedly different-sized emperor and Adélie penguins are the most southerly breeding penguin species and are the two that are the most properly associated with Antarctica. (Photo courtesy of David Ainley)

penguins have bare areas on their faces during breeding, although these areas are feathered post-molt, when the species spend extended periods at sea. The high-latitude, polar Adélie penguins' beaks are covered by feathers almost to the end all year round.

Penguin Evolution: Radiation into Vacant Niches

One should not develop the conception that body/appendage size in penguins is all about issues of heat generation and loss, despite the tie between penguin evolution and climate, as discussed above. Body size also affects the diving capacity of modern penguins and the size of the prey captured (in part due to bill size; Chaps. 4, 5, 6, 7, and 8). Moreover, body size is related to what nesting habitat is accessible and, in turn, which stretches of ocean they will be exploiting (Chaps. 2 and 3). Ultimately, this likely relates to which penguin species would be compatible for foraging in the same region (Chaps. 3 and 5).

Comparing Fig. 1.4 with Table 1.2 and Fig. 1.7, it is interesting that the largest penguins are among the older lineages and that the smallest penguins are among the youngest. This is perhaps the result of the climate regime in which recent islands, with difficult access, have emerged, bringing forth several "new" penguin species,

Fig. 1.7 Today's penguin species arranged by body size

as noted above, mostly in the warmer north. However, is this just geological serendipity? As Antarctica and the southern land masses separated, rift zones with their volcanic activity lay in between and along these is where many volcanic islands appeared. Transitioning north from the cold Southern Ocean is a series of increasingly warmer climatic zones (see Chap. 3). Within these, the shores of existing and even new volcanic islands are mostly steep, and the large, lumbering, older-lineage king and emperor penguins had already taken the easy terrain. The implications are explored below.

Most of this book will clearly be about what penguins do at sea, presented in great detail. What we present here will be the basis for the short summary and proposal that follows (also Chap. 2). In the context of the DNA-derived lineage that has been recently enumerated, mostly related to the physical changes in the Southern Ocean and surrounding islands and continents, as discussed above, we would like to propose some ecological/biological factors that would have likely contributed to penguin evolution and species radiation. Nothing more complex than the maxim, "survival of the fittest" or better "existence by being forced to do something else and doing so better than other similar species." This is more akin to the analyses of Darwin's finches of the Galápagos [25, 31], which involves the radiation of one species to fill numerous niches that were "waiting for occupation." Ecologists talk about "niche space," which is the collection of habitat factors that a species evolved in order to use. The use of niche space requires ecological, morphological, physiological, and behavioral aspects. Our proposed scenario, goes like this:

1. The oldest penguin lineage of large, lumbering (on land) *Aptenodytes* took up and occupied all the easily attainable breeding habitats, such as gradual beaches and fast ice, in the process of occupying ice-free versus sea ice-covered oceans, respectively (Chaps. 2 and 3). Being large led to/correlated with their extremely long breeding cycles (longer in large animals [68]). The one egg clutch or the one chick brood of kings, most of which nested on oceanic islands, was required owing to them foraging in pelagic, oligotrophic (prey-depleted) waters, and, those of emperors was needed in order to accomplish breeding in just 1 year, due to the ephemeral nature of fast ice (the only breeding habitat that they were able to access). Differences in the length of the breeding cycle also related to the high energy-dense prey available to the emperors (they could not have had a 1-year cycle otherwise) and the more depleted and more distant forage species available to the kings (Chaps. 4 and 5). The deep diving of both opened up a larger oceanic volume in which to find prey, which would be an advantage in the prey-searching issue (Chaps. 10 and 11).
2. Some long time after (see above, Fig. 1.4) came the *Pygoscelis*, *Spheniscus*, and *Eudyptula* species, which had to have more agility than emperors/kings in order to colonize the steeper-sloped terrain bordering the coastal ocean that was left available to them. The *Aptenodytes* species gave them no option, but there was plenty of that vacant nesting niche ready to be used. There was, for sure, a temperature gradation in the zones occupied, varying from cold to cool to warm for *Pygoscelis*, *Spheniscus*, and *Eudyptula*, respectively (see above). Except for *Eudyptula*, the other two genera populated highly productive waters (south of the Southern Boundary of the Antarctic Circumpolar Current (SBACC)) on the part of *Pygoscelis* and upwelling currents on the part of *Spheniscus* (Chap. 3). The northern gentoos perhaps were a bit of an exception, too, though they foraged over the more productive insular shelves (as did the yellow-eyed and little penguins) rather than out in the more depleted

pelagic realm (Chap. 3). Being smaller than emperors/kings, these three genera could accomplish an annual breeding cycle, in which they had a clutch of two eggs and often raised two chicks. The oceanic habitat available to these smaller species, i.e., coastal waters, was rich enough for their life history strategy.

3. Finally, remaining to be colonized by penguins were oceanic islands, a number of which have appeared relatively recently in "penguin time" (see above) but by nature were surrounded by pelagic, oligotrophic waters – abyssal depths begin immediately off shore in many cases. That is, little in the way of the more productive continental and insular shelves were still available for exploitation. The *Eudyptes* took up the challenge, so to speak, and as described above, radiated according to island appearance/availability, in differing oceanic climates [9–12, 65]. These species had to be the terrestrial acrobats of the penguin world, able to climb rocky slopes that smart humans would not attempt without climbing equipment and, in a flash, access the rocky shore between waves (see below). Moreover, the penguins could do it without the use of hands! They were an array of small, highly agile penguins. However, having the capacity to access these islands, owing to the relatively depleted ocean remaining for them, the crested penguins diverged from the other small, more coastal penguins and evolved their brood reduction strategy, their marked range in body–bill size, their marked sexual dimorphism and breeding duty dimorphism, and a staggering range of breeding phenology (timing) (August to November, depending on the species; Chap. 2). All these factors, involved in species radiation, facilitated a reduction in both intraspecific and interspecific trophic competition, needed for the oligotrophic oceanic habitat available to them (Chaps. 3, 4 and 5).

Conclusions

When all is said and done, the evolutionary history of penguins has left us today with about 20 species, ranging in size from 1 to 30 kg, inhabiting the world's southern oceans where water temperatures range from −1 to 18 °C. They range in dress code from the flamboyant (crested penguins) to the dowdy (little penguins) and have capacities to acquire sea creatures that have only evolved in the ocean, governed by the rule of "big fish eat little fish" (Chaps. 4 and 12). Thus, these prey seek to evade capture using a suite of tricks, including descending to great depths, far from the free oxygen that penguins and other air-breathing predators require. That penguins have evolved to be spectacularly adept at finding and catching their endlessly elusive prey, as well as eluding their own predators, despite having evolved from aerial precursors, is surely cause for celebration, and we shall do just that in the remainder of this book.

References

1. Ainley DG (2002) The Adélie penguin: bellwether of climate change. Columbia University Press, New York
2. Baker AJ, Pereira SL, Haddrath OP, Edge KA (2006) Multiple gene evidence for expansion of extant penguins out of Antarctica due to global cooling. Proc R Soc B Biol Sci 273:11–17
3. Bertelli S, Giannini NP (2005) A phylogeny of extant penguins (Aves: sphenisciformes) combining morphology and mitochondrial sequences. Cladistics 21(3):209–239
4. Boersma PD, Frere E, Kane O, Pozzi LM, Pütz K, Rey AA, Rebstock GA, Simeone A, Smith J, Van Buren A, Yorio P, Borboroglu PG (2013) Magellanic penguin (*Spheniscus magellanicus*). In: Borboroglu PG, Boersma PD (eds) Penguins: natural history and conservation. University of Washington Press, Seattle, pp 233–263
5. Boersma PD, Steinfurth A, Merlen G, Jimenez-Uzcategul G, Hernen Vargus F, Parker PG (2013) Galapagos penguin (*Spheniscus mendiculus*). In: Borboroglu PG, Boersma PD (eds) Penguins: natural history and conservation. University of Washington Press, Seattle, pp 285–302
6. Boessenkool S, Austin JJ, Worthy TH, Scofield P, Cooper A, Seddon PJ, Waters JM (2009) Relict or colonizer? Extinction and range expansion of penguins in southern New Zealand. Proc R Soc B Biol Sci 276:815–821
7. Bost C-A, Delord K, Barbraud C, Cherel Y, Pütz K, Cotté C, Péron C, Weimerskirch H (2013) King penguin (*Aptenodytes patagonicus*). In: Borboroglu PG, Boersma PD (eds) Penguins: natural history and conservation. University of Washington Press, Seattle, pp 7–21
8. Cairns DK, Gaston AJ, Huettmann F (2008) Endothermy, ectothermy and the global structure of marine vertebrate communities. Mar Ecol Prog Ser 356:239–250
9. Cole TL, Dutoit L, Dussex N, Hart T, Alexander A, Younger JL, Clucas GV, Frugone MJ, Cherel Y, Cuthbert R, Ellenberg U (2019) Receding ice drove parallel expansions in Southern Ocean penguins. Proc Natl Acad Sci U S A 116(52):26690–26696
10. Cole TL, Ksepka DT, Mitchell KJ, Tennyson AJD, Thomas DB, Pan H, Zhang G, Rawlence NJ, Wood JR, Bover P, Bouzat JL, Cooper A, Fiddaman SR, Hart T, Miller G, Ryan PG, Shepherd LD, Wilmshurst JM, Waters JM (2019) Mitogenomes uncover extinct penguin taxa and reveal island formation as a key driver of speciation. Mol Biol Evol 36(4):784–797
11. Cole TL, Rawlence NJ, Dussex N, Ellenberg U, Houston DM, Mattern T, Miskelly CM, Morrison KW, Scofield RP, Tennyson AJ (2019) Ancient DNA of crested penguins: testing for temporal genetic shifts in the world's most diverse penguin clade. Mol Phylogenet Evol 131:72–79
12. Cole TL, Zhou C, Fang M, Pan H, Ksepka DT, Fiddaman SR, Emerling CA, Thomas DB, Bi X, Fang Q, Ellegaard MR, Feng S, Smith AL, Heath TA, Tennyson AJD, Borboroglu PG, Wood JR, Hadden P, Grosser S, Bost C-A, Cherel Y, Mattern T, Hart T, Sinding M-HS, Shepherd LD, Phillips RA, Quillfeldt P, Masello JF, Bouzat JL, Ryan PG, Thompson D, Ellenberg U, Dann P, Miller G, Boersma PD, Zhao R, Thomas M, Gilbert P, Yang H, Zhang D-X, Zhang G (2022) Genomic insights into the secondary aquatic transition of penguins. Nat Commun 13(1):1–13
13. Crawford RJM, Altwegg R, Barham BJ, Barham PJ, Durant JM, Dyer BM, Geldenhuys D, Makhado AB, Pichergru L, Ryan PG, Underhill LG, Upfold L, Visagie J, Waller LJ, Whittington PA (2011) Collapse of South Africa's penguins in the early 21st century. Afr J Mar Sci 33(1):139–156
14. Croll DA, Gaston AJ, Burger AE, Konnoff D (1992) Foraging behavior and physiological adaptation for diving in thick-billed murres. Ecology 73:344–356
15. Crossin GT, Trathan PN, Crawford RJM (2013) Macaroni penguin (*Eudyptes chrysolophus*), royal penguin (*Eudyptes schlegeli*). In: Borboroglu PG, Boersma PD (eds) Penguins: natural history and conservation. University of Washington Press, Seattle, pp 185–208
16. Cuthbert RJ (2013) Northern rockhopper penguin (*Eudyptes moseleyi*). In: Borboroglu PG, Boersma PD (eds) Penguins: natural history and conservation. University of Washington Press, Seattle, pp 131–143

17. Davis LS (2013) Erect-crested penguin (*Eudyptes sclateri*). In: Borboroglu PG, Boersma PD (eds) Penguins: natural history and conservation. University of Washington Press, Seattle, pp 145–151
18. De La Puente S, Bussalleu A, Cardeña M, Valdés-Velásquez A, Maluf P, Simeone A (2013) Humboldt penguin (*Spheniscus humboldti*). In: Borboroglu PG, Boersma PD (eds) Penguins: natural history and conservation. University of Washington Press, Seattle, pp 265–283
19. Earle SA, Gover LK (2009) Ocean, an illustrated atlas. National Geographic Society, Washington, DC
20. Elliott KH, Davoren GK, Gaston AJ (2007) The influence of buoyancy and drag on the dive behaviour of an Arctic seabird, the thick-billed murre. Can J Zool 85:352–361
21. Elliott KH, Ricklefs RE, Gaston AJ, Hatch SA, Speakman JR, Davoren GK (2013) High flight costs, but low dive costs, in auks support the biomechanical hypothesis for flightlessness in penguins. Proc Natl Acad Sci 110:9380–9384
22. Fretwell PT, Trathan PN (2020) Discovery of new colonies by Sentinel2 reveals good and bad news for emperor penguins. Remote Sens Ecol Conserv 7(2):139–153
23. Galasso S (2014) When the last of the great auks died, it was by the crush of a fisherman's boot. Smithsonian Magazine, 10 July 2014
24. Gavryushkina A, Heath TA, Ksepka DT, Stadler T, Welch D, Drummond AJ (2017) Bayesian total evidence dating reveals the recent crown radiation of penguins. Syst Biol 66(1):57–73
25. Grant PR, Grant BR (2008) How and why species multiply: the radiation of Darwin's finches. Princeton University Press, Princeton
26. Grémillet D, Ponchon A, Paleczny M, Palomares M-LD, Karpouzi V, Pauly D (2018) Persisting worldwide seabird-fishery competition despite seabird community decline. Curr Biol 28:4009–4013
27. Harrison P, Perrow M, Larsson H (2021) Seabirds: the new identification guide. Lynx Edicions, Barcelona
28. Ksepka D, Ando T (2011) Penguins past, present, and future: trends in the evolution of the sphenisciformes. In: Dyke G, Kaiser G (eds) Living dinosaurs: the evolutionary history of modern birds. Wiley, New York, pp 155–186
29. Ksepka D, Thomas DB (2012) Multiple cenozoic invasions of Africa by penguins (Aves, Sphenisciformes). Proc R Soc B Biol Sci 279(1730):1027–1032
30. Kuroda N (1954) On the classification and phylogeny of the order Tubinares, particularly the shearwaters (Puffinus): with special considarations [sic] on their osteology and habit differentiation. Kuroda (self published), Tokyo
31. Lack DL (1947) Darwin's finches. Cambridge University Press, Cambridge
32. Lustick S (1984) Thermoregulation in adult seabirds. In: Whittow GC, Rahn H (eds) Seabird energetics. Plenum Press, New York, pp 183–201
33. Lynch HJ (2013) Gentoo penguin (*Pygoscelis papua*). In: Borboroglu PG, Boersma PD (eds) Penguins: natural history and conservation. University of Washington Press, Seattle, pp 73–88
34. Lynch HJ, Larue MA (2014) First global census of the Adélie penguin. Auk 131:457–466
35. Lynch W (2007) Penguins of the world. Firefly Books, Buffalo
36. Marples BJ (1953) Fossil penguins from the mid-tertiary of Seymour Island. HMSO, London
37. Mattern T (2013) Fiordland penguin (*Eudyptes pachyrhychus*). In: Borboroglu PG, Boersma PD (eds) Penguins: natural history and conservation. University of Washington Press, Seattle, pp 153–167
38. Mattern T (2013) Snares penguin (*Eudyptes robustus*). In: Borboroglu PG, Boersma PD (eds) Penguins: natural history and conservation. University of Washington Press, Seattle, pp 169–183
39. Mattern T, Wilson KJ (2018) New Zealand penguins – current knowledge and research priorities: birds New Zealand. West Coast Penguin Trust, NZ
40. Mayr G, De Pietri VL, Love L, Mannering AA, Scofield RP (2017) A well-preserved new mid-paleocene penguin (Aves, Sphenisciformes) from the Waipara Greensand in New Zealand. J Vertebr Paleontol 37(6):e1398169

41. Mitchell KJ, Llamas B, Soubrier J, Rawlence NJ, Worthy TH, Wood J, Lee MSY, Cooper A (2014) Ancient DNA reveals elephant birds and kiwi are sister taxa and clarifies ratite bird evolution. Science 344(6186):898–900
42. Obiol J F, James H, Chesser R, Bretagnolle V, Gonzales-Solis J, Rozas J, Welch A, Riutort M (2020) Paleoceanographic changes in the late Pliocene promoted rapid diversification in pelagic seabirds. Authorea, October 12
43. Olson SL (1973) Evolution of the rails of the South Atlantic islands (Aves: Rallidae). Smithsonian Contrib Zool 152:1–53
44. Olson SL, James HF (1982) Prodromus of the fossil avifauna of the Hawaiian Islands. Smithsonian Contrib Zool 365:1–59
45. Paleczny M (2008) An analysis of temporal and spatial patterns in global seabird abundance during the modern industrial era, 1950-2010, and the relationship between global seabird cecline and marine fisheries catch. MSc thesis: University of British Columbia
46. Paleczny M, Hammill E, Karpouzi V, Pauly D (2015) Population trend of the world's monitored seabirds, 1950–2010. PLoS One 10(6):e0129342
47. Pan H, Cole TL, Bi X, Fang M, Zhou C, Yang Z, Ksepka DT, Hart T, Bouzat JL, Argilla LS, Bertelsen MF (2019) High-coverage genomes to elucidate the evolution of penguins. GigaScience 8(9):giz117
48. Pennycuick CJ (1987) Flight of seabirds. In: Croxall JP (ed) Seabirds: feeding biology and role in marine ecosystems. Cambridge University Press, Cambridge, pp 43–62
49. Pennycuick CJ, Croxall JP, Prince PA (1984) Scaling of foraging radius and growth rate in petrels and albatrosses. Ornis Scand 15:145–154
50. Peterson RT (1979) Penguins. Houghton Mifflin, Boston
51. Pratt HD (2005) The Hawaiian honeycreepers. Oxford University Press, Oxford
52. Pütz K, Rey AA, Otley H (2013) Southern rockhopper penguin (*Eudyptes chrysocome*). In: Borboroglu PG, Boersma PD (eds) Penguins: natural history and conservation. University of Washington Press, Seattle, pp 113–129
53. Sato K, Shiomi K, Watanabe Y, Watanuki Y, Takahashi A, Ponganis PJ (2010) Scaling of swim speed and stroke frequency in geometrically similar penguins: they swim optimally to minimize cost of transport. Proc R Soc B Biol Sci 277(1682):707–714
54. Seddon PJ, Ellenberg U, Van Heezik Y (2013) Yellow-eyed penguin (*Megadyptes antipodes*). In: Borboroglu PG, Boersma PD (eds) Penguins: natural history and conservation. University of Washington Press, Seattle, pp 91–110
55. Simeone A, Anguita C, Daigre M, Arce P, Vega R, Luna-Jorquera G, Portflitt-Toro M, Suazo CG, Miranda-Urbina D, Ulloa M (2021) Spatial and temporal patterns of beached seabirds along the Chilean coast: linking mortalities with commercial fisheries. Biol Conserv 256:109026
56. Simpson GG (1946) Fossil penguins. American Museum of Natural History, New York
57. Simpson GG (1975) Fossil penguins. In: Stonehouse B (ed) The biology of penguins. Macmillan Press, New York, pp 19–41
58. Sparks J, Soper T (1967) Penguins. David & Charles Publishers Limited, Devon
59. Stephenson J, Budd GM, Manning J, Hansbro P (2005) Major eruption-induced changes to the McDonald Islands, southern Indian Ocean. Antarct Sci 17(2):259–266
60. Stonehouse B (1967) The general biology and thermal balance of penguins. In: Cragg JB (ed) Advances in ecological research. Academic, London, pp 131–196
61. Storer RW (1966) Sexual dimorphism and food habits in three North American accipiters. Auk 83(3):423–436
62. Strycker N, Wethington M, Borowicz A, Forrest S, Witharana C, Hart T, Lynch HJ (2020) A global population assessment of the Chinstrap penguin (*Pygoscelis antarctic*a). Sci Rep 10:19474
63. Thomas DB, Ksepka DT, Fordyce RE (2011) Penguin heat-retention structures evolved in a greenhouse Earth. Biol Lett 7(3):461–464

64. Tyler J, Bonfitto MT, Clucas GV, Reddy S, Younger JL (2020) Morphometric and genetic evidence for four species of gentoo penguin. Ecol Evol 10(24):13836–13846
65. Vianna JA, Fernandes FA, Frugone MJ, Figueiró HV, Pertierra LR, Noll D, Bi K, Wang-Claypool CY, Lowther A, Parker P, Le Bohec C (2020) Genome-wide analysis reveals drivers of penguin diversification. Proc Natl Acad Sci U S A 117(36):22303–22310
66. Webster T (2018) The pathway ahead for hoiho. Yellow-Eyed Penguin Trust, Dunedin
67. Williams TD (1995) The penguins, sphenisidae. Oxford University Press, Oxford
68. Williams TD (2018) Avian reproduction – overview (wild birds). In: Encyclopedia of reproduction. Elsevier, New York, pp 595–601
69. Wilson RP, Hustler K, Ryan PG, Burger AE, Noldeke EC (1992) Diving birds in cold water: do Archimedes and Boyle determine energetic costs? Am Nat 140(2):179–200
70. Wong K (2002) Diminutive dinosaur from China sheds light on bird evolution. Sci Am 273(1582):11–17

Chapter 2
Land Ahoy: A Tiresome Business

> "Penguins coming ashore to rest, breed and molt are out of the element which shaped them, and they are handicapped in several ways. Weightless at sea, they suddenly become ponderous and awkward on stepping ashore. From swimming effortlessly in three dimensions they must trot short-leggedly in two. Ill-poised, probably short-sighted, suffering climate fluctuations unknown in the water, open to attack by land predators and parasites,....it is not surprising that they seem to spend as little time on land as possible. While ashore they are completely separated from food, and living on a diminishing store of body fat. The length of time they can spend away from the sea depends on their food reserves, and largely determines their patterns of activity on land". (B. Stonehouse [35], p. 84)

Or

> "...the Penguins and Manchots appear to form the shade between birds and fish. Instead of wings they have little pinions, which might be said to be covered with scales rather than feathers, and which serve as fins; their body is large, compact, and cylindrical, behind which are attached two broad oars, rather than two legs: the impossibility of advancing far into the land, the fatigue even of remaining there, otherwise than by lying; the necessity, the habit of being almost always at sea, their whole economy of life, mark the analogy between the aquatic animals and these shapeless birds, strangers to the region of air, and almost equally exiled from those of the land". Count de Buffon (1812) *Natural History* Vol 20, as presented in Stahel and Gales [34], p. 13.

It is important to add an element of perspective, before diving into our promised treatise on the aquatic lives of penguins. In this chapter, we will briefly review the process of penguins going to and coming from the sea and then get into the relevant aspects of their land lives, i.e., why they come to land, with which most penguin enthusiasts are at least generally familiar. Indeed, all existing books on penguins exhaustively cover the land part of their lives or at least the breeding and nesting biology portion. We present just a short review here. At the same time, the reader needs to keep reminding themselves that some penguins spend about 90% of their lives at sea, when, as an example, for the Adélie penguins, one considers their juvenile years (1–3), pre-breeding years (3–5), and breeding years (5–20 [1]). Yearlings and juveniles do not visit land (occasional, extremely short visits being an exception among 1-year-olds) and spend their lives completely at sea, pre-breeders spend a few days to 2 months investigating the colony with the remainder of the time at sea

D. G. Ainley, R. P. Wilson, *The Aquatic World of Penguins*, Fascinating Life Sciences, https://doi.org/10.1007/978-3-031-33990-5_2

(a few trips between the sea and land), and adults spend 4–5 months associated with their colonies each year, although they spend half that time foraging to recover their condition or feed their chicks. After breeding, and some reconditioning, penguins also molt all their feathers, which are extremely important to them (Chaps. 7 and 14). This is the pattern for most penguin species that, to varying degrees, give up the "central-based foraging mode" of the nesting season (Chap. 5) and disperse away from their colonies (where most of them will return to molt before the ultimate dispersal; see below). This would not be the case for the sedentary species, most of which visit land daily (though mostly at night) and even their nest sites for much of the year [16] (see below).

Crossing the Land–Ocean Interface Is Affected by Body Size

In Chap. 1, we discussed much about the agility of small penguins on land, i.e., the ones that have appeared most recently in the penguin lineage – the larger species having had taken the easily accessible terrain for breeding. Here, we need to discuss a bit more about their agility before we touch on the reason why penguins bother to come to land in the first place. They do so for breeding and molting, despite being so incredibly adept at existing in their watery world (later chapters). In this chapter, we describe the actual, often acrobatic, process of coming ashore, as it does relate to penguin species radiation, as covered in Chap. 1.

Indeed, one clear problem (other than the lack of prey-searching capacity) for a seabird unable to fly (in air) is the transiting between the ocean and the land. Of course, when the water is calm, the matter is relatively trivial. In those conditions, penguins swim toward the shore until they reach a depth at which they can wade and then walk up the beach slope, or vice versa. This is the preferred mode for species such as king, yellow-eyed, and little penguins (Figs. 2.1, 2.2, 2.3 and 2.4). On the other hand, some penguins that have to negotiate a steep shore may gather incredible speed underwater and then leap out, traveling ballistically, as do *Pygoscelis* penguins and emperor penguins alighting onto flat ice (we will discuss the far more

Fig. 2.1 Adélie penguins exhibiting two techniques for entering the sea. (Photos courtesy of David Ainley)

Fig. 2.2 Having made it successfully off the rocky shore, these Magellanic penguins will dive into the approaching wave, just like human surfers would do as they paddle out. (Photo courtesy of David Ainley)

Fig. 2.3 Australian little penguins and surf conditions (or the lack thereof) that they seek when finding an appropriate colony location. (Photo courtesy of Eric Woehler)

Fig. 2.4 King penguins in surf conditions to their liking; the water is shallow enough that if these penguins were coming ashore rather than leaving, they could probably stand up, water "waist" high to wade the rest of the way to the shore. (Photo courtesy of Eric Woehler)

spectacular leaps of the smaller *Pygoscelis* when we consider predator avoidance; Chap. 12).

Two issues are relevant, especially for small-bodied penguins that regularly access steep, rocky shores in areas with large waves, which is the norm for oceanic islands: (i) how to deal with the often highly energetic and potentially destructive surf zone and (ii) how to exit efficiently, which depends on their body size and the terrain at the interface. These factors, of course, are not independent since being thrown onto a gently sloping sandy beach is clearly less damaging than being thrown onto jagged rocks.

Wind-generated waves travel at speeds ranging from 2 to 15 m/s, with 5 m/s being a reasonable norm (https://en.wikipedia.org/wiki/Wind_wave). Given that most penguins' maximum speeds lie in this range (see Chap. 6), being thrown onto land at any of these speeds and certainly being thrown at maximum speeds is not a trivial issue, even for birds as fast, and as "tough" (with significant amount of padding and having an extremely robust body), as penguins. Penguins are, however, judicious in their landings and will come in immediately behind (on or within) a wave and pop onto land just after the wave breaks against it. If, during this process, they have misjudged something, however, then they will turn and dive into the next wave, swimming out rapidly underwater perpendicular to the shore. They then will keep away from the water surface where the fluid movement pushing them onto land is the highest. Once at a safe distance though, and yes penguins also need to be mindful of lurking predators (see Chap. 12), they will turn and reattempt the landing.

The penguin species, however, are not equal when it comes to crossing the sea-to-land boundary, and this is another penguin attribute related to body size (Chap. 1). First, the two little penguin species, the smallest of penguins, have the lowest maximum swim speeds, with no reliable estimates exceeding 3.5 m/s (Chap. 6).

Fig. 2.5 African penguins waiting for the current set of large waves to pass before diving in. (Photo courtesy of Rory Wilson)

This puts them at a considerable disadvantage compared to the other species when they are trying to swim through a fast wave and may explain why these species tend to not have colonies associated with more challenging surf conditions (see Figs. 2.5, 2.6 and 2.7). On the other hand, being small does have some advantages in rabid surf zones, and *Eudyptes* penguins appear to take advantage of this. In the worst case scenario, when a penguin is taken by a wave and cast against a rock, what is called the work–energy principle is in operation (see Box 2.1).

Box 2.1
The work–energy principle. The change in kinetic energy is equal to the net work done on the animal. If the final velocity is 0 m/s, and the wave is traveling at 5 m/s, then the work done is 0.5 × mass × velocity of the wave squared. In other words, a 1-kg little penguin, subjected to this, has net work done of 12.5 joules, whereas a 28-kg emperor has 350 joules. It should be noted how the squared term of the wave velocity makes it important to try and come ashore when waves are traveling slowly, particularly for larger penguins.

Eleanor Pettingill [25] had some telling descriptions of northern rockhopper penguins coming to land (p. 109):

> "Far out the water boiled as a horde of porpoising birds approached a flat rock that slanted into the sea. A wave crashed in and suddenly about fifty penguins stood on the rock; another

Fig. 2.6 Southern rockhoppers on the Falkland/Malvinas Islands exemplifying the type of habitat available to *Eudyptes* penguins after their easy-access habitat was taken by the earlier evolving penguin species. (Photo courtesy of Virginia Morandini)

> wave, and dozens more. The first arrivals hippity-hopped up the rock….Each wave reminded me of the arrival of a commuters' train at a suburban station." Also (p. 136).
>
> "They porpoised toward their rock by the hundreds and sprang from the surf. Some of them made it the first time; others missed, were swept back and had to try again….They were always exuberant, as if battling the sea were exhilarating." And (p. 141).
>
> "The breakers sent up towers of spray shot with rainbows. Between the crests the troughs were brilliant turquoise. The rockhoppers hurled their solid little bodies, sleek and glistening, right toward us onto the landing rock, and hopped swiftly ashore. Often a giant wave collapsed and washed over the rock, sweeping hundreds of birds out to sea again. Struggling, they porpoised into view once more, got another foothold, and climbed to safety above the surf".

With regard to being cast against a rock (Figs. 2.5, 2.6 and 2.7), the stopping force applied to an object (our penguin) in a wave is what in particular determines the potential for detriment, i.e., with what force is the penguin going to come in contact with the land substrate? This force is calculated by simply dividing the work done (see Box 2.1) by the distance over which the bird stops. So, a medium-sized penguin, such as a 4-kg macaroni penguin, being thrown on a sandy beach by a 5-m/s wave, coming to a stop over 50 cm of sand experiences an average force of 100 N (equivalent to a thump of about 10-kg force). If the same bird is forced against a rock over a distance of 5 cm (rocks are notoriously unyielding), then it will experience a stopping force of about 100 kg! Definitely, this would not be its choice! The equivalent figures for little and emperor penguins would be 2.5 and 70 kg for

Fig. 2.7 The types of sea conditions that the highly agile *Eudyptes* penguins typically encounter in the process of landing or exiting the rocky shores where their nesting colonies are located; shown here are southern rockhopper penguins on the Falkland/Malvinas Islands. (Photo courtesy of Virginia Morandini)

the sandy beach and 25 and 700 kg for the rocks, respectively! This should make it abundantly clear why large penguins, especially, should not be getting out of the water onto difficult rocky shorelines where there are tumultuous waves. It explains why the *Aptenodytes* first "chose" the easy landing places – shallow beaches or low freeboard sea ice.

There are, though, two more disadvantages of being a large penguin attempting to scramble up a rocky slope when the next big wave is bearing down. Larger animals, in general, have lower metabolic rates in relation to their masses than do smaller animals (Chap. 7). Therefore, for any work required for quickly moving up a slope (before the next wave), they have to expend energy to account for the potential energy (PE) increase associated with the height and distance to climb according to PE = mass × height × 9.81 – this is a higher proportion of their normal resting metabolic rate compared to small animals. In addition, the force that can be generated by a muscle (e.g., used for scrambling up a rocky slope) is proportional to the cross-sectional area of that muscle (in other words, a squared term (the area of a circle is π X radius squared)), whereas that force has to be applied to a body mass moving up a slope that increases according to an approximately cubed term (the volume of a box is length X width X height). Thus, although larger animals are absolutely stronger, they are, relative to their mass, weaker, which explains why chipmunks or squirrels can race up a tree but gorillas can only ascend slowly. In penguin terms, this helps explain why small species like rockhoppers can move rapidly up slopes between waves on steep shores to get out of the surf zone quickly, whereas king (and emperor) penguins would think twice about doing so.

Why and How Often Do Penguins Come Ashore?

Overall, considering the extremes, from the courtship to the chick fledging period, king and emperor penguins make just 12–14 trips to sea, although periods at sea or on land last for multiple days. Indeed, their large size means that they are able to bring a lot of food back to their colony, whereupon they fast and/or incrementally feed their chicks small meals while their partner forages [8]. In contrast, Adélie penguins go to sea 38 or more times during this part of the breeding season (depending on colony size – at small colonies, foraging trips are shorter and more numerous [1]). As to be discussed in Chap. 3, neither of the two high-latitude Antarctic species visits the colony outside the breeding period, and king penguins molt on land but then spend months at sea. Crested penguins exhibit a similar pattern. Those penguin species that are sedentary and frequent their breeding sites year round, i.e., yellow-eyed, Fiordland, and little penguins (Chap. 3), make a trip to the sea daily or almost daily, returning for the night [16, 29]. Therefore, from courtship through chick fledging, they may make in the order of 50–60 trips to/from the sea and 100 trips over the course of a year (see also Stahel and Gales [34]). For penguins to visit the sea versus come to land requires a major transition in their behavior and mindset, as will hopefully become apparent in this book. As it is, for purposes of this book, we are interested in their aquatic mindset.

Vertebrates that interact with the ocean might be broadly grouped into those that never leave the sea and those that depend on the ocean but have some interaction with the land. Of those that never leave the water, one group, fish, the first vertebrates to appear on Earth, evolved in the oceans and never left [22]. Another group, marine mammals such as whales, dolphins, and sea otters, however, reentered the oceans after evolving from a terrestrial ancestor, much later in time from the appearance of fish [39], and now never come to land unless stressed or injured. Land interaction is minimal in sea turtles and in an unusual small, slender Californian fish called the Grunion – both species, under the protection of night, visit beaches for seconds to hours just to lay eggs, which then hatch on their own (although male turtles never come ashore). Then, there are those ocean-associated animals that have more interaction with land. Polar bears, for example, reside in snow caves on land, hibernating and raising cubs from approximately October into March, with some variation (~5 months), but then are generally "at sea" (in the pack ice) continuously during late spring to early autumn. Seals and fur seals, depending on the species, spend a few months on land per year during the pupping/breeding season (departing periodically to forage) as well as to molt; to varying degrees they are "capital breeders," which subsist on body fat for long periods. They then become pelagic. Sea lions have a similar schedule but also haul out repeatedly on coastal rocks during the remainder of the year, diving in to forage now and again. In short, a number of marine taxa are still tied to the land because they have to reproduce there. Penguins, and all other seabirds, are in this camp, given that they lay eggs that, unlike fish eggs, will not hatch if they are in the water.

Importantly though, all penguins are closely wedded to the sea, not least because they catch all of their prey in the ocean – live prey, not by scavenging dead ones (Chap. 4) – for which they are supremely adapted (as this book will describe). Consequently, any period on land, which may last days to weeks, involves them fasting, which is hardly ideal, although the sea is not a perfect refuge for penguins either. Not being the so-called "top predators," the ocean can be a dangerous place for them (see Chap. 12), though, except for being caught in our fishing nets, not as dangerous as the land for most species (they have evolved to form colonies where there are no mammalian predators, which has become an issue during the Anthropocene because humans have introduced many mammals alien to certain habitats). As pointed out by George Simpson [32], penguins would much rather deal with predators on land than with those at sea. On land, if not on a land-predator-free island, penguins nest in out-of-the-way places or in burrows and come to land at night. Otherwise, catching prey and avoiding predators is what penguins do at sea, quite effectively. So, in essence, the marine ecology of penguins is inextricably tied to what they do on land, and thus it is appropriate to provide some basic knowledge of their respective annual cycles. Here, we provide a brief review of the penguin relationship with land. Some of what we discuss below should make more sense once the remainder of this book is digested by readers.

Before we cover relevant penguin natural history common to most species, we will mention some outliers in the land visitation issue. First, the two penguin species most different from the remaining, namely, emperor and king penguins, represent the oldest penguin lineage among the existing species (Chap. 1, Fig. 1.4). Clearly, both species have been highly "successful" in an evolutionary sense (present for an extremely long time) and, as we noted, have occupied all the easily accessible breeding habitats, thus leaving the steep slopes to the more recently evolved species. Indicated in existing summaries of penguin breeding biology [8, 16, 43, 44], both of the *Aptenodytes* penguins form incredibly dense colonies but neither species builds, nor defends, a nest, existing essentially shoulder to shoulder, also a departure from all other penguin species. Members of mated pairs move to the vicinity in which they last left their mate to find one another in the congested mass, by voice [4]. On the basis of having 10 times the density of hair cells in their ears, compared to us (and other mammals; S. Emslie, pers. comm.), penguins are better able to distinguish notes and recognize one another. The single, large egg is positioned on top of the incubator's feet, tucked into a unique, large fold of the skin. Both species take an extremely long time to complete their breeding–molt cycle, consistent with the general pattern that larger animals generally take longer than do smaller ones [45]. The major difference between the two is derived from the extreme, highly seasonal environment in the Antarctic, where emperors exclusively reside, and from the much less extreme, quasi-seasonal environment of the subantarctic, where kings reside. Emperors can breed annually using their extraordinary adaptation characteristics to deal with extreme climate and by availing high energy-dense prey in close proximity to their colonies. King penguins, on the other hand, do not have to deal with such extreme and seasonal environments. However, they lack close proximity to energy-dense prey and, in fact, at most, breed in 2 out of 3 years due to the long time needed by them to adequately provide for and raise their young.

Emperor penguins require about 9 months to successfully raise their young and another month for molting (see below). They then have to recover from that, and thus they lack any significant temporal cushion between breeding attempts. They form colonies not on land but rather on stable, thick sea ice, floating on the sea but held in place by the physical features of the coast (called fast ice, i.e., held fast; Fig. 2.8). Lacking much agility (other than in the sea), emperors can access only the low freeboard of sea ice, which they accomplish by a vigorous hurdle from the sea with a belly flop on landing. They court in April–May and lay their large egg in May–June to then fledge their chick in December; only the male incubates (with a total fasting of 115 days, including 62–67 days of incubation!) and the female, having fasted for 40 days during courtship and laying, returns to brood the chick from the time of hatching for the next few weeks while the male goes to sea to recover. They then both forage for themselves and the chick over a 5-month period. All of this has to be completed before the fast ice disappears, which it does for a couple of months (January–February) – the sea ice breeding habitat is ephemeral to say the least! Their high-quality, energy-dense prey, mostly Antarctic silverfish (Chap. 4), is available in close proximity, all of which facilitates rapid chick growth and eventual fledging. When at sea, emperor penguins remain within the pack ice or its margins; as the sea ice retreats seasonally so too do emperor penguins. While for the

Fig. 2.8 Emperor penguins can only access locations to form colonies on the flat sea ice that is anchored in place along the shore, for instance here at Cape Crozier. (Photo courtesy of David Ainley)

most part, emperor penguins spend their time standing or lying on ice floes or walking/tobogganing over consolidated sea ice, the time that they are at sea is otherwise spent in a number of ways such as diving to obtain the food to accumulate a thick layer of fat needed during incubation–brooding fasts, catching prey to feed their young, recovering body mass lost from raising their young (~25%), building mass in order to fast during the molt, or recovering mass lost in that process. Their extremely large mass (a pre-breeding emperor male weighs ~37 kg, female 28 kg [28]; Table 1.2) and lack of a territory are necessary in order to withstand the intense cold, facilitated by huddling. They also molt while on ice floes, hunkered down behind pressure ridges in the ice to avoid the full force of the cold wind. An emperor penguin, despite having an incredible diving capability (see Chaps. 6, 7, 8, 9, 10 and 11), probably spends the least amount of time actually in water compared to any other penguin species, though technically it does spend all of its time at sea. It accomplishes this ironic feat by standing/lying for long periods on frozen water – ice floating on the sea surface! Only juveniles, in their first year after fledging, actually spend continuous time in water, north of the sea ice-covered portion, before returning to icy regions for the remainder of their lives [20, 21].

Like the emperors, king penguins also have a large body and mass (11–16 kg, depending on their sex [8]; Table 1.2). They too lack much agility on land, requiring a gently sloped beach to reach potential or occupied nesting habitats (Fig. 2.9). If successful at raising a chick, individuals do not breed again that year – the only penguin (and one of the few avian species) that takes longer than a year to court, tend to eggs, and raise a chick to independence. In fact, they require 14–15 months, with large chicks sometimes going untended or unfed for up to 5 months during winter, a time when adults find it difficult to obtain high-quality prey close to their colony. The adults are then a long way at sea from their colony. Adding the time

Fig. 2.9 A king penguin colony at Salisbury Plain, South Georgia, located in the only habitat that they can access – flat to gently sloping terrain that requires little in the way of agility to navigate. (Photo courtesy of Jean Pennycook)

needed for molt, an adult king penguin requires 18 month to complete its full annual cycle. At any one time at a colony, one can find pairs in all stages of the nesting cycle, including new breeders or failed breeders of the previous year laying eggs (in the summer) in the midst of successful pairs still commuting to feed and tend to their chicks and some adults in the molt stage of the cycle. This extended breeding (supra-annual) cycle arises because, unlike their congener, king penguins lack extremely high-quality prey in close proximity. Much of a king penguin's diet is low-caloric squid, available through its deep diving (Chap. 4). Prey are also farther away for breeding king penguins than for many other penguin species. Sometimes they travel as far as 600 km beyond the continental/insular shelf, which is a distance that significantly increases foraging effort and time and reduces chick provisioning rate. Thus, it is hardly surprising that chicks grow unusually slowly for a penguin. As with emperor penguins, although king penguin chicks are not fed all that often, when a parent arrives, it does so with a significant amount of food that gets pumped incrementally into the chick. Those individuals that are not breeding during a given year are at sea, residing in the open ocean far from land.

Deprived of any easy access to a potential breeding habitat by the *Aptenodytes* penguins, the remaining penguin species nevertheless appeared in the penguin family tree, given that there still remained much ocean to explore and exploit at shallower depths (Chap. 1). However, in order to get to a breeding habitat, they needed to possess greater land agility than the emperors and kings to make the climbs up steeper terrain to colonies or nest sites. All build nests (unlike the two nest-less *Aptenodytes*) on land, either in large colonies in the open (Fig. 2.10) or in smaller

Fig. 2.10 The dense concentration of Adélie penguins in a surface nesting colony at Cape Royds, typical of many of the smaller penguin species. (Photo courtesy of David Ainley)

Fig. 2.11 A yellow-eyed penguin standing near its burrow entrance, with chicks hiding under the tussock grass, among loosely associated neighboring nesting pairs on Campbell Island. (Photo courtesy of Louise Chilvers)

ones in which nest territories are under bushes, are among tussocks or in burrows/cavities (Fig. 2.11). All breed annually and have clutches of two eggs (again, unlike *Aptenodytes*; see pertinent chapters in Davis and Renner [16] and Williams [44]). Among these penguin species, some travel widely between breeding seasons, disappearing from their colonies, whereas other species are largely sedentary, lacking much of any post-nesting dispersion (Chaps. 3 and 5). Included in the latter category are the yellow-eyed, Fiordland, gentoo, and little penguins, whose foraging habitats are close to their colonies, thus enabling them to spend more time on land compared to other penguins [14, 16]. In accordance, during the chick phase of nesting, parents in these sedentary species feed their chicks daily more often than not. They form much smaller but dispersed colonies than do the other penguin species, nesting in burrows or beneath bushes or tussock grass and visit land at times throughout the year, including their territory. Thus, without territorial display, they exercise a somewhat passive mode of territory guarding, with possession being the rule of law. These sedentary penguins generally forage more or less within a few porpoising hours of their nest locality, exploiting the more spatially limited, though rich waters of the continental or insular shelf (Chap. 5). The farther-ranging penguins, such as Adélie or chinstrap penguins, nest in large colonies, numbering thousands to millions of pairs. They become totally pelagic going far away both within and outside of their annual breeding period, and chick-feeding foraging trips in the cases of large colonies are measured in days. The idea that colony size and foraging habitat extent are strongly connected, a subject in Chap. 5, was long ago established by R.W. Storer [36], comparing the alcid genera, *Cepphus* (black/pigeon guillemots), which form small, loose colonies and forage in coastal reefs, in contrast to *Uria* (common/thick-billed murres), which form huge, dense colonies and forage far at sea over the continental shelf (aka "penguins of the north" but see Chap. 14). This concept was generalized even more by P. Ashmole [3], who investigated seabirds nesting on tropical, oceanic islands, which involved species like sooty terns that forage over pelagic waters hundreds of kilometers to sea in all directions – hence termed "Ashmole's halo."

Besides the Adélie and chinstrap, the more far-ranging penguins also include three of the four species of the genus *Spheniscus* and the four of the five species of *Eudyptes*. All conduct their breeding and molt over 4 months and then remain at sea for the remainder of the year, although African penguins occasionally, and seemingly unpredictably, pop back to visit their nest sites during their nonbreeding periods. The Galápagos penguins seem to be an exception to this among the *Spheniscus* species. They have nowhere to go, living on an oceanic island, surrounded by productive waters but not far away from a warm, unproductive ocean full of predators (sharks, sea lions, etc.). The exception to the *Eudyptes* species is the Fiordland penguins, but the reason why they stay close home is not clear. The marine ecology of the two *Pygoscelis* species has been investigated far more extensively, and more thoroughly, than those of the other penguins, owing to their proximity to Antarctic national research bases and the requirement that any facilities be used for scientific research under the Antarctic Treaty. In these species' annual cycle, which exhibits little variation in timing in a markedly seasonal environment (egg laying in October–November, depending on the latitude [1]), there is a near-equal division of nesting duty between partners in pairs, with frequent nest changes. Initially, one bird has to brood the chicks for a few days, whereas the other forages for them, but, then, once the chicks are large enough to be thermally independent and "look after themselves" (i.e., resist aerial land predators), the two parents forage simultaneously. *Spheniscus, Megadyptes*, and *Eudyptula* penguins, somewhat less widely studied, are also involved in nesting for about 4 months each year, although the timing of breeding varies with ocean/food web conditions, usually associated with the onset of coastal upwelling (see below). They nest in burrows or far back under bushes (almost a cavity). Partners in pairs put in equal effort, quite similar in nature to the *Pygoscelis* species.

Finally, penguins belonging the genus *Eudyptes* are the most varied and exhibit an interesting range in the nesting cycle strategy. They have been the most recent to appear from an evolutionary perspective, and what resources remained to them were the steep terrain of oceanic islands in mostly oligotrophic (depleted), though climate-varying, oceans. Some crested penguins like macaroni, rockhopper, and erect-crested nest in the open in large colonies, whereas Snares and Fiordland penguins nest in small colonies, with nests loosely spaced under trees and boulders. As noted above (including Chap. 1), all require great agility for accessing their nesting areas. All nest annually, though timing can vary, given that most colonies are at lower latitudes in warmer, more seasonally variable waters compared to penguins in subpolar/polar seas where seasons are predictable. Egg laying by Fiordland penguins occurs in August, northern rockhopper penguins in September, erect-crested and Snares penguins in October, and southern rockhopper penguins in November.

The staggering of annual cycles could work to alleviate interspecific competition in their somewhat depleted oceanic habitat (Chap. 5). In addition, crested penguins are more sexually dimorphic than are the other penguin species (Table 1.2), in part related to their nesting duties (large males are the guardians) but likely also related to reducing intraspecific competition for food on the basis of body–bill size (Chaps. 4 and 5). All lay a clutch of two eggs, but the first-laid egg is significantly smaller than the second and is part of a brood reduction strategy, again an adaptation to a depleted ocean. Therefore, pairs rarely, if ever, raise two chicks to fledging, depending on the species [41]. In a significant departure from the other penguin species, and perhaps still one more means to reduce intraspecific competition for food, members of the pair rarely forage at the same time, depending on how they divide nesting duties. For instance, male macaroni penguins only feed chicks in the later portion of the chick period [6].

Tying Land Life to the At-Sea Life of Fish-Birds: Foraging and Breeding Success Vary with Prey Availability

Virtually all marine vertebrate species reproduce by laying eggs, but since fish-birds are actually birds, rather than shedding their eggs into the ocean, penguins lay eggs on land so that they remain dry and warm in order to hatch. This, and molting, is their main reason for coming to land. In accordance with the range of breeding strategies, body and bill size, colony size, prevalence of marine predators, and distance to a productive prey habitat, the foraging behavior of breeding penguins also varies, i.e., their traveling and diving behavior, which are the subjects of much of the remainder of this book. This variation has consequences for breeding success. While countless studies provide details of variation in breeding success among various penguin species, few researchers have been blessed by having access to information on the actual preyscape (prey abundance, its depth and patchiness, etc.). In those, we can perceive the degree to which penguins alter their at-sea behavior in compensation or not. Some of this information is discussed in greater detail further on, but as long as we are considering penguins coming ashore, a tie between that aspect of their lives and their at-sea lives will supply some perspective or context – a prelude to the chapters that follow.

What we are mostly considering here is the efficiency with which penguins go and come from the sea, generally involving mates or parents returning at a rate that is timely enough to relieve their partners (which are fasting) or provide food for their chicks, and how much/what type of food they provide. For penguins, as the following chapters will elucidate, it is all about maintaining their high-energy lifestyle and foraging efficiency. That is what impacts their eventual breeding success. Efficiency can be affected by factors that alter (1) the time and distance (including depth) to adequate prey patches, (2) the abundance of prey, and (3) the quality of prey. Affecting their provisioning efficiency also involves penguins' capability of compensating for changes in these factors, which they can do well within limits, but essentially that is the subject of Chaps. 6, 7, 8, 9, 10 and 11.

How provisioning efficiency is affected by prey availability, i.e., the abundance of prey within the foraging "range," is best demonstrated by the "experimental" alteration of the marine protected area (MPA) boundaries to push forage fish extraction by the purse seine fleet farther away from African penguin colonies [26, 27]. With expansion of the MPA, thus limiting overexploitation by purse seiners, the penguins reduced their foraging range by 25–30%, from a 70-to-50-km maximum distance. As they were catching the same, high-energy-dense prey (also targeted by fisheries) regardless of MPA size, neither diet nor foraging depth changed. Adult mass and chick growth rates were positively related to MPA size (i.e., more prey closer to the colony led to fatter penguins). In a "natural experiment" that emphasizes the importance of prey quality, once competing emperor penguins and Weddell seals were chased away by killer whales, Adélie penguins, whereas not changing their foraging range nor depth, brought back more high-energy fish rather than krill. The result was that their chicks grew faster and reached heavier fledging mass, leading to higher probability of post-fledging survival [2, 30]. In another "natural experiment," in years of changed oceanic conditions, the availability of little penguins' preferred high-energy-dense prey was reduced, leading to greater foraging effort and lower reproductive success and chick growth [9, 10]. In the Antarctic, in a series of natural experiments, in years when increased sea ice cover forced penguins to walk rather than swim, thus affecting the time needed for foraging and commuting, Adélie and emperor penguins exhibited decreased breeding success [5, 17, 18]. In years when acoustically measured krill biomass was lower, chinstrap penguins had to make longer foraging trips, and dive deeper, with the increased effort resulting in lower breeding success [31]. There are many more examples cited in subsequent chapters, especially in Chaps. 5 and 13, though, in most cases, only proxies for prey availability were used, e.g., phytoplankton concentration (more on that to follow).

Molt: A Necessary, Brief Respite from the Sea

The evolutionary change of the petrel-like avian species into penguins required some astounding physical, physiological, and behavioral changes, as reviewed in Chap. 1. Among these changes were the development of special feathers (Chap. 7) as well as the methods by which these feathers are maintained (Chap. 14) and subsequently replaced. Penguins are in trouble unless their feather "coat" is in perfect working order (Chap. 13). Penguins are among the very few avian species that replace their feathers all at one time (except for face feathers among a select few subadults – Chap. 14). They have to do this because they need the magical properties of their special feathers, as they cannot deal with the ocean with old, worn ones.

Penguins molt after fledging their chicks or otherwise ceasing colony activities. This can be confusing in some *Spheniscus* species (African, Galápagos) that in some years breed more than once (depending on a favorable food web), and so may appear to molt before breeding (D. Boersma, pers. comm.). In all penguin species, after finishing with the breeding season, individuals intensively forage to recover from raising chicks, guarding territories or other nesting duties, i.e., a stint of "hyperphagia." Then, in most cases, they return to the colony area or vicinity to molt, during which time they do not enter the water and thus live off fat reserves. In addition, their new feathers are derived from stored tissue resources. The molting process takes about 3 weeks to accomplish in smaller penguins, e.g., *Pygoscelis*, and 4 weeks for larger ones, e.g., *Aptenodytes* [37, 24, 12, 19]. Thus, they fast and their metabolic rate is elevated for the process of growing a full coat of new feathers. They lose about 45% of their premolt mass during this process [11, 12]. Although it does not appear to be the case for the polar and subpolar species, which are blessed with the elevated food resources that come with cooler oceans, among the warm-water species (coping with the vagaries of the Subtropical Front), such as Galápagos, yellow-eyed, and little penguins, molt can be taxing enough that mortality can occur among unprepared individuals [7, 15, 40].

During molting, which most of the time occurs in groups, it is not like the penguins position themselves in the tight spacing that they exhibit when nesting, or gather en masse on the beach or at sea, but rather they space out so that they are not touching one another. Their skin becomes impregnated with a dense array of blood vessels, becoming highly sensitive, and the old feather cost begins to look "puffy" (Fig. 2.12). If they are not the forest- or tussock-nesting species, which easily find shelter, then they find somewhere that offers protection from the wind and other elements. Their thermoregulation is not normal either. Their existing coat of feathers is pushed out by the feathers growing beneath. They become highly

Fig. 2.12 Top to bottom, molting individuals of Adélie, gentoo, and king penguins, initially exhibiting puffing out as new feathers begin to grow beneath old ones, followed by the shedding of extensive feather patches. (Photos courtesy of Jean Pennycook)

Fig. 2.13 Chinstrap penguins molting at Baily Head, Deception Island. Their spacing can be observed so that not even their flippers have a chance of touching. That is not snow upon which the penguins are standing but rather mats of white feathers! Thus, the molt was pretty far along when this photograph was taken (March). (Photo courtesy of Jean Pennycook)

ragged-looking, as a pile of old feathers accumulates around where each one of them stands (Fig. 2.13). The wind blows feathers everywhere. Generally, the yearlings molt first, toward the end of the period when successful adults are still tending to their chicks. Then, the unsuccessful adults molt, finally followed by the successful ones after fledging their chick(s) and undergoing hyperphagia.

The exception to molting on land, generally in the vicinity of where they nest, is exhibited by the two highest-latitude penguin species, the Adélie and emperor. All emperor penguins, owing to their clumsiness when climbing from the ocean and up the slopes, molt while riding on ice floes or even better on fast ice. They have to choose large floes that they know will not break during the 4-week period while they lose their old feathers and grow new ones. Such floes are often multiyear, which is harder and thicker ice, and tend to accumulate in several areas around the Antarctic coast. For instance, along the coast of the western Weddell (east coast of the Antarctic Peninsula), western Amundsen Sea/eastern Ross Sea (Marie Byrd Land), Bellingshausen Sea, and coastal waters of eastern Adélie Land. Thus, technically, emperor penguins are "at sea" for their entire lives, whether or not the sea is frozen. In the case of Adélie penguins, especially the highest-latitude breeding populations, they move north after nesting to forage in prey hotspots (see Chap. 3) and then ride on ice floes for a few weeks while molting. The lowest-latitude Adélie penguins, mostly those of the western Antarctic Peninsula and Scotia Sea but also in some cases those of East Antarctica, in many cases, actually return to their colonies to molt [23, 33, 38, 13]. This was the case on Anvers Island back when there was still plenty of sea ice in the surrounding waters, which apparently made the penguins feel more "at home." However, in recent decades, as sea ice has disappeared along the western coast of the northern Antarctic Peninsula (Chap. 13), these penguins now travel south to where there is sea ice in order to molt and also experience some hours of light each day for the premolt hyperphagia [38, 42].

References

1. Ainley DG (2002) The Adélie penguin: bellwether of climate change. Columbia University Press, New York
2. Ainley DG, Dugger KM, La Mesa M, Ballard G, Barton KJ, Jennings S, Karl BJ, Lescroël A, Lyver POB, Schmidt A, Wilson P (2018) Post-fledging survival of Adélie penguins at multiple colonies: chicks raised on fish do well. Mar Ecol Prog Ser 601:239–251
3. Ashmole NP (1963) The regulation of numbers of tropical oceanic birds. Ibis 103(3):458–473
4. Aubin T, Jouventin P, Hildebrand C (2000) Penguins use the two-voice system to recognize each other. Proc R Soc Lond B 267:1081–1087
5. Barbraud C, Delord K, Weimerskirch H (2015) Extreme ecological response of a seabird community to unprecedented sea ice cover. R Soc Open Sci 2:140456
6. Barlow KE, Croxall JP (2002) Provisioning behaviour of macaroni penguins *Eudyptes chrysolophus*. Ibis 144(2):248–258
7. Boersma P (1978) Breeding patterns of Galapagos penguins as an indicator of oceanographic conditions. Science 200(4349):1481–1483
8. Bost C-A, Delord K, Barbraud C, Cherel Y, Pütz K, Cotté C, Péron C, Weimerskirch H (2013) King penguin (*Aptenodytes patagonicus*). In: Borboroglu PG, Boersma PD (eds) Penguins: natural history and conservation. University of Washington Press, Seattle, pp 7–21
9. Cannell B, Thomson P, Schoepf V, Pattiaratchi C, Fraser M (2019) Impacts of marine heatwaves. In: Techera E, Winter G (eds) Marine extremes: ocean safety, marine health and the blue economy. Routledge, New York. Chapter 10
10. Cannell BL, Chambers LE, Wooller RD, Bradley JS (2012) Poorer breeding by little penguins near Perth, Western Australia is correlated with above average sea surface temperatures and a stronger Leeuwin Current. Mar Freshw Res 63(10):914–925
11. Chappell MA, Janes DN, Shoemaker VH, Bucher TL, Maloney SK (1993) Reproductive effort in Adélie penguins. Behav Ecol Sociobiol 33:173–182
12. Cherel Y, Charrassin JB, Challet E (1994) Energy and protein requirements for molt in the king penguin *Aptenodytes patagonicus*. Am J Physiol 266:R1182–R1188
13. Clarke J, Kerry K, Fowler C, Lawless R, Eberhard S, Murphy R (2003) Post-fledging and winter migration of Adélie penguins *Pygoscelis adeliae* in the Mawson region of East Antarctica. Mar Ecol Prog Ser 248:267–278
14. Croxall JP, Davis LS (1999) Penguins: paradoxes and patterns. Mar Ornithol 27:1–12
15. Dann P, Cullen JM, Thoday R, Jessop R (1992) Movements and patterns of mortality at sea of Little Penguins *Eudyptula minor* from Phillip Island, Victoria. Emu 91:278–286
16. Davis LS, Renner M (2003) Penguins. Yale University Press, New Haven
17. Dugger KM, Ballard G, Ainley DG, Lyver POB, Schine C (2014) Adélie penguins coping with environmental change: results from a natural experiment at the edge of their breeding range. Front Ecol Evol 2:68
18. Emmerson L, Southwell C (2008) Sea ice cover and its influence on Adélie penguin reproductive performance. Ecology 89:2096–2102
19. Kooyman GL, Hunke EC, Ackley SF, Van Dam RP, Robertson G (2000) Moult of the emperor penguin: travel, location, and habitat selection. Mar Ecol Prog Ser 204:269–277

20. Kooyman GL, Ponganis P (2008) The initial journey of juvenile emperor penguins. Aquat Conserv Mar Freshw Ecosyst 17(S1):S37–S43
21. Labrousse S, Orgeret F, Solow AR, Barbraud C, Bost CA, Sallée J-B, Weimerskirch H, Jenouvrier S (2019) First odyssey beneath the sea ice of juvenile emperor penguins in East Antarctica. Mar Ecol Prog Ser 609:1–16
22. Nelson JS (1984) Fishes of the world. Wiley, New York
23. Parmelee DF, Fraser WR, Neilson DR (1977) Birds of the Palmer Station area. Antarct J U S 12:14–21
24. Penney RL (1967) Molt in the Adélie penguin. Auk 84:61–71
25. Pettingill ER (1960) Penguin summer. Clarkson N. Potter Inc, New York
26. Pichegru L, Grémillet D, Crawford RJM, Ryan PG (2010) Marine no-take zone rapidly benefit threatened penguin. Biology Letters 6(4):498–501
27. Pichegru L, Ryan PG, Van Eeden R, Reid T, Grémillet D, Wanless R (2012) Industrial fishing, no-take zones and endangered penguins. Biol Conserv 156:117–125
28. Prévost J (1961) Ecologie des manchots empereur. Actualities Scientifique et Industrielles 1270
29. Richdale LE (1951) Sexual behavior in penguins. University of Kansas Press, Lawrence
30. Saenz BT, Ainley DG, Daly KL, Ballard G, Conlisk E, Elrod ML, Kim SL (2020) Drivers of concentrated predation in an Antarctic marginal-ice-zone food web. Sci Rep 10:7282
31. Salmerón N (2022) Warming precludes lower biomass of Antarctic krill in the South Shetland Islands: consequences for foraging and breeding success of a chinstrap penguin population. MSc Thesis. Ghent: University of Ghent
32. Simpson GG (1946) Fossil penguins. American Museum of Natural History, New York
33. Sladen WJL (1958) The pygoscelid penguins. I. Methods of study. II. The Adélie penguin *Pygoscelis adeliae* (Hombron and Jacquinot). London
34. Stahel C, Gales R (1987) Little penguin. New South Wales University Press, Kensington
35. Stonehouse B (1968) Penguins. Golden Press, New York
36. Storer RW (1952) A comparison of variation, behavior and evolution in the seabird genera Uria and Cepphus. Univ Calif Publ Zool 52:121–222
37. Taylor RH (1962) The Adélie penguin *Pygoscelis adeliae* at Cape Royds. Ibis 104:176–204
38. Trivelpiece WZ, Fraser WR (1996) The breeding biology and distribution of Adélie penguins: adaptations to environmental variability. In: Ross RM, Hofmann EE, Quetin LB (eds) Foundations for ecological research west of the Antarctic Peninsula. American Geophysical Union, Washington, DC, pp 273–285
39. Uhen MD (2007) Evolution of marine mammals: back to the sea after 300 million years. Anat Rec 290(6):514–522
40. Van Heezik Y, Davis L (1990) Effects of food variability on growth rates, fledging sizes and reproductive success in the yellow-eyed penguin *Megadyptes antipodes*. Ibis 132:354–365
41. Warham J (1975) The crested penguins. In: Stonehouse B (ed) The biology of penguins. Macmillan, London
42. Warwick-Evans V, Downie R, Santos M, Trathan PN (2019) Habitat preferences of Adélie *Pygoscelis adeliae* and Chinstrap Penguins *Pygoscelis antarctica* during pre-moult in the Weddell Sea (Southern Ocean). Polar Biol 42:703–714
43. Wienecke B, Kooyman G, Le Maho Y (2013) Emperor penguin (*Aptenodytes forsteri*). In: Borboroglu PG, Boersma PD (eds) Penguins: natural history and conservation. University of Washington Press, Seattle, pp 23–34
44. Williams TD (1995) The penguins, Sphenisidae. Oxford University Press, Oxford
45. Williams TD (2018) Avian reproduction – overview (wild birds). In: Encyclopedia of reproduction. Elsevier, New York, pp 595–601

Part II
Penguin Marine Haunts and Food Habits

Chapter 3
Fish-Birds at Home in Their Oceanic Habitats

"Penguin distribution, like that of other seabirds, is not random but is closely related to cold water currents and food availability. The highest density of penguins occurs near convergences and upwellings. Convergences are areas of ocean where different water masses meet, thereby creating extensive turbulence, and upwelling are areas where cold ocean currents that are rich in nutrients rise and replace surface waters. These water characteristics result in areas of high marine productivity that support extensive food webs, high levels of nutrients result in increased plankton abundance, which in turn supports krill, fish and squid – the diet of penguins." (Colin Stahel and Rosemary Gales [191]; p. 14)

Oceanographic Fronts and Water Masses Are Important to Penguins: General Discussion

Back in the day when seabird biologists participated in numerous oceanographic cruises, i.e., in the 1960–1990s (a history is summarized in Ainley et al. [14]), they found that water mass boundaries or fronts were often seabird "hotspots." They also found that seabird density and species composition correlated with ocean productivity, at least on the broad scale, beginning with the classic review by R.C. Murphy [144], who addressed the oceanography around South America; this subject was subsequently further quantified by others [158] (see below). It was also found that, due to increased productivity and prey availability, underwater diving in pursuit of active prey among birds, for a number of reasons, is confined to temperate and polar waters, regardless of the hemisphere (auks in the north, penguins and diving petrels in the south [4, 6]; see Figs. 3.1 and 3.2). Therefore, ocean climate significantly affects where certain species occur, particularly penguins with their high-energy lifestyle.

These days, with research vessels becoming fewer owing to the increasing costs for their maintenance and operation, oceanographers have turned to robots, i.e., buoyancy gliders, satellite-linked buoys, and sensors on satellites, to reveal and monitor ocean climate and structure. At the same time, with another development of

D. G. Ainley, R. P. Wilson, *The Aquatic World of Penguins*, Fascinating Life Sciences, https://doi.org/10.1007/978-3-031-33990-5_3

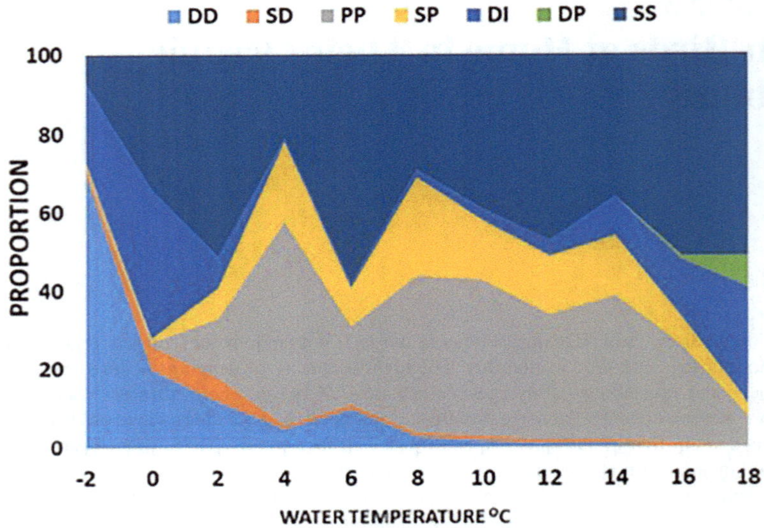

Fig. 3.1 Change in the proportion of the seabird species' main feeding method exhibited as a function of water temperature (and thus approximate productivity), from the subtropical equator to polar waters during at-sea surveys on six cruises from just below the equator to Antarctica in the Pacific region. (Data from Ainley and Boekelheide [6]). DD, deep pursuit diving (penguins); SD, shallow pursuit diving (diving petrels); PP, pursuit plunge (shearwaters); SP, shallow plunge (snow petrels, terns); DI, dipping (gulls); DP, deep plunge (gannets); SS, surface seize (albatross)

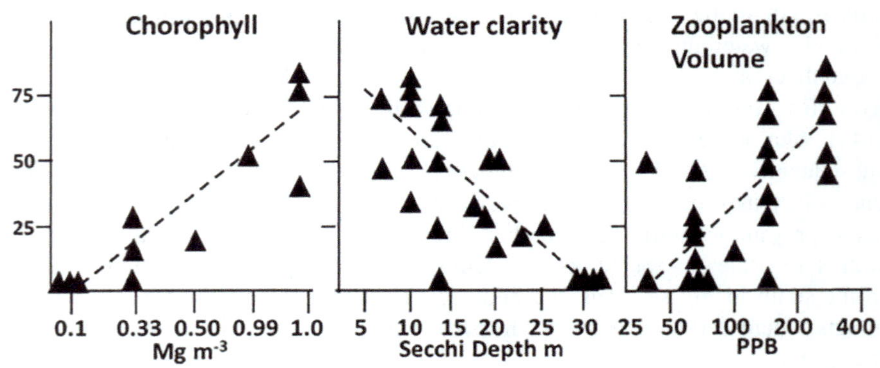

Fig. 3.2 Proportion of species that forage by pursuit diving in 11–23 multispecies breeding avifaunas relative to a few ocean properties (depending on data availability), with data points ranging from the Arctic to the Antarctic in the eastern Pacific. (Data from Ainley [4])

technology, seabird researchers have turned to applying microelectronic sensors to the backs or heads of birds (by suturing, gluing, or using tape). In addition to tracking the movements of the seabirds, the electronics in these biologgers are miniature enough to also allow quantification of their foraging behavior (especially, 3D movement, depth, and behavior over time) as well as some ocean properties (like temperature and light), a subject to be covered thoroughly elsewhere in this book (see Chaps. 5, 6, 7, 8, 9, 10 and 11). With regard to where penguins forage, the biologging data have confirmed what previous at-sea surveys had revealed: the association of seabirds and penguins (and other species) with ocean fronts and the change in species composition as one moves from one water mass to the next (Fig. 3.1).

Ocean fronts attract seabirds due to increased prey availability associated with the mixing of water types, convergences, small-scale eddies, etc. as well as species bumping against, so to speak, the boundaries of their preferred habitat. That is, fronts separate water masses, with ocean properties changing greatly (and sometimes extremely rapidly) across them. Fronts are particularly rich in nutrients and plankton, mixed with surface layers, as investigated in the southern subtropical to Antarctic waters [69, 73, 74, 82, 96]. An increased change in seabird species composition, as well as increased abundance, has been observed at the Subantarctic Front (SAF), the Antarctic Polar Front (APF), and in waters south of the Southern Boundary of the Antarctic Circumpolar Current (SBACC) [6, 38, 81, 106, 170]. The waters south of the SBACC are the most productive of the Southern Ocean, given, among other things, the stretch of ocean from which more than a million whales were extracted during the 1930s–60s [146, 207]. In the well-studied Benguela Current off the west coast of southern Africa, increased seabird abundance has been linked to sea surface temperature and primary productivity [1, 93]; this was also observed and investigated in the Peru/Humboldt Current off Chile [181] and in the Scotia Sea [12, 107, 144]. Investigations of spatial variation in fronts on a smaller scale, particularly with respect to penguins, have been made around Tasmania and southern New Zealand (NZ) [3, 79, 129, 157].

Penguins Require High-Productivity Water Masses

The Antarctic Circumpolar Current (ACC) flows eastward around Antarctica, driven by the strong west winds characteristic of southern polar latitudes, especially from about 65° S southward [69, 70, 110, 149] (Fig. 3.3). The ACC is composed of a series of mostly parallel circumpolar "belts." The southernmost, along the Antarctic coast, is the westward-flowing East Wind Drift Current or coastal current, which is actually a segment of a series of gyres, namely, the Ross, Weddell, and Kerguelen Gyres, whose "bases" are along the continental shelf break, flowing west, but whose "tops," in accordance with their clockwise circulation, are part of the Southern Boundary of the Antarctic Circumpolar Current (SBACC), flowing east. Farther south in large "bays," such as Prydz Bay and the Ross Sea, compensatory, counter circulation is apparent. These waters are cold and highly saline due to the freezing of surface water into sea ice close to the continent, in latent heat polynyas (to freeze seawater, salt needs to be rejected to "purify" it– sea ice is less salty than the water it floats upon). The cold, saline water sinks to become the Antarctic bottom water,

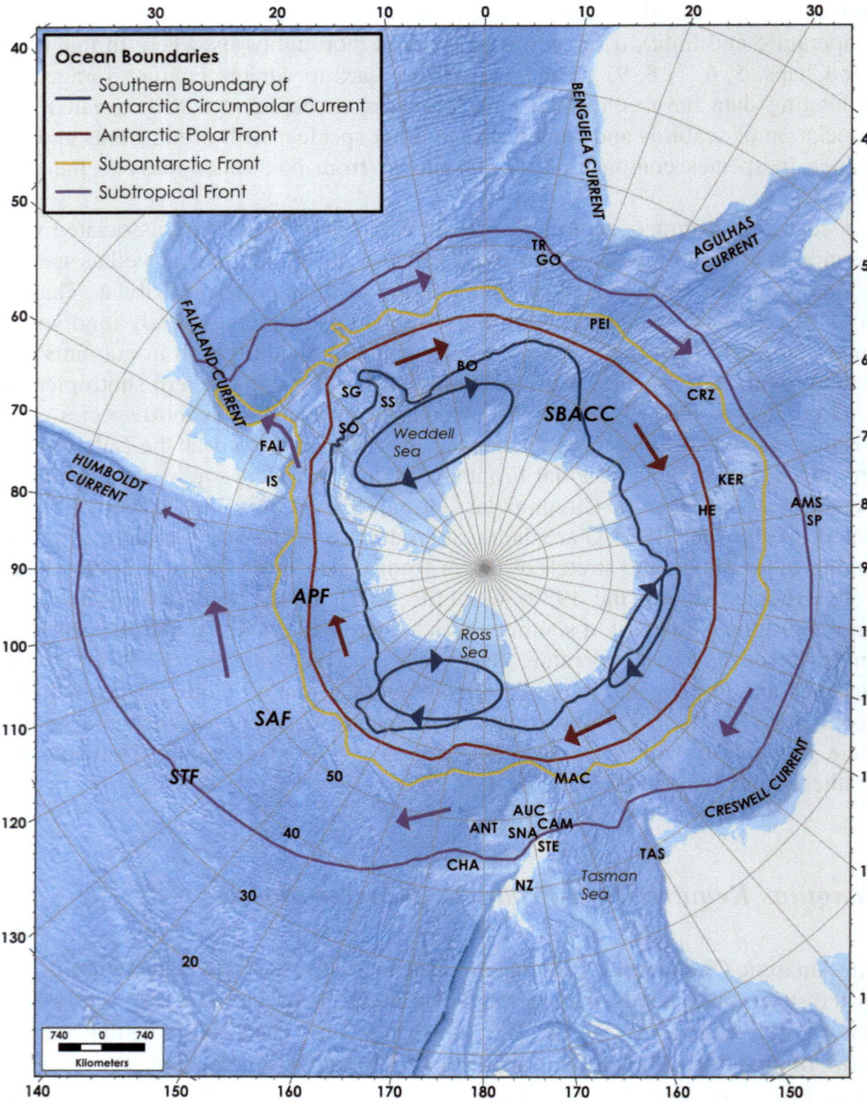

Fig. 3.3 The oceanic islands upon which penguins nest, showing their proximity to the major oceanographic boundaries and current systems. *STF* Subtropical Front, *SAF* Subantarctic Front, *APF* Antarctic Polar Front, *SBACC* Southern Boundary of Antarctic Circumpolar Current; the Ross, Weddell, and Kerguelen Gyres also shown within the southern boundary waters. The islands are *FAL* Falkland Islands/Islas Malvinas, *IS* Isla do los Estados, *SG* South Georgia, *SO* South Orkneys, *SS* South Sandwich, *BO* Bouvetøya, *PEI* Prince Edward/Marion, *CRZ* Crozet, *KER* Kerguelen, *HE* Heard, *AMS* Amsterdam, *SP* St. Paul, *TAS* Tasmania, *MAC* Macquarie, *CAM* Campbell, *SNA* Snares, *AUC* Auckland, *ANT* Antipodes, *STE* Stewart, *CHA* Chatham; and *NZ* New Zealand

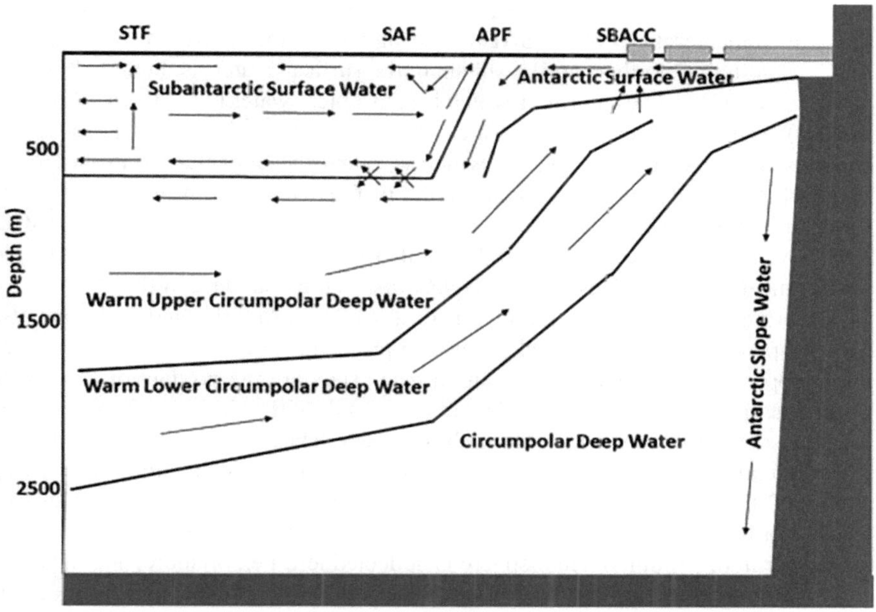

Fig. 3.4 A schematic of the array of water types and fronts along a hypothetical section in the Pacific sector of the Southern Ocean (adapted from Anderson [17] and ultimately from the American Geographical Society); acronyms for the fronts are similar to those used in Fig. 3.3. Antarctic slope water is also known as Antarctic bottom water, which is extremely cold and highly saline

which flows along the bottom, reaching circulation systems even into the Northern Hemisphere (see Fig. 3.4).

Some distance north of the SBACC is the Antarctic Polar Front (APF), which has been recently designated by the Geographical Society to be the northern boundary of the Southern Ocean, now the seventh official ocean. The SBACC is the only one of the fronts that does not represent a change in the characteristics of surface water, in that the Antarctic surface water extends from the Antarctic continent out to the APF. What the SBACC does represent is the location where water less than the freezing temperature of sea water (< −1.7 °C) reaches the surface, having traveled subsurface under the subantarctic and Antarctic surface water (this is called the upper Circumpolar Deep Water (CDW); there is a lower CDW, which continues to flow south to the Antarctic shelf break). Water north of the SBACC is warmer. Thus, only occasionally, depending on the strength and persistence of the wind, does sea ice extend north of the SBACC (but not very far; see below, Fig. 3.4).

The APF is the northern boundary of Antarctic surface water (<3.9 °C) and is actually a convergence zone in which the 2 °C isotherm of the Antarctic surface water sinks below the warmer overlying ocean (to reach a ~ 200-m depth). On the north side of the APF is subantarctic surface water (4° to ~13–14 °C), within which lies the Subantarctic Front (SAF). The latter is the northern edge of the ACC. Continuing farther north, subantarctic surface water is bounded by the Subtropical Front (STF), north of which is subtropical water (13–21.9 °C; Figs. 3.3 and 3.4). The subantarctic surface water mass, between the APF and STF, as noted, contains two weaker fronts, one of which is called the Subantarctic Front (SAF). The SAF represents the sinking and mixing of fresher surface water beneath salty water to its north. This is another convergence. These four major fronts– SBACC, APF, SAF and STF– occur south of the southern continents, including Australia, as well as mostly south of New Zealand (NZ brushes the boundary of the STF).

The currents along the southern west coasts of South America and Africa are also cool-temperate but are not actual northward projections of the ACC water, as they might appear on some maps of ocean climate. They are the product of cold, deep water that upwells, to then trend offshore, as a result of the Coriolis effect driven by the persistent northward winds along these coasts. These winds are the product of what are called Hadley cells, which represent air rising at the equator, then sinking at about 30 °S. In accordance, the counter winds in these cells blow these surface waters north as the respective Humboldt and Benguela Currents. The moving surface water in the Southern Hemisphere is deflected left, i.e., offshore, and the deep, cold, nutrient-rich water flows upward to replace it, close to the coasts. Without the upwelling, waters off these coasts would be subtropical, which may be the case during the non-upwelling season. These areas are known as eastern boundary currents (the eastern boundaries of the major oceanic gyres of the South Pacific and South Atlantic) and are among the most productive stretches of water on the planet, contributing a huge portion to total human fishery consumption [88] much to the expense of penguins (see later in this book). There are three eastern boundary currents in the Northern Hemisphere (North Pacific and Atlantic and the Indian Ocean) but no penguins dwell within them (though flightless seabirds once did!). The Humboldt or Peru Current, upon reaching northern Peru, curves westward, thus becoming the seasonally present Cromwell Current, bathing the Galápagos Islands at the equator with cold waters. The Cromwell Current is enhanced, and cooled, by upwelling along the west coasts of the westernmost islands of that archipelago.

On the east coast of South America, somewhat contrary to the flow of the South Atlantic (ocean scale) Gyre farther offshore in the central basin, the Falkland Current is a countercurrent, sweeping northward around Cape Horn, bringing cool subantarctic water over the shelf as far as northern Argentina. It pushes the SAF and STF north along the coast and displaces subtropical waters offshore. Finally, in completing the ocean world of penguins, along the southern coast of Australia, flowing east to west over the narrow and relatively deep shelf, including around Tasmania, the cool Creswell Current occurs just inshore of the warm, subtropical Leeuwin Current, flowing west to east along the shelf break of Australia. In a sense, the STF is the inshore boundary between the Leeuwin and Creswell Currents. These currents, and their connections to penguins, will be discussed in more detail below.

Seasonal At-Sea Movement of Penguins: Is It Migration or Dispersion?

Later in this chapter, we will review the at-sea ranges of penguins, which include their occurrence through their entire annual cycle. In a nutshell, for most penguins, these ranges are largely "mere" extensions or expansions of the waters they occupy when they are exercising their central place foraging (CPF) mode during the breeding season. In this context, we will consider their ranges within the types of seasonal movements that have been identified among seabirds as follows [122]:

Residency: The population is sedentary, remaining within the same area and water type the entire year.

Dispersion: The population spreads or fans out following the breeding season, with the area spread depending on the species but remaining in the same water type or mass.

Partial migration: A fraction of the population is sedentary, whereas another fraction moves, e.g., young, nonbreeders remain in the nonbreeding area, whereas the adults move to a colony area for nesting; if a specific portion of the population is involved, i.e., sorted by age or sex, then this is termed "differential" movement [195];

Migration: The entire population completely deserts their breeding grounds and surrounding waters, with movement to a nonbreeding habitat on a precise schedule, passing over habitats with which it normally would not associate.

Critically, the movement of an animal population from one site to another in a true migration results from clear advantages in occupying two (or more) productive regions that are geographically separated while it is disadvantageous being in the space between regions. For example, the areas/regions may have (season-linked) enhanced feeding conditions but no food between them. The distance between such sites, the speed at which animals can travel between them (which relates to the time spent traveling), as well as the energy they expend doing so, in tandem with the extent of their own body reserves to underpin the between-site movement, will all be relevant in determining the extent of migrations that can be undertaken. In short, slow-moving large whales, for example, can undertake global-scale migrations because they have massive body reserves that can be used during the long periods of fasting needed for traveling between calving and foraging areas. For instance, the humpback whales of the Southern Hemisphere travel back and forth between seven subtropical calving regions (A–G) and foraging areas to the south along the Antarctic shelf break, taking weeks to travel between both [40]. On the other hand, fast-moving aerial seabirds can also globe trot but much faster; they can do the humpback trip in a day or two. So, where in the movement continuum (above) do penguins sit?

For any penguin species, seasonal movement is quite restricted compared to a number of aerial seabird species, many of which actually "migrate" using their extraordinary powers of flight during fast, long-range movements (discussed by others [86]). For instance, mottled petrels' annual movement cycle goes like this: they nest in New Zealand, and during that time, their confinement to central place foraging can take them as far as the edge of the Southern Ocean sea ice several hundred kilometers away (SBACC; Figs. 3.3 and 3.4; like the humpbacks but carried out in a few days); released from nesting, they fly north to the Bering Sea, at a distance of ~15,000 km, where they spend the Austral winter, and, then, they return south to NZ the following early (Austral) spring to nest again [9, 10]. The same is true for short-tailed and sooty shearwaters, which nest in Tasmania or New Zealand and, in the nonbreeding part of their cycle, frequent the California Current, Gulf of Alaska, and Bering Sea in the Northern Hemisphere [188, 190]. A similar pattern is true for the sooty shearwaters that nest in Patagonia and spend their nonbreeding season in the Peru/Humboldt and California Currents [188, 189] or the North Atlantic Ocean [147]. In addition, there are several volant seabird species (petrels, albatross) that nest in New Zealand but spend their nonbreeding periods on the other side of the South Pacific Ocean, in the Peru/Humboldt Currents [186]. In all these cases, these species are mobile enough to make use of the extremely productive stretches of the ocean elsewhere when food availability diminishes or conditions become stormy in waters within their breeding season CPF ranges. In large measure, they use the energy that is contained in the wind to propel them through the required long distances [152, 187, 188]. Being able to use wind for energy, they can quickly cross great expanses of the "ocean desert," e.g., the central portions of the South and North Pacific and South Atlantic gyres, even parts of the Southern Ocean, taking just a week or a few days without having to stop to forage. Truly, they exhibit classic migratory behavior.

There is no penguin species that can equal the movement of these true seabird migrants, especially given that penguins cannot use wind for energy and neither can they cover ocean distances anywhere nearly as quickly, crossing unproductive waters (for instance, the equator; see also below). Later chapters in this book explain the high-energy lifestyle of penguins, which includes the immense amount of energy that penguins must marshal in order to dive and the high costs of transport ("distance unit per gallon" equivalent) compared to flying birds. Needing to stay in highly productive waters, unlike volant seabirds, they don't possess the physical capabilities needed to trot the globe, so to speak, from cool, productive subantarctic waters, overflying unproductive waters, to reach cool, productive subarctic waters (or the like). The latter, as noted above, is what many subantarctic volant seabirds undertake annually [122].

Flying seabirds of penguin size pass through their medium (air) much faster than do penguins through theirs (water): 7–14 $m.s^{-1}$ in crosswinds (volant seabirds' preferred wind angle), depending on the species [187], compared to 2–3 $m.s^{-1}$ for penguins [41, 171]. The differences in "ranging speed" (flight progression over water) is even more disparate – in seven species of penguin-sized petrels and albatross traveling at a speed ranged 4.9 to 6.9 $m.s^{-1}$ [153], compared to 1.2 $m.s^{-1}$ measured in Magellanic penguins [218]. Even an emperor penguin's fastest speed on ice, which is typically a great expanse to negotiate, is just 0.7 $m.s^{-1}$ by tobogganing (slower by walking [112]), though certainly its ranging speed in directional swimming would be double that. According to C. Pennycuick [152], a macaroni penguin, which has the same wing aspect ratio of a giant petrel and the same shape though much shorter wings, would have to beat its wings like a hummingbird in order to achieve the "ranging speed" of a giant petrel. That is hard to imagine Consider, too, in this comparison, the Southern Ocean is the smallest of the seven oceans, despite seeming so large in certain geographic map projections (see Fig. 3.3). Thus, what might appear to be a long distance in the penguin realm, as displayed on a map, is not really all that long compared to the travels of many volant seabird species.

A number of penguin species are pretty much "resident," exploiting the same productive, coastal waters year round. They visit their nesting areas or even their nests year round, though somewhat less frequently during the nonbreeding season. We have briefly mentioned this in Chap. 2. The penguin species that are in this group comprise Galápagos, Humboldt, yellow-eyed, little, and, to some extent, gentoo penguins. No one can make the case that these penguins migrate. Considering this case, however, Culik and Luna-Jorquera [62, 63]), in their analysis of post-breeding movement, entitled one of their papers, "The Humboldt Penguin *Spheniscus humboldti:* A migratory bird?" Although they concluded that they are not as sedentary as advertised, i.e., as some penguin enthusiasts had claimed, using satellite tags at just their study colony, they found that most penguins remained within 5–90 km (the central place foraging radius of nesting). One, however, went 600 km away, though remaining in the same water type (Peru/Humboldt Current). Their study colony was midway in the species' nesting range along the Chile–Peru coast. Essentially, these penguins are just spreading out, i.e., dispersing, when they don't have to be central place foragers while in the throes of colony duties. By no means would a researcher of land birds or waterfowl define such movement as migration – the distance moved is so short that even in their slow traveling speed these penguins could be back to their nesting areas in a couple of days, remaining in the same water type all the while, and foraging along the way.

The post-central place foraging during nesting among most other penguin species is also "dispersal," though the distances of foraging area expansion are much greater than, for instance, those of the resident/sedentary species. Dispersal reduces the intraspecific competition that is heightened by central place foraging (Chap. 5). Take for example, the other penguins of the *Spheniscus* genus, which broaden their north–south distribution somewhat along the respective continental shelves, though some portions of their total populations still frequent a good portion of the waters exploited during breeding. In the case of African penguins, the range of nonbreeders, which remain at sea, spreads farther but overlaps that of breeders [59, 182]. If food availability remains high, the adults may lay eggs again after fledging their chicks of the usually timed breeding season, i.e., double brooding, and thus they are actually in a "sedentary" mode. In the case of Magellanic penguins, the population shifts from occupying coastal waters stretching from Patagonia to northern Argentina to coastal waters stretching from southern Argentina to central Brazil, though some individuals even remain in the waters off Patagonia [31]. On the basis of long-term banding at a southern Argentinian colony, the distribution of recoveries of juveniles, subadults, and breeding adults almost entirely overlapped spreading along the entire breeding range, except for an extension of adults ~400 km north in southern Brazil during nonbreeding [192]. This distance is just a ~ 12% extension of the species' nesting range along that coast of ~3200 km (see below). Thus, there is almost complete overlap in the breeding and nonbreeding ranges.

Perhaps the best case for actual penguin migration might be that exhibited by the king penguins of the Kerguelen and Heard Islands, where parents leave their chicks in their colony, famished and existing on residual fat, while they forage along the outer edge of the Antarctic pack ice for a few months, though even that is just a few to several hundred kilometers away [37, 139]. It is possible that these breeders found the waters in which the subadults occur, but no at-sea surveys exist in documentation. This then might be a case of "differential migration." Another candidate for migration might be macaroni penguins that nest on South Georgia but which, in the nonbreeding season, disperse west to almost reach the Falkland Islands/Islas Malvinas [164]. In this case, too, the distance is only a few hundred kilometers. Again, though, the distribution of subadults is not known. Macaroni penguins that nest on the Kerguelen Islands disperse eastward about the same distance during the interbreeding period, staying at sea for a few months in a movement that looks like an expansion of the foraging range used during nesting; some researchers [38] have called this movement "dispersion," but others [200] have called it "migration." The point is, though, that these penguins, as do a few other species, vacate their colonies entirely in the nonbreeding season, with the population remaining at sea but spread out seemingly to reduce intraspecific competition, i.e., being more dispersed than when in the central place foraging mode. The macaroni penguins of Heard Island appear to behave in the same way [71]. These examples and others, as well as the competition angle, are discussed in more detail in Chap. 5.

In the case of emperor penguins, there might be what could be considered actual "migration," though this too would be "differential migration" because it appears that only juveniles do it. Upon fledging, independent of adults, they travel to the Antarctic Polar Front, where they spend a couple of months. They then head south to the pack ice where they will spend the next few years among adults, whose presence will ebb and flow depending on the breeding stage [113, 114, 199]. Some authors refer to this juvenal behavior as "dispersal."

The most interesting case is that of the Adélie penguin, perhaps only because biologging of breeding birds at colonies has been conducted far more than any other species, ranging around the Antarctic continent (Chap. 5), and there also have been a number of at-sea surveys of seabird distribution that cover an appreciable portion of the biologging regions. Among all its various populations, Adélie penguins demonstrate a range of seasonal movement patterns. At northern colonies (62–64° S), adult Adélie penguins remain in the general vicinity of their colonies, visiting now and again year round [150]; this was the case back when sea ice was present in winter (Chap. 13). They are thus sedentary or "resident." These days with less winter sea ice, Adélies of this coast of the western Antarctic Peninsula shift slightly southward to remain where there is sea ice, and those of northern-most Antarctic Peninsula shift to the sea ice that is east of the peninsula tip [84, 95]. Nesting and nonbreeding waters overlap or at least abut. These populations now are trending toward "dispersing." The situation on Signy Island, Scotia Sea, where both Adélie and chinstrap penguins nest, is additionally instructive in that Adélies disperse but chinstraps remain in the vicinity [208]; the two species are true to their habitat preferences, i.e., sea ice cover versus ice-free ocean. Off East Antarctica (68° S), fledgling, subadult, and adult Adelie penguins spread out in the same waters extending outward from colony areas, with their travels subsequently being a function, to a great degree, of sea ice movement in the Kerguelen Gyre [52] (Fig. 3.3). These populations should be clearly considered as "dispersing" following the nesting season. Finally, the most southern-nesting populations of the species in the Ross Sea (76–77 °S) for the nonbreeding season shift north to remain where there is sea ice and enough hours of sun to navigate (Chap. 14), i.e., 700 or more kilometers farther north. However, what really happens, best seen by comparing two studies [10, 22], is that after breeding, adults leave their colonies and expand their presence northward to occupy the waters where young nonbreeders (1–3-year-olds), who do not visit the colonies, had remained during the entire summer. All age classes are then combined with the sea ice movement of the Ross Gyre during the nonbreeding period (Fig. 3.3). Looked at in reverse, it appears that penguins in these southernmost populations (colonies) exhibit "partial migration," in which they occupy a pack ice area year round, but a segment (the adults) departs south to nesting colonies, leaving the juveniles and subadults in the species' typical pack ice habitat (Fig. 3.5). This, thus, would be "differential partial" migration.

Comparison of the results of the latter two studies [10, 22] point out the dangers of basing concepts about penguin movements entirely on those of colony occupiers (breeders, nonbreeders) and entirely on biologging (e.g., satellite or even GLS/GPS) devices. Biologging instruments are never attached to individuals that are not present at colonies, or in many cases even to nonbreeders, as the researchers fear of never getting the transmitter back (for downloading the data) since these penguins are minimally attached to a nest site. Why apply the instrument to a bird that one has little chance of seeing again (they cost thousands of dollars each)? However, past at-sea surveys from oceanographic vessels, which these days unfortunately have become extremely rare, have shown where the subadult Adélie penguins hang out in summer. Really, all the adults do is spread out to reoccupy the species' at-sea "range" already defined by where the subadults locate year round (Fig. 3.5).

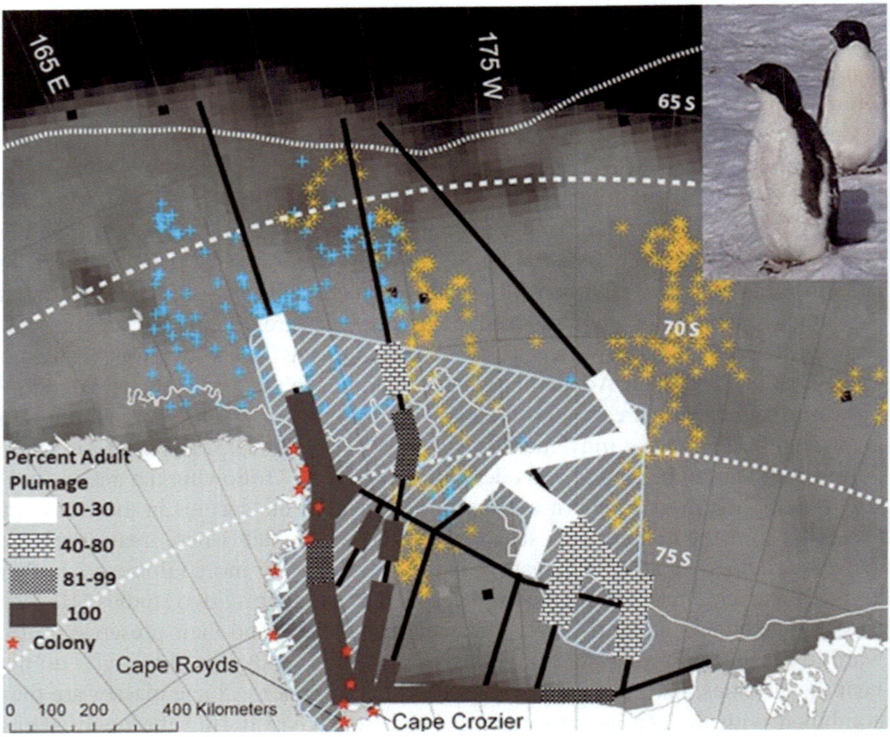

Fig. 3.5 An example of differential migration in penguins – Adélie penguins, nesting at high latitudes, showing Austral winter and summer distributions. The entire population is in the pack ice during the nonbreeding season, until adults to undertake nesting duties leave pre-breeders in the pack ice. This figure shows the Ross Sea region with the positions of adult Adélie penguins, from Cape Royds (blue asterisks) and Cape Crozier (gold asterisks), to which GLS loggers were applied, in August (2003–05), which is the month when large-scale pack ice is typically at its maximum (fuzzy gray extending from the coast out to 65° S [22]). Superimposed are icebreaker cruise tracks (rectangles) in December (1976–79) and the reduced pack ice that typically occurs then (the northern edge having retracted south and open water having developed over the continental shelf [10]). The reduced December ice extent is shown by diagonal barring, as determined from the ice-breaking ship (before the era of satellite imagery). Along those tracks is shown the proportion of adult-plumaged penguins, distinguishable from yearlings by their black throats (insert). These proportions are an index of the prevalence of 1–3-year-olds that have yet to visit any colony [5]. Colonies shown by red stars and extending out from those, the prevalence of adult-plumaged birds is 100%, representing colony-based foraging areas. Note that the number of Ross Sea Adélie penguins roughly doubled between the late 1970s and the early 2000s

Large-Scale Edge: The At-Sea Distributions of Penguins and Oceanographic Boundaries

The Southern Boundary of the Antarctic Circumpolar Current and the Large-Scale Sea–Ice Edge

Perhaps the most famous of ocean boundaries, when it comes to penguins, is the large-scale edge of the Antarctic pack ice. The latter (pack ice) is composed of the spread of ice floes covering the ocean surface and produced from the breakup of the ocean's frozen surface close to land. Sometimes the floes are spread out and sometimes they are packed together, depending on wind direction and other factors. As noted above, when surface sea temperatures reach lower than −1.7 °C, the ocean surface freezes. The ocean freezes near to the Antarctic coast, in a sheet that becomes broken by waves and swells (Fig. 3.6). The floes are advected or blown northward, driven off-continent by gale force winds and the Coriolis effect, especially once the sea ice reaches the region where the circumpolar winds are from the west. Thus, the floes are forced to head northward in their movement east, to be replaced by ice from the south in a continual process, at least during the late autumn to early spring ice formation season (air temperatures are cold, e.g., -35 °C, freezing the sea surface). Without wind, the Antarctic pack ice would not occur as far north as it does

Fig. 3.6 A sheet of frozen ocean being broken into floes by sea swells coming in from the right. (Photo courtesy of D. Ainley)

[119, 205]. Southern Ocean winds have been increasing (except very recently), leading to sea ice extending farther from the coast (Chap. 13). In the Southern Ocean, once air temperatures reach high enough that surface waters no longer freeze, by late summer, most of the pack ice disappears, growing again to reach the maximum extent sometime between July and October, depending on the year and location (Weddell Sea vs. Ross Sea vs. East Antarctica). At maximum, the northern sea ice boundary closely corresponds to the Southern Boundary of the Antarctic Circumpolar Current (Figs. 3.3 and 3.4). The SBACC represents the line at which the upper CDW is mixed with surface water, and surface temperatures are slightly warmer (Fig. 3.4).

Ice edges generally exhibit increased productivity compared to the surrounding ocean due to the encouragement of diatomic (single-cell phytoplankton) growth by a somewhat fresh, meltwater lens associated with the edge [185]. The lens forms a stratum (somewhat akin to an oil slick floating on the sea surface) that prevents, due to wind-generated turbulence, the downward mixing of surface water with deeper water. This process would take the light-dependent diatoms away from light. The diatoms are grazed by penguin prey (zooplankton), and the zooplankton predators (fish) are then preyed upon by penguins (see Chap. 4). A multiyear, multi-season, multi-ship, and multi-investigator effort, in which one of us was involved, researched the Antarctic marginal ice zone (MIZ) of the Scotia–Weddell Confluence, during the 1980s (AMERIEZ: Antarctic Marine Ecosystem Research at the Ice Edge Zone). Quantification of what R.C. Murphy [144] had earlier described qualitatively was the avoidance of pack ice by some penguin species that occur south of the APF (chinstrap, gentoo, macaroni, king) and the association with pack ice by others (emperor, Adélie). In fact, the avifauna associated with the Antarctic pack ice is the least varying assemblage known among seabird communities. This was deduced using several at-sea surveys, looking to quantify seabird community composition, extending from the equator to the Antarctic coast in the eastern Pacific sector [169]. Since then, this transition in the avifauna, including penguins, and related to the presence of sea ice, has been seen in other locations around Antarctica [170].

The shift in penguin species composition along the MIZ occurred despite the fact that when the pack ice edge retreated southward, owing to change in wind direction, this allowed southward movement of the ice-avoiding species (they do roost on icebergs). However, in the newly exposed, sea ice-free waters, they ate many of the same prey as did the ice-loving penguins and other seabird species that were present just a week earlier [13, 83]. Clearly, this demonstrated the importance of physical habitat characteristics, rather than there being entirely a food-type-related factor, to explain penguin species occurrence. This pattern would be akin to the more familiar, to most people, association of different song bird species with grasslands adjacent to forests, with all eating seeds and/or insects. Cut down the forest, and grassland species replace the forest species. This explains why, as sea ice becomes reduced in the northern part of the Antarctic Peninsula due to climate change, gentoo and chinstrap penguins are replacing Adélie penguins [178, 179] (Chap. 13), though they all have closely similar diets (Chap. 4). Historically it explains why Adélie penguin populations in that region have come and gone with long-term fluctuation in the presence of sea ice, for example, fluctuations associated with the Medieval Warm Period (less sea ice) or Little Ice Age (more sea ice) [75–78, 183] (see Chap. 13).

In the more recent era of biologging, fledgling emperor penguins have been tracked from their spring-summer nesting colonies to move at times to near the Polar Front or at least to the outer edge of the large-scale pack ice in portions of the Southern Ocean where the SBACC is relatively close to the continent of East Antarctica [222] (Fig. 3.3). In areas where the SBACC is north of the pack ice edge, fledgling emperor penguins travel to the SBACC before heading south to the MIZ [114, 121]. In areas where the sea ice belt is narrow, Adélie penguins have been tracked to spend a good portion of their winter in the vicinity of the ice edge [95], and as well in the case of those from East Antarctica riding floes in the Kerguelen Gyre [52](Fig. 3.3). Similarly, in winter, king penguins from colonies on islands in the low-latitude Southern Ocean, a long way from sea ice, have been tracked southward to the large-scale ice edge [37, 139]. Where the pack ice extent is extremely broad, as in the Ross Sea sector, Adélie penguins mostly spend their time in the interior of the pack overlying those especially productive waters, riding floes as they circle in the Ross Gyre [22](Fig. 3.3). In years of extremely extensive ice, some penguins may find themselves in the unproductive waters north of the SBACC [217].

In the large-scale perspective, the SBACC demarcates the boundary between the at-sea distribution of king and emperor penguins (Fig. 3.7) and Adélie and chinstrap penguins (Fig. 3.8). All four of these species exhibit minimal genetic variation in their populations despite their circumpolar distributions, other than the unique genomes of emperor and Adélie in the Ross Sea [54, 221]. This indicates high levels of gene flow among the populations despite the spread of islands and coastal colonies, consistent with their homogeneous ocean habitat in zones encircling the continent. Emperor and Adélie penguins, as noted, are confined to the sea ice zone, and it is only when the sea ice has retreated southward in summer that king and chinstrap penguins begin to visit the productive waters south of the SBACC. Hence, though an apparent overlap occurs in the respective at-sea distributions of king with emperor and Adélie with chinstrap, it's a spatial and not temporal overlap. In fact, there is little overlap among these ice-obligate (pagophilic) vs ice-avoiding species when the sea ice extends far from the coast. The SBACC, owing to its loose correspondence with the outer edge of the Antarctic ice pack, is also the southern boundary of the gentoo penguin's range (Fig. 3.9). This penguin species has a range that also extends northward into temperate subantarctic waters, and, so, it appears to be the most widely spread of all penguin species (in terms of latitude). While the spread of the chinstrap penguin colonies looks somewhat similar to that of the gentoo, longitudinally from about 90 °E through 180–120 °W, they are very few and far between. Thus, the SBACC constitutes the sort of species-limiting ocean boundary referred to by some authors [54, 143], though they did not recognize this specific habitat feature in their analyses. North of the SBACC, they did note that there are genetically distinct clusters of gentoo penguins north and south of the APF, and, in fact, some researchers recognize a northern gentoo (*P. p. papua*) (Falkland Islands/

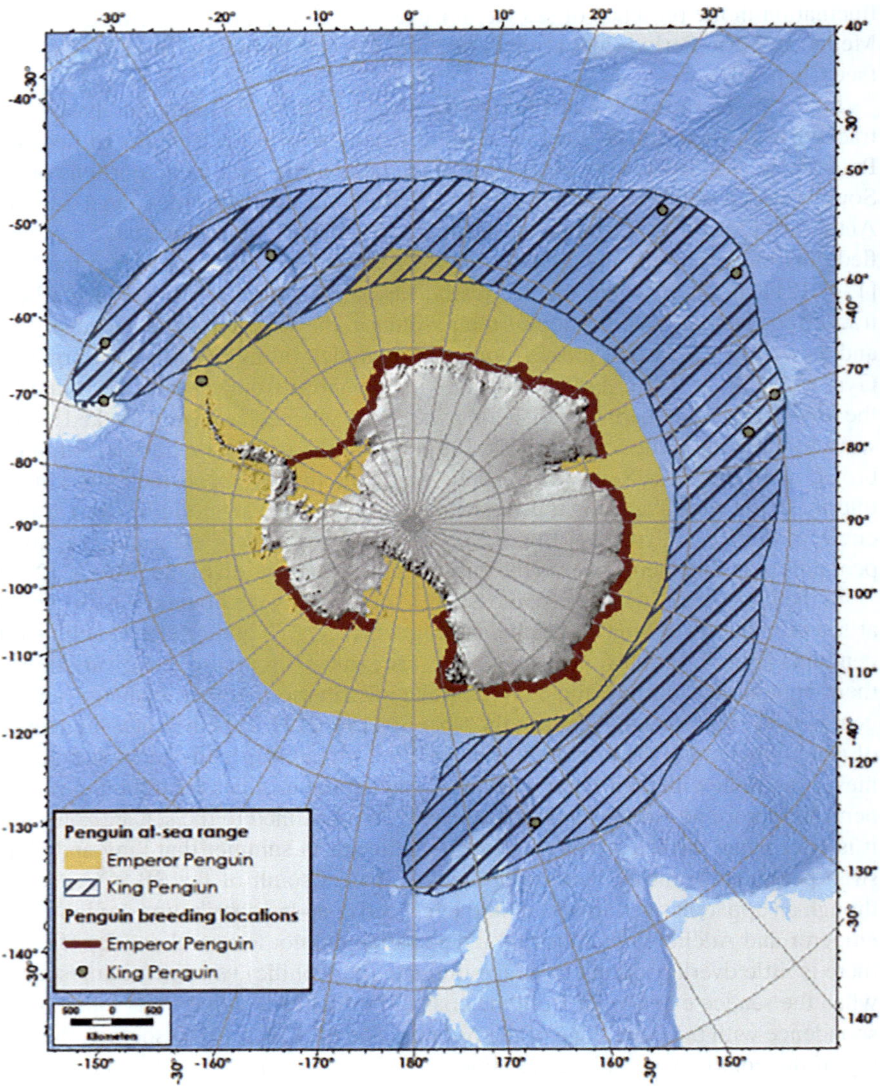

Fig. 3.7 Location of breeding colonies and the respective at-sea ranges of emperor and king penguins, drawn from several sources [36, 85, 172, 216]. King penguins venture south to the large-scale ice edge in winter, which corresponds to the Southern Boundary of the ACC (see Fig. 3.3). At the edge, the king penguins can exploit the highly productive waters that lie south of the SBACC

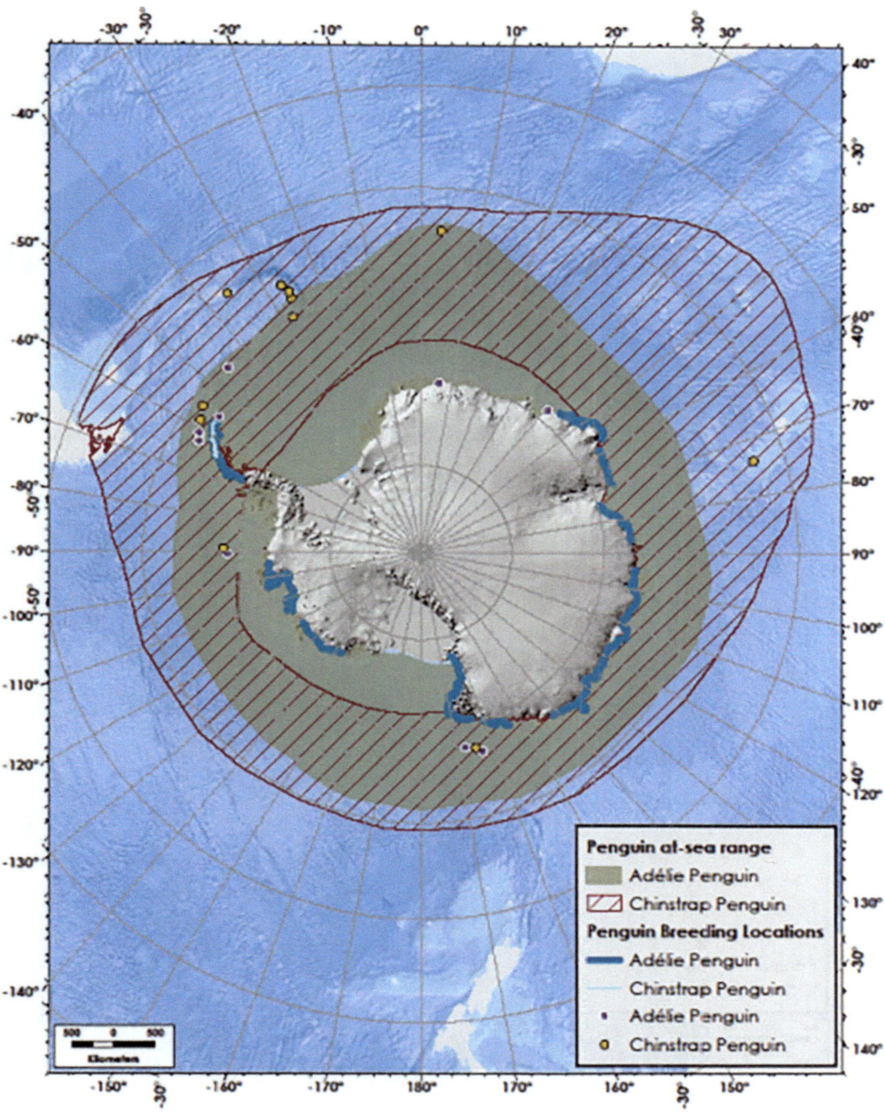

Fig. 3.8 Location of breeding colonies and the respective at-sea ranges of Adélie and chinstrap penguins, drawn from several sources [126, 172, 193, 216]. To a large extent, chinstrap penguins venture south of the SBACC (Fig. 3.3; the northernmost limit of the Adélie penguin range) only when the sea ice has seasonally disappeared and Adélie penguins have retreated south

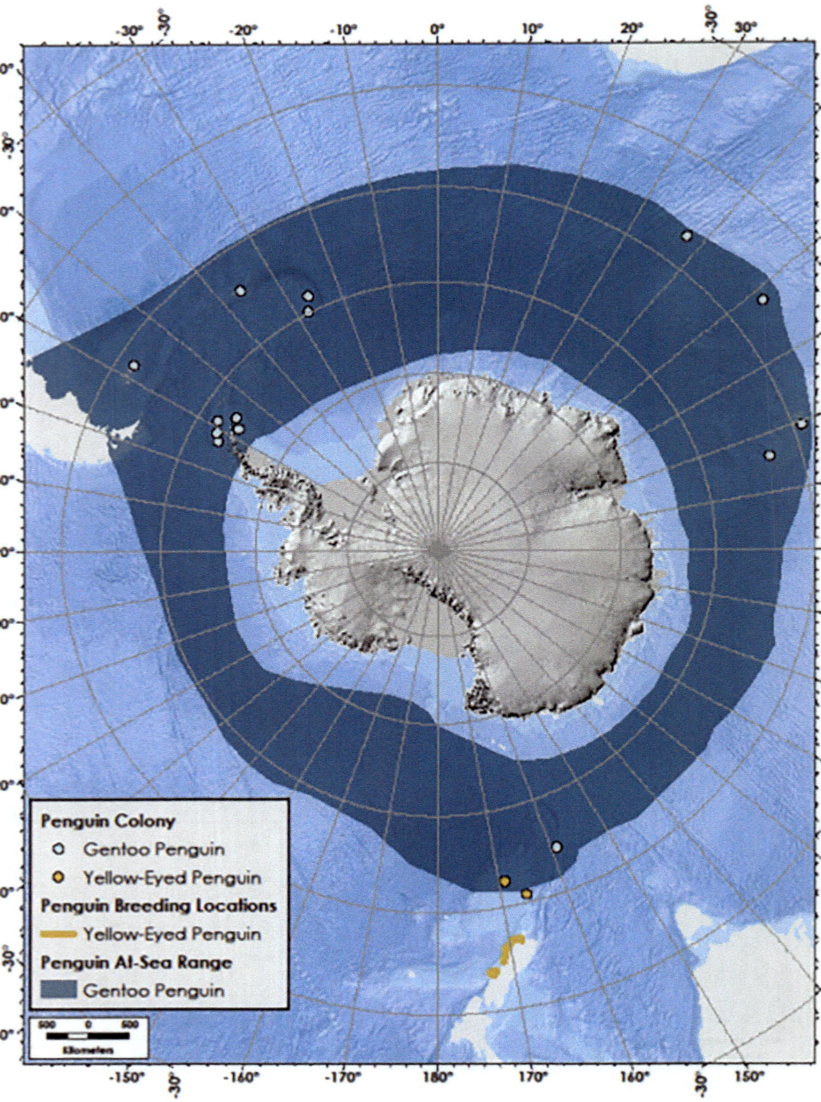

Fig. 3.9 Location of breeding colonies and the respective at-sea ranges of gentoo and yellow-eyed penguins, drawn from several sources [125, 172, 216]. Gentoo penguins, in general, do not venture south of the SBACC, other than along the west coast of the northern Antarctic Peninsula (where sea ice is present only for a short period of the year)

Islas Malvinas, Kerguelen, Heard and Macquarie Islands) and a southern gentoo (*P. p. ellsworthi*) (South Georgia, South Shetlands) [206].

The Antarctic Polar Front

North of the SBACC, the APF also encircles the Southern Ocean and represents the zone in which cold, saline Antarctic surface water, being nudged northward in its eastward journey by the Coriolis effect, sinks beneath the less dense subantarctic surface water (Figs. 3.3 and 3.4). The temperature of the surface waters within the zone, occurring at about 59–62 °S, would be in the 3–4 °C range. The abrupt juxtaposition of cold with warm waters in many locations along the front brings about highly persistent fog, as well as a line of flotsam and jetsam, e.g., fronds of kelp (and also plastic these days) released from subantarctic islands, and prevented from traveling south by the Coriolis effect of the west-to-east blowing winds and surface flows.

As assessed by biologging, king penguins of the Crozet, Kerguelen, and Heard Islands in the Indian Ocean sector and of the Falkland Islands/Islas Malvinas in the Atlantic sector, during summer when making their long foraging trips (for feeding themselves rather than their chicks) forage south to reach, and especially forage within, the APF where foraging success is deemed to be the highest. Annual variation in the terminations of their trips tracked any north–south interannual excursions of the front [39, 139, 156, 159, 163]. When foraging within the front, mesoscale eddies appear to be important, with such eddies likely driving prey to shallower depths [56, 177]. Only when released from constant foraging to simultaneously feed chicks and themselves, i.e., during winter, do they venture even farther south to the SBACC and pack ice edge (see above). Even more indicative of the importance of the APF to this species is the fact that from colonies that are *south* of the APF, e.g., South Georgia, king penguins travel *north* to feed in that front [159, 203]. It is within the APF, and particularly in the thermocline/halocline transition at the bottom of the surface mixed layer (SML), that king penguins have best success at acquiring the energy-dense, mesopelagic myctophid fishes, thus exercising their deep diving capacity [116, 177] (see Chaps. 4, 6, 7, 8, 9, 10 and 11). It is within these transitional waters that king penguins from South Georgia change their movement patterns, from linear north–south as they swim to/from the zone to short, multidirectional excursions once foraging begins– the so-called area-restricted search prevalent in the biologging literature [203].

Except for macaroni penguins, by and large, the APF, and perhaps more so the SAF, is the approximate southern boundary of the ranges of crested penguins (compare Figs. 3.3 and 3.10). Mostly, these penguins are subantarctic temperate-water species, encircling the southern, cooler portion of the globe, limited only in their longitudinal extent by the existence of the islands on which they nest. Macaroni penguins, on the other hand, venture significantly farther south than do other crested penguins and, to some degree, to the northern edge of the pack ice zone (SBACC)

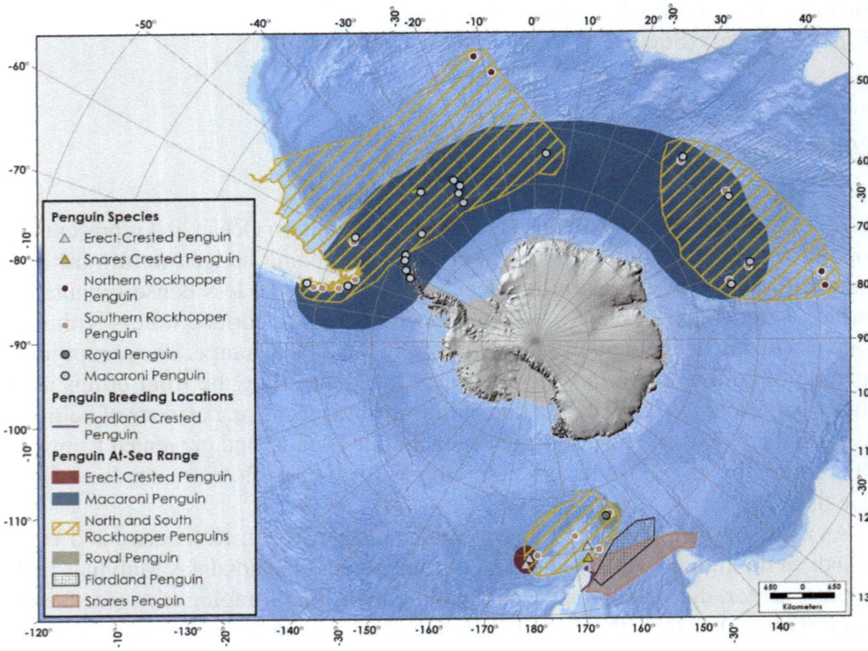

Fig. 3.10 Location of the breeding colonies and the known respective at-sea ranges of the seven crested penguin species, drawn from several sources [60, 64, 66, 90, 130, 131, 161, 172, 201, 216]

at least during the nonbreeding season. They are much larger than the other crested penguins, which may impart some thermoregulatory capacity not available to the other, smaller crested species, which remain in warmer temperate waters (Chap. 7). Thus, too, as noted above, the northern gentoos would be sympatric with most of the crested penguin species, whereas the southern gentoos would be sympatric only with the macaroni penguins (of the crested penguins).

Macaroni penguin foraging has been investigated using biologging at a number of locations. On the Marion and Crozet Islands, which sit pretty much on the SAF, macaroni penguins traveled along the front, slightly to the south, but mostly eastward and, thus, downstream [198]. Following breeding, and before the molt, macaroni penguins from Crozet and Kerguelen frequented waters between the SAF and APF to the south; their routes during this season, and the winter, were also downstream of the respective islands and overlapped very little [197, 198, 200]. Those two fronts in this region, especially Kerguelen, constitute a narrow corridor being only about 5° of latitude apart (~500 km). The penguins from Kerguelen ventured east for about 50° of longitude, or about 3000 km. Thus, the foraging/wintering area was relatively narrow but stretched a long ways away, relatively speaking.

On Heard Island, in the southern portion of the Kerguelen Plateau and extremely close to the APF, macaroni penguins when feeding chicks spread in all directions, fairly close but, again, downstream of the island [26, 71, 92]. On Macquarie Island, extremely close to the SAF, the very closely related royal penguins forage to the east and southward into the front, fairly near the island (as do southern rockhoppers [102, 103]. The APF is known to meander somewhat in this eastern Indian Ocean region, and, thus, one might expect penguin patterns to differ somewhat from year to year (E. Woehler, pers. comm.).

At South Georgia during chick provisioning, post-breeding, and before the molt, macaroni penguins forage from the island north to the APF and closely occurring SAF, infrequently passing farther north [23, 91, 98, 164]. At Bouvetøya, which is located close to the SBACC (and the APF), macaroni penguins (and chinstrap penguins), during the brood guard stage of breeding, forage to the west of the island, i.e., to some degree along the SBACC, but not very far from the island (10–40 km [29]. The APF is only about 300 km north of the island (Fig. 3.10).

The Subantarctic and Subtropical Fronts

These two fronts also encircle Antarctica to the north of the APF, with the northernmost Subtropical Front (STF) marking the northern reach of subantarctic waters [149] (Figs. 3.3 and 3.4). It is at the STF that upper Circumpolar Deep Water (CDW) is formed and begins its descent beneath surface waters, progressing southward, eventually reaching the SBACC. Lower CDW reaches even farther, to the shelf break and canyons of Antarctica at a >200-m depth (Fig. 3.4). At the STF, the surface temperature sharply changes by 4–5 °C, with its northern side being highly salty and having a temperature >13.0 °C [69, 149] (Table 3.1).

These two fronts are the oceanographic boundaries for most of the crested penguins. Southern rockhopper penguins from the eastern islands of the Falkland Islands/Islas Malvinas mostly venture eastward, with the curl northward of the respective foraging areas mirroring the northward tongue of subantarctic water in the vicinity of the Falkland Islands/Islas Malvinas [164](Fig. 3.3). Southern rockhoppers from the western islands of the Falkland Islands/Islas Malvinas, on the other hand, occupied the subantarctic waters to the west and north, somewhat offshore the Patagonian coast, extending north to the SAF [127, 164]. Thus, they followed the northward projection of cooler waters along the coast and foraged downstream in the Falkland Current.

Table 3.1 Approximate ranges of sea surface temperatures and salinities across a broad oceanographic zone [6]. These ranges do vary somewhat in different regions of the world oceans

	Sea surface temperature (°C)	Sea surface salinity (‰)
Antarctic	−1.8 to 3.9	34.8–33.8
Subantarctic	4.0–13.0	34.8–33.8
Subtropical	13.0–21.9	35.8–35.8
Tropical	≥22.0	36.2–29.0

These patterns of southern rockhopper penguins at the southern tip of South America are particularly interesting for several reasons, irrespective of whether there are interspecific overlaps or not. During the breeding season, while sympatrically nesting Magellanic penguins foraged in waters over the continental shelf between Isla de los Estados and the mainland (the tip of Patagonia), rockhopper penguins foraged farther south into deeper subantarctic waters, thus with not a lot of overlap with the Magellanics [173]. Of added interest, following the breeding season, the Isla de los Estados rockhoppers dispersed along the Argentinian coast north and south generally within ~1000 km of the island, again within the northward projection of subantarctic water (Fig. 3.3); a few dispersed a bit farther away [162]. As noted above, those nesting in the western Falkland Islands/Islas Malvinas dispersed northward, downstream in the Falkland Current, and to the east of the Isla do los Estados population, staying mostly within that tongue of subantarctic water bordered on the north and west by the STF [127, 160]. Thus, their wintering area abutted that of the rockhoppers from Isla de los Estados, with little overlap. Moreover, near-coast waters, occupied densely by Magellanic penguins, are devoid of these other populations (see Fig. 1 in Ratcliffe et al. [164]).

Considering the three rockhopper populations of Patagonia (Isla do los Estados and West and East Falkland Islands/Islas Malvinas), they each exhibited abutting and minimally overlapping wintering areas, thus avoiding each other, as well as more coastal Magellanic penguins; see above [162, 164, 166]. The only crested penguins whose wintering area they overlapped were those foraging east and thus overlapping the winter range of the much larger (in body size and mass) macaroni penguins, venturing west along the SAF from South Georgia. The penguins of Patagonia and Falkland Islands/Islas Malvinas provide a good example of penguin populations dealing with intraspecific competition by avoidance, especially among similarly sized species (see Chap. 5).

The east coast of southern South America is the principle location of colonies of Magellanic penguins, though 150,000 pairs in a couple dozen small colonies do occur on the Peru/Humboldt Current side (north to ~41 °N; see below). The species' main breeding colonies are on the southern east coast of South America, extending from the tip of Patagonia and Falkland Islands/Islas Malvinas north to northern Argentina (41.5 °N; Fig. 3.11). These coastal waters are dominated by the temperate Falkland Current, which is a countercurrent inshore of the western side of the South Atlantic Gyre (Fig. 3.3). As a consequence, the SAF and the STF bend northward between the Gyre and the coast (as noted above). During the nonbreeding season, Magellanic penguins remain in these waters, with a large portion moving north along the coast as far as southern Brazil, that is, downstream in the Falkland Current. Investigations on Martillo Island, Magellan Strait (55 °S), indicate that while some remain over the shelf to the east of their island, many disperse north over the shelf as far as 42 °S, with females going farther than males [72]. Magellanic penguins nesting farther north (44 °S) are the ones who winter as far north as Brazil [168]. As noted above, this species' at-sea density along the Patagonian and central coast of Argentina seems to preclude penguins from other populations from using those coastal waters.

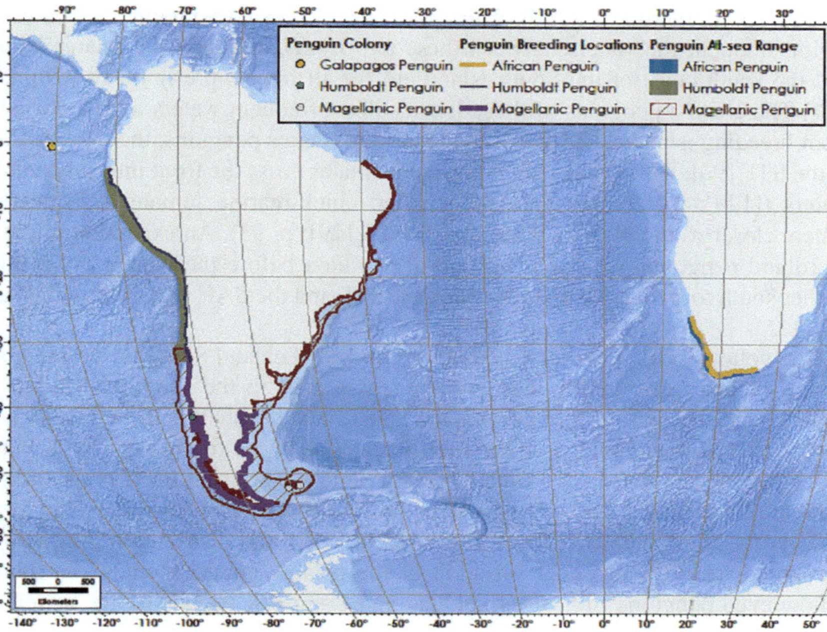

Fig. 3.11 Location of breeding colonies and the respective at-sea ranges of the *Spheniscus* penguins, drawn from several sources [31, 34, 58, 59, 67, 216]

Eastern rockhopper penguins nesting on Crozet and Kerguelen during the interbreeding period foraged to the east (downstream) of those islands and essentially along the SAF [198]. However, on the Marion/Prince Edward islands (to the northwest of Crozet), eastern rockhopper penguins often foraged to the south toward the APF [213]. Northern rockhopper penguins, nesting on Amsterdam Island, which is just north of the STF (Figs. 3.3 and 3.10), also foraged in winter to the east (downstream) but southward over a couple thousand kilometers and thus toward cooler, more productive waters of the front [198].

The behavior of the SAF and STF in the region of New Zealand is highly complex, owing to the bathymetric ridges and plateaus, now submerged, which were once part of the continent of Zealandia and which help direct current flow in the region. For instance, the Chatham Rise, which extends from the northern coast of the South Island of New Zealand eastward to the Chatham Islands may force the STF to be north of the islands [68, 94]. On the west coast of New Zealand, the STF bends northward partially into the Tasman Sea and, then to the east, it bows south to be south of Stewart Island, before turning north to be offshore the Chatham Islands; the SAF occurs south of the various islands that emerge south of New Zealand (Fig. 3.3) [134].

The temperature-based delineation between subtropical and subantarctic water changes with seasons and has been considered to be around 10 °C in winter and around 15 °C in summer [94]. Thus, the various crested penguin species of the region frequent the waters between these two fronts, although the Snares Islands and the south coast of the South Island are for all intents pretty much within the STF. The crested penguins of those islands don't frequent waters all that far away from breeding islands [129, 220]. In the case of Snares penguins, they forage close to the STF [90]. During the incubation stage, males cross the front into subantarctic waters ([129]: p. 69), but then the smaller, chick-rearing females seek warmer waters closer to the New Zealand mainland ([129]: p. 93). An exception might be Fiordland penguins, which in their quest to fatten before the annual molt, travel farther south to frequent waters between the STF and the SAF, particularly the latter [90, 135].

The yellow-eyed penguins breeding on the New Zealand South Island mainland forage exclusively over the narrow continental shelf to the southeast (SE) [133, 140]. The shelf edge itself represents the STF. Since the shelf is merely 10–15-km wide, these penguins forage on the subtropical side of the front (~12–16 °C water, depending on the season; T. Mattern pers. comm.). The front itself plays only an indirect role in their foraging, as the penguins are almost exclusively benthic foragers over the shelf and, as such, their prey availability and distribution is somewhat decoupled from the seasonal variation of the pelagic environment. Even though yellow-eyed penguins from the Auckland and Campbell Islands forage in subantarctic waters, they still forage to a great degree on the seafloor, although foraging in the water column seems to dominate in some seasons and years [142]. When foraging in the water column among the Auckland Islands, the foraging area is larger than with just benthic foraging [141]. With benthic foraging in their behavioral repertoire, they would also be less affected by the seasonality of ocean conditions affording them their sedentary lifestyle at the offshore islands. In general, at least around the New Zealand coast, productivity is largely driven by nutrient influx from the Tasman Sea and the mainland river systems, and this could be the reason that Snares penguins tend to forage closer to the mainland. Their foraging from the Auckland Islands, or at least from Enderby Island of the group where their foraging has been investigated, is downstream of the island and, moreover, in the vicinity where a wake would be pronounced (off the northern end of the Auckland Islands; see below for discussion on island wakes).

Due to temporal vagaries in the strength of the STF in waters around New Zealand, the timing of egg laying of little penguins is far more spread out than that of the little penguins in Australia (see below). Eggs in New Zealand can be laid from the beginning of May through September, with the schedule correlated with the ocean temperature; lower SSTs bring about earlier laying [3, 46, 136]. Little penguins, in some years, also double brood [3]. Such variation in phenology means the intensity and perhaps the location of the STF are at play, and indeed there are locations along the eastern shores of Northland (South Island) and in places along the South Island's west coast where there is local upwelling of cool water. At least on the SW coast of the South Island, the penguins appear to forage within the STF, which is cooler water than inshore; on the southeast coast, they forage inshore of the STF [3]. In Australia, egg laying of little penguins lasts over a period of just 2 months when the STF is the weakest (see below). In the Tasman Sea waters off the east coast of Tasmania, little penguins are found where the SST is 13–15 °C, which essentially is the STF [80].

Eastern Boundary Currents

The Peru/Humboldt and Benguela Currents, due to intensive upwelling of cold, nutrient-rich water, are responsible for the existence of penguins northward in what would otherwise be subtropical and tropical waters along the west coast of South America (Chile and Peru) and southern Africa (Namibia, South Africa) (Figs. 3.3 and 3.11). The *Spheniscus* penguins are the denizens of these currents and, in fact, nest north to the equator (Galápagos penguins). These currents are temperate, with subtropical waters running along their seaward boundaries. It has been found that African penguins, especially once they finish nesting, exploit waters that are 14.5–17.5 °C, which, as noted above, characterize the STF [182]. The Cromwell Current extension of the Peru/Humboldt Current allows Galápagos penguins to exist in the Galápagos archipelago, especially on the western islands that are the closest to the upwelling of cool water. Equally characteristic of these currents are the high-energy density anchovies and sardines that these penguins target during their foraging (see Chap. 4).

The Leeuwin Current, which is technically an eastern boundary current, poses an anomaly in explaining why little penguins breed from a bit northward along the Western Australia coast, to about Perth eastward at locations along the southern coast of Australia to Tasmania, and a bit northward (Fig. 3.12). Unlike the eastern boundary currents of South America and southern Africa, which flow north and are cooled by the extensive upwelling of cold water due to the Coriolis effect, the Leeuwin Current, which flows along the eastern boundary of the Indian Ocean, is subtropical/tropical and flows south along Western Australia and then eastward along the southern coast east to Tasmania [61]. Working in their favor, the penguins are molting and between breeding seasons from February to June, when the current is strongest and the waters warmest. It is the weakest in summer, i.e., from September to January, the peak of little penguin nesting. It represents the Subtropical Front. However, it flows mainly seaward of the shelf break, across which these penguins apparently rarely venture as they do not travel very far and only frequent waters of the shelf (Appendix Table A3). There are large meanders and eddies that draw the warm water away from the coast, to be replaced by cool water (sourced from cool Flinders and Leeuwin undercurrents). Most importantly, during summer along the southern Australia coast, the cooler Creswell Current flows over the shelf, east to west [16], and the cooler Capes Current flows west and north around the southwestern cape of Australia [219] (Fig. 3.3).

The timing of the range of little penguin breeding and foraging in Australia fits with and indicates the seasonal change in oceanography. They lay eggs within a fairly short window, from mid-September to mid-November, when the Leeuwin

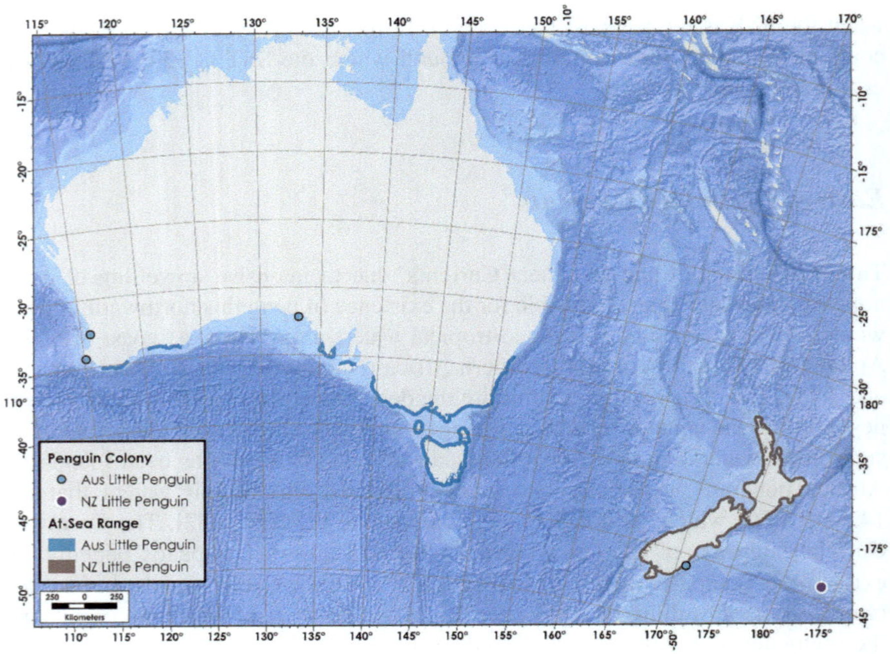

Fig. 3.12 Location of breeding colonies and the respective at-sea ranges of Australian and New Zealand little penguins, drawn from two sources [65, 216]

Current is weakest and competing factors are stronger [65]. Moreover, they rarely travel more than ~15 km from the shore (Appendix Table A3), i.e., they remain over the shelf [55, 151, 210, 211] and, during the nonbreeding season, disperse alongshore rather than offshore [137]. Little penguin nesting on the SE coast of Tasmania essentially forage within the STF in cool, subtropical waters, at 13–15 °C, but do so in windy areas in which the water column stratification is disrupted, allowing mixing of cooler, deep waters [79, 87, 157]. The penguins fare best in their breeding when water temperatures are on the cool side, i.e., <14 °C [43, 45].

Meso- and Small-Scale Factors: Regional and Local Ocean Processes that Facilitate the At-Sea Foraging of Penguins

The macroscale factors that affect where the penguin species occur have been reviewed above. Now we will look at the smaller scales of the biophysical factors that affect where penguins forage in the neighborhood of their colonies. In many cases, penguin colonies are located where there is almost no continental or insular shelf, which is the case for many oceanic islands (Fig. 3.3), on the west coasts of Africa and South America, and on the coast of southern Australia. Otherwise, there is a substantial extent of shelf. We will first consider the penguin movements, mainly in the nonbreeding season but also in the breeding season, reviewed above, noting that many extend downstream of their colonies, i.e., island and headland wakes. Then, we will consider the factors related to insular or continental shelves that affect penguin foraging location when they are in the "central place foraging" mode, e.g., especially while provisioning chicks. Finally, we will consider another largely shelf or coastal factor, open water (or polynyas), a factor that significantly affects where colonies are located in the Antarctic.

Island Wakes

Noted several times above in the discussion of the major frontal zones that influence penguin biogeography was the tendency of many penguin populations nesting on islands, especially oceanic ones– surrounded by deep oceans rather than along a mainland coast– to forage downstream in the general ocean flow or current, e.g., macaroni and other crested penguins from the Crozet and Kerguelen Islands [197, 198, 200], royal penguins from the Heard and Macquarie Islands [26, 71, 92, 102, 103], and yellow-eyed penguins from Enderby, Auckland Islands [141]. In the eastward-flowing ACC, which flows around a majority of penguins' oceanic nesting islands (Fig. 3.1), this means that penguins tend to frequent waters to the east of the breeding island, downstream. It is well known that islands in fast-flowing currents have "wakes" [24, 138], i.e., a zone of turbulence and eddies extending out from their lee shore at extremely long distances, especially from the larger islands, such as Crozet, Kerguelen, Macquarie, etc. Most people know wakes as the turbulence that one sees behind a moving boat – recall the popular saying, especially among runners, of bettering someone or something by leaving them in one's wake. Only, in the case of island wakes, it is the ocean that is moving past the stationary island; a boat anchored in a river would also generate a wake. In these wakes, productivity is enhanced by the mixing brought about by eddies and the fine-scale turbulence, which actually draw deeper water to the surface, enhancing all levels of the food web. While this phenomenon has been well-researched in the subtropical and tropical oceans, owing, for instance, to where one finds tuna, e.g., East Indies, Wake, Johnston, Hawaii, Madeira, Canary Islands, etc. [18, 30, 100, 105, 214], it has received less attention in south polar seas. In one case, a richer ocean was found downstream as opposed to upstream of the Kerguelen Islands [28, 104]; a similar pattern is associated with the Prince Edward, Bouvetøya, and South Sandwich Islands [154, 155]. Nevertheless, much of the Southern Ocean is of low productivity (oligotrophic), and, thus, these wakes, as indicated by penguins, appear to be significantly important for raising the regional productivity level and providing more food.

In a twist, the penguins nesting on the western and northern shores of the Falkland Islands/Islas Malvinas – gentoos, Magellanic, and southern rockhoppers– also frequent waters downstream in the major current flow. However, in their cases, this involves the northward-flowing Falkland Current and so they forage to the north of their nesting islands [127]. In another twist, king, macaroni, and southern rockhopper penguins from the Prince Edward Islands forage "upstream" in the ACC to an area of generally oligotrophic (depleted) waters where there is a ridge of seamounts. Through upwelling related to ACC flow, a local, productive food web is generated as the ACC flows in its "downstream" direction around these seamounts [180]. These penguins thus forage between their nesting islands and the seamounts, and this explains some of the patterns detailed by other authors [199]. In the final twist, the large population of macaroni penguins of South Georgia, investigated by Ratcliffe et al. [164], also forage downstream, to the west of the island toward the Falkland Islands/Islas Malvinas, owing to an eddy that projects to the west, despite the general eastward flow of the ACC [19].

Headland Wakes, Eddies, and Upwelling Plumes

Similar to islands (or seamounts), any land projecting a significant way out or into ocean currents, e.g., a cape or a subsea pinnacle, generates turbulence and mixing as the ocean flows past [108, 209]. Mesoscale (measured in tens of kilometers) plumes and eddies are generated, which increase productivity and prey availability. With regard to penguins, perhaps the best known of these, on a fairly large scale, would be Cape Horn and the tip of the Antarctic Peninsula. Indeed, Isla de los Estados at the easternmost tip of Patagonia (Cape Horn) is in the ACC wake that initiates the coastal Falkland Current that is so important to Magellanic and rockhopper penguins (see above). The northward-flowing Falkland Current initiates a wake from the Falkland Islands/Islas Malvinas. Evidence that productivity and prey availability is increased in this neighborhood is the noteworthy presence of large industrial fishing efforts, including effects on the penguins [160], in this wake on the Patagonian shelf. Farther north in Argentina, several coastal capes project out into the Atlantic, generating turbulence and fronts [2]. At well-studied Punta Tombo, Magellanic penguins forage downstream (northward) and, with little doubt, in the plumes and eddies generated by those capes [32, 33]. Off the central coast of Chile, one of the largest collection of colonies of Humboldt penguins occur between Punta Lengua de Vaca (30 °S) and Cabo Bascuñán (29 °S) [67], and the penguins are mostly concentrated close to the shore, downstream of the point in an area of intense upwelling [212].

At the tip of the Antarctic Peninsula, flowing east is the Scotia–Weddell Confluence, where the Scotia Sea water of the ACC converges with the Weddell Gyre, flowing counterclockwise (Fig. 3.3). This would be a large headland wake. The food web in the Confluence is rich and has been fairly well-investigated, with eddies strong enough to drive mesopelagic (deep living) organisms to the surface, which are then consumed as seabird prey [11, 15, 97, 165, 202]. A secondary result of the enhanced production is one of the world's largest concentrations of Adélie penguins situated among the Danger Islands at the northern tip of the Peninsula

[35]. An equally large concentration of Adélie penguins occurs at the northern tip of Victoria Land, at Cape Adare and associated islands [126], where the northerly winds and resulting current actually form the large gravel bar that is the Cape and play a huge role in generating the Ross Gyre (Fig. 3.3). This is another large headland wake.

River Outflow Plumes

Among the penguin species that frequent waters close to the coast, river plumes and the salinity fronts that occur where the fresh/brackish water meets ocean water can offer important foraging habitats. Of course, the vast majority of penguin species occur along desert coasts and oceanic islands where rivers are all but absent (see above). Nevertheless, there are examples of river plumes providing important foraging habitats. In Port Phillip Bay, Victoria, with a major outflow from the Yarra River, little penguins forage in the outer portion of the plume, which are more turbid, warmer, and less saline waters, having a higher chlorophyll concentration than waters outside the plume, where the penguins do not forage [117, 118]. The referenced study provides possible reasons for the attraction – perhaps a different community of fish, more fish, and, with more turbid water, perhaps it is easier for the penguins to approach. As well, or alternatively, the radical change in salinity may provide an oceanographic "surface" against which fish concentrate or can be trapped (see below).

Another example of this phenomenon is the river plumes along the east coast of New Zealand's South Island, e.g., the Rangitata River, which is an important foraging habitat for yellow-eyed penguins, especially juveniles [132]. Finally, the coast of Argentina has tidal plumes and salinity fronts emanating from several major rivers [2]. Some of these salinity fronts are utilized by Magellanic penguins, especially the extremely extensive Rio de la Plata plume [168]. The tidal salinity front that stretches along the coast offshore of the outflow from the Strait of Magellan is an area rich in anchovies [2] and is another important habitat to penguins and other seabirds (DGA, pers. obs.).

Shelves and Banks

In addition to the continental shelves of South America, South Africa, and Australia, along which penguins nest, extensive shelves or plateaus occur in proximity to a number of islands where penguins breed. Notable are the Kerguelen Plateau in the southern Indian Ocean (Kerguelen and Heard Islands), the Campbell Plateau and Chatham Rise east and south of New Zealand (Chatham, Bounty, Antipodes, Auckland, and Campbell Islands), and the Burdwood Bank (southeast of the

Falkland Islands/Islas Malvinas). The ocean is <200-m deep over these banks, high enough that some amount of light can reach their respective benthic communities, but also involved are the upwelling and mixing of waters along their flanks. Thus, they are highly productive, and major fishing industries are associated with all of them.

A number of penguins are associated with oceanographic shelves/banks, confining much of their breeding season foraging to their vicinity: yellow-eyed, gentoo, little, and all the *Spheniscus* and *Aptenodytes* species. Yellow-eyed penguins forage on the bottom in shelf waters out to the shelf break but concentrate their foraging inshore from that (see above) [132]; on Enderby Island, among the Auckland Islands, when foraging in the water column, yellow-eyed penguins forage on the insular shelf [141]. A similar foraging strategy is exhibited by many gentoo penguins, for instance, those on the Kerguelen Islands [124] and on the Marion [44], Macquarie [171], Falkland Islands/Islas Malvinas [25], South Georgia [194], and on the western coast of the northern Antarctic Peninsula [115]. In other words, they use shelf waters by venturing away from the coast and their colonies only to the brink of the shelf break. Approximately 95% of little penguins are found within 4 km of the shore, and bottom composition is important [148]. This could be true for all penguin species foraging over shelves, as bottom type likely influences prey type and prevalence/distribution and abundance.

The continental shelf on the east coast of southern South America is extremely wide, and the entire Argentinian portion of the Magellanic penguin breeding population coincides with it. This shelf extends from Isla de los Estados (Cape Horn, eastern tip of Patagonia) to the east of the Falkland Islands/Islas Malvinas (with the Burdwood Bank a short distance south) and then turns back west toward the mainland coast, narrowing as it extends toward the north to about 30 °S (Laguna Patos), where the shelf break reaches close to the coast [2]. Magellanic penguins frequent waters of the shelf out to its edge [32, 33]. The length of Magellanic penguin foraging ranges when feeding chicks are in accordance with the breadth of the shelf, being short in the south (25–30 km) and lengthening midway where the shelf is the broadest (70–110 km [2]; Appendix Table A3, no data for more northern colonies). On the west coast of South America, the shelf is extremely narrow, with Magellanic and Humboldt penguins foraging no farther than 15–35 km from their colonies (Appendix Table A3). We will have more to say about this when we discuss shelf-break fronts (below).

A significant portion of king penguins' foraging activity, of those nesting on the eastern Falkland Islands/Islas Malvinas, is associated with the Burdwood Bank just to the south of the islands [159]. King penguins nesting on the eastern Kerguelen Islands concentrate their foraging along the shelf break of the Kerguelen Plateau to the southeast [49]; those nesting on Heard Island, farther southeast along the Kerguelen Plateau, also appear to frequent this habitat [139]. However, the best data on how shelf topography affects penguin foraging come from emperor penguins in the Ross Sea [111]. Studied over a 22-year period, penguins from Cape Washington (the largest emperor penguin colony) headed east, across the Drygalski Trough, to continue foraging over Crary Bank. It was only over the bank that the penguins made their deepest dives, seemingly to access the energy-dense Antarctic silverfish found there. The outer portions of Crary Bank are shallow enough to ground icebergs and to allow sunlight to influence the benthic community.

Submarine Canyons

Submarine canyons that cross continental shelves are what remains from glacial periods. This was when the sea level was 200 m lower and runoff from rivers or glacial streams, extending from what would otherwise be continental river valleys, gouged troughs in the adjacent dry shelf. In these canyons, these days, deeper, more productive water upwells and extends toward the coast closer than it would otherwise. Around Antarctica, the colonies of emperor and *Pygoscelis* penguins are associated with such canyons [176], with which latent heat polynyas (formed by wind blowing the pack ice aside) coincide due to the katabatic winds that blow down the glacial valleys. Circumpolar Deep Water (CDW), which is warmer than surface water, upwells in these canyons, and this leads to sensible heat polynyas [170]. A major, multidisciplinary, long-term ecological effort has been researching the association of Adélie and gentoo penguins and their prey with such a canyon, the Palmer Deep, on the western side of the Antarctic Peninsula [50, 51, 145, 174]. In that particular canyon, and presumably others, the movement of water forms eddies that enhance prey availability to the penguins by retaining zooplankton, such as krill [101]. In the canyons and narrow fiords, even tides have important effects on penguin foraging [27]. Another sensible heat polynya is formed by upwelled CDW at Cape Adare, Victoria Land, with this polynya affording Adélie penguins to nest there even during glacial maxima [196]. As far as anyone knows, Adélie penguins were excluded from elsewhere along the continental margin by all the glacial ice that overlaid the coast and extended across continental shelves during glacial maxima [17].

Shelf-Break Fronts

Continental shelves generally tend be overlain by water whose oceanographic characteristics differ from waters seaward of their outer edge, known as a shelf break. For one, shelf waters incorporate runoff from coastal rivers, thus reducing salinity, but there are other factors at play as well, such as those affecting temperature. A shelf break, i.e., a drop-off of a shelf into abyssal depths, is often highly steep, and various oceanic layers abut against it (Fig. 3.4). In the eastern boundary currents, and elsewhere such as Antarctica, deep water from off the shelf upwells to replace the water being pushed or advected offshore by the winds (Coriolis effect; see Fig. 3.4). Where this new water meets waters adjacent to the shelf, at the shelf break, there is a front (discontinuity in water types; see above). This is the shelf-break front, and it is generally at least a few kilometers wide, i.e., a zone. More than just being a discontinuity in seabird species composition between water types, but also owing to the mixing of waters and increased prey prevalence, a number of seabirds (and marine mammals and fish) typically occur in higher abundance along the front on either side.

Antarctic-wide, distance to the shelf-break front negatively affects the abundance of both Adélie and chinstrap penguins, i.e., the closer to the front, the higher is penguin abundance [170]. The shelf-break front, in this case, is a product of the upwelled Circumpolar Deep Water mixed with the surface waters of the shelf [7, 8]. This pattern was particularly revealing in the case of Adélie penguins because especially in winter, their sea ice habitat extended from the coast far to the sea. Very few penguins were found to occur between the coast and the shelf break, where overlying waters were covered by concentrated sea ice. The increase in penguin numbers at the front was especially graphic as it exceeded the numbers at the large-scale marginal ice zone farther yet to the sea, where there was another increase in abundance (see above). Foraging over the continental slope, essentially the shelf-break front, by Adélie penguins has also been seen in smaller regions, Amundsen and Bellingshausen Seas, in late summer [8]. A similar pattern is exhibited off the coast of East Antarctica in summer near Mawson [53], at Pointe Géologie [57], and generally anywhere along that coast [167].

Returning to chinstrap penguins, along the western side of the northern Antarctic Peninsula, the SBACC (see above; Fig. 3.3) is so close to the shore that it pretty much overlaps or coincides with what would be a shelf-break front. Not surprisingly, along the shelf-break/SBACC front is where krill are concentrated, and, again, not surprisingly, chinstrap penguins are there in the highest numbers foraging upon them [204]. It's also a region where krill trawlers concentrate (Chap. 5).

One of the better demonstrations of increased penguin abundance as a result of the enhanced production and food availability in a shelf-break front is from the Ross Sea (Fig. 3.13). The shelf break of the Ross Sea is where the ocean deepens, i.e., from the 800-m to the 3000-m isobath, and against that slope, as noted, Circumpolar Deep Water upwells, providing additional nutrients to enhance the food web [7]. The emperor penguins whose occurrence patterns are depicted in the figure mostly breed in colonies along the coast of Victoria Land, fronting the western side of the Ross Sea [176], but the period shown (mid-December to mid-January) is when these penguins forage to recover from breeding and before fasting for their annual molt. The best place to forage and fatten is in the productive waters of the shelf-break front. In the eastern Ross Sea, there is also some pack ice at that time, and upon it is where they molt. While adults, intending to breed, then move back toward colony locations, as shown by biologging, subadults remain at the front for period of time (post-molt), into April, and then become entrained in the sea ice movements of the Ross Gyre, taking them farther into the sea [89]. Foraging in waters of the shelf-break front, i.e., the continental slope, in order to fatten during the premolt is also exhibited by emperor penguins in East Antarctica [167].

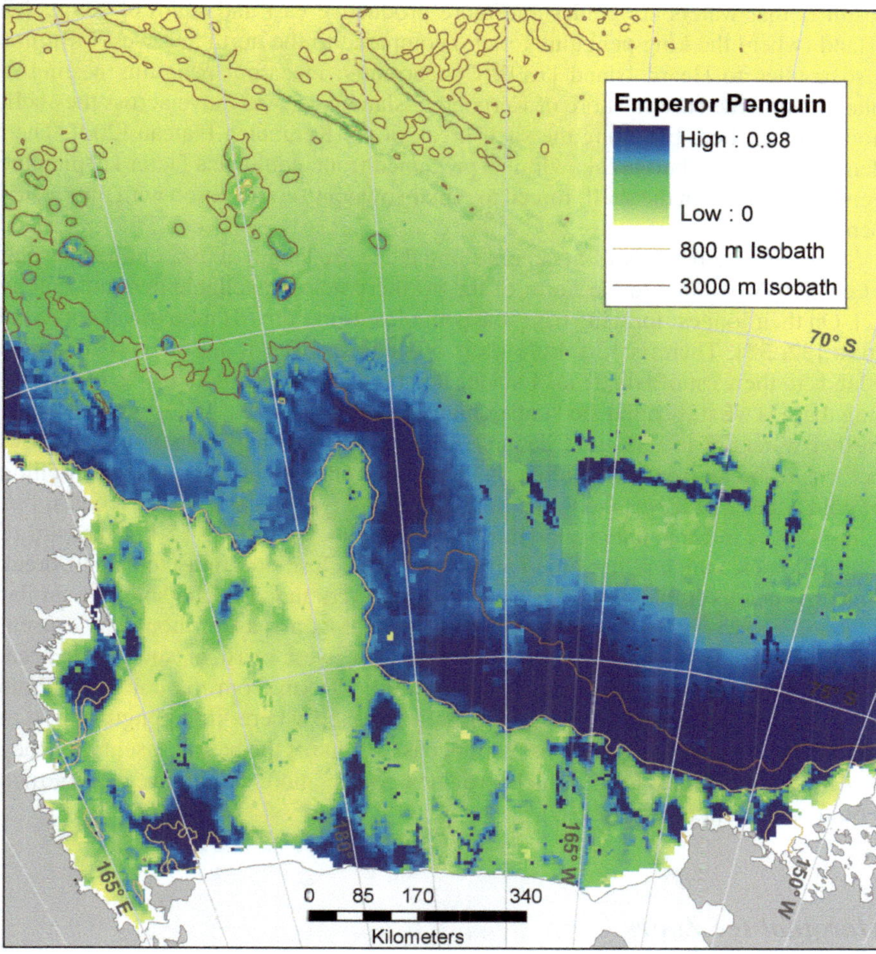

Fig. 3.13 Using a Maxent model [20], the distribution of emperor penguins in the Ross Sea during cruises that occurred mid-December to mid-January, 1976–1981 [10]. (From Elsevier, by permission)

The association of penguin foraging with shelf-break fronts elsewhere has been well-studied in only a few locations. On Kerguelen, those gentoo penguins that forage the farthest from the shore appear to do so in waters overlying the upper portion of the Kerguelen Plateau slope [42, 124]. Thus, an analogous front could be important. Being the leeward side of the archipelago, upwelling should occur along the front, with the presence of turbulence leading to developing an island wake. Indeed, temperature sensors deployed on king penguins on the Kerguelen Island have detected a tongue of cold, nutrient-rich water intruding from the south in the deeper portion of the shelf break, as the surface water is forced to the east [49]. The overall result is that waters are colder and more productive east and south of Kerguelen Island (where the king penguins forage) compared to the north and west [48]; this also applies to Heard Island [139]. Concurrently, macaroni penguins nesting at places along the eastern shore of Kerguelen Island appear to move across the shelf, not foraging there but along the shelf break of the Kerguelen Plateau [200]. Their density along the shelf break, within a presumed front, could be a factor keeping the gentoo penguins on the shelf, forced to remain owing to competition with the crested penguins.

A marked shelf-break front occurs along the entire length of the continental shelf of eastern Argentina, to the north of the northernmost Magellanic penguin colony [2]. On their longest foraging trips, the penguins from Punta Tombo travel out to the front [32, 33]. There is very little shelf offshore the Tierra del Fuego/Patagonian islands to the west of Isla de los Estados and then north to Chiloe, Chile, but occurring there is what is called the "Patagonian cold estuarine zone." This is a mix of the fresher waters of the Beagle Channel. To the north, the shelf is extremely narrow, but a shelf-break front exists along its length associated with the intense upwelling of cold water that drives the Peru/Humboldt Current [2]. Along the west coast, foraging ranges of Magellanic and Humboldt penguins are just 15–35 km seaward of their colonies, which reflects the narrow shelf and close proximity to a shelf-break front (Appendix Table A3). The frontal zone covers much of the shelf. A similar situation occurs in the Benguela Current, an analogous eastern boundary current flowing along an extremely narrow shelf off the west coast of southern Africa. African penguin foraging excursions reach only 20 km from the colonies (Table 5.1). Finally, and similarly, there is virtually no shelf where little penguins nest at colonies in Victoria along the Bass Strait coast. Right off the beach, essentially, they feed in a shelf-break front, which provides cooler, more productive waters than would be present otherwise [99].

Marginal Ice Zones

Important to penguin occurrence and prevalence on the macroscale (see above), marginal ice zones (MIZs) on a smaller scale, especially those of coastal polynyas, are where Adélie and emperor penguins concentrate a good deal of their foraging (Fig. 3.14). In fact, close proximity to polynyas is a major factor in assessing where Adélie and emperor penguin colonies occur [176]. It is the pack ice and its ephemeral openings, around the edges of the polynyas, and not the open water toward the middle, that are important for penguin foraging, in other words, the MIZ [120]. The food web built from diatoms, rather than less grazable phytoplankton of central polynya waters, occurs in these MIZs [184]. Where they are along the gradation of ice cover from concentrated ice to open water in these MIZs an affect the penguins' diet [128]. Biologging of emperor penguins has shown an affinity toward MIZs in the Ross Sea and along the East Antarctica coast [111, 120, 215, 222]. In the case of Adélie penguins in the western Ross Sea, biologging shows that they prefer to

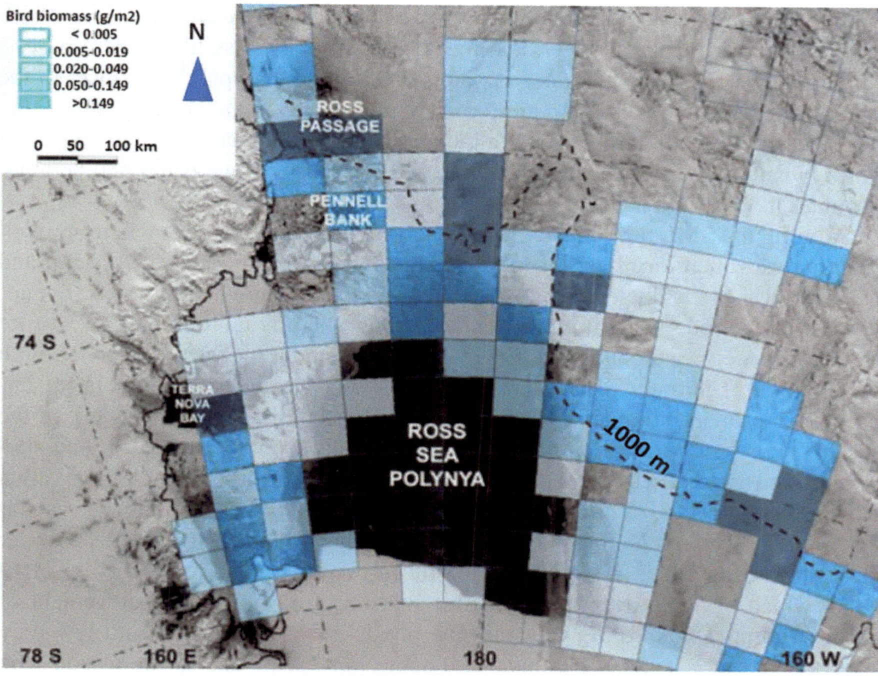

Fig. 3.14 The concentration of seabird biomass, including penguins, in the marginal ice zones bordering polynyas in the Ross Sea. Cruise tracks crisscrossed the entire area [10] (see also Fig. 3.5). (From Karnovsky et al. [109] and from Elsevier, by permission)

forage where the sea ice cover is just 6–10%, which is characteristic of the outer edge of the Ross Sea polynya's MIZ [123].

Thermoclines/Haloclines

Thus far, apart from where penguins foraged near the bottom or within the water column, we have only considered two-dimensional aspects of penguin foraging habitats, i.e., distance from colony and direction. Of course, there is a third dimension in the ocean, which is depth. As repeatedly noted above, the ocean is constructed of layers whose attributes differ (Fig. 3.4). The most important to most penguins foraging in the water column is what is called the surface mixed layer (SML), the lower boundary of which is a thermocline and a halocline, usually approximately coinciding, which represents a sharp transition in values. The temperature or salinity changes dramatically within just 10–20 m or so, with the depth of the transition and intensity varying by region and season. The SML is called thus because the wind is able to disturb the water, i.e., mix it, above the clines, with little influence below, depending on how sharp is the transition. The cline forms a boundary, like a layer of oil on water, and the sharper it is, the less likely there will be mixing between surface and deeper waters when wind disrupts the surface. In the colder portion of the penguin realm, salinity mostly drives the layering, given that temperature does not have much of a range, regardless of depth. As one progresses north, where higher surface temperatures and a greater range occurs, temperature becomes the layering factor in the SML.

Altering the SML, i.e., causing it to become a weak boundary, is where some of the processes mentioned earlier in this chapter become important, i.e., island and headland wakes that include eddies, upwellings, etc. When/where the SML boundary is weak, or eddy activity is extremely strong, there is more mixing of surface water with deep water, and vice versa. This brings about higher productivity, with the addition of nutrient-rich water to the surface layer, more influence from sunlight, high primary production, and higher prey availability. This is the story of the eastern boundary currents and shelf-break/upwelling fronts. One reason that subtropical and tropical seas are relatively unproductive, compared to polar seas, is that the thermocline is too strong and deep (≥150 m) for the winds to influence very far down in the water column, allowing little mixing of nutrient-rich, deep water upward. Thus, surface water is depleted of nutrients. This is also the story of El Niño, which brings a deeper thermocline that, in turn, inhibits the movement of deep, nutrient-rich water upward into the photic zone. Otherwise, with few prey species in surface waters of tropical and subtropical seas, this is one reason why penguins, whose lifestyle requires ample food, are absent. Thus far, the only connection we have considered with regard to the SML and its depth is with respect to marginal ice zones (MIZs). There, a strong, fresher water surface layer (from melting sea ice) inhibits downward mixing of diatoms by the wind below the photic zone, leading to diatomic growth, more grazers, and more food for penguins. Diatoms need to stay high in the water column as they are not adapted to low light levels.

Shallow, sharp thermoclines that structure the water column, a temperature change that penguins likely may be able to detect, may help them relocate prey by giving them a reference point by which to compare how deep are the better prey patches. Thus, it's possible that diving efficiency would increase with less search time, but this has been minimally researched. In the stratified water of the MIZ of southwestern Ross Sea, the forging efforts of Adélie penguins was the most intense where stratification (as deduced by loggers worn by the penguins) was the strongest [21]. However, it is not known whether this was a result of variation in prey availability (the halocline/thermocline affected the prey directly) or it was an aid by which penguins relocated prey. In McMurdo Sound, Ross Sea, where water column properties were investigated independently of the penguins, and bottom depths were ~1000 m, Adélie penguins foraged at the lower edge of the intense surface phytoplankton bloom, which was advected (along with fish and krill) from the Ross Sea- and within the bloom, visibility was near to zero [175]. In the shallower waters off Anvers Island (~400 m), Adélie and gentoo penguins foraged down to the greater depths where krill were more concentrated- the krill actually were balancing their need to occur high enough in the water column to find phytoplankton for their own foraging but not too near the surface to be subjected to predation by the penguins [50]. The krill were most dense at the bottom of the SML (~20 m), which was that shallow due to tidal mixing in this coastal, fiord system and is where chlorophyll maximum occurred as well. This was a much shallower SML than in the two Ross Sea examples above, where penguins were diving to 40–50 m on average.

In a more oceanic context, king penguins from the Crozet Island, which lies between the Subantarctic Front and the Antarctic Polar Front (Fig. 3.3), traveled to the APF in summer but in winter traveled farther south to the SBACC [47]. At the APF, the SML ranged to a 80–140-m depth compared to 100–150 m at the SBACC. The penguins maximized their effort within and just below the thermocline, with little effort in waters of the actual SML. This same preoccupation with the thermocline was also evident in king penguins from Kerguelen also foraging in the APF [116] – within the thermocline, the penguins showed their highest prey capture rates [177]. These penguins mostly sought the high-energy-dense myctophid fish.

Unfortunately, there is a lack of studies on the foraging of other penguin species relative to the water column structure. This would be especially interesting with regard to *Eudyptula*, *Spheniscus*, *Eudyptes*, and chinstrap penguins, all of which mostly forage high in the water column where the SML is a major factor driving prey availability. Especially telling would be investigations of *Spheniscus* penguin foraging in the eastern boundary currents subjected to dramatic changes in the SML depth occurring between El Niño and La Niña conditions (Chap. 13). The results would shed more light on their actual foraging efficiency, i.e., energy gained vs energy expended.

References

1. Abrams RW (1985) Environmental determinants of pelagic seabird distribution in the African sector of the southern ocean. J Biogeogr 12:473–492
2. Acha EM, Mianzan HW, Guerrero M, Favero M, Bava J (2004) Marine fronts at the continental shelves of austral South America: physical and ecological processes. J Mar Syst 44:83–105
3. Agnew P (2014) Demographic parameters, foraging and responses to environmental variation of little penguins (*Eudyptula minor*). PhD dissertation, Dunedin, New Zealand: University of Otago
4. Ainley DG (1977) Feeding methods in seabirds: a comparison of polar and tropical communities in the eastern Pacific Ocean. In: Llano GA (ed) Adaptations within Antarctic ecosystems. Gulf Publishing Company, Houston, pp 669–685
5. Ainley DG (2002) The Adélie penguin: bellwether of climate change. Columbia University Press, New York
6. Ainley DG, Boekelheide RJ (1983) An ecological comparison of oceanic seabird communities of the South Pacific Ocean. Stud Avian Biol 8:2–23
7. Ainley DG, Jacobs SS (1981) Affinity of seabirds for ocean and ice boundaries in the Antarctic. Deep-Sea Res A 28:1173–1185

8. Ainley DG, Jacobs SS, Ribic CA, Gaffney I (1998) Seabird distribution and oceanic features of the Amundsen and southern Bellingshausen seas. Antarct Sci 10:111–123
9. Ainley DG, Manolis W (1979) First records of mottled petrels in California. Western Birds 10:113–123
10. Ainley DG, O'Connor EF, Boekelheide RJ (1984) Ecology of seabirds in the Ross Sea, Antarctica. Ornithol Monogr 32:iii–97
11. Ainley DG, Ribic CA, Fraser WR (1992) Does prey preference affect habitat choice in Antarctic seabirds? Mar Ecol Prog Ser 90:207–221
12. Ainley DG, Ribic CA, Fraser WR (1994) Ecological structure among migrant and resident seabirds of the Scotia-Weddell Confluence region. J Anim Ecol 63:347–364
13. Ainley DG, Ribic CA, Spear LB (1993) Species-habitat relationships among Antarctic seabirds: a function of physical and biological factors? Condor 95:806–816
14. Ainley DG, Ribic CA, Woehler EJ (2012) Adding the ocean to the study of seabirds: a brief history of at-sea seabird research. Mar Ecol Prog Ser 451:231–243
15. Ainley DG, Smith WO, Sullivan CW, Torres JJ, Hopkins TL (1983) Antarctic mesopelagic micronekton: evidence from seabirds that pack ice affects community structure. Science 232:847–850
16. Akhir MF, Pattiaratchi C, Meuleners M (2020) Dynamics and seasonality of the Leeuwin current and the surrounding counter-current system in the region south of Western Australia. J Mar Sci Eng 8(8):552
17. Anderson JB (1999) Antarctic Marine geology. Cambridge University Press, Cambridge
18. Arístegui J, Tett P, Hernández-Guerra A, Basterretxea G, Montero MF, Wild K, Sangrá P, Hernández-Leon S, Canton M, Pacheco M, Barton ED (1997) The influence of Island-generated eddies on chlorophyll distribution: a study of mesoscale variation around Gran Canaria. Deep-Sea Res I Oceanogr Res Pap 44(1):71–96
19. Atkinson A, Whitehouse MJ, Priddle J, Cripps GC, Ward P, Brandon MA (2001) South Georgia, Antarctica: a productive, cold water, pelagic ecosystem. Mar Ecol Prog Ser 216:279–308
20. Ballard G, Jongsomjit D, Veloz SD, Ainley DG (2012) Coexistence of mesopredators in an intact polar ocean ecosystem: the basis for defining a Ross Sea marine protected area. Biol Conserv 156:72–82
21. Ballard G, Schmidt AE, Toniolo V, Veloz S, Jongsomjit D, Arrigo KR, Ainley DG (2019) Fine-scale oceanographic features characterizing successful Adélie penguin foraging in the SW Ross Sea. Mar Ecol Prog Ser 608:263–277
22. Ballard G, Toniolo V, Ainley DG, Parkinson CL, Arrigo KR, Trathan PN (2010) Responding to climate change: Adélie penguins confront astronomical and ocean boundaries. Ecology 91:2056–2069
23. Barlow KE, Croxall JP (2002) Seasonal and interannual variation in foraging range and habitat of macaroni penguins *Eudyptes chrysolophus* at South Georgia. Mar Ecol Prog Ser 232:291–304
24. Barton ED (2001) Island wakes. In: Steele JH (ed) Encyclopedia of ocean sciences. Academic, pp 1397–1403
25. Baylis AMM, Tierney M, Orben RA, De La Peña DG, Brickle P (2021) Non-breeding movements of Gentoo penguins at The Falkland Islands. Ibis 163:507–518
26. Bedford M, Melbourne-Thomas J, Corney S, Jarvis T, Kelly N, Constable A (2015) Prey-field use by a Southern Ocean top predator: enhanced understanding using integrated datasets. Mar Ecol Prog Ser 526:169–181
27. Bernard KS, Steinberg DK (2013) Krill biomass and aggregation structure in relation to tidal cycle in a penguin foraging region off the Western Antarctic Peninsula. ICES J Mar Sci 70(4):834–849
28. Blain ST, Tréguer P, Belviso S, Bucciarelli E, Denis M, Desabre S, Fiala M, Jezehque VM, Le Fevre J, Mayzaud P, Marty JC, Razouls S (2001) A biogeochemical study of the Island

mass effect in the context of the iron hypothesis: Kerguelen Islands, Southern Ocean. Deep Sea Res II 48:163–187
29. Blanchet MA, Biuw M, Hofmeyr GG, De Bruyn PN, Lydersen C, Kovacs KM (2013) At-sea behaviour of three krill predators breeding at Bouvetøya—Antarctic fur seals, macaroni penguins and chinstrap penguins. Mar Ecol Prog Ser 477:285–302
30. Boehlert GW, Watson W, Sun LC (1992) Horizontal and vertical distributions of larval fishes around an isolated oceanic Island in the tropical Pacific. Deep Sea Res A Oceanogr Res Pap 39(3–4):439–466
31. Boersma PD, Frere E, Kane O, Pozzi LM, Pütz K, Rey AA, Rebstock GA, Simeone A, Smith J, Van Buren A, Yorio P, Borboroglu PG (2013) Magellanic penguin (*Spheniscus magellanicus*). In: Borboroglu PG, Boersma PD (eds) Penguins: natural history and conservation. University of Washington Press, Seattle, pp 233–263
32. Boersma PD, Rebstock GA (2009) Foraging distance affects reproductive success in Magellanic penguins. Mar Ecol Prog Ser 375:263–275
33. Boersma PD, Rebstock GA, Frere E, Moore SE (2009) Following the fish: penguins and productivity in the South Atlantic. Ecol Monogr 79(1):59–76
34. Boersma PD, Steinfurth A, Merlen G, Jimenez-Uzcategul G, Hernen Vargus F, Parker PG (2013) Galapagos penguin (*Spheniscus mendiculus*). In: Borboroglu PG, Boersma PD (eds) Penguins: natural history and conservation. University of Washington Press, Seattle, pp 285–302
35. Borowicz A, Mcdowall P, Youngflesh C, Sayre-Mccord T, Clucas G, Herman R, Forrest S, Rider M, Schwaller M, Hart T, Jenouvrier S, Polito MJ, Singh H, Lynch HJ (2018) Multi-modal survey of Adélie penguin mega-colonies reveals the Danger Islands as a seabird hotspot. Sci Rep 8:3926
36. Bost C-A, Delord K, Barbraud C, Cherel Y, Pütz K, Cotté C, Péron C, Weimerskirch H (2013) King penguin (*Aptenodytes patagonicus*). In: Borboroglu PG, Boersma PD (eds) Penguins: natural history and conservation. University of Washington Press, Seattle, pp 7–21
37. Bost CA, Charrassin JB, Clerquin Y, Ropert-Coudert Y, Le Maho Y (2004) Exploitation of distant marginal ice zones by king penguins during winter. Mar Ecol Prog Ser 83:293–297
38. Bost CA, Cotté C, Bailleul F, Cherel Y, Charrassin JB, Guinet C, Ainley DG, Weimerskirch H (2009) The importance of oceanographic fronts to marine birds and mammals of the southern oceans. J Mar Syst 78(3):363–376
39. Bost CA, Cotté C, Terray P, Barbraud C, Bon C, Delord K, Gimenez O, Handrich Y, Naito Y, Guinet C, Weimerskirch H (2015) Large-scale climatic anomalies affect marine predator foraging behaviour and demography. Nat Commun 6:8220
40. Branch TA (2011) Humpback whale abundance south of 60°S from three complete circumpolar sets of surveys. J Cetacean Res Manag Special Issue 3:53–69
41. Brown CR (1987) Traveling speed and foraging range of macaroni and rockhopper penguins at Marion Island. J Field Ornithol 58:118–125
42. Camprasse E, Cherel Y, Bustamante P, Arnould J, Bost CA (2017) Intra and inter-individual variation in the foraging ecology of a generalist subantarctic seabird, the gentoo penguin. Mar Ecol Prog Ser 578:227–242
43. Cannell BL, Chambers LE, Wooller RD, Bradley JS (2012) Poorer breeding by little penguins near Perth, Western Australia is correlated with above average sea surface temperatures and a stronger Leeuwin current. Mar Freshw Res 63(10):914–925
44. Carpenter-Kling T, Handley JM, Green DB, Reisinger RR, Makhado AB, Crawford RJM, Pistorius PA (2017) A novel foraging strategy in gentoo penguins breeding at sub-Antarctic Marion Island. Mar Biol 164:33
45. Chambers LE (2004) Delayed breeding in little penguins – evidence of climate change? Aust Meteorol Mag 53(1):13–19
46. Chambers LE, Dann P, Cannell BL, Woehler EJ (2014) Climate as a driver of phenological change in southern seabirds. Int J Biometeorol 58(4):603–612

47. Charrassin JB, Bost CA (2001) Utilisation of the oceanic habitat by king penguins over the annual cycle. Mar Ecol Prog Ser 221:285–297
48. Charrassin JB, Park Y-H, Le Maho Y, Bost CA (2004) Fine resolution 3D temperature fields off Kerguelen from instrumented penguins. Deep-Sea Res I Oceanogr Res Pap 51(12):2091–2103
49. Charrassin JB, Park Y-H, Le Maho YL, Bost CA (2002) Penguins as oceanographers unravel hidden mechanisms of marine productivity. Ecol Lett 5(3):317–319
50. Cimino MA, Moline MA, Fraser WR, Patterson-Fraser DL, Oliver MJ (2016) Climate-driven sympatry may not lead to foraging competition between congeneric top-predators. Sci Rep 6:18820
51. Cimino MA, Patterson-Fraser DL, Stammerjohn S, Fraser WR (2019) The interaction between Island geomorphology and environmental parameters drives Adélie penguin breeding phenology on neighboring islands near Palmer Station, Antarctica. Ecol Evol 9(16):9334–9349
52. Clarke J, Kerry K, Fowler C, Lawless R, Eberhard S, Murphy R (2003) Post-fledging and winter migration of Adélie penguins *Pygoscelis adeliae* in the Mawson region of East Antarctica. Mar Ecol Prog Ser 248:267–278
53. Clarke J, Manly B, Kerry K, Gardner H, Franchi E, Corsolini S, Focardi S (1998) Sex differences in Adélie penguin foraging strategies. Polar Biol 20(4):248–258
54. Clucas GV, Younger JL, Kao D, Emmerson L, Southwell C, Wienecke B, Rogers AD, Bost CA, Miller GD, Polito MJ, Lelliott P, Handley J, Crofts S, Phillips RA, Dunn MJ, Miller KJ, Hart T (2018) Comparative population genomics reveals key barriers to dispersal in Southern Ocean penguins. Mol Ecol 27(23):4680–4697
55. Collins M, Cullen JM, Dann P (1999) Seasonal and annual foraging movements of little penguins from Phillip Island, Victoria. Wildl Res 26(6):705–721
56. Cotté C, Park YH, Guinet C, Bost CA (2007) Movements of foraging king penguins through marine mesoscale eddies. Proc R Soc B Biol Sci 274(1624):2385–2391
57. Cottin M, Raymond B, Kato A, Amélineau F, Le Maho Y, Raclot T, Galton-Fenzi B, Meijers A, Ropert-Coudert Y (2012) Foraging strategies of male Adélie penguins during their first incubation trip in relation to environmental conditions. Mar Biol 159(8):1843–1852
58. Crawford RJM, Altwegg R, Barham BJ, Barham PJ, Durant JM, Dyer BM, Geldenhuys D, Makhado AB, Pichegru L, Ryan PG, Underhill LG, Upfold L, Visagie J, Waller LJ, Whittington PA (2011) Collapse of South Africa's penguins in the early 21st century. Afr J Mar Sci 33(1):139–156
59. Crawford RJM, Kemper J, Underhill LG (2013) African penguin (*Spheniscus demersus*). In: Garcia-Borboroglu PG, Boersma PD (eds) Penguins: natural history and conservation. University of Washington Press, Seattle/Washington, pp 211–231
60. Crossin GT, Trathan PN, Crawford RJM (2013) Macaroni penguin (*Eudyptes chrysolophus*), Royal Penguin (*Eudyptes schlegeli*). In: Borboroglu PG, Boersma PD (eds) Penguins: natural history and conservation. University of Washington Press, Seattle, pp 185–208
61. Csiro Marine Research (2001) The Leeuwin current – life of the west. CSIRO Marine Research – Understanding Our Oceans, Fact Sheet No. 44
62. Culik BM, Luna-Jorquera G (1997) The Humboldt penguin *Spheniscus humboldti*: a migratory bird? J Ornithol 138:325–329
63. Culik BM, Luna-Jorquera G (1997) Satellite tracking of Humboldt penguins (*Spheniscus humboldti*) in northern Chile. Mar Biol 128:547–556
64. Cuthbert RJ (2013) Northern Rockhopper penguin (*Eudyptes moseleyi*). In: Borboroglu PG, Boersma PD (eds) Penguins: natural history and conservation. University of Washington Press, Seattle, pp 131–143
65. Dann P (2013) Little Penguin (*Eudyptula minor*). In: Borboroglu PG, Boersma PD (eds) Penguins: natural history and conservation. University of Washington Press, Seattle, pp 305–320

66. Davis LS (2013) Erect-crested penguin (*Eudyptes sclateri*). In: Borboroglu PG, Boersma PD (eds) Penguins: natural history and conservation. University of Washington Press, Seattle, pp 145–151
67. De La Puente S, Bussalleu A, Cardeña M, Valdés-Velásquez A, Maluf P, Simeone A (2013) Humboldt Penguin (*Spheniscus humboldti*). In: Borboroglu PG, Boersma PD (eds) Penguins: natural history and conservation. University of Washington Press, Seattle, pp 265–283
68. Deacon G (1937) The hydrology of the Southern Ocean. Cambridge University Press, Cambridge
69. Deacon GER (1982) Physical and biological zonation in the Southern Ocean. Deep Sea Res A Oceanogr Res Pap 29(1):1–15
70. Deacon GER (1984) The Antarctic Circumpolar Ocean. Cambridge University Press, Cambridge
71. Deagle BE, Gales NJ, Hindell MA (2008) Variability in foraging behaviour of chick-rearing macaroni penguins *Eudyptes chrysolophus* and its relation to diet. Mar Ecol Prog Ser 359:295–309
72. Dodino S, Lois NA, Riccialdelli L, Polito MJ, Pütz K, Rey AR (2021) Sex-specific spatial use of the winter foraging areas by Magellanic penguins and assessment of potential conflicts with fisheries during winter dispersal. PLoS One 16(8):e0256339
73. El-Sayed SZ (1970) On the productivity of the Southern Ocean. In: Holdgate MW (ed) Antarctic ecology. Academic, London, pp 119–135
74. Emery KO, Milliman JD, Uchupi E (1973) Physical properties and suspended matter of surface waters in the southeastern Atlantic Ocean. J Sediment Res 43(3):822–837
75. Emslie SD (2001) Radiocarbon dates from abandoned penguin colonies in the Antarctic Peninsula region. Antarct Sci 13(3):289–295
76. Emslie SD, Berkman PA, Ainley DG, Coats L, Polito M (2003) Late-Holocene initiation of ice-free ecosystems in the southern Ross Sea, Antarctica. Mar Ecol Prog Ser 262:19–25
77. Emslie SD, Coats L, Licht K (2007) A 45,000 yr record of Adélie penguins and climate change in the Ross Sea, Antarctica. Geology 35(1):61–64
78. Emslie SD, Fraser W, Smith RC, Walker W (1998) Abandoned penguin colonies and environmental change in the Palmer station area, Anvers Island, Antarctic Peninsula. Antarct Sci 10(3):257–268
79. Evans R, Hindell M, Kato A, Phillips LR, Ropert-Coudert Y, Wotherspoon S, Lea MA (2020) Habitat utilization of a mesopredator linked to lower sea-surface temperatures and prey abundance in a region of rapid warming. Deep-Sea Res II Top Stud Oceanogr 175:104634
80. Evans R, Lea MA, Hindell MA (2021) Predicting the distribution of foraging seabirds during a period of heightened environmental variability. Ecol Appl 31(5):e02343
81. Force MP, Santora JA, Reiss CS, Loeb VJ (2015) Seabird species assemblages reflect hydrographic and biogeographic zones within Drake Passage. Polar Biol 38(3):381–392
82. Foxton P (1956) The distribution of the standing crop of zooplankton in the Southern Ocean. Discov Rep 28:191–236
83. Fraser WB, Ainley DG (1986) Ice edges and seabird occurrence in Antarctica. Bioscience 36:258–263
84. Fraser WR, Trivelpiece WZ (1996) Factors controlling the distribution of seabirds: winter-summer heterogeneity in the distribution of Adélie penguin populations. Found Ecol Res West Antarctic Peninsula 70:257–272
85. Fretwell PT, Trathan PN (2020) Discovery of new colonies by Sentinel2 reveals good and bad news for emperor penguins. Remote Sens Ecol Conserv 7(2):139–153
86. Gaston AJ (2004) Seabirds: a natural history. Yale University Press, New Haven
87. Gibbs CF (1991) Oceanography of Bass Strait: implications for the food supply of little penguins *Eudyptula minor*. Emu 91(5):395–401
88. Glantz MH, Thompson JD (eds) (1981) Resource management and environmental uncertainty: lessons from coastal upwelling fisheries. Wiley, New York

89. Goetz KT, Mcdonald BI, Kooyman GL (2018) Habitat preference and dive behavior of non-breeding emperor penguins in the eastern Ross Sea, Antarctica. Mar Ecol Prog Ser 593:155–171
90. Green CP, Ratcliffe N, Mattern T, Thompson D, Lea MA, Wotherspoon S, Borboroglu PG, Ellenberg U, Morrison KW, Pütz K, Sagar PM, Seddon PJ, Torres LG, Hindell MA (2022) The role of allochrony in influencing interspecific differences in foraging distribution during the non-breeding season between two congeneric crested penguin species. PLoS One 17(2):e0262901
91. Green JA, Wilson RP, Boyd IL, Woakes AJ, Green CJ, Butler PJ (2009) Tracking macaroni penguins during long foraging trips using behavioural geolocation. Polar Biol 32(4):645–653
92. Green K, Williams R, Green MG (1998) Foraging ecology and diving behaviour of macaroni penguins *Eudyptes chrysolophus* at Heard Island. Mar Ornithol 26:27–34
93. Grémillet D, Pichergru L, Kuntz G, Woakes AG, Wilkinson S, Crawford RJM, Ryan PG (2008) A junk-food hypothesis for gannets feeding on fishery waste. Proc R Soc B Biol Sci 275(1639):1149–1156
94. Heath RA (1985) A review of the physical oceanography of the seas around New Zealand – 1982. N Z J Mar Freshw Res 19(1):79–124
95. Hinke JT, Polito MJ, Goebel ME, Jarvis S, Reiss CS, Thorrold SR, Trivelpiece WZ, Watters GM (2015) Spatial and isotopic niche partitioning during winter in chinstrap and Adélie penguins from the South Shetland Islands. Ecosphere 6(7):125
96. Holm-Hansen O, El-Sayed S Z, Francerschini GA, Cuhel RL (1977) Primary production and the factors controlling phytoplankton growth in the Southern Ocean. In: Proceedings of the third SCAR symposium on Antarctic biology, Washington, DC
97. Hopkins TL, Ainley DG, Torres JJ, Lancraft TM (1993) Trophic structure in open waters of the marginal ice zone in the Scotia-Weddell confluence region during spring (1983). Polar Biol 13(6):389–397
98. Horswill C, Matthiopoulos J, Ratcliffe N, Green JA, Trathan PN, Mcgill RAR, Phillips RA, O'Connell TC (2016) Drivers of intrapopulation variation in resource use in a generalist predator, the macaroni penguin. Mar Ecol Prog Ser 548:233–247
99. Hoskins AJ, Dann P, Ropert-Coudert Y, Kato A, Chiaradia A, Costa DP, Arnould JPY (2008) Foraging behaviour and habitat selection of the little penguin *Eudyptula minor* during early chick rearing in Bass Strait, Australia. Mar Ecol Prog Ser 366:293–303
100. Hsu TY, Chang Y, Lee MA, Wu RF, Hsiao SC (2021) Predicting skipjack tuna fishing grounds in the western and Central Pacific Ocean based on high-spatial-temporal-resolution satellite data. Remote Sens 13(5):861
101. Hudson K, Oliver MJ, Kohut J, Dinniman MS, Klinck JM, Cimino MA, Bernard KS, Statscewich H, Fraser W (2022) Subsurface eddy increases quantities and delivery of simulated krill to penguin foraging regions. Mar Ecol Prog Ser. https://doi.org/10.3354/meps14211
102. Hull CL (1999) The foraging zones of breeding royal (*Eudyptes schlegeli*) and rockhopper (*E. chrysocome*) penguins: an assessment of techniques and species comparison. Wildl Res 26(6):789–803
103. Hull CL, Hindell MA, Michael K (1997) Foraging zones of royal penguins during the breeding season, and their association with oceanographic features. Mar Ecol Prog Ser 153:217–228
104. Hunt BPV, Espinasse B, Pakhomov EA, Cherel Y, Cotté C, Delegrange A, Henschke N (2021) Pelagic food web structure in high nutrient low chlorophyll (HNLC) and naturally iron fertilized waters in the Kerguelen Islands region, Southern Ocean. J Mar Syst 224:103625
105. Hyrenbach KD, Forney KA, Dayton PK (2000) Marine protected areas and ocean basin management. Aquat Conserv Mar Freshwat Ecosyst 10(6):437–458
106. Hyrenbach KD, Veit RR, Weimerskirch H, Metzl N, Hunt GL Jr (2007) Community structure across a large-scale ocean productivity gradient: marine bird assemblages of the Southern Indian Ocean. Deep-Sea Res I Oceanogr Res Pap 54(7):1129–1145
107. Joiris CR (2017) Seabird and marine mammal "hotspots" in polar seas. Lap Lambert Scientific Publishing

108. Kai ET, Rossi V, Sudre J, Weimerskirch H, Lopez C, Hernandez-Garcia E, Marsac F, Garcon V (2009) Top marine predators track Lagrangian coherent structures. Proc Natl Acad Sci 106(20):8245–8250
109. Karnovsky N, Ainley DG, Lee P (2007) The impact and importance of production in polynyas to top-trophic predators: three case histories. In: Smith WO Jr, Barber DG (eds) Polynyas. Elsevier, London, pp 391–410
110. Klinck J, Nowlin WD Jr (2001) Southern Ocean-Antarctic circumpolar current. Academic/CCPO Publications
111. Kooyman GL, Goetz K, Williams CL, Ponganis PJ, Sato K, Eckert S, Horning M, Thorson PT, Van Dam RP (2020) Crary bank: a deep foraging habitat for emperor penguins in the western Ross Sea. Polar Biol 43(7):801–811
112. Kooyman GL, Kooyman TG (1995) Diving behavior of emperor penguins nurturing chicks at Coulman Island, Antarctica. Condor 97:536–549
113. Kooyman GL, Kooyman TG, Horning M, Kooyman CA (1996) Penguin dispersal after fledging. Nature 383:397
114. Kooyman GL, Ponganis P (2008) The initial journey of juvenile emperor penguins. Aquat Conserv Mar Freshwat Ecosyst 17(S1):S37–S43
115. Korczak-Abshire M, Hinke JT, Milinevsky G, Juáres MA, Watters GM (2021) Coastal regions of the northern Antarctic peninsula are key for gentoo populations. Biol Lett 17(1):20200708
116. Koudil M, Charrassin JB, Le Maho Y, Bost CA (2000) Seabirds as monitors of upper-ocean thermal structure. King penguins at the Antarctic polar front, east of Kerguelen sector. Comptes Rendus de l'Académie des Sciences-Series III-Sciences de la Vie 323(4):377–384
117. Kowalczyk ND, Reina RD, Preston TJ, Chiaradia A (2015) Environmental variability drives shifts in the foraging behaviour and reproductive success of an inshore seabird. Oecologia 178(4):967–979
118. Kowalczyk ND, Reina RD, Preston TJ, Chiaradia A (2015) Selective foraging within estuarine plume fronts by an inshore resident seabird. Front Mar Sci 2:42
119. Kwok R, Comiso JC (2002) Southern Ocean climate and sea ice anomalies associated with the Southern Oscillation. J Clim 15(5):487–501
120. Labrousse S, Fraser AD, Sumner M, Tamura T, Pinaud D, Wienecke B, Kirkwood R, Ropert-Coudert Y, Reisinger R, Jonsen I, Porter-Smith R, Barbraud C, Bost C-A, Ji R, Jenouvrier S (2019) Dynamic fine-scale sea icescape shapes adult emperor penguin foraging habitat in East Antarctica. Geophys Res Lett 46(20):11206–11218
121. Labrousse S, Orgeret F, Solow AR, Barbraud C, Bost CA, Sallée J-B, Weimerskirch H, Jenouvrier S (2019) First odyssey beneath the sea ice of juvenile emperor penguins in East Antarctica. Mar Ecol Prog Ser 609:1–16
122. Lambert C, Fort J (2022) No evidence that seasonal changes in large-scale environmental conditions drive migration in seabirds. J Anim Ecol. https://doi.org/10.1111/1365-2656.13759
123. Lescroël A, Ballard G, Grémillet D, Authier M, Ainley DG (2014) Antarctic climate change: extreme events disrupt plastic phenotypic response in Adélie penguins. PLoS One 9(1):e85291
124. Lescroël A, Bost CA (2005) Foraging under contrasting oceanographic conditions: the gentoo penguin at Kerguelen Archipelago. Mar Ecol Prog Ser 302:245–261
125. Lynch HJ (2013) Gentoo Penguin (*Pygoscelis papua*). In: Borboroglu PG, Boersma PD (eds) Penguins: natural history and conservation. University of Washington Press, Seattle, pp 73–88
126. Lynch HJ, Larue MA (2014) First global census of the Adélie penguin. Auk 131:457–466
127. Masello JF, Mundry R, Poisbleau M, Demongin L, Voigt CC, Wikelski M, Quillfeldt P (2010) Diving seabirds share foraging space and time within and among species. Ecosphere 1(6):1–28
128. Massaro M, Ainley DG, Santora JA, Quillfeldt P, Lescroël A, Whitehead A, Varsani A, Ballard G, Lyver POB (2020) Diet segregation in Adélie penguins: some individuals attempt to overcome colony-induced and annual foraging challenges. Mar Ecol Prog Ser 645:205–218

129. Mattern T (2006) Marine ecology of offshore and inshore foraging penguins: The snares penguin *Eudyptes robustus* and yellow-eyed penguin *Megadyptes antipodes*. PhD dissertation. Dunedin, New Zealand: University of Otago
130. Mattern T (2013) Fiordland penguin (*Eudyptes pachyrhychus*). In: Borboroglu PG, Boersma PD (eds) Penguins: natural history and conservation. University of Washington Press, Seattle, pp 153–167
131. Mattern T (2013) Snares penguin (*Eudyptes robustus*). In: Borboroglu PG, Boersma PD (eds) Penguins: natural history and conservation. University of Washington Press, Seattle, pp 169–183
132. Mattern T (2020) Modelling marine habitat utilisation by yellow-eyed penguins along their mainland distribution: baseline information. Fisheries New Zealand, Wellington
133. Mattern T, Ellenberg U, Houston DM, Davis LS (2007) Consistent foraging routes and benthic foraging behaviour in yellow-eyed penguins. Mar Ecol Prog Ser 343:295–306
134. Mattern T, Houston DM, Lalas C, Setiwan AN, Davis LS (2009) Diet composition, continuity in prey availability and marine habitat – keystones to population stability in the snares penguin (*Eudyptes robustus*). Emu 109(3):204–213
135. Mattern T, Pütz K, Garcia-Borboroglu P, Ellenberg U, Houston DM, Long R, Lüthi B, Seddon PJ (2018) Marathon penguins – reasons and consequences of long-range dispersal in Fiordland penguins/Tawaki during the pre-moult period. PLoS One 13(8):e0198688
136. Mattern T, Wilson KJ (2018) New Zealand penguins – current knowledge and research priorities. Birds New Zealand
137. Mccutcheon C, Dann P, Salton M, Renwick L, Hoskins AJ, Gormley AM, Arnould JPY (2011) The foraging range of little penguins (*Eudyptula minor*) during winter. Emu 111(4):321–329
138. Merrifield ST, Colin PL, Cook T, Garcia-Moreno C, Mackinnon JA, Otero M, Schramek TA, Siegelman M, Simmons HL, Terrill EJ (2019) Island wakes observed from high-frequency current mapping radar. Oceanography 32(4):92–101
139. Moore GJ, Wienecke B, Robertson G (1999) Seasonal change in foraging areas and dive depths of breeding king penguins at Heard Island. Polar Biol 21(6):376–384
140. Moore PJ (1999) Foraging range of the yellow-eyed penguin *Megadyptes antipodes*. Mar Ornithol 27:49–58
141. Muller CG, Chilvers BL, Chiaradia A, French RK, Kato A, Ropert-Coudert Y, Battley PF (2021) Foraging areas and plasticity of yellow-eyed penguins *Megadyptes antipodes* in their subantarctic range. Mar Ecol Prog Ser 679:149–162
142. Muller CG, Chilvers BL, French RK, Battley PF (2020) Diving plasticity in the ancestral range of the yellow-eyed penguin *Megadyptes antipodes*, an endangered marine predator. Mar Ecol Prog Ser 648:191–205
143. Munro J, Burg TM (2017) A review of historical and contemporary processes affecting population genetic structure of Southern Ocean seabirds. Emu-Austral Ornithol 117(1):4–18
144. Murphy RC (1936) Oceanic birds of South America. Macmillan Company, New York
145. Nardelli SC, Cimino MA, Conroy JA, Fraser WR, Steinberg DK, Schofield O (2021) Krill availability in adjacent Adélie and gentoo penguin foraging regions near Palmer Station, Antarctica. Limnol Oceanogr 66(6):2234–2250
146. Nicol S, Pauly T, Bindoff NL, Wright S, Thiele D, Hosie GW, Strutton PG, Woehler E (2000) Ocean circulation off East Antarctica affects ecosystem structure and sea-ice extent. Nature 406(6795):504–507
147. Nisbet ICT, Veit RR, Auer SA, White TP (2013) Marine birds of the eastern United States and the Bay of Fundy. Nuttall Ornithological Monographs, No. 29. Cambridge, MA
148. Norman FI (1991) Counts of little penguins Eudyptula minor in port Phillip Bay and off southern Phillip Island, Victoria, 1986–1988. Emu 91:287–301
149. Orsi AH, Whitworth LT III, Nowlin WD Jr (1995) On the meridional extent and fronts of the Antarctic circumpolar current. Deep-Sea Res I Oceanogr Res Pap 42(5):641–673
150. Parmelee DF, Fraser WR, Neilson DR (1977) Birds of the Palmer Station area. Antarct J US 12:14–21

151. Pelletier L, Chiaradia A, Kato A, Ropert-Coudert Y (2014) Fine-scale spatial age segregation in the limited foraging area of an inshore seabird species, the little penguin. Oecologia 176(2):399–408
152. Pennycuick CJ (1987) Flight of seabirds. In: Croxall JP (ed) Seabirds: feeding biology and role in marine ecosystems. Cambridge University Press, pp 43–62
153. Pennycuick CJ, Croxall JP, Prince PA (1984) Scaling of foraging radius and growth rate in petrels and albatrosses. Ornis Scand 15:145–154
154. Perissinotto R, Laubscher RK, Mcquaid CD (1992) Marine productivity enhancement around Bouvet and the South Sandwich Islands (Southern Ocean). Mar Ecol Prog Ser 88:41–53
155. Perissinotto R, Lutjeharms JRE, Van Ballegooyen RC (2000) Biological–physical interactions and pelagic productivity at the Prince Edward islands, Southern Ocean. J Mar Syst 24:327–341
156. Péron C, Weimerskirch H, Bost CA (2012) Projected poleward shift of king penguins' (*Aptenodytes patagonicus*) foraging range at the Crozet Islands, southern Indian Ocean. Proc R Soc B Biol Sci 279(1738):2515–2523
157. Phillips LR, Hindell M, Hobday AJ, Lea MA (2019) Variability in at-sea foraging behaviour of little penguins *Eudyptula minor* in response to fine-scale environmental features. Mar Ecol Prog Ser 627:141–154
158. Pocklington R (1979) An oceanographic interpretation of seabird distributions in the Indian Ocean. Mar Biol 51(1):9–21
159. Pütz K (2002) Spatial and temporal variability in the foraging areas of breeding king penguins. Condor 104(3):528–538
160. Pütz K, Ingham RJ, Smith JG, Lüthi BH (2002) Winter dispersal of rockhopper penguins *Eudyptes chrysocome* from The Falkland Islands and its implications for conservation. Mar Ecol Prog Ser 240:273–284
161. Pütz K, Rey AA, Otley H (2013) Southern Rockhopper penguin (*Eudyptes chrysocome*). In: Borboroglu PG, Boersma PD (eds) Penguins: natural history and conservation. University of Washington Press, Seattle, pp 113–129
162. Pütz K, Rey AR, Huin N, Schiavini A, Pütz A, Lüthi BH (2006) Diving characteristics of southern rockhopper penguins (*Eudyptes c. chrysocome*) in the Southwest Atlantic. Mar Biol 149(2):125–137
163. Pütz K, Ropert-Coudert Y, Charrassin JB, Wilson RP (1999) Foraging areas of king penguins *Aptenodytes patagonicus* breeding at Possession Island, southern Indian Ocean. Mar Ornithol 27:77–84
164. Ratcliffe N, Crofts S, Brown R, Baylis AMM, Adlard S, Horswill C, Venables H, Taylor P, Trathan PN, Staniland IJ (2014) Love thy neighbour or opposites attract? Patterns of spatial segregation and association among crested penguin populations during winter. J Biogeogr 41(6):1183–1192
165. Rau GH, Ainley DG, Bengston JL, Torres JJ, Hopkins TL (1992) $^{15}N/^{14}N$ and $^{13}C/^{12}C$ in Weddell Sea birds, seals, and fish: implications for diet and trophic structure. Mar Ecol Prog Ser 84:1–8
166. Raya Rey A, Trathan P, Pütz K, Schiavini A (2007) Effect of oceanographic conditions on the winter movements of rockhopper penguins *Eudyptes chrysocome chrysocome* from Staten Island, Argentina. Mar Ecol Prog Ser 330:285–295
167. Raymond B, Lea MA, Patterson T, Andrews-Goff V, Sharples R, Charrassin JB, Cottin M, Emmerson L, Gales N, Gales R, Goldsworthy SD, Harcourt R, Kato A, Kirkwood R, Lawton K, Ropert-Coudert Y, Southwell C, Van Den Hoff J, Wienecke B, Woehler E, Wotherspoon S, Hindell MA (2015) Important marine habitat off East Antarctica revealed by two decades of multi-species predator tracking. Ecography 38(2):121–129
168. Rebstock GA, Boersma PD (2018) Oceanographic conditions in wintering grounds affect arrival date and body condition in breeding female Magellanic penguins. Mar Ecol Prog Ser 601:253–267

169. Ribic CA, Ainley DG (1989) Constancy of seabird species assemblages: an exploratory look. Biol Oceanogr 6(2):175–202
170. Ribic CA, Ainley DG, Ford RG, Fraser WR, Tynan CT, Woehler EJ (2011) Water masses, ocean fronts, and the structure of Antarctic seabird communities: putting the eastern Bellingshausen Sea in perspective. Deep-Sea Res II Top Stud Oceanogr 58(13–16):1695–1709
171. Robinson SA, Hindell MA (1996) Foraging ecology of gentoo penguins *Pygoscelis pupua* at Macquarie Island during the period of chick care. Ibis 138(4):722–731
172. Ropert-Coudert Y, Hindell MA, Phillips R, Charrassin JB, Trudelle L, Raymond B (2014) Biogeographic patterns of birds and mammals. In: De Broyer C, Koubbi P, Griffiths HJ, Raymond B, Udekem d'Acoz C, Van de Putte AP, Danis B, David B, Grant S, Gutt J, Held C, Hosie G, Huettmann F, Post A, Ropert-Coudert Y (eds) Biogeographic atlas of the Southern Ocean. Scientific Committee on Antarctic Research, Cambridge, pp 364–387
173. Rosciano NG, Polito MJ, Rey AR (2016) Do penguins share? Evidence of foraging niche segregation between but not within two sympatric, central-place foragers. Mar Ecol Prog Ser 548:249–262
174. Saba GK, Fraser WR, Saba VS, Iannuzzi RA, Coleman KE, Doney SC, Ducklow HW, Martinson DG, Miles TN, Patterson-Fraser DL, Stammerjohn S, Steinberg DK, Schofield OM (2014) Winter and spring controls on the summer food web of the coastal West Antarctic Peninsula. Nat Commun 5:1–8
175. Saenz BT, Ainley DG, Daly KL, Ballard G, Conlisk E, Elrod ML, Kim SL (2020) Drivers of concentrated predation in an Antarctic marginal-ice-zone food web. Sci Rep 10:7282
176. Santora JA, Larue MA, Ainley DG (2020) Geographic structuring of Antarctic penguin populations. Glob Ecol Biogeogr 29(10):1716–1728
177. Scheffer A, Trathan PN, Edmonston JG, Bost C-A (2016) Combined influence of meso-scale circulation and bathymetry on the foraging behaviour of a diving predator, the king penguin (*Aptenodytes patagonicus*). Prog Oceanogr 141:1–16
178. Schofield O, Ducklow H, Bernard K, Doney S, Patterson-Fraser D, Gormon K, Martinson D, Meredith M, Saba G, Stammerjohn S, Steinberg D (2013) Penguin biogeography along the West Antarctic Peninsula: testing the canyon hypothesis with Palmer LTER observations. Oceanography 26(3):204–206
179. Schofield O, Ducklow HW, Martinson DG, Meredith MP, Moline MA, Fraser WR (2010) How do polar marine ecosystems respond to rapid climate change? Science 328:1520–1523
180. Sergi S, Baudena A, Cotté C, Ardyna M, Blain S, D'ovidio F (2020) Interaction of the Antarctic circumpolar current with seamounts fuels moderate blooms but vast foraging grounds for multiple marine predators. Front Mar Sci 7:416
181. Serratosa J, Hyrenbach KD, Miranda-Urbina D, Portflitt-Toro M, Luna N, Luna-Jorquera G (2020) Environmental drivers of seabird at-sea distribution in the eastern South Pacific Ocean: assemblage composition across a longitudinal productivity gradient. Front Mar Sci 6:838
182. Sherley RB, Ludynia K, Dyer BM, Lamont T, Makhado AB, Roux JP, Scales KL, Underhill LG, Votier SC (2017) Metapopulation tracking juvenile penguins reveals an ecosystem-wide ecological trap. Curr Biol 27(4):563–568
183. Smith RC, Domack E, Emslie S, Fraser WR, Ainley DG, Baker K, Kennett J, Leventer A, Mosley-Thompson E, Stammerjohn S, Vernet M (1999) Marine ecosystem sensitivity to historical climate change: Antarctic Peninsula. Bioscience 49:393–404
184. Smith WO Jr, Ainley DG, Cattaneo-Vietti R, Hofmann EE (2012) The Ross Sea continental shelf: regional biogeochemical cycles, trophic interactions, and potential future changes. In: Rogers AD, Johnston NM, Murphy EJ, Clarke A (eds) Antarctic ecosystems: an extreme environment in a changing world. Blackwell Publishing, London, pp 213–242
185. Smith WO Jr, Nelson DM (1985) Phytoplankton bloom produced by a receding ice edge in the Ross Sea: spatial coherence with the density field. Science 227(4683):163–166
186. Spear LB, Ainley DG, Webb SW (2003) Distribution, abundance and behaviour of Buller's, Chatham Island and Salvin's albatrosses off Chile and Peru. Ibis 145:253–269

187. Spear LB, Ainley DG (1997) Flight speed of seabirds in relation to wind speed and direction. Ibis 139:234–251
188. Spear LB, Ainley DG (1999) Migration routes of sooty shearwaters in the Pacific Ocean. Condor 101:205–218
189. Spear LB, Ainley DG (2008) The seabird community of the Peru current, 1980-1995, with comparisons to other eastern boundary currents. Mar Ornithol 36:125–144
190. Springer AM, Van Vliet GB, Bool N, Crowley M, Fullagar P, Lea M-A, Mcnash R, Price C, Vertigan C, Woehler EJ (2018) Transhemispheric ecosystem disservices of pink salmon in a Pacific Ocean macrosystem. Proc Natl Acad Sci U S A 115(22):E5038–E5045
191. Stahel C, Gales R (1987) Little penguin. New South Wales University Press, Kensington
192. Stokes DL, Boersma PD, De Casenave JL, Garcia-Borboroglu P (2014) Conservation of migratory Magellanic penguins requires marine zoning. Biol Conserv 170:151–161
193. Strycker N, Wethington M, Borowicz A, Forrest S, Witharana C, Hart T, Lynch HJ (2020) A global population assessment of the Chinstrap penguin (*Pygoscelis antarctica*). Sci Rep 10:19474
194. Tanton JL, Reid K, Croxall JP, Trathan PN (2004) Winter distribution and behaviour of gentoo penguins *Pygoscelis papua* at South Georgia. Polar Biol 27(5):299–303
195. Terrill SB, Able KP (1988) Bird migration terminology. The Auk 105:205–206
196. Thatje S, Hillenbrand CD, Mackensen A, Larter R (2008) Life hung by a thread: endurance of Antarctic fauna in glacial periods. Ecology 89(3):682–692
197. Thiebot JB, Cherel Y, Acqueberge M, Prudor A, Trathan PN, Bost CA (2014) Adjustment of pre-moult foraging strategies in macaroni penguins *Eudyptes chrysolophus* according to locality, sex and breeding status. Ibis 156(3):511–522
198. Thiebot JB, Cherel Y, Crawford RJM, Makhado AB, Trathan PN, Pinaud D, Bost CA (2013) A space oddity: geographic and specific modulation of migration in *Eudyptes* penguins. PLoS One 8(8):e71429
199. Thiebot JB, Lescroël A, Barbraud C, Bost C-A (2013) Three-dimensional use of marine habitats by juvenile emperor penguins *Aptenodytes forsteri* during post-natal dispersal. Antarct Sci 25:536–544
200. Thiebot JB, Lescroël A, Pinaud D, Trathan PN, Bost CA (2011) Larger foraging range but similar habitat selection in non-breeding versus breeding sub-Antarctic penguins. Antarct Sci 23(2):117–126
201. Thompson D (2016) Penguins reveal unknown swimming talents. NIWA Press Release. https://www.niwa.co.nz/news/penguins-revealunknown-swimming-talents
202. Torres JJ, Aarset AV, Donnelly J, Hopkins TL, Lancraft TM, Ainley DG (1994) Metabolism of Antarctic micronektonic crustacea as a function of depth of occurrence and season. Mar Ecol Prog Ser 113:207–219
203. Trathan PN, Bishop C, Maclean G, Brown P, Fleming A, Collins MA (2008) Linear tracks and restricted temperature ranges characterise penguin foraging pathways. Mar Ecol Prog Ser 370:285–294
204. Trathan PN, Warwick-Evans V, Hinke JT, Young EF, Murphy EJ, Carneiro APB, Dias MP, Kovacs KM, Lowther AD, Godø OR, Kokunun N, Kim JH, Takahashi A, Santos M (2018) Managing fishery development in sensitive ecosystems: identifying penguin habitat use to direct management in Antarctica. Ecosphere 9(8):e02392
205. Turner J, Comiso JC, Marshall GJ, Lachlan-Cope TA, Bracegirdle T, Maksym T, Meredith MP, Wang Z, Orr A (2009) Non-annular atmospheric circulation change induced by stratospheric ozone depletion and its role in the recent increase of Antarctic Sea ice extent. Geophys Res Lett 36:L08502
206. Tyler J, Bonfitto MT, Clucas GV, Reddy S, Younger JL (2020) Morphometric and genetic evidence for four species of gentoo penguin. Ecol Evol 10(24):13836–13846
207. Tynan CT (1998) Ecological importance of the Southern Boundary of the Antarctic circumpolar current. Nature 392:708–710

208. Warwick-Evans V, Downie R, Santos M, Trathan PN (2019) Habitat preferences of Adélie *Pygoscelis adeliae* and chinstrap penguins *Pygoscelis Antarctica* during pre-moult in the Weddell Sea (Southern Ocean). Polar Biol 42:703–714
209. Watson JR, Fuller EC, Castruccio FS, Samhouri JF (2018) Fishermen follow fine-scale physical ocean features for finance. Front Mar Sci 5:46
210. Weavers BW (1992) Seasonal foraging ranges and travels at sea of little penguins *Eudyptula minor*, determined by radiotracking. Emu 91:302–317
211. Weavers BW, Dowling B, Smout AD (1991) Radio-tracking the little penguin (*Eudyptula minor*) off Phillip Island, South-Eastern Australia: techniques and field data. Arthur Rylah Institute for Environmental Research
212. Weichler T, Garthe S, Luna-Jorquera GMJ (2004) Seabird distribution on the Humboldt Current in northern Chile in relation to hydrography, productivity, and fisheries. ICES J Mar Sci 61(1):148–154
213. Whitehead TO, Connan M, Ropert-Coudert Y, Ryan PG (2017) Subtle but significant segregation in the feeding ecology of sympatric penguins during the critical pre-moult period. Mar Ecol Prog Ser 565:227–236
214. Whitney JL, Gove JM, Mcmanus MA, Smith KA, Lecky J, Neubauer P, Phipps JE, Contreras EA, Kobayashi DR, Asner GP (2021) Surface slicks are pelagic nurseries for diverse ocean fauna. Sci Rep 11:3197
215. Wienecke BC, Robertson G (1997) Foraging space of emperor penguins *Aptenodytes forsteri* in Antarctic shelf waters in winter. Mar Ecol Prog Ser 159:249–263
216. Williams TD (1995) The penguins, Sphenisidae. Oxford University Press, Oxford
217. Wilson PR, Ainley DG, Nur N, Jacobs SS, Barton KJ, Ballard G, Comiso JC (2001) Adélie penguin population change in the Pacific sector of Antarctica: relation to sea-ice extent and the Antarctic circumpolar current. Mar Ecol Prog Ser 213:301–309
218. Wilson RP (1985) The jackass penguin, *Spheniscus dernersus*, as a pelagic predator. Mar Ecol Prog Ser 25:219–227
219. Woo M, Pattiaratchi C (2010) How the Capes current ends: an investigation of a west Australian coastal current using an autonomous ocean glider
220. Xavier JC, Cherel Y, Ceia FR, Queirós JP, Guimarães B, Rosa R, Cunningham DM, Moors PJ, Thompson DR (2018) Eastern rockhopper penguins *Eudyptes filholi* as biological samplers of juvenile and sub-adult cephalopods around Campbell Island, New Zealand. Polar Biol 41(10):1937–1949
221. Younger JL, Clucas GV, Kao D, Rogers AD, Gharb K, Hart T, Miller KJ (2017) The challenges of detecting subtle population structure and its importance for the conservation of emperor penguins. Mol Ecol 26(15):3883–3897
222. Zimmer I, Wilson RP, Gilbert C, Beaulieu M, Ancel A, Plötz J (2008) Foraging movements of emperor penguins at pointe Géologie, Antarctica. Polar Biol 31(2):229–243

Chapter 4
Sea Food: The Fish-Bird Menu

> "Food has a culture. It has a history. It has a story." Winona LaDuke (March 2012, TED talk: "Seeds of our ancestors, seeds of life")

In 1987, J. Croxall and G. Lishman published a summary of the diet of penguins that derived its information from 31 studies presented in a table spanning 2 pages [29]. They had the following to say in summary (pp. 103–105):

> "There are few records of penguins taking prey other than fish, squid or crustaceans. The smaller penguins…south of the Antarctic Convergence [APF] feed mainly on crustaceans…almost exclusively…krill; fish are taken….only by Adélie penguins and…Gentoo Penguins. In contrast, the large…penguins in the same area take fish (and squid)…For penguins….at or north of the [APF}, but south of the [STF], there is no systematic patterns to the importance of…dietary categories but squid and/or fish are….at least as important as crustaceans…Little is known of the fish and squid prey. Further north the Little Penguin and *Spheniscus* species…prey chiefly on abundant shoaling fish…"

Since then, we have gained significant knowledge about what penguins eat and how location, colony size, gender, and season affect their diet. Some of this is covered in the next chapter (Chap. 5), but in all we cover information derived from more than 140 studies, with the results summarized in a table spanning 10 pages (Appendix Table A2).

General Considerations

The study of diet in seabirds, and, in our case, penguins, has a long history owing to the insights gained into better understanding seabirds' natural history by knowing what they eat and from where/how they obtain their prey. Not only do we learn about the fruits of penguins' foraging efforts, that is, foraging efficiency (energy expended for energy gained), but, often times, we learn more about the prey's natural history that was previously unknown. That is, a penguin is a "sampling device." We also learn, ecologically speaking, where a penguin resides in the food web or in

D. G. Ainley, R. P. Wilson, *The Aquatic World of Penguins*, Fascinating Life Sciences, https://doi.org/10.1007/978-3-031-33990-5_4

the larger marine community, and acquire the basis for investigating or just understanding what its actions are in that role as well as the effects that it might have on its prey. Finally, we learn who penguins' competitors are, be they commercial fishing, other upper food web predators such as whales, or other penguins themselves, including between species and also within the same penguin species or even the same colony. In other words, how does inter- and intraspecific competition for food play a role in penguin population trends and colony distribution, among other things? Indeed, the life history of any wild organism, including penguins, can be seen as a game of risk, which involves eating but not getting eaten, living long and effectively enough to reproduce viable young, and doing so in the face of ecological issues. We cannot really gather insights into these issues if we do not know what penguins eat and from where they obtain their food.

Research efforts have been devoted by fisheries to learn about fish that, as it turns out, are not only a human-exploited resource but are also the prey of seabirds/penguins and other creatures such as marine mammals. In cases in which ichthyology (fish research) is underway or has been undertaken, we have a head start in knowing more about penguins by virtue of knowing their diet. If a prey species is extremely well-known in terms of where its various life stages occur, the degree of its abundance, and how those variables change seasonally, year to year or even decade by decade, then we can gain insights into some of the major factors that affect a penguin's life history strategy. For instance, specific prey species are not found everywhere in a penguin colony's region at all times of the year (e.g., Van Heezik [74])– if the prey is a species that dwells on or near the bottom, then we know that the penguin had to dive to demersal depths to obtain it, or, in contrast, if the prey occurs only in the upper depths of the water column, then the penguin puts in less diving effort, and, finally, if it is a prey species that occurs only in waters of the continental shelf, as opposed to the deep ocean beyond the shelf, then we get some idea about how far the penguin traveled to obtain it. If during one season, year, or decade, the penguin had to make many foraging trips, compared to fewer trips at other times, then we know that the availability, or even the abundance, of that prey is varied [26, 56, 57, 68]. Again, this is an example of a penguin being a sampling device. Varying prey abundance could have consequences for penguins, which are seabirds that cannot search as easily or as widely for food as can aerial species (the subject of other chapters), and less abundance could lead to prey switching, perhaps to a more difficult-to-catch prey or to a prey of less quality, e.g., those with less fat and lower energy content or density (for example, Ludynia et al. [45]).

To add to our understanding of diet, the practice has recently been perfected, or at least newly applied to penguins, in which the determination of isotopic ratios of carbon and nitrogen in an animal's blood (or other tissues) can tell us the trophic level of its prey. In the case of penguins in which there is a range of different types of prey at the trophic level, say fish versus crustaceans, using what are called "isotopic mixing models," we can assess whether a penguin ate more fish than krill (because fish occur at a higher trophic level) and/or caught prey in waters over the continental shelf as opposed to beyond the continental shelf (variation in carbon richness, more so in shelf waters; for example, Forero et al. [33] and Polito et al. [58]). Unfortunately, such a method cannot tell us the species composition of the diet, unless it is fairly simple, i.e., one krill and one fish species at different trophic levels (e.g., Ainley et al. [3]), nor can it compensate for secondary prey (i.e., prey in the stomachs of penguins' prey). Therefore, generally, what can be learned from stable isotope analysis is confined to the general foraging niche or whether it differs among co-occurring penguins [35, 59]. Additionally, the practice of biologging and use of electronic trackers has been perfected to better provide information on penguins' foraging efforts (distance of trips, depth of diving, frequency of diving, etc.; for example, Kooyman [40] and Wilson and Peters [80]).

While definitely providing new insights, almost always such isotopic and biologging information is gathered without any context to why penguins forage in the way they do, what species they actually eat, and with whom they share their food. As well, to what degree does energy expended during foraging achieve an adequate energy gained? These questions can only be answered by knowing what prey penguins actually catch, how available that prey actually is (measured independently), and what its nutritive value is. When a penguin consumes a commercially valuable prey species, then this is usually an ideal seabird research situation owing to the research that fishery scientists are conducting on fish, which is often quite extensive for some prey species. Fishery scientists are tasked with this research, or monitoring, because, theoretically, they are responsible for determining how much penguin prey can be extracted, without harming the penguin (and other predator) populations. This is a subject discussed in Chap. 5.

Diet Quality: Survival in Cold Water

Energy Density of Prey

The "quality" of a penguin's diet, as in all seabirds, depends on the accessibility, abundance, and energetic aspects of its prey [8, 18, 20, 45, 62]. The quality or nutritional value of prey is largely determined by lipid content, which varies significantly between species ([10, 42]; Appendix Table A4); a high-energy diet, for example, one based on fish rather than krill, results in elevated chick growth rates and increased digestive and metabolic efficiency and survival to fledging [8, 20, 65]. In some seabird species, adults preferentially feed their chick(s) fish rather than invertebrates, such as krill or squid, because fish provide higher-calorific, protein and calcium content, lower salt loads, and are easier to digest, thus leading to more rapid growth [18, 33, 74]. Penguins may also actively enhance chick survival by increasing the frequency at which chicks are provisioned and by bringing back larger meals [19, 20, 65, 74]. Nutritional stress in penguins, possibly with reduced breeding success, can result from absolute reductions in the abundance of their preferred prey on local or ecosystem-wide scales, changes in the composition of prey communities, resulting in an increase in the abundance of species deficient in energetic content or one or more nutrients, and reductions in the energy content of commonly taken prey [21, 22, 45, 68]. Observed population declines in some seabird and penguin species have been attributed to changes in the structure of marine communities that lead to a switch to prey with relatively low lipid densities or energy content [45, 50]. Even if abundant, prey with a low nutritional value may not be adequate to meet the energetic requirements of growing chicks [20].

Prey Size May or May Not Differ by Penguin Size

Then, we consider the size of prey taken by penguins (Table 4.1). It has long been known that body/bill size in birds strongly affects the size of the prey taken, with prey size being an important component of a species' foraging niche [14, 15, 67, 69, 70]. In addition, by maximizing both the size of the prey and its energy density, a seabird gains optimal foraging efficiency, i.e., maximizing "bang for the buck." In the absence of one of these alternatives, a seabird somehow has to compensate in its prey switching [54, 75]. It comes as no surprise, therefore, that the largest penguins, the emperor penguins, are able to and do take the largest prey (see both mean size and range of sizes in Table 4.1). This species makes the deepest dives, and, so, it

Table 4.1 A summary of the prey size among the penguin species, as indicated by fish total length (TL)

	All fish		
	Mean size	All fish Range	Main fish Mean size
	TL ± SD	TL	TL ± SD
Emperor penguins	21.4 ± 10.1	1–38	12.5 ± 4.5
King penguins	11.3 ± 6.5	2–33	6.6 ± 1.9
Adélie penguins	14.4 ± 3.6	1–24	9.4
Chinstrap penguins	17.1 ± 10.2	1–52	9.6
Gentoo penguins	12.5 ± 7.0	1–45	12.0 ± 6.0
Yellow-eyed penguins	10.9 ± 6.2	1–32	10.1 ± 5.5
NZ little penguins	7.9 ± 4.8	1–17	7.9 ± 4.8
Australian little penguins	6.7 ± 3.0	1–12	7.7 ± 1.2
African penguins	8.3 ± 3.0	2–26	8.8 ± 3.5
Humboldt penguins	12.5 ± 4.6	5–21	11.7 ± 4.1
Galápagos penguins	6.7 ± 3.2	3–10	9.3
Magellanic penguins	8.4 ± 4.2	2–14	8.4 ± 4.2
Royal penguins	2.4		2.4
Macaroni penguins	6.9 ± 2.8	1–21	6.9 ± 2.8
Northern rockhopper penguins	6.0		6.0
Southern rockhopper penguins	5.3 ± 1.8	1–19	5.3 ± 1.8
Fiordland penguins	3.2	1–14	3.2
Snares penguins	7.3 ± 4.0	1–12	7.3 ± 4.0

In studies that reported standard length, these were converted; details can be found in Appendix Table A4. *NZ* New Zealand, *SD* standard deviation pays dividends to find large, energy-dense prey to make its dives more worthwhile, i.e., it pays back the energetic costs of making those dives in the first place. King penguins also make deep dives (though not as deep as the emperors), and, without having access to large prey (half the size of those taken by the emperors), they take the most energy-dense fish, those of the family Myctophidae. To be sure, the fish caught by the emperors are also toward the upper end of the energy density range exhibited by penguin prey (Appendix Table A4). King penguins consume larger prey in the form of squid, but these are of relatively low energy density. Perhaps there is some compensation extended by their size? Interestingly, the other higher-latitude penguin species, specifically those of the genus *Pygoscelis* (Adélie, chinstrap, gentoo), also tend to consume larger-sized, more energy-dense fish compared to the remainder of the penguin species. In the case of chinstrap penguins, there is not a lot of information on the size of fish it consumes, given that it is a krill specialist (see below). Hence, the rather disparate size is exhibited between all fish consumed versus the most important of those consumed (in turn, a victim of small sample size with a few species affecting the result).

Interestingly, these larger penguins that consume a wider range of prey sizes are also the oldest in the penguin lineage (see Chap. 1). In fact, the skeletal and muscular structure of *Aptenodytes*, especially of the head and bill, clearly differs from that of smaller penguins and are conducive to consuming larger prey [83]. The more recent penguins, in the resource competition that is a major force in penguin evolution, seemingly had to evolve to mostly resort to consuming smaller prey. Being able to consume a wide range in prey size, which only large penguins (or seabirds in general) can do, does allow a cushion in the event of variation in prey availability [53]. Was this the reason for the earlier (in geological time) success of these lineages? The size of fish taken by yellow-eyed penguins, a relatively large species, is similar to what is taken by gentoo penguins, and both forage close to or on the bottom of the shelf habitat (see below); both have rather similar cranial structures [83]. Then, there are the *Eudyptula* (both species of little penguins) and the *Spheniscus* species (African, Humboldt, Galápagos, and Magellanic), which have similar cranial structures [83] and concentrate on the long, fusiform, schooling clupeid fish (anchovies, sardines, sprats) of, surprisingly, fairly similar size [82]. In this regard, little penguins are "fighting above their weight" in consuming these fish despite having a much smaller body and bill. The prey size if these clupeid-feeders would appear to be even more homogeneous but there seems to have been little interest among researchers in measuring prey size in the diet studies of these penguins. Perhaps given little variation in anchovy–sardine–sprat size, prey size is not all that inspiring to be measured? Finally, the *Eudyptes* penguins (rockhoppers, crested), which tend to consume plenty of pelagic crustaceans (i.e., krill), which are small, follow a diet that includes a lot of small myctophids, making up for the low energy density of krill. There is a lot of overlap in the size of adult krill and myctophids consumed by penguins.

Prey Availability

Finally, there is the question of the availability of prey that can alter diet, not only determined by the natural history cycles of the prey (for example, Van Heezik et al. [74]) but also by competition between penguins and other mesopredators or among penguins themselves. However, we cannot substantiate these possible interactions, if we do not know what the penguins, their competitors, and their facilitators are eating. Fisheries can compete with penguins (Chap. 5), as, for example, for anchovies off South Africa [27, 57, 68] and Peru [31] or for krill off the northwestern Antarctic Peninsula [72, 76]– humans taking too much of the resource can negatively affect penguin populations [27, 31, 57, 68, 77]. Baleen whales, constituting a "fishery" in their own right, are well-known to affect or have affected the availability of prey common to their own and penguins' diets (Chap. 5). The results include short-term changes in penguins' diet [5, 6, 64] or long-term changes that eventually affect penguins' population size [4, 12, 46]. Finally, penguins from a large colony can affect prey abundance in their foraging areas, necessitating their own increased foraging efforts [13] or precluding growth of neighboring, smaller colonies by consuming a disproportionate amount of the prey available within the reach of their colonies [33, 34, 44, 66]. Interestingly, on the basis of biochemical analysis of sediments beneath the sea ice where the Amanda Bay emperor penguin colony occurs, a change in the stable isotopes of nitrogen (N_2) (trophic level index) indicated that as the colony increased in size during the Little Ice Age, the penguins had to resort to consuming less fish and more krill [38] (see also Chap. 13). The penguins sought fish in preference to krill. The same phenomenon is exhibited today in the southern Ross Sea, where a large, expanding colony of Adélie penguins provides less fish to chicks than does a small colony in the region; chicks at the large colony do not grow as large [8] (Chap. 5).

Penguin Diet Composition: A Comparison

In the sections that follow, we review the diet of the different penguin species based on the results of a large body of research (Appendix Table A2). In fact, other than perhaps the common and thick-billed murres of the northern polar and subpolar latitudes, which live close to most of the world's human population and marine research stations and have further been directly exploited by humans, we probably know more about the diet of penguins than that of any other seabird and even most land bird species. This is mainly the result of the ease of approaching penguins, their large size, and thus the ease in conducting the sampling needed to determine their diet, as noted above (either directly or by biochemical means). This overview of diet feeds into, not really a pun, the other components of penguins' marine lives discussed elsewhere in this book.

Overall, there are a few emerging patterns that seemingly reveal why all penguin species do not appear in the same place or at the same time. That is, their respective distributions, by ecotype, may reflect competition for foraging space, both vertical and horizontal, and prey. The overall patterns are more complex than merely dividing penguins into "inshore" versus "offshore" foragers [28, 30], reflecting more, at least in part, on the evolutionary history of penguins (Chaps. 1 and 2). These patterns are as follows:

First, the diet of most penguin species is dominated by fish, and not krill, which perhaps dispels the generally held misconception of penguin diet by the public. The latter is possibly the result of the large number of national research stations that are present in the Scotia Sea/Antarctic Peninsula sector of the Southern Ocean, leading to numerous foraging studies on penguins (see below) and where krill are the most important. Fish, by far, are more energetically valuable than are crustaceans (see Appendix Table A4), important for predators, including penguins, which are warm-blooded and have a high-energy lifestyle (discussed at length elsewhere in this book). Unlike aerial birds, penguins cannot use the wind as a source of energy but must instead cope with the viscous qualities of their watery environment to search for and acquire food (discussed elsewhere in this book). Their only option is to obtain energy from their food. At times, when on land, they could resort to basking in the warmth of the Sun's rays, but we are not concerned with their land lives in this book.

Second, there appears to be only two penguin ecotypes that could be categorized as being prey specialists. Chinstrap penguins and macaroni penguins prey heavily on krill, with the chinstraps in particular consuming Antarctic krill almost exclusively; both species are largely confined, or have their largest populations, in that part of the Southern Ocean where Antarctic krill predominate, which is the Scotia Sea region [11, 48]. A few small populations occur on other islands near the Antarctic Polar Front (APF). Where krill are scarcer, the macaroni penguins prey on myctophid fish, with these penguins being somewhat complementary to the almost indistinguishable and closely related royal penguins confined to Macquarie Island (with more myctophids than krill). In contrast, the large proportion of other penguin species are more circumpolar, spread more evenly among subantarctic and temperate islands. Krill are particularly abundant in the Scotia/Antarctic Peninsula region, densely swarm and are slow, and occur in surface waters abundant enough to satisfy penguins. Curiously (?), the chinstrap and macaroni penguins divide the Scotia Sea into west and east and south and north, respectively, with virtually no nesting (and therefore foraging) overlap. The other penguins that are close to being specialists constitute a group of six species: the four species of the genus *Spheniscus* (African, Magellanic, Humboldt, and Galápagos penguins) and the two of the genus *Eudyptula* (Australian and New Zealand (NZ) little penguins). As noted above, they forage almost exclusively on fast-moving, dense schooling fish, characteristic of continental-shelf/continental-slope habitats, such as anchovies, sardines, and similar clupeid fish (e.g., sprats). All these are extremely high-energy density fish and are an important human fishery resource. Seemingly, these penguins would be able to catch sufficient quantities of such prey only by cooperative foraging, i.e., a flock of penguins encircling a fish school to then pick off fish at the periphery or to dash through the middle [79, 82]. The school, however, needs to be cornered first. The dramatic, black-and-white striping of their plumage may help in this process [81], which is also true for the raucous calling among the flock encircling the prey school [52] (see Chap. 14).

Third, except for the temperate coasts of the southern continents (home of the *Spheniscus* and *Eudyptula* penguins), there are two penguin species that divide between them, with little spatial overlap, all of the southern subpolar and polar oceans, namely, the emperor and king penguins (of the genus *Aptenodytes*). Both are capable of exploiting the deep depths inaccessible to all the other penguin species, thus invoking a vertical partitioning of their foraging habitat. Both follow diets dominated by high-energy fish, the Antarctic silverfish in the case of emperor penguins and various fish species of the family Myctophidae in the case of king penguins. Their secondary prey are squid. As noted above, the larger bill size of these penguins allows them to also consume a larger range in prey size than the other penguin species.

Fourth, foraging in surface waters at depths overlying those of the deep-diving *Aptenodytes* penguins are the several penguins of the genus *Eudyptes* in the subantarctic region and two more of the polar species of *Pygoscelis* (gentoo and Adélie). Gentoos have a range that extensively overlaps with the *Eudyptes* and king penguins, but the Adélies overlap extensively only with emperor penguins. These species follow a diet that extensively overlaps in composition with the deep-diving penguins in their respective regions, thus, in a sense, as noted, dividing the water column, i.e., exploiting different foraging niches by depth. Gentoos forage near the bottom of continental shelves (relatively shallow as penguin foraging depth goes), deeper than the *Eudyptes* or for that matter the Adélies, where these two *Pygoscelis* species overlap (not extensively).

Finally, somewhat alone are the yellow-eyed penguins of southern New Zealand. They are of an ecotype similar to the gentoos (demersal or lower water column foraging over the shelf), and, perhaps not surprisingly, the ranges of these two penguins do not overlap. Seemingly, since the gentoos occur on the Falkland Islands, they should also occupy New Zealand locations (similar climate), but they do not. Is that because the yellow-eyed penguins are present in NZ? To a great degree, both species acquire the juveniles of large fish, mainly of demersal character, which are the target of human fishery exploitation. Yellow-eyed penguin are in jeopardy because the industry leaves them with only larger prey of relatively low energy density [50, 77] (Chap. 5).

Polar/Subpolar, Mesopelagic Foraging Penguins

The two species of penguins of the genus *Aptenodytes*, emperor and king, owing largely to their size, are capable of using the entire water column from the surface to the extremely deep depths (to 600 m), far beyond the capabilities of all other penguin species. Both are circumpolar in their distribution, together covering the entire ocean from temperate to extreme high-latitude Antarctic coasts. That their distributions overlap only very little latitudinally, argues for a separation brought about by trophic competition (Chap. 3). The emperor penguin breeding population, numbering about 265,000 pairs, spreads among ~60 breeding sites around the Antarctic continent (table in Chap. 1), the even spread of which is believed to be the result of trophic competition among colonies [66]. Their breeding season starts in the dead of winter, allowing chicks to fledge in early summer seven or so months later. They can only do this by consuming high-energy density prey. King penguins have a much larger breeding population, totaling 1.6 million pairs found among all the subantarctic and some temperate islands (table in Chap. 1); the number of colonies must total several dozen. Their breeding, contrary to that of the emperors, is spread over 14–15 months, meaning that successful pairs breed every other year. This is probably necessary because their diet includes relatively low-energy density squid. Despite the challenges posed to researchers by these two disparate breeding strategies, the diets of both species have been fairly well-studied.

Emperor Penguins

Emperor penguins forage within the sea ice that surrounds Antarctica for several months of the year and along the edges of coastal polynyas, flaw leads, or larger-scale ice edges. They reach most depths available in coastal waters, particularly seeking submarine banks [40] and the shelf break (see Chap. 2), which actually in a sense is a bank, as the oceanic floor is much shallower than waters adjacent to the coast (owing to the isostatic depression of the continent by the mass of the Ice Cap). The shelf break, in many cases, is shallow enough for icebergs to ground (~200 m), with depths adjacent to the coast in places up to 1000+ m. In most locations, fish, particularly energy-dense Antarctic silverfish, dominate the diet from winter to summer; less prevalent in the diet is the glacial squid (Appendix Table A2.1). Both prey species are closely associated with sea ice [41]. The common occurrence in the emperor penguin diet consisting of bald rockcod, which inhabits cavities on the underside of sea ice, is testament to their foraging throughout the water column, on the surface and at the bottom, which only they can do among all other penguin species, at least when ice is overhead [60]. On occasion, they also consume Antarctic krill, though fish have much higher energetic value and are thus seemingly more appropriate for the penguins to cope with the intense thermoregulation and deep diving required for their lifestyle. During the nonbreeding season, in which diet actually has been far less intensively studied, emperor penguins occur in the pack ice along the productive shelf break and off the continental shelf, where the species dines on mesopelagic fish and squid. In this case, the myctophid species is often Antarctic lanternfish ([9].

King Penguins

King penguins frequent the type of oceanic habitat frequented by emperor penguins during their nonbreeding period, that is, open ocean or at least the more productive oceanic fronts of the open ocean (Chap. 2). Their diet is composed of mesopelagic fish, particularly high-energy content myctophids as well as squid (of lesser energetic value; Appendix Table A2.1). The species of prey consumed are similar in almost the entire circumference of their range, in particular, three species of myctophid fish (rhombic, Tension's, and electron subantarctic lanternfish) and two of squid (giant warty and greater hooked squid). King penguins concentrate their effort where eddies and gyres related to water mass boundaries tend to concentrate prey, for instance, the Antarctic Polar Front (APF) [16, 25, 71] (Chap. 2). The abovementioned prey in the diet of king penguins breeding on the Falkland Islands would indicate that they forage to the south in vicinity of the Antarctic Polar Front. The fact that the myctophids, Antarctic lanternfish, have not been found in their diet indicates that king penguins do not travel much to the waters south of the Southern Boundary of the Antarctic Circumpolar Current (Chap. 2).

Subpolar, Full Water Column/Demersal/Benthic, Continental-Shelf Foraging Penguins

The two species in this group, like in the previous group, exhibit disjunct breeding distributions, i.e., nonoverlapping, though both occur in temperate waters. That is, oceanic climate is similar between, for example, the Falkland Islands and southern New Zealand. The presence of yellow-eyed penguins in New Zealand perhaps, as noted above, precludes the gentoo penguins from nesting there owing to exploiting a similar foraging niche, through they exhibit similar foraging behavior and diet.

Yellow-Eyed Penguins

Yellow-eyed penguins are a temperate species present year round at their regionally restricted breeding areas, which include the southeast coast of the South Island, NZ, and offshore Stewart/Codfish, Auckland, and Campbell Islands [47] (Chap. 2). Apparently, they forage during the day, leaving at dawn and returning at dusk with some variation depending on whether they need to feed their chicks [63]. The total current population, much reduced in the present era, and continuing to decrease, is estimated at 1670 pairs, which breed in isolation or in small groups of just a few pairs each (Table 1.1); for example, they come and go from 42 sites on mainland New Zealand, 172 on Campbell Island [47], and 19 on Stewart Island [49]. Therefore, it is likely, or because of the dispersed distribution of small breeding assemblies (i.e., colonies), that not much intraspecific competition for food exists among them (see the "Penguin Diet Competition: A Comparison" section).

The diet of yellow-eyed penguins reveals that they mostly forage on prey <200 mm in length, found not far above or on/near the bottom (that is, demersal species), in waters over the continental or insular shelves [18, 55, 73, 74] (Appendix Table A3). Such can be surmised from their once-a-day foraging trips and the prey caught (Appendix Table A2.2). They feed on fish, including more than 20 species making up ~90% of their diet, and arrow squid, a frequent but sparse prey species in their overall diet. Even though krill are found in surface waters over the shelf in their region, these penguins do not forage on them [74]. Judging from this behavior, and the predominance of just a few fish species in their diet, yellow-eyed penguins appear to be somewhat selective in their prey choice. Primary among these is the blue cod, a shallow, rocky bottom fish that is caught as adults by recreational and commercial fisheries. Among the fish species, only a few densely school, such as sprats; the remainder, including blue cod, at best occur in loose shoals. Thus, it appears that prey are captured and consumed one or a few at a time, as confirmed by penguin-mounted video cameras [51]. That fisheries are mostly leaving behind large blue cod for the penguins is causing issues with their efforts to raise chicks, owing to prey that are too large for chicks to swallow and brought back with reduced frequency [50, 77]. The squid that they catch do occur above the bottom, within the water column.

Gentoo Penguins

The diet of gentoo penguins is one of the most frequently studied of the penguin order. This species is found for most of the year on islands and coasts in a circumpolar, temperate, and subpolar distribution. There are an estimated 387,000 pairs found in a few hundred colonies in 12 general locations or subantarctic island groups (Table 1.1). "Colonies," are generally small and include no more than a few thousand pairs or fewer. Therefore, like yellow-eyed penguins, it is not likely that there is much in the way of intraspecific competition for food within or among neighboring colonies, possibly a reason for being as spread out as they are [66]. They mostly forage during the day, which at least during summer, includes most of the 24-h period at their relatively high latitude.

To a large degree, gentoos forage on fish, particularly the deep-dwelling, demersal species (near the bottom), especially during the breeding season and chick-feeding period (Appendix Table A2.3). This is particularly true for "colonies," i.e., for the majority that are in coastal bays that tend to have broad shelves [23, 43, 61]. They also include an appreciable amount of cephalopods, including octopus, in their diet. Many of the fish prey consumed are the juveniles of larger species, for instance, nototheniid fish [1, 2]. Exceptions are the colonies of the Scotia Sea and northwestern Antarctic Peninsula where gentoo penguins forage most extensively on krill, which occur in the water column, in part because the demersal fish species of the region were fished to commercial extinction during the 1960–1970s [39] and subsequent diet study occurred after these fish disappeared [7]. Indeed, perhaps an artifact of the dearth in available fish at present, recent studies have indicated a fish size that is larger than among gentoo penguins nesting elsewhere: 17.9 ± 4.3 cm vs. 12.0 ± 6.0 cm TL (Karnovsky, unpubl.; Appendix Table A2.3). Otherwise, in this area, males catch slightly more fish than do females [35] (see the section "Intraspecific Competition" in Chap. 5). On the open coasts of oceanic islands, such as Marion, Crozet, and Kerguelen, at colonies adjacent to a narrow shelf, these penguins' diets include a significant proportion of mid-water myctophid fish and more euphausiid crustaceans, which are typically found off the shelf [37, 43]. These prey are also extensively consumed during the nonbreeding portion of the year [1, 17, 36].

When foraging off the shelf, there is a remarkable similarity in diet composition, composed of a few euphausiid crustaceans and myctophid fish, through the entire range of subpolar penguins. These crustaceans include *Euphausia valentini* and *Nauticaris marionis*, whereas the myctophids include electron subantarctic and Nichol's, Tenison's, and rhombic lanternfish or at least other representatives of these genera. Only the prey consumed from the insular shelves differ much from one island group to the next, although juveniles of various nototheniids are usually represented. Only at the Falkland Islands are schooling fish consumed, namely, the Falkland sprat; in certain places, yellow-eyed penguins also consume a close sprat relative.

Temperate, Upper Water Column, Continental-/Insular-Shelf Foraging Penguins

This penguin ecotype includes the four members of the *Spheniscus* genus (African, Magellanic, Humboldt, and Galápagos penguins) and the two members of *Eudyptula* (Australian and NZ little penguins). These are the northernmost occurring penguins, frequenting shallow, coastal locations (Chap. 2). It is possible that the two genera are closely aligned in the evolution of the penguin avifauna (Chap. 1).

Spheniscus Penguins

A remarkable association exists between the four *Spheniscus* penguin species and a community of mid-water, densely schooling, small-sized fishes that occur in boundary currents, overlying the shelf and slope waters of western and eastern South America and western South Africa/Namibia (Chap. 2). These fish, which dominate these penguins' diet, include the clupeids and similar schooling species (anchovy, sardine, saury, jack mackerel, and hake; Appendix Table A2.4). In the case of the last two fish species, it is the juveniles that the penguins consume, as the adults are far too large. In waters on both coasts of Patagonia, including the Falkland/Malvinas Islands, i.e., the southern extensions of the Magellanic penguins' range, these fish are replaced in the diet by other, small-sized schooling fish, namely, sprat, and by the schooling species of squid, e.g., arrow squid. Crustaceans are a highly infrequent prey item, although when included it is a dense schooling species, i.e., squat lobster. Owing to their abundance and dense schooling, these prey species maintain enormous industrial fisheries, though not around the Galápagos Islands.

One of the three least abundant of all penguin species, the Galápagos penguins (numbering 3000 pairs in just a few colonies), have, by far, the least studied diet, and the diet of the once abundant, but now uncommon Humboldt penguins (24,000 pairs now in just a few locations), has also been sparsely investigated. On the other hand, given the association of Peru/Humboldt Current penguins with the densely schooling fish listed above (and in the table below), the prey species of these penguins are known through fishery research but not so much because of their relative contribution to the diet. In addition, however, little is known of a seasonal diet change, if any. Humboldt penguins tend to forage in association with the boobies and cormorants that are well-known to forage on anchovies [78]. The diet of the once abundant African penguins (now numbering just 25,000 pairs) has been better studied, including in earlier years when it was abundant and, more recently, at those locations where the species remains. In contrast, the diet of Magellanic penguins is one of the most frequently studied among all penguin species.

Australian and New Zealand Little Penguins

Whereas the *Spheniscus* penguin species populate the temperate oceanic coasts of the eastern Pacific and Atlantic (east and west), the little penguins are their analogues in the western Pacific and Indian Oceans – that is, the southern temperate coasts of Australia and New Zealand. The Australian little penguins in particular, like the *Spheniscus* species, extensively subsist on anchovies, sardines, and similar surface-schooling, near-shore fish species (Appendix Table A2.5). Little penguins, not having those fish species for a choice, extensively feed on sprat, a similar fish ecotype, and the shoaling arrow squid. A co-generic relative of sprat (sandy sprat) and that of squid is a minor item in the diet of little penguins in Australian waters. Crustaceans are only an extremely minor item in the diet of either penguin species, similar to the *Spheniscus* penguins.

Polar, Upper Water Column, Continental-Shelf/ Continental-Slope Foraging Penguins

Just one penguin species qualifies as foraging on the Antarctic continental (and insular) shelf and slope in the upper portion of the water column, the Adélie penguins. Thus, to a great degree, they are equivalent to the *Spheniscus* and *Eudyptula* and, to a lesser degree, to the *Eudyptes* species of temperate and northern subantarctic waters in terms of foraging habitat. An Adélie penguin is one of just two true pagophilic penguins (the other is the emperor penguin). Thus, its association with sea ice sets it apart from the temperate, continental-shelf/continental-slope species of a similar ecotype. The Adélies rank among the top six most abundant penguins, with all having breeding populations exceeding one million pairs (Table 1.1). The Adélie population, exceeding 3.7 million pairs, nests at 189 sites in clusters of colonies, with the smaller ones being associated with one large colony. Such a distribution is an indication of trophic competition, with the large colony controlling the size of its neighboring colonies [66].

The diet of Adélie penguins is among the best studied of all penguin species at least during summer and the chick-feeding period. True to the surface depths of shelf waters where they mostly forage, their main prey include high-energy fish, Antarctic silverfish, and a krill species, crystal krill (Appendix Table A2.6). Both are also pagophilic species, like the penguins. When foraging in waters of the continental slope, crystal krill are replaced by Antarctic krill. Thus, this species' diet is fairly simple, dominated by fish, similar to its analogues foraging in waters of temperate continental shelves. The Adélies do much better in raising large, more viable chicks when their diet is dominated by energy-dense fish [8, 20]. Their diet thus greatly overlaps in the horizontal space with the deep-diving emperor penguins. However, the smaller Adélies consume much smaller prey than do the emperors (Table 3.2). During winter, when Adélies are present in the pack ice away from the coast, Antarctic lanternfish, an energy-dense myctophid, become the fish of choice. Squid are also consumed (Appendix Table A2) [32]. Winter diet, however, has not been well-studied.

Subpolar, Upper Water Column, Continental-Slope/Pelagic Foraging Penguins

This penguin ecotype includes five species, all of them belonging to the genus *Eudyptes*. Some are widespread in subantarctic waters, nesting at multiple island groups and sites, such as the two rockhopper penguins, whereas others are highly limited in distribution: Fiordland (southwest coast New Zealand) and Snares (Snares Islands) penguins and royal penguins (Macquarie Island) (Chap. 3). The rockhoppers are abundant, from >300,000 to 1.2 million pairs. Crested penguins, except for the royal penguins, however, have small populations of <100,000 pairs and all are listed as "threatened" or "endangered" (Chap. 1). The population of royals is ~500,000 pairs, but, like the crested penguins, they are found in very few, i.e., one, locations. The breeding locations of these two crested penguins overlap a little if at all with any other penguins. The larger macaroni penguins consume prey that average larger than those consumed by the other crested penguins, except for Snares crested penguins, with which they do not overlap (Chap. 2, Table 3.2). While the breeding sites of the three rockhopper penguins do not overlap one another, they do overlap with other *Eudyptes* species, but consume smaller prey. The *Eudyptes* species' respective ranges extensively overlap with those of the larger king and gentoo penguins, which consume much larger prey and dive deeper.

As a group, the *Eudyptes* penguins are a shallow-diving species in whose diet crustaceans dominate, mainly different species of open-ocean krill and amphipods (Appendix Table A2.7). When and where fish are important, the species consumed tend to be off-shelf myctophids, and, in some cases, off-shelf squid are also important. In general, it appears that these penguins have a similar diet to that of king penguins but acquire their prey at much shallower depths. Where *Eudyptes* species co-occur, for instance, southern rockhopper and macaroni/royal penguins at four locations (Marion, Possession, Heard, Macquarie Islands), the rockhopper diet by mass is composed of 70–91% crustaceans and 8–29% fish, whereas that of the macaroni/royal is 51–77% crustaceans and 19–46% fish [24]. Thus, the larger *Eudyptes* species consume larger prey, essentially more fish, than do the smaller *Eudyptes*.

Polar, Upper Water Column, Continental-Slope/Pelagic, Foraging Penguins

Two penguin species qualify as foraging mainly in the waters of the continental slope and farther to sea in the polar region, that is, at or south of the Antarctic Polar Front: the chinstrap and macaroni penguins (Chap. 2). These penguins feed in surface waters and, by far, are the most tropically specialized of any penguin species, foraging almost entirely on krill, especially Antarctic krill (Appendix Table A2.8). Any fish consumed are of the family Myctophidae or other mesopelagic species that rise to surface waters during night and are found seaward of the continental or insular shelf breaks. The small populations of macaroni penguins that nest elsewhere than the Scotia Sea predominantly prey on the other species of euphausiids.

Chinstrap Penguins

This species mostly occurs in the ice-free waters of the eastern Scotia Sea and on the northwest coast of the Antarctic Peninsula. To avoid sea ice, with Antarctic sea ice retreating southward each spring, it nests a month later than any Adélie penguins who might overlap the chinstrap range. Even though its range is mostly confined to the Scotia Sea sector, its breeding population is close to being the second largest for a penguin, about 3.4 million pairs spread among greater than 300 breeding sites (table in Chap. 1). Three tiny colonies occur outside the Scotia Sea region, downstream at the Bouvetøya, Heard, and Balleny Islands, with each having just a few hundred pairs. Chinstraps do consume the occasional myctophid fish and males do more so than females [35] (see Chap. 5).

Macaroni Penguins

This species occurs in ice-free waters with the huge majority of the population nesting on South Georgia and, in opposition to the chinstrap penguins, frequents waters of the western Scotia Sea (Chap. 2). Like the chinstraps, small populations nest on a few other islands that occur close to the Antarctic Polar Front. It does not occur on Macquarie Island, where the very closely related royal penguins occur. Macaroni penguins are the world's most abundant penguin species, with more than six million breeding pairs (Chap. 1). As with the chinstraps, they dine almost entirely on Antarctic krill and other large pelagic crustacean species and/or to some degree on myctophids.

References

1. Adams NJ, Klages NT (1989) Temporal variation in the diet of the gentoo penguin *Pygoscelis papua* at sub-Antarctic Marion Island. Colon Waterbirds 12:30–36
2. Adams NJ, Wilson MP (1987) Foraging parameters of gentoo penguins *Pygoscelis papua* at Marion Island. Polar Biol 7(1):51–56
3. Ainley DG, Ballard G, Barton KJ, Karl BJ, Rau GH, Ribic CA, Wilson PR (2003) Spatial and temporal variation of diet composition and quality within a presumed metapopulation of Adélie penguins. Condor 105(1):95–106
4. Ainley DG, Ballard G, Blight LK, Ackley S, Emslie SD, Lescroël A, Olmastroni S, Townsend SE, Tynan CT, Wilson P, Woehler E (2010) Impacts of cetaceans on the structure of southern ocean food webs. Mar Mamm Sci 26(2):482–489
5. Ainley DG, Ballard G, Dugger KM (2006) Competition among penguins and cetaceans reveals trophic cascades in the Ross Sea, Antarctica. Ecology 87(8):2080–2093
6. Ainley DG, Ballard G, Jones RM, Jongsomjit D, Pierce SD, Smith WO Jr, Veloz S (2015) Trophic cascades in the western Ross Sea, Antarctica: revisited. Mar Ecol Prog Ser 534:1–16
7. Ainley DG, Blight LK (2009) Ecological repercussions of historical fish extraction from the Southern Ocean. Fish Fish 10(1):13–38
8. Ainley DG, Dugger KM, La Mesa M, Ballard G, Barton KJ, Jennings S, Karl BJ, Lescroël A, Lyver POB, Schmidt A, Wilson P (2018) Post-fledging survival of Adélie penguins at multiple colonies: chicks raised on fish do well. Mar Ecol Prog Ser 601:239–251
9. Ainley DG, Ribic CA, Fraser WR (1992) Does prey preference affect habitat choice in Antarctic seabirds? Mar Ecol Prog Ser 90:207–221
10. Ainley DG, Wilson PR, Barton KJ, Ballard G, Nur N, Karl B (1998) Diet and foraging effort of Adélie penguins in relation to pack-ice conditions in the southern Ross Sea. Polar Biol 20(5):311–319
11. Atkinson A, Siegel V, Pakhomov EA, Rothery P, Loeb V, Ross RM, Quetin LB, Schmidt K, Fretwell P, Murphy EJ, Tarling GA, Fleming AH (2008) Oceanic circumpolar habitats of Antarctic krill. Mar Ecol Prog Ser 362:1–23
12. Ballance L, Pitman RL, Hewitt RP, Siniff DB, Trivelpiece WZ, Clapham PJ, Brownell RL Jr (2006) The removal of large whales from the Southern Ocean: evidence for long-term ecosystem effects? In: Estes JA, Demaster DP, Doak DF, Williams TE, Brownell RL Jr (eds) Whales, whaling and ocean ecosystems. University of California Press, Berkeley, pp 215–230
13. Ballance LT, Ainley DG, Ballard G, Barton K (2009) An energetic correlate between colony size and foraging effort in seabirds, an example of the Adélie penguin *Pygoscelis adeliae*. J Avian Biol 40(3):279–288

14. Bedard J (1967) Adaptive radiation in the Alcidae. Ibis 111(2):189–198
15. Bedard J (1969) Feeding of the least, crested, and parakeet auklets around St. Lawrence Island, Alaska. Can J Zool 47(5):1025–1050
16. Bost CA, Cotté C, Terray P, Barbraud C, Bon C, Delord K, Gimenez O, Handrich Y, Naito Y, Guinet C, Weimerskirch H (2015) Large-scale climatic anomalies affect marine predator foraging behaviour and demography. Nat Commun 6:8220
17. Bost CA, Koubbi P, Genevois F, Ruchon L, Ridoux V (1994) Gentoo penguin *Pygoscelis papua* diet as an indicator of planktonic availability in the Kerguelen Islands. Polar Biol 14(3):147–153
18. Browne T, Lalas C, Mattern T, Van Heezik Y (2011) Chick starvation in yellow-eyed penguins: evidence for poor diet quality and selective provisioning of chicks from conventional diet analysis and stable isotopes. Austral Ecol 36(1):99–108
19. Chapman EW, Hofmann EE, Patterson DL, Fraser WR (2010) The effects of variability in Antarctic krill (*Euphausia superba*) spawning behavior and sex/maturity stage distribution on Adélie penguin (*Pygoscelis adeliae*) chick growth: a modeling study. Deep-Sea Res II Top Stud Oceanogr 57(7-8):543–558
20. Chapman EW, Hofmann EE, Patterson DL, Ribic CA, Fraser WR (2011) Marine and terrestrial factors affecting Adélie penguin *Pygoscelis adeliae* chick growth and recruitment off the western Antarctic Peninsula. Mar Ecol Prog Ser 436:273–289
21. Cimino MA, Fraser WR, Irwin AJ, Oliver MJ (2013) Satellite data identify decadal trends in the quality of *Pygoscelis* penguin chick-rearing habitat. Glob Chang Biol 19(1):136–148
22. Clarke J, Kerry K, Irvine L, Phillips B (2002) Chick provisioning and breeding success of Adélie penguins at Béchervaise Island over eight successive seasons. Polar Biol 25(1):21–30
23. Clausen AP, Arkhipkin AI, Laptikhovsky VV, Huin N (2005) What is out there: diversity in feeding of gentoo penguins (*Pygoscelis papua*) around the Falkland Islands (Southwest Atlantic). Polar Biol 28(9):653–662
24. Cooper J, Brown CR, Gales RP, Hindell MA, Klages NT, Moors PJ, Van Heezik Y (1990) Diets and dietary segregation of crested penguins (*Eudyptes*). In: Davis LS, Darby JT (eds) Penguin biology. Academic Press, San Diego, pp 131–156
25. Cotté C, Park YH, Guinet C, Bost CA (2007) Movements of foraging king penguins through marine mesoscale eddies. Proc R Soc B Biol Sci 274(1624):2385–2391
26. Crawford RJM, Barham PJ, Underhill LG, Shannon LJ, Coetzee JC, Dyer BM, Leshoro TM, Upfold L (2006) The influence of food availability on breeding success of African penguins *Spheniscus demersus* at Robben Island, South Africa. Biol Conserv 132(1):119–125
27. Crawford RJM, Underhill LG, Upfold L, Dyer BM (2007) An altered carrying capacity of the Benguela upwelling ecosystem for African penguins (*Spheniscus demersus*). ICES J Mar Sci 64(3):570–576
28. Croxall JP, Davis LS (1999) Penguins: paradoxes and patterns. Mar Ornithol 27:1–12
29. Croxall JP, Lishman GS (1987) The food and feeding ecology of penguins. In: Croxall JP (ed) Seabirds: feeding ecology and role in marine ecosystems. Cambridge University Press, London, pp 101–134
30. Davis LS, Renner M (2003) Penguins. Yale University Press, New Haven
31. Duffy DC, Hays C, Plenge M (1984) The conservation status of Peruvian seabirds. In: Croxall JP, Evans PGH, Schreiber RW (eds) Status and conservation of the world's seabirds. Cambridge, pp 245–260
32. Emslie SD, Woehler EJ (2005) A 9000-yr record of Adélie penguin occupation and diet at the Windmill Islands, East Antarctica. Antarct Sci 17:57–66
33. Forero MG, Hobson KA, Bortolotti GR, Donázar JA, Bertellotti M, Blanco G (2002) Food resource utilization by the Magellanic penguin evaluated through stable-isotope analysis: segregation by sex and age and influence on offspring quality. Mar Ecol Prog Ser 234:289–299
34. Furness RW, Birkhead TR (1984) Seabird colony distributions suggest competition for food supplies during the breeding season. Nature 311:655–656

35. Gorman KB, Williams TD, Fraser WR (2014) Ecological sexual dimorphism and environmental variability within a community of Antarctic penguins (Genus *Pygoscelis*). PLoS One 9(3):e90081
36. Hindell M (1989) The diet of gentoo penguin *Pygoscelis papua* at Macquarie Island: winter and early breeding season. Emu 89(2):71–78
37. Hindell M, Robertson GG, Williams R (1995) Resource partitioning in four species of sympatrically breeding penguins. In: Dann P, Norman I, Reilly P (eds) The penguins: ecology and management. Surrey Beatty, Chipping Norton, pp 196–215
38. Huang T, Yang L, Chu Z, Sun L, Yin X (2016) Geochemical record of high emperor penguin populations during the Little Ice Age at Amanda Bay, Antarctica. Sci Total Environ 565:1185–1191
39. Kock K-H (1992) Antarctic fish and fisheries. Cambridge University Press, Cambridge
40. Kooyman GL, Goetz K, Williams CL, Ponganis PJ, Sato K, Eckert S, Horning M, Thorson PT, Van Dam RP (2020) Crary bank: a deep foraging habitat for emperor penguins in the western Ross Sea. Polar Biol 43(7):801–811
41. La Mesa M, Eastman JT (2012) Antarctic silverfish: life strategies of a key species in the high-Antarctic ecosystem. Fish Fish 13(3):241–266
42. Lenky C, Eisert R, Oftedal OT, Metcalf V (2012) Proximate composition and energy density of nototheniid and myctophid fish in McMurdo Sound and the Ross Sea, Antarctica. Polar Biol 35(5):717–724
43. Lescroël A, Ridoux V, Bost CA (2004) Spatial and temporal variation in the diet of the gentoo penguin (*Pygoscelis papua*) at Kerguelen Islands. Polar Biol 27(4):206–216
44. Lewis ST, Sherratt N, Hamer KC, Wanless S (2001) Evidence of intra-specific competition for food in a pelagic seabird. Nature 412(6849):816–819
45. Ludynia K, Roux JP, Jones R, Kemper J, Underhill LG (2010) Surviving off junk: low-energy prey dominates the diet of African penguins *Spheniscus demersus* at Mercury Island, Namibia, between 1996 and 2009. Afr J Mar Sci 32(3):563–572
46. Lynch HJ, Naveen R, Trathan PN, Fagan WF (2012) Spatially integrated assessment reveals widespread changes in penguin populations on the Antarctic Peninsula. Ecology 93(6):1367–1377
47. Marchant S, Higgins P (1990) Handbook of Australian, New Zealand and Antarctic birds. Oxford University Press, Melbourne
48. Marr JWS (1962) The natural history and geography of the Antarctic krill (*Euphausia superba* Dana). Discov Rep 32:33–464
49. Massaro M, Blair D (2003) Comparison of population numbers of yellow-eyed penguins, *Megadyptes antipodes*, on Stewart Island and on adjacent cat free islands. N Z J Ecol 27:107–113
50. Mattern T, Ellenberg U (2018) Yellow-eyed penguin diet and indirect effects affecting prey composition – collation of biological information. Department of Conservation, Wellington
51. Mattern T, Mcpherson MD, Ellenberg U, Van Heezik Y, Seddon PJ (2018) High definition video loggers provide new insights into behaviour, physiology, and the oceanic habitat of a marine predator, the yellow-eyed penguin. PeerJ 6:e5459
52. Mcinnes AM, Thiebault A, Cloete T, Pichergru L, Aubin T, Mcgeorge C, Pistorius PA (2020) Social context and prey composition are associated with calling behaviour in a diving seabird. Ibis 162(3):1047–1059
53. Montevecchi WA, Benvenuti S, Garthe S, Davoren GK, Fifield D (2009) Flexible foraging tactics by a large opportunistic seabird preying on forage- and large pelagic fishes. Mar Ecol Prog Ser 385:295–306
54. Montevecchi WA, Gerrow K, Buren AD, Davoren GK, Lewis KP, Montevecchi MW, Regular PM (2019) Pursuit-diving seabird endures regime shift involving a three-decade decline in forage fish mass and abundance. Mar Ecol Prog Ser 627:171–178
55. Moore PJ, Wakelin MD (1997) Diet of the yellow-eyed penguin *Megadyptes antipodes*, South Island, New Zealand, 1991–1993. Mar Ornithol 25:17–29

56. Pichegru L, Grémillet D, Crawford RJM, Ryan PG (2010) Marine no-take zone rapidly benefit threatened penguin. Biol Lett 6(4):498–501
57. Pichegru L, Ryan PG, Van Eeden R, Reid T, Grémillet D, Wanless R (2012) Industrial fishing, no-take zones and endangered penguins. Biol Conserv 156:117–125
58. Polito MJ, Trivelpiece WZ, Karnovsky NJ, Ng E, Patterson WP, Emslie SD (2011) Integrating stomach content and stable isotope analyses to quantify the diets of pygoscelic penguins. PLoS One 6(10):e26642
59. Polito MJ, Trivelpiece WZ, Patterson WP, Karnovsky NJ, Reiss CS, Emslie SD (2015) Contrasting specialist and generalist patterns facilitate foraging niche partitioning in sympatric populations of *Pygoscelis* penguins. Mar Ecol Prog Ser 519:221–237
60. Ponganis PJ, Van Dam RP, Marshall G, Knower T, Levenson DH (2000) Sub-ice foraging behavior of emperor penguins. J Exp Biol 203(21):3275–3278
61. Pütz K, Ingham RJ, Smith JG, Croxall JP (2001) Population trends, breeding success and diet composition of gentoo *Pygoscelis papua*, Magellanic *Spheniscus magellanicus* and rockhopper *Eudyptes chrysocome* penguins in the Falkland Islands. Polar Biol 24(11):793–807
62. Randall RM, Randall BM (1986) The diet of jackass penguins *Spheniscus demersus* in Algoa Bay, South Africa, and its bearing on population declines elsewhere. Biol Conserv 37(2):119–134
63. Richdale LE (1957) A population study of penguins. Oxford University Press, Oxford
64. Saenz BT, Ainley DG, Daly KL, Ballard G, Conlisk E, Elrod ML, Kim SL (2020) Drivers of concentrated predation in an Antarctic marginal-ice-zone food web. Sci Rep 10:7282
65. Salihoglu B, Fraser WR, Hofmann EE (2001) Factors affecting fledging weight of Adélie penguin (*Pygoscelis adeliae*) chicks: a modeling study. Polar Biol 24(5):328–337
66. Santora JA, Larue MA, Ainley DG (2020) Geographic structuring of Antarctic penguin populations. Glob Ecol Biogeogr 29(10):1716–1728
67. Selander RK (1966) Sexual dimorphism and differential niche utilization in birds. Condor 68(2):113–151
68. Sherley RB, Ludynia K, Dyer BM, Lamont T, Makhado AB, Roux JP, Scales KL, Underhill LG, Votier SC (2017) Metapopulation tracking juvenile penguins reveals an ecosystem-wide ecological trap. Curr Biol 27(4):563–568
69. Spear LB, Ainley DG, Walker WA (2007) Trophic relationships of seabirds in the eastern Pacific Ocean. Stud Avian Biol 35:1–99
70. Storer RW (1966) Sexual dimorphism and food habits in three North American accipiters. Auk 83(3):423–436
71. Trathan PN, Bishop C, Maclean G, Brown P, Fleming A, Collins MA (2008) Linear tracks and restricted temperature ranges characterise penguin foraging pathways. Mar Ecol Prog Ser 370:285–294
72. Trivelpiece WZ, Hinke JT, Miller AK, Reiss CS, Trivelpiece SG, Watters GM (2011) Variability in krill biomass links harvesting and climate warming to penguin populations in Antarctica. Proc Natl Acad Sci 108(18):7625–7628
73. Van Heezik Y (1990) Diets of yellow-eyed, Fiordland crested, and little blue penguins breeding sympatrically on Codfish Island, New Zealand. N Z J Ecol 17(4):543–548
74. Van Heezik Y (1990) Seasonal, geographical, and age-related variations in the diet of the yellow-eyed penguin (*Megadyptes antipodes*). N Z J Ecol 17(2):201–212
75. Warzybok P, Santora JA, Ainley DG, Bradley RW, Field JC, Capitolo PJ, Carle RD, Elliott M, Beck JN, Mcchesney GJ, Hester MM, Jahncke J (2018) Prey switching and consumption by seabirds in the central California Current upwelling ecosystem: implications for forage fish management. J Mar Syst 185:25–39
76. Watters GM, Hinke JT, Reiss CS (2020) Long-term observations from Antarctica demonstrate that mismatched scales of fisheries management and predator-prey interaction lead to erroneous conclusions about precaution. Sci Rep 10:1–9
77. Webster T (2018) The pathway ahead for hoiho. Yellow-Eyed Penguin Trust, Dunedin

78. Weichler T, Garthe S, Luna-Jorquera G, Moraga J (2004) Seabird distribution on the Humboldt Current in northern Chile in relation to hydrography, productivity, and fisheries. ICES J Mar Sci 61(1):148–154
79. Wilson RP (1985) The jackass penguin, *Spheniscus dernersus*, as a pelagic predator. Mar Ecol Prog Ser 25:219–227
80. Wilson RP, Peters G (1999) Foraging behaviour of the chinstrap penguin *Pygoscelis Antarctica* at Ardley Island, Antarctica. Mar Ornithol 27:85–95
81. Wilson RP, Ryan PG, James A, Wilson MP (1987) Conspicuous coloration may enhance prey capture in some piscivores. Anim Behav 35:1558–1560
82. Wilson RP, Wilson MP (1990) Foraging ecology of breeding *Spheniscus* penguins. In: Davis L, Darby J (eds) Penguins. Academic Press, Orlando, pp 181–206
83. Zusi RL (1975) An interpretation of skull structure in penguins. In: Stonehouse B (ed) The biology of penguins. MacMillan Press, London, pp 59–84

Chapter 5
Ecological Consequences of Diet Composition

> "Species sharing the same habitat do not generally compete for the same food, however, because otherwise one would tend to oust the others through inherent differences in hunting efficiency. [Though various penguins] may breed together and seek their food in the same areas, each species is probably adapted to taking a different proportion of the various kinds of food available……The competition between the different species of the smaller penguins in the Antarctic zone, and between penguins and whales, for the vast but not inexhaustible supplies of krill may have vital bearing upon penguin numbers." (Sparks and Soper [163]; pp. 112–113)

Food, i.e., prey such as fish or krill, is a resource, so, as is true for all resources, its availability can be limiting to various species' needs depending on how much there is, how far away it is, and the number of predators trying to exploit it. If predators of the same, or different species, are interested in acquiring the same prey, at the same time and place, then this situation is ripe for competition to occur. Whether competition occurs is dependent on the size/availability of the resource and the number of exploiters; if relatively small predator populations are involved, and plenty of food is easily available, then the likelihood is that no competition occurs. The converse is true if for one reason or another, there is a limiting factor involved.

There are two main types of competitions. The first one is called "exploitative competition," which occurs when a predator reduces the overall abundance and availability of a prey stock depriving others of what would be best for them. This is what happens when commercial fisheries or abundant, major predators such as baleen whales "forage" on penguin prey in areas where penguins forage, thus taking far more than their "fair share" (from a penguin's perspective!). This would be a large-scale interaction among predators, having to do with regional predator abundance and the overall prevalence of the exploited prey stock. For seabirds, this was first shown among cormorants that foraged along the bottom of the water column for individual fish hiding in the substrate [28]. Such a situation would be ripe for depletion, given the territorial nature of the fish being extracted and the ensuing delay caused by other fish moving in to take the place of those captured. This

D. G. Ainley, R. P. Wilson, *The Aquatic World of Penguins*, Fascinating Life Sciences, https://doi.org/10.1007/978-3-031-33990-5_5

phenomenon has since been shown to occur in situations in which intensive foraging by seabirds, including penguins, deplete the abundance of their surface and mid-water prey around their breeding colonies [8] (see below).

The other type of competition is called "interference competition," which occurs when other predators cause penguins to increase their search efforts by disrupting the preyscape (3D distribution and abundance of prey) but without necessarily reducing the overall, regional amount of the prey resource. This is a more local interaction among predators. One example would be the foraging behavior of one predator forcing the prey deeper, out of the range of other predators, without actually changing the abundance of the prey. Another example would be a flock of penguins foraging on a school of fish, but, then, a baleen whale arrives and in one gulp consumes the entire school. This forces the penguins to search for another fish school, thus increasing their effort; if this happens with high frequency, it reduces the penguins' foraging efficiency by increasing their search efforts. Foraging efficiency, i.e., the ratio of energy gained on the energy spent while foraging, is the "name of the game." Of course, if whales consume too many prey schools, then this becomes exploitative competition. Otherwise, the reader can recall their own similar experiences when shopping, especially for a product of limited availability! One has to find a store that has the product, but it might not be worth crossing the entire country to get it. This has been shown among gannets and kittiwakes that, by the intensity of their foraging, drive their prey deeper, hence making the fish less accessible to these surface-foraging birds [8, 116]. The birds have to find schools that are still near the surface.

Penguins (and people) are particularly prone to negative consequences of food resource competition due to two main factors: (1) compared to other marine predators, especially other seabirds, penguins have reduced powers to search for prey and (2), at least during certain portions of their life cycle (mainly breeding or, in some species/populations, foraging to fatten for molt), penguins (and people?) operate as a "central place forager," with the central place that they forage from being a colony (home in the case of humans) or, in some cases, even the water surface. This means that there is some factor that dictates that they forage as close to their colony (or point of origin) as possible. The first factor is exhaustively explored elsewhere in this book (Chaps. 6, 7, 8, 9 and 10) and the second more so here.

Although over their entire life span, individuals of a number of penguin species spend most of their time at sea, i.e., 95% [2] (Chap. 2), for critical periods, they are confined to the land, from which they act like yo-yo's and travel away for food and then return to the same point of land, usually their breeding site. During this yo-yoing, an individual penguin wants to obtain its food as early as possible in its foraging trip, and thus begins to search for food as close as possible to its colony. Therefore, the most easily accessible, usually closest prey are the first to be depleted. Eventually, as it becomes necessary for adult penguins to venture farther and farther away from their central place in search of food, chick growth may be compromised by the decreased feeding frequency or pair synchrony might be disrupted if one parent or the other is away too long and the other, fasting individual has reached the end of its reserves. Such a need for close-by food increases the competition, both exploitative and interference, among members of the colony or land-based group.

Below, we explore examples showing how penguins have coped, or not, with competition from other predators, including other penguins (intraspecific competition) as well as whales and fisheries (interspecific competition).

Intraspecific Competition Among Penguins

Foraging Range Is Key: General Considerations

Since the earliest days of the modern era of ecology, it has been known that large seabird colonies have large foraging areas, with the converse being true for small colonies (Box 5.1). This could have to do with the amount of foraging habitats available as hypothesized by Robert Storer [166] in his studies on penguin analogues in coastal, north polar seas, namely, the auks. In other words, once a colony reaches a certain size relative to foraging habitat availability, or once a saturated density level is reached (no more space, including 3D, for more birds), competition becomes intense and other foragers of a given species cannot be added. The excess birds need to find another colony/roosting site, one that has a suitable foraging habitat within range. The colony size–foraging area size relationship, if the habitat is not restricted, could also be due to exploiting an increasingly larger area as competition forces birds to travel farther to find food. This was the conclusion of biologists who worked on tropical oceanic islands, completely surrounded by potential foraging habitats stretching from the islands to beyond the horizon [17, 65]. In other words, there was no limit to the amount of foraging habitats. Birds from small colonies fed close to the islands, and those from larger colonies fed farther and farther away depending on the colony size and on their ability to bring food back to their colony at a frequency or trip duration that did not compromise their chicks' growth. The radius of foraging birds stretching out from a colony became known as "Ashmole's halo," despite the fact that the phenomenon was described, with less fanfare, 10 years earlier [166] (see Fig. 5.1).

This was the level of understanding that stood into the 1980s, when researchers showed that large seabird colonies, through intraspecific competition, can limit the size of neighboring colonies that forage in the same area [77]. In other words, there is a limit to the density or even volume of central place foragers in a given oceanic area, including members of all colonies within the foraging range of the numerically dominant colony. These researchers, too, considered auks in their analysis. Subsequently, this phenomenon has been shown to be the rule of thumb on the basis of studies on dozens of seabird species and as summarized by others [31, 98]. Among the 39 seabird species that these authors covered, studies of 6 penguin species were included, which we will review below. We will also consider information obtained about all other penguin species, which, as a rule, also support the colony size–foraging area size relationship as a result of competition. We will look at both two- (2D) and three-dimensional (3D) patterns, considering the depth of foraging, which, for most seabird species, is not an issue.

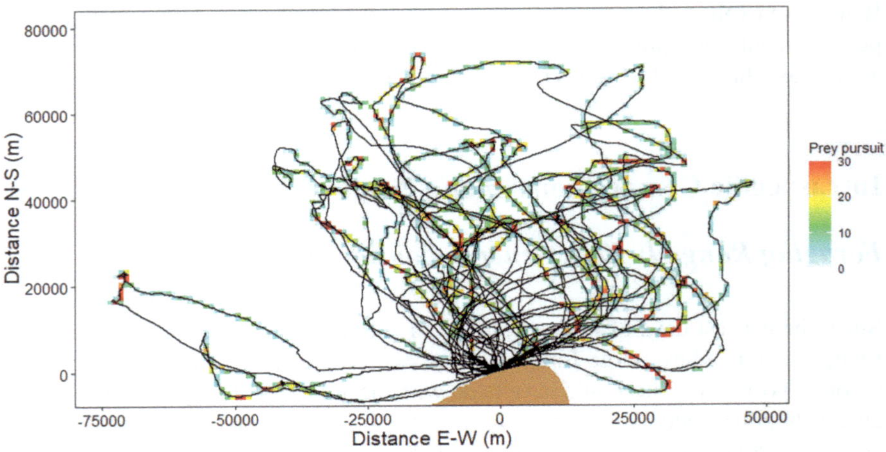

Fig. 5.1 The partial Storer–Ashmole "halo," actually a fan, exhibited by foraging Magellanic penguins at Peninsula Valdez, Argentina. The colony, which occurs in segments along a ~5-km stretch of the coast, contains about 350,000 pairs, and the data were gathered by attaching biologgers to parents during the first week after chick hatching. The loggers recorded the location and frequency of diving. The red dots indicate actual, intensive diving to catch prey (anchovies); it should be noted that there are relatively few such spots close to the colony, in part due to earlier prey depletion. As chicks grow older, and their demands for food increase, one would expect the foraging area to expand even more. (Data and graphics were obtained from Richard Gunner)

Box 5.1: How Does One Define a Penguin Colony?
During the first days of polar exploration and penguin studies, i.e., in the early 1900s, concentrations of breeding penguins were known as "rookeries," a name that is also and still applied to seals and sea lions. Pauline Reilly, then, in her book, *The Fairy Penguin and Earthy People* [150], which was a summary of what was known about Australian little penguins (called Fairy penguins up to the early 1980s) and the people investigating them, argued that the term "colony" be used rather than rookery. Doing so would be consistent with the research conducted and writing on aerial species of seabirds. Up through the 1980s, and to some degree even now, a colony has been defined as any concentration of breeding penguins that had access to a landing beach. Thus, penguins could breed on adjacent islets, with a narrow channel in between, and be called separate colonies [119, 167]. Lance Richdale [153] got around this in his studies on yellow-eyed penguins by referring to "breeding areas," which entailed a geographical unit (bay, cape) having one or more nesting-pair assemblages. However, based on banding penguins and, subsequently, searching for them beyond their natal location, it become apparent to him that penguins on nearby islets show much interchange and, so, in effect, the various groups are part of one "colony," as also found elsewhere [2, 158]. The latter authors suggest, as a rule of thumb, that penguins, as demonstrated in

(continued)

Adélie penguins, nesting within an 8-km (5 nautical miles) radius be considered a colony (possibly composed of different "breeding sites," as per Richdale [153], and as presented in Fig. 5.1 for Magellanic penguins. In the following text, when we refer to a "colony" as a known breeding site combination or as an isolated location, we will use that term. If the study was conducted at a breeding site, without reference to whether other penguins nest close by in another assemblage of nests, we also will use the word "colony".

Ultimately, the effect of seabirds depleting prey near colonies, thus requiring a larger area to allow a larger foraging population, has been shown directly, i.e., measurement of prey abundance as a function of distance from a colony (more prey farther away). This has happened in a few seabird species, including penguins. As noted above, interference competition has been shown to be at play in reducing prey availability among northern gannets [116] and black-legged kittiwakes [13], through persistent harrying of prey schools that leads to fish swimming deeper, making them harder to obtain by these near-surface-foraging seabirds. Exploitative competition, i.e., actual depletion of prey, has been shown to occur in benthic foraging cormorants, as also previously noted [28], as well as in mid-water column-foraging Adélie penguins [8]. As penguins deplete and alter the prey patches near their colonies, they are forced to forage farther away and deeper to find high-quality patches to thus maintain their foraging efficiency [75, 115, 138] (Box 5.2). Greater foraging efficiency translates to better nesting coordination and provisioning of young, leading to higher breeding success and, in turn, population regulation consequences [78]. This has actually been experimentally shown in African penguins: as boundaries of marine protected areas expanded or contracted to exclude the fishing industry, accordingly, prey availability and penguins' foraging efforts and efficiency also altered [140, 141]. With the exclusion of fishing, penguins did not have to go as far to find food.

Thus far, the foraging range of most penguin species has been investigated using biologging at least at one location or, in other words, by attaching devices that record where penguins are going or where they have been (Appendix Table A3). Such devices communicate with navigation satellites providing time and place data, sometimes needing to recapture the penguin to download the data or transmit the data in real time to a home-based computer via a satellite. Of these investigated penguin species, the foraging ranges of four (Fiordland, royal, northern rockhopper, and Galápagos penguins) have been investigated at only one colony each, thus directly assessing whether the presence of a colony size–foraging range size relationship is problematic. For two others, yellow-eyed and little penguins, which only occur in small colonies, the foraging range is not extensive and varies little from colony to colony. Their current colonies are too small to elicit trophic competition, with colony size having been negatively affected by land-based predators. It could be, too, that these species' strategy is to nest in low numbers in many locations to thus reduce competition for food. This leaves six penguin species for which to evaluate the relationship and evidence of intraspecific competition, and all show a colony size–foraging area size pattern.

Box 5.2: Why Consider Trophic Competition During the Crèche Period?
Evidence of intraspecific competition in penguins is most apparent during the crèche period, when not only are both parents foraging simultaneously, thus almost doubling the foraging population from earlier in the breeding period, but they are also foraging for the hungry chick population. Before the crèche period, members of each pair took their turns foraging, one at a time. Thus, with just half the population foraging at any one time, pressure to go and come repeatedly and frequently is somewhat relaxed, as is competition for the closest available food. This is not the case during the crèche period, when chicks' are left without parental protection at the colony so that both parents can forage simultaneously. It has been shown that feeding frequency has strong effects on penguin chick growth [44, 22]. During the incubation and guard stage, which in many penguin species are periods in which parental duties are dominated by only one member of the pair (incubating eggs, guarding young chick(s)), as noted, individuals whose turn is to forage at sea experience much reduced competitive pressure. In species such as emperor, king, Adélie, royal, macaroni, and Magellanic penguins, biologging has been intensive enough that all breeding cycle periods have been investigated. Foraging ranges tend to be much greater during the incubation period, when parents have the luxury of spreading out and to some degree "taking their time," than during crèche, as well as them being highly variable [66] (Appendix Table A3). Thus, below, we will mostly be discussing about the foraging range during the crèche period, unless otherwise noted, when central place foraging exerts its maximum pressure.

Penguin Species' Central Place Foraging Range Patterns

The most intensively studied penguin species in terms of foraging range, as affected by colony size, are the Adélie penguins. Indeed, the results of the comparisons made between different colonies in 19 studies conducted at 12 colonies showed that colony size differed but in some cases has significantly changed over the past decade or more. In our review of this species, colony size is determined by the 8-km-rule discussed above [158] (Box 5.1). Moreover, the entire colony size range of these penguins was included in these studies, from the largest to the smallest (Fig. 5.2), which is not the case for several penguin species. Clearly, as an Adélie penguin colony grows in size (i.e., more breeding pairs), or in the comparison of large with small colonies, more penguins force each other to forage farther from the central place. Thus, the size of a colony's foraging area is affected by the size of the colony,

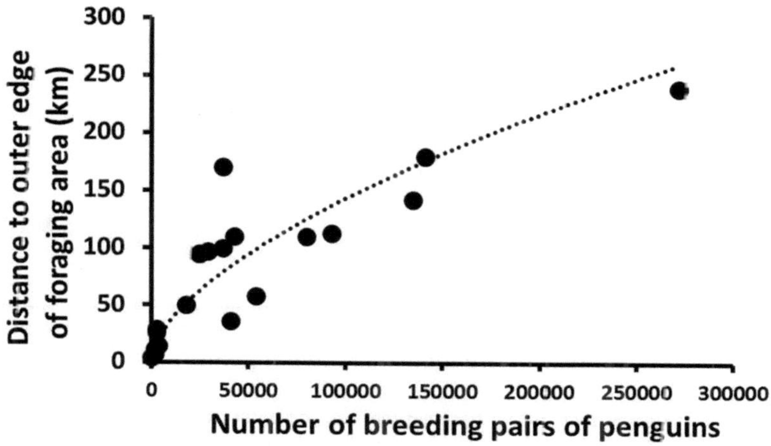

Fig. 5.2 The statistical relationship between the size of an Adélie penguin colony (number of pairs; Box 5.1) and the distance to the outer edge of its foraging area (in kilometers) during the chick crèche period when both parents are foraging simultaneously to cope with needs of their growing chick(s) (Box 5.2). When both parents are foraging, the abundance and thus the density of the foraging birds is at its highest as central place foragers

and this has to do with intraspecific competition for food. While there is a 2D pattern, there appears to be no relationship with the depth of foraging among colonies, i.e., a 3D pattern, although this is evident at one of the largest colonies where foraging has been investigated, at Cape Crozier, Ross Island (with >275,000 pairs of Adélie penguins). In this large colony, not just the foraging range but also the foraging depth increases as the chick-raising period progresses. Thus, the foraging volume increases. The increasing average depth of foraging coincides with an increase in the proportion of fish in the diet, with fish occurring deeper than krill [8] (as discussed above). The energy density of fish is much greater than that of krill (Chap. 4, Appendix Table A4), which works to maintain a level of foraging efficiency (energy gained relative to energy expended).

The phenomenon of increasing the foraging range with colony size has also been detected in gentoo and chinstrap penguins, which have received a significant amount of biologging attention as well (Appendix Table A3, Fig. 5.3). Investigations have been made at 10 and 8 breeding sites, i.e., "colonies," respectively. However, and rather important for our purposes, the foraging range has been investigated only at small "colonies," that is, up to about 5500 pairs in gentoo penguins and 20,000 pairs in chinstraps. Gentoo penguins exist where populations are double that number, and chinstraps exist in assemblages that are more than an order of magnitude larger [158, 167]. It should be noted, again, that the results that we summarize here are in terms of foraging area relative to how many penguins are nesting at a given site, i.e., a "colony." As it is, several colonies can occur among these species in a relatively confined region, with many close to one another, thus further making "colony size" a problematic term, which leads to the question, how actually large are the effective

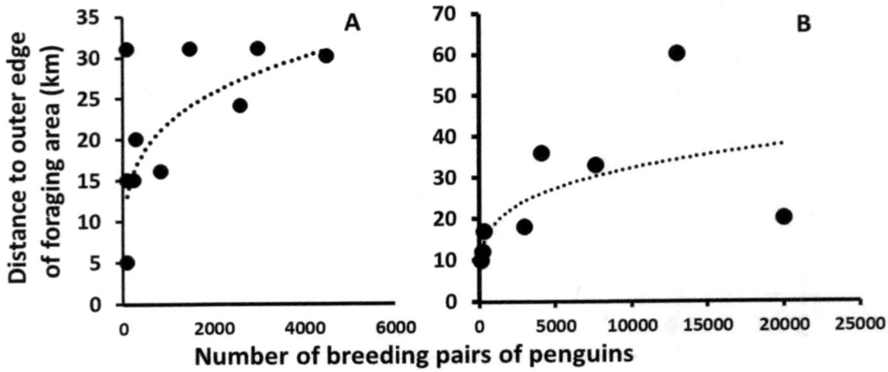

Fig. 5.3 The statistical relationship between the size of gentoo (**a**) and chinstrap (**b**) colonies (number of pairs; Box 5.1) and the distance to the outer edge of their foraging area (in kilometers) during the chick crèche period when both parents are foraging simultaneously to cope with needs of their growing chick(s) (Box 5.2)

"colonies" that contribute to foraging density, in which penguins compete (see maps in Bearhop et al. [26] and Santora et al. [157)? Thus, while the population–foraging area size trend appears to be similar to that evident among Adélie penguins at the low colony size end of their data distribution, much work is needed in these two other *Pygoscelis* species, i.e., at larger "colonies" (and perhaps in the more specific use of the term "colony" [158]), before a valid comparison can be made and we can better assess the effects of colony size on foraging behavior.

Biologging to determine foraging range and behavior repeatedly has been directed at the two *Aptenodytes* species, at six colonies for emperor and five for king penguins (Appendix Table A3). In the case of emperors, but, perhaps, not in kings, population size does refer to an actual, independent colony. This is because the distributions of emperor colonies are very regularly spaced at an average of about 220 km apart; they do not occur in clumps, and 220 km is at the periphery of the measured foraging ranges. Such even spacing of emperor colonies, with the distance in accordance with the foraging range, is believed to be a result of intraspecific competition for food [158]. That is to say, like the same poles of two magnets, emperor penguin colonies repel one another. This repulsion can happen because at the high latitudes where emperor penguins breed, there is minimal limit to the presence of fast ice on which their colonies form (Chap. 3). Therefore, it might well be fairly "easy" for them to adjust their location to minimize competition with neighboring colonies, assuring the presence of more food in the process.

It is not known what effect "colony" size may have on foraging in the case of king penguins at locations where biologging has occurred. Problematic for the penguin species, especially the emperor penguins, is that although biologging is implemented at all stages of their reproductive cycle, it is done so by often using a tiny sample size, which reduces our ability to perform inter-colonial comparisons. Nevertheless, larger populations do appear to have larger foraging ranges in emperors but not in kings (Appendix Table A3). For instance, during the emperor penguin chick guard stage (with one member of each pair foraging), the foraging range at Cape Washington (20,000–26,000 pairs) has been measured to be 160–200 km, at smaller Auster (13,400 pairs) 133 km, and at the much smaller Point Géologie (3000 pairs) 92 km. In the case of king penguins, there appears to be no population–foraging range relationship but, too, the size of study "colonies," as reported, may not represent a colony but rather only the number of penguins at the specific study site. The apparent importance of a targeted foraging habitat for these penguins, as reviewed elsewhere in this book, such as ocean current boundaries for kings and banks for emperors (Chap. 3), may well override any effect that population size has on foraging range. In keeping with this point, the foraging depth of these two species shows no population effect. In the case of emperor penguins, for sure, the maximum foraging depth appears to be greatly affected by the depth of the bank toward which their efforts are directed rather than other factors [105].

Magellanic and African penguins, as well as macaroni penguins, have also received an appreciable amount of biologging attention among researchers, with the Magellanic at eight and the African and macaroni at four breeding sites each (Appendix Table A3). Again, we do not know much about the demographic structure of these populations and the degree to which the population at a single breeding site can be defined as a full "colony" (other than the colony in Fig. 5.1). Nevertheless, biologging has been conducted at "colonies" with a range of sizes In all three species, it appears that there is a relationship between "colony" size and foraging range, though comparison is a bit challenged by data obtained from different stages of the breeding cycle. With regard to the crèche stage, the foraging range of African penguins nesting in Algoa Bay decreased from 40 to 14–25 km as the population (at two sites) decreased from ~20,000 to ~10,000 pairs over a couple of decades; the foraging range of this species among the other, smaller "colonies", in accordance, was even less than these. With regard to the Magellanic penguins, for which most data were obtained from the guard stage, the foraging range decreased from ~80 to 25 km comparing a population of 225,000 to one of 5500 pairs; sites having <2000 pairs showed no further reduction, with a foraging range of 24–30 km. Adding to the Magellanic penguin story, by performing a stable isotope analysis of their diet, it was found that as the breeding population increases, in an investigation covering "colony" size from 47 to 175,000 pairs, foraging on high-quality prey decreases (as anchovy/sprat are depleted) and the penguins begin to forage closer to the shore on lower-quality prey [76]. Finally, macaroni penguins, though not consistently sampled by the breeding stage, during the crèche period, exhibited a whopping 398-km range for a "colony" of 405,000 pairs compared to 48 km for one of 1100 pairs.

Many of the remaining penguin species, mostly those endangered or threatened, were studied where breeding populations, not surprisingly, are small. Accordingly, they exhibited short foraging ranges. That is, for all "colonies" with <3000 pairs, Fiordland, yellow-eyed, little, Galápagos, and Humboldt penguins have maximum foraging ranges of <35 km. The foraging range of yellow-eyed penguins has been investigated at a number of locations over the years (10 locations; Appendix Table A3). Where the number of pairs are 10 or less, the foraging range tends to be

<10 km, but where there are ~50 or more penguins, the foraging range can be 30–50 km. In the case of southern rockhopper penguins, during the guard stage, a population of 13,000 pairs had a foraging range of 30–36 km compared to 16 km for a population of 5000 pairs. For northern rockhoppers, a population of 25,000 pairs foraged within 6 km. Such small foraging ranges might possibly be the result, too, of temporal segregation between the sexes (one member forages at a time), thus reducing the number and density of penguins foraging at any one time, until the very end of the chick period [175]. On the other hand, where there are immense numbers, for example, in a "colony" of 500,000 pairs, royal penguins on Macquarie Island foraged as far as >700 km away during the chick guard stage.

In somewhat of an instructive twist to the patterns discussed thus far, with regard to Australian little penguins, among all of the 28 "colonies" that were arrayed along the coast of Victoria, Australia, ranging in population from 3 to 17,500 pairs (hence, generally, small as penguin "colonies" go), as "colony" size increased, chick fledging mass decreased as a function of the density of birds foraging [63] (Fig. 5.4). This was possible because, as far as we know, no foraging range extended more than 30 km from any "colony," possibly constrained by the less productive subtropical water farther offshore (Chap. 3). Thus, the ensuing increase in foraging density would increase the foraging effort of individual penguins. In turn, this pattern would explain why chick fledging mass decreased as the "colony" size and foraging density in adjacent waters increased. That is, with increased foraging density, it was more difficult for parents to bring back food to their chicks in a timely manner, owing to increased intraspecific competition, and, therefore, they brought back less food. In the sample of "colonies" in this study, colony size had a lot to do with island size and also the prevalence of predators (many colonies were on the mainland coast rather than on islands). As noted, the intention behind these penguins to reduce competition might also be to nest in concentrations of low numbers spread widely.

On the other hand, and in comparison, this is where the following example becomes especially instructive. In the case of an Adélie penguin colony at Cape Crozier that was 2 orders of magnitude larger than all 28 Australian little penguin "colonies" combined (41,850 breeding pairs), as illustrated in Fig. 5.4 [63] as the

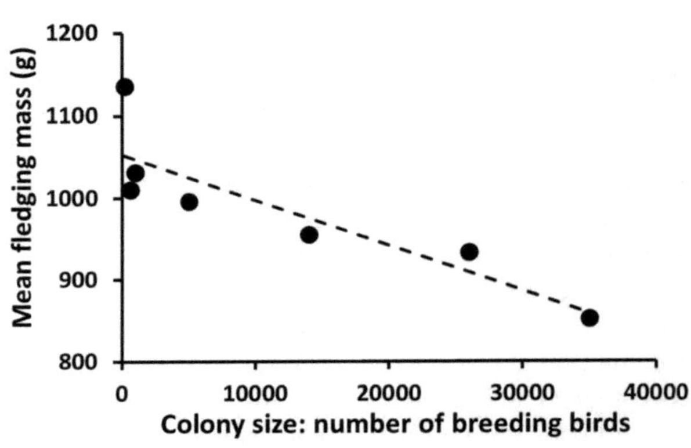

Fig. 5.4 Trends in the fledging mass of little penguins in Victoria, Australia, as a function of "colony" size (number of breeding birds, not pairs; data from Dann and Norman [63])

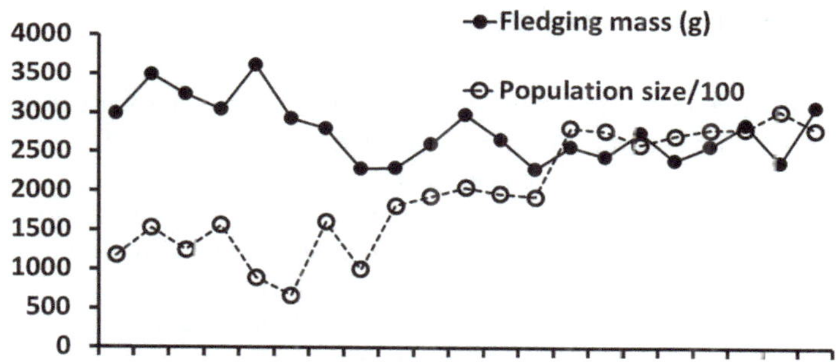

Fig. 5.5 The change in fledging mass (in grams) as the Cape Crozier Adélie penguin colony grew (number of pairs/100), 1996–2016. (Data from Ainley et al. [11])

Fig. 5.6 Shown here are robust chicks that are raised at the small Cape Royds colony where parents need to put in only minimal foraging effort. (Photo courtesy of D. Ainley)

colony grew from ~150,000 to >300,000 pairs over a 20-year period, fledging mass also decreased [11] (Figs. 5.5 and 5.6). This trend was in stark contrast to the fledging mass at smaller Adélie penguin colonies in the Crozier region over the same period, some of which grew but to a size less than those of the total little penguin colonies. Consistent and in support of Dann and Norman's findings for little penguins, another study of Adelie penguins, including other colonies in the Crozier region, found that energy expenditure increased as both colony size and its foraging range increased [21].

Sex Differences in Foraging

It appears that the degree to which gender differences in foraging behavior occur, that is, intraspecific competition on a smallest scale, can also be affected by colony size. In other words, if the research pertains to small colonies, then no gender differences in forging are likely to be observed. However, mainly from numerous biologging studies, especially those conducted at the Adélie penguin colonies reviewed above, we know that colony size is important. At the extremely large Cape Crozier colony on Ross Island (~275,000 pairs), as the chick-feeding period progresses, although members of both sexes increase their foraging range, the increase is more evident among females as males begin to forage increasingly deeper [112, 113]. In so doing, the males bring back a larger proportion of prey in the form of energy-rich fish, as fish occur deeper than krill [8, 123]. Foraging in different depths of the strata could explain why the two sexes bring back different-sized prey, with the males finding larger prey, rather than this being due to the males being larger and thus able to catch larger prey [12]. Silverfish do occur deeper in the water column as they age and grow [108]. At a somewhat smaller, but still large colony (only 2000 pairs on Béchervaise Island but with >100,000 pairs in the Mawson Bay neighborhood), a similar pattern is apparent, though more so during the guard than crèche stage [49, 50]. Apparently, although the average foraging depth does not differ much between sexes in the chick phase [151], the males nevertheless bring back more fish as the crèche period progresses [8, 11]. The same pattern is evident at another large "colony" (Hope Bay, >104,000 pairs), where females and males forage farther as chick feeding progresses, with females farther still but with males bringing back more fish [53]. At a medium-sized colony on Ross Island (Cape Bird, 58,000 pairs), both males and females forage increasingly farther away but no depth change occurs on the part of males [114]. Although not statistically different, at another smaller but nevertheless medium-sized colony (Pt. Géologie, 39,000 pairs), females also tended to forage farther away than did males [16]. Finally, at a small Ross Island colony (Cape Royds, <3000 pairs), there was no difference or change in either foraging range or depth by either sex as the chick-feeding period progressed [156]. Similarly, neither gender differences nor increased foraging range was detected at two other small colonies of this species, Anvers Island, Bellingshausen Sea (<3000 pairs [48, 79]) and Edmonson Point, Ross Sea (<2500 pairs) [49]. When a colony in the vicinity of the Anvers Island had 8000 pairs, in 1991–1992, no difference was evident, at least in terms of diving depth [45].

There have been a number of studies that have considered differences related to sex in foraging among other *Pygoscelis* penguins. When both diving behavior and diet of gentoo penguins were investigated, males, being larger than females, in most locations, such as South Georgia [26, 186], South Shetland Islands [79, 99, 142]), and Kerguelen Islands [42], foraged deeper and consumed more fish than did females. For chinstraps, the results have been a bit mixed, perhaps due to differing years of study and size of study "colonies" in the region of the Antarctic Peninsula.

Gorman et al. [79] found that male chinstraps consumed more fish than did females chinstraps, but other researchers [99, 142] found no difference in their diets.

Among other penguin species, gender differences involved in intraspecific competition have been observed to be sparser. In species such as emperor and the various *Eudyptes* penguins, the marked seasonal partitioning of nesting duties between the two sexes, e.g., in emperor penguins, only the males incubate, makes comparison of foraging behavior between the two sexes pretty much impossible, at least from a contemporaneous perspective. This is because for most of the egg and early chick guard stages, members of each pair do not forage at the same time and oceanic conditions and prey availability likely changes over time.

Biologging studies on emperor penguins are not instructive as to gender differences owing to the small sample sizes investigated thus far and the lack of simultaneous foraging on the part of the two sexes. All that can be said is that males, consistent with being larger, tend to dive deeper than do females and, as a result, tend to bring back slightly more fish [105, 180, 187]. In one study available for king penguins, the opposite was found: older females dived deeper than did the younger females and males and brought back more fish [111].

Among *Eudyptes* penguins, understanding of gender differences in foraging are pretty much confined to macaroni penguins, the crested species most often studied, but mostly just in the late crèche stage when both members of pairs forage to feed their chicks. Males do not provision chicks until then, and, therefore, most of the studies of this species' foraging have been confined to just females. However, the fact that females do most of the foraging for chicks is in itself a gender difference and serves to reduce the pressure on the prey resource, i.e., less intraspecific competition [175]. In addition, the much larger bill of male *Eudyptes* penguins results in them catching larger prey than females, which is another strategy toward reducing competition for food [51] (Table 1.2). Males beginning to provision chicks only late in the chick period is a factor that greatly affects the size to which chicks will grow, i.e., it is toward the end of growth that they receive more food more often [22]; during that period, at South Georgia, females' trips are apparently longer, which is opposite to the pattern during incubation [23]. However, consistent with a larger body size, males dive deeper and longer than do females [82] (Table 1.2). The diet followed by the two sexes of macaroni penguins at South Georgia was found not to differ, though stable isotope analysis indicated slightly more fish intake by males [26]. Likewise, on the Prince Edward Islands, no sex difference in the diet of macaroni penguins was apparent, but, in the case of southern rockhopper penguins, the males consumed slightly more fish than did the females [179].

Findings among other penguin species often revealed consistency in foraging behavior with body size, with the larger size usually, but in some cases, not necessarily being of males (Table 1.2) [42]. For little penguins, studies indicated little difference in foraging behavior between the two sexes, although males appeared to forage slightly closer to the shore [127, 136] and, although the sample size was small, males may forage deeper [27, 94]. Among African penguins, the larger males dived significantly deeper and longer than did females, who performed more dives per hour than did males on foraging trips; and females exploited larger areas than did males, especially at the larger of two colonies studied [139]. Finally, among Magellanic penguins, the larger males dove slightly deeper and also foraged closer to the breeding sites than did the females. In general, these differences translated to more fish in the diet, especially a higher proportion of schooling forage fish such as anchovies on the part of males [30, 76]. The latter diet aspect varied by colony, and, indeed, at the tiny colony studied at Isla de Los Estados (1600 pairs), no sexual differences in diet or foraging behavior were found [155].

Interspecific Competition Involving Penguins

Penguins Eat a Lot!

In order for competition for a resource to mean much, or indeed anything, a population of a species has to consume an ecologically significant amount of the commodity, in this case, food, to have any effect on populations. For penguins, how much prey they catch is not easy to calculate, but just how much food do people think that penguins consume? One thing is for certain and that is, it is "a lot." This can get us to questions that some people ask, "Do we really need penguins, what do they do for us?" (Chap. 15).

Being highly active, but small (compared to seals, for example), subsurface fast movers living in cold seas, penguins play a high-energy game (see Chap. 7), in which the "currency" is paid through food consumed. In particular, penguins spend most of their time immersed in sea water (even when resting because they sit low in the water compared to other pelagic seabirds; see, e.g., Chap. 7, Fig. 7.1) and expend plenty of energy traveling, especially when they are feeding hungry chicks [20]; see, e.g., Chap. 9, the Snares penguin in Fig. 9.1). So, on top of their metabolism that helps keep them warm, they have energy needs for their movement – traveling, diving for prey etc. (Chaps. 6 and 7). In fact, most penguins are on the move for appreciable parts of their annual cycle. For example, foraging distances and intervals during breeding are one thing, as just noted (Appendix Table A3), but they do not just sit around during the nonbreeding season either. For example, gentoo penguins, which stay relatively close to breeding locations, have been found to travel up to 800 km during their nonbreeding exploits, whereas macaroni penguins travel as much as 13,000 km during theirs [169]. This is just the horizontal distances traveled! Actually, of course, they operate in a 3D world and have to eat, a lot, as they go, traveling downward in a viscous medium just to catch their dinner! Beyond that, unlike aerial seabirds, they cannot use the wind to provide lift to help them save energy as they move.

The problem with trying to calculate how much a particular penguin population may eat, and worse still, how much all the world's penguins eat, is that many assumptions have to be made. This is because we do not know enough about the energetics and activities of penguins at sea. This information also has to be dovetailed with metabolic rates (and time budgets) of birds on land, including how much food they give to their chicks. Nor indeed, do we have particularly accurate population estimates, and they are dynamic (Table 1.1; Chap. 13). On top of all that, exactly how much penguins have to eat to pay for the overall metabolic rate of their lifestyle depends on what they eat, how much energy that prey type consists of (and that may vary over time – it is not just farm animals that get fat for the holidays), and how much of that energy can be derived from their foodstuffs (their assimilation efficiency), which varies according to prey type (see Chap. 14). Some of these topics will be discussed in various chapters of this book, but not that many attempts have been made to fit them together to come up with the "harvest" capacity of a penguin colony or group of colonies. However, we do know that penguins must be big eaters in the marine ecosystems which they inhabit based on at least two factors: large populations and high biomass of individuals. A summary of the attempts to quantify some of this is provided in Table 5.1. This shows that, although estimates vary widely, penguins clearly consume a lot of food in a year.

The earliest attempt at estimating how much food a penguin population consumes was made by studying African penguins during 1954–1956, when an estimated 103,000 birds nested at Cape Province, South Africa, and not much was known of the various metabolic rates listed above. Based on the average amount of

Table 5.1 A summary of the estimates of the amount of food, in tonnes, which the breeding adults of various penguin populations consume in a given year

Location, year	Penguin species	Population size ($\times 10^3$)	Fish	Squid	Crustaceans (krill)	Total ($\times 10^3$)	Source
Cape Province, South Africa, 1950s	African	103	7			7	[147]
Saldana Bay, South Africa, 1981	African	122	16,800			16,800	[132]
St. Croix, 2010	African	7	1			1	[141]
South Georgia, late 1970s	King	686	19	45		64	[61]
	Gentoo	220	29		61	90	
	Chinstrap	8			2.5	2.5	
	Macaroni	8000	79		3160	3952	
1991	Macaroni	3000			8080	8080	[34]
South Shetland Islands, 1980s	Chinstrap	1600			7800	7800	[59]
	Adélie	1800			3100	3100	
	Gentoo	1300			1200	1200	
Heard and McDonald Islands, late 1980s	King	7.6	0.4	5.8		6.2	[84, 184]
	Gentoo	33.1	7.1	0.3	3.3	10.7	
	Southern rockhopper	2	0.09	0.01	0.1	0.1	
	Macaroni	4000	93.6	0.4	309	403	
Argentina	Magellanic	1800				1500	[157]

food (380 g) in the stomachs of 247 penguins collected at sea, assuming that they filled their stomachs once per day, it was estimated that during a year those penguins extracted 7000 tonnes of food [147]. The calculation also assumed that these penguins spent 185 days at sea during the year. Most of their prey were sardines and anchovies (Appendix Table A2, below and Chap. 13). A comparison of this rough estimate with the more recent ones in which the various metabolic rates were known shows that the study severely underestimated African penguin consumption.

Using many of the factors of penguin natural history in the above list (for calculating prey consumption), an attempt was made to estimate the prey consumed annually by the breeding penguins at South Georgia during the late 1970s [61]. However, even that attempt, not having all the factors to consider, apparently underestimated penguin consumption. For instance, there were 8 million macaroni penguins at that time, and they were estimated to consume 3.9 million tonnes of krill. Yet with more known about what affects prey consumption, in an exercise completed several years later [34], with the population having decreased by almost two-thirds (see below, as fur seals increased), almost two-thirds more krill were estimated to have been consumed by the macaroni penguins of South Georgia. Regardless, both series of estimates were impressive! Putting this in perspective, the four species of penguins investigated earlier consumed slightly more than 50% of the total consumption of the entire 21-species seabird nesting community (the intake of aerial species was calculated using the same variables as used for the penguins)! It is no wonder that aerial birds “choose” to forage farther at sea than do penguins, as the penguins may catch most of their prey closer to the shore (Chap. 1; also below).

Likewise, on the Heard and McDonald Islands during the late 1980s, the consumption of prey by breeding penguins on an annual basis was also estimated [184] using the same modeling procedure used at South Georgia [61]. Also impressive, but more so, the penguins consumed 98% of what the entire 15-species nesting seabird community caught from the surrounding ocean!

We believe that it would be unwise for us to try and calculate the consumption of any population or species of penguins or even try to cover the consumption of the world population of penguins, though maybe some of the new data provided in this book might help someone try. Hey, we encourage people to try! That being said, 20 years ago, there was a noble attempt to estimate the amount of food consumed by the breeding populations of the penguin species – emperor, king, Adélie, chinstrap, gentoo, macaroni, and southern rockhopper penguins – that inhabit the entire “Southern Ocean ecosystem” (Antarctic Polar Front and to the south; Chap. 3) [183]. That, of course, at the very least, required completion of the herculean job of compiling all information existing then on the abundance of these species at all their respective nesting sites, a task excellently undertaken and completed by Eric Woehler just a few years earlier [182]. The result of the food consumption estimate – hold on to your hats – was 3.7 million tonnes of fish, 13.9 million tonnes of crustaceans (mostly krill of various species), and 0.54 million tonnes of squid! Breaking that total down by penguin species, for the Southern Ocean: emperor 447,870 tonnes; king 1,611,000 tonnes; Adélie 1,632,600 tonnes; chinstrap 4,878,220 tonnes; gentoo 156,655 tonnes; macaroni 8,535,630 tonnes; and southern rockhopper 192,930 tonnes. Of course, as detailed in Chap. 4, these species differed in terms of the actual dietary composition of their intake. Given the new appreciation for the energy-demanding lifestyle of penguins (the subject of this book), compared to what was known in 1995, the grand total, 18 million tonnes annually consumed by penguins in the Southern Ocean, is more than likely several times over a gross underestimate.

We will reiterate, however, that because penguins are numerous, large (compared to most seabirds), and lead the high-energy lifestyle required in cold water, they certainly do eat "a lot," in fact "a helluva lot" of food. This gives them a special role, especially in the coastal oceanic ecosystems that they exploit and especially when raising their offspring. However, then, penguins are all about being special, as this book will continually point out (leading to Chap. 15).

Competition Between Penguin Species

In the avian family Spheniscidae, there are 19–21 species of penguins (or distinct genomes, as molecular taxonomists would call them), depending on the degree to which a researcher tends to "lump" versus "split" the species (see Chap. 1). In the discussion of penguin diet (Chap. 4), we divided the species into six "ecotype" groups, each of which contained members that forage in a similar way related to region and foraging habitat/method, with consequences for similarity of diet: (1) polar/subpolar, Mesopelagic (emperor, king); (2) polar, continental shelf/slope, upper water column (Adélie); (3) polar, continental–insular slope/pelagic, upper water column (chinstrap, macaroni); (4) subpolar, continental–insular shelf, full water column/demersal/benthic (gentoo, yellow-eyed, plus a recently extirpated relative; see Chap. 1); (5) subpolar, continental slope/pelagic, upper water column (northern and southern rockhoppers, erect-crested, Snares, royal, plus others further split; Chap. 1); and (6) temperate, continental–insular shelf, upper water column (African, Magellanic, Humboldt, Galápagos, and little). Supposedly, the emperor–king lineage is the oldest among the existing penguin species, followed more recently by the gentoo (and then other *Pygoscelis*), with the others apparently radiating (becoming species) still more recently sometime near two million years ago (Chap. 1). This progression in appearance likely had influence on the size and agility (and thus on the choice of nesting habitat), which affected the subsequent appearance of various penguin ecotypes/genera. It may have also had an influence on what oceanographic niche became occupied/exploited.

The present distribution of species within the penguin family is likely influenced by competition for food, either interference or exploitative, or, perhaps, more appropriately, by avoiding the competition that initially led to speciation. This avoidance occurs as a result of penguins not finding much to eat, with what there is having been taken by the current occupants of a niche, and so they go elsewhere or adapt. Current latitudinal and ecological distribution of penguins is pretty much the same as the fossil record, although many more penguin species have gone extinct than exist today [107]. Extinctions presumably occurred due to environmental changes, making possessed adaptations no longer useful, but also including increased prevalence of competitors (e.g., seals, dolphins, even other penguins) over the last 40 million years (Chap. 1).

The factors that lead to speciation are variable, but the following scenario could be correct, and is close to what has been described for "Darwin's finches" and the honeycreepers of Hawaii, which are impressive arrays of land bird species that reside in the Galápagos and Hawaiian Islands, respectively, with each likely originating from one or just a few invasions by a "lost" flock of birds [80, 144]. Essentially, a group/flock of a particular species finds its way to unexploited habitats in which conditions, while allowing their survival, are different enough that by evolutionary adjustment (selection of phenotypes giving advantages/performing better in this new environment), their progeny, with each generation, become increasingly better-suited to exploit, rendering them into forms that differ from the original species, thus eventually being unable interbreed with it anymore. This is called "species adaptive radiation," that is, through evolution, the number of species with a common ancestor rapidly increases through the adaptation of organisms to new ecological contexts. For instance, it has been shown [51, 52] (see Chap. 1) that radiation in modern penguins accompanied the emergence of islands (breeding locations), many of volcanic origin, and relatively young, geologically speaking (and thus so are the penguin species involved). The oceanic environment in the region around each of these islands was different than from where the invaders originated. For instance, there was little in the way of insular (or continental) shelf, should the pioneering penguins have come from a continental coastal habitat (that includes mainland New Zealand (NZ) and Australia). Facilitating penguins' progression is their highly social nature – penguins do not go anywhere alone but always travel in a flock [1] (Chaps. 12 and 14). Thus, potential partners abound upon landing on a foreign shore.

On the other hand, species can coexist if they do things differently, i.e., be diverged enough that they have different life histories and/or ecologies than the others in their presence. They exploit different ecological niches (the assembly of different habitat features in which different species exist). Such is true among members of the six penguin ecotypes, and, besides foraging habitat differences, they also differ in body/bill size. Thus, prey size also plays a major role in their coexistence (compare Table 1.2 in Chap. 1 with Table 4.1 in Chap. 4). The geographic scale, however, i.e., the ocean in the southern third of the world globe, is far greater than that of the finches on the Galápagos Islands. They have much to divvy up among themselves.

- Among the two polar, mesopelagic foragers, both being circumpolar in distribution (Chap. 3), one occurs only in the sea ice-free waters of the Southern Ocean (kings), whereas the other occurs within the farther south sea ice zone waters (emperors). Hence, as long as there is sea ice, they do not overlap. However, both are large as penguins go and forage by diving to deep depths, deeper than any other penguin species (Chaps. 6, 7, 8, 9 and 10). Being large, they also catch prey in a wider range of sizes than do all the smaller penguin species (Chap. 4).

- By exploiting deeper depths, and being much larger in size, the circumpolar continental-shelf/continental-slope emperors can coexist with the circumpolar, shallow depth-foraging (surface waters) Adélie penguins (Chap. 3). The two are the only true sea ice-seeking penguin species (pagophilic), with both occurring in the Antarctic sea ice zone. There is much overlap in the prey species that they catch (see Chap. 4), but being larger and diving deeper, the emperors can also catch larger individuals (as noted above in the discussion of Adélie sex differences).
- Kings coexist with several smaller, shallow depth-foraging, subpolar species as well as macaroni penguins. The latter are technically "polar" (they nest and forage south of the Antarctic Polar Front), but their range overlaps only slightly with that of Adélie penguins (Chap. 3) because macaroni penguins, like kings, avoid sea ice.
- In their circumpolar distribution, king penguins overlap in many locations with almost all of the subpolar, continental-slope/pelagic *Eudyptes* (crested) penguins but do so by foraging deep, whereas the others forage in waters much closer to the surface. Among the several crested penguins (6–9 species depending on how they are split; Chap. 1), ranges are nonoverlapping for many of those nesting and foraging around isolated island groups: Snares, erect-crested, Fiordland, Warham's, royal, and, more or less, macaroni.
- The various rockhopper penguins overlap with other crested penguins to some degree but are decidedly smaller in body size (Table 1.2 in Chap. 1), with implications related to differences in the foraging depth, prey size, and diet [55].
- As for gentoo penguins, though they exploit the entire water column, they forage greatly on demersal fish of continental and insular shelves. Therefore, they coexist in many locations with various crested penguins and king penguins, which forage much farther above the oceanic floor but generally in waters off the shelf.
- The same full water column foraging is true for yellow-eyed penguins, which have a similar body size to the gentoos, and, although they do not overlap with the gentoos, they do overlap with a few other species. In particular, yellow-eyed penguins overlap with the much smaller, shallow depth-foraging New Zealand little penguins. The latter is disjunct from the similar Australian little penguins, which overlap with no other penguin species.
- Finally, the other members of the temperate, shallow depth-foraging, shelf penguin group, i.e., those of the genus *Spheniscus*, essentially do not overlap with each other (slight overlap of Humboldt with Magellanic in southern Chile) and only rarely with any other penguin species. The exception occurs on the Falkland Islands, which have king, southern rockhopper, gentoo, and Magellanic penguins. Given what we have reviewed above, they all get along owing to limited ecotype overlap.

These patterns are what can be inferred by visual comparison of the overall ranges and consideration of other natural historical aspects, such as body size (and its ecological implications; Chaps. 1 and 7). Unfortunately, there have been few studies in which, as "natural experiments," the foraging ecology of co-occurring penguin species has been investigated simultaneously to directly learn about the consequences of interspecific competition. A "natural experiment" would be a situation that changes the present scene, such as a dramatic variation in oceanic climate that changes food availability, invasion (even if just seasonally) of a competitor that changes the preyscape, and others, but the key aspect would be that a researcher is present and is aware enough to observe/collect data during the change. Of course, for the several endangered penguin species, finding a large colony in which even intraspecific competition would occur is no longer possible (see discussion on foraging range, above). We have lost any inkling of that in their natural historical patterns. Therefore, the degree of actual, direct competition for prey or habitat among these penguin species remains little known, that is, what are the ecological aspects by which the penguin species have diverged in their lives at sea? Moreover, in most instances where a comparison is made for non-endangered penguins, the research effort involves tiny to small colonies, which, as is apparent in the review of intraspecific (within colony) competition (above), exhibits either zero or minimal effects of competition. This is illustrated in a comparison between two small (<3000 pairs) colonies, 12 km apart, of gentoo and Adélie penguins on the west coast of the northern Antarctic Peninsula (southern edge of the gentoo subpolar range, near-northern edge of the Adélie polar range). While ecological segregation is somewhat achieved by a month's difference in the timing of breeding – gentoos avoid sea ice, so they wait for it to seasonally disappear before laying eggs – it was found, not surprisingly, that there was little overlap in the foraging areas of the two small colonies [48]. However, on occasions when there was slight, actual overlap of where penguins of the two species were foraging on a given day, the larger gentoo penguins foraged deeper than did the Adélie penguins. In doing so, the gentoos caught different size/age classes of krill. Thus, in a "natural experiment," these two species expressed their ecotypes even more clearly.

A few foraging studies have paired gentoos with other co-occurring penguin species, although not in real-time, coincident, overlapped foraging, i.e., a "natural experiment." Along the west coast of the northern Antarctic Peninsula, in accordance with gentoos usually foraging deeper than most of the smaller penguin species, they were found to feed much more on fish, and demersal ones at that, than did Adélie and chinstrap penguins [99]. The chinstraps, by foraging more off the shelf than Adélies from the respective colonies, followed a diet in which the minimal amount of fish that they consumed was composed of myctophids (a pelagic, i.e., off-shelf fish). Thus, for the latter two penguin species, the results were consistent with the ecotypes, one is a shelf/slope forager and the other a slope/pelagic forager. Similarly, at other locations of the northern Antarctic Peninsula, and using stable isotope analysis, findings supported the idea that gentoos caught more fish and over the shelf, diving deeper, than chinstraps, which fed mostly on krill off the shelf. That is, the gentoos had an enriched nitrogen (N_2) isotope and the chinstraps had a more enriched carbon (C) isotope in the analysis of their blood and feathers [67, 143]. The same pattern was evident when comparing gentoos to macaroni penguins on the

Kerguelen Island, Indian Ocean sector, and South Georgia, Scotia Sea [26, 60, 169, 174]. Getting back to the northern Antarctic Peninsula, during the nonbreeding season, the Adélies disperse east to remain within pack ice (see Chap. 3), whereas the chinstraps disperse west to occupy the open water adjacent to the pack ice in that direction [89] – this could be a pattern related to reducing competition and/or occupying a typical habitat for each of these species [15].

The foraging of gentoo penguins has also been compared to co-occurring Magellanic and rockhopper penguins, in this case on New Island, Falkland Islands/Islas Malvinas [122, 146]. The results were consistent with the respective ecotypes. Given the small size of the study colonies (two of gentoo and Magellanic and one of rockhopper; all small, <13,000 pairs), not surprisingly, little overlap in the foraging range was found among them. Thus, it cannot really be deduced whether these penguins segregate foraging areas through interspecific competition. Intraspecifically, they all foraged closer to their respective colonies at the start of chick-feeding stage compared to later, which is consistent with what has been observed in general among penguins with regard to the foraging range, as the pressure to find food increases (see above). Magellanics fed farther away and fed more on surface fish/squid than did gentoos (demersal fish); rockhoppers fed more on euphausiids. These findings were consistent with the ecotype, but there were no instances in which foraging behavior was compared simultaneously, i.e., in a "natural experiment."

King penguins also have colonies in the Falklands, but as noted above, they feed much farther away, i.e., along the Antarctic Polar Front to the south of the islands (and of course deeper [145]). In a study comparing king with gentoo penguins on Heard Island, on the Kerguelen Plateau in the southern Indian Ocean, although fish dominated the diet of both species, the fish species consumed by the gentoos were dominated by the coastal, demersal species (especially the Nototheniidae), whereas those consumed by the kings were dominated by mid-water myctophids from waters at the edge of or off the shelf, confirming once again the respective ecotypes [103]. In a study using stable isotopes of C and N_2 on Possession Island, part of the Crozet Islands (the size and proximity of study "colonies" are unknown), in accordance with the ecotype, in spring, king penguins ate significantly more fish, far from the island (myctophids), whereas southern rockhopper and macaroni penguins consumed mostly euphausiids much nearer to their "colonies" [47].

While on the subject of king penguins, and intraspecific competition, it does appear that those nesting on the east coast of the Kerguelen Islands, during summer, forage to the south along the outer shelf and shelf break of the Kerguelen Plateau [46, 145] (Fig. 5.7). In doing so, however, they do not appear to travel far enough south to overlap their foraging area much with the king penguins of Heard Island, which forage in a similar habitat and mostly to the south of their island [129]. If the king penguins of Heard Island foraged to the north along the shelf break, then they would entirely overlap the area used by the Kerguelen birds. At times, a few actually do forage to the northeast, but what the Kerguelen penguins may or may not do at the time in response is not known. This comparison is clearly somewhat tenuous owing to different trackers being used and differences in the seasonal timing of investigation (only slight temporal overlap, years and months) and is thus not an

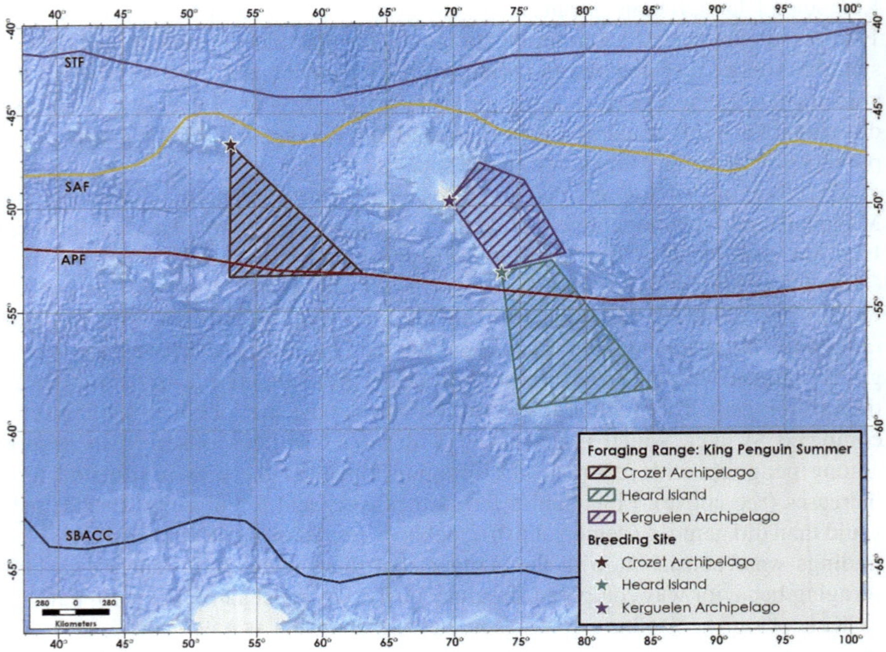

Fig. 5.7 Approximate foraging range of king penguins nesting on islands in the southern Indian Ocean sector of the Southern Ocean during summer. (Data for various sources: Heard Island [129], Kerguelen Islands [46, 145], and Crozet [86, 137, 145])

ideal "natural experiment." Although, perhaps, the distance between islands could be too great for consideration, king penguins nesting on Possession Island forage to the south and east but without overlapping the foraging areas of those nesting on Kerguelen, at least those on the east coast of Kerguelen [137]. Where the king penguins that nest on the west coast of Kerguelen forage during summer has not been investigated. A juxtaposition of foraging areas on a similar scale also exists among the king penguins nesting on the east coast of the Falkland/Malvinas Islands and those on the central north coast of South Georgia, i.e., not foraging far enough toward one another to achieve an overlap [145].

Somewhat of a "natural experiment" occurred on Prince Edward Island between macaroni and southern (eastern) rockhopper penguins when researchers simultaneously compared foraging ranges during premolt for enough years to include one of reduced overall ocean productivity [179]. Both species fed south of the Antarctic Polar Front, but in the year of reduced ocean productivity (and likely reduced prey availability), overlap in the foraging region was markedly less, with macaroni penguins traveling farther south than in other years; stable isotope analysis indicated that the macaroni penguins also ate more fish than usual compared to the rockhoppers. Otherwise, southern rockhopper and macaroni penguins, during the nesting season, exhibited similar foraging patterns and consumed similar prey species on

Marion Island, which was not too surprising as only two small, neighboring "colonies" were investigated [39, 40]. Thus, it was not really a good test of niche partitioning. On Heard Island, where macaroni penguins are two orders of magnitude more abundant than southern rockhoppers (10,000 pairs), diets were similar but the macaroni penguins consumed larger fish than did the smaller rockhoppers [102]. On both Marion and Heard islands, the rockhopper diet contains appreciably more crustaceans than does the diet of the macaroni [55]. In a comparison of macaroni and chinstrap penguins on Bouvetøya, both had similar diets but the larger macaroni penguins foraged deeper [29].

Comparing foraging ranges can be instructive when assessing the degree of compatibility among different penguin species in the same space and at the same time. On the Kerguelen, Amsterdam, and Heard islands, all in the same general portion of the Indian Ocean, during the postbreeding/late breeding stage of the penguin annual cycle, the rockhopper species do not overlap much, although they do with the much larger-bodied macaroni (Fig. 5.8). As noted in Chap. 4, and above, macaroni penguins with their much larger bill consume much larger prey than do the other crested penguin species, which could be a result of them dealing with competition.

On Macquarie Island in the Tasman Sea portion of the Southern Ocean, the closely related macaroni and royal penguins exhibited broad overlap in their diets

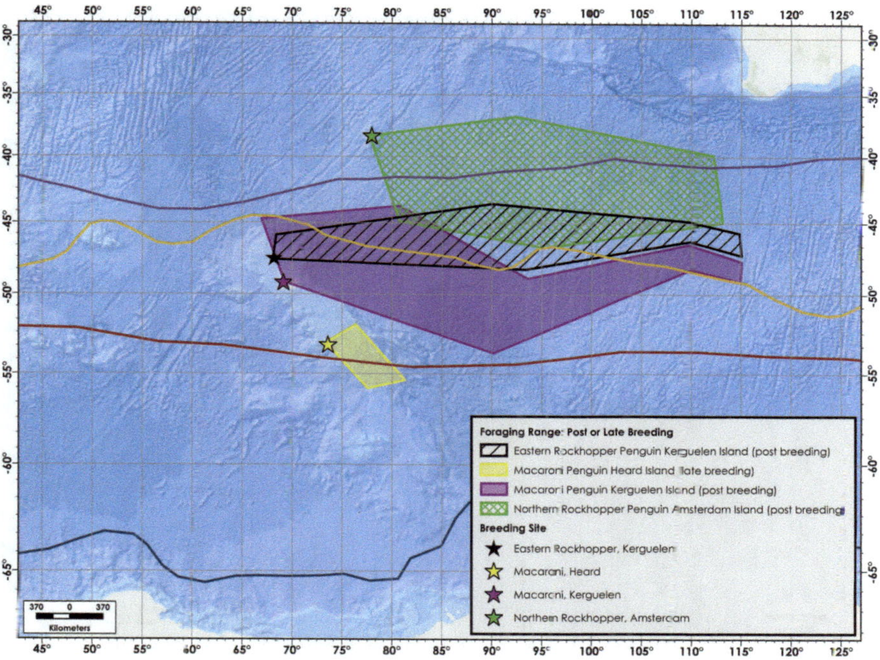

Fig. 5.8 Foraging ranges of *Eudyptes* penguins nesting in the eastern Indian Ocean portion of the Southern Ocean, just after or at the end of their nesting seasons. (Maps are reconstructed from several sources [32, 96, 97, 168, 169])

during the premolt and late chick periods, but the macaroni penguins foraged outward from their "colonies" on the east coast, as opposed to the royal penguins foraging out from their west coast colonies [92]. Hence, geography likely had an important role in the difference in diet between the two species. While these two, similar-sized, large *Eudyptes* penguins had little foraging area overlap, they did significantly overlap with the range of the much smaller-bodied Fiordland penguins, whose foraging extends southwest from the South Island of New Zealand (Fig. 5.9). Therefore, it was highly likely that their diet differed on the basis of body size, which it does: larger penguins consume larger prey (Table 4.1). Later in the year, during the post-molt nonbreeding season, the very closely related, similar-sized Snares and Fiordland penguins also exhibit minimal overlap in respective foraging areas, with both extending their foraging southwestward from the west coast of New Zealand's South Island [81]. However, most of the Snares' foraging area lies north of that of Fiordland penguins' area and also extends much farther west, even west of Tasmania [170]. This pattern is close to that exhibited during the premolt foraging period on the part of the Fiordland penguins. These penguins apparently avoid spending much time in the coastal waters that are occupied by yellow-eyed and little penguins (Fig. 5.9). It also appears that they avoid traveling east of NZ into waters where there are other *Eudyptes* penguins foraging from a number of islands (but

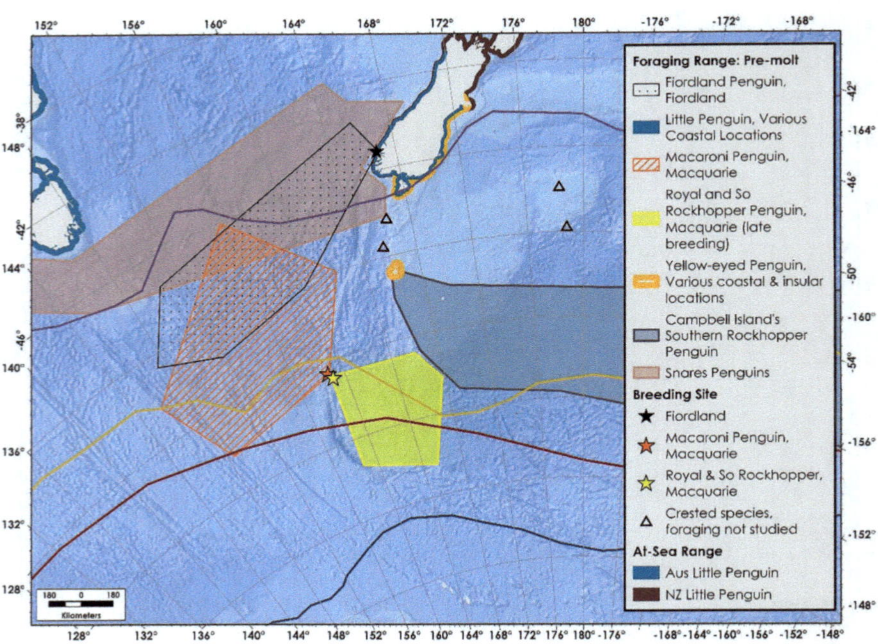

Fig. 5.9 Postbreeding foraging ranges of *Eudyptes, Megadyptes*, and *Eudyptula* penguins nesting in the Tasman Sea portion of these penguins' global ranges. (Maps are reconstructed from several sources (Green [81, 83, 96, 97, 124–126, 170]) and Fig. 3.12)

their foraging remains to be investigated). In fact, the southern rockhopper penguins from Campbell Island occupy much of that stretch of ocean [170] (Fig. 5.9).

Among the Falkland Islands, the same pattern of interaction occurs (Fig. 5.10). Two large colonies of southern rockhopper penguins (>100,000 pairs each), on either side of the archipelago, frequent foraging areas during winter and exhibit little overlap; their foraging areas also did not overlap much with the southern rockhoppers nesting on Isa de los Estados, Argentina, to the south. This was in contrast to the southern rockhoppers on the east coast of the Falklands, in which winter foraging area extended a long distance to the east, overlapping entirely with that of the much larger-sized macaroni penguins from colonies on the west end of South Georgia and foraging to the west. That is, they overlapped foraging in the waters stretching between the two islands. In other words, penguins from colonies of the same species, of course of similar body size, failed to overlap the foraging range between large colonies, but those closely related species of significantly differing body/bill size did overlap extensively [148]. Their diet and foraging behavior, other than location, was not investigated concomitantly, though we know from other studies that the macaroni penguins consume much larger prey (Table 4.1). On Isla de los Estados, tracking and C^{13} stable isotope data indicated little overlap in the foraging areas of southern rockhopper and Magellanic penguins, with the rockhoppers (two

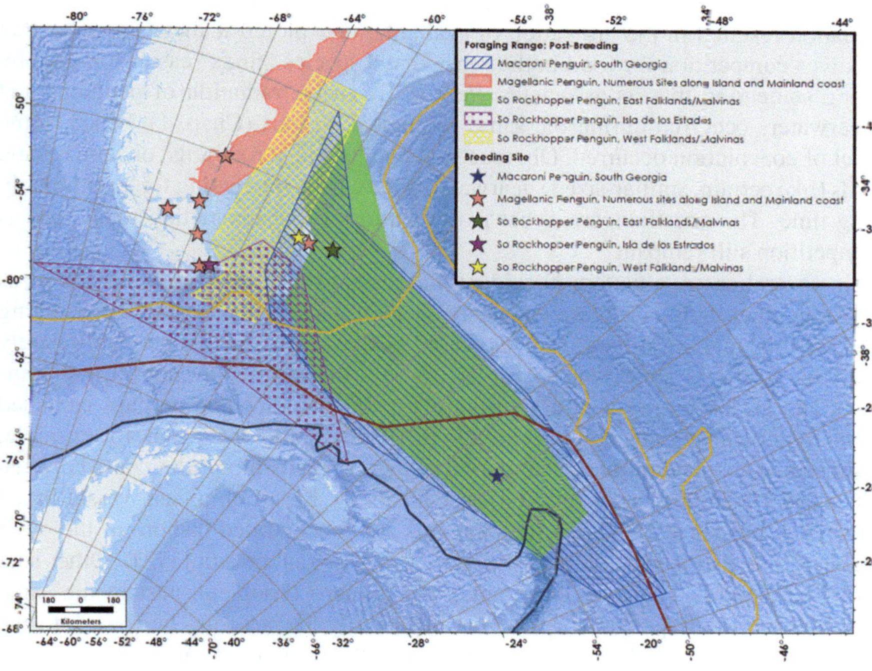

Fig. 5.10 Postbreeding foraging ranges of *Eudyptes* and *Spheniscus* penguins in the southwestern Atlantic region of the Southern Ocean. (Maps are reconstructed from several sources [148, 154, 165])

orders of magnitude more abundant) foraging farther to sea and deeper than the other species; N^{15} results were consistent with other studies, having the Magellanics feeding on more fish than the rockhoppers [154]. Magellanic penguins nest on a number of locations in this area, but during the postbreeding period, Magellanic distribution shifted north, staying in waters over the inner and central continental shelf of Argentina. The rockhopper penguins that nest and winter in the region occupied slope and deeper waters, adjacent to those occupied by the Magellanics. Finally, not surprisingly, the foraging areas of the much larger king penguins that nest on the Falkland/Malvinas Islands completely overlapped the foraging areas of all these *Eudyptes* penguins, but of course the king penguins foraged at depths below those exploited by these others (see Pütz [145]).

Competition Between Penguins and Other Seabirds

Noted in Chap. 1 is the evolutionary origin of the penguins, which, many millions of years ago, were derived from aquatic members of the petrel–albatross avian group. Even today, certain petrels and shearwaters use their wings to propel themselves underwater in pursuit of prey, though, eventually, penguins evolved a capacity to do this much more efficiently (Chap. 6). Nevertheless, penguin species radiation resulted in the utilization of available, unexplored niches, a process that involves competition for food in the sense that a species "finds" easier foraging by doing something that its associates are not. Extensive radiation of penguins and shearwaters occurred during the same geological periods (Chap. 1), hence, some level of coevolution occurred. Obviously, penguins and aerial birds, or semiaquatic birds (like certain, similar-sized shearwaters), have been interacting for an extremely long time. The questions are, what is the nature of that interaction today, and is competition still rampant?

In several cases, a pattern is evident that leads to the question, are aerial birds avoiding inner shelf, neritic waters where penguins have concentrated their foraging and are thus making it harder for the aerial birds to find shallow prey sufficiently available to them? Unfortunately, at-sea surveys of seabirds are not all that common, and are becoming less so, but there are some studies that perhaps can shed some light on answering that question. In the Peru/Humboldt Current, for example, a substantial concentration of Humboldt penguin breeds along the coast between 30° and 29°S in northern Chile [64] (see Chap. 3). Based on at-sea, vessel-based surveys, it has been found that the penguins are concentrated inshore of the other seabirds that also feed mainly on the same fish prey (anchoveta [178]). Other seabirds, particularly guanay (cormorants) and piquero (boobies), nest in the same localities as do the penguins (penguins historically, in fact, dug burrows into the "guano birds" guano deposits) but apparently fly over the foraging penguins to reach prey schools. In the outer part of the penguins' foraging area, plenty of flocks include all three species, but farther to sea, few if any penguins occur and aerial species are abundant [164] (Chap. 3). A somewhat similar pattern occurs in the

Galápagos with the penguins there feeding close inshore relative to a tendency of the aerial species to be feeding farther from the shore [128]. The Galápagos (flightless) cormorants, which also feed close inshore, cope with the penguins by feeding on or near the seafloor, rather than in the water column where the Galápagos penguins forage. In the waters off East Antarctica and elsewhere in the Southern Ocean, Adélie penguins tend to feed over the shelf and upper shelf break (Chap. 3), inshore from large numbers of foraging petrels and shearwaters [14, 133, 152, 185]. In the waters off the South Island of New Zealand, yellow-eyed penguins and little penguins forage over the shelf (Chap. 3), seaward of which high densities of sooty shearwaters forage from the same or near the same breeding localities; the shearwaters only rarely forage over the shelf [25, 120]. As a mild confirmation that some direct species interaction is going on between the diving and pursuit diving seabirds, in the California Current, where the two most abundant seabird species are sooty shearwaters and common murres ("penguins of the north"), the shearwaters traveled thousands of kilometers to spend their nonbreeding period foraging seaward of the murres. The irony is that the resident murres can exploit the entire water column over the shelf (like penguins), but the shallower-foraging shearwaters are shifted to forage in deeper waters on essentially the same prey [10].

This sort of pattern, although intriguing, is not necessarily universal. In waters off Tasmania, little penguins, which are nowhere highly concentrated in their nesting, and thus in their foraging density, appear to coexist with the aerial birds, such as various shearwater species [149]. On the other hand, as discussed in more detail in Chap. 13, marine heat waves off Tasmania affected the food web but not the little penguins all that much compared with shearwaters that tend to feed a bit more offshore [74].

A different slant on the interaction between penguins and other seabirds has been observed in waters off South Georgia. Like the multispecies seabird foraging flocks in the Peru/Humboldt Current, which include Humboldt penguins (noted above), such flocks also occur in the Scotia Sea [87]. Macaroni and gentoo penguins are found there, and it appears that aerial birds (petrels, albatross) depend on the penguins, along with Antarctic fur seals, to drive Antarctic krill closer to the surface. This makes the krill more accessible to the aerial birds. Presumably, where there is enough krill to interest a passing petrel, the penguins would not have had negative effects on prey patch quality (?). In waters around the Galápagos Islands, multispecies foraging flocks are of longer duration if penguins are within them, indicating that the penguins are improving the foraging of the petrels and boobies there [128].

Competition Between Penguins and Marine Mammals

The discussion in Chap. 1 includes the influence of the appearance of dolphins and pinnipeds (seals, sea lions, fur seals) millions of years ago on the disappearance of a number of penguin species, particularly the larger ones. Marine mammals continue to have an ecological influence on penguins, beyond these mammals merely preying upon them (covered in Chap. 12). One of the hallmark periods in the history of human presence in the Southern Ocean, and exploitation of its resources, was the near extirpation during the 1800–1900s of krill-eating Antarctic fur seals and then baleen whales (several species), mostly in the Scotia Sea and Antarctic Peninsula region, though elsewhere as well. More than 1.2 million fur seals were taken from the Scotia Sea region and the species was nearly extirpated by the 1820s. Likewise, >1 million baleen whales were taken from this area, especially from the 1930s to early 1970s [90, 171]. Such overexploitation led to what became known as the "Krill Surplus Hypothesis," in which loss of whales and fur seals is believed to have led to a dramatic release of prey for other krill-eating predators [90, 162]. Indeed, it is believed that a five-fold decrease occurred in the amount of krill that whales consumed once whale populations were decimated [181]. However, some estimates of consumption prior to whaling reach 430 million tonnes of krill consumed annually by whales in the Southern Ocean [159]. This is a significant amount of food to have been consumed by other predators, although the critical issue is how much became available within the foraging areas of the affected competitors! To determine whether or not such an ecological release really happened and had consequences, investigators subsequently sought to determine how various potentially competing predators responded with population change or altered reproductive strategy, including penguins, minke whales, and crabeater seals. For the whales and seals, researchers sought evidence that the age of recruitment perhaps became earlier (recorded in teeth and ear bone growth rings). A final detailed analysis revealed no such evidence among the mammals and that only the chinstrap penguin population grew in accordance with more food becoming available to them [20]. As noted above (Chap. 4), the chinstrap penguins are the closest there is to a dietary specialist among penguins, with Antarctic krill being pretty much their exclusive prey of choice at least during summer. Therefore, given the specialization, the potential for whale–penguin interaction is further amplified.

Nowadays, while researchers are doing their best to enlist climate change as the cause for penguin decrease [119, 167], and indeed a decrease in the seasonal prevalence of sea ice in that region has apparently led to reduced overall krill abundance [18, 19], the chinstrap penguin populations of the Antarctic Peninsula/Scotia Sea have been dramatically decreasing in the last couple of decades. During this period, baleen whale numbers in that area have been recovering in an equally dramatic manner [35–37]. Ignoring the likely competition for krill would seem to be off base, given the historical picture. While humans currently extract krill from this region at just ~1% of the supposed krill biomass, certainly a lot less than consumed by whales, the fishing by whales, humans, and penguins is narrowly concentrated along the shelf break [176]. The point is that the whale–penguin interaction is far more spatially and temporally intense than a regional analysis would reveal, a subject discussed in more detail below.

Simultaneously, upon recovering from previous near extirpation, Antarctic fur seal population growth in the region has slowed down or even reversed [95, 106], with the fur seals competing for krill with both whales and penguins [117]. Although there are complexities and logistical difficulties in figuring out what each predator eats and where [117], the inescapable truths among these supposed krill specialists are that baleen whales have greatly recovered, fur seals spectacularly so but then began to decrease, and chinstrap penguins have begun to decrease as well. For sure, if overall there is less krill now than in the mid-1900s, owing to climate change (and fewer whales? [159]), then competition for it would be further intensified. At least, as judged by stable isotopic signatures in colony soils dating back 38,000 years, the diet of *Pygoscelis* penguins changed to include more krill, and less fish, in accordance with the near extirpation of krill-consuming whales and fur seals in the Scotia Sea region during the last 200 years [72].

More direct information shows the influence of competition for food among penguins and cetaceans. A "natural experiment" is repeated each austral summer in the southern Ross Sea to show the trophic interaction between baleen whales and Adélie penguins. At the three penguin colonies on Ross Island, the penguins mostly feed their chicks crystal krill with short foraging trips until minke whales show up. Then, with the whales foraging, the penguins increase the length and duration of their foraging trips and change their diet to prey more on silverfish [7]. Using the acoustic sensors of buoyancy gliders and remotely operated vehicles (ROVs), researchers showed that in fact the shallow krill do disappear with the arrival of minke whales, with the penguins themselves aiding that disappearance close to the large Cape Crozier but not the small Cape Royds colonies [8, 156]. This penguin–whale interaction helps explain the increase in the number of Ross Sea Adélie penguins with the removal of thousands of minke whales in the 1970 to early 1980s and then the penguin numbers decreased as the whale numbers recovered following prohibition of commercial whaling in the early 1980s [3, 5, 6, 20].

A similar penguin–mammal story is apparent in the case of macaroni penguin populations in the Scotia Sea, where at South Georgia, their numbers have been closely monitored and have decreased from ~5.4 million penguins during the 1970s to ~1 million currently [172]. The macaroni penguins in this region are also pretty much krill specialists, at least during modern times, especially of Antarctic krill (see Chap. 4). The cause of change, in large part, is increased competition due to recovery from human exploitation (for their fur) of the krill-eating Antarctic fur seals (especially on the part of female fur seals [93, 172]). Only tens of fur seal individuals were present in the 1950s, before the first macaroni penguin counts were made (mid-1970s [172]). Clearly, once the fur seals were gone, there was no commercial incentive to continue the hunt (for their fur), and, in the 1970s, they became officially protected under the Antarctic Treaty [90]. Fur seal numbers grew to reach >1.5 million seals by the 1990s [33]. If climate change (above) and a krill fishery further reduced krill availability, both would serve to intensify competition for krill between the penguins and fur seals by reducing krill availability [172]. As noted above, macaroni penguins and fur seals do participate at times in the same foraging aggregations [87]. Male fur seals are also predators of penguins to some degree (see Chap. 12).

Another fur seal species, the South African or Cape fur seals, has also been dramatically recovering from near extirpation (also depleted for their fur during the 1700–1800s) and, in doing so, may be inhibiting the recovery of the African penguin population [58, 101]. Both species are dietary specialists; in this case, both forage greatly on anchovies and sardines (see Chap. 4). The penguin population has become greatly reduced, and it now being considered "Endangered" by the International Union for Conservation of Nature (IUCN), due to extensive oil pollution and competition with fisheries (Chap. 1). Also involved could be an effect of predation by fur seals, especially males, of the penguins (Chap. 12) as well as competition for island space [101]. Likewise, the recovery of subantarctic fur seals on Tristan da Cunha/Gough Islands appears to be involved in the decrease of northern rockhopper penguins [62]. On the Falkland Islands, penguins avoid occupancy of portions of the coast where South American fur seals are especially abundant [122]. This could be to avoid trophic competition as well as competition for space.

Other seals, such as Weddell seals, through competition for food, appear to be able to affect the location and size of Adélie and emperor penguin colonies [109, 110]. All three species forage intensively on Antarctic silverfish (and rockcod of several species in the case of emperors and seals). Although, on average, seals consume slightly larger silverfish than do penguins, there is much overlap. The interaction is especially important for the penguins and seals during overlapping periods when adults need to recover mass from what is lost during breeding and initial independence of their offspring. Only small concentrations of seals occur in proximity to large emperor penguin colonies, and vice versa, with both species seeking stable fast ice on which to form their breeding colonies/assemblages. Seal prevalence is lower as a function of the size of Adélie penguin colonies in the immediate area. In McMurdo Sound, Ross Sea, Adélie penguins at the small Cape Royds colony (<3000 pairs), during the guard stage starting in mid- to late December, feed their chicks crystal krill, until minke whales show up. Then, they feed much more on silverfish as the minke whales deplete the krill and the Weddell seals (and emperor penguins) depart from the area with the arrival of killer whales (for fear of being eaten). Mostly, it is the fish-eating killer whales (type C), but, apparently, the seals and emperor penguins cannot afford to spend time distinguishing them from the seal- and penguin-feeding killer whales (type B). In the end, departure of the seals and emperor penguins free up the silverfish along the fast ice edge, and, hence there is an increased proportion of fish in the Adélie penguins' diet [7, 156]. Type B killer whales do not prey on Adélie penguins but do prey on the seals and emperor penguins [4] (see Chap. 12).

Competition Between Penguins and Industrial Fisheries

Their high-energy lifestyle (largely coping with the dense medium of water), and their reduced capacity to search for prey (compared to aerial birds), has been the bane of penguins' existence in their need to cope with modern-day industrial fishing. Many seabird populations the world over have been affected, usually negatively, by competition with humans [85, 134, 135], and a noticeable array of penguins are included. As noted, penguins require high-quality (energy-dense) prey relatively close to their colonies so that nesting pairs can adequately provision their chicks (Chap. 2). As reviewed below, there is little doubt that extraction to the point of overexploitation of forage fish is a major factor behind the endangered status of African, Humboldt, and yellow-eyed penguins (see also Chap. 1). This would be a case of exploitative competition on the part of the fisheries. There is immediate concern that if Antarctic krill are overexploited, then macaroni and chinstrap penguin populations will be negatively impacted in the Scotia Sea–northern Antarctic Peninsula region (see below). Other penguin species have already been affected by fishing in other ways.

The endangered yellow-eyed penguins of southern New Zealand and associated islands, appear to have had their food web altered. Historical numbers are difficult to ascertain, but a gradual decrease in the number of yellow-eyed penguins occurred from 1981 to 2007, with only that time series currently available, leaving the present world population at <2000 pairs [160] (Chap. 1). While the species has had difficulties with mammalian predators introduced to the mainland of New Zealand, fisheries have also been involved in the population decrease. The prey species most important to yellow-eyed penguins are targeted by commercial fisheries, with indications of overfishing first appearing in the mid-twentieth century [177]. Unfortunately, there has been no easily available, high-quality prey replacement for these near-bottom foraging penguins [100, 177]. They have lost the exploitative competition with the fisheries. Adding to the problem, the bottom has been seriously altered by shellfish dredging, which has destroyed the habitat of numerous fish [71]. These penguins are left with picking through 43 prey species, many of poor quality, with blue cod being consumed more so among the 7 species most prevalent in the diet [41, 125, 130] (see Chap. 4). However, with a dearth of juvenile fish, the available blue cod tend to be too large to be swallowed or fed to chicks. As a result, when they dominate a diet, starvation of adults and chicks becomes prevalent in colonies [126, 173].

Gentoo penguins are of the same penguin ecotype as are yellow-eyed penguins (Chap. 3). The prey available to them have been affected by fisheries, but other factors have brought in a suitable substitute. The diet of the gentoos in the Scotia Sea–Antarctic Peninsula region was affected by overexploitation of ground fish during the 1960–1970s to the point that it was no longer commercially feasible to continue the fishery. With no protest from the industry, one of the first actions by the cognizant fishery agency, CCAMLR (Commission for the Conservation of Antarctic Marine Living Resources), upon coming into existence in the early 1980s, was to close the finfish fishery [104]. As noted in Chap. 4, the gentoo penguins turned to foraging almost exclusively on Antarctic krill in that area, with krill having been made more available by the decimation of baleen whales (noted above; also Ainley and Blight [9] and Emslie et al. [73]). Thus, the penguins exhibited no population repercussions, unlike the co-occurring, demersal fish-eating imperial shag, which significantly decreased in number in the region [43]. Elsewhere than that region, the diet of gentoo penguins includes a large proportion of the juveniles of the type of demersal fish species acquired by the penguins and the shag. In fact, the gentoo penguin population of the west coast of the northern Antarctic Peninsula, though small, is increasing, as sea ice retreats southward (gentoo penguins prefer seas free of sea ice [68] (Chaps. 2 and 13) and also as the ground fish stocks have been slowly recovering [24, 121].

A similar story of population decrease with the advent of fisheries is apparent in the population trends of Humboldt and African penguins. As noted in Chap. 4, they are specialists on densely schooling fish of the continental shelf and slope, such as anchovies and sardines. However, they have lost the exploitative competition with massive trawl fisheries targeting these prey, as detailed below [56, 57, 69, 70, 161].

Unfortunately, there is little in the way of quantitative, historical data to estimate what was the size of the Humboldt penguin population in the mid-eighteenth century, before the appearance of industrial fishing. The exception is the writings of R.C. Murphy [131], who noted that they nested abundantly on the approximately hundreds of islands and islets that occur along the Peruvian and Chilean coast, digging burrows in the soft layer of seabird guano that covered them. Then, as the industrial revolution unfolded, these islands were stripped of the guano for its value as a nitrate source, leaving rocky habitats. The penguins decreased in number and were forced to live in sea caves. Then came the anchoveta fishery in the 1950s, with the stock overfished and soon decreasing dramatically; the stock was estimated to have been reduced to 15 million metric tonnes by 1971, falling to 4 million tonnes by 1981 [69]. It has not recovered. Now, there are just 24,000 pairs of Humboldt penguins, scattered in small "colonies" among a few coastal islets, where penguin numbers are possibly increasing very slightly if at all [64] (Chap. 1).

In the case of the African penguins, the species may have numbered ~1 million breeding pairs in the 1920s, falling to ~150,000 pairs by the mid-1950s, and eventually to 25,000 pairs in the first decade of the twenty-first century [56] (Chap. 1). This trend tracked the capture of sardines and anchovies using a purse seine fishery of the Benguela Current [58]. Benguela Current catches of sardines reached their peak in the late 1960s (~1.4 million tonnes per year, in a stock estimated at 12 million tonnes), then crashed to <0.1 million tonnes by the late 1970s (combination of heavy fishing, as oceanic conditions changed). Sardines were replaced by anchovies, with the two species exhibiting temporal complementarity – as one decreases, the other increases in natural cycles in accordance with the decadal swings in oceanic climate. With sardines gone, the fishery turned to anchovies as a replacement. The anchovy stock grew to about 8 million tonnes by the late 1990s (catch at ~1 million tonnes), before crashing. Both fish species, subsequently, have failed to recover much, and their distribution has changed, too, moving away from where the penguins had colonized. Changes in the diet of the penguins has occurred in accordance with these fish stock changes, and, recently, more and more low-quality prey have been consumed, i.e., pelagic goby [118]. Moreover, the penguins are tricked into sensing where anchovy schools should be located by the smell of phytoplankton patches, but, then, upon locating the phytoplankton patch, they find no fish [161]. Therefore, while the penguins' foraging areas are protected from fishing for part of the year [140, 141], such is not the case once they disperse after breeding.

Experience with the anchovy and sardine sagas has led to concern by CCAMLR managers about the fishery that has arisen for Antarctic krill, especially in the Scotia

Sea region [38]. CCAMLR came into being out of concern that humans might disproportionately compete with baleen whales for krill and thus compromise the whales' recovery from former decimation [90, 91]. While CCAMLR views the take as "precautionary" [54], determining the actual stock size has been elusive, mainly due to problematic logistics (too remote for concerted scientific sampling). Most importantly, the fishery appears to be overlapping with the foraging areas of penguins, such as chinstrap penguins, and also fur seals [88, 117, 176]. In this case, it could be more the case of interference competition, given that krill trawlers, whales, and fur seals, but especially the central place foraging penguins, seek the highest-quality patches where krill are most easily caught, mostly along the shelf break of the South Shetland Islands and South Georgia. However, while krill concentrate there, in total, they are spread widely [19], so it may not be a reduction in total biomass that is important. Rather it might be smaller, more focused-scale competition among the predators, including industrial fishing (e.g., [88, 117]). The whales and the immense nets towed by the trawlers have the capacity to remove entire krill swarms, thus forcing the penguins to search elsewhere for another swarm. This results in a need for increasing their foraging efforts and, in the process, decreasing their foraging efficiency. As noted above, as the whales and fur seals are recovering their numbers, the chinstrap penguins are decreasing. This is a model for how penguins of other species may be responding to the recovery of their competitors.

References

1. Ainley DG (1972) Flocking of Adélie penguins. Ibis 114:388–390
2. Ainley DG (2002) The Adélie penguin: bellwether of climate change. Columbia University Press, New York
3. Ainley DG (2010) A history of the exploitation of the Ross Sea, Antarctica. Polar Rec 46(3):233–243
4. Ainley DG, Ballard G (2012) Trophic interactions and the decrease in killer whale (*Orcinus orca*) prevalence with reduced availability of large fish in the southern Ross Sea. Aquat Mamm 38:153–160
5. Ainley DG, Ballard G, Ackley S, Blight LK, Eastman JT, Emslie SD, Lescroël A, Olmastroni S, Townsend SE, Tynan CT, Wilson P, Woehler E (2007) Paradigm lost, or, is top-down forcing no longer significant in the Antarctic marine ecosystem? Antarct Sci 19(3):283–290
6. Ainley DG, Ballard G, Blight LK, Ackley S, Emslie SD, Lescroël A, Olmastroni S, Townsend SE, Tynan CT, Wilson P, Woehler E (2010) Impacts of cetaceans on the structure of southern ocean food webs. Mar Mamm Sci 26(2):482–489
7. Ainley DG, Ballard G, Dugger KM (2006) Competition among penguins and cetaceans reveals trophic cascades in the Ross Sea, Antarctica. Ecology 87(8):2080–2093
8. Ainley DG, Ballard G, Jones RM, Jongsomjit D, Pierce SD, Smith WO Jr, Veloz S (2015) Trophic cascades in the western Ross Sea, Antarctica: revisited. Mar Ecol Prog Ser 534:1–16
9. Ainley DG, Blight LK (2009) Ecological repercussions of historical fish extraction from the Southern Ocean. Fish Fish 10(1):13–38
10. Ainley DG, Dugger KD, Ford RG, Pierce SD, Reese DC, Brodeur RD, Tynan CT, Barth JA (2009) The spatial association of predators and prey at frontal features in the northern California Current: competition, facilitation, or merely co-occurrence? Mar Ecol Prog Ser 389:271–294

11. Ainley DG, Dugger KM, La Mesa M, Ballard G, Barton KJ, Jennings S, Karl BJ, Lescroël A, Lyver POB, Schmidt A, Wilson P (2018) Post-fledging survival of Adélie penguins at multiple colonies: chicks raised on fish do well. Mar Ecol Prog Ser 601:239–251
12. Ainley DG, Emison WB (1972) Sexual size dimorphism in Adélie penguins. Ibis 114(2):267–271
13. Ainley DG, Ford RG, Brown ED, Suryan RM, Irons DB (2003) Prey availability, interference competition, and the geographic structure of seabird colonies: a study of black-legged kittiwakes and forage fish in Prince William Sound, Alaska. Ecology 84:709–723
14. Ainley DG, O'connor EF, Boekelheide RJ (1984) Ecology of seabirds in the Ross Sea, Antarctica. Ornithol Monogr 32:iii–97
15. Ainley DG, Ribic CA, Spear LB (1993) Species-habitat relationships among Antarctic seabirds: a function of physical and biological factors? Condor 95:806–816
16. Angelier F, Bost CA, Giraudeau M, Bouteloup G, Dano S, Chastel O (2008) Corticosterone and foraging behavior in a diving seabird: the Adélie penguin, *Pygoscelis adeliae*. Gen Comp Endocrinol 156(1):134–144
17. Ashmole NP (1963) The regulation of numbers of tropical oceanic birds. Ibis 103(3):458–473
18. Atkinson A, Siegel V, Pakhomov EA, Rothery P (2004) Longterm decline in krill stock and increase in salps within the Southern Ocean. Nature 432:100–103
19. Atkinson A, Siegel V, Pakhomov EA, Rothery P, Loeb V, Ross RM, Quetin LB, Schmidt K, Fretwell P, Murphy EJ, Tarling GA, Fleming AH (2008) Oceanic circumpolar habitats of Antarctic krill. Mar Ecol Prog Ser 362:1–23
20. Ballance L, Pitman RL, Hewitt RP, Siniff DB, Trivelpiece WZ, Clapham PJ, Brownell RL Jr (2006) The removal of large whales from the Southern Ocean: evidence for long-term ecosystem effects? In: Estes JA, Demaster DP, Doak DF, Williams TE, Brownell RL Jr (eds) Whales, whaling and ocean ecosystems. University of California Press, Berkeley, pp 215–230
21. Ballance LT, Ainley DG, Ballard G, Barton K (2009) An energetic correlate between colony size and foraging effort in seabirds, an example of the Adélie penguin *Pygoscelis adeliae*. J Avian Biol 40(3):279–288
22. Barlow KE, Croxall JP (2002) Provisioning behaviour of macaroni penguins *Eudyptes chrysolophus*. Ibis 144(2):248–258
23. Barlow KE, Croxall JP (2002) Seasonal and interannual variation in foraging range and habitat of macaroni penguins *Eudyptes chrysolophus* at South Georgia. Mar Ecol Prog Ser 232:291–304
24. Barrera-Oro E, Marschoff E, Ainley DG (2017) Changing status of three notothenioid fish at the South Shetland Islands (1983–2016) after impacts of the 1970–80s commercial fishery. Polar Biol 40(10):2047–2054
25. Bartle SA (1974) Seabirds of eastern Cook Strait, New Zealand, in autumn. Notornis 21:135–166
26. Bearhop S, Phillips RA, Mcgill R, Cherel Y, Dawson DA, Croxall JP (2006) Stable isotopes indicate sex-specific and long-term individual foraging specialization in diving seabirds. Mar Ecol Prog Ser 311:157–164
27. Bethge P, Nicol S, Culik BM, Wilson RP (1997) Diving behaviour and energetics in breeding little penguins (*Eudyptua minor*). J Zool 242(3):483–502
28. Birt VL, Birt TP, Goulet D, Cairns DK, Montevecchi WA (1987) Ashmole's halo: direct evidence for prey depletion by a seabird. Mar Ecol Prog Ser 40(3):205–208
29. Blanchet MA, Biuw M, Hofmeyr GG, De Bruyn PN, Lydersen C, Kovacs KM (2013) At-sea behaviour of three krill predators breeding at Bouvetøya—Antarctic fur seals, macaroni penguins and chinstrap penguins. Mar Ecol Prog Ser 477:285–302
30. Boersma PD, Rebstock GA (2009) Foraging distance affects reproductive success in Magellanic penguins. Mar Ecol Prog Ser 375:263–275
31. Bolton M, Conolly G, Carroll M, Wakefield ED, Caldow R (2019) A review of the occurrence of inter-colony segregation of seabird foraging areas and the implications for marine environmental impact assessment. Ibis 161(2):241–259

32. Bost C-A, Thiebot J-B, Pinaud D, Cherel Y, Trathan PN (2009) Where do penguins go during the inter-breeding period? Using geolocation to track the winter dispersion of the macaroni penguin. Biol Lett 5:473–476
33. Boyd IL (1993) Pup production and distribution of breeding Antarctic fur seals (*Arctocephalus gazella*) at South Georgia. Antarct Sci 5(1):17–24
34. Boyd IL (2002) Estimating food consumption of marine predators: antarctic fur seals and macaroni penguins. J Appl Ecol 39:103–119
35. Branch TA (2006) Abundance estimates for Antarctic minke whales from three completed circumpolar sets of surveys, 1978/79 to 2003/04. International Whaling Commission, Cambridge
36. Branch TA (2011) Humpback whale abundance south of 60°S from three complete circumpolar sets of surveys. J Cetacean Res Manag Special Issue 3:53–69
37. Branch TA, Stafford KM, Palacios DM, Allison C, Bannister JL, Burton CLK, Cabrera E, Carlson CA, Vernazzani BG, Gill PC, Hucke-Gaete R (2007) Past and present distribution, densities and movements of blue whales *Balaenoptera musculus* in the Southern Hemisphere and northern Indian Ocean. Mammal Rev 37(2):116–175
38. Brooks CM, Ainley DG (2016) Fishing the bottom of the Earth: the political challenges of ecosystem-based management. In: Dodds K, Hemmings AD, Roberts P (eds) Handbook on the politics of Antarctica. Edward Elgar Publishing, Northampton, pp 422–438
39. Brown CR (1987) Traveling speed and foraging range of macaroni and rockhopper penguins at Marion Island. J Field Ornithol 58:118–125
40. Brown CR, Klages NT (1987) Seasonal and annual variation in diets of macaroni (*Eudyptes chrysolophus*) and southern rockhopper (*Eudyptes crestatus*) penguins at sub-Antarctic Marion Island. J Zool 212(1):7–28
41. Browne T, Lalas C, Mattern T, Van Heezik Y (2011) Chick starvation in yellow-eyed penguins: evidence for poor diet quality and selective provisioning of chicks from conventional diet analysis and stable isotopes. Austral Ecol 36(1):99–108
42. Camprasse E, Cherel Y, Bustamante P, Arnould J, Bost CA (2017) Intra and inter-individual variation in the foraging ecology of a generalist subantarctic seabird, the gentoo penguin. Mar Ecol Prog Ser 578:227–242
43. Casaux R, Barrera-Oro E (2006) Shags in Antarctica: their feeding behaviour and ecological role in the marine food web. Antarct Sci 18(1):3–14
44. Chapman EW, Hofmann EE, Patterson DL, Ribic CA, Fraser WR (2011) Marine and terrestrial factors affecting Adélie penguin *Pygoscelis adeliae* chick growth and recruitment off the western Antarctic Peninsula. Mar Ecol Prog Ser 436:273–289
45. Chappell MA, Shoemaker VH, Janes DN, Bucher TL, Maloney SK (1993) Diving behavior during foraging in breeding Adélie penguins. Ecology 74:1204–1215
46. Charrassin JB, Park Y-H, Le Maho YL, Bost CA (2002) Penguins as oceanographers unravel hidden mechanisms of marine productivity. Ecol Lett 5(3):317–319
47. Cherel Y, Hobson KA, Guinet C, Vanpe C (2007) Stable isotopes document seasonal changes in trophic niches and winter foraging individual specialization in diving predators from the Southern Ocean. J Anim Ecol 76(4):826–836
48. Cimino MA, Moline MA, Fraser WR, Patterson-Fraser DL, Oliver MJ (2016) Climate-driven sympatry may not lead to foraging competition between congeneric top-predators. Sci Rep 6:18820
49. Clarke J, Manly B, Kerry K, Gardner H, Franchi E, Corsolini S, Focardi S (1998) Sex differences in Adélie penguin foraging strategies. Polar Biol 20(4):248–258
50. Clarke JR (2001) Partitioning of foraging effort by Adélie penguins at Béchervaise Island. Polar Biol 24(1):16–20
51. Cole TL, Dutoit L, Dussex N, Hart T, Alexander A, Younger JL, Clucas GV, Frugone MJ, Cherel Y, Cuthbert R, Ellenberg U (2019) Receding ice drove parallel expansions in Southern Ocean penguins. Proc Natl Acad Sci U S A 116(52):26690–26696

52. Cole TL, Ksepka DT, Mitchell KJ, Tennyson AJD, Thomas DB, Pan H, Zhang G, Rawlence NJ, Wood JR, Bover P, Bouzat JL, Cooper A, Fiddaman SR, Hart T, Miller G, Ryan PG, Shepherd LD, Wilmshurst JM, Waters JM (2019) Mitogenomes uncover extinct penguin taxa and reveal island formation as a key driver of speciation. Mol Biol Evol 36(4):784–797
53. Colominas-Ciuró R, Santos M, Coria N, Barbosa A (2018) Sex-specific foraging strategies of Adélie penguins (*Pygoscelis adeliae*): females forage further and on more krill than males in the Antarctic Peninsula. Polar Biol 41(12):2635–2641
54. Constable AJ, de la Mare WK, Agnew DJ, Everson I, Miller D (2000) Managing fisheries to conserve the Antarctic marine ecosystem: practical implementation of the Convention on the Conservation of Antarctic Marine Living Resources (CCAMLR). ICES J Mar Sci 57:778–791
55. Cooper J, Brown CR, Gales RP, Hindell MA, Klages NT, Moors PJ, Van Heezik Y (1990) Diets and dietary segregation of crested penguins (*Eudyptes*). In: Davis LS, Darby JT (eds) Penguin biology. Academic, San Diego, pp 131–156
56. Crawford RJM, Altwegg R, Barham BJ, Barham PJ, Durant JM, Dyer BM, Geldenhuys D, Makhado AB, Pichergru L, Ryan PG, Underhill LG, Upfold L, Visagie J, Waller LJ, Whittington PA (2011) Collapse of South Africa's penguins in the early 21st century. Afr J Mar Sci 33(1):139–156
57. Crawford RJM, Kemper J, Underhill LG (2013) African penguin (*Spheniscus demersus*). In: Garcia-Borboroglu P, Boersma PD (eds) Penguins: natural history and conservation. University of Washington Press, Seattle, pp 211–231
58. Crawford RJM, Underhill LG, Upfold L, Dyer BM (2007) An altered carrying capacity of the Benguela upwelling ecosystem for African penguins (*Spheniscus demersus*). ICES J Mar Sci 64(3):570–576
59. Croll DA, Tershey BR (1998) Penguins, fur seals, and fshing: prey requirements and potential competition in the South Shetland Islands, Antarctica. Polar Biol 19:365–374
60. Croxall JP, Davis RW, O'connell MJ (1988) Diving patterns in relation to diet of gentoo and macaroni penguins at South Georgia. Condor 90(1):157–167
61. Croxall JP, Rickets C, Prince PA (1984) Impact of seabirds on marine resources, especially krill, of South Georgia. In: Whittow GC, Rahn H (eds) Seabird energetics. Plenum Press, New York, pp 285–316
62. Cuthbert R, Cooper J, Burle MH, Glass CJ, Glass JP, Glass S, Glass T, Hilton GM, Sommer ES, Wanless RM, Ryan PG (2009) Population trends and conservation status of the northern rockhopper penguin *Eudyptes moseleyi* at Tristan da Cunha and Gough Island. Bird Conserv Int 19(1):109–120
63. Dann P, Norman FI (2006) Population regulation in little penguins (*Eudyptula minor*): the role of intraspecific competition for nesting sites and food during breeding. Emu 106(4):289–296
64. De La Puente S, Bussalleu A, Cardeña M, Valdés-Velásquez A, Maluf P, Simeone A (2013) Humboldt penguin (*Spheniscus humboldti*). In: Borboroglu PG, Boersma PD (eds) Penguins: natural history and conservation. University of Washington Press, Seattle, pp 265–283
65. Diamond AW (1978) Feeding strategies and population size in tropical seabirds. Am Nat 112:215–223
66. Dias MP, Warwick-Evans V, Carneiro APB, Harris C, Lascelles BG, Clewlow HL, Manco F, Ratcliffe N, Trathan PN (2019) Using habitat models to identify marine important bird and biodiversity areas for chinstrap penguins *Pygoscelis antarcticus* in the South Orkney Islands. Polar Biol 42(1):17–25
67. Dimitrijević D, Paiva VH, Ramos JA, Seco J, Ceia FR, Chipev N, Valente T, Barbosa A, Xavier JC (2018) Isotopic niches of sympatric gentoo and chinstrap penguins: evidence of competition for Antarctic krill? Polar Biol 41(9):1655–1669
68. Ducklow HW, Baker K, Martinson DG, Quetin LB, Ross RM, Smith RC, Stammerjohn SE, Vernet M, Fraser W (2007) Marine pelagic ecosystems: the west Antarctic Peninsula. Philos Trans R Soc B Biol Sci 362:67–94
69. Duffy DC (1983) Environmental uncertainty and commercial fishing: effects on Peruvian guano birds. Biol Conserv 26(3):227–238

70. Duffy DC, Hays C, Plenge M (1984) The conservation status of Peruvian seabirds. In: Croxall JP, Evans PGH, Schreiber RW (eds) Status and conservation of the world's seabirds. International Council of Bird Preservation Technical Publ. No. 2, Cambridge, pp 245–260
71. Elley T, Mattern T, Ellenberg U, Young MJ, Hickcox RP, Van Heezik Y, Seddon PJ (2022) Consistent site-specific foraging behaviours of yellow-eyed penguins/hoiho breeding on Stewart Island, New Zealand. Biology 11(6):844
72. Emslie SD, Patterson WP (2007) Abrupt recent shift in δ13C and δ15N values in Adélie penguin eggshell in Antarctica. Proc Natl Acad Sci 104(28):11666–11669
73. Emslie SD, Polito MJ, Patterson WP (2013) Stable isotope analysis of ancient and modern gentoo penguin egg membrane and the krill surplus hypothesis in Antarctica. Antarct Sci 25(2):213–218
74. Evans R, Lea MA, Hindell MA (2021) Predicting the distribution of foraging seabirds during a period of heightened environmental variability. Ecol Appl 31(5):e02343
75. Ford RG, Ainley DG, Lescroël A, Lyver POB, Toniolo V, Ballard G (2015) Testing assumptions of central place foraging theory: a study of Adélie penguins *Pygoscelis adeliae* in the Ross Sea. J Avian Biol 46(2):193–205
76. Forero MG, Hobson KA, Bortolotti GR, Donázar JA, Bertellotti M, Blanco G (2002) Food resource utilization by the Magellanic penguin evaluated through stable-isotope analysis: segregation by sex and age and influence on offspring quality. Mar Ecol Prog Ser 234:289–299
77. Furness RW, Birkhead TR (1984) Seabird colony distributions suggest competition for food supplies during the breeding season. Nature 311:655–656
78. Gaston AJ, Ydenberg RC, Smith GEJ (2007) Ashmole's halo and population regulation in seabirds. Mar Ornithol 35:119–126
79. Gorman KB, Williams TD, Fraser WR (2014) Ecological sexual dimorphism and environmental variability within a community of Antarctic penguins (Genus *Pygoscelis*). PLoS One 9(3):e90081
80. Grant PR, Grant BR (2008) How and why species multiply: the radiation of Darwin's finches. Princeton University Press, Princeton
81. Green CP, Ratcliffe N, Mattern T, Thompson D, Lea MA, Wotherspoon S, Borboroglu PG, Ellenberg U, Morrison KW, Pütz K, Sagar PM, Seddon PJ, Torres LG, Hindell MA (2022) The role of allochrony in influencing interspecific differences in foraging distribution during the non-breeding season between two congeneric crested penguin species. PLoS One 17(2):e0262901
82. Green JA, Boyd IL, Woakes AJ, Warren NL, Butler PJ (2005) Behavioural flexibility during year-round foraging in macaroni penguins. Mar Ecol Prog Ser 296:183–196
83. Green JA, Wilson RP, Boyd IL, Woakes AJ, Green CJ, Butler PJ (2009) Tracking macaroni penguins during long foraging trips using 'behavioural geolocation'. Polar Biol 32(4):645–653
84. Green K, Williams R, Green MG (1998) Foraging ecology and diving behaviour of macaroni penguins *Eudyptes chrysolophus* at Heard Island. Mar Ornithol 26:27–34
85. Grémillet D, Ponchon A, Paleczny M, Palomares M-L D, Karpouzi V, Pauly D (2018) Persisting worldwide seabird-fishery competition despite seabird community decline. Curr Biol 28:4009–4013
86. Guinet C, Koudil M, Bost C-A, Durbec JP, Georges JY, Mouchot MC, Jouventin P (1997) Foraging behaviour of satellite-tracked king penguins in relation to sea-surface temperatures obtained by satellite telemetry at Crozet Archipelago, a study during three austral summers. Mar Ecol Prog Ser 150:11–20
87. Harrison NM, Whitehouse MJ, Heinemann D, Prince PA, Hunt GLJ, Veit RR (1991) Observations of multispecies seabird flocks around South Georgia. Auk 108(4):801–810
88. Hinke JT, Cossio AM, Goebel ME, Reiss CS, Trivelpiece WZ, Watters GM (2017) Identifying risk: concurrent overlap of the Antarctic Krill fishery with krill-dependent predators in the Scotia Sea. PLoS One 12(1):e0170132

89. Hinke JT, Polito MJ, Goebel ME, Jarvis S, Reiss CS, Thorrold SR, Trivelpiece WZ, Watters GM (2015) Spatial and isotopic niche partitioning during winter in chinstrap and Adélie penguins from the South Shetland Islands. Ecosphere 6(7):125
90. Hofman RJ (2017) Sealing, whaling and krill fishing in the Southern Ocean: past and possible future effects on catch regulations. Polar Rec 53(1):88–99
91. Hofman RJ (2019) Stopping overexploitation of living resources on the high seas. Mar Policy 103:91–100
92. Horne RSC (1985) Diet of royal and rockhopper penguins at Macquarie Island. Emu 85(3):150–156
93. Horswill C, Ratcliffe N, Green JA, Phillips RA, Trathan PN, Matthiopoulos J (2020) Unravelling the relative roles of top-down and bottom-up forces driving population change in an oceanic predator. Ecology 97(8):1919–1928
94. Hoskins AJ, Dann P, Ropert-Coudert Y, Kato A, Chiaradia A, Costa DP, Arnould JPY (2008) Foraging behaviour and habitat selection of the little penguin *Eudyptula minor* during early chick rearing in Bass Strait, Australia. Mar Ecol Prog Ser 366:293–303
95. Hucke-Gaete R, Osman LP, Moreno CA, Torres D (2004) Examining natural population growth from near extinction: the case of the Antarctic fur seal at the South Shetlands, Antarctica. Polar Biol 27(5):304–311
96. Hull CL (1999) The foraging zones of breeding royal (*Eudyptes schlegeli*) and rockhopper (*E. chrysocome*) penguins: an assessment of techniques and species comparison. Wildl Res 26(6):789–803
97. Hull CL, Hindell MA, Michael K (1997) Foraging zones of royal penguins during the breeding season, and their association with oceanographic features. Mar Ecol Prog Ser 153:217–228
98. Jovani R, Lascelles M, Garamszegi LZ, Mavor R, Thaxter CB, Oro D (2015) Colony size and foraging range in seabirds. Oikos 125(7):968–974
99. Karnovsky NJ (1997) The fish component of *Pygoscelis* penguin diets. MSc thesis. Montana State University, Bozeman
100. King SD, Harper GA, Wright JB, Mcinnes JC, Van Der Lubbe JE, Dobbins ML, Murray SJ (2012) Site-specific reproductive failure and decline of a population of the endangered yellow-eyed penguin: a case for foraging habitat quality. Mar Ecol Prog Ser 467:233–244
101. Kirkman SP (2009) Evaluating seal–seabird interactions in southern Africa: a critical review. Afr J Mar Sci 31(1):1–18
102. Klages NTW, Gales RP, Pemberton D (1989) Dietary segregation of macaroni and rockhopper penguins at Heard Island. Wildl Res 16(6):599–604
103. Klages NTW, Pemberton D, Gales RP (1990) The diets of king and gentoo penguins at Heard Island. Wildl Res 17(1):53–60
104. Kock K-H (1992) Antarctic fish and fisheries. Cambridge University Press, Cambridge
105. Kooyman GL, Goetz K, Williams CL, Ponganis PJ, Sato K, Eckert S, Horning M, Thorson PT, Van Dam RP (2020) Crary bank: a deep foraging habitat for emperor penguins in the western Ross Sea. Polar Biol 43(7):801–811
106. Krause DJ, Hinke JT (2021) Finally within reach: a drone census of an important, but practically inaccessible, Antarctic fur seal colony. Aquat Mamm 47(4):349–354
107. Ksepka D, Ando T (2011) Penguins past, present, and future: trends in the evolution of the sphenisciformes. In: Dyke G, Kaiser G (eds) Living dinosaurs: the evolutionary history of modern birds. Wiley, New York, pp 155–186
108. La Mesa M, Eastman JT (2012) Antarctic silverfish: life strategies of a key species in the high-Antarctic ecosystem. Fish Fish 13(3):241–266
109. Larue MA, Salas L, Nur N, Ainley DG, Stammerjohn S, Barrington L, Stamatiou K, Pennycook J, Dozier M, Saints J, Nakamura H (2019) Physical and ecological factors explain the distribution of Ross Sea Weddell seals during the breeding season. Mar Ecol Prog Ser 612:193–208

110. Larue MA, Salas L, Nur N, Ainley DG, Stammerjohn S, Pennycook J, Dozier M, Saints J, Stamatiou K, Barrington L, Rotella J (2021) Insights from the first global population estimate of Weddell seals in Antarctica. Sci Adv 7(39):eabh3674
111. Le Vaillant M, Le Bohec C, Prud'homme O, Wienecke B, Le Maho Y, Kato A, Ropert-Coudert Y (2013) How age and sex drive the foraging behaviour in the king penguin. Mar Biol 160(5):1147–1156
112. Lescroël A, Ballard G, Grémillet D, Authier M, Ainley DG (2014) Antarctic climate change: extreme events disrupt plastic phenotypic response in Adélie penguins. PLoS One 9(1):e85291
113. Lescroël A, Ballard G, Toniolo V, Barton KJ, Wilson PR, Lyver POB, Ainley DG (2010) Working less to gain more: when breeding quality relates to foraging efficiency. Ecology 91(7):2044–2055
114. Lescroël A, Lyver POB, Jongsomjit D, Veloz S, Dugger KM, Kappes P, Karl BJ, Whitehead AL, Pech R, Cole TL, Ballard G (2020) Inter-individual differences in the foraging behavior of breeding Adélie penguins are driven by individual quality and sex. Mar Ecol Prog Ser 636:189–205
115. Lescroël A, Schmidt A, Elrod M, Ainley DG, Ballard G (2021) Foraging dive frequency predicts body mass gain in the Adélie penguin. Sci Rep 11(1):1–10
116. Lewis ST, Sherratt N, Hamer KC, Wanless S (2001) Evidence of intra-specific competition for food in a pelagic seabird. Nature 412(6849):816–819
117. Lowther AD, Staniland I, Lydersen C, Kovacs KM (2020) Male Antarctic fur seals: neglected food competitors of bioindicator species in the context of an increasing Antarctic krill fishery. Sci Rep 10(1):1–12
118. Ludynia K, Roux JP, Jones R, Kemper J, Underhill LG (2010) Surviving off junk: low-energy prey dominates the diet of African penguins *Spheniscus demersus* at Mercury Island, Namibia, between 1996 and 2009. Afr J Mar Sci 32(3):563–572
119. Lynch HJ, Naveen R, Trathan PN, Fagan WF (2012) Spatially integrated assessment reveals widespread changes in penguin populations on the Antarctic Peninsula. Ecology 93(6):1367–1377
120. Marchant S, Higgins P (1990) Handbook of Australian, New Zealand and Antarctic birds. Oxford University Press, Melbourne
121. Marschoff ER, Barrera-Oro ER, Alescio NS, Ainley DG (2012) Slow recovery of previously depleted demersal fish at the South Shetland Islands, 1983–2010. Fish Res 125:206–213
122. Masello JF, Mundry R, Poisbleau M, Demongin L, Voigt CC, Wikelski M, Quillfeldt P (2010) Diving seabirds share foraging space and time within and among species. Ecosphere 1(6):1–28
123. Massaro M, Ainley DG, Santora JA, Quillfeldt P, Lescroël A, Whitehead A, Varsani A, Ballard G, Lyver POB (2020) Diet segregation in Adélie penguins: some individuals attempt to overcome colony-induced and annual foraging challenges. Mar Ecol Prog Ser 645:205–218
124. Mattern T (2020) Modelling marine habitat utilisation by yellow-eyed penguins along their mainland distribution: baseline information. Fisheries New Zealand, Wellington
125. Mattern T, Ellenberg U, Houston DM, Davis LS (2007) Consistent foraging routes and benthic foraging behaviour in yellow-eyed penguins. Mar Ecol Prog Ser 343:295–306
126. Mattern T, Ellenberg U (2018) Yellow-eyed penguin diet and indirect effects affecting prey composition – collation of biological information. NZ Department of Conservation, Wellington
127. Mccutcheon C, Dann P, Salton M, Renwick L, Hoskins AJ, Gormley AM, Arnould JPY (2011) The foraging range of little penguins (*Eudyptula minor*) during winter. Emu 111(4):321–329
128. Mills KL (1998) Multispecies seabird feeding flocks in the Galápagos Islands. Condor 100(2):277–285
129. Moore GJ, Wienecke B, Robertson G (1999) Seasonal change in foraging areas and dive depths of breeding king penguins at Heard Island. Polar Biol 21(6):376–384

130. Moore PJ, Wakelin MD (1997) Diet of the yellow-eyed penguin *Megadyptes antipodes*, South Island, New Zealand, 1991–1993. Mar Ornithol 25:17–29
131. Murphy RC (1936) Oceanic birds of South America. Macmillan Company, New York
132. Nagy KA, Siegfried WR, Wilson RP (1984) Energy utilization by free-ranging Jackass penguins, *Spheniscus demersus*. Ecology 65:1648–1655
133. Pakhomov EA, Mcquaid CD (1996) Distribution of surface zooplankton and seabirds across the Southern Ocean. Polar Biol 16(4):271–286
134. Paleczny M (2008) An analysis of temporal and spatial patterns in global seabird abundance during the modern industrial era, 1950–2010, and the relationship between global seabird cecline and marine fisheries catch. MSc thesis, University of British Columbia
135. Paleczny M, Hammill E, Karpouzi V, Pauly D (2015) Population trend of the world's monitored seabirds, 1950–2010. PLoS One 10(6):e0129342
136. Pelletier L, Chiaradia A, Kato A, Ropert-Coudert Y (2014) Fine-scale spatial age segregation in the limited foraging area of an inshore seabird species, the little penguin. Oecologia 176(2):399–408
137. Péron C, Weimerskirch H, Bost CA (2012) Projected poleward shift of king penguins' (*Aptenodytes patagonicus*) foraging range at the Crozet Islands, southern Indian Ocean. Proc R Soc B Biol Sci 279(1738):2515–2523
138. Phillips JA, Fayet AL, Guilford T, Manco F, Warwick-Evans V, Trathan P (2021) Foraging conditions for breeding penguins improve with distance from colony and progression of the breeding season at the south Orkney Islands. Mov Ecol 9:22
139. Pichegru L, Cook T, Handley J, Voogt N, Watermeyer J, Nupen L, Mcquaid CD (2013) Sex-specific foraging behaviour and a field sexing technique for endangered African penguins. Endanger Species Res 19(3):255–264
140. Pichegru L, Grémillet D, Crawford RJM, Ryan PG (2010) Marine no-take zone rapidly benefit threatened penguin. Biol Lett 6(4):498–501
141. Pichegru L, Ryan PG, Van Eeden R, Reid T, Grémillet D, Wanless R (2012) Industrial fishing, no-take zones and endangered penguins. Biol Conserv 156:117–125
142. Polito MJ, Brasso RL, Trivelpiece WZ, Karnovsky N, Patterson WP, Emslie SD (2016) Differing foraging strategies influence mercury (Hg) exposure in an Antarctic penguin community. Environ Pollut 218:196–206
143. Polito MJ, Trivelpiece WZ, Patterson WP, Karnovsky NJ, Reiss CS, Emslie SD (2015) Contrasting specialist and generalist patterns facilitate foraging niche partitioning in sympatric populations of *Pygoscelis* penguins. Mar Ecol Prog Ser 519:221–237
144. Pratt HD (2005) The Hawaiian honeycreepers. Oxford University Press, Oxford
145. Pütz K (2002) Spatial and temporal variability in the foraging areas of breeding king penguins. Condor 104(3):528–538
146. Pütz K, Ingham RJ, Smith JG, Croxall JP (2001) Population trends, breeding success and diet composition of gentoo *Pygoscelis papua*, Magellanic *Spheniscus magellanicus* and rockhopper *Eudyptes chrysocome* penguins in the Falkland Islands. Polar Biol 24(11):793–807
147. Rand RW (1960) The biology of guano-producing seabirds. The distribution, abundance and feeding habits of the Cape penguin, *Spheniscus demersus*, off the southwestern coast of the Cape Province. Investl Rep Div Fish S Afr 41:1–28
148. Ratcliffe N, Crofts S, Brown R, Baylis AMM, Adlard S, Horswill C, Venables H, Taylor P, Trathan PN, Staniland IJ (2014) Love thy neighbour or opposites attract? Patterns of spatial segregation and association among crested penguin populations during winter. J Biogeogr 41(6):1183–1192
149. Reid TA, Hindell MA, Eades DW, Newman M (2002) Seabird Atlas of South-Eastern Australian Waters, vol 4. Birds Australia Monograph, Victoria
150. Reilly P (1983) Fairy penguins and earthy people. Lothian, Maryborough
151. Riaz J, Bestley S, Wotherspoon S, Freyer J, Emmerson L (2020) From trips to bouts to dives: temporal patterns in the diving behaviour of chick-rearing Adélie penguins, East Antarctica. Mar Ecol Prog Ser 654:177–194

152. Ribic CA, Ainley DG, Ford RG, Fraser WR, Tynan CT, Woehler EJ (2011) Water masses, ocean fronts, and the structure of Antarctic seabird communities: putting the eastern Bellingshausen Sea in perspective. Deep Sea Res Part II Top Stud Oceanogr 58(13–16):1695–1709
153. Richdale LE (1949) A study of a group of penguins of known age. Biological Monographs 1, Dunedin
154. Rosciano NG, Polito MJ, Rey AR (2016) Do penguins share? Evidence of foraging niche segregation between but not within two sympatric, central-place foragers. Mar Ecol Prog Ser 548:249–262
155. Rosciano NG, Pütz K, Polito MJ, Rey AR (2018) Foraging behaviour of Magellanic penguins during the early chick-rearing period at Isla de los Estados, Argentina. Ibis 160(2):327–341
156. Saenz BT, Ainley DG, Daly KL, Ballard G, Conlisk E, Elrod ML, Kim SL (2020) Drivers of concentrated predation in an Antarctic marginal-ice-zone food web. Sci Rep 10:7282
157. Sala Je WR, Quintana F (2012) How much is too much? Assessment of prey consumption by Magellanic Penguins in Patagonian colonies. PLoS One 7(12):e51487. https://doi.org/10.1371/journal.pone.0051487
158. Santora JA, Larue MA, Ainley DG (2020) Geographic structuring of Antarctic penguin populations. Glob Ecol Biogeogr 29(10):1716–1728
159. Savoca MS, Czapanskiy MF, Kahane-Rapport SR, Gough WT, Fahlbusch JA, Bierlich KC, Segre PS, Di Clemente J, Penry GS, Wiley DN, Calambokidis J, Nowacek DP, Johnston DW, Pyenson ND, Friedlaende AS, Hazen EL, Goldbogen JA (2020) Baleen whale prey consumption based on high-resolution foraging measurements. Nature 599:85–90
160. Seddon PJ, Ellenberg U, Van Heezik Y (2013) Yellow-eyed penguin (*Megadyptes antipodes*). In: Borboroglu PG, Boersma PD (eds) Penguins: natural history and conservation. University of Washington Press, Seattle, pp 91–110
161. Sherley RB, Ludynia K, Dyer BM, Lamont T, Makhado AB, Roux JP, Scales KL, Underhill LG, Votier SC (2017) Metapopulation tracking juvenile penguins reveals an ecosystem-wide ecological trap. Curr Biol 27(4):563–568
162. Sladen WJL (1964) The distribution of the Adélie and chinstrap penguins. In: Carrick R, Holdgate MW, Prévost J (eds) Biologie antarctique. Hermann, Paris, pp 359–365
163. Sparks J, Soper T (1967) Penguins. David & Charles Publishers Limited, Devon
164. Spear LB, Ainley DG (2008) The seabird community of the Peru Current, 1980–1995, with comparisons to other eastern boundary currents. Mar Ornithol 36:125–144
165. Stokes DL, Boersma PD, De Casenave JL, Garcia-Borboroglu P (2014) Conservation of migratory Magellanic penguins requires marine zoning. Biol Conserv 170:151–161
166. Storer RW (1952) A comparison of variation, behavior and evolution in the seabird genera Uria and Cepphus. Univ Calif Publ Zool 52:121–222
167. Strycker N, Wethington M, Borowicz A, Forrest S, Witharana C, Hart T, Lynch HJ (2020) A global population assessment of the Chinstrap penguin (*Pygoscelis antarctica*). Sci Rep 10:19474
168. Thiebot JB, Cherel Y, Crawford RJM, Makhado AB, Trathan PN, Pinaud D, Bost CA (2013) A space oddity: geographic and specific modulation of migration in *Eudyptes* penguins. PLoS One 8(8):e71429
169. Thiebot JB, Lescroël A, Pinaud D, Trathan PN, Bost CA (2011) Larger foraging range but similar habitat selection in non-breeding versus breeding sub-Antarctic penguins. Antarct Sci 23(2):117–126
170. Thompson D (2016) Penguins reveal unknown swimming talents. NIWA Press Release. https://www.niwa.co.nz/news/penguins-revealunknown-swimming-talents
171. Tønnessen JN, Johnsen AO (1982) The history of modern whaling. C. Hurst and Co., London
172. Trathan PN, Ratcliffe N, Masden EA (2012) Ecological drivers of change at South Georgia: the krill surplus, or climate variability. Ecography 35(11):983–993
173. Van Heezik Y (1990) Seasonal, geographical, and age-related variations in the diet of the yellow-eyed penguin (*Megadyptes antipodes*). New Zealand J Ecol 17(2):201–212

174. Waluda CM, Hill SL, Peat HJ, Trathan PN (2017) Long-term variability in the diet and reproductive performance of penguins at Bird Island, South Georgia. Mar Biol 164:39
175. Warham J (1975) The crested penguins. In: Stonehouse B (ed) The biology of penguins. London, Macmillan
176. Watters GM, Hinke JT, Reiss CS (2020) Long-term observations from Antarctica demonstrate that mismatched scales of fisheries management and predator-prey interaction lead to erroneous conclusions about precaution. Sci Rep 10:1–9
177. Webster T (2018) The pathway ahead for hoiho. Yellow-Eyed Penguin Trust, Dunedin
178. Weichler T, Garthe S, Luna-Jorquera G, Morage J (2004) Seabird distribution on the Humboldt Current in northern Chile in relation to hydrography, productivity, and fisheries. ICES J Mar Sci 61(1):148–154
179. Whitehead TO, Connan M, Ropert-Coudert Y, Ryan PG (2017) Subtle but significant segregation in the feeding ecology of sympatric penguins during the critical pre-moult period. Mar Ecol Prog Ser 565:227–236
180. Wienecke BC, Robertson G (1997) Foraging space of emperor penguins *Aptenodytes forsteri* in Antarctic shelf waters in winter. Mar Ecol Prog Ser 159:249–263
181. Willis J (2007) Could whales have maintained a high abundance of krill? Evol Ecol Res 9:651–662
182. Woehler EJ (1993) The distribution and abundance of Antarctic and Subantarctic penguins. Scientific Committee for Antarctic Research, Scott Polar Research Institute, Cambridge
183. Woehler EJ (1995) Consumption of Southern Ocean marine resources by penguins. In: Dann P, Norman I, Reilly P (eds) Penguins: ecology and management. Chipping Norton, Surrey Beatty & Sons Pty Limitied, pp 266–296
184. Woehler Ej GK (1992) Consumption of marine resources by seabirds and seals at Heard Island and McDonald Islands. Polar Biol 12:659–665
185. Woehler EJ, Raymond B, Boyle A, Stafford A (2010) Seabird assemblages observed during the BROKE-West survey of the Antarctic coastline (301E–801E), January – March 2006. Deep Sea Res Part II Top Stud Oceanogr 57:982–991
186. Xavier JC, Trathan PN, Ceia FR, Tarling GA, Adlard S, Fox D, Edwards EWJ, Vieira RP, Medeiros R, De Broyer C, Cherel Y (2017) Sexual and individual foraging segregation in gentoo penguins *Pygoscelis papua* from the Southern Ocean during an abnormal winter. PLoS One 12(3):e0174850
187. Zimmer I, Wilson RP, Gilbert C, Beaulieu M, Ancel A, Plötz J (2008) Foraging movements of emperor penguins at Pointe Géologie, Antarctica. Polar Biol 31(2):229–243

Part III
The Hardware of a Fish-Bird

Chapter 6
The Slippery Shape, Hot Air, and the Powerhouse: How Fish-Birds Swim

If there is one thing that penguins do extremely well, it is swim. There have been many texts that document exactly that. For example:

> "If they [Gentoo fledglings upon first swim] rested too close to the water, a wave was sure to sweep them seaward....No matter how they got into the water [others walked in] they were instantly at home. They had met the sea and recognized their element. Here they would live for the rest of their lives." (E. Pettingill [58])

Actually, on the basis of our own experience, we find that fledgling penguins are often not that good in water, requiring some days to show "respectable competence." This occurs some days even after they rid themselves of tufts of down that act like life jackets, inhibiting any diving. We have seen down-covered emperor penguin fledglings departing for sea, flailing across the oceanic surface like steamer ducks, unable to shed their life jackets. Adults are quite a different matter, however. Despite having seen penguins in the water at zoos many times, we (specifically Rory) were overwhelmed upon seeing an African penguin swimming in the wild. The bird was floating in a largish rock pool, perhaps a meter deep, 15-m long but just 1.5-m wide. It was sculling around lazily, preening when Rory arrived, but, on seeing him, it immediately dived to then shoot up the length of the canal, leaving tiny dancing bubbles that rose to the surface in its wake. This "spirit," that had become little more than a perception as demonstrated by its speed, flickered left and right before, somehow, turning instantaneously with the touch of a flipper, to come fizzing back, swerving close to him to disappear back up the length of the pool again. The bird leapt out of the water at the far end, a projectile, still hugged tightly by a thin layer of water, not, it seemed, wishing to let it go, and then waddled off. Penguins at zoos never do this sort of thing, and few penguin researchers have had the honor and luck to witness such athleticism. However, to juxtapose the sublime being that slipped between waters, epitomizing fluidity and penguins' otherworldly ether, with the inept rocking gait of a popular clown was profound. If nothing else, we hope that we can convey to the readers of this book, at least to some extent, the metamorphosis of water – as opposed to the land version of penguins because to

D. G. Ainley, R. P. Wilson, *The Aquatic World of Penguins*, Fascinating Life Sciences, https://doi.org/10.1007/978-3-031-33990-5_6

understand the aquatic prowess of fish-birds is to come closer to what they really are (see the quote at the start of Chap. 1, stated by the first Europeans to encounter them, failing to figure out what they were).

Water: A Hard Taskmaster

To begin dissecting the energetics and some of the biomechanics of penguin swimming, we need to define a few of their physical characteristics within the context of seawater, and, although it seems otherwise for those who have watched swimming penguins being embraced by their liquid surroundings, water is actually a hard taskmaster. It is dense, viscous, and has a high heat-absorbing capacity. This translates into a medium that attempts to force penguins to the surface (i.e., buoyancy) while opposing their every move to swim in any direction and, at the same time, have a frightening capacity to suck heat from their bodies. The shivering that usually accompanies this can be avoided with a fat layer and/or with air trapped among the feathers, but both these factors encourage greater buoyancy. Before moving onto the complexities of heat, it is germane to consider the suite of physical forces that penguins have evolved to use or overcome.

The Four Forces Relevant to Penguins

At any time that penguins are moving in water, they experience four main mechanical forces – their weight, buoyancy (upthrust), thrust, and drag (Fig. 6.1). These are basic forces that govern the motion of all bodies anywhere in the universe, but the way these play out for fish-birds depends on their weight (see Box 6.1) and volume, the depth at which they are swimming, and their speed and swim angle with respect to the water (Earth's) surface.

Box 6.1
What is the difference between the weight and mass of a penguin? Answering this is a bit of a distraction for us earthlings. The weight and mass of something on Earth are basically often treated as equivalent. Weight is the force exerted on something by gravity so that it changes depending on which planet you are on or whether you are drifting about in space (where you are weightless). The mass tells you how much stuff (matter) there is in an object/body– it does not change if you move from one planet to another. In fact, mass is a measure of the tendency of an object to resist motion if a force is exerted on it. It gets complicated, though. The way that we deal with it is to refer to "weight" when talking about how much something actually weighs (a chinstrap, for example; Table 1.2) but to use "mass" when we are talking about any formula that uses mass for inertia etc.

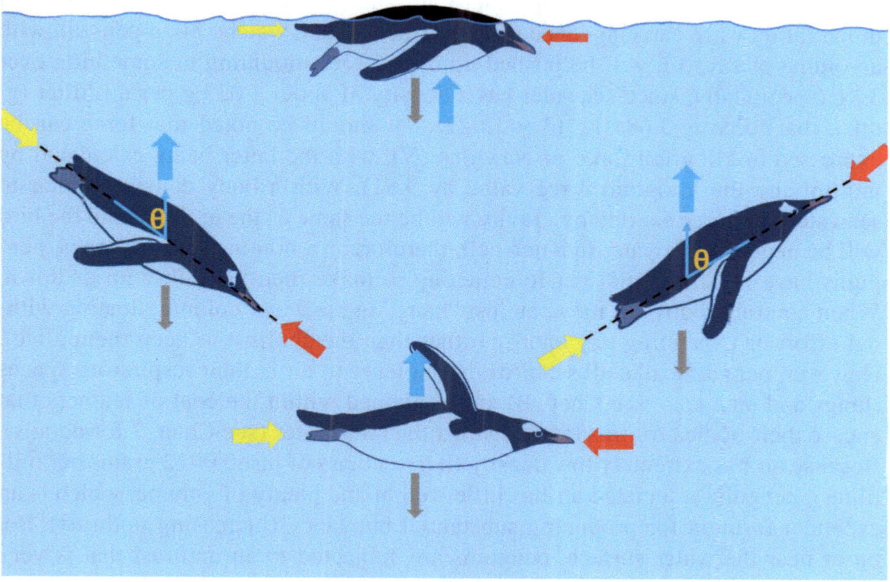

Fig. 6.1 There are four forces acting on a moving penguin, but, in water, some become more critical than in air. Two forces act perpendicularly to Earth's surface – penguins have a **weight** (gray arrow), which exerts a direct downward force (from gravity), usually less than the force of **buoyancy** (upthrust) (blue arrow). The latter is due to the amount of water displaced by the penguin's volume and exerts a force directed vertically upward, thereby tending to push the penguin toward the surface. During swimming, using their flippers, the penguins exert a **thrust** force (yellow arrow) along their longitudinal axis (action), which is directly opposed by the **drag** force (red arrow; reaction). This thrust (and therefore drag) is much less in penguins floating on the surface because birds do not swim fast in that situation. The extent to which buoyancy helps or hinders motion underwater depends on the angle of the penguin with respect to the surface (Θ)– in other words, relative to the angle of the upward and downward forces (see text). The dense medium of the ocean hinders any form of travel far more than does air

Vertical Forces: Weight and Upthrust

Whether penguins are in water or out of it, they have a constant weight due to gravity (see Box 6.1). The strength of this force stems from the high density of their body tissues (e.g., muscles, solid bones) and their volume (ignoring air spaces for the moment). Although this weight exerts the same downward-acting force in water as it does in air, in water, it is largely counterbalanced by a buoyancy force imparted by the water due to the penguin's overall density. This is called upthrust. Specifically, a penguin (or anything) underwater is pushed upward by this force, which is equivalent to the amount (mass) of the water that has been displaced by the body (thanks to Archimedes for figuring that out in the bath). If, for the moment, we exclude any air associated with a penguin (primarily respiratory and in the feathers), penguins have a body density that is roughly the same as seawater [86] (although there are slight variations depending on how much fat they are carrying since fat is less dense than water). So, a penguin with a volume of say, 3 l, will be pushed up by a force amounting to something over 3 kg (specifically, since seawater has a density of about 1.027 g per milliliter (g/mL), that force is 3.081 kg (3 × 1.027) – it should be noted that force can be expressed in kilogram force or Newtons (N), with the latter being calculated by multiplying the kilogram force value by 9.81). With a body density similar to seawater, the force exerted by gravity will be the same as the upthrust and the bird will be neutrally buoyant. In a nutshell, therefore, for non-air body volumes, penguins have no substantial net force acting to make them go either up or down. When neutrally buoyant, they can just "hang" in the water column, floating without effort or expending any energy (other than metabolism to keep them alive). However, penguins, like all seabirds, have plenty of air in their respiratory spaces (lungs and air sacs – see Chap. 8) and contained within the coat of feathers that encase their bodies for insulation against the cold water (see Chap. 7 for details). Because air has extremely low density (it has a mass of just 0.0012 grams per milliliter), penguin-associated air has little weight and plenty of volume, which is an excellent formula for producing substantial buoyancy (increasing upthrust). So, on or near the water surface, penguins are subjected to an upthrust that is very much greater than their "downthrust" (the force of gravity). Penguins on the sea surface benefit from this upthrust, allowing them to float if they want – which, for example, makes sleeping and preening easy. They float like corks, though low in the water. However, upthrust is also important for penguins when they dive because it constantly tends to push them toward the surface (Fig. 6.1).

Upthrust changes, though, as birds go deeper – in part because the amount of air that penguins inhale increases as they anticipate deeper dives (see later) – but upthrust would change with depth even if they did not. Penguin body material, i.e., muscles and bones, are incompressible but the penguin air space within and around these is not. Thus, the pressure stemming from the mass of water above a diving penguin (called hydrostatic pressure) causes the volume of the air to be compressed (according to Boyle's Law; Box 6.2) as birds travel deeper (Fig. 6.2).

Box 6.2

Boyle's Law (or the Boyle–Mariotte Law) states that the absolute pressure exerted by a given mass of ideal gas is inversely proportional to its volume (at constant temperature). In the case of volumes of air at depth, this translates to $P1.V1 = P2.V2$, where $P1$ and $P2$ are the absolute pressures at two different depths (in atmospheres) and $V1$ and $V2$ are the corresponding gas volumes at those depths.

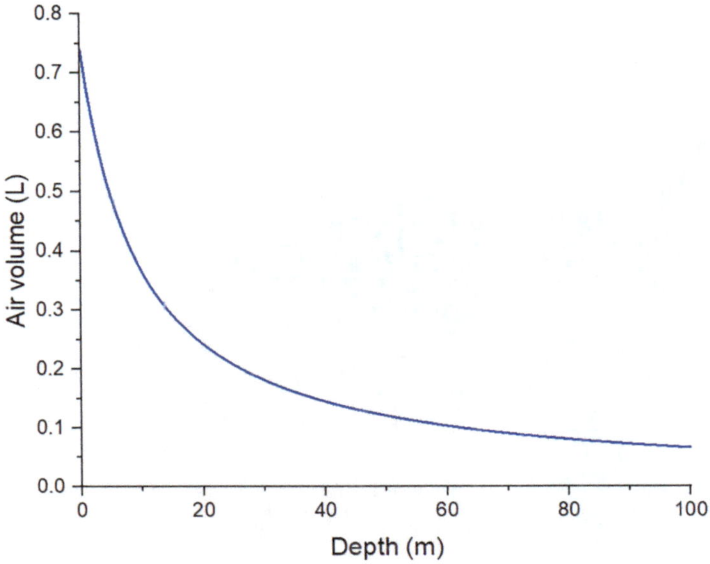

Fig. 6.2 The change in the volume of 0.74 L of air, measured on the surface, as a function of water depth. This volume corresponds to the amount of air that a medium-sized (4-kg) penguin would be expected to take down with it (in its respiratory spaces and feathers – see text). Air volume is important in diving penguins because the buoyancy (upthrust) to which penguins are subjected to is directly proportional to it and tends to push the penguin toward the water surface

Since the upthrust directly equates with their volume, penguins become less buoyant as they go deeper (their volume is squeezed by water pressure that increases with depth) and, consequently, are less required to battle with upthrust pushing them back to the surface. This eases the energy used to descend but makes ascent more onerous. Exactly to what degree does buoyancy hinder or enhance penguin swimming depends on the amount of air that is associated with them as well as their depth and swimming angle with respect to the water surface (Fig. 6.3).

Although penguins underwater actually lose a small amount of air when they initiate their dives (see Chap. 14 for more details on this) and during high acceleration or fast turns (Fig. 6.4), our measurements of the bodies of two African penguins (that were killed by a land predator) show that the amounts of air lost are trivial and that, in dead penguins at least (which may not have the same amount of air in their plumage and respiratory spaces as live birds do but still have the same anatomical features for air retention), buoyancy basically reduces and increases with depth as predicted by conventional physics (i.e., the laws of the universe; e.g., Box 6.1). We measured this by attaching a spring balance to the cadavers and then by looking at its readings while we descended the water column using scuba gear. Observations of penguins underwater also indicate that some air is lost when they travel at high speeds, most notably during their rapid exit from water (see Chap. 1, Fig. 1.2). This, and the general air loss over time in water, is examined in Chap. 14. Importantly, the

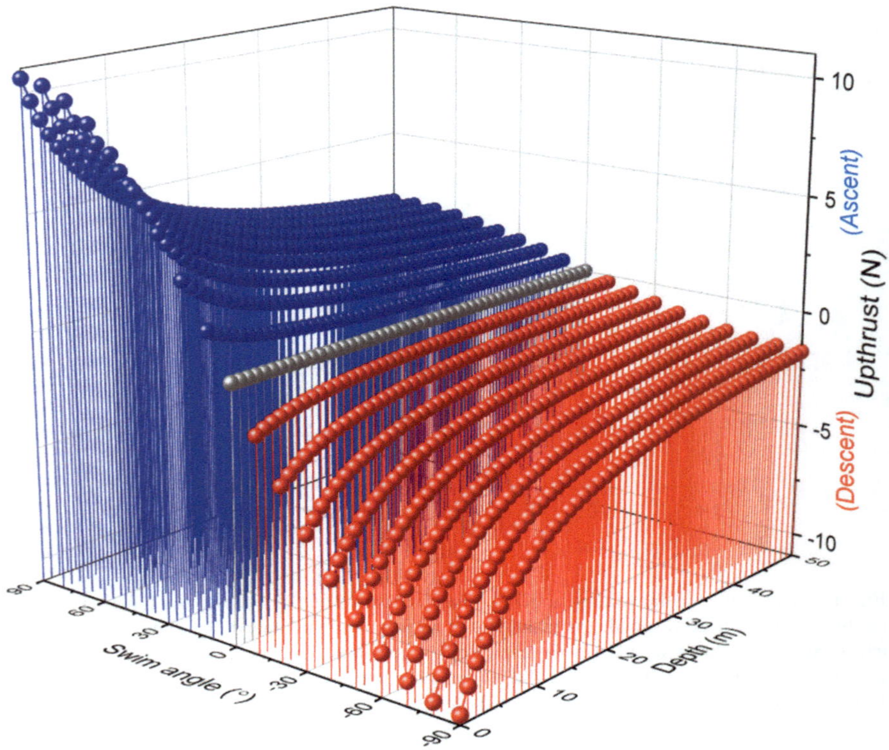

Fig. 6.3 The extent to which upthrust (measured in Newtons, where 1 kg force = 9.81 N; Isaac Newton first formulated the force of gravity after supposedly contemplating an apple of all things) is predicted to hinder (red points) or aid (blue points) penguin swimming (we have used a Fiordland penguin in this example) according to its depth and swim angle- negative angles are the bird pitching down and positive angles are the bird pitching up. The representation is based on the estimates of total body air (see below), Boyle's Law (see Box 6.2), and simple physics where forces are treated vectorially (i.e., combining direction and magnitude of forces acting in tandem) and the contribution of the upthrust force to swimming depends on the sine (a geometric measure) of the swim angle (Θ) (see Fig. 6.1). It should be noted that there is about a 10-N force opposing a penguin swimming directly down (at 90°) near the surface and that it decreases to <2 N at 50 m, with the reverse being the case for a bird swimming directly up

critical function of air as insulation for penguins against the heat-sapping colder water means that there must always be an appreciable amount of air in their feathers (see Chap. 7) and that this leads to buoyancy (Fig. 6.3).

In short, therefore, the upthrust forces primarily stemming from air carried by penguins are substantial and can greatly exceed those of thrust and drag (see below). Thus, they are a major element modulating the swimming and diving energetics of penguins, i.e., they lay down the ground (sea?) rules. One of these is that it is impossible for penguins to swim slowly underwater near the water surface without angling their bodies down (pitched down) considerably, and this is often observed in zoos.

Fig. 6.4 An Adélie penguin swimming just beneath the water surface, releasing air from its respiratory spaces (see bubbles emanating from the beak) and losing some air from its plumage (most obvious on the back between the flippers), both of which will decrease its buoyancy. (Photo courtesy of John Weller)

At that time, the birds simply generate an appropriate thrust with their flippers to counteract their buoyancy. Otherwise, in order to swim horizontally at shallow depths, penguins have to swim relatively fast so as to generate negative lift (stemming from their body moving through the water – see later) to counteract their buoyancy.

Buoyancy: How Much Air Do Penguins Hold?

As trivial as it may seem, there is still considerable debate about how much air penguins take down with them when they dive– not least because air has both respiratory- and feather-based components– i.e., the amount of air in their lungs/air sacs held in place by their scale-like, cupped feathers.

Perhaps the most informative estimates for respiratory (but not feather-based) air come from the work by Ponganis et al. [59], who used sophisticated computed tomography (CT) scans on live emperor, king, and Adélie penguins to determine that they probably dive with approximately 25 mL of air per kilogram body mass [81]. These figures actually echoed previously made estimates in a rather seminal study on many bird species [40]. The matter is somewhat complicated by other studies that have provided convincing evidence that penguins actually "know" to inhale the right amount of air so that the pressure at their target depth results in them operating with minimized buoyancy [64, 88]. The theory is that birds swimming nearer to the surface can save energy by not having to combat excessive upthrust if they just inhale a little air (cf. Fig. 6.3). However, if birds are diving deeper, then the upthrust is minimized because the respiratory air volume is greatly reduced due to the pressure (though the amount of oxygen is not). In this way, penguins can easily cruise at depths without having to waste energy on correct overexaggerated buoyancy. Our Magellanic penguin data support this wholeheartedly because birds use their air volume-dependent buoyancy to glide on the surface, beginning to glide at greater depths if they have dived to greater depths (Fig. 6.5).

How much air is contained by the uniquely cupped feathers of diving penguins is less known. An early study on cadavers in which penguins (and other waterbirds) were immersed in water to determine their volume by the volume of water displaced – following Archimedes – before plucking all their feathers (no easy feat) and repeating the process, indicated that rockhopper and African penguins had 250 mL and 210 mL of air held by their feathers (against their skin) per kilogram body mass, respectively [86]. Our experiments with penguins, with us

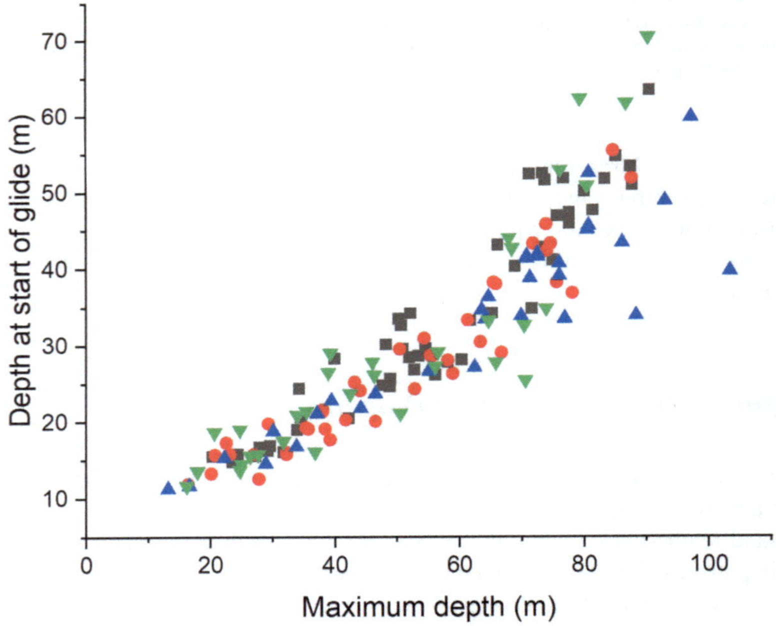

Fig. 6.5 The depth at which Magellanic penguins begin to glide as they make their way back to the surface in relation to the maximum depth reached during the dive (different symbols show different individuals, and each point denotes a single dive). It should be noted that the deeper the dive, the deeper the depth at which gliding begins, implying that birds inhale more air for deeper dives. The (linear) line of best fit is: gliding depth = 0.58 maximum depth + 0.35 ($r^2 = 0.84$)

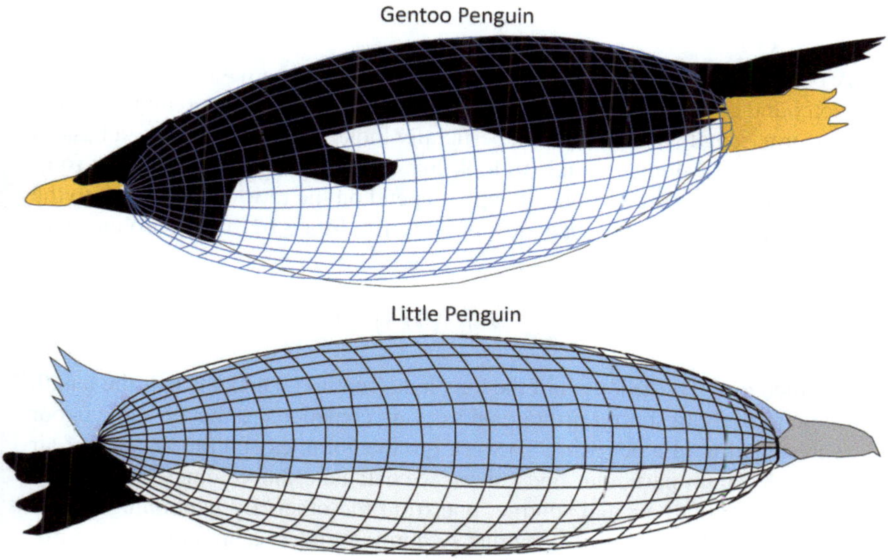

Fig. 6.6 The basic penguin shape during swimming can be approximated by a prolate spheroid, effectively a sphere that is stretched along one axis but which accommodates well the shape of a penguin swimming underwater. This is indicated by two extreme examples in terms of body lengths and widths – gentoo and little penguins. The surface area and volume of prolate spheroids can be readily calculated using spheroid-based formulae (see text)

donning scuba gear (see above), produced similar figures, but it is impossible to be sure whether the air-holding capacity of a dead penguin's feathers approximates that of a live bird. Sato et al. [64] used data on the gliding behavior of penguins at depth to calculate that Adélie penguins only had 17 mL/kg body mass in their feathers. We consider this value a considerable underestimate because we can use this value to calculate the mean air layer thickness over the penguin's body by approximating its shape to that of a prolate spheroid – something that seems to approximate the shape of penguins swimming underwater pretty well (Fig. 6.6).

Indeed, the specific advantage of using a prolate spheroid means that parameters such as volume and surface area can be easily calculated (Box 6.3).

A mean body weight of 4.575 kg (Chap. 1), a body density (without air) of 1.027 g/mL, and a prolate spheroid semimajor axis of 204.3 mm (and conditions as detailed in Box 6.3), translates to a mean overall air layer thickness of less than 0.5 mm for an Adélie penguin on the surface. This would dramatically reduce with depth (cf. Fig. 6.2), as during diving water pressure would compress the feathers. Given that the primary function of the air layer must be insulation, it seems inconceivable that this value is realistic.

In an attempt to gain more insight into the volumes of air that might be contained within penguin feathers, we measured penguin feather air layer thickness using a

Box 6.3

A prolate spheroid is a fancy word referring to a shape having circular symmetry and two semidiameters (half the diameters across the two major axes of the form, length, and girth). It superficially looks like a sphere, e.g., a basketball, stretched along one axis to look like a rugby ball or an American football. The precise dimensions, though, depend on the ratio of the length to the diameter of the shape at its widest point (girth; Fig. 6.6). It has a volume calculated by the formula:

$$\text{Vol} = (4/3)\pi ab^2,$$

in which *a* is half the length of the long axis (semiaxis) and *b* is half the length of the girth (both shorter orthogonal axes are considered to be the same in our representation of a penguin). To work out the volumes and thicknesses of air associated with particular body sizes, we calculate the volumes of the body with and without air and subtract one from the other. By examining photographs of eight species of penguins swimming underwater, and by superimposing prolate spheroids onto them to maximize the coherence between the penguin body form and the superimposition of the prolate spheroid (e.g., Fig. 6.6), we determined that the mean ratio of length to diameter of penguin prolate spheroids was 2.79 ± standard deviation (SD) 0.01.

"featherometer" on a number of stuffed penguins in scientific collections (these were penguins that had once been housed alive in zoos and aquaria). The effort was based on the noted linear relationship between the thickness of the air layer in the breast feathers of waterbirds and the feather-associated air determined using water displacement techniques (see above) [86]. A featherometer is a contraption that we invented, which can be used to apply a known pressure (to simulate water pressure) to bird feathers before measuring the distance from the skin to the outer surface of the feather layer (Box 6.4).

The major results from our ministrations with the featherometer and a respectable number of stuffed penguins are detailed in the next chapter (Chap. 7), where air layer thickness (and how it changes with depth) is considered with the insulative capacity of penguin feathers in mind. However, for now, and to help with our buoyancy considerations, it seems that penguins have two major regions of feather thickness on their body: (i) the bits that are above the water when they float on the surface and (ii) the bits that are below the waterline when they float on the surface (Table 6.1 – but see Chap. 7 and Figs.7.2, 7.11, and 7.13 for more details). To that, we can add the head region, which has a particularly thin air layer in the feathers, presumably because penguins, unlike highland cattle, like to be able to see things out of their eyes! However, for our purposes here, this part is a relatively small fraction of the total surface area of the body and we shall ignore it.

Box 6.4
A "featherometer" was developed for measuring the thickness of air layers in feathers at defined points (see the photograph below). A brass rod (diameter 3 mm × 10 cm long) has its tip lathed down until it has a diameter of 0.5 mm. This rod sits inside an elongated collar of brass that emulates seawater pressure at a depth of 0.5 m (51.35 g/cm^2) by virtue of its weight and surface area in contact with the feathers. To use this instrument, the experimenter holds the bird so that the feather surface to be measured faces upward. The rod is gently pushed perpendicularly through the feathers until it meets the skin. The brass collar is then allowed to slip down until it rests on the feathers. The distance between the tip of the rod and the resting face of the collar indicates the feather thickness at that point at a water pressure corresponding to a depth of 0.5 m. This can be converted to the thickness of the feather layer at any depth or standardized to be on the surface (= 1 atm pressure), using Boyle's Law (Box 6.2).

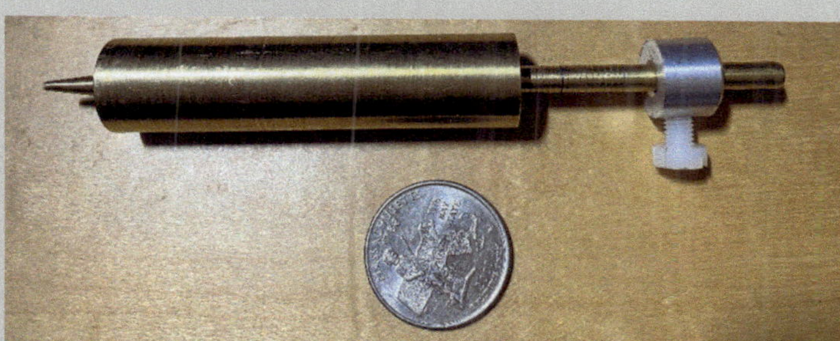

The discrepancy in air layer thickness between breast and back feathers is to be expected. Penguins, like all birds, rely heavily on air for insulation [32, 47] but always occur in water that is much colder than their body temperature, more so the

Table 6.1 Air layer thickness (in millimeters, mean ± standard error (SE), calculated at a 0-m depth) within the feathers of various penguin species according to the area of the body. This information was obtained using a featherometer (see Box 6.4) to measure the penguin skins in museums. More details are provided in Chap. 7 on the precise sites used (three measurements are made per region)

Species	Head	Breast	Back	*N*	Breast/back ratio
Adélie	4.3 ± 0.3	9.5 ± 0.7	6.7 ± 0.5	18	1.41
Emperor	3.8 ± 0.4	14.5 ± 1.2	12.2 ± 1.0	28	1.42
Galápagos	2.0 ± 0.1	6.2 ± 0.3	4.3 ± 0.2	15	1.18
Humboldt	2.2 ± 0.3	7.6 ± 0.8	4.7 ± 0.5	16	1.59
Little	2.8 ± 0.3	7.6 ± 0.8	4.7 ± 0.5	15	1.60
Southern rockhopper	2.8 ± 0.3	8.8 ± 0.7	6.1 ± 0.5	16	1.44
Yellow-eyed	3.1 ± 0.3	10.9 ± 1.3	5.4 ± 0.3	11	2.02

polar species compared to the subtropical ones (see Chap. 3). Thus, the ocean is, in a sense, always trying to rob the penguins of their body heat. The capacity of air to restrict heat loss directly depends on the thickness of the cupped spaces between feather layers (Chap. 7). Thus, to minimize heat loss in cold water, penguins (and other seabirds) should ideally have a thick layer of air. In fact, the thicker the air layer, the less the overall density of the bird and the higher it floats on the water surface – exposing the bird body to less of the heat-sapping water (Fig. 6.7).

However, larger amounts of air result in proportionately greater buoyancy, making underwater swimming energetically expensive, especially near the surface when the air volume is at its greatest (Figs. 6.2 and 6.3). So, with seeming irony, it is actually easier for pursuit-diving seabirds with larger amounts of air in their feathers (e.g., the common murre, an auk, has 330 mL/kg [86]) to swim at greater depths than in the surface waters. Indeed, this may be the reason why auks swim deeper than do penguins once their maximum dive depths have been corrected for bird mass [78] (Fig. 6.8) – especially noting that larger animals tend to dive deeper (see Chap. 8 for a discussion on this).

Fig. 6.7 The amount of air in the body feathers of seabirds determines their overall density and, therefore, via the upthrust from the water, how high their body floats on the water surface. Auks of the Northern Hemisphere, often likened to penguins ("penguins of the north," but see Chap. 14), have much more air trapped by their feather layers per unit body mass and so float much higher, much like the cruise ship shown here. This means that a smaller proportion of their body surface area is exposed to the heat-sapping properties of water compared to penguins (see Chap. 7). This is well-exemplified by the comparison of an ancient murrelet, an auk (top left), with a Snares penguin (top right). However, large amounts of air render descent during diving and level swimming energetically onerous. Therefore, the penguin solution is to have less insulating air as a response to that (similar to the submarine shown here). (Photos: auk from Wikipedia; penguin courtesy of Peter Ryan)

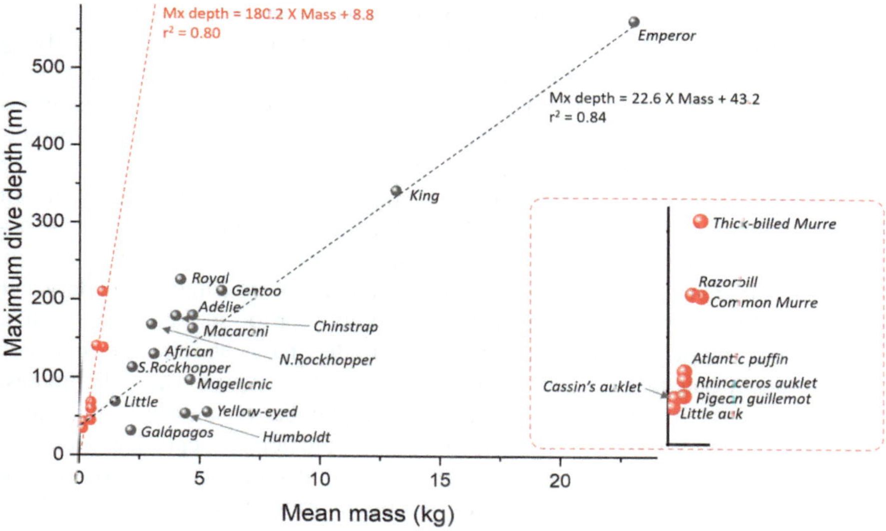

Fig. 6.8 The maximum dive depth of penguins and auks as a function of their body mass (data from Penguiness.net). The inset on the right-hand side shows the auk data with associated names. It should be noted that the auks seemingly "outperform" the penguins, although this may be, in part, because these birds have to dive deep to minimize the high upthrust that they experience at shallow depths due to the large volumes of air in their feathers

There is clearly more to it than this, though, since many highly buoyant diving ducks dive shallowly [44]. One thing is clear, however, and that is that, contrary to the situation for humans who have little body-associated air, it is shallow diving that is the most energetically challenging for seabirds. Since penguins have the least amount of air in their feathers of any seabird family, except for the cormorants [86], they are able to utilize both shallow and deeper waters – surely a substantial benefit for having given up aerial flight (Chap. 1).

Buoyancy and Bergmann's Rule Revisited

The effect of air trapped by the feathers, however, is bird mass-dependent. As discussed in Chap. 1, Bergmann's Rule [13, 46] predicts that warm-blooded creatures (homeotherms) tend to be larger in cooler environments. This is because larger animals have a smaller surface area relative to their volume and so have a greater potential to produce heat and less to lose it. Penguins appear to abide by this rule. Indeed, this has been suggested to explain how penguin species size varies with latitude (Chaps. 1 and 3). Body size, however, takes on a somewhat different role when applied to penguin diving and the amount of air contained within the feathers per kilogram body mass.

We can use our prolate spheroid approximation for penguin body shapes (Fig. 6.6) to examine how bird mass interacts with upthrust/buoyancy by applying different air layer thicknesses in the feathers – even creating penguins that do not exist anymore. This shows that the ratio of the body surface area to its volume is responsible for dramatically increasing the mass-specific volume of feather-associated air and, therefore, buoyancy in smaller penguins (Fig. 6.9).

This again highlights the problems of being small and operating as a pursuit diver in cold water (see Chap. 3, Fig. 3.1). Air is needed in the feathers for insulation, and the thicker the better, but air is responsible for disproportionately large (even becoming huge) buoyancy (upthrust) in small birds, which makes underwater swimming difficult or perhaps untenable, especially near the surface. The idea that birds should "waste" the upper water column by having to traverse the first, say, 10 m of depth quickly in order to minimize time where power use is the highest seems ludicrous. However, the contradictory interplay between air layer thickness, water temperature, heat loss, and buoyancy as a function of depth has produced penguin (and auk) air layer thicknesses and behavioral solutions that have to work according to each species' respective various environmental constraints. What

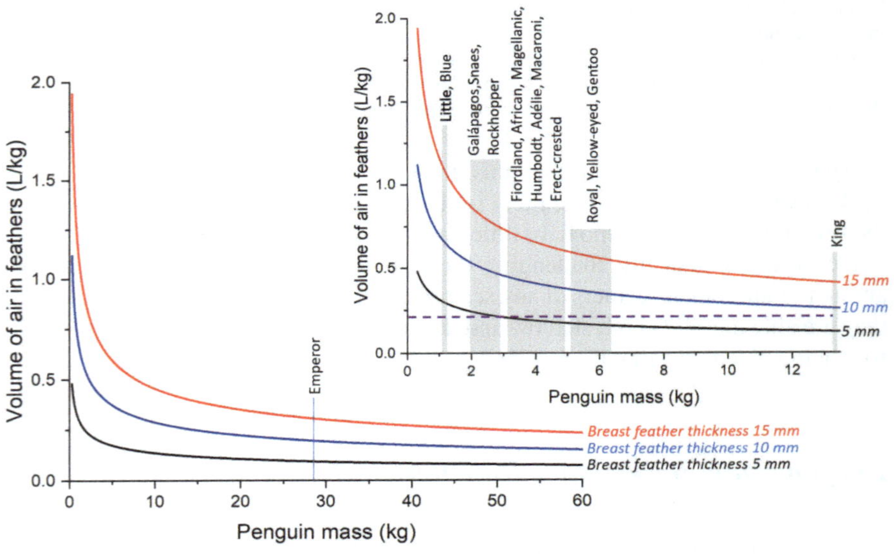

Fig. 6.9 Predicted volume of air on the surface in penguin feathers as a function of penguin body mass using a prolate spheroid to approximate their shape and using different air layer thicknesses for the breast area (indicated in millimeters and color coded to the lines – cf. Table 6.1) scaled up to the whole of the body surface area. For this, we assume that the air layer thickness in the breast feathers is 1.5 times that of the back (Table 6.1) and that the back and breast areas each account for 50% of the penguin body surface area). The gray bars show the approximate mass range of the various penguin species and provide the estimated and approximate volume of air in the feathers according to breast feather thickness. The purple dashed line shows the value obtained for African and rockhopper penguins from water displacement studies (see text), which would indicate a breast feather air layer thickness of around 6 mm

seems clear is that, of the various penguin species, little/blue penguins (and Galápagos penguins) are going to be the least able to cope with colder water because the thicker layers of feather-bound air needed for insulation increase the upthrust to an unacceptable degree (Fig. 6.9). Indeed, this may be the reason why little penguins increase their resting metabolic rate (RMR) in water so dramatically with decreasing temperature [10, 73]. It is also quite possible that this is the reason that little/blue penguins are at home only in the vicinity of the Subtropical Front (the same is true for Galápagos penguins; Chap. 3).

Horizontal Forces: Drag

Drag is a force that opposes penguin motion (Fig. 6.1) [42, 43]. Really fast race cars, and jet airplanes, deploy a "drogue chute" to provide drag, thus to slow themselves down, as brakes are not enough if they need to stop quickly. Drag is highly relevant to penguins, too, because these remarkable birds habitually swim fast (they have to in order to cover the oceanic space in search of prey and catch it when they find it or to avoid predators; Chaps. 4 and 12). Drag increases with the square of velocity (Box 6.5). The drag imposed on a penguin traveling through water varies with the bird's cross-sectional area, the density of the water, and something called the drag coefficient (Box 6.5). This is a dimensionless number that indicates how easily a particular body moves through the liquid surrounding it, usually varying between about 0.04 and 1.2. For example, a cube traveling directly through water has a drag coefficient of 1.05, a sphere has a drag coefficient of 0.47, and a streamlined body, with the classic torpedo shape, has a drag coefficient of 0.04. Penguins are shaped like a torpedo, but, unlike all other aquatic creatures or contraptions, they can change shape to fit the situation.

Box 6.5
Drag (conventionally given the term "F_d") opposes penguin motion and is mathematically characterized by the formula:

$$F_d = 0.5 \rho u^2 A C_d,$$

where ρ is the density of seawater, u is the penguin velocity A is the penguin cross-sectional area at its point of maximum girth, and C_d is the drag coefficient of the penguin.

Various authors have attempted to work out the drag coefficient of penguins (e.g., [10, 12, 33, 42, 49], with the earliest study being the one conducted by Clark and Bemis [19]. They analyzed the video footage of gentoo penguins swimming around in a large tank in Detroit and, by studying the rate at which birds gliding underwater slowed down (cf. [76]), calculated that penguins had a drag coefficient of 0.07. This was similar to estimations made by Nachtigall and Bilo [49] of 0.05–0.06. Following this, we helped in work performed by Rudolph Bannasch [9] in the 1990s in a huge water flume (5 m wide X 1 m deep) in Berlin, normally used to examine the hydrodynamics of model ships before their construction is started in earnest. Rudi had created a beautiful life-sized gentoo penguin, which was placed in the flume together with measuring gear while a team of engineers – from their "Star Trek-esque" control bridge – turned up the water speed and watched. After the first run, they declared that there was something wrong because the drag coefficient, at 0.03–0.04, seemed to them to be impossibly low. Indeed, they considered it inconceivable that a simple bird could outperform what the best theoretical minds believed were the most streamlined shapes. Incredibly, it turned out that the penguin values were lower than those of the most streamlined body ever created by man! Many runs later, becoming suitably impressed, they conceded that nothing was wrong, having to conclude that penguins were fascinating (something we already knew; see Chap. 15!).

Quite why penguin drag coefficients are so low is not clear, although some have suggested that the orientation of the feathers around the beak may play a role [53] and that the curious body form, with a slight recess for the neck – which is not used in man-made ideal shapes – may be important [9]. Another option is that the scale-like body feathers (Fig. 6.10) may constitute what is called a "compliant surface" [8, 17], which acts like grease to reduce drag.

Such surfaces act a bit like grease, doing their magic by allowing water to slip past the outer penguin surface particularly easily. Rory must certainly be amongst the very few people, perhaps the only person, to have handled a penguin underwater

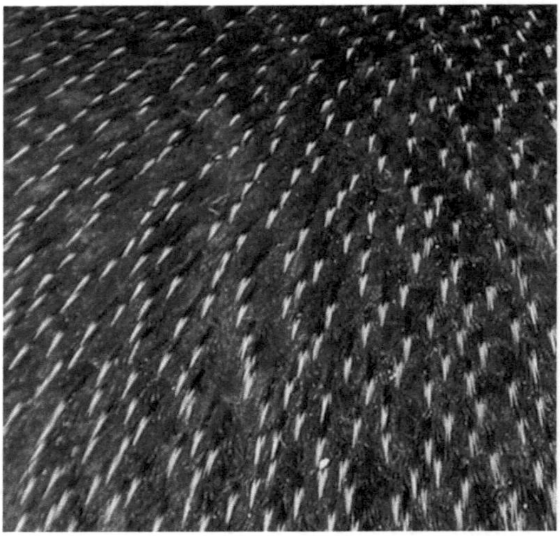

Fig. 6.10 Penguin feathers are unique in the bird world (these are Adélie penguin back feathers). Compared to the size of the birds, these feathers are tiny and are remarkably homogeneous, covering the body in a dense, compliant layer. A number of authors have speculated that they may play a particular role in reducing drag, thereby allowing the birds to move through water with reduced effort. (Photo courtesy of Jean Pennycook)

to observe how feathers react to hydrostatic pressure and how this might relate to compliancy. Rory was afforded this opportunity because African penguins on Marcus Island, South Africa, occasionally had to be caught if they had oil spots in their plumage (to bring them into rehabilitation centers before rerelease; Chap. 13). In addition, because such birds give up breeding and simply hang around on the shoreline, they have to be caught by trickery. Rory would don a wet suit and swim along the shore parallel to them, which would cause the assembled beach flock to move nervously up the beach, no doubt thinking he was a fur seal, known to predate these birds (Chap. 12). The birds realized their mistake when Rory emerged from the surf, and they would rush into the water where, in the kelp-ridden shallows, the oiled birds could be caught. It was at this time that he noted that the penguin feathers, which, out of the water, feel much like fur in superficial texture, became locked together the instant that they were submerged. Indeed, running the hand over the smoothed lattice underwater would cause the feathers to move to a new position, but they would lock in that new position rather than bounce back. So, there are hidden elements in penguin featheration, and it would seem that this locking mechanism is one of them. Certainly, it likely plays its role as a compliant surface, although it may also be part of the penguin box of tricks for keeping air securely within their feathers (see Chaps. 7 and 14). The particularities of feathers for reducing drag have also been noted in flying birds [48], though not the actual mechanism of how this is achieved, as well as the clever way that flying birds and some diving birds modify the shape of their bodies to reduce drag [41, 54]. Therefore, it is perhaps expected, but still cool, that penguins seem to do all this too – underwater. As it is, these days, racing sailboat hulls have a textured surface similar to the scale-like feather surface matrix of penguins (and fish), thanks to what the German engineers had learned. However, they have some way to go before they get their ideal yacht shapes to morph, like penguins, according to circumstance!

Times and technology have changed since the first attempts at determining penguin drag. These days, it is possible to make an informed measurement of the drag of wild, free-swimming penguins, unconstrained by the vagaries of aquaria and not represented by mere models, with their artificial surfaces and immobile shapes. All that is needed is what is called an accelerometer that, attached to a penguin, measures its change in velocity in three dimensions (forward, vertically, and sideways). Armed with such loggers, we attempt this below and, like Clark and Bemis did more than 40 years ago, we will look at the gliding phases of penguins. As mentioned above, the buoyancy (upthrust), primarily due to the air in the penguin respiratory spaces and trapped by their feathers, helps propel them toward the surface (cf. Fig. 6.1). At a respectable depth, the air is so compressed that the buoyancy force is not enough to drive penguins upward, and, therefore, birds wishing to surface have to initiate and supplement the upthrust with help from the odd flipper beat. However, as they ascend and the air expands – increasing the upthrust– there comes a place in the water column at which the buoyancy alone is enough to propel the penguins upward. At this point, they begin gliding (Fig. 6.11) [64, 79].

During the ascent at the end of dives, penguins have approximately constant velocity (at least over any short interval such as 2 seconds). Thus, the angular component of the upthrust (which depends on the penguin's return-to-the-surface angle)

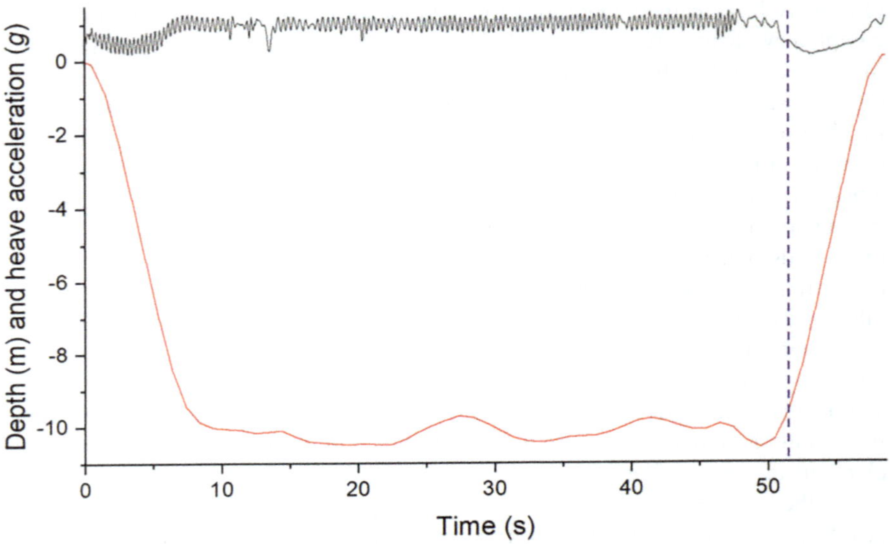

Fig. 6.11 The dive profile of a little penguin (red line) showing, in the waveform above it, the heave (dorsoventral) acceleration (black line) indicating flipper beating. It should be noted how the flippers beat regularly on the descent, become somewhat irregular at the bottom of a dive, probably due to prey pursuit, but cease completely at a certain point (between 9- and 10-m depth) during the ascent (dashed line). At that time and depth, the upthrust force (buoyancy) alone is enough to propel the penguin to the surface (Data courtesy of Belinda Cannell)

is equivalent to the opposing drag (Newton reminds us that at constant velocity, the forces acting on the body are balanced – action vs. reaction). Therefore, knowing the rate of change of depth and swim angle from appropriate sensors in a penguin-attached tag (including a pressure sensor and accelerometers) and combining them with our best estimates of body air ([86] and see above), we can estimate the full body drag (Box 6.6).

This approach indicates that Magellanic penguins, for example, have a total body drag of around 1.5 N at ca. 2 m/s (Fig. 6.12). This ostensibly ties in well with measurements made by Rudi Bannasch on a carefully made plastic model of a gentoo penguin in the swim channel in Berlin. There, he obtained a drag of 1.97 N at 2.2 m/s [7] – it should be noted, however, that gentoos are bigger than Magellanic penguins (Chap. 1) and so will have a higher drag. It is also worth noting that Rudi's model had no wings, whereas our Magellanics obviously did and that this should increase the drag considerably. Thus, taken together, our data on wild Magellanic penguins operating normally (and, likely importantly, also having to deal with the increased drag of an attached device; Chap. 14) seem to have an extraordinarily low drag. Putting this information into the drag formula (Box 6.5), gives us a mean drag coefficient of about 0.028 (± standard deviation (SD) 0.003) across all depth ranges. This again compares well with Rudi Bannasch's speculation that penguin drag coefficients may be as low as 0.025. It seems that Rudi was right.

Box 6.6

Determination of drag in a free-swimming wild penguin as it glides to the surface propelled by its upthrust, first requires the mean penguin body pitch to work out the mean swim angle of the surfacing bird over a defined 2-second period. This includes correcting for imperfect tag placement by rotating the smoothed acceleration data so that the bird on the surface is allocated zero pitch. For that 2-s period, we then determine the change in depth to calculate the vertical velocity at that time before using simple trigonometry to derive absolute penguin velocity. We then use our estimates of penguin body air on the surface (see earlier – e.g., a Magellanic penguin has 0.16 L/kg of air per body mass and inhales 0.025 L /kg body mass), before using Boyle's Law (see Box 6.2) to estimate the volume of air at the depth considered and thereby the force of the upthrust. The relevant vector of the upthrust for powering the passive penguin ascent, however, is the sine of the ascent angle. This can then be multiplied by the total upthrust to derive the thrust powering the penguin upward, which is equaled by the drag. This process can be repeated for various depths and dives as long as the penguin is ascending passively. It should be noted, however, that our process is critically dependent on our estimates of body air and that must be borne in mind when considering our results.

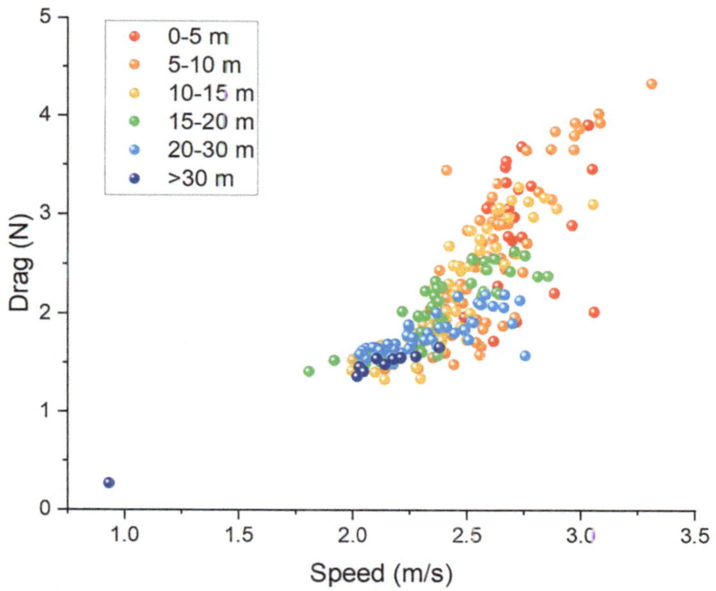

Fig. 6.12 The calculated drag of a Magellanic penguin according to depth (see color coding) over 2-second periods as the bird glides upward in the terminal phases of (31) dives (cf. Fig. 6.11). It should be noted how the overall shape shows dramatically increasing drag with speed as predicted from the drag equation (Box 6.5) but that there is much greater variation in drag at shallower depths

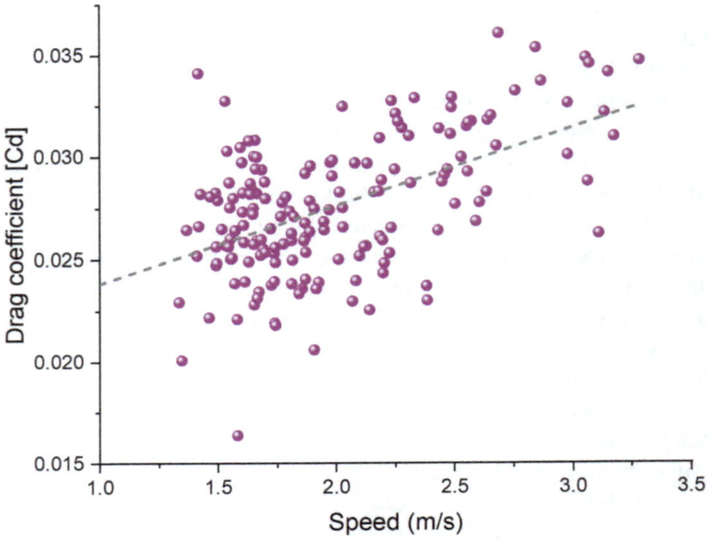

Fig. 6.13 The calculated change in the drag coefficient of a Magellanic penguin as a function of speed as it glides toward the surface at the end of dives. These values were derived using the data shown in Fig. 6.12 for all depths greater than 10 m because as penguins approach the surface, they deliberately increase their drag (see below)

As the drag formula specifies, though, drag increases disproportionately with increasing speed, which is what our example data clearly show (Fig. 6.12). Rudi Bannasch also suggested that the drag coefficient would change with speed [7], and the data in Fig. 6.12 indicate that he is right (again). We can convert each data point into a drag coefficient value using the drag formula (Box 6.6), assuming that Magellanic penguins have a cross-sectional area (including the flippers) of 0.0247 m^2 to show this explicitly (Fig. 6.13).

The Interplay Between Upthrust and Drag in Gliding Penguins

Our method for determining drag uses the gliding, surfacing phase of a penguin dive, a phase when forces on the birds are changing radically all the time. Mainly, though, water pressure decreases as depth becomes less. So, specifically, as penguins ascend, their associated air volume increases – owing to decreased hydrostatic pressure squeezing their feather layer less and – creating greater upthrust (more buoyancy). Buoyancy is what drives the penguin ascent in the first place if it is not beating its flippers. So, how do penguins react to this changing scenario and what are the consequences for their eventual surfacing behavior?

First, and with little doubt, it is to the penguin's energetic advantage to glide as much as possible rather than to use energy to "power up" – wild animals are notoriously good at being miserly with their energy, and penguins are no exception. To achieve this, penguins glide almost as soon as they can (as long as the glide allows them to continue moving at an "appropriate" pace – e.g., it is no good moving upward at a snail's pace since they are limited by their breath-holding capacity). To initiate gliding as early as possible means to do so as deep as possible. However, deeper water has higher pressure, compressing the feathers, resulting in less penguin body air and minimal upthrust. In order to capitalize on this, penguins have to have their return-to-surface angles close to vertical so that the paltry upthrust at depth essentially pushes them directly up (Fig. 6.14).

However, as penguins get closer to the surface and are subjected to lower hydrostatic pressure, their feather-trapped air volume (and the air in their respiratory spaces) increases dramatically (Fig. 6.2) and so does their buoyancy. If penguins maintained their steep return-to-surface angle, the high vertical component of the upthrust would accelerate them at extremely high speeds [63]. Indeed, Katsufumi Sato and colleagues note a particular case of an emperor penguin wanting to leap out onto the ice and needing high speed to do so. This bird accelerated to a speed of 3.0 m/s by simply increasing its body angle (to 58°), even though it did not beat its flippers for the full final 7 s of the trajectory at all [65]! Similar behavior is exhibited by some other penguin species coming ashore, most notably by the three *Pygoscelis*

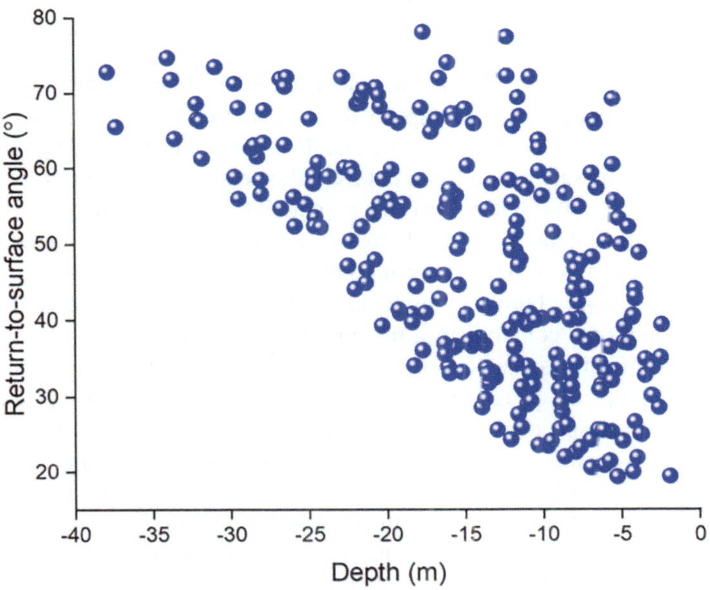

Fig. 6.14 The body pitch angle for a Magellanic penguin gliding to the surface for 2-s periods during (31) dives. It should be noted that body angles with respect to the sea surface are always steep at deeper depths but, overall, tend to decrease as the bird approaches the surface

Fig. 6.15 Some penguin species, such as these Adélie penguins, use their buoyancy to attain high vertical speeds to propel themselves out of the water to reach land. Otherwise, they would have difficulty landing, when dealing with such an overhang in the beach ice. The overhang is far enough from the shore that the water is deep enough to allow this maneuver. (Photo courtesy of David Ainley)

penguins when wanting to "alight" onto ice (Fig. 6.15). Importantly, as pointed out by Katsufumi, although their great speed allows them to project themselves up to 2 m in air (we have had gentoo penguins even land on our heads by standing too close to ice floe edges –who said penguins do not fly!), this feat may involve little locomotive effort (flipper beating) if the water is deep enough [75]. This is also apparent if the footage of the birds underwater just before emergence is closely examined.

Figure 6.14 makes it clear that penguins respond to the increasing thrust provided by their growing buoyancy by reducing their return-to-surface angles as they rise. This reduces the component of the upthrust powering them to a fraction of its potential (specifically to the sine of the return-to-surface angle multiplied by the upthrust) and helps the penguins maintain their speeds within limits. Even so, they do tend to increase speed at shallower depths (Fig. 6.16), as has also been shown in auks [63, 79]. Near the surface, they do need to be more vigilant, and more in control, should predators be present in the area. Most of those predators are air breathers too and do their "lurking" pretty much only near the surface (Chap. 12).

Reduced return-to-surface angles are not necessarily always advantageous to penguins from a foraging perspective, though. If the birds continue in a straight line at shallow return-to-surface angles, then this would result in them being taken far

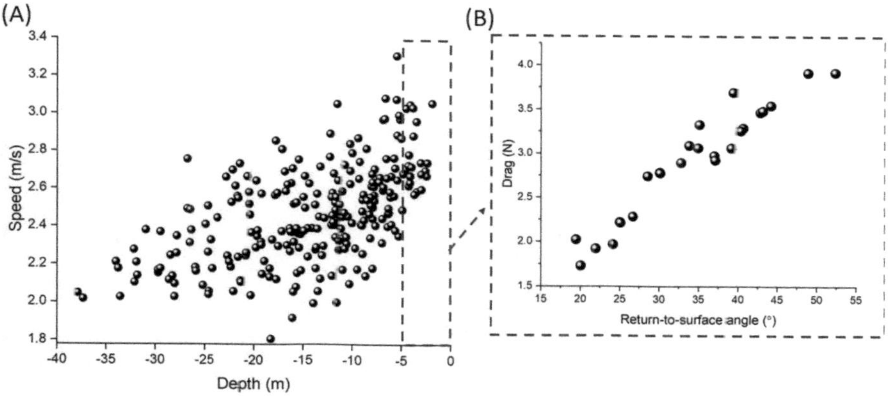

Fig. 6.16 (**a**) The absolute speed of a Magellanic penguin in 2-second periods as it glides to the surface during (31) dives. The overall increasing speed as the bird nears the surface should be noted. (**b**) The interplay between drag and return-to-surface angle during glides occurring within 5 m of the water surface. At this time, there is no relationship between speed and the return-to-surface angle (the mean speed for the data shown here is 2.75 m/s (± SD 0.18)) because the steeper angles (with a consequent greater thrust component from buoyancy) require more dramatic braking by the bird before it reaches the surface. Failure to do this would have the penguin shoot out of the water, something it would not normally want to do

from the place where they might have just foraged and could return [63]. So, another, complementary, approach adopted by penguins to mitigate this is to increase their drag by, e.g., lowering their feet, akin to a drag chute (cf. Sato et al. [64]) (Fig. 6.17). That this occurs is apparent in Fig. 6.10, which shows that gliding at greater depths shows minimal variation in drag (in tandem with little variation in the return-to-surface angle; Fig. 6.14). However, as penguins approach the surface, the variation in drag increases as birds juxtapose drag with return-to-surface angles to control their speed. Over the shallowest few meters of the water column, the proportionate increase in body air volume is the highest so that the birds that do not want to leap out onto an ice floe or ledge have to brake their ascent dramatically, especially if their return-to-surface angle is steep (Fig. 6.15b).

So, faced with a "free" powerhouse, surfacing penguins can control both their absolute speed and their vertical velocity (the rate at which they move upward) within limits during their passive ascents. Why would penguins want to control their speed within specific limits? Aside from the prodigious leap or being wary of predators discussed above, other reasons include simple control, thus minimizing heat loss because higher speeds result in more body heat being lost (see the "wind chill effect" in Chap. 7). There is also the possibility that measured ascent may mitigate the bends (see Chap. 8), being comparable to decompression "stops" [64].

Fig. 6.17 A Galápagos penguin in the foreground apparently doing a backward loop-the-loop, as the one in the background applies gentle brakes by slightly lowering its feet, perhaps to observe the show? Actually both, being ever vigilant, probably had their attention on fish until interrupted by the approach of the photographer. (Photo courtesy of Tui De Roy)

The Drag Devil Lies in the Detail

The formulation of our estimates of drag, and therefore drag coefficients (Box 6.6), is explicit in terms of the elements used to derive our final estimates, but it is worth noting that errors in our evaluations of penguin-associated air volume, in particular, can make a substantial difference in our "answers." In this context, it is also worth noting that we expect penguin-associated air to vary over time. For a start, the volume of lungs and air sacs that make up the respiratory volume are likely to reduce as soon as penguins start to eat appreciable amounts [59]. This is because the penguin thorax is essentially immovable, except, perhaps, for some distension that can protrude between the legs – a full penguin on land has an obvious "between-the-legs bulge"! In this state, they waddle (even) more! As such, large amounts of food must begin to take up the volume that would otherwise be allocated to respiratory air. In addition to this, there is good evidence that the volume of air in the penguin plumage varies over time, decreasing during serial dives to depth because some air bubbles escape as birds approach the surface after deep dives. This air loss is replaced during extended time on the surface, most particularly after a "wash." This topic is examined in detail in Chap. 14.

The Penguin Powerhouse

How Penguins Swim

Given that penguins evolved from flying birds, their apparatus for driving them through water – their wing in the form of a flipper– has dramatically changed from the conventional bird wing [6] (see Chap. 1). Most obviously, flippers are tiny compared to wings (something that hugely helps in underwater locomotion in the dense medium of water compared to air [25]). These flippers are composed of a series of flattened bones that can effectively lock tightly together with a minimally associated musculature (Fig. 6.18). In addition, penguin flipper bones are stronger (as demonstrated by tests on the mass-specific humeral strength) than those of flighted birds [28]. This is easy to believe if you have ever been whacked on the shins by the flippers of an irate penguin! The flipper bones are covered by tissue (mostly tendons) with the skin having tiny, short, stiff feathers (Fig. 6.18).

Flippers cannot be used unless the bones within are in the locked position, which creates a complex blade that can be, otherwise, bent. In fact, one of the critical factors that determines whether a fledgling penguin can go to sea is the hardening of its wing bones. Early on, fledging penguins acquire fully developed foot and leg bones

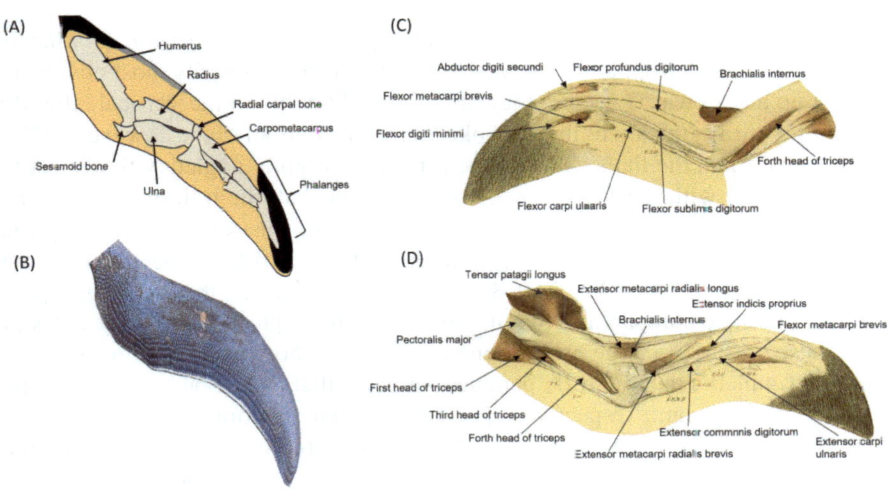

Fig. 6.18 (**a**) A schematic diagram of a penguin flipper showing the tight arrangement of the bones locked together by their shape and tendons to create a paddle with the sesamoid bone. (**b**) The dorsal surface of an emperor penguin flipper (from a dead specimen), highlighting the tight feather covering (the flipper is covered by the skin having minute feathers at the leading edge, transitioning to longer feathers lying parallel on the trailing edge). (**c**) The ventral and (**d**) the dorsal surface in the drawings of the musculature in a southern rockhopper flipper taken from Nares [51]. The minimal musculature within the flipper should be noted

in order to chase their parents for food. In contrast, these same chicks have soft, floppy flippers for much of their pre-fledging period. However, as they begin to lose their down and acquire their typical penguin featheration (albeit the juvenile color scheme), their wing bones solidify. Specifically, it is the maturation of the movement of a bone in the elbow (the sesamoid bone; see Fig. 6.18) to between the humerus and the radius that creates the locking mechanism. In fact, while following the growth of penguin chicks up to the point when they would leave for the sea, we used to massage this bone into place to see if the flippers would lock. If they did not, there were still a few days before fledging could happen.

Penguin flippers are far more than a simple paddle [6]. They react to pressures as the penguin moves through water by deforming along their length in a sophisticated manner that enhances their ability to provide thrust – in other words, to power the bird forward [29]. This bending, though, is not analogous to what is evident in flying birds, which can extend their wingspan variously depending on flight conditions for flight control [31]. Instead, penguins exert control by elaborate rotation maneuvers at the articulation of the flipper with the body – the "caput humeri." Rudi Bannasch [7] writes: "…examination of [the specialised shoulder] structure and kinematics reveals a two-axis-eccentric motion of the head of the humerus conducted by a strong ligament and an articulation surface of the bone shaped like a logarithmic spiral" (cf. [4, 5])! Suffice it to say that it is complex, but, along with changing body shape, this gives penguins the extraordinary control that we observe as they dash and swerve beneath the sea surface.

The actual generation of a force that propels penguins forward is what is termed "lift-based propulsion," whereby the flippers are primarily most obviously moved up and down. However, they are at a slight angle to the vertical, slightly further forward at the top of the upstroke than at the bottom of the downstroke [7]. During that time, the flat surface of the flippers is slightly angled with respect to the horizontal, with the leading edge being higher than the trailing edge on the upstroke and lower on the downstroke [29]. The downstroke is driven by the extremely large "pectoralis" muscle, whereas the upstroke is powered by the "supracoracoideus" muscle [4, 69, 80]. The latter is only half the size of the "pectoralis" but huge compared to that in most flying birds [1, 35]. The large sizes of both these muscles (a large proportion of a penguin's body weight – for example, Belinda Cannell's data on little penguins indicate that these constitute 8.3% and 3.5% of body weight, respectively [16]) make their function apparent, namely, that penguins exert considerable force during both the downstroke and the upstroke of their flippers. This is what drives them forward for both phases of the flapping cycle [6, 10, 19, 80]. Doing that is in stark contrast to most flying birds (with the exception being hummingbirds). Anyone who has been whacked in the shins by a charging, territorial penguin, as noted above, knows their strength fully well, especially since penguins turn their flippers on the side so that they can hit you with the "sharp part of their wing." Thus, it is like being hit by the thin edge of a ruler multiple times per second! However, underwater, the angled flippers also generate "lift" (force in the vertical dimension) as part of this process [7], something that causes the birds to move up and down slightly in the water column

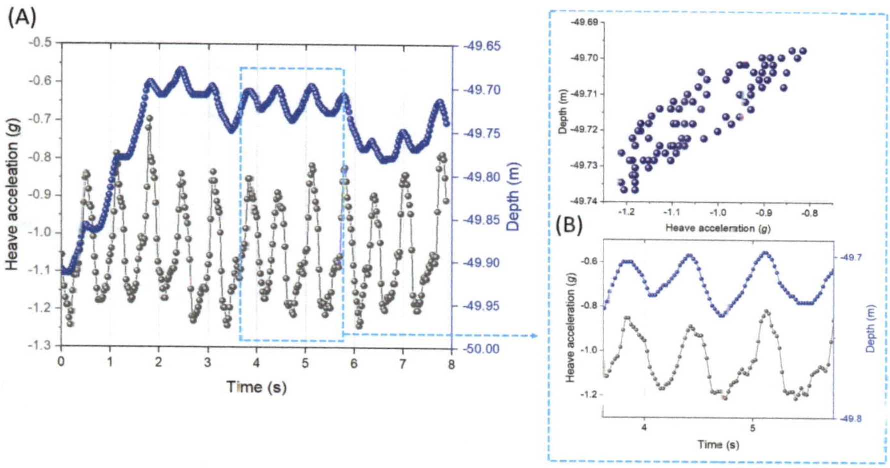

Fig. 6.19 (**a**) The (dorsoventral) heave acceleration exhibited by a Magellanic penguin during level swimming at a 70-m depth graphed together with the change in depth due to its flipper motion (superimposed on the normal changes in the penguin's depth). The lift-based propulsion system used by penguins means that, as the flippers move down and up, they cause the body to rise and fall slightly in the water column. A section of panel (**a**) where the bird did not change its overall depth at all is shown in panel (**b**) to clarify the detail, both in terms of time (lower pane) and with respect to how the heave acceleration relates to the depth change – effectively provided by the gradient of the line of best fit (upper pane). Overall, the change in depth per flipper cycle is extremely small (about 2.3 cm). Therefore, at a nominal swim speed (2.1 m/s) during one flipper cycle, the bird will have traveled 4.6 cm up and down but about 1.4 m forward. Thus, the relative efficiency in terms of movement is ca. 97%!

as they progress (Fig. 6.19). This is clearly visible for anyone who has watched penguins swimming side on and may help emphasize the lateral stripe in *Spheniscus* penguins to help them catch prey (see details in Chap. 11). This vertical wobble, though, is a waste of energy from a movement efficiency perspective.

Much seminal work on the kinematics of penguin swimming underwater has been carried out using the filming techniques of birds in captivity (e.g., [19]). However, as noted above, the relatively recent development and deployment of high sample rate accelerometers on free-living penguins has now provided us with an extraordinary amount of high-quality data, even as birds swim at their impressive depths and speeds far from land. The clever part is that the flippers are used to propel the body and, thus, by measurement of the propulsion of the body by accelerometers (specifically as the body is accelerated up and down, forward and back, and even left and right), we can deduce what the flippers are doing and examine how well they do it. If we examine the flipper beating pattern of a typical penguin swimming horizontally by looking at the acceleration in the body (Fig. 6.19), the flipper beat cycles are extremely obvious, as is the clear variation in the vertical (heave) axis of the bird. Fast, accurate sampling of acceleration in tandem with penguin depth reveals the extent of the inefficiencies in the forward with respect to the

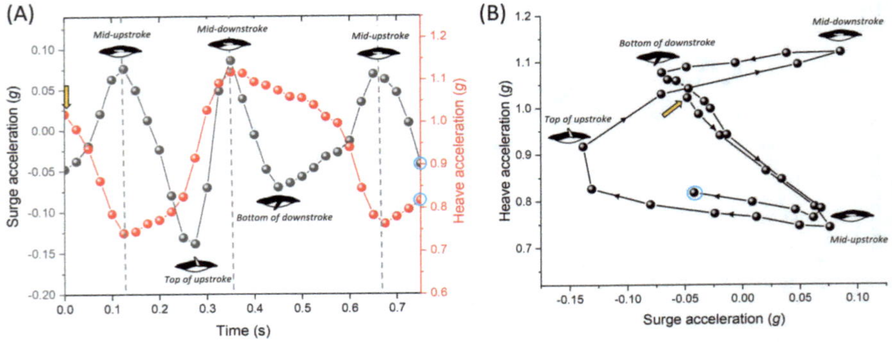

Fig. 6.20 Details of the raw surge and heave acceleration during (a bit more than) one flipper beat cycle highlighted in the dashed box in Fig. 6.19. (**a**) Surge and heave over time and (**b**) surge plotted against heave. In both panes, the yellow arrow shows the start of the cycle and the blue circle the end. In the bi-acceleration plot in panel (**b**), it should be noted how part of the repeating cycle is overlaid on the first cycle. A predominant, apparently negative surge (which would imply that the bird was swimming backward) is due to the static acceleration component stemming from the angle of the penguin body. This is corrected in Fig. 6.21, below

vertical motion. In Magellanic penguins, the ratio of vertical to horizontal distance traveled during a wingbeat cycle is about 1:30 (Fig. 6.19).

However, penguins are famed for their speed, and this means that there must be a surging acceleration component in the body generated by the movement of the flippers. Closer examination of both the heave and surge accelerations within a little more than a single flipper beat, both with respect to time and with respect to each other (Fig. 6.20), indicates how the peak and trough in the heave acceleration occur at the times when the surge acceleration is maximum. This happens when the flippers are roughly horizontal, at which point they can apply the highest force, with the downbeat resulting in upward body acceleration and the upbeat resulting in downward body acceleration. There are also two occasions per flipper beat cycle when the surge acceleration is negative – i.e., the bird experiences a braking force – with the one after the flipper upbeat being more extreme than the one after the downbeat. Both of these occur when the heave acceleration is around zero, i.e., at the top and at the bottom of the flipper beats, respectively (Fig. 6.20).

What is obvious from the surge versus heave bi-plots (Fig. 6.20b) is that there is effectively some deceleration (braking) for some of the cycle. This happens when the flippers are turning from their points of maximum upstroke and downstroke and the angling of the flippers switch from one position to another. This shows the inevitable inefficiencies within the lift-based propulsive system as it flips from the two phases of the full stroke cycle. To examine it more precisely requires proper correction for penguin body pitch (see below) since the surge component is particularly sensitive to the gravity component of the acceleration.

The Effect of Upthrust and Body Angle on Penguin Thrust and Lift Forces

Accelerometers measure, unsurprisingly, change in velocity (aka acceleration), but if we know the mass of the penguin carrying an accelerometer, then we can determine the thrust and lift forces generated by the flippers during swimming. Newton said, "Force equals Mass X Acceleration." So, applying this, and removing any forces due to gravity, we can reconstruct the forces generated by the penguin flippers to examine whether there is any indication that penguins might have elements within their flipper kinematics that compensate for their positive buoyancy, especially on the surface, where it is the highest. The answer, as suggested by Bannasch years ago [7], is clearly "no" (Fig. 6.21). In short, this approach helps separate the surge from the heave forces generated by the penguin during flipper beating. While there is a clear positive surge force, which is what is needed to push the penguin effectively through water, there is no substantive net heave force that might help mitigate the effects of buoyancy (Fig. 6.21).

In the event of the heave force helping correct buoyancy issues, we would expect there to be a marked preponderance of negative heave forces (equating to negative lift) – in other words, tending to push the penguin down. How then, do penguins counteract the considerable force of buoyancy acting to push them back to the surface?

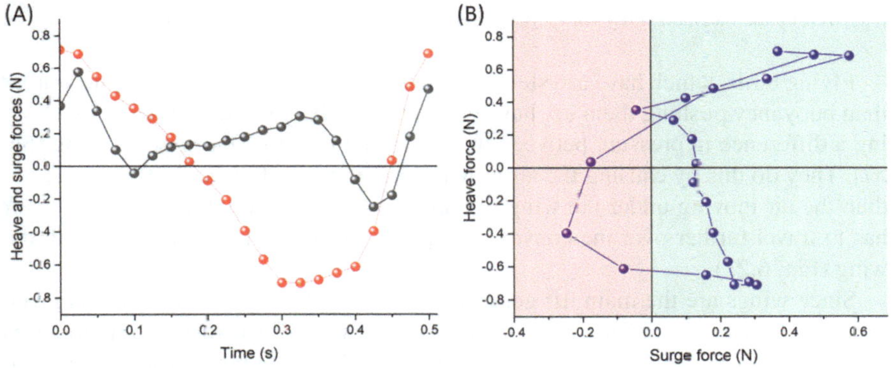

Fig. 6.21 The forces generated by a single flipper beat (from a Magellanic penguin swimming horizontally at a 11-m depth), derived by transforming the acceleration (cf. Fig. 6.20) and removing forces due to gravity. Shown in panel (**a**) are heave (red) and surge (black) forces over time and in panel (**b**) heave vs. surge forces. In both diagrams, it should be noted that the surge forces are predominantly positive (mean = 0.166 N) despite the two braking periods as the flipper changes direction (cf. Fig. 6.20). This is why the penguin moves forward overall. Against this, the heave forces are evenly balanced between positive and negative lift (mean = −0.027 N). Thus, there is no substantive net tendency for the bird to move up or down the water column due to the movement of its flippers

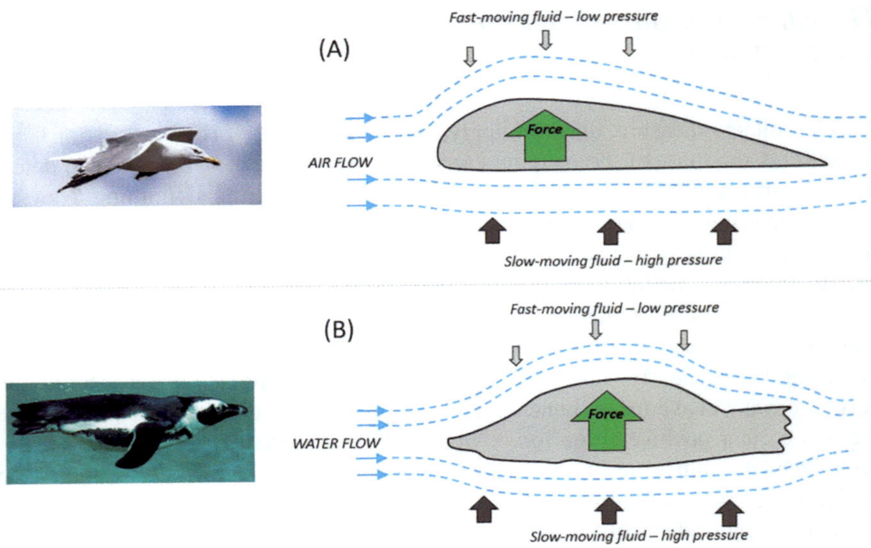

Fig. 6.22 (**a**) A schematic diagram of a flying bird's wing, showing how the lift force to keep the bird in air is generated: faster-moving air over the convex top surface of the wing creates lower pressure compared to the under surface of the wing. The inequality acts to push the wing (and therefore the bird) up. (**b**) If a swimming (African) penguin is inverted and presented as a cross section, then it too has a markedly more convex side, which presumably also results in a force being generated perpendicular to its long axis. In reality, however, the force acts downward and may well be the mechanism by which penguins compensate for their positive buoyancy

Flying birds, which have a system to counteract gravity pulling them down rather than buoyancy pushing them up, have wing profiles that provide lift based on creating a difference in pressure between the upper and lower surfaces of the wing [56, 57]. They do this by causing the air moving over the top of the wing to move faster than the air moving under the wing because, within the same time interval, the air has to travel further over the convex surface to "meet" the air flowing beneath the wing (Fig. 6.22).

Since wings are the main lift generators for aerial birds, little attention is given to the potential effect of the body for providing lift [55]. However, some treatises such as that by Cofta [20] show that flying bird bodies have more or less equivalent shapes above and below their midline and, therefore, lift forces are not expected from this. In penguins, the flippers are tiny compared to a volant bird's wing, so they provide little potential for direct negative lift. In addition, the flipper's cross-sectional geometry shows that the upper and lower surfaces are symmetrical [7]. Thus, there seems little chance that the flipper construction per se helps in producing the negative force needed to counteract buoyancy. Although it is often suggested that penguins steer with their feet, examination of them swimming does not indicate that they are used in the same way as wing elevators in planes. Instead, we suggest that the whole penguin body shape may effectively act as a "reverse wing" to thus produce negative lift and it is this that would counteract the buoyancy in penguins

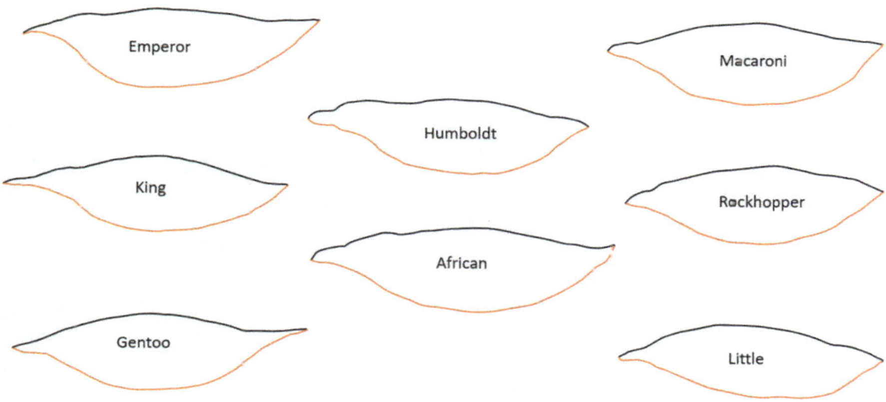

Fig. 6.23 Profiles of the penguin species swimming horizontally (all traveling left) derived from the photographs with the feet and flippers ignored. The much greater length of the ventral distance (in orange) as it goes round the longitudinal axis of the body compared to the dorsal distance (black) should be noted. All species have been normalized to have approximately the same size

swimming horizontally. As with wings in flying birds, the cross-sectional profile needs to be asymmetrical, with the lower surface being more convex than the upper. This is exactly the way this seems to be in penguins (Figs. 6.22 and 6.23).

To underpin this, we examined the ratio of the lengths of the upper and lower longitudinal body profiles in eight penguin species based on photographs. The result was a ratio of 1:1.07 (± SD 0.038) (Fig. 6.23). As with flighted wings, the extent of (negative) lift in penguins will depend on speed, with higher speeds producing greater negative forces. An expected corollary of this is that, with all other things being equal, horizontally swimming penguins nearer the surface should swim faster than deeper birds. What this also means is that penguins can benefit from generation of this force, specifically because they swim fast anyway. Sadly, no amount of cleverly designed (by evolution) body shape will allow slower-swimming species, such as diving ducks [15, 30], to counteract upthrust by any other method than angling their body down and driving against it with their feet.

When one watches penguins in aquaria, it becomes clear that horizontally swimming birds that slow down tend to rise toward the surface. Precise control of the degree of negative lift (e.g., with varying depths) could be exerted by positioning of the head and neck, which would change the path length of the water flowing from the tip of the beak to the tail across the upper and lower surfaces. Rudi Bannasch [7] suggests that the head and neck in penguins could be important in general steering and we concur with this, for both pitch and directionality (Fig. 6.24). Certainly, observations of penguins underwater show that they travel in the direction that their head points. Belinda Cannell's PhD thesis [16] reports that little penguins commonly use a combination of their head, feet, and tail to assist in a turn. Specifically, she said that the beak was pointed in the direction of the turn, the outside foot was abducted to the vertical position and splayed while the tail was initially raised and

Fig. 6.24 A Galápagos penguin turning in the direction toward which its head is pointing, arching its body sideways in that direction at the same time. These actions change the way that water flows over its body. Whether or not its feet always assist in the turning is open to question. The penguins in the background with their feet variously deployed, including as brakes, should be noted. (Photo courtesy of Tui De Roy)

then lowered. The flippers on the outside of the turn were dropped below the horizontal line of the body if the turn was unpowered.

Production of negative lift by the penguin body would also explain another interesting difference between penguins when they turn underwater compared to birds in air. Flying birds bank with their belly out, whereas penguins turn with their belly inward (the penguin in Fig. 6.17, doing a back flip, maybe to show off, is the exception). The flighted situation makes sense because the lift generated by the upper wing surface helps counteract the outward, centrifugal force (specifically the horizontal component of the lift matches the centrifugal force). To benefit in a similar manner, penguins should have their lift-generating surfaces inward too – in other words, they should travel with their belly inward, which is what they do. An intriguing question is what do auks do? Kyle Elliot, an auk pundit, tells us that they bank outward, like aerial birds, when they fly, and inward, like penguins when underwater. This is a good reason to mix and match we guess.

The issue of producing negative lift is less problematic when penguins descend because, by pitching their body down, they can use the vertical component of the surge force to do this, presumably balancing it with body-generated negative lift. Penguins swimming slowly, particularly if they are picking up stones or even fish from the bottom, have to do what diving ducks and auks do: by angling their bodies down, they use the surge force to entirely counteract the upthrust. What this also means is that penguins cannot swim slowly and horizontally at the same time, as we have noted already, unless they are deep enough for buoyancy to be largely negated (cf. [61]).

Top Speeds: Power and Upthrust

There is huge public interest in how fast penguins can swim. Given how impressive penguin porpoising seems, or perhaps more particularly, how breathtaking their brief "aerial flights" are as they catapult themselves out of the water to alight on land (Fig. 6.12), this is little surprising. However, the simple question of how fast a penguin can swim does not have a simple answer. Moving shiftily from foot to foot (like a politician), the sensible scientist will say, "that depends…"! In fact, it depends on five factors: depth, swim angle, amount of air inhaled on the surface, how recently the penguin has molted, and, finally, how recently it has fed.

Penguin molt is important because the condition of a penguin's "coat" affects top speeds dramatically (Chap. 2). In the 1980s, one of us (Rory) put African penguins into the water at one end of a long (15 m), canal-shaped (with a depth of about 1 m and width ca. 1.5 m) rock pool and timed how long it took for them to cover a 10-m stretch in the middle. This allowed him to calculate their speed [83]. What came out of the calculation was dramatic. First, molting penguins cannot really swim, at least in penguin terms. At the first stage of molting, when all the old feathers lift to sit closer perpendicularly to the body, birds achieved a maximum speed of about 4 km/h (1.1 m/s). Many publications report that molting penguins do not go to sea because their insulation is compromised [26]. This may be true, but it is also of little use to go into the water when the top speed is a fraction of that needed to catch prey (Chap. 11) or evade predators (Chap. 12) and when the costs of transport skyrocket (see below). Interestingly, in the African penguins that had been put into the rock pool, as the old feathers dropped off and the new feathers constituted the ever greater proportion of the birds' surface areas, the top speed rose to about 10 km/h (2.8 m/s). Despite the new, slick featheration, which should have given the birds a reduced drag due to optimized surface compliance (see above), this is less than the top speed of penguins just premolt, which was about 13.5 km/h (3.75 m/s). Although the feathers of premolt penguins are in poor condition (inspection of the feather tips shows them to be worn and the color of the black feathers starts to fade to brown), at least they have large muscles to power their flippers effectively and optimal body shape, which minimizes body drag. This contrasts the condition in post-molt birds that have fasted for ~21 days and have used their body protein stores to synthesize new feathers and help pay metabolic costs. They consequently look as if they are wearing baggy pajamas as their skin hangs down their breasts, latterly full of powerful muscles. So, even with a new surface coat, the engine room is weakened and the shape compromised. Fast forward to 3 weeks later, the same penguins have been to the sea and fattened up. Then, they benefit from their muscle power, optimal shape, and fabulous feathers and reach their maximum speed of ~20 km/h (>5 m/s). This is hugely impressive to watch!

The conditions under which we examined the maximum speed of African penguins were particular with respect to the other factors that determine how fast a penguin can swim. Unbeknown to us, we had corrected for the effects of upthrust and depth, which, theoretically at least, will make a substantial difference to penguin maximum swim speed. As we have considered above, the upthrust increases disproportionately for penguins nearer the surface (Fig. 6.2). This means that birds swimming downward have to use power to counteract buoyancy, reducing their top speed, especially as their dive angle becomes steeper (Fig. 6.3). The reverse is true for penguins swimming upward. The situation in both descending and ascending penguins is exacerbated if the birds have inhaled air to their maximum capacity (cf. Wilson et al. [86]).

To determine what this amounts to, we have to consider how the drag of a penguin increases with speed (Fig. 6.12) (accepting that our calculations on a bird with stationary flippers (gliding) will be less than those on a powering penguin). Then, we have to consider how metabolic power scales with drag (see below), followed by estimating the maximum power that a bird can generate (e.g., about 100 W/kg if we use estimates made for larger auks [25]), and then insert the relevant upthrust forces. Cutting to the chase, using the back of an envelope and accepting that we will have some degree of error, we calculate that a "standard penguin," the size of a chinstrap, powering vertically down or up at a 2-m water depth (assuming that it could have these conditions for a theoretical period long enough for it to have reached terminal velocity) would have top speeds of 17.4 km/h (4.83 m/s) while going down and 21.8 km/h (6.07 m/s) while going up. Swimming horizontally, the same bird would swim at a (theoretical) maximum speed of 20.2 km/h (5.6 m/s). Top speeds at given angles are likely to be a couple of kilometers per hour faster in larger penguin species (we have recorded gentoo penguins reaching 6.6 m/s (23 km/h), and Kooyman et al. clocked an emperor traveling at 7.1 m/s (25.6 km/h) [39]) and slower in the smallest species (Bethge et al. documented the fastest speed by a little penguin as 3.3 m/s (11.9 km/h) [11]), but this gives some general idea of what is involved.

The Energy Costs of Swimming

General Considerations

A critical element of penguin swimming is what it all costs in terms of energy. Ultimately, this affects performance and, in the end, penguin ecology and well-being. Energy expenditure, aptly termed the "fire of life" [36], affects virtually everything in the lives of all animals, not just penguins. Its analogy with money is quite compelling. Unless you have huge initial resources, if you do not earn (or somehow acquire!) at least the same amount of money as you spend, you have no long-term prospects. Judicious expenditure, especially with respect to gain, is a strategy that works with money for people and with energy for animals (and, incidentally, with people, too).

The extraordinary, environmentally variable water world of penguins, as we began to review above, makes determining the energetics of their swimming no easy task, which is why it took until 1984 before anyone had a go at it. Then, it was Ken Nagy [50], who used the "doubly labeled water" method for penguins in an attempt to figure out how much it costs African penguins to swim. The (exceedingly simplistic and quick) explanation of the technique is that it provides scientists with values for the total amount of carbon dioxide (CO_2) produced by a free-living animal over two points in time (typically a day or two). While oxygen is used to break down food during cellular respiration, water, carbon dioxide, and energy are produced, and it is the easiest to measure the CO_2. In fact, Ken was at Rory's research base (on Marcus Island) in South Africa to do this and caused some initial surprise by asking (then) impossible questions such as, "How much water does a penguin drink at sea?" and the like. However, Ken is a magnificent fellow, and we worked well together. In his publication, he provided some highly useful base data about the energetics of African penguins, such as the fact that, during a 2-day chick brooding and foraging cycle, they expended about 3890 kJ, which equates to 22.5 W per bird (1 watt = 1 joule expended per second) or (weighing in at 3.17 kg [50]) 7.1 W per kilogram of penguin. The difficulty is that this average energy expenditure incorporated three main activities: sitting quietly on the nest in warm air, floating quietly at sea in cold water, and, finally, swimming (including diving) in cold water. A major issue is how to separate these elements to get to the costs of "swimming" (Chap. 7). What was easy, though, was to determine the costs of sitting on the nest because some birds did only that, and Ken's work determined that individuals doing that expended an average of 4.4 W (1.38 W/kg). Knowing the amount of time spent on the nest and the amount of time spent at sea, allowed Ken and colleagues to surmise that the costs of sitting on land could be subtracted from the total costs of being on land "and" at sea to give the costs of being at sea. This turned out to be 57.4 W (18.1 W/kg). So, it is clear that life at sea for African penguins foraging for their chicks is vastly more costly than sitting on the nest. The seemingly clever thing that Ken did to access the costs of swimming was to use some of our autoradiographic speed/distance meters [85] (primitive bird-attached gadgets that we invented for the specific purpose of finding out more about penguins at sea) to find out how far his doubly labeled water study birds had swum. Then, he plotted the distance swum against the energy used, reasoning that the gradient of the line would indicate the cost of swimming per kilometer and, by knowing the speed, time underwater etc., the cost of swimming per second could be obtained. The results showed that African penguins used 27.3 W/kg of their body mass to swim at normal speeds (Table 6.2) if the energy expended is equated with the time spent underwater (penguins only swim seriously underwater, of course). This accorded well with the graphed data, which produced 26 W/kg/km (Table 6.2).

The method was an exciting but crude tool by which to determine the complexities of penguin power underwater. This was especially so, given that swimming penguins can choose to do anything from dilly-dallying along to shooting up and down the depths with immense athleticism, rather like aquatic swallows. A key feature, therefore, is how power relates to speed. This is not just relevant for overall

Table 6.2 The metabolic rates (MRs) of swimming among several penguin species (blue values show non-normal speeds)

Species	Activity	MR (W/kg)	Source
African	Swimming underwater	27.3	[50]
Humboldt	Swimming at 1.25 m/s	14.8	[34]
Humboldt	Swimming at 0.6 m/s	8	[45]
Humboldt	Swimming at 2.2 m/s	23.1	[45]
Little	Swimming at 0.62 m/s	7.4	[10]
Little	Swimming at 0.72 m/s	8.4	[10]
Little	Swimming at 1.8 m/s	20	[11]
Adélie	Swimming at 2.0 m/s	14.7	[21]
Adélie	Swimming at ca. 2.0 m/s	15.8	[21]
Adélie	Swimming at 2.1 m/s	15.7	[22]
Chinstrap	Swimming at 2.0 m/s	15.2	[21]
Chinstrap	Swimming at 2.5 m/s	10	[22]
Gentoo	Swimming at 2.0 m/s	17	[21]
Gentoo	Swimming at 2.3 m/s	16.1	[21]
King	Swimming at ca. 2 m/s	10.3	[23]
King	Swimming at 1.0 m/s at (5 °C)	5.46	[37]
Macaroni	Normal diving	9	[27]
Emperor	Swimming under ice	2.07	[81]
Emperor	Swimming under ice	2.27	[82]

penguin energetics, but their energy use also equates to their oxygen use and that determines how long they can spend underwater.

The first serious attempt to measure penguin power in relation to speed was made by Clifford Hui in 1983 [34]. He put a Humboldt penguin in a submerged cage with a breathe hood over it next to his boat at sea (in California). In this way, he could measure carbon dioxide production (and therefore calculate energy expenditure) by analyzing the respiratory gases of the bird – a technique called gas respirometry or indirect calorimetry – when it surfaced after a dive. Of course, the cage did not allow the penguin to go anywhere, but the clever part was that Clifford could then move the boat at various speeds with the penguin mirroring those speeds within the cage [34]. Actually, the protocol had its problems, not least of which was that pushing the cage through the water caused unusual turbulence, which would have affected swim performance. In addition, Clifford could not get the system up to speeds higher than 1.26 m/s (3.6 km/h), which is well below the normal cruising speed of penguins of around 2 m/s (7.2 km/h) [66, 67]. All that apart, he determined that his two test birds used about 14.8 W at 1.25 m/s (Table 6.2) [34].

One way that people in the process of determining the energy used by aquatic animals to swim is to put them in a water flume, rather like the one used by Rudi Bannasch in Berlin (see earlier) – though obviously smaller – and have them swim at defined speeds while breathing between submergences into a special dome where the respiratory gases can be analyzed. This approach was used to determine that the power consumption of little penguins underwater was about 8.4 W/kg for speeds of

0.7 m/s [10]. The latter figure was interesting because it showed how much more efficient the penguins were when they swam underwater than on the surface (comparable values for surface swimming were 12.5 W/kg). Again, though, the speeds were too low to represent normal traveling speeds of penguins, even for little penguins that cruise at about 1.8 m/s [11].

Following this, we mulled over how to get a swim flume to Antarctica to determine the swimming costs of the medium-sized *Pygoscelis* penguins but, importantly, at the speeds they normally use. The first projections revealed that we had to have a flume custom-made, and, second, it had to work for our quite big (see Chap. 1) and fast penguins. Therefore, the cost was going to be almost a million dollars! Then, as we morosely sipped our coffee, we remembered the speed trials conducted on African penguins on Marcus Island, where the birds swam underwater up and down a natural canal (see above) and we hit on the idea of replicating that in Antarctica. So, after an immense amount of planning, that was what we did. In this scenario, the birds breathed at either end of a 21-m channel into respiratory domes where the carbon dioxide could be analyzed. Of course, this system could not entice (or force!) the penguins to swim at particular speeds like the water flume. However, we could, at least, let them swim as they wished, note down their speeds, and then correlate those speeds with energy expenditure. So, a chap named Boris Culik, who did all the real work in designing and setting up the swim channel system, spent many hours standing on top of a tall ladder watching penguins swim up and down his channel while he talked into a recorder to say exactly where they were within the channel. This enabled him to calculate swim speeds and ultimately equate them to energy expenditure. In fact, the approach was so successful that it was subsequently used by a number of other researchers around the Southern Hemisphere to tell us more about the swimming energetics of several penguin species. This, and subsequent work, indicated that penguins swimming at "normal speeds" of around 2 m/s use energy at a rate between about 10 and 23 J/kg/s (or 14–23 W/kg) [11, 21–23] (Table 6.2). We have ignored the notably higher figure for African penguins of 27.3 W/kg (Table 6.2) because these birds were wearing a harness to carry their speed/distance meters and this would have increased swim costs considerably. We noted that among birds swimming at "normal" speeds (of 9, 10, and 10.3 W/kg), the three lowest estimates did not obviously come from the largest species (macaroni, chinstrap, and king penguins, respectively) [22–24, 27] (Table 6.2). This raises a question: why was there such apparent variability among species? First, given the universally slick hydrodynamics and fairly standardized propulsion system of penguins, it seems unlikely that the variability of estimates shown in Table 6.2 is genuine. This is despite efforts to get it right, having major implications for calculating food consumption of penguin populations, understanding and quantifying foraging success and efficiency [50], and understanding how long these birds can spend underwater using just their oxygen stores (e.g., [27]). Rather, the range of values simply indicates the difficulties of doing something as esoteric as attempting to calculate the swimming and diving metabolic rates (MRs) of penguins, the birds of wild oceans, plying their trade in places and at extraordinary depths that cannot be easily replicated.

Specific Considerations

We expect there to be considerable variability in the costs of penguin movement, depending primarily on the bird dive or return-to-surface angle, speed, and depth for the reasons explained above. Thus, without taking these into account, our estimates of energy expenditure – more specifically "power" (the rate at which energy is used per second) – are liable to be coarse.

There is another way that we can estimate these activity-specific costs, and that is using data from the technology that we attach to birds. In particular, accelerometers (for measuring acceleration!) and depth can be combined with a little (basic) Newtonian physics to estimate the biomechanical power of swimming (Box 6.7).

Box 6.7
Bird-attached accelerometers measure change in speed in all three axes of movement, two of which are important for penguin swimming – the surge and the heave axes (Figs. 6.19 and 6.20). These can be converted to force (F) using the relationship (referred to previously) $F = m.a$, where m = mass and a = acceleration (Fig. 6.21). The acceleration is given by the tag and the mass is that of the penguin. The work done (WD – in joules) by a moving system is given by WD = $F.D$, where D is the distance traveled during the application of a particular force. Some penguin tags have speed sensors on them, in which case, speed, important for the next step, is measured directly. Otherwise, accurate depth sensors in the tag allow computation of the rate of change of depth (d) over time ($\Delta d/dt$), which is also equal to the vertical velocity) at any moment during the dive. The dive angle, which is also the penguin body pitch angle, i.e., the swim angle with respect to the surface (p), can be calculated from the smoothed surge axis. Using the geometry worked out long ago by Pythagoras, true penguin speed (u) can be calculated with the formula, $u = (\Delta d/dt)/sine\ p$, as long as the penguin pitch (swim) angle is not too close to zero. Once speed is known, distance can be calculated ($D = u.t$) to get the work done per unit distance. Since Newton stated that biomechanical power (P – in watts or joules per second) is the rate at which the energy (in joules) is being expended, the work done can be divided by time to obtain power.

This biomechanical power, which just examines the motion associated with the body (and ignores, for example, the marginal, but extra, heave costs because the penguin body moves up and down slightly with every flipper beat; see above, Fig. 6.19), is not the same as "chemical power," which is effectively the energy that the bird has to expend in order to produce the biomechanical power. How much chemical power is needed for the biomechanical power depends on the efficiency of the biological system for doing what the physicists call "work." However, the relationship between the chemical and biomechanical power for animals is reasonably well-known, with the process being about 20% efficient (though it varies according to what the animals are doing). Values for penguins are estimated to lie between about 0.17 [66] and 0.25 [7], although efficiency likely varies with activity and the optimal contraction frequency of each muscle fiber. Therefore, biomechanical values need to be corrected (multiplied) to produce chemical power, which is what the bird pays in energy (and it can use its fat reserves, for example, for doing this) for its activities. In addition, if we wish to determine the chemical power of penguins doing something, we also need to add the chemical power of "resting" – i.e., the costs of the bird being alive, to which the costs of movement have to be added (Box 6.8). There are actually other things that can be relevant, depending on what the penguin is doing, an example being the energetic costs of digestion. We shall ignore these for convenience.

Box 6.8
We can convert biomechanical power to chemical power by multiplying it by about 5 [7, 66] and then adding the metabolic cost of resting. To this, we need to add the cost of the resting metabolic rate (RMR). It would seem logical to add the RMR in water since the birds are in water – a good value for our example species (the Magellanic penguins) being that measured for the Humboldt penguin of 5.95 W/kg [45]. However, as discussed in the next chapter, this is likely to be wrong because a large fraction of the costs of resting in water come from thermoregulation. Swimming penguins can provide the heat normally lost by resting penguins in water without substantial extra metabolic cost because about 80% of their chemical energy is produced as heat (Chap. 7). Therefore, we suggest taking the RMR of penguins on land, a good value being provided by congeneric African penguins of 1.4 W/kg [50]. So, we suggest, and adopt, adding 1.4 W/kg to any estimated chemical energy power cost of active swimming.

Actually, the energetic costs of movement in a fluid medium have been well-studied and are nicely summarized by the drag equation, which basically states that costs should go up with the cube of the velocity (Box 6.9 – we apologize for all the boxes! They are particularly nasty in flocks!).

So, do our estimates of penguin swim speed increase in a roughly similar manner? The approach described in Boxes 6.7 and 6.8 can be used to estimate the cost of swimming (the chemical power) at different speeds for penguins equipped with accelerometers. However, ultimately, due to variable buoyancy with depth, we have to do this within the prescribed dive angles (and we should do it for varying depths and maximum depth reached, too, because penguins inhale more when they dive deeper [64]- but the effect is secondary, so we will ignore it here). Doing this nicely shows how Magellanic penguin power to swim increases with speed, and does so, as predicted, with higher speeds costing disproportionately more energy (Fig. 6.25).

Box 6.9

The power (P) to overcome drag (or the rate at which energy is expended per second) is mathematically characterized by the formula:

$$P = 0.5\rho u^3 AC_d,$$

where ρ is the density of seawater, u is the penguin velocity, A is the penguin cross-sectional area at its point of maximum girth, and C_d is the drag coefficient of the penguin (cf. Box 6.5). It should be noted that, since ρ and A are constants, and C_d is increased by the addition of an external tag, it is C_d and the swim velocity that determine the rate at which a swimming penguin uses energy.

These figures compare well with those provided by other authors in Table 6.2 but are perhaps a touch higher, probably because our birds are swimming down against buoyancy, whereas those measured in Table 6.2 are swimming horizontally.

So how does swim angle fit into this? Using our approach, and holding speed constant (instead of angle; Fig. 6.25) in our dataset, we see that penguin allocation

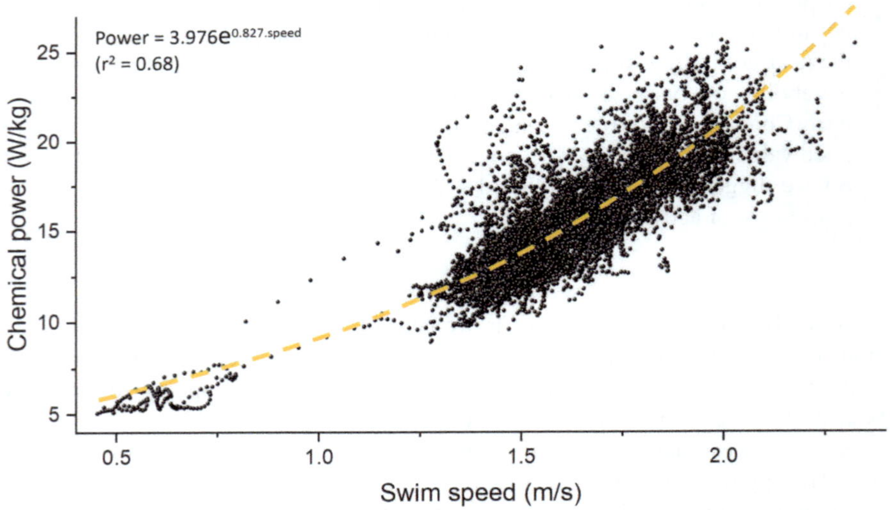

Fig. 6.25 The calculated power to swim as a function of speed for Magellanic penguins based on the approach presented in Boxes 6.7 and 6.8 for birds only, using dive angles restricted to 30–50° (because swim angle is predicted to affect power). Variation around the line of best fit will be due to many factors, including swim angle, maximum depth reached during the dive, and the instantaneous depth at which the measurement was made. (It should be noted that speed is used in the calculation of power (Box 6.7) to derive the distance traveled. Thus, there is enhanced correlation although if the graph is plotted of vectorial force against swim speed (which is less helpful for our purposes)], it is still highly significant)

to biomechanical power when they swim down is greater than when they swim nearer the surface. Allocation also increases with increasingly steep dive angles, although the effect decreases with depth (Fig. 6.26). This is presumably due to them having to "pay for" counteracting their increasing buoyancy at shallower depths and pay more if they swim more directly against buoyancy by descending at steep angles. We can use the data in Fig. 6.26 to define the power to swim at a chosen speed, descent angle, and depth more precisely in order to model it (Box 6.10).

The formulation of power use against performance (Box 6.10) is more important than simply mathematical fun because it allows us to determine the total mechanical energy costs of particular dive types (see Chap. 9), and we will do this in later chapters. For the moment, it is enough to see how penguins have to invest energy according to what they do (descent angle and speed) and how this varies with depth. In order to compare our results derived from our Magellanic penguin acceleration

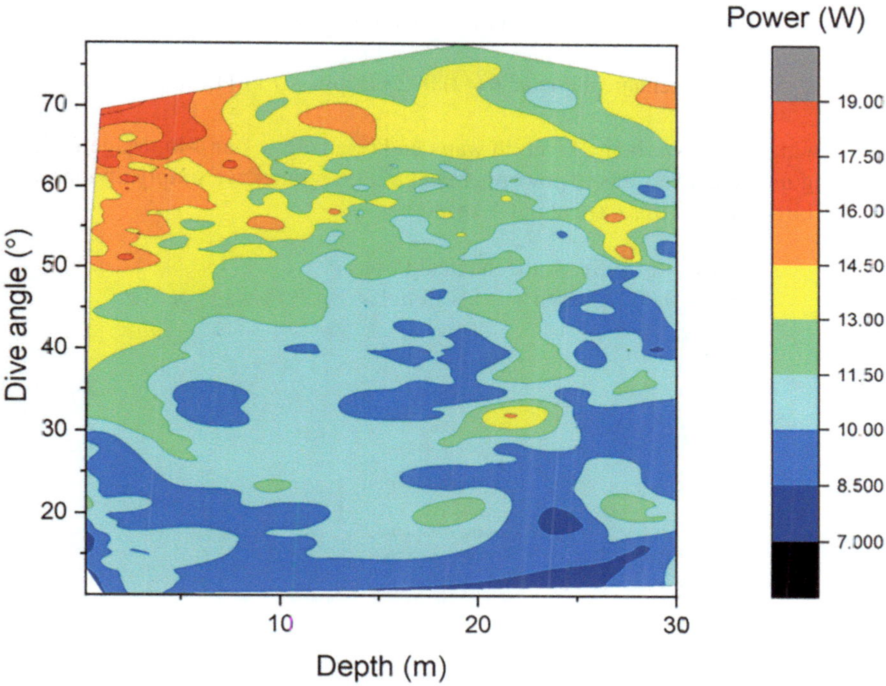

Fig. 6.26 The biomechanical power estimated (see Box 6.7) for a Magellanic penguin to swim at a constant downward speed (≈1.6 m/s – data limits 1.5–1.7 m/s) as a function of its dive angle and depth. This contour plot was generated using data from 36 dives, all of which had the descent uninterrupted by prey capture. Magellanic penguins dive deeper than 30 m, of course, but to get appropriate point density for a chosen approximate constant speed and a well-defined dive angle, we curtailed the depth to 30 m. it should be noted how steeper descent angles use higher power, especially near the surface where the upthrust due to buoyancy is the highest. It should also be noted that as the pitch angle approaches zero, the power costs are approximately constant irrespective of depth. This also underpins some clever mechanism, such as body shape, providing a downward acting force that penguins use to deal with upthrust (see above)

data to those from the literature, most of which are derived from penguins swimming at about a 0.5-m depth in a swim channel (see Table 6.2), we need to perform the simple operations described in Box 6.8.

Box 6.10

If we derive regressions for "transects" across Fig. 6.26, with the lines drawn parallel to the *y*-axis (pitch) within defined depth categories, then what we get are best fits that are linear across depths (Fig. 6.27a). The gradients are clearly dependent on swim depth (Fig. 6.27a), though the intercepts are not (all aggregating around 8 W – with mean = 8.12 W). We can statistically compare these gradients against swim depth (the midpoint of the ranges given in Fig. 6.27a) to derive a relationship between swim depth and gradient (Fig. 6.27b). We then use this to derive biomechanical power as a function of bird descent angle (pitch) and swim depth for the defined swim speed (1.6 m/s). For this instance, this amounts to:

$$\text{Biomechanical Power} = \left(-0.0023.\text{Depth} + 0.114\right).\text{Pitch} + 8.12,$$

with the answer being given in watts and referring to our 4-kg Magellanic penguin. To convert this to power per penguin kilograms, we simply divide by the bird mass (in this case about 4 kg) to get:

$$\text{Mass} - \text{specific biomechanical Power}$$
$$= [(-0.0023.\text{Depth} + 0.114).\text{Pitch} + 8.12] /4 \text{ watts per kg.}$$

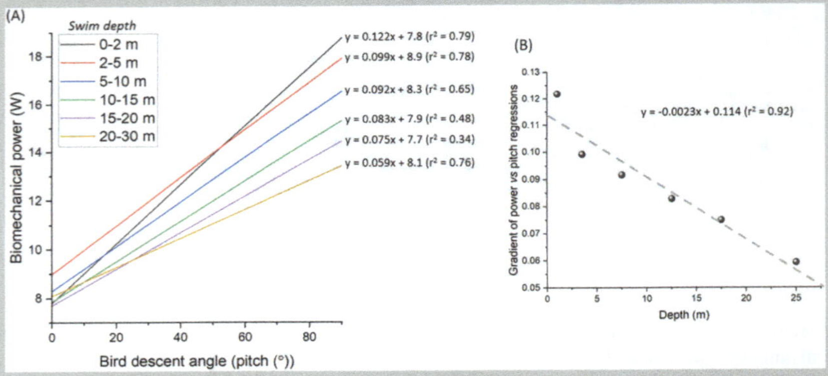

Fig. 6.27 (**a**) The relationship between biomechanical power and penguin descent angle for specified depth ranges for the Magellanic penguin data shown in Fig. 6.26. (**b**) If the gradients of the regressions in panel (**a**) are plotted against depth (midpoint of the depth ranges), then there is a clear relationship to indicate that the biomechanical power to swim at a given speed (here, 1.6 m/s) at defined descent angles and depths can be accessed by formulation

These operations show that Magellanic penguins use approximately 11.4 W/kg of chemical power to swim horizontally at 1.6 m/s at any depth. This compares favorably to the ~14 W/kg measured for Humboldt penguins swimming in a water channel [45]. That our value is lower by about 20% will be considered in the next chapter.

"Sensible" Swim Strategies and Costs of Transport

A commonly used metric to examine how animals maximize the "value" of their travel is "cost of transport" (generally shortened to COT [68]). COT provides a measure of how much energy animals need in order to travel a given distance and is generally specified as joules per meter. It can be derived by dividing the power to move by the speed of movement. People apply this principle to cars but generally using the inverse term of "miles per gallon." A thoughtful car driver gets plenty of miles per gallon, whereas an animal moving judiciously in the same way expends little energy to go a given distance. The nub of the matter for penguins, with all other things being equal, is that they benefit by moving with a low COT. Going the farthest for a given amount of energy is important for penguins traveling the sometimes large distances between their breeding and foraging sites, as discussed in Chap. 5, but it is also important for maximizing their efficiency underwater. Traveling at minimized costs (COT_{min}) means that birds travel the greatest possible distance underwater during any single dive – and this is important because they need to cover the distance to find prey (see Chaps. 5 and 10). However, travel at COT_{min} also helps penguins use their oxygen stores appropriately (oxygen use equates with energy expenditure) and effectively allows them to maximize their dive durations for a given distance moved. Doing so gives them time to descend repeatedly to where they had found prey, sometimes at phenomenal foraging depths.

Our above calculations show, for example, that, because power consumption for swimming increases with increasing descent angle for any given speed and decreases with increasing instantaneous depth (Fig. 6.26), both dive angle and instantaneous depth are relevant in any consideration of penguin COT, i.e., swimming efficiency (cf. Sato et al. [66]) (Fig. 6.28).

Application of COT, however, is also more complex in penguins than it is for many other animals. Conventional calculations simply divide the power used by the distance traveled per second. However, penguins have to optimize distance in two dimensions, vertically and horizontally, because of the way that their prey distribute themselves in the water column (see Chaps. 5 and 10, Appendix Table 3). It is thus germane to consider both vertical and horizontal COTs, derived by dividing the power to swim by the vertical and horizontal distances, respectively. Using the formulation provided in Box 6.10 to describe the COT for our model Magellanic penguin swimming at 1.6 m/s, we can see that COT varies greatly depending on whether traveling vertical, horizontal or absolute distance is the most important (Fig. 6.29).

So, in our examination of descent angles chosen by penguins for their movement down the water column, considerations have to reflect on what is optimum: (i)

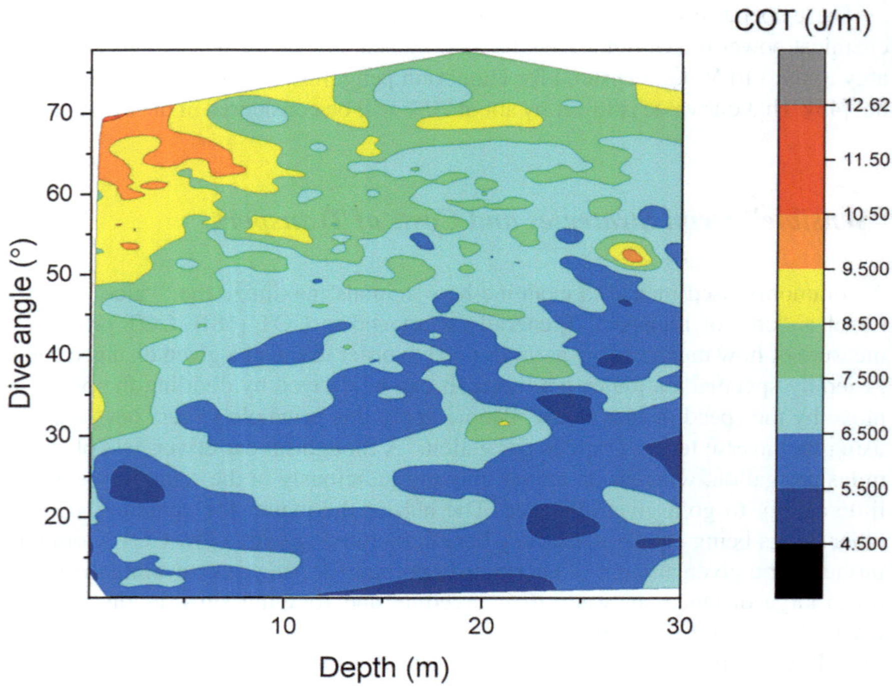

Fig. 6.28 The biomechanical cost of transport (COT) estimated (using the approach described in Box 6.7) for a Magellanic penguin to swim at constant speed (≈1.6 m/s – data limits 1.5–1.7 m/s) down the water column as a function of dive angle (pitch) and depth. This contour plot was generated using data from 36 dives, all of which had the descent uninterrupted by prey capture. It should be noted how this graph replicates the power graph (Fig. 6.26) because swim speeds are approximately constant. We keep in mind that steeper descent angles have higher costs of transport, especially near the surface where the upthrust due to buoyancy is the highest and, as the pitch angle approaches zero, the costs of transport are approximately constant irrespective of depth

whether it is better to descend efficiently to a preferred foraging depth (as in U- or W-dives; see Chap. 9 for dive definitions), thus minimizing vertical COT (Fig. 6.29b), (ii) whether penguins should move horizontally most efficiently (e.g., traveling dives) (Fig. 6.29c), or (iii) whether they might do better hedging their bets by doing a bit of both (e.g., in parabolic dives) (Fig. 6.29a). Clearly, to utilize a prey patch in the water column at depths directly below them, penguins should minimize vertical COT and should use dive angles closer to 90° (Fig. 6.29b), although it is interesting that they never descend vertically (e.g., [84]). This refusal to dive at, or near, 90° contrasts with data from seabirds that feed on benthic prey, such as most cormorants [60, 87] and some auks [80]. Rather, the maximum descent angle for penguins is about 70°, which actually coincides with a clear COT_{min} in Fig. 6.29b – a consequence of the way power to swim interacts with vertical velocity. We also note that, since penguin body shape likely generates negative lift (see earlier), traveling too

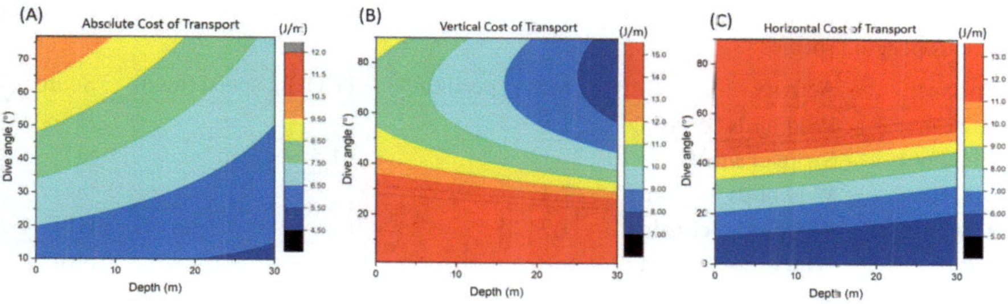

Fig. 6.29 The biomechanical costs of transport (COTs) for a Magellanic penguin descending the water column according to whether COT refers to (**a**) the absolute distance traveled, (**b**) the vertical distance traveled, or (**c**) the horizontal distance traveled. Lowest COTs (cold colors) represent the most energetically efficient travel. These contour plots were generated using a model based on real data (Box 6.10), the general applicability of which can be verified by comparing panel (**a**) in this figure to Fig. 6.28

close to the absolute vertical might instigate forces that would tend to pull them upside down!

It is also relevant that penguins traveling horizontally have both minimized COTs for absolute and horizontal distance regardless of depth (Fig. 6.29a, c). This is reason enough to have descent angles tending toward 0° generally (something that may explain the preponderance of dive types that display this pattern (see Chap. 9)).

Cruising Speed, Cost of Transport, and Beyond

Swim speed is a critical parameter that effectively defines how animals may deal with distance. Faster-swimming animals move through distance more rapidly than do slower-swimming animals, which has repercussions in terms of foraging range and the rate at which prey can be encountered. Here again, penguins are exceptional. However, in breath-holding divers, there is a generally accepted relationship between animal mass and swim speed [77], seabirds in general, and penguins in particular, swim at cruising speeds (Table 6.3) that are appreciably faster than the general trend (than they should) for their size [77]. This also holds if penguins are compared to fish that habitually swim fast (e.g., Blank et al. [14], Sepulveda and Dickson [70], and Yuen [89]). In a wide-ranging analysis of this, Yuuki Watanabe and colleagues [77] broadly found marine mammals to cruise at somewhat less than 2 m/s, despite their size, whereas most penguins cruised at, or faster than, 2 m/s (Fig. 6.30) (other seabirds swam between about 1 and 1.85 m/s).

The overarching question as to why, was not resolved in this publication, particularly given that the study highlighted that higher speed was a previously unrecognized advantage of greater size in breath-holding divers, although the authors suggested that heat flux might somehow play a role. So, penguins, flashing around

Table 6.3 Normal underwater swim speeds and minimum costs of transport speed measured for various penguin species

Species	Activity	Mean speed (m/s)	COT_{min} speed (m/s)	Source
African	Wild birds "cruising"	1.8		[38]
Humboldt	Swim channel		1.5	[45]
Magellanic	Wild birds "cruising"	2.1		[88]
Magellanic	Wild birds "cruising"	1.7–2.8		[18]
Little	Wild birds "cruising"	1.8		[66]
Little	Wild birds "cruising"	0.7		[38]
Little	Wild birds "cruising"	1.8	1.8	[11]
Little	Wild birds "cruising"	1.85		[80]
Adélie	Wild birds "cruising"	2.1		[66]
Adélie	Wild birds "cruising"	2		[38]
Adélie	Wild birds "cruising"	2.2		[21]
Adélie	Swim channel		2.1	[22]
Chinstrap	Wild birds "cruising"	2.3		[66]
Chinstrap	Wild birds "cruising"	2.4		[21]
Chinstrap	Swim channel		2.5	[21]
Gentoo	Wild birds "cruising"	2.3		[66]
Gentoo	Wild birds "cruising"	1.8		[21]
Gentoo	Swim channel		2.3	[21]
King	Wild birds "cruising"	2.1		[66]
King	Swim channel		2.2	[23]
Emperor	Wild birds "cruising"	2.2		[66]
Emperor	Wild birds "cruising"	2.1		[38]
Macaroni	Wild birds "cruising"	2		[66]

in the depths, buck the aquatic swimmers trend and thus appear to be the high-speed underwater "swallows" or "swifts" in their subsurface realm. Why though?

What was clear from the above study was that most animals considered swam at, or close to, their lowest cost of transport (in terms of the absolute traveled distance rather than horizontal or vertical distance). Indeed, the most widely discussed COT modulator in animals in general is speed, and this is particularly the case in aquatic animals since, as we have said, the power to swim increases with the cube of the velocity (Box 6.9). So, at what speed should penguins normally travel (cruise), and how does this compare to reality [66]? Sensible swim speeds will obviously depend on circumstance. Thus, for example, speeds should sometimes be high to outrun predators (Chap. 12) and capture some prey (Chap. 11), but during normal "cruising," achieving judicious energy expenditure is a strategy that serves all animals, not just penguins [68].

Penguin COT with respect to speed can be usefully compared to cars. A car idling in a queue, perhaps slowly creeping forward, uses plenty of fuel for the minimal distance covered. This is comparable to a car speeding down the motorway using a disproportionate amount of fuel per unit distance traveled. In fact, most cars, these days, have the sweet spot for COT_{min} at around 50 mph (80 km/h). The

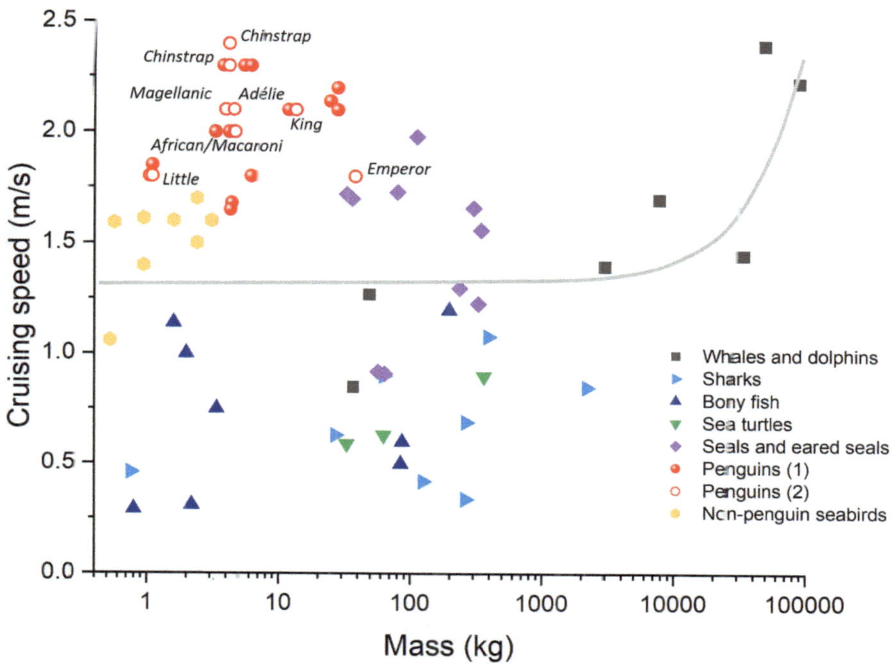

Fig. 6.30 The mean speeds of various swimming marine species in relation to their mass (each point refers to one species, and, in this graph, taxon is denoted by symbol type and color using data from [77] for all filled symbols). The unfilled symbols (Penguins (2)) show additional data provided in Table 6.3 (below), except for one anomalous point for the little penguin from [38]. The dark line shows the overall best fit between speed and mass for all species taken from [77]. It should be noted how the cruising speed of most penguins lies well above that of most other species, but especially when size is considered, making a strong case for penguins being the fastest cruising animals in the seas. For a full list of species, see [77]

principle is exactly the same for penguins and, since the derivation of COT is the rate at which energy is used divided by the speed (Joules/second divided by meter/second, which converts to Joules/meter), slow-traveling birds have high costs of travel and so do birds that swim too fast. The penguin sweet spot for COT_{min}, at least for horizontal swimming, is at about 2 m/s (Fig. 6.31), and this is the speed at which most of them seem to travel (Table 6.3), irrespective of their size [74].

Specific studies on penguin speed report that, in Adélie penguins at least, the swim speed for the descent phase of dives is approximately constant, around what is projected to be COT_{min}, irrespective of depth [62], although Fig. 6.28 implies that COT may be minimized at different speeds with depth to some degree due to buoyancy changes. However, speeds are much more variable during the bottom phase of dives, including sometimes dropping to extremely low values (<0.5 m/s), and these seem to be associated with prey capture (for more details on this, see Chap. 11).

Obviously, the mean speed values hide any natural variability in penguin speed that occurs due to, e.g., prey capture, but, nonetheless, over a given foraging trip for

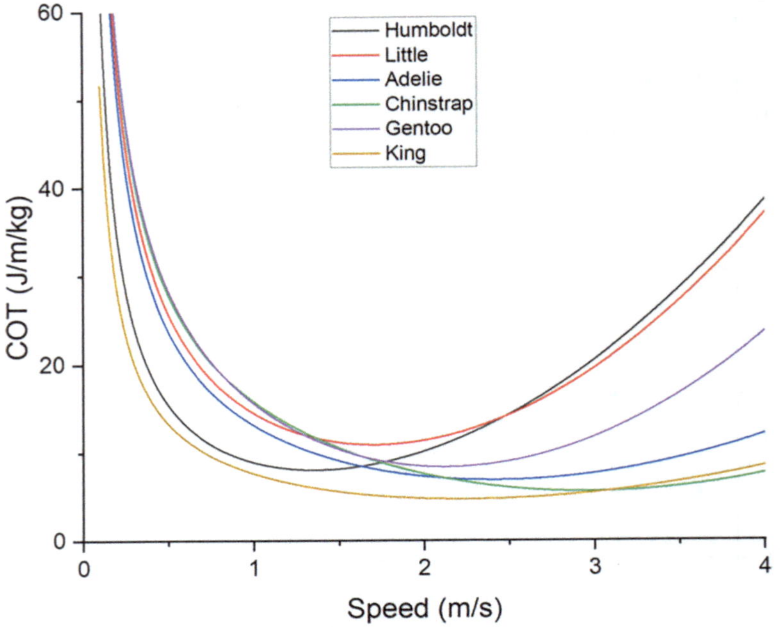

Fig. 6.31 Cost of transport (COT) as a function of swim speed for six species of penguins using best-fit regression lines provided by the literature, with these being derived from swim channel experiments. It should be noted how COT is exceedingly high at extremely low speeds due to the costs of the resting metabolic rate with little advance in distance but then decreases to a minimum before rising, with increasing speed, due to the disproportionate power costs (proportional to the velocity cubed) at high speeds. Nonetheless, the COT for every species has a fairly extended trough, thus demonstrating that the energetic penalties for deviating slightly from the minimum costs of transport are not too onerous

their chicks, penguin swim speed is surprisingly constrained with the swim speed peaks close to COT_{min} (Fig. 6.32).

The reasons given for adopting COT_{min} all make sense but, returning to the high-speed penguin issue above, the central question for penguins is then why do their own particular COT_{min} values occur at such high speeds? A clue is perhaps provided in Magellanic penguin data because this species, at least, travels faster in colder waters [18]. The obvious reasoning is the same as for persons cycling. Dressed in cycling gear, people can travel slowly when it is warm but get chilled if the temperature drops and so have to pedal more to stay warm because their muscular activity will generate more heat.

Based on this line of thinking, we suggest that heat production and loss may be a critical element that helps explain why penguins swim so fast for their size. Penguins resting in their normal seawater temperatures use between some two and five times as much energy as resting in air at the same temperature (covered in detail in Chap. 7). The increased energy expenditure in water is needed to balance the

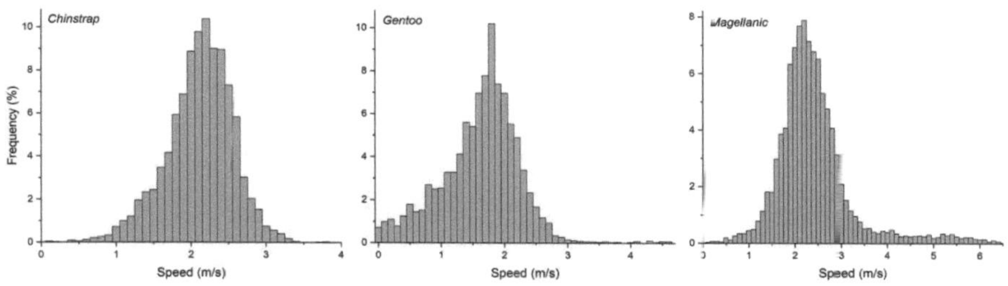

Fig. 6.32 Distribution of swim speeds for three species of penguins showing a surprisingly narrow range of preferred speeds with, however, clear interspecific differences

increased heat loss so that the penguin body temperature does not cool. When the penguin starts to swim, it incurs the biomechanical costs of the actual swimming, which increase dramatically with speed (see Box 6.9), and a substantial associated heat production (see earlier). However, by moving, the bird also has increased heat loss – the water equivalent of the "wind chill" factor. Putting these two together can help us compare how heat produced during swimming compares with heat loss. Producing a model based on deriving biomechanical power to swim for a Magellanic penguin (formulation in Box 6.9) and using resting metabolic rates in water for Humboldt penguins (the closest that we could get to Magellanics) (from [45]) provided us with the heat production estimates of Magellanic penguins swimming at different speeds. Against this, we could use a model produced specifically for heat loss in Magellanic penguins [18] to look at energy lost as heat. A graph of both of these (Fig. 6.33) shows that at swim speeds of less than about 1.7 m/s, Magellanic penguins start to lose out – they would have to invest extra energy (beyond that used in swimming) to maintain body temperatures. Conversely, at speeds in excess of 1.7 m/s, the birds produce enough muscle-generated heat to cope with any further cooling ("wind chill") factors.

This is a good reason for Magellanic penguins to swim at speeds in excess of around 1.7 m/s and, more generally, would nicely explain why penguins swim faster than they "should," for their size. The next chapter provides more detail on heat flux, how it is expected to change with depth, and how heat generated by the muscles of fast-swimming penguins can be stored to help them save even more energy.

So, energy saving due to heat production and loss can help explain much of the high-speed life of penguins, but, on top of this, the extraordinarily low drag coefficients of penguins will also tend to push the COT_{min} point into the higher-speed regions (Fig. 6.34): World-beating low drag coefficients lead naturally to world-beating optimal cruising speeds!

Finally, powered swimming apart, such extraordinary low drag coefficients coupled with appreciable buoyancy (see earlier) will also tend to help increase the speed of penguins moving up to the water surface, even if they glide passively – a torpedo-shaped balloon released at depth will move up the water column faster than a conventionally shaped balloon.

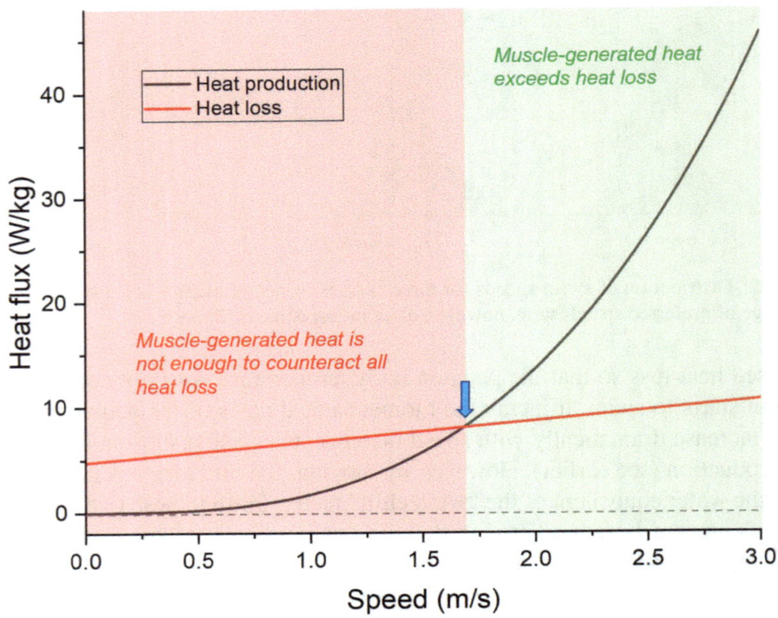

Fig. 6.33 Estimates of the heat produced by a Magellanic penguin swimming (above that produced during resting on the water surface) as a function of swim speed (black line) in relation to heat lost (which increases due to the "wind chill" factor – see text) (red line). Up to speeds of about 1.7 m/s (red zone), the muscles used for swimming do not generate enough heat to compensate for that lost, so if penguins swim at such low speeds, they have to provide more energy to keep their bodies warm –the little time that Magellanic penguins spend at these speeds shown in Fig. 6.32 should be noted. Above that speed (green zone – the blue arrow marks the point of equivalence), like a thinly clad cyclist in the Alps, their activity is enough to keep them warm. (To derive this figure, we used the formulation for power provided in Box 6.9, using a cross-sectional area of 0.0293 m^2 for Magellanic penguins, assuming a resting metabolic rate of 5.95 W/kg of the birds in water [45] at a temperature of 14 °C [18] and a 20% muscle efficiency (see earlier). We used the approach provided in [18] for heat loss)

So, there seem to be multiple reasons why penguins cruise at such high speeds compared to other water users, all of which ultimately boil down to their high body temperature (which tends to increase the rate of heat loss in cold environments), small size (which, because of their increased surface area-to-volume ratio, increases mass-specific heat loss compared to larger marine mammals), and their stunningly low drag. It is the specific combination that makes penguins unique! Moreover, it is these high cruise speeds that help them move fast through their ocean, increasing the rate at which they encounter prey while extending their foraging ranges and reducing the time it takes for them to shuttle between breeding and foraging sites. All of this is not a bad thing, particularly when one considers penguin evolution and how the penguin-specific features that result in them having most efficient travel at

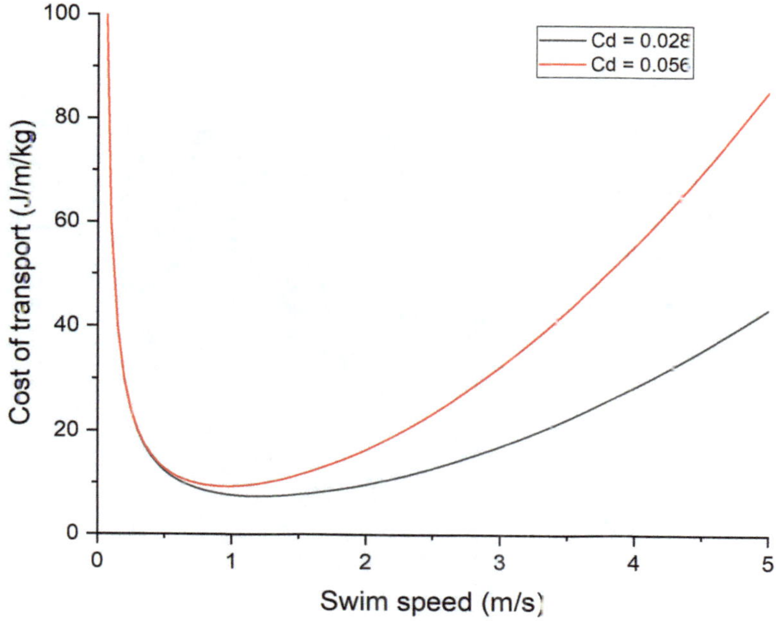

Fig. 6.34 The theoretical cost of transport of a penguin (based on the Magellanic penguin) as a function of swim speed and inserting two different drag coefficients (Cd). The lower value is that derived by us for Magellanic penguins (cf. Fig. 6.13), whereas the higher value, although double that, is still a reasonably low drag coefficient that may be expected for penguins and similar marine animals [77]. It should be noted how the minimum cost of transport (the lowest part of the trough) occurs at a higher speed in the case of the lower drag coefficient and is accompanied by a flatter trough, extending the "operational speed range" with lower energetic penalties. To derive this figure, we used the same parameters as used in Fig. 6.33

unusually high speeds will have played a role in their competition with smaller marine mammals (Chap. 1).

Penguin swimming is not all about just saving energy though. There are other important reasons why "cruising" penguins should modulate their swim speed to be other than at COT_{min}. For example, king penguins reduce their swim speed when ambient light conditions decrease, and this is believed to be because they are compensating for the decreased resolution of vision caused by low light levels [72]. In essence, though, the speeds that penguins select should depend on the "currency" that they are optimizing. Optimizing energy is a compelling case for traveling at COT_{min}, but birds may also be attempting to minimize the chick-feeding intervals to promote the chick's growth rate and thus travel faster than predicted by judicious energy use alone [see 52, 71]. In the process of doing this, they lose body condition, devoting their fat stores to chick provisioning rather than their own well-being [3]. Certainly, speed sensors on Magellanic penguins foraging for their chicks and commuting between the colony and the foraging area show marked variability (Fig. 6.35).

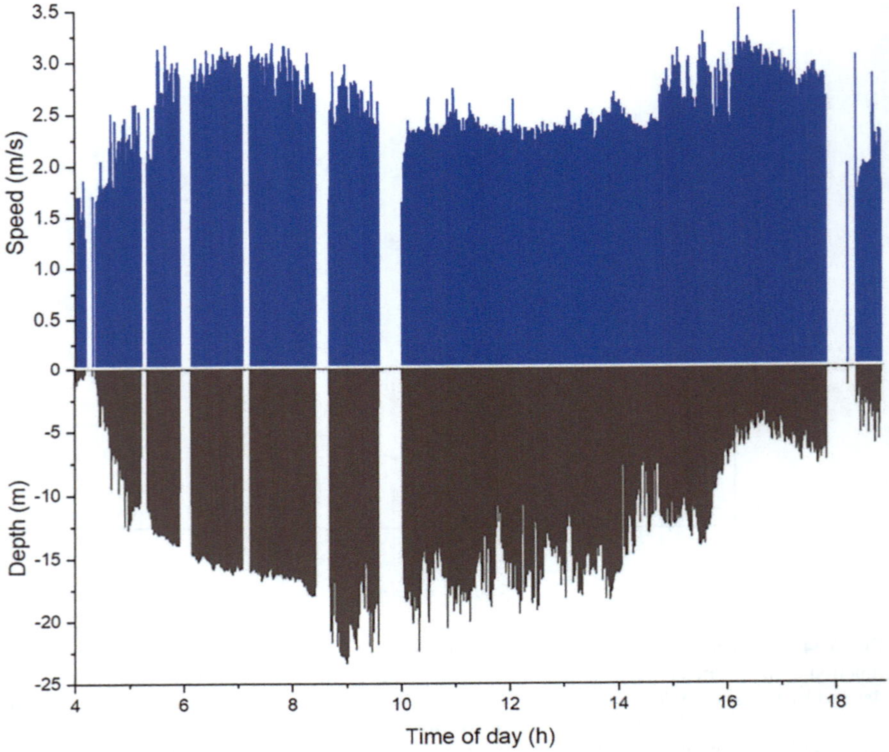

Fig. 6.35 Notably variable swim speeds and dive depths of a Magellanic penguin during a foraging trip from Cabo Virgenes, Argentina. Despite much penguin literature adhering to a speed that equates to the minimum cost of transport, this bird makes it clear that there are many factors that may modulate speed, one of which may be light intensity. In this regard, it should be noted how the increasing dive depth as light intensity increases around dawn is paralleled by increasing swim speed – might birds operating at the limits of their visual acuity have to slow down to enhance prey perception?

In addition, Adélie penguins use disproportionately more energy when they have to go on long compared to short foraging trips, and the energetic balance appears to limit how far they can go while provisioning their chicks [2].

Overall, therefore, although we are beginning to understand about the physical processes that affect the energetics of penguin swimming, we would be foolish to expect them to adhere to our projections about idealized behavior if we consider elements in isolation. Penguin lives are extraordinarily colorful and varied, with many factors affecting the viability and sensibility of the behaviors they adopt in their lives as fish-birds. It is nonetheless exciting to have pieced together a few aspects of the energetics puzzle in our quest to resolve the whole picture as best we can. In the meantime, keeping an open mind to things we have not considered also seems to be a sensible option.

References

1. Baldwin J (1988) Predicting the swimming and diving behaviour of penguins from muscle biochemistry. Hydrobiologia 165(1):255–261
2. Ballance LT, Ainley DG, Ballard G, Barton K (2009) An energetic correlate between colony size and foraging effort in seabirds, an example of the Adélie penguin *Pygoscelis adeliae*. J Avian Biol 40(3):279–288
3. Ballard G, Dugger KM, Nur N, Ainley DG (2010) Foraging strategies of Adélie penguins: adjusting body condition to cope with environmental variability. Mar Ecol Prog Ser 405:287–302
4. Bannasch R (1986) Morphologisch-funktionelle Untersuchung am Lokomotionsapparat der Pinguine als Grundlage für ein allgemeines Bewegungsmodell des Unterwasserfluges. I. Jahrbuch für Morphologie und mikroskopische Anatomie. 1. Abteilung, Gegenbaurs morphologisches Jahrbuch 132(5):645–679
5. Bannasch R (1987) Morphologisch-funktionelle Untersuchung am Lokomotionsapparat der Pinguine als Grundlage für ein allgemeines Bewegungsmodell des «Unterwasserfluges». Jahrbuch für Morphologie und mikroskopische Anatomie. 1. Abteilung, Gegenbaurs morphologisches Jahrbuch 133(1):39–59
6. Bannasch R (1994) Functional anatomy of the 'flight'apparatus in penguins. Mech Physiol Anim Swim 99:163–192
7. Bannasch R (1995) Hydrodynamics of penguins-an experimental approach. In: The penguins; ecology and management, pp 141–176
8. Bannasch R (1997) Experimental studies on the boundary layer development in penguins: mechanisms of turbulence control and their applicability to engineering
9. Bannasch R (2013) Widerstandsverminderung bei Meeres-tieren-was und wie Ingenieure von der. Bionik: Ökologische Technik nach dem Vorbild der Natur?: 187
10. Baudinette R, Gill P (1985) The energetics of 'flying'and 'paddling'in water: locomotion in penguins and ducks. J Comp Physiol B 155(3):373–380
11. Bethge P, Nicol S, Culik B, Wilson R (1997) Diving behaviour and energetics in breeding little penguins (*Eudyptula minor*). J Zool 242(3):483–502
12. Bilo D, Nachtigall W (1980) A simple method to determine drag coefficients in aquatic animals. J Exp Biol 87(1):357–359
13. Blackburn TM, Gaston KJ, Loder N (1999) Geographic gradients in body size: a clarification of Bergmann's rule. Divers Distrib 5(4):165–174
14. Blank JM, Farwell CJ, Morrissette JM, Schallert RJ, Block BA (2007) Influence of swimming speed on metabolic rates of juvenile Pacific bluefin tuna and yellowfin tuna. Physiol Biochem Zool 80(2):167–177
15. Butler PJ (2000) Energetic costs of surface swimming and diving of birds. Physiol Biochem Zool 73(6):699–705
16. Cannell B (1994) The feeding behaviour of Little Penguins, *Eudyptula minor* (Forster). Moansh University
17. Choi H, Park H, Sagong W, Lee S, Biomimetic flow control based on morphological features of living creatures. Physics of Fluids 24(12):1–20
18. Ciancio JE, Quintana F, Sala JE, Wilson RP (2016) Cold birds under pressure: can thermal substitution ease heat loss in diving penguins? Mar Biol 163(2):43
19. Clark BD, Bemis W (1979) Kinematics of swimming of penguins at the Detroit Zoo. J Zool 188(3):411–428
20. Cofta T (2021) Flight identification of European passerines and select Landbirds: an illustrated and photographic guide. Princeton University Press
21. Culik B, Wilson R, Bannasch R (1994) Underwater swimming at low energetic cost by pygoscelid penguins. J Exp Biol 197(1):65–78
22. Culik B, Wilson RP (1991) Swimming energetics and performance of instrumented Adélie penguins (*Pygoscelis adeliae*). J Exp Biol 158(1):355–368

23. Culik BM, Pütz K, Wilson R, Allers D, Lage J, Bost C, Le Maho Y (1996) Diving energetics in king penguins (*Aptenodytes patagonicus*). J Exp Biol 199(4):973–983
24. Culik BM, Wilson RP, Dannfeld R, Adelung D, Spairani HJ, Coria NC (1991) Pygoscelid penguins in a swim canal. Polar Biol 11(4):277–282
25. Elliott KH, Ricklefs RE, Gaston AJ, Hatch SA, Speakman JR, Davoren GK (2013) High flight costs, but low dive costs, in auks support the biomechanical hypothesis for flightlessness in penguins. Proc Natl Acad Sci 110(23):9380–9384
26. Erasmus T, Randall R, Randall B (1981) Oil pollution, insulation and body temperatures in the jackass penguin *Spheniscus demersus*. Comp Biochem Physiol A Physiol 69(1):169–171
27. Green J, Butler P, Woakes A, Boyd I (2003) Energetics of diving in macaroni penguins. J Exp Biol 206(1):43–57
28. Habib M (2010) The structural mechanics and evolution of aquaflying birds. Biol J Linn Soc 99(4):687–698
29. Harada N, Oura T, Maeda M, Shen Y, Kikuchi DM, Tanaka H (2021) Kinematics and hydrodynamics analyses of swimming penguins: wing bending improves propulsion performance. J Exp Biol 224(21):jeb242140
30. Heath JP, Gilchrist HG, Ydenberg RC (2006) Regulation of stroke pattern and swim speed across a range of current velocities: diving by common eiders wintering in polynyas in the Canadian Arctic. J Exp Biol 209(20):3974–3983
31. Henderson CL (2008) Birds in flight: the art and science of how birds fly. Voyageur Press, Minneapolis
32. Herreid Ii CF, Kessel B (1967) Thermal conductance in birds and mammals. Comp Biochem Physiol 21(2):405–414
33. Hui CA (1988) Penguin swimming. I. Hydrodynamics. Physiol Zool 61(4):333–343
34. Hui CA (1988) Penguin swimming. II. Energetics AND BEHAVIOR. Physiol Zool 61(4):344–350
35. Kaiser GW (2010) The inner bird: anatomy and evolution. UBC Press
36. Kleiber M (1961) The fire of life. An introduction to animal energetics. Wiley, New York
37. Kooyman G, Cherel Y, Maho YL, Croxall J, Thorson P, Ridoux V, Kooyman C (1992) Diving behavior and energetics during foraging cycles in king penguins. Ecol Monogr 62(1):143–163
38. Kooyman GL, Lynch W (2013) Penguins: the animal answer guide. John Hopkins University Press, Baltimore
39. Kooyman GL, Ponganis PJ, Castellini M, Ponganis E, Ponganis K, Thorson P, Eckert S, Lemaho Y (1992) Heart rates and swim speeds of emperor penguins diving under sea ice. J Exp Biol 165(1):161–180
40. Lasiewski RC, Calder WA Jr (1971) A preliminary allometric analysis of respiratory variables in resting birds. Respir Physiol 11(2):152–166
41. Lovvorn J, Liggins G (2002) Interactions of body shape, body size and stroke-acceleration patterns in costs of underwater swimming by birds. Funct Ecol 16(1):106–112
42. Lovvorn J, Liggins GA, Borstad MH, Calisal SM, Mikkelsen J (2001) Hydrodynamic drag of diving birds: effects of body size, body shape and feathers at steady speeds. J Exp Biol 204(9):1547–1557
43. Lovvorn JR (2001) Upstroke thrust, drag effects, and stroke-glide cycles in wing-propelled swimming by birds. Am Zool 41(2):154–165
44. Lovvorn JR, Jones DR (1991) Effects of body size, body fat, and change in pressure with depth on buoyancy and costs of diving in ducks (*Aythya* spp.). Can J Zool 69(11):2879–2887
45. Luna-Jorquera G, Culik BM (2000) Metabolic rates of swimming Humboldt penguins. Mar Ecol Prog Ser 203:301–309
46. Mcnab BK (1971) On the ecological significance of Bergmann's rule. Ecology 52(5):845–854
47. Metwally S, Comesaña SM, Zarzyka M, Szewczyk PK, Karbowniczek JE, Stachewicz U (2019) Thermal insulation design bioinspired by microstructure study of penguin feather and polar bear hair. Acta Biomater 91:270–283
48. Nachtigall W (1998) Starlings and starling models in wind tunnels. J Avian Biol 29:478–484

49. Nachtigall W, Bilo D (1980) Strömungsanpassung des Pinguins beim schwimmen unter Wasser. J Comp Physiol 137(1):17–26
50. Nagy KA, Siegfried WR, Wilson RP (1984) Energy utilization by free-ranging jackass penguins, *Spheniscus demersus*. Ecology 65(5):1648–1655
51. Nares CGS (1876) Report on the scientific results of the voyage of H. M. S. challenger during the years 1873–76 under the Commnand of Captain George S. Nares and the late Captain Frank Tourle Thomson. Edinburgh, Scotland
52. Norberg RA (1981) Optimal flight speed in birds when feeding young. J Anim Ecol 50:473–477
53. Parfitt AR, Vincent JF (2005) Drag reduction in a swimming humboldt penguin, *Spheniscus humboldti*, when the boundary layer is turbulent. J Bionic Eng 2(2):57–62
54. Pennycuick C, Klaassen M, Kvist A, Lindström Å (1996) Wingbeat frequency and the body drag anomaly: wind-tunnel observations on a thrush nightingale (*Luscinia luscinia*) and a teal (Anas crecca). J Exp Biol 199(12):2757–2765
55. Pennycuick CJ (1968) A wind-tunnel study of gliding flight in the pigeon Columba livia. J Exp Biol 49(3):509–526
56. Pennycuick CJ (1989) Bird flight performance: a practical calculation manual. Oxford University Press, Oxford
57. Pennycuick CJ (2008) Modelling the fying bird. Elsevier, London
58. Pettingill ER (1960) Penguin Summer. Clarkson N. Potter Inc. New York
59. Ponganis P, St Leger J, Scadeng M (2015) Penguin lungs and air sacs: implications for baroprotection, oxygen stores and buoyancy. J Exp Biol 218(5):720–730
60. Ropert-Coudert Y, Kato A, Baudat J, Bost C-A, Le Maho Y, Naito Y (2001) Time/depth usage of Adélie penguins: an approach based on dive angles. Polar Biol 24(6):467–470
61. Ropert-Coudert Y, Kato A, Bost C-A, Rodary D, Sato K, Le Maho Y, Naito Y (2002) Do Adélie penguins modify their foraging behaviour in pursuit of different prey? Mar Biol 140(3):647–652
62. Ropert-Coudert Y, Kato A, Sato K, Naito Y, Baudat J, Bost A, Le Maho Y (2002) Swim speed of free-ranging Adélie penguins *Pygoscelis adeliae* and its relation to the maximum depth of dives. J Avian Biol 33(1):94–99
63. Sato K, Charrassin J-BT, Bost C-A, Naito Y (2004) Why do macaroni penguins choose shallow body angles that result in longer descent and ascent durations? J Exp Biol 207(23):4057–4065
64. Sato K, Naito Y, Kato A, Niizuma Y, Watanuki Y, Charrassin J, Bost C-A, Handrich Y, Le Maho Y (2002) Buoyancy and maximal diving depth in penguins: do they control inhaling air volume? J Exp Biol 205(9):1189–1197
65. Sato K, Ponganis PJ, Habara Y, Naito Y (2005) Emperor penguins adjust swim speed according to the above-water height of ice holes through which they exit. J Exp Biol 208(13):2549–2554
66. Sato K, Shiomi K, Watanabe Y, Watanuki Y, Takahashi A, Ponganis PJ (2010) Scaling of swim speed and stroke frequency in geometrically similar penguins: they swim optimally to minimize cost of transport. Proc R Soc B Biol Sci 277(1682):707–714
67. Sato K, Watanuki Y, Takahashi A, Miller PJ, Tanaka H, Kawabe R, Ponganis PJ, Handrich Y, Akamatsu T, Watanabe Y (2007) Stroke frequency, but not swimming speed, is related to body size in free-ranging seabirds, pinnipeds and cetaceans. Proc R Soc B Biol Sci 274(1609):471–477
68. Schmidt-Nielsen K (1972) Locomotion: energy cost of swimming, flying, and running. Science 177(4045):222–228
69. Schroeder KL, Sylvain NJ, Kirkpatrick LJ, Rosser BW (2015) Fibre types in primary 'flight'muscles of the African Penguin (*Spheniscus demersus*). Acta Zool 96(4):510–518
70. Sepulveda C, Dickson KA (2000) Maximum sustainable speeds and cost of swimming in juvenile kawakawa tuna (*Euthynnus affinis*) and chub mackerel (*Scomber japonicus*). J Exp Biol 203(20):3089–3101
71. Shepard EL, Wilson RP, Quintana F, Gomez Laich A, Forman DW (2009) Pushed for time or saving on fuel: fine-scale energy budgets shed light on currencies in a diving bird. Proc R Soc B Biol Sci 276(1670):3149–3155

72. Shiomi K, Sato K, Handrich Y, Bost CA (2016) Diel shift of king penguin swim speeds in relation to light intensity changes. Mar Ecol Prog Ser 561:233–243
73. Stahel C, Nicol S (1982) Temperature regulation in the little penguin, *Eudyptula minor*, in air and water. J Comp Physiol 148(1):93–100
74. Takahashi A, Costa DP, Bost CA, Trathan PN (2007) Stroke frequency, but not swimming speed, is related to body size in free-ranging seabirds, pinnipeds and cetaceans. Proc R Soc B 274:471–477
75. Van Dam R, Ponganis P, Ponganis K, Levenson D, Marshall G (2002) Stroke frequencies of emperor penguins diving under sea ice. J Exp Biol 205(24):3769–3774
76. Watanabe Y, Baranov EA, Sato K, Naito Y, Miyazaki N (2006) Body density affects stroke patterns in Baikal seals. J Exp Biol 209(17):3269–3280
77. Watanabe YY, Sato K, Watanuki Y, Takahashi A, Mitani Y, Amano M, Aoki K, Narazaki T, Iwata T, Minamikawa S (2011) Scaling of swim speed in breath-hold divers. J Anim Ecol 80(1):57–68
78. Watanuki Y, Burger AE (1999) Body mass and dive duration in alcids and penguins. Can J Zool 77(11):1838–1842
79. Watanuki Y, Niizuma Y, Geir WG, Sato K, Naito Y (2003) Stroke and glide of wing–propelled divers: deep diving seabirds adjust surge frequency to buoyancy change with depth. Proc R Soc Lond Ser B Biol Sci 270(1514):483–488
80. Watanuki Y, Wanless S, Harris M, Lovvorn JR, Miyazaki M, Tanaka H, Sato K (2006) Swim speeds and stroke patterns in wing-propelled divers: a comparison among alcids and a penguin. J Exp Biol 209(7):1217–1230
81. Williams CL, Czapanskiy MF, John JS, St Leger J, Scadeng M, Ponganis PJ (2021) Cervical air sac oxygen profiles in diving emperor penguins: parabronchial ventilation and the respiratory oxygen store. J Exp Biol 224(1):jeb230219
82. Williams CL, Meir JU, Ponganis PJ (2011) What triggers the aerobic dive limit? Patterns of muscle oxygen depletion during dives of emperor penguins. J Exp Biol 214(11):1802–1812
83. Wilson R (1985) The Jackass Penguin (*Spheniscus demersus*) as a pelagic predator. Mar Ecol Prog Ser 25:219–227
84. Wilson R, Culik B, Peters G, Bannasch R (1996) Diving behaviour of Gentoo penguins, *Pygoscelis papua*; factors keeping dive profiles in shape. Mar Biol 126(1):153–162
85. Wilson RP, Bain CA (1984) An inexpensive speed meter for penguins at sea. J Wildl Manag 48:1360–1364
86. Wilson RP, Hustler K, Ryan PG, Burger AE, Noldeke EC (1992) Diving birds in cold water: do Archimedes and Boyle determine energetic costs? Am Nat 140(2):179–200
87. Wilson RP, Wilson M-PT (1988) Foraging behaviour in four sympatric cormorants. J Anim Ecol 57:943–955
88. Wilson RP, Zimmer I (2004) Inspiration by Magellanic penguins: reduced swimming effort when under pressure. Mar Ecol Prog Ser 278:303–307
89. Yuen HS (1966) Swimming speeds of yellowfin and skipjack tuna. Trans Am Fish Soc 95(2):203–209

Chapter 7
Hot Penguins: Cold Water

Penguins, like all birds, are "hot-blooded" ("homeothermic" is the scientific term), and this makes their internal chemistry "cook" fast, i.e., their physiological rates are rapid, requiring elevated energy compared to other creatures. In turn, they possess extraordinary athletic powers (speed, acceleration, maneuverability, etc.), which also require energy. Importantly, their hot-bloodedness allows them to exist in relatively cold water, where food is more available than in warm waters, i.e., polar vs. tropical (Chap. 3). It is an irony: they need plenty of food just to exist in cold water, but it is in cold water where they find sufficient food for their high-energy lifestyle (Chaps. 3 and 6). Incidentally, this also applies to the penguin species that live in "hot" places – most notably the Galápagos, Humboldt, and African penguins – because they exploit the cold, coastal waters that upwell from the deep (Chap. 3). Being supercharged as penguins are, having high body-operating temperatures also allows them to outperform their "cold-blooded" (ectothermic) prey (Chaps. 4 and 5), which have to make do with capacities primarily dictated by the temperature of the water – the colder it gets, the slower they are (cf. [9]). However, there is a severe constraint to being supercharged, especially for hot-blooded animals operating in cold water. Not only do they need to capture a large amount of high-energy-density food but they also experience a form of "heat tax" – the loss of heat energy to the water– and this becomes more severe as the sea temperatures drop.

The whole "heat taxation" process is dictated, at the least, by the properties of water and the physics of heat transfer. Heat energy always moves from places (or bodies) of high temperature to those of lower temperature (causing the temperature of the hotter body to decrease), and it does so at a rate that depends on the temperature difference between the two entities. In other words, relatively speaking, the hotter the penguin and the colder the water, the more the bird needs to be concerned about "heat tax." To some degree, the large majority of all bird species have to deal with this, since birds have higher metabolic rates (MRs) and body temperatures than do mammals (and certainly higher than those of reptiles and amphibians), and air is generally cooler than most birds, especially when flying at an altitude. However,

D. G. Ainley, R. P. Wilson, *The Aquatic World of Penguins*, Fascinating Life Sciences, https://doi.org/10.1007/978-3-031-33990-5_7

water is also a more demanding heat taskmaster than air because it has a frighteningly high "volumetric heat capacity" – i.e., the heat energy required to cause an increase in temperature of a substance of one unit volume, e.g., a cubic centimeter, by one degree (units are J/°C.mL). In fact, it takes 3200 times more energy to heat a given volume of water by one degree than the same volume of air. Water also requires more energy to heat per gram than almost any other substance (the specific heat capacity (SHC) of water is 4.18 J/°C.g). In practical penguin terms, this means that a stationary layer of cold water surrounding a penguin (or a human) heats up extremely slowly compared to air. As such, the temperature gradient in water is extremely steep and the rate of heat transfer from a penguin (or a person) to the water is high. This is also the reason why people in water at, say, 10 °C, find it so much colder than air at the same temperature. As an aside, this is also why the ocean, rather than the atmosphere, has absorbed 90% of the excess heat that humans have generated over the recent millennia, with repercussions for penguins (see Chap. 13).

In this chapter, we first will consider heat generation and retention among stationary or floating penguins. We will then consider these subjects as they change with active, diving penguins.

Resting or Floating Penguins

Irrespective of whether they like it or not, penguins continuously lose heat to both water and air simultaneously when they float on the surface (sleeping, preening, resting, etc.) with their upper surface exposed to the air and their lower surface underwater, and this affects their heat loss correspondingly. What is notable about penguins is that they sit lower in the water than virtually any other seabird (see cormorants [26, 70]) and, therefore, have a particularly large part of their body surface area exposed to the water with its high heat tax (Fig. 7.1). The low profile of penguins comes about because air, which generates the buoyancy for most seabirds to sit "high and dry" on the water, exerts a high energy cost for the diving species as, underwater, they have to work harder against this upthrust (Chap. 6). It is a case of "win on the roundabout and lose on the swings"! The highest-latitude penguins, such as emperors and Adélies (Chap. 3), at least in winter, find that it is warmer to be in water (>1.7 °C) than in air (many degrees below zero). Nevertheless, they cannot escape the "heat tax."

However, just like with people dealing with real taxmen, penguin bodies are strategically built to help minimize the rate at which they pay both their "heat tax" and their "underwater buoyancy tax." Given that air trapped by the feathers is a major insulator against heat loss in penguins (air is an excellent insulator because its molecules resist heat transfer owing to their rarity), it should theoretically benefit penguins to have thicker feather "coats" for those parts of their bodies that are

Fig. 7.1 A seabird resting on the sea surface loses heat at a rate that depends on the temperature of the water and the proportion of its body that is underwater – the greater the proportion of the surface area underwater, the greater the heat loss. This figure illustrates the huge difference between a penguin and a gull in that respect. Upper photographs: left, an erect-crested penguin. (Courtesy of Peter Ryan) and, right, a Sabine's gull. (Photo courtesy of Ómar Runólfsson, Wikipedia)

mostly underwater (especially their underside). They should have thinner coats for those body parts that spend more time in air – the head, neck, and back. In this way, their overall buoyancy underwater is not overly important.

In fact, this is exactly what our measurements with the featherometer (see Chap. 6 for a description of how this works) show, for example, the air layer thickness on the breast is almost twice that of the back (Fig. 7.2).

Although the difference in heat loss from the upper and lower body surfaces of floating penguins is important, it is hard to separate the contribution of the two elements to the overall heat loss experimentally or otherwise. Instead, researchers who have studied heat loss in penguins have simply dealt with the energy expenditure of "whole-body" floating penguins, and this is what we shall consider here. These studies also deal with metabolic rate, i.e., the heat-generating capacity of a penguin.

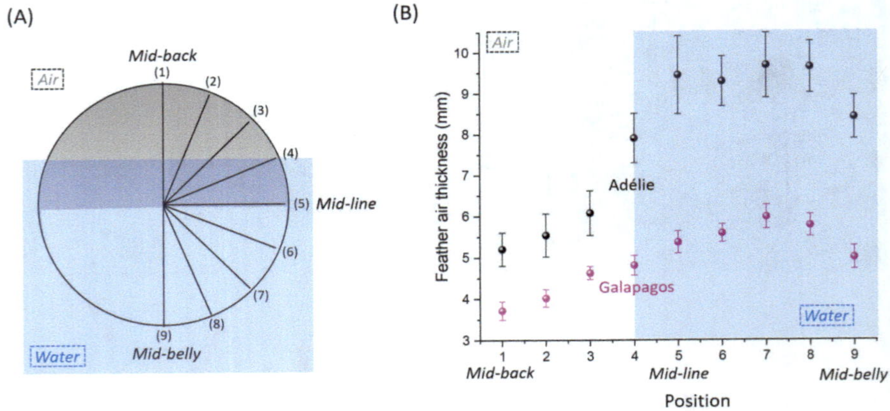

Fig. 7.2 (**a**) A schematic diagram of a cross section of a penguin at the point of its widest girth, indicating the nine points where feather air layer thickness measurements were made with a "feath-erometer" (see Box 6.3, Chap. 6). (**b**) The mean air layer thicknesses (bars show standard error (SE)) at the nine points indicated in panel (**a**) from measurements on 18 Adélie and 28 Galápagos penguin carcasses. The parts of the body that are underwater during normal surface floating are indicated by the blue box. Note how these parts also have the thickest air layers, which help reduce heat loss in cold water and that the Adélie penguins have noticeably thicker feathers than do the Galápagos penguins across the full cross section due to the Adélies living in much colder climes (see also below)

The Metabolic Rate of Floating Penguins

The best information that we have on the costs of maintaining body temperature in floating, inactive penguins comes from gas respirometry studies – in which penguins are placed in water in a chamber while their oxygen consumption and carbon dioxide production are measured (see Chap. 6). Simple conversions of gas-use data can indicate how much it costs a penguin to float in water in terms of energy lost as heat, and, for the really dogged researchers, this can be carried out at different temperatures. In this, in the typical manner of much game-changing biological work conducted in the 1970s and 1980s, Stahel and Nicol [63] published their seminal work on the metabolic rate of Australian little penguins resting both in air and floating on water at temperatures ranging from about 0 °C to 35 °C (Fig. 7.3). The clarity and attention with which this work was conducted means that the results can be used to describe trends in how penguins are generally expected to deal with varying temperatures in water and air, although, of course, there will be interspecific differences in the details (see later).

This work showed the huge difference that the environment – water or air– makes in affecting the resting metabolic rates of penguins. In short, the rate at which little penguins used energy (primarily to keep warm since they were resting) was much the same for both air and water at around 33 °C. However, for temperatures below 33 °C, the allocation of energy to "heat tax" increased massively and

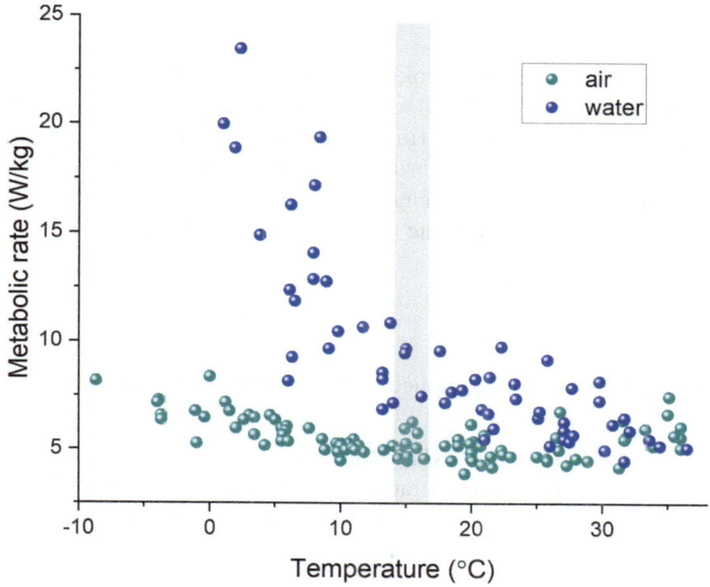

Fig. 7.3 The metabolic rate of little penguins resting in air and floating on water at different temperatures. (Data are transcribed from [63]). Note the huge difference in energy expenditure for birds in water and for those in air at the same temperature. The normal sea temperature range experienced by these birds [53] is indicated by the gray bar

disproportionately as the temperature dropped when the penguins were in water. It should be noted that both species of little penguins inhabit subtropical waters (Chap. 3), where water temperatures, at least on the surface, generally range from 13 °C to 22 °C (cooler at depth), with air temperatures often much higher. Regardless, their high heat tax is due to the high volumetric heat capacity of water compared to that of air (see above) – water effectively "sucks" heat energy out of the penguins much faster than does air and, to keep body temperatures hot and constant especially when in water, penguins have to generate that energy by increasing their metabolic rate. The rub, though, as the equation in Box 7.1 shows, critically depends on the difference in temperature between the penguins and the water.

Accordingly, we have replotted the Stahel and Nicol data to show how the difference in environmental (air or water) temperature and little penguin core body temperature affects the metabolic rate (Fig. 7.4).

As Stahel and Nicol point out, what is shown in this figure indicates that little penguins in air have a thermoneutral zone- where temperature changes make no difference to their metabolic rate- that ranges from about 10 °C to 30 °C. It should be noted that Fig. 7.4 shows that there is no substantial change in the position of the green circles within this range. Above and below that, the metabolic rate climbs as the birds use energy to either keep cool or keep warm. This is in stark contrast to the same birds in water that have to increase their metabolic rate steadily (but

Box 7.1
The rate at which heat is transferred from a hot penguin to cold water ($\Delta E/dt$ – in J/s or W) depends upon the (i) temperature of the penguin ($T_{penguin}$), (ii) the temperature of the water (T_{water}), (iii) the heat transfer coefficient of the insulation between the penguin's hot insides and the environment (h – in W/°C/m^2), (iv) the surface area over the penguin ($A_{penguin}$), and (v) the thickness of the insulator (d). Specifically, the rate of energy loss is given by:

$$\frac{\Delta E}{dt} = \left[h.A_{\text{penguin}} \cdot \left(T_{\text{penguin}} - T_{\text{water}} \right) \right] / d$$

A similar process can be applied to penguins resting in air.

substantially) from a minimum, where the water temperature is just a few degrees cooler than they are, up to a temperature difference of about 30 °C. At that point, their energy costs soar. We consider, at this point (which corresponds to the sea

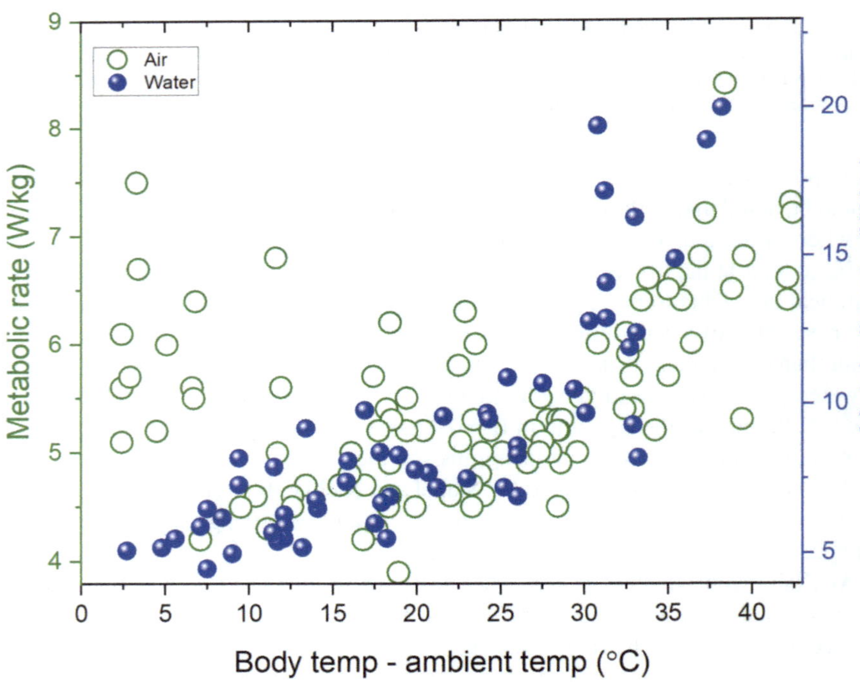

Fig. 7.4 The metabolic rate of Australian little penguins resting in air (green circles) and floating on water (blue spheres) at different temperatures. (Data are transcribed from [63] and reanalyzed) as a function of the difference in temperature between the penguins and their environment (the penguin mean body temperatures were 38.4 °C and 39.2 °C for air and water, respectively)

temperature being around 9 °C), that the birds were no longer able to deal with the low temperatures at all. Tellingly, Stahel and Nicol also noted that penguin body temperatures dropped at this time. The normal operating sea temperatures for little penguins are actually not much higher than this, at about 15–17 °C [53], indicating that these birds operate close to their physiological "edge." Overall, the data also make it clear that it is impossible for little penguins to rest in the water using metabolic rates comparable to those in air (Fig. 7.4).

This book almost exclusively deals with the aquatic lives of penguins, and it seems relevant that the Stahel and Nicol data mentioned above imply that the heat flux and energy expenditure of penguins on land are trivial by comparison to the at-sea situation. In fact, this is an oversimplification because land-bound penguins can get cold, notably during molt (see Chap. 2) when all their feathers simultaneously drop. In this situation, they lose heat more easily than the properly "clothed" versions – and have to have higher energy expenditure to mitigate that as a consequence [25]. Conversely, penguins can get too hot on land, most particularly the Galápagos, Humboldt, and African penguins, living, as they do in, sometimes, searingly hot places with high sun intensities [20, 62]. This can cause higher metabolic rates as birds try to lose heat, at least until they decide to abandon land and pop into the ocean to cool off (actually not a good idea for birds brooding or shading small chicks). Interestingly, the same situation can occur in any penguin species because, as we will show below, all species are insulated only for their own conditions of nominal water and air temperatures and, as such, they can all get too hot on land under the right conditions.

Patterns of Heat Loss to the Sea

The great work on little penguins was possible, in part, because these birds are small, and it is obviously easier to manipulate smaller animals, including varying their environmental temperatures, than large ones. However, there is more to being a "little" penguin than that. It is simplistic to talk about environmental temperatures and insulation in hot penguins without taking allometry – the consequences of animal size- into account. In essence, the volume of the penguin body (which is proportional to the mass) is the powerhouse, the size of the "engine" producing heat, whereas the surface area of that body is what loses it. When exploring this, we will adopt a standard protocol for estimating the surface area of a penguin based on its mass (Box 7.2).

This approach precisely shows how the relative amount of surface area decreases with increasing body mass (Fig. 7.5).

Thus, bigger penguins should be better equipped to deal with cold water than the smaller ones- they have much lower surface areas compared to their mass (see the red line in Fig. 7.5). To put this into a quantitative context, we can plot the overall metabolic rate of different penguin species (resting in water that approximates the temperatures to which they are normally exposed), which range in size (Table 1.2,

Box 7.2

Penguins resting on the sea surface have an extremely different body shape (rather duck-like) compared to those swimming underwater (whose body shape is closer to a torpedo (see Chap. 6)). Therefore, for floating penguins, it seems appropriate to use a body surface area approximation that stems from "standard" birds. Stahel and Nicol used a formula for surface area of:

$$A = k.M^{0.667}$$

where M is the mass of the bird (in kilograms), A is measured in meter squared, and k is a constant that describes the bird body shape. In this, $k = 0.006$ for a cube and $k = 0.048$ for a sphere (Schmidt-Nielsen (1984), cited in [17]). The latter suggested using $k = 0.1$ for penguins, although precisely which value of k is used makes an appreciable difference [42]. For simplicity, we adopt Drent and Stonehouse's suggestion for our mass-related comparisons.

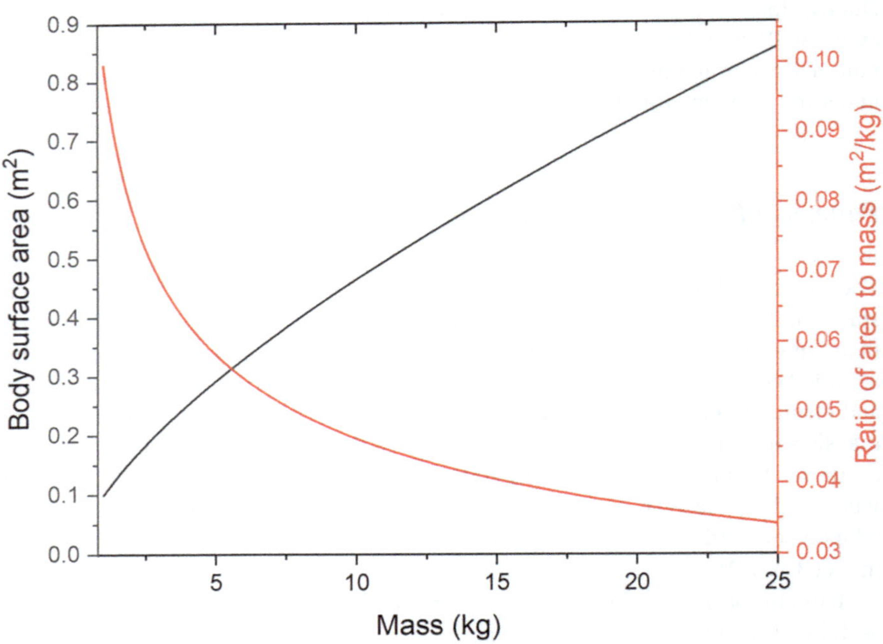

Fig. 7.5 The theoretical relationship between penguin body mass and the surface area of the body (for details, see Box 7.2) showing how bigger species have a proportionately less surface area

Fig. 1.7). against their surface area (Fig. 7.6). Interestingly, and against what we might expect, this shows that the energy lost as heat per unit surface area is remarkably consistent across the different-sized species operating in their different sea temperature regimes (Table 1.2). Specifically, this is because the slope of the graph shown in Fig. 7.6 is constant, indicating a surface area-specific metabolic rate of about 108 W/m^2. Thus, penguins of all sizes and operating in waters of different temperatures all seem to lose the same amount of heat per unit body surface area. This is pretty remarkable!

This consistency between the different penguin species seems surprising because, superficially, we might expect there to be substantial interspecific differences, depending on their size and the temperature of the water in which they live. Thus, there would be a connection between body size and a penguins' geographic range. There actually is a connection to a slight degree for the extremes of body size and

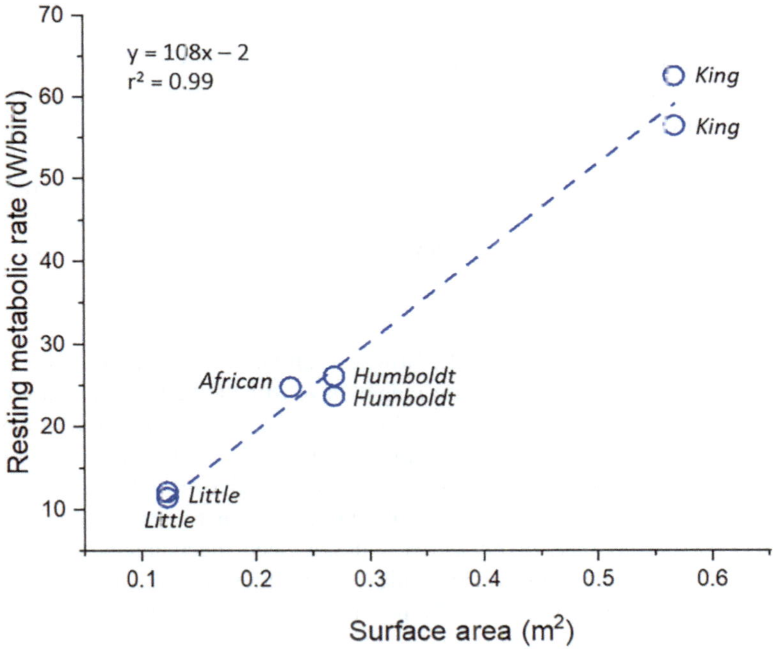

Fig. 7.6 The overall metabolic rate of four different species of penguins (for data, see Table 7.1) resting in water (at temperatures close to those experienced in the wild) as a function of their estimated surface area (see Box 7.2 for the method and using the mass data provided in Table 1.2). Although there are other estimates of the "resting metabolic rate" of penguins in water (see Table 7.1), the values presented here only refer to those in which the birds were resting for prolonged periods and not those in which the metabolic rate of "not swimming" was calculated for penguins in a swim channel. These latter birds generally only "rested" for short periods between active swimming bouts and so would have had higher metabolic rates anyway (in a manner analogous to runners who stop but continue panting for some time afterward). The apparent linear relationship indicates that, irrespective of their masses, all the species in the relationship lose roughly the same amount of heat per unit of body surface area (see line equation)

Table 7.1 The resting metabolic rate (MR) of penguin in water and air

		Water	Air	
Species	Temperature/notes	MR (W/kg)	MR (W/kg)	Source
Little penguins	10 °C	8.5		[4]
Little penguins	15 °C	≈9		[63]
Little penguins	At thermoneutrality		3.26	[4]
Little penguins	At thermoneutrality		4.93	[63]
African penguins	≈13 °C	7.1		[44]
African penguins	At thermoneutrality		1.42	[44]
King penguins	In air		3.3	[35]
King penguins	5 °C	4.2		[35]
King penguins	5 °C (sleeping)	3		[35]
King penguins	At thermoneutrality		2.15	[35]
King penguins	9 °C	4.65	3.3	[35]
Humboldt penguins	19 °C	5.95		[41]
Humboldt penguins	18 °C	5.4		[8]
Humboldt penguins	At thermoneutrality		3.8	[42]
Humboldt penguins	At thermoneutrality		2.83	[17]
Adélie penguins	4 °C	8.4		[12]
Adélie penguins	2–6 °C	8.4		[13]
Chinstrap penguins	4 °C	8.75		[12]
Gentoo penguins	4 °C	8.2		[12]
Macaroni penguins	At thermoneutrality		6.27	[23, 24]
Emperor penguins	1.5–6.1 °C	2.08–2.24		[38]

range (Chaps. 1, 2 and 3, Table 1.2). However, all animals maintaining body temperatures within narrow temperature ranges have to balance heat generation with heat loss, wherever they live – and well-insulated animals living in cold environments can be hot too. Thus, while hot-blooded animals living in cold water might be expected to have highly efficient insulation to minimize heat loss, the situation changes when they are active. Since heat generation constitutes about 80% of the cost of muscular activity, highly active animals need mechanisms to be able to release this heat. In penguins, much of this can be achieved by enhanced blood flow to the flippers and feet [68] (Fig. 7.7), but it may not be enough when their metabolic rate is high.

We suggest that the consistent rate of energy loss per unit surface area across species accords with a heat conductance value in water that caters to overheating but also comes with a cost when birds rest at sea. Put in another way, penguins may be equipped with a "body suit" (a feather coat) that is a compromise between one that works for both lethargy on the sea surface and huge athleticism underwater. However, the consistent rate of heat loss per unit body surface area also means that the different species have different body insulations because they operate in water at different temperatures (see below).

Fig. 7.7 An Adélie penguin, fresh from the sea, engaged in the ritual of preening that immediately follows on arrival to a beach [1]. In part, this activity restores air to their plumage. The pink tone to the undersides of its flippers is due to the increased flow of warm blood to flipper capillaries as a means to cool the bird from its exertion. (Photo courtesy of D. Ainley)

Overall Body Insulation/Conductance

The overall similarity in the rate at which heat is lost per unit surface area of penguins hides the reduced mass-specific metabolic rates in larger species [64] and does not take into account the sea temperatures to which the different species are exposed. Based on the expectation that the rate of heat loss critically depends on the difference in temperature between the penguin and the water (Box 7.1), we would expect penguins operating in colder waters to have better insulation, which is why so many authors are interested in "conductance" (the opposite of insulation). Such a metric specifically details how much heat energy is lost per unit body surface area and per degree temperature difference between the penguin and the sea. This heat energy loss depends on the properties of a penguin's surface, i e., the thickness of the insulating air layer as well as the insulating properties of the penguin skin and any subcutaneous fat. Conductance, then, may be the factor behind the difference in geographic range, as it relates to water temperature, for the array of penguin species mentioned above (see Chaps. 1 and 3).

The values for the whole-body thermal conductance of penguins in the literature are bewildering. This is partly because various authors use different relationships between penguin body mass and surface area. However, values for penguins are also compared, within and between various reports, both in air and water with, sometimes, little apparent appreciation for the fact that, at any given temperature, these two media are likely to have wildly different thermal conductance metrics (see above for the metabolic rate consequences of this). Thus, Drent and Stonehouse [17] calculated the mean conductance (below the thermoneutral zone) of 1.93 W/m^2/°C (standard deviation (SD) 0.15, $n = 37$) for a Humboldt penguin, a value that is less than the 2.6 W/m^2/°C obtained by others [42] for the same species. Subsequently, Stahel and Nicol [63], working with the smallest penguin, obtained a surprisingly similar value of 1.72 W/m^2/°C. In contrast, a suite of authors obtained between 1.31 and 1.56 W/m^2/°C for the largest penguin, the emperors [39, 46]. Values of 1.8 W/m^2/°C were quoted for midsized macaroni penguins [3]. In a theoretical treatise on heat loss, Dawson et al. [16] calculated a thermal conductivity of 2.38 W/m^2/°C, a value that they said accorded closely with a measured value of 1.93 W/m^2/°C.

In an attempt to normalize things, we assumed that virtually all of the energy expenditure of penguins resting in water could be allocated to heat production [63]. This was before examining the published data on the metabolic rate of various penguin species resting in water (Fig. 7.6). We then standardized the calculation of species-specific surface areas using the method detailed in Box 7.2. This was before combining this information with data on the experimental water temperatures (so that the difference between the penguin body and water temperature could be incorporated) to estimate the standardized thermal conductance. This approach indicated that the thermal conductance of the outer layer of various penguin species matches the water temperature very well (Fig. 7.8). This ties in with much of what is reported in the literature, including the older papers in which conductance was measured across penguin pelts [37].

Specifically, the lower the water temperatures experienced by penguins in the wild, the better their insulation (the lower their conductance). This is to be expected because hot-blooded animals in cold environments need to be better protected. It also ties in with the fact that, irrespective of mass, penguins have a constant rate of heat loss per unit surface area (Fig. 7.6).

When performing our standardized calculations, we found no obvious relationship between penguin mass and thermal conductance (contrary to some accounts [42]), presumably because the temperature effects of cold water are more important than the mass-specific metabolic rate in the range of sizes displayed by extant penguins (Table 1.2). However, assuming that our regression of conductance relative to water temperature for resting penguins holds across penguin species (shown in Fig. 7.8), we can estimate the resting metabolic rate of all penguin species in the water temperatures that they usually inhabit (Box 7.3).

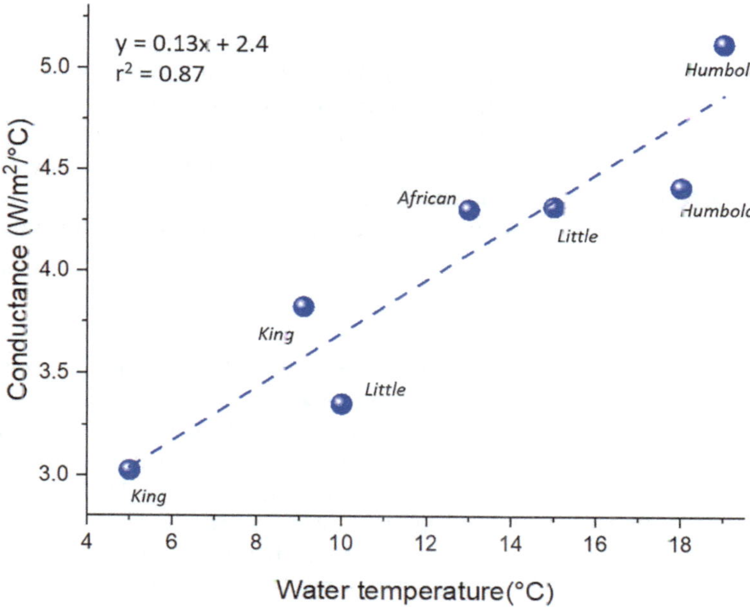

Fig. 7.8 The calculated thermal conductance of the plumage, skin, and subcutaneous fat layers of various penguin species in relation to the experimental water temperatures when they matched those experienced by these birds in the wild

Box 7.3

Estimation of the resting metabolic rates of penguins in water: For this, we assume that the relationship between the thermal conductance of penguin plumage and the approximately "normal" sea temperatures for those species holds across all penguin species and can be described by our linear fit (Fig. 7.7). We take penguin species' mass estimates as the midpoint of the minima and maxima presented in Table 1.2 (Chap. 1) and Meeh's formula for the relationship between penguin surface area and mass as described in Box 7.2 to estimate the surface areas of the different species. We can use these relationships to estimate the "standard conductance" of each species for the "standard sea temperature" to which they are exposed by using the midpoint of the temperature ranges provided in Table 1.2. These "standard conductance" values (in $W/m^2.K$) provide a standardized figure for the insulating properties of the various penguin body surfaces under their species-specific standard temperature conditions. These can be scaled up to per bird "standard conductance" values (in *W/bird.K*) by multiplying the surface area-specific

(continued)

values by the species-specific surface areas. To derive the metabolic rate across the full temperature range experienced by each species (provided in Chap. 1, Table 1.2), we subtract the appropriate sea temperatures from 38 °C (taken to be the standard body temperature of penguins resting in water [23, 37, 63]) and multiply this by the conductance per bird per degree difference between body temperature and water temperature. Finally, this can be divided by the species masses to derive mass-specific metabolic rate as a function of temperature.

Our approach indicates that penguin resting metabolic rate (their generation of heat) in water changes linearly with water temperature, which is exactly what has been shown, as early as 1976, for Adélie penguins [37]. We note that larger penguins tend to have lower resting metabolic rates in water and that these species also seem to frequent generally lower sea temperatures anyway (Fig. 7.9, Table 1.2).

Importantly, the species having the highest mass-specific metabolic rates are most sensitive to temperature change (provided by the steeper gradients of the lines in Fig. 7.9). Given that climate change will bring about shifts in sea temperatures, it is likely that the penguins' response will be related to their overall energy expenditure at sea (Chap. 13). Relevant to this are three factors: the extent to which the resting metabolic rate in water is higher than that in air (see above), the percentage of the year that the different species spend at sea (Chap. 2), and how much of this time is spent at different water temperatures across the annual cycle (see Chap. 3 for penguin ranges). Within this, the literature is clear that, across species and oceanic environments, penguins resting at sea have metabolic rates in water that are typically two to three times than those in air at similar temperatures (Table 7.1), despite a report indicating an increase of 3.5 times for Adélie penguins [37].

So, the energetic costs required in the marine life of penguins are high, presumably even when birds are sleeping at sea, and this affects the balance of energetics according to whether penguins return to their colonies between foraging periods or not. This applies to penguins on a daily basis but also likely determines whether or not they disperse or migrate widely outside the breeding season. The warmer-water species (little, Galápagos, Humboldt, African, yellow-eyed, and, to some degree, gentoo penguins) do not disperse; the remaining species disperse (even migrate to some degree) widely, especially the polar penguins (Chap. 3). We know that, for penguins, the metabolic cost of resting on land is much less than that of resting at sea. Therefore, if the distance between foraging sites and land is small enough, then it is (energetically) worthwhile returning to the colony to rest, despite the heightened travel costs [41] (Fig. 7.10). However, the critical distance at which penguins would do better to stay at sea rather than return to land depends on water temperature (as well as predation avoidance; Chap. 12), with lower water temperatures making the return to land more profitable for any given distance between foraging and

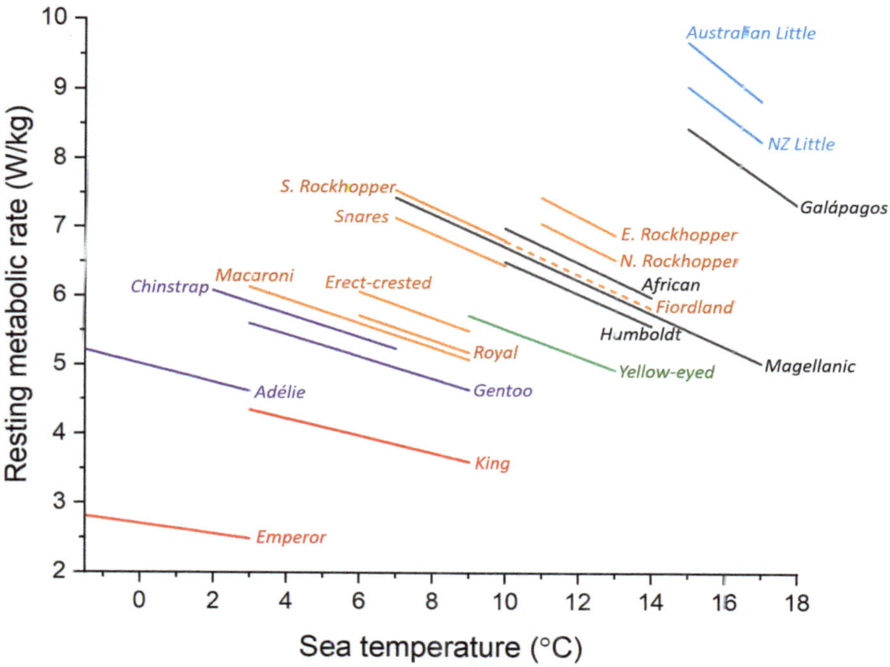

Fig. 7.9 The estimated resting metabolic rate of penguins in water in relation to the different sea temperatures to which each species is exposed (see Box 7.3 for details). The different genera are color coded to highlight trends; the Fiordland and southern rockhopper lines are differentiated by dashed and continuous lines, respectively. Note how the different genera tend to occupy specific spaces and that the species inhabiting warmer waters show a greater change in metabolic rate with temperature variation than the colder-water species (see Table 1.2)

sleeping/roosting areas (Fig. 7.10). Although all breeding birds have to return to land to provision their chicks, warmer distant waters with plenty of high-energy prey (Appendix 4) make a compelling case for a mobile lifestyle with reduced energy expenditure, even when the costs of traveling are subtracted. For example, Magellanic penguins breeding along the coast of Argentina are exposed to water temperatures between about 6 and 14 °C [10] but, after breeding, many move north to inhabit waters as far as those off Brazil [15], where sea temperatures can be as high as 17 °C (Table 1.2). We assume that Magellanic penguins have plumage conductance that can be predicted by the equation in Fig. 7.6 for a "normal" water temperature of, say, 10 °C. Thus, a simple calculation predicts that they would have at-sea resting metabolic rates that vary between 30.7 W/bird compared to just 20.8 W/bird in the in the southern and northern portions of their annual range, respectively.

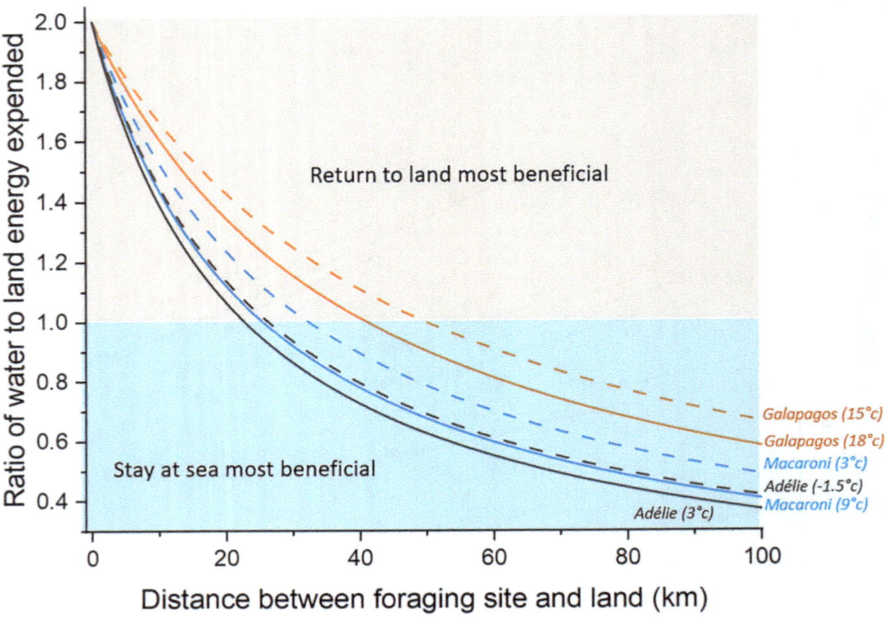

Fig. 7.10 The energetic consequences for nonbreeding penguins at different water temperatures according to whether they return to land or stay at sea daily as a function of distance between the foraging site and land. These data are modeled for three species living in waters of quite different temperatures (Fig. 7.9). For convenience, we assume (i) a 24-h foraging rhythm, (ii) that the at-sea metabolic rate as a function of temperature is shown in Fig. 7.9, (iii) that the land metabolic rate is half that (cf. Table 7.1), and (iv) that swim speeds and costs are 2 m/s and 14 W/kg, respectively. When the ratio of energy expended (energy expended by staying at sea/energy expended by returning to land) is greater than 1.0, the penguins benefit more by returning to land at night, whereas values less than 1.0 indicate that it is energetically better to stay at sea. Although only considered for a 24-h cycle here, the same principles apply for breeding versus nonbreeding birds, with the distance between lucrative foraging sites and land influencing whether penguins outside the breeding season should be migratory or not. Among the species presented, the Galápagos penguins are the only ones that show no postbreeding dispersal (Chap. 3)

The Nature of Penguin Insulation

The main element affecting the energy expenditure of resting penguins at sea is the insulation that they have surrounding their hot bodies. Although limited work has been conducted on this, the general belief is that penguins primarily depend on the air in their plumage for insulation. In the 1970s, investigators [33, 39] estimated that the feathers of emperor penguins account for 86% of their total insulation. For Adélie penguins, another polar species, it was found that their insulating surface layer was up to 1.8-cm thick, including both the skin and feathers [37]. So, air would seem important [47], and there are some fascinating features of penguin feather construction that might capitalize on air's insulating capacity [65]. Included is the fact that the barbules of the feathers have a relatively small radius and a particular geometric structure, both of which serve to reduce heat loss [18]. Air is generally a good insulator for animals anyway because it is light and its spread-out molecular configuration resists heat transfer. Indeed, air is universally used by birds and many mammals to reduce heat loss [6], and the thicker the air layer, i.e., the thickness of their "coat" of fur or feathers, the better it insulates (see the formula in Box 7.1). Based on this, we can predict that penguins operating in warmer water should have thinner air layers in their plumage, especially if they do not dive deeply where the air is compressed (reducing the plumage's insulative capacity). Warmer-water penguins also tend to lose feathers around their face once nesting begins, which is assumed to be something that helps them lose excess heat in high air temperatures or when they are subjected to excessive insolation [20]. This subject of feather coat thickness, as insulation, takes on added significance in the case of oil pollution, with coated feathers offering little heat loss protection (Chap. 13).

Our featherometer results, which allow us to estimate the thickness of the air layer in feathers (see Box 6.3, Chap. 6), certainly underpin the importance of air in penguin insulation. For example, measurement of feather air layer thickness at standardized spots on penguin bodies (Fig. 7.11) show that feather air layer thickness increases across all regions of the body in example species from warm- to- temperate to cold-water regions (Fig. 7.11).

Interspecific comparisons, though, are likely to be complicated by the mass effect (Fig. 7.5). Thus, it is useful to include this by comparing sea temperature to mass to feather air layer thickness for a standardized site on the penguin body across

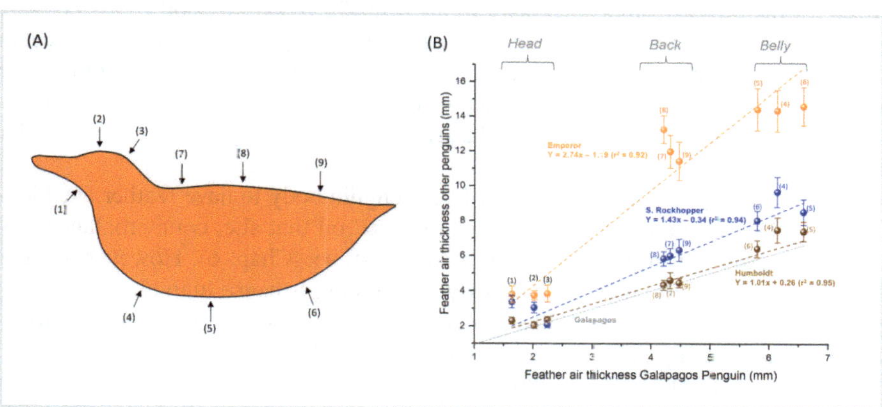

Fig. 7.11 (**a**) Spots used to measure for feather air layer thickness using a "featherometer" (see Box 6.3, Chap. 6) on museum penguin bodies. In panel (**b**), the mean air thickness per site (±SE) is plotted for a cold-water species (the emperor; $n = 15$), a temperate-water species (the southern rockhopper; $n = 16$), and a warm-water species (the Humboldt; $n = 16$) against values for the most tropical penguin, the Galapagos penguin ($n = 28$). Note how the best-fit line shows an increasing gradient with decreasing water temperatures, indicating that the colder the water, the more air the penguins have in their feathers and that this holds true across the whole body. The gray line shows the line of parity for the Galápagos penguin for comparison

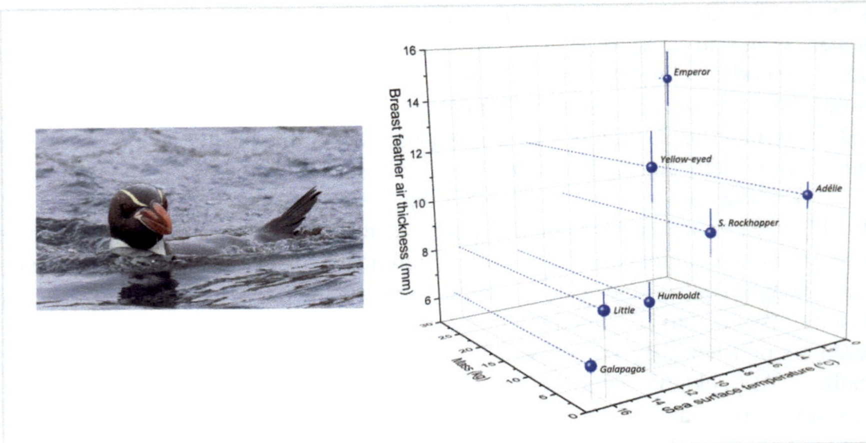

Fig. 7.12 Whether penguins are diving or floating on the surface – as exemplified by this Snares penguin – their breast feathers are almost always underwater and so are particularly important in mitigating heat loss. This becomes especially critical as water temperature decreases. The graph on the right shows the mean breast feather air layer thickness (bars show SE) in seven species of penguins of differing masses (Chap. 1) and occupying different water temperatures (Chap. 3). Note how warmer-water species tend to have less air, as do smaller species. The sample sizes for these penguin species are Adélie: 17, emperor: 15, Galápagos: 28, Humboldt: 16, little: 15, southern rockhopper: 16, and yellow-eyed: 11. (Photo courtesy of Peter Ryan)

a number of penguin species. For this, we managed to obtain appreciable sample sizes for the breast feather thicknesses (site 4 in Fig. 7.11 – a good spot since it is permanently immersed when birds are at sea) for seven penguin species. This again highlights not only the importance of air as an insulator as water temperature decreases but also that as mass increases, so too does the thickness of the air layer in the feathers (Fig. 7.12) (although larger species also tend to be found in colder regions; Table 1.2, Chap. 3).

We would expect the specific site of the penguin body to have feather air thicknesses reflecting the thermal challenges imposed on that site, counterbalanced by the energetic implications of buoyancy (upthrust; see Chap. 6). However, feather thickness also has to allow normal body functioning because high air layer thicknesses can compromise the ability of locomotory and sensory systems to function – as any images of Highland cattle will show (Fig. 7.13)!

Consideration of feather air thickness in relation to measurement site shows a general trend of the thinnest air layers on the head, followed by the back, with, universally, the thickest feather layers on the belly (Fig. 7.14).

This accords with the belly being permanently exposed to the water, with its high capacity to "suck" heat out of the birds, the back sometimes being awash, and the head being generally held clear of the water when penguins are on the surface. The belly is therefore the most thermally challenging interface for a penguin, whereas the head is the least, although both sites are subjected to the same conditions when

Fig. 7.13 The thickness of warm-blooded animal insulation typically varies across the body surface. Not having to catch fast-moving prey by sight, this Highland cow can afford to go around having severely compromised visual capacities because grass is not that fast! Because penguins have a critical need for acute vision to perceive and catch their food, they have short feathers on their head, especially around their eyes. The price they pay for this is increased heat loss at this site. (Photo courtesy of D. Valdemaras on Unsplash)

the birds dive. It is also notable that the interspecific variation in feather air thickness is least in the head (Fig. 7.14), presumably because excessively long feathers in the head would compromise vision.

The consistency of the thickness of the head feathers can be used to examine relative changes in feather air layer thickness across the body in different species. This is because we might expect relative thicknesses of back versus belly feathers to reflect the temperature differences of the air compared to water. In essence, the colder the air, relative to the water, and the more time, overall, that birds spend in cold air (including on land or ice), the more we would expect the back feather air layer thicknesses to approach that of the belly. To examine this, we divided air layer thicknesses of the various sites by the mean head feather thickness of each given species to obtain a ratio, with high values showing proportionately thicker feather air layers (Fig. 7.15).

These ratios show that the temperate/tropical species have remarkable consistency in the patterns displayed, seemingly irrespective of their mass and water temperature. The belly section has a feather air layer thickness that is three to four times that of the head, whereas the back is around two times thicker (Fig. 7.15). In contrast, the two polar species, Adélie and emperor penguins, have species-specific back and belly ratios that are much more similar to each other (the Adélies have a

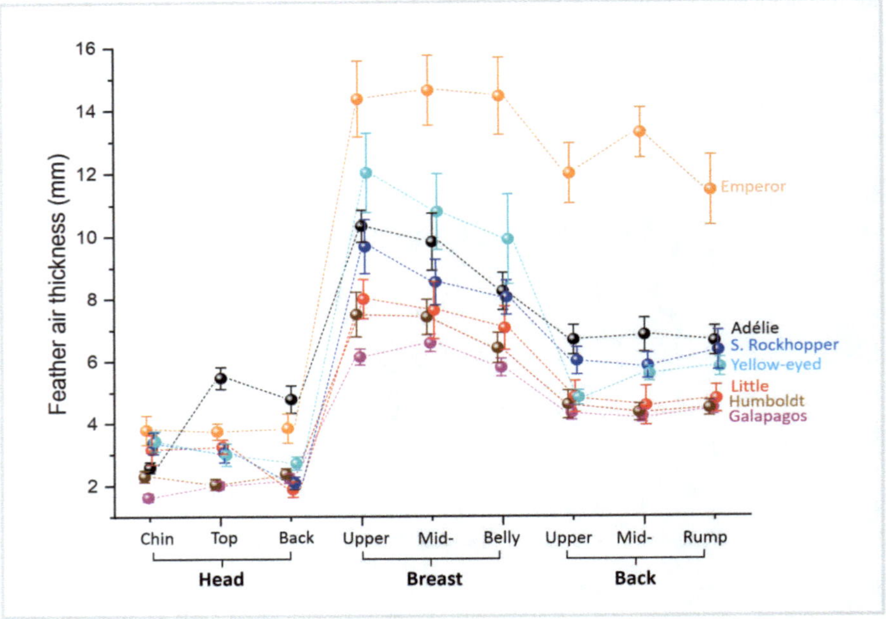

Fig. 7.14 The mean feather air thickness (bars show SE) as a function of site on the body for seven species of penguins (the sample sizes are shown in Fig. 7.11). Note that colder-water species tend to have thicker plumage (cf. Fig. 7.11). Also note that the thinnest plumage universally is on the head, followed by the back, with the thickest plumage on the breast, which is permanently underwater when birds are at sea

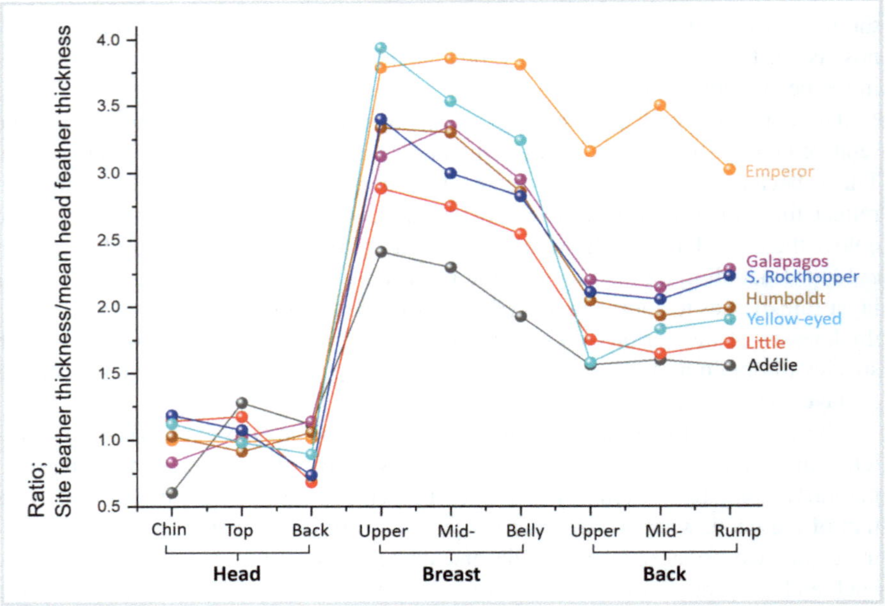

Fig. 7.15 The relative feather air layer thicknesses of various sites on penguin bodies compared to the mean thickness of the head feathers. The points show the mean feather air layer thickness of each site divided by the mean of the head feathers for seven penguin species (the sample sizes are as shown in Fig. 7.12)

belly/head ratio of about 2.2 and a back/head ratio of 1.7, whereas the emperors have a belly/head ratio of ca. 3.6 and a back/head ratio of around 3.2). The implication is that the potential for heat loss through the back is higher in these two species compared to the others and that they mitigate for this by having better dorsal insulation. This ties in with what is known about both species. Although they may swim in water at temperatures down to −1.7 °C, they are exposed to air temperatures that are much lower and, in the case of the emperor penguins, down to −45 °C [48]. In addition, both species spend long periods in cold air standing on ice or ice floes, even during the nonbreeding period, and we suggest that the unusually high amounts of air in the backs of these two species are a reflection of this.

Active Penguins

Activity Produces Heat

Kooyman et al. [36] noted that even the largest penguins, emperors, resting in water can shiver violently; we have also observed this in Adélie penguins in swim canals (see Chap. 6), and Guillermo Luna-Jorquera tells us that he has also seen it in wild Humboldt penguins. Shivering generates heat, which can help pay for heat loss, but it also highlights the value of movement as a heat-generating behavior (Fig. 7.16).

Indeed, after a short time shivering in the resting chamber, our swim canal Adélie penguins would always reinitiate swimming, shooting up and down the length of the canal for extended periods before undertaking more short rests (less than 1 min), at the end of which the birds would begin shivering again. It would seem, therefore, that the act of swimming generates enough heat to mitigate any heat lost to the water, sometimes in excess (Fig. 7.7). The details of the heat generation and loss process are important, however, in helping penguins save energy in their thermally challenging environments. A critical element of this is the lability of their body temperature, i.e., not a constant temperature [20]. Indeed, penguins maintain their body temperature in somewhat broad limits (stenothermy), despite being considered "hot-blooded" (homeothermic) like all birds. Our swim canal *Pygoscelis* penguins (Adélies, gentoos, and chinstraps) demonstrated this clearly, with the body temperature (measured in the stomach by the pills that they swallowed, which contained mini-temperature sensors) rising during swimming and dropping during rest. We observed the same phenomenon in all three *Pygoscelis* penguin species in the wild when they entered the water, swimming away from their breeding island to forage (Fig. 7.17), with the initial temperature rising by 2 °C. This was apparent before feeding and/or drinking confused the picture. A similar pattern has been noted in African penguins [69]. Therefore, it appears that penguin body temperature in general increases during swimming. Indeed, to some degree, the cold water may help them to not overheat!

Fig. 7.16 Although shivering is assumed to occur at sea, it can readily be observed on land as well. These Adélie penguins have had their nests inundated by glacial melt water but persist with incubation, shivering to maintain their body temperature. (Photos courtesy of D. Ainley)

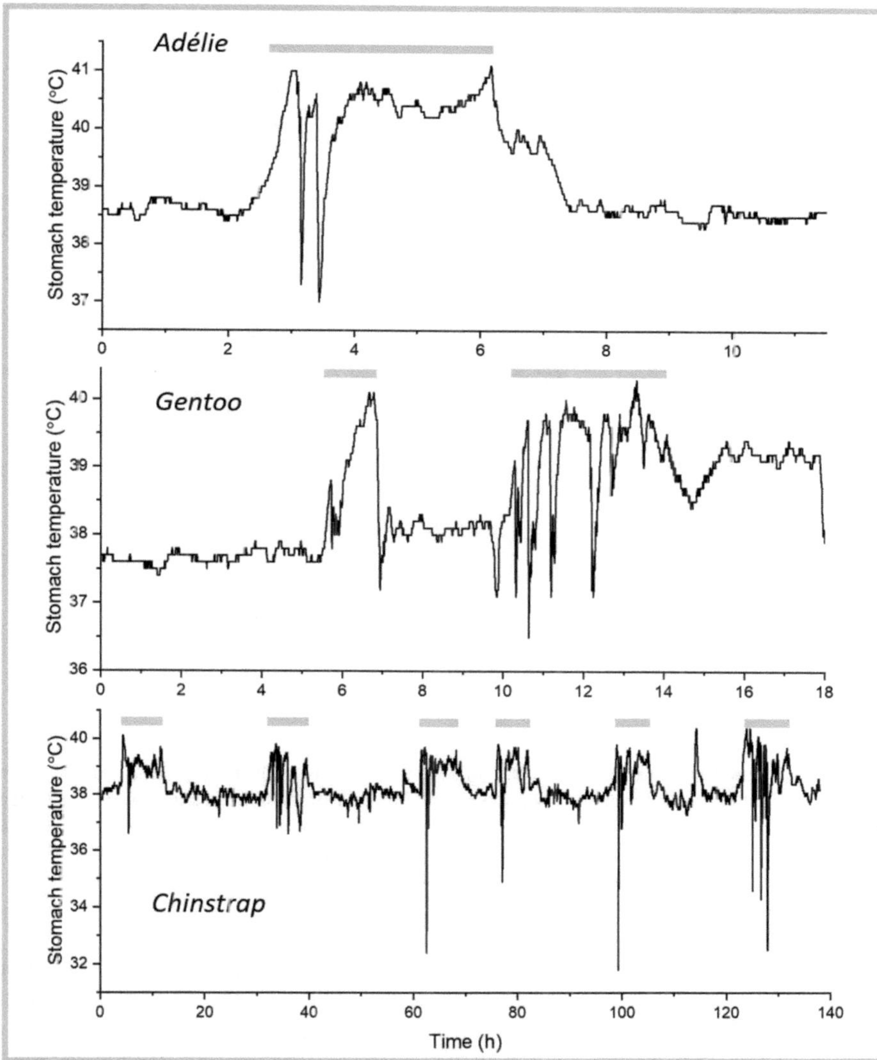

Fig. 7.17 Body temperature variability of *Pygoscelis* penguins in water and air as measured by stomach temperature sensors, swallowed as pills, showing a clear, almost square wave, higher temperatures when swimming (gray bars), and generally lower temperatures when on land. The graphs illustrate different time intervals – the Adélie penguins with one foraging trip of ca. 4.5 h in an 11-h monitoring period, the gentoo penguins undertaking two trips to sea over a total time of 18 h, and the chinstrap penguins with six substantive periods at sea over 140 h of monitoring. Rapid, but transient, temperature drops during the periods at sea are due to ingestion of (ectothermic) prey or water

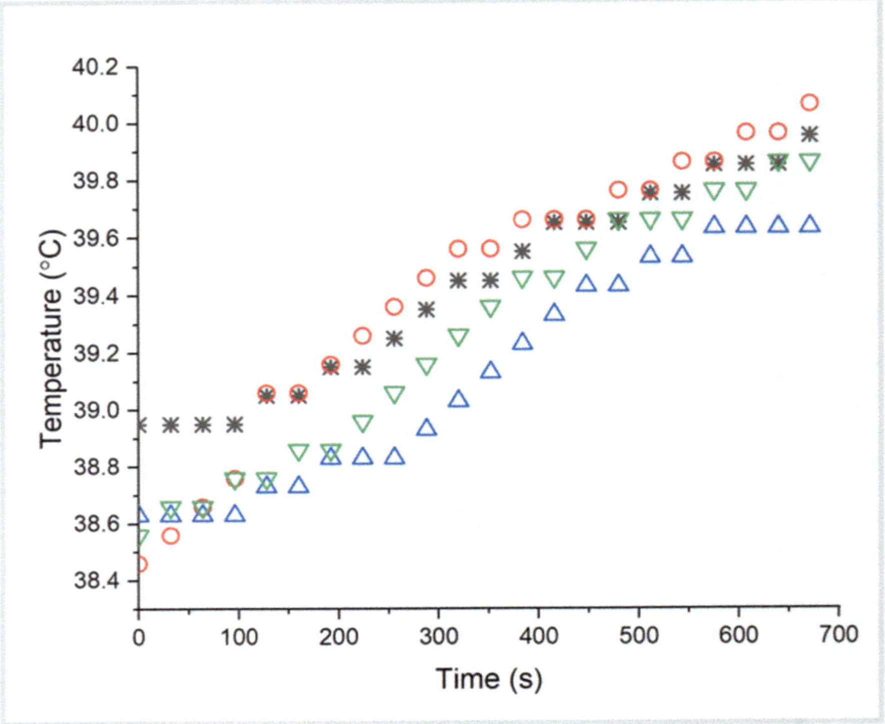

Fig. 7.18 The stomach temperatures of four Adélie penguins (only four of nine examples are shown for clarity), which, immediately upon entering the sea after extended periods on land, began continuous swimming. The data show a linear increase in temperature with some variation in the gradient (see text), presumably primarily due to differences in swim speed and the length of their brief rests. The temperature data were only resolved to the nearest 0.1 °C and taken every 32 s, and hence the presence of the apparent "step function" in increases

The details of body temperature elevation when penguins start swimming after spending time on land are relevant because they allow us to estimate how much energy is being used to heat the birds from the activity of swimming. This comes from the basic understanding that if the temperature rises, then the heat production is higher than the heat loss, and vice versa. As part of this, and as an example of how this plays out in penguins, we looked at the initial temperature rises documented by our temperature pills in Adélie penguins on King George Island, Antarctica. We were specifically looking for instances in which the birds dived into the water and swam continuously for a period without drinking or feeding (because that messes up the recordings of body temperature as the cold material goes into the stomach). In fact, of 14 birds monitored, some of which went to sea multiple times, we managed to find 9 examples of what we had hoped. These showed linear increases in temperature over the first 10 min (or more) of their swimming (Fig. 7.18) with the mean rate of increase being 0.0013 °C/s (SD 0.00067).

Assuming that an average Adélie penguin weighs 4.65 kg (Chap. 1, Table 1.2) and that the specific heat capacity of penguins is 3.47 J/g/°C [42], there is a surplus of heat energy amounting to about 21 W per bird being provided to heat the penguins at this time (the energy provided for this increase = mass X specific heat capacity X temperature rise per second). The linear nature of the rise indicates that active heat loss mechanisms, such as shunting blood to the flippers [68], are not brought into play until upper critical temperatures are reached (which seem to be around 40–41 °C; Fig. 7.17). Therefore, the heat energy produced during swimming is effectively stored by the birds. This can theoretically save them energy when they eventually come to rest because they can allow their heat losses to be paid for by their stored energy until their body temperature cools to a lower critical level. This is that point they have to counteract heat loss by increasing their metabolic rate [69]. The specific advantage to penguins of their easily changeable body temperature, i.e., lability, becomes clear if we consider the energetic costs that the birds would have if they had no such flexibility. Here, the penguin has to pay the heightened metabolic costs of resting in water (approximately twice those in air; see Table 7.1) the instant it stops swimming, despite already paying the metabolic cost of that swimming. The difference in energy costs between the two activity modes, which highlight the theoretical benefits of having them, can be modeled with a few assumptions (Box 7.4).

This modeling approach shows that, for Adélie penguins swimming for long periods near the surface (the relevance of which is explained below) and resting for short periods, as well as doing the reverse, there is no difference in the rate of energy

Box 7.4

We assume that our model penguin has a resting metabolic rate in water (RMR_{water}) that is twice that in air (RMR_{air}) (see earlier due to the "obligate thermal cost" of being in water (colder than body temperature) so that $RMR_{water} = 2\ RMR_{air}$. We further assume that the penguin has a metabolic rate during swimming (MR_{swim}) that produces heat at a rate that is four times the biomechanical power (Chap. 6) so that the heat production during swimming is $0.8\ MR_{swim}$.

The power needed for heating up the penguin (Heat $gain_{swim}$) is provided by the heat produced by the penguin minus the heat lost to the environment. The heat loss is via convection (noting that there are two types of convection: "free" (ambient, where the object losing heat does not move) and "forced" (which relates to the rate of flow of water over an object)).

The heat gain minus the heat loss is given by:

$$\text{Heat gain}_{swim} = 0.8\text{MR}_{swim} - \text{heat loss}_{swim} \tag{7.1}$$

(continued)

whereby the heat loss during swimming can be conveniently represented by the resting metabolic rate in water plus some speed-dependent increment (equivalent to the wind chill factor [45]) so that:

$$\text{heat loss}_{\text{swim}} = \text{RMR}_{\text{water}} + \text{k1.speed} \quad (7.2)$$

where k1 is the linear gradient describing heat loss in water as a function of speed (see [43]). Thus, the overall heat gain can be given by:

$$\text{Heat gain}_{\text{swim}} = 0.8\text{MR}_{\text{swim}} - \left(\text{RMR}_{\text{water}} + \text{k1.speed}\right) \quad (7.3)$$

The metabolic rate of a penguin swimming is a function of its velocity cubed (Chap. 6) so that

$$\text{MR}_{\text{swim}} = \text{k2.v}^3 \quad (7.4)$$

where k2 is a constant (cf. the drag equation in Chap. 6, Box 6.4).

Nominally, for any given swim speed adopted by a penguin starting to swim after resting at sea (where its metabolic rate at $t = 0$ is described by $\text{RMR}_{\text{water}}$), we might expect MR_{swim} to be added to $\text{RMR}_{\text{water}}$. However, in reality, it seems likely that once the heat production component of the metabolic cost of swimming rises, it can be used to substitute for the increased costs of $\text{RMR}_{\text{water}}$ (above RMR_{air}), plus any speed-linked increment described in Eq. (7.2). Therefore, the expectation is that the rate of energy expenditure of a penguin that starts swimming after resting represents a decreasing obligate thermal cost for the left-hand term in Eq. (7.2). Ignoring this for convenience, we concentrate on steady-state swimming in which, if the heat produced is greater than the heat lost, then the temperature of the penguin can rise. Specifically, this occurs when

$$0.8\left(\text{k2.v}^3\right) > \text{RMR}_{\text{water}} + \text{k1.speed} \quad (7.5)$$

When this is true, the rate at which energy is provided to heat up the penguin is given by combining Eqs. (7.2), (7.3), and (7.4):

$$\text{Heat gain}_{\text{swim}} = 0.8\left(\text{k2.v}^3\right) - \left(\text{RMR}_{\text{water}} + \text{k1.speed}\right) \quad (7.6)$$

Luna-Jorquera et al. [42] suggest using a specific heat capacity (SHC) for penguins of 3.47 J/g/°C, so the rate at which the body should heat up ($\Delta T/dt$) is given by the mass of the penguin (M) multiplied by its specific heat capacity ($\text{SHC}_{\text{penguin}}$) divided by the heat gained during swimming according to Eq. (7.6).

(continued)

$$\frac{\Delta T}{dt} = \left(\text{SHC}_{\text{penguin}} M\right) / \left[0.8\left(\text{k2.v}^3\right) - \left(\text{RMR}_{\text{water}} + \text{k1.speed}\right)\right] \quad (7.7)$$

Approximate values for all the terms in Eq. (7.7) are provided in the literature for Adélie penguins as our example species, except for k1. However, we can access this by looking at the rate at which these penguins heat up once they start swimming after a period of rest (Fig. 7.17) in order to calculate the heat energy surplus to provoke the observed body temperature rises (which we estimate to be about 21 W – see text). For this, we can rearrange Eq. (7.7) and insert values of RMR_{water} of 4.9 W/kg (≈22.8 W/bird if the mean mass is taken to be 4.65 kg (Chap. 1, Table 1.1)) as well as the costs of swimming at "cruising speeds" of 2.2 m/s as 71.9 W/bird [12]. This yields a heat loss component due to speed (k1) of 6.27 W/bird/m/s. We can then use this and all other parameters from the literature (see above) to feed into Eq. (7.6) for swimming alternating with non-swimming (resting) periods for Adélie penguins at sea to examine how having a flexible body temperature can save them energy. For this, we will assume minimum and maximum body temperatures of 38 °C and 40.5 °C, respectively (Fig. 7.12); have a heat-up rate of 0.0013 °C/s/bird during swimming and a cool down rate of 0.00071 °C/s/bird during resting (the latter is derived by subtracting RMR_{air} from RMR_{water}; and expressing the difference in °C/s/bird by reference to the SHC of penguins and the assumed body mass of Adélies). When resting birds have a body temperature of 38 °C, we assume that they have paid RMR_{water} costs, but when they are resting with a body temperature > 38 °C, they pay RMR_{air} costs. Swimming birds at temperatures of <40.5 °C accumulate heat in their bodies until they reach 40.5 °C, at which point they shift blood to their flippers and feet to prohibit further rises.

expenditure between thermally flexible and inflexible birds (Fig. 7.19). However, there are swim periods versus rest periods in which penguins with labile body temperatures can save an appreciable amount of energy (Fig. 7.19).

The process by which heat produced by activity can be used to mitigate heat loss costs during other activities has been called "thermal substitution" [40] and has been discussed for diving ducks [34]. Otherwise, the subject has been little considered for its potential to explain activity patterns in hot-blooded diving animals in cold water in general and in penguins in particular. Our consideration of thermal substitution in penguins thus far, though, has been limited to birds swimming near the surface. Its significance changes once penguins dive to appreciable depths.

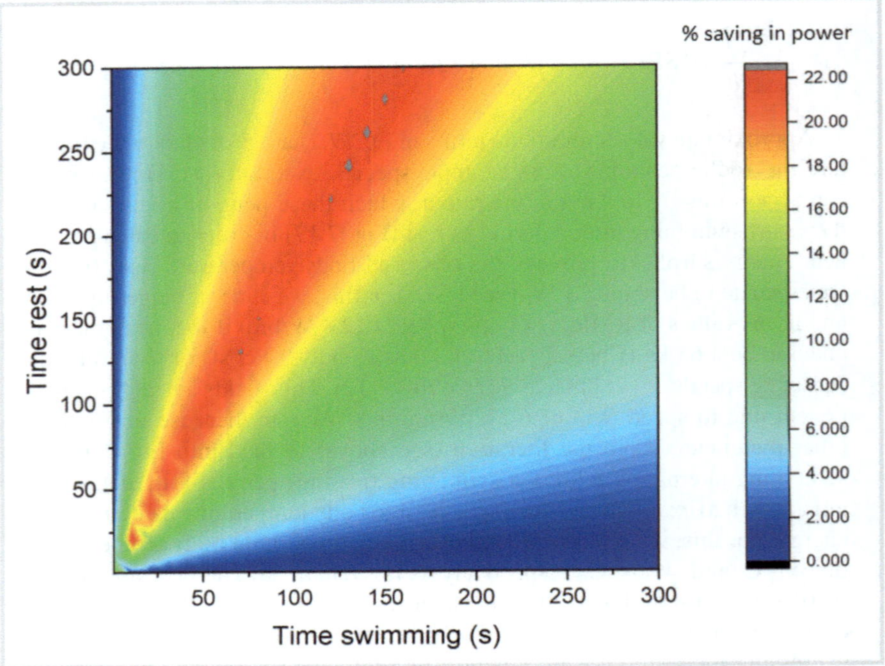

Fig. 7.19 A model predicting the energy savings (expressed as a percentage) made by Adélie penguins by having labile body temperatures compared to birds that have a fixed body temperature for different continuous surface swimming versus resting periods. Since the rate of heat loss per unit body surface area appears to be roughly constant for all penguin species, we expect this basic pattern to hold for penguins other than just Adélies

Greater Depths Impose a Higher Heat Tax

Plumage and respiratory air affect the buoyancy of diving penguins, as Chap. 6 points out, i.e., how air volume, which provides penguins with substantive upthrust near the surface, decreases with increasing depth. However, if air is the main insulator for penguins, the reduction in plumage air volume with depth will result in a proportionately thinner insulating layer and increased heat loss (the same for an oil-coated penguin; Chap. 13). In short, deeper-diving penguins should be losing heat quite quickly compared to when they are on the surface. This is obviously not an area that is easy to research empirically but Gerry Kooyman, a true pioneer in the diving physiology of penguins, attempted to address this matter some time ago [37]. He had Adélie penguins "resting" in water (actually they were held onto a board in a water bath, so probably they were "fidgeting" a lot) at normal pressures before doubling the atmospheric pressure – effectively taking the birds down to 10 m (by having them in a pressure chamber), although they could still breathe. In both instances, he measured the rate of oxygen uptake, a process that directly relates to the rate of energy expenditure and so could simulate the effect of birds "resting" on the surface compared to "resting" at about 10-m depth. In tandem with these experiments, Gerry measured the conductance of penguin pelts in water both on the surface and at 10 m. What his work clearly showed was that heat loss with depth is important, although precisely how important was equivocal. The total conductance of Adélie and gentoo penguin pelts increased from the surface (0 m) to a simulated 10 m by about 2.8 and 4.6 times, respectively, whereas the metabolic rate of penguins subjected to conditions that simulated 10-m water depth was ostensibly higher than birds at 0 m (a relevant graph seems to show this [37]), although not statistically so (this is difficult to check because the raw data are not available). Gerry, musing over this apparent discrepancy, pointed out that pelts are unlikely to behave in the same way as would a live penguin's outer feather layers when subject to pressure. Possibly, there would be water ingression to their inner layers, which is not something that real penguins experience. In addition to this, if the "resting" penguins were, in fact, straining at their restraining systems (and, knowing penguins, it is hard to imagine they would not be doing this), they could have been producing much more metabolic heat than actual resting birds. This could thermally substitute for the varying heat losses, blurring depth-dependent patterns.

For us, there is no question that the air layer in the coats of specimens was compromised by water ingression at the simulated 10-m depth. Theory predicts that the rate of heat loss is directly proportional to the thickness of the insulator (see Box 7.1). Indeed, using this standard formulation would lead to the prediction that the pelts at 10-m depth would lose about twice as much heat energy as those on the surface (because the pressure is doubled and the air thickness halved). Both Adélie and gentoo penguins had values that were much higher than this, a difference that would indicate that heat loss could not be explained by insulation layer thickness changes alone. So, how do we try and approximate how heat flux varies with depth in an attempt to understand how hot penguins may deal behaviorally and physiologically with cold water under varying pressures?

In an attempt to do this, we consider both heat production and heat loss, essentially subtracting one from the other to predict what might happen to fluctuating penguin body temperatures during dives (Box 7.5). To understand how energy expenditure for swimming varies with depth and dive angle, we refer back to Chap. 6, where we had a formulation for power used by Magellanic penguins (Box 6.7). This relationship can be minimally modified to make it applicable to our Adélie penguin model, not least because both species are likely to have comparable conductance values for their specific water temperatures (cf. Fig. 7.8). For the heat loss, we can use the approach described in Box 7.5, modified by estimates of how the insulating properties of penguin outer layers change with depth.

Box 7.5

Heat production with depth: Heat production for swimming at given speeds near the surface (conditions under which metabolic rates of penguin swimming usually have been derived) is provided by Eqs. (7.4), (7.5), and (7.6) in Box 7.4. This formulation, however, needs to be modified by depth and swim angle because increasing the depth reduces the power to swim, whereas increasing the descent angle increases it (Box 6.7). The equation for this in descending (Magellanic) penguins swimming at 1.6 m/s is provided in Box 6.7 for mass-specific biomechanical power. Given that the chemical power for penguin swimming is much the same across similar mass species (Table 6.1), we make the Magellanic penguin data approximately applicable to Adélie penguins by multiplying the mass-specific metabolic rate by the mean mass of Adélies (of 4.65 kg – Chap. 1, Table 1.2). The heat production associated with such biomechanical energy is usually ascribed to be four times that value, and we use this here. Surfacing penguins glide during a substantial proportion of the upward trajectory (Chap. 6; [59]) with little or no muscular effort. Thus, for simplicity, we have restricted our treatment of heat flux in our model penguin to dives to 40 m and assumed that penguins glide between 40 m and the surface and have metabolic rates equivalent to resting in water. Actually, 40–50 m is the depth at which much Adélie penguin foraging occurs [2, 56].

Heat loss with depth: This requires us to calculate how air volume decreases with depth (given in Box 6.1), assuming that the air layer thickness varies in direct proportion to this so that the insulating properties of the air layer also change in direct proportion (Box 7.1). Although, in their model of heat flux in diving Magellanic penguins, Ciancio et al. [10] assumed that air is the only insulator, we consider that it is more realistic to add some insulating effect of the penguin skin to this. Kooyman et al. [37] consider that their tests using penguin pelts under pressure resulted in water influx onto the skin so that conductance values measured under those conditions should provide some measure of the insulative capacity of the skin alone. One should keep in mind that the dried skin (as in a stuffed, penguin specimen) is presumably less effective than the "normal" skin. Kooyman et al. provide an insulation of $0.11\ W^{-1}.m^2.°C$, whereas Le Maho et al. [39] suggest that about 15% of penguin insulation comes from the skin. We use this value, taking its measure from RMR_{water}, and assume that it does not change with depth. This is, therefore, a component of the air-based conductance that changes due to the decreasing air layer with increasing depth based on using heat loss values given for surface-swimming penguins (calculated in Box 7.4). Note that our approach here uses RMR_{water}, rather than RMR_{air}, for the metabolic rate of penguins gliding on the surface. This reduces the value of thermal substitution and decreases the calculated temperature drops, so our theoretical values of heat loss are probably conservative.

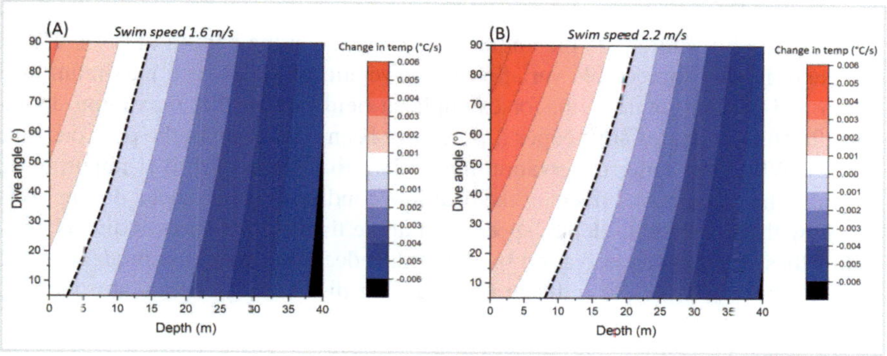

Fig. 7.20 Predicted heat flux (overall heat gain over the whole penguin body (red areas) or heat loss (blue areas)) for an Adélie penguin descending the water column as a function of dive angle and depth. The dashed black line shows the condition under which heat production = heat loss. In both panels (**a**) and (**b**), there is greater net heat loss with depth because the insulating capability of the air layer is compromised by its compression and because birds produce less heat, requiring less power to swim against the decreased buoyancy (cf. Chap. 6, Fig. 6.26). Modeling the effect of speed on heat flux indicates that because the power to swim increases as the cube of the velocity, whereas heat loss only increases linearly, faster-swimming penguins can dive deeper before they start incurring net body heat loss. The net heat loss is also reduced if they descend at steeper angles

Not at all surprising, our approach indicates that the deeper the penguins dive, the more heat they lose, although they can produce more heat to mitigate this loss by stroking faster (increased muscle activity) and/or by diving at steeper angles (Fig. 7.20).

We assume that penguins, with their changeable body temperatures, simply allow their body temperatures to vary within limits, and, indeed, this has been shown in a number of studies that have used implanted temperature sensors [14, 23, 29]. This thermal flexibility means that birds can substitute the excess heat produced during the initial part of the descent phase of the dive (vigorous stroking), to partially cover the heat losses that occur later in the dive (more gliding). As with surface swimming, if penguins did not have this flexibility in body temperature, they would have had to correct net losses immediately with metabolic heat production, thus increasing the energy expended during the dive.

Thermal substitution, coupled with a labile body temperature, is a reminder that it is naïve to consider that penguin-diving strategies might be about optimization of just one currency, such as time or energy in the pursuit of prey. For example, the literature documents how penguins have steeper dive angles with increasing maximum depth of the dive, (e.g., Hanuise et al. [30], Irving [32], and Sato et al. [58]), with conventional wisdom advocating that penguins do this to balance their horizontal search distance with their vertical search distance as part of their foraging strategy. Also involved is a penguin "decision" to consider an appropriate breath-holding time relative to the depth it intends to achieve. However, steeper dive angles also result in a greater rate of metabolic heat production (more stroking), which could be stored as higher body temperatures. Therefore, steeper dive angles could also partially pay for the higher rates of heat loss at greater depths (when they will be likely to glide more, i.e., the undulations common at the bottom of dives; Chap. 9). There is also appreciable variation in the relationship between maximum dive depth and descent angle with, for example, it being noted that maximum depth-specific descent angles are steeper if prey have been consumed in the previous dive [7, 52]. While this latter observation makes intuitive sense [50], it is important to consider the cumulative effects of the heat flux conditions experienced during dives and how these vary with depth, speed, and angle through the various dive phases. All of these factors can be varied by the bird. Indeed, our heat flux model can use specific dive examples to indicate how varying dive characteristics are likely to affect body temperature (Fig. 7.21).

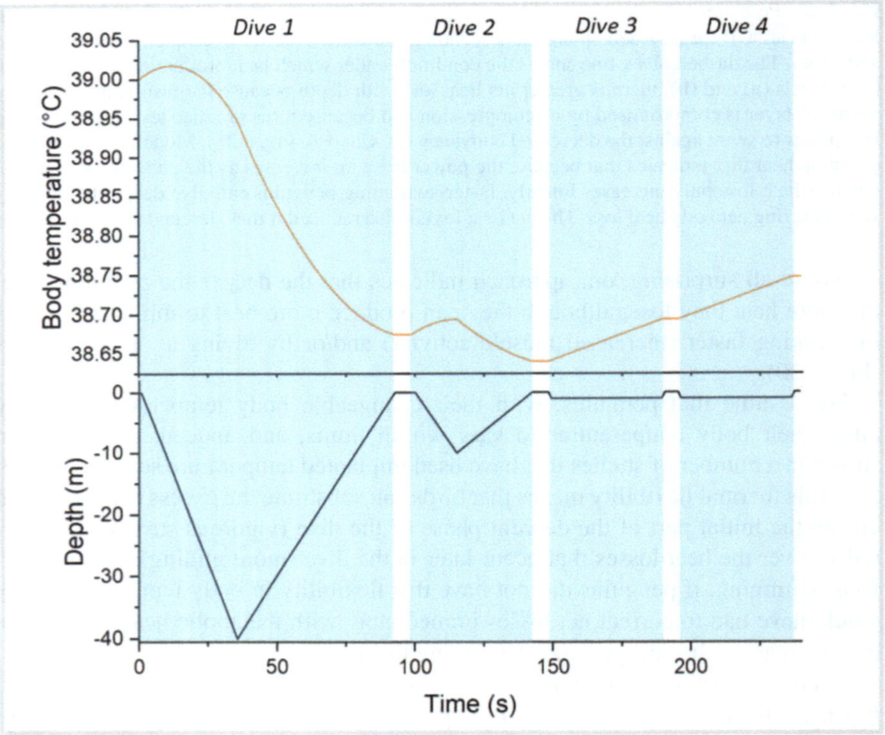

Fig. 7.21 Modeled body temperature changes (see Box 7.5; using constant swim speeds of 2.2 m/s and depth-dependent dive and return-to-surface angles derived from data provided in Ropert-Coudert et al. [52]) in an Adélie penguin executing four dives, two V-dives (bounce dives), to 40 m then 10 m, followed by two surface-swimming dives (dive types in Chap. 9). Note how the initial part of all dives results in an increase in the body temperature which, in the V-dives, is followed by a decreasing body temperature as the insulative capacity of the plumage is compromised and the heat production during swimming is reduced, both due to decreasing air volume with increasing depth. This effect is most pronounced in the deeper dive. Against this, surface swimming (dives 3 and 4) generates a net heat increment. It is conceivable that penguins may sometimes interspace shallow with deeper dives in order to minimize their body temperature drops that occur during deeper dives

Fig. 7.22 Penguin species feeding on slower-swimming prey, like these snares penguins, which predominantly prey on small crustaceans, will generate less muscular heat during prey acquisition than will fish eaters, such as Humboldt penguins. Whether this affects the air layer thickness of the plumage remains to be seen. (Photo courtesy of Peter Ryan)

Clearly, the precise details of each individual dive affect heat flux (Fig. 7.21). The most important element is dive depth, but, for example, extending bottom durations at depth (U-dives – see Chap. 9) will lead to continued overall heat loss just as slow, shallow-angled gliding will result in returns to the surface. Against this, the penguin species that engage in high-speed chases to capture their fast-swimming prey, such as the *Spheniscus* species [71] and little penguins [54] (Chaps. 4 and 10), can presumably partially mitigate heat loss because they produce more heat during their high-power movements. In contrast, penguins foraging on crustaceans generally slow down during prey capture [71] (Fig. 7.22), and these will incur increased rates of temperature drop.

Extrapolating dive characteristics over extended time indicates that a series of deeper dives will result in a considerable amount of heat lost by the body. Consideration of what dive types and characteristics occur at which points in a dive may provide clues as to how penguins might deal with excessive cumulative heat loss. They may, for example, increase heat generation by increasing their swim speed (more stroking, particularly in the first stages of descent [55]), noting that Magellanic penguins, for example, have been shown to swim faster in colder water [10]. Otherwise, "overcooled" penguins may engage in a series of shallow dives in which the plumage has its maximum insulation and swim power can generate heat efficiently – like "star jumps" undertaken by army troops in cold weather (Fig. 7.21). In this respect, it has been noted that many penguin species, most notably the deep-diving king penguins, do engage in shallow dives between bouts of deeper ones [51]. Such behavior might be a hint that the sequence of maximum depths may be related, in some way, to judicious heat modulation.

Consuming Prey Imposes a Heat Tax

As if it is not enough for penguins to have to deal with water "sucking" heat from them from the outside, they also have to deal with heat being "sucked" out of them from the inside. This happens when they ingest their prey, which have the temperature of the water from which they were extracted. This "ice cream effect" is substantial, and no amount of insulation helps. In essence, penguins, with their nominal body temperature at 38 °C (similar to most birds), have to swallow prey at (water) temperatures varying between −1.7 °C and 18 °C, depending on the species and locality (Chap. 1, Table 1.2). Penguins have no other option other than to heat this ingesta, and because water, which constitutes such a high percentage of their prey, has such a high specific heat capacity (see above, the highest of almost any substance), the process is costly, taking a significant amount of heat to warm the prey to ease digestion. The formulation behind this is provided in Chap. 10 (Box 10.7), but it basically amounts to penguins having to invest between about 69 and 138 joules (depending on water temperature) to heat every single gram of prey from ambient to "penguin temperature." This can amount to 8% of the daily energy budget of Adélie penguins [66], but, more importantly, for our consideration of heat flux, it can either push the body temperature down or has to be counteracted by an increase in energy expenditure by the penguins. Translated into body temperature terms, this would mean that if we ignored any other options, then penguins ingesting 30% of their body mass over, say, 8 h (Sala et al. [57] and the references therein) would incur body temperature drops between 6 °C and 13 °C at rates varying between 0.75 °C and 1.6 °C per hour (for Galápagos and Adélie penguins, respectively). If our example of an Adélie diving to 40 m is taken to represent the scenario of heat loss in a bird serially diving to this depth (with the dive pause ratio provided by Wilson (2010) [67]), then this all adds up to potential body temperature drops of 8.9 °C per hour. The 1.6 °C drop from ingesting food should also be added to this. Maintenance of this rate of drop over continuous periods over hours is clearly impossible and may explain why Adélie penguins do not spend hours diving deeply without breaks. However, it also points to the limits of allowing body temperatures to go down. They do dive repeatedly if they encounter a rich prey patch, e.g., 10–102 dives per bout of successive dives and 32–237 bouts per feeding trip (when feeding chicks) depending on a number of factors, such as intraspecific competition for food [19]. The only way for them to forestall serious decreases in body temperature is to ensure that their metabolic rates are appropriate to produce the heat needed to stop overcooling. There are a number of ways for doing that (see above), including actively generating heat (thermolysis) using the brood patch [27, 60] as well as pausing longer between dives and increasing the metabolic rate on the sea surface [28] when the oxygen needed to underpin this metabolism is freely available.

Embracing the Fish in the "Fish-Bird"

We have pointed to fluctuating penguin body temperatures, noting that it would seem, in birds on the surface at least, that the lowest temperatures are at around 38 °C [60], although they may rise to more than 40 °C during surface swimming. However, the lower limits change once birds begin a series of deeper dives. Although some studies had reported small body temperature drops in diving seals [31, 61], the first to demonstrate how dramatic this process could be in penguins was the groundbreaking publication in 1997 by Yves Handrich et al. [29]. These authors showed that abdominal temperatures in king penguins could fall to as low as 18 °C during sustained deep diving. These temperature drops were the most extreme in the deepest dives, as predicted by our heat loss model. In addition, since these abdominal temperatures were lower than those in the stomach, it appears that they were primarily the result of peripheral, rather than stomach-based, cooling. Similar results have since been reported by other groups [23], who provide a graph showing how the abdominal temperature of a macaroni penguin dropped from about 38 °C to 28 °C over the course of an 8-h period of diving. Subsequently, the same team [23] provided a relationship between the temperature drop and the number of dives in a diving bout for this species (the decrease in abdominal temperature was best approximated by 0.04 × the number of dives in a bout +0.91). There is notable variation in this regression, presumably partly due to the differing depths of dives, with varying swim speeds, angles, etc. (see above). In fact, it turns out that dramatic cooling of body tissues also occurs in diving seabirds other than penguins, with the imperial shag being an excellent example [5]. Therefore, in diving birds, the phenomenon of body tissue cooling may be more common than we think. The truth is that we have not really looked for it in many diving species due to the difficulty of measuring such phenomena, mainly because this normally involves surgical implantation of the temperature-sensing tags. Actually, it may turn out that lowered body temperatures are more widespread even in non-diving birds, especially in species that have to operate in cold climes (e.g., geese that migrate over the Himalayan mountains).

There is plenty of literature that shows how some bird species undergo torpor during their inactive periods, with concomitant dramatic reduction in their body temperature [21, 22]. However, what happens to the Andean condor's naked head as it flies at 70 km/h in subzero air temperatures? If wind chill is taken into account, then an ambient temperature of 0 °C equates to an effective temperature of about −8 °C, so how can a naked piece of tissue the size of a grapefruit not suffer frostbite? We suggest that, rather than being supplied with huge amounts of energy-expensive heat to counteract what would be a massive heat loss, the head is allowed to cool, being perfused with blood periodically to keep temperatures at an "appropriate" level. The advantage to both penguins and (possibly) condors of having temperature lability in large parts of their bodies is clear because by doing so, they can save energy on heat production. Importantly, the lower the operational temperatures of these heat-fluctuating portions of these birds, the less heat is lost. This is because the rate of heat transfer depends on the difference in temperature between the bird and environment. In a nutshell, by allowing lower body temperatures, penguins save energy.

Although our modeling of heat loss in penguins has assumed that the temperature of the whole body drops homogeneously, this is simplistic. It is well accepted that the extremities of animals living in cold environments can be at temperatures that are much lower than the body core. Reindeers' legs [32] and penguins' flippers and feet [68] are good examples, but it would appear, in penguins at least, that some of their inner body layers may considerably drop their temperatures. This is most obviously the case in the abdomen near the stomach where both the cold stomach contents and the proximity to the thin penguin brood patch, itself a minimally feathered interface with the seawater, will tend to remove heat from the penguin [27, 29]. More detailed studies on king penguins, since the original work of Handrich et al. (1997), have even shown that the pectoral muscles, important heat-generating elements in swimming penguins, may drop by a few degrees during deep diving [60]. We do note, however, that this phenomenon is not observed in emperor penguins [47].

The picture that emerges then is that when penguins dive deep, they can adopt an unequal distribution of temperatures within their body, with the heart and brain likely to have the highest temperatures, whereas other regions, such as the lower abdomen, have the lowest. There are presumably clever blood shunting and other mechanisms that allow this heterogeneity (cf. Schmidt et al. [60]). Physiologists rightly point to the value of the operational capacity of organs being dependent on temperature, with high body temperatures being a critical part of bird (and mammal) performance compared to reptiles, amphibia, and fish. However, it is perhaps time to appreciate that selective drops in temperature in hot-blooded animals living in cold environments may still allow high-quality operation, whether that is in penguins chasing prey 100 m underwater or in condors flying above the Andes. After all, reptiles can operate extremely well with brain and muscle temperatures of 30 °C, so why should birds not do the same?

The other notable phenomenon about the drops in penguin body temperature during deep dives is that there are small rises when the birds resurface [29]. This can be accompanied by tachycardia [23] and a general resuming of digestive activity (indicated by stomach churning), which is also curtailed during diving (Chap. 14). Thus, the process of dive recovery, reoxygenation, reheating, and other oxygen-heavy processes are reserved for the time when the birds are on the surface. There, they can use the unlimited supplies of oxygen provided by air at high rates.

In fact, oxygen use is a critical part of the penguin strategy of having thermolability because it plays a role in enhancing their dive performance– the energy used by living tissues is a function of its temperature, with more energy used in hotter tissues [49]. Moreover, in normal aerobic (oxygen-facilitated) metabolism, the higher the metabolic rate, the more oxygen is used. So, by allowing their body temperatures to drop, penguins not only save energy but also use less oxygen, and this has been used to explain how their dive durations can be so extraordinarily long [29].

We started this chapter by pointing out that the high body temperature of penguins is what gives them the edge over their cold-blooded prey, conferring great speed, agility, and reactive capacity and making them formidable aquatic predators [9]. Indeed, being hot-blooded is widely acknowledged as being a large part of the success of mammals and birds, as they evolved from their cold-blooded ancestors [11]. It is somewhat ironic, therefore, that as evolution pushed penguins to ever greater deviations from the general bird plan as they adapted more fully to a life aquatic (Chap. 1), they would, in a temperature sense at least, become more like the prey they so successfully acquire. If ever there was a case for referring to penguins as "fish-birds," it is this.

References

1. Ainley DG (1974) The comfort behaviour of Adélie and other penguins. Behaviour 43:16–51
2. Ainley DG, Ballard G, Jones RM, Jongsomjit D, Pierce SD, Smith WO Jr, Veloz S (2015) Trophic cascades in the western Ross Sea, Antarctica: revisited. Mar Ecol Prog Ser 534:1–16
3. Barre H, Roussel B (1986) Thermal and metabolic adaptation to first cold-water immersion in juvenile penguins. Am J Phys Regul Integr Comp Phys 251(3):R456–R462
4. Bethge P, Nicol S, Culik B, Wilson R (1997) Diving behaviour and energetics in breeding little penguins (*Eudyptula minor*). J Zool 242(3):483–502
5. Bevan R, Boyd I, Butler P, Reid K, Woakes A, Croxall J (1997) Heart rates and abdominal temperatures of free-ranging South Georgian shags, *Phalacrocorax georgianus*. J Exp Biol 200(4):661–675
6. Blix AS (2016) Adaptations to polar life in mammals and birds. J Exp Biol 219(8):1093–1105
7. Bost C-A, Handrich Y, Butler P, Fahlman A, Halsey L, Woakes A, Ropert-Coudert Y (2007) Changes in dive profiles as an indicator of feeding success in king and Adélie penguins. Deep-Sea Res II Top Stud Oceanogr 54(3–4):248–255
8. Butler P, Woakes A (1984) Heart rate and aerobic metabolism in Humboldt penguins, *Spheniscus humboldti*, during voluntary dives. J Exp Biol 108(1):419–428
9. Cairns DK, Gaston AJ, Huettmann F (2008) Endothermy, ectothermy and the global structure of marine vertebrate communities. Mar Ecol Prog Ser 356:239–250
10. Ciancio JE, Quintana F, Sala JE, Wilson RP (2016) Cold birds under pressure: can thermal substitution ease heat loss in diving penguins? Mar Biol 163(2):43
11. Clarke A, Pörtner HO (2010) Temperature, metabolic power and the evolution of endothermy. Biol Rev 85(4):703–727
12. Culik B, Wilson R, Bannasch R (1994) Underwater swimming at low energetic cost by pygoscelid penguins. J Exp Biol 197(1):65–78
13. Culik B, Wilson RP (1991) Swimming energetics and performance of instrumented Adélie penguins (*Pygoscelis adeliae*). J Exp Biol 158(1):355–368
14. Culik BM, Pütz K, Wilson R, Allers D, Lage J, Bost C, Le Maho Y (1996) Diving energetics in king penguins (*Aptenodytes patagonicus*). J Exp Biol 199(4):973–983
15. Dantas GP, Almeida VS, Maracini P, Serra SD, Chame M, Labarthe N, Kolesnikovas C, Siciliano S, Matias CA, Moura JF (2013) Evidence for northward extension of the winter range of Magellanic Penguins along the Brazilian coast. Mar Ornithol 41:195–197
16. Dawson C, Vincent JF, Jeronimidis G, Rice G, Forshaw P (1999) Heat transfer through penguin feathers. J Theor Biol 199(3):291–295
17. Drent R, Stonehouse B (1971) Thermoregulatory responses of the Peruvian penguin, *Spheniscus humboldti*. Comp Biochem Physiol A Physiol 40(3):689–710

18. Du N, Fan J, Wu H, Chen S, Liu Y (2007) An improved model of heat transfer through penguin feathers and down. J Theor Biol 248(4):727–735
19. Ford RG, Ainley DG, Lescroël A, Lyver POB, Toniolo V, Ballard G (2015) Testing assumptions of central place foraging theory: a study of Adélie penguins *Pygoscelis adeliae* in the Ross Sea. J Avian Biol 46(2):193–205
20. Frost P, Siegfried W, Burger A (1976) Behavioural adaptations of the Jackass penguin, *Spheniscus demersus* to a hot, arid environment. J Zool 179(2):165–187
21. Geiser F (1988) Reduction of metabolism during hibernation and daily torpor in mammals and birds: temperature effect or physiological inhibition? J Comp Physiol B 158(1):25–37
22. Geiser F (2004) Metabolic rate and body temperature reduction during hibernation and daily torpor. Annu Rev Physiol 66:239–274
23. Green J, Butler P, Woakes A, Boyd I (2003) Energetics of diving in macaroni penguins. J Exp Biol 206(1):43–57
24. Green JA, Butler PJ, Woakes AJ, Boyd I (2002) Energy requirements of female macaroni penguins breeding at South Georgia. Funct Ecol 16:671–681
25. Green JA, Butler PJ, Woakes AJ, Boyd IL (2004) Energetics of the moult fast in female macaroni penguins *Eudyptes chrysolophus*. J Avian Biol 35(2):153–161
26. Grémillet D, Chauvin C, Wilson RP, Le Maho Y, Wanless S (2005) Unusual feather structure allows partial plumage wettability in diving great cormorants *Phalacrocorax carbo*. J Avian Biol 36(1):57–63
27. Grémillet D, Meslin L, Lescroël A (2012) Heat dissipation limit theory and the evolution of avian functional traits in a warming world. Funct Ecol 26(5):1001–1006
28. Halsey L, Handrich Y, Fahlman A, Schmidt A, Bost C-A, Holder R, Woakes A, Butler P (2007) Fine-scale analyses of diving energetics in king penguins *Aptenodytes patagonicus*: how behaviour affects costs of a foraging dive. Mar Ecol Prog Ser 344:299–309
29. Handrich Y, Bevan R, Charrassin J-B, Butler P, Pütz K, Woakes A, Lage J, Maho YL (1997) Hypothermia in foraging king penguins. Nature 388(6637):64–67
30. Hanuise N, Bost CA, Handrich Y (2013) Optimization of transit strategies while diving in foraging king penguins. J Zool 290(3):181–191
31. Hill RD, Schneider RC, Liggins GC, Schuette AH, Elliott RL, Guppy M, Hochachka PW, Qvist J, Falke KJ, Zapol WM (1987) Heart rate and body temperature during free diving of Weddell seals. Am J Phys Regul Integr Comp Phys 253(2):R344–R351
32. Irving L (1972) Heterothermic operation of homeotherms. Chapter 10 in Arctic life of birds and mammals. Springer, Heidelberg, pp 137–162
33. Jarman M (1973) Experiments on the emperor penguin, *Aptenodytes forsteri*, in various thermal environments. Bull Brit Antarct Surv 33(34):57–63
34. Kaseloo PA, Lovvorn JR (2005) Effects of surface activity patterns and dive depth on thermal substitution in fasted and fed lesser scaup (*Aythya affinis*) ducks. Can J Zool 83(2):301–311
35. Kooyman G, Cherel Y, Maho YL, Croxall J, Thorson P, Ridoux V, Kooyman C (1992) Diving behavior and energetics during foraging cycles in king penguins. Ecol Monogr 62(1):143–163
36. Kooyman G, Drabek C, Elsner R, Campbell W (1971) Diving behavior of the emperor penguin, *Aptenodytes forsteri*. Auk 88:775–795
37. Kooyman G, Gentry R, Bergman W, Hammel H (1976) Heat loss in penguins during immersion and compression. Comp Biochem Physiol A Physiol 54(1):75–80
38. Kooyman G, Ponganis P (1994) Emperor penguin oxygen consumption, heart rate and plasma lactate levels during graded swimming exercise. J Exp Biol 195(1):199–209
39. Le Maho Y, Delclitte P, Chatonnet J (1976) Thermoregulation in fasting emperor penguins under natural conditions. Am J Physiol Leg Content 231(3):913–922
40. Lovvorn J (2007) Thermal substitution and aerobic efficiency: measuring and predicting effects of heat balance on endotherm diving energetics. Philos Trans R Soc B Biol Sci 362(1487):2079–2093
41. Luna-Jorquera G, Culik BM (2000) Metabolic rates of swimming Humboldt penguins. Mar Ecol Prog Ser 203:301–309

42. Luna-Jorquera G, Wilson R, Culik B, Aguilar R, Guerra C (1997) Observations on the thermal conductance of Adélie (*Pygoscelis adeliae*) and Humboldt (*Spheniscus humboldti*) penguins. Polar Biol 17(1):69–73
43. Mccafferty D, Moncrieff J, Taylor I (1997) The effect of wind speed and wetting on thermal resistance of the barn owl (*Tyto alba*). I: total heat loss, boundary layer and total resistance. J Therm Biol 22(4–5):253–264
44. Nagy KA, Siegfried WR, Wilson RP (1984) Energy utilization by free-ranging jackass penguins, *Spheniscus demersus*. Ecology 65(5):1648–1655
45. Osczevski RJ (1995) The basis of wind chill. Arctic 48:372–382
46. Pinshow B, Fedak MA, Battles DR, Schmidt-Nielsen K (1976) Energy expenditure for thermoregulation and locomotion in emperor penguins. Am J Physiol Leg Content 231(3):903–912
47. Ponganis P, Van Dam R, Levenson D, Knower T, Ponganis K, Marshall G (2003) Regional heterothermy and conservation of core temperature in emperor penguins diving under sea ice. Comp Biochem Physiol A Mol Integr Physiol 135(3):477–487
48. Prévost J (1961) Écologie du manchot empereur *Aptenodytes forsteri* Gray. Hermann, Paris
49. Rao KP, Bullock TH (1954) Q10 as a function of size and habitat temperature in poikilotherms. Am Nat 88(838):33–44
50. Ribak G, Strod T, Weihs D, Arad Z (2007) Optimal descent angles for shallow-diving cormorants. Can J Zool 85(4):561–573
51. Ropert-Coudert Y, Bost C-A, Handrich Y, Bevan RM, Butler PJ, Woakes AJ, Le Maho Y (2000) Impact of externally attached loggers on the diving behaviour of the king penguin. Physiol Biochem Zool 73(4):438–444
52. Ropert-Coudert Y, Kato A, Baudat J, Bost C-A, Le Maho Y, Naito Y (2001) Time/depth usage of Adélie penguins: an approach based on dive angles. Polar Biol 24(6):467–470
53. Ropert-Coudert Y, Kato A, Chiaradia A (2009) Impact of small-scale environmental perturbations on local marine food resources: a case study of a predator, the little penguin. Proc R Soc B Biol Sci 276(1676):4105–4109
54. Ropert-Coudert Y, Kato A, Wilson RP, Cannell B (2006) Foraging strategies and prey encounter rate of free-ranging Little Penguins. Mar Biol 149(2):139–148
55. Ropert-Coudert Y, Kato A, Sato K, Naito Y, Baudat J, Bost A, Le Maho Y (2002) Swim speed of free-ranging Adélie penguins *Pygoscelis adeliae* and its relation to the maximum depth of dives. J Avian Biol 33(1):94–99
56. Saenz BT, Ainley DG, Daly KL, Ballard G, Conlisk E, Elrod ML, Kim SL (2020) Drivers of concentrated predation in an Antarctic marginal-ice-zone food web. Sci Rep 10(1):1–17
57. Sala JE, Wilson RP, Quintana F (2012) How much is too much? Assessment of prey consumption by Magellanic penguins in Patagonian colonies. PLoS One 7(12):e51487
58. Sato K, Charrassin J-BT, Bost C-A, Naito Y (2004) Why do macaroni penguins choose shallow body angles that result in longer descent and ascent durations? J Exp Biol 207(23):4057–4065
59. Sato K, Naito Y, Kato A, Niizuma Y, Watanuki Y, Charrassin J, Bost C-A, Handrich Y, Le Maho Y (2002) Buoyancy and maximal diving depth in penguins: do they control inhaling air volume? J Exp Biol 205(9):1189–1197
60. Schmidt A, Alard F, Handrich Y (2006) Changes in body temperatures in king penguins at sea: the result of fine adjustments in peripheral heat loss? Am J Phys Regul Integr Comp Phys 291(3):R608–R618
61. Scholander P, Irving L, Grinnell SW (1942) On the temperature and metabolism of the seal during diving. J Cell Comp Physiol 19(1):67–78
62. Simeone A, Luna-Jorquera G, Wilson RP (2004) Seasonal variations in the behavioural thermoregulation of roosting Humboldt penguins (*Spheniscus humboldti*) in north-central Chile. J Ornithol 145(1):35–40
63. Stahel C, Nicol S (1982) Temperature regulation in the little penguin, *Eudyptula minor*, in air and water. J Comp Physiol 148(1):93–100
64. White CR, Marshall DJ, Alton LA, Arnold PA, Beaman JE, Bywater CL, Condon C, Crispin TS, Janetzki A, Pirtle E (2019) The origin and maintenance of metabolic allometry in animals. Nat Ecol Evol 3(4):598–603

65. Williams CL, Hagelin JC, Kooyman GL (2015) Hidden keys to survival: the type, density, pattern and functional role of emperor penguin body feathers. Proc R Soc B Biol Sci 282(1817):20152033
66. Wilson R, Culik B (1991) The cost of a hot meal: facultative specific dynamic action may ensure temperature homeostasis in post-ingestive endotherms. Comp Biochem Physiol A Comp Physiol 100(1):151–154
67. Wilson RP (2010) Resource partitioning and niche hyper-volume overlap in free-living Pygoscelid penguins. Funct Ecol 24(3):646–657
68. Wilson RP, Adelung D, Latorre L (1998) Radiative heat loss in gentoo penguin (*Pygoscelis papua*) adults and chicks and the importance of warm feet. Physiol Zool 71(5):524–533
69. Wilson RP, Grémillet D (1996) Body temperatures of free-living African penguins (*Spheniscus demersus*) and bank cormorants (*Phalacrocorax neglectus*). J Exp Biol 199(10):2215–2223
70. Wilson RP, Hustler K, Ryan PG, Burger AE, Noldeke EC (1992) Diving birds in cold water: do Archimedes and Boyle determine energetic costs? Am Nat 140(2):179–200
71. Wilson RP, Ropert-Coudert Y, Kato A (2002) Rush and grab strategies in foraging marine endotherms: the case for haste in penguins. Anim Behav 63(1):85–95

Chapter 8
Fish-Birds: The Inside Story

It is extraordinary to contemplate that a bird, i.e., a creature conventionally considered to make effective use of the highly rarefied air, could even entertain the idea of surging downward underwater to astonishing depths, where there is no air, for unreasonable periods (especially in the case of *Aptenodytes* penguins). One should just think of the minutes that they must hold their breath underwater and consider the immense water pressures that they will experience at depths in excess of 500 m (pressures of more than 56 kg for every square centimeter of the body surface (i.e., 796 lbs./inch^2), which is enough to easily crush an aluminum beer can) – and just to get lunch! They do this repeatedly, for long periods at a time!

Chapter 6 explains why the process of swimming downward, at the high speeds that penguins use, is so energetically taxing. However, there are plenty of birds that do energetically taxing things for long periods – e.g., hummingbirds hovering in front of flowers or geese flying over the Himalayas during migration. However, high-power lifestyles need oxygen (O_2) to release stored energy aerobically and birds have flow-through breathing systems to help them with this, i.e., they have a near continual access to oxygen. Critically for penguins, however, they carry out important activities underwater, but, there, they cannot renew their oxygen. In a nutshell, having food (their energy source) and oxygen so far apart from each other in space (and time) is not just an inconvenience. It is also hugely constraining and potentially life-threatening, especially as they need the oxygen to produce the energy required to make their muscles work. It is little wonder, but no less fascinating, that penguins have developed a suite of internal features in their hardware that help them, and they integrate these with behavioral decisions about when, where, and how to move during foraging (detailed in Chap. 10), to make them efficient at getting their "lunch." Indeed, most extraordinary is the suite of physical and physiological adaptations that penguins have undertaken, through evolution (Chap. 1), in order to achieve this, and as detailed in the remainder of this chapter. This is truly incredible!

D. G. Ainley, R. P. Wilson, *The Aquatic World of Penguins*, Fascinating Life Sciences, https://doi.org/10.1007/978-3-031-33990-5_8

Surface Issues: Uptake of Oxygen

What better place to start examining the mechanisms that fish-birds employ to maximize their efficiency underwater than on the surface? This is where oxygen is taken from the air and loaded into the body, whereas carbon dioxide (CO_2) is expelled. CO_2 is the by-product of using oxygen. The avian respiratory system is covered in detail by Frank Powell [75], but, in essence, even in penguins, it consists of a trachea that leads into a rather complex system of thin-walled air sacs that can be broadly subdivided into the anterior (thoracic) and posterior (abdominal) sacs with the bird lungs lying in between them (Fig. 8.1). It is much more efficient than in mammals such as us, for instance, as we have to expel what is in our lungs in order to refill them with new air. There is always some residual, carbon dioxide-saturated air preventing an entire exchange. This is not the case in birds, as detailed below.

The air sacs are just what they look like – sacs in which air is stored but there is no gas exchange between them and the blood. The exchange occurs in the lungs. The bird lungs, in general, comprise about 10% of the total volume of their respiratory system [76], and penguins are no different [59]. Penguins' lungs, unlike the air sacs, are considered to have a rather fixed volume that has air flowing through in one

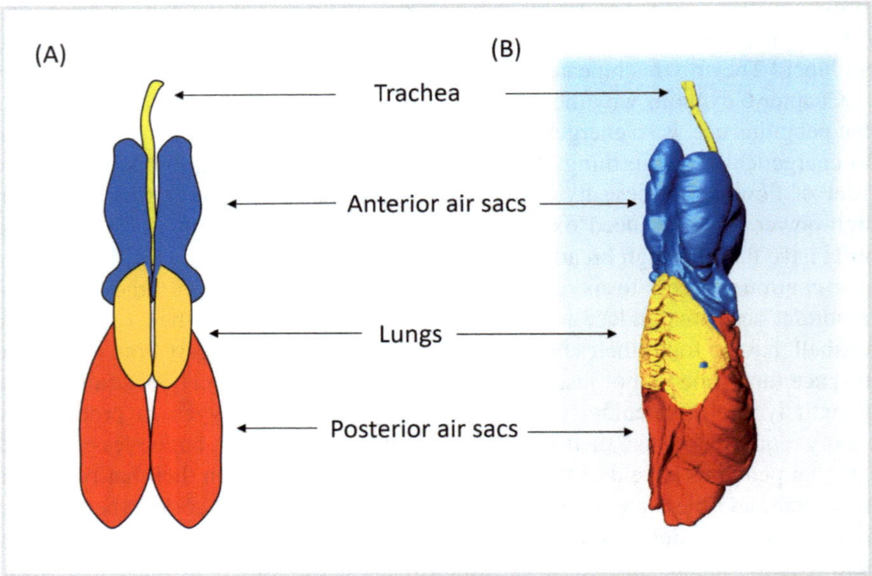

Fig. 8.1 (**a**) A schematic diagram of a penguin respiratory system (dorsal view) and (**b**) an oblique view of the respiratory system of an emperor penguin adapted from computerized tomographic reconstructions (by Dr. Miriam Scadeng in Ponganis et al. [59]) conducted on a live bird. The lungs are shown in yellow, the anterior air sacs (cervical, interclavicular, and anterior thoracic) are shown in blue, whereas the posterior air sacs (posterior thoracic and abdominal) are shown in red. (Image courtesy of Paul Ponganis)

direction. The order of a complete respiration cycle seen from the outside looks superficially just like the typical mammalian inhale-and-exhale cycle with the bird body diameter increasing and decreasing, just as in it does in us [99], whereas the respiratory system is filled with air and is emptied. It actually involves an inhalation of the outside air into the expanding posterior air sacs and the lungs while, cleverly, the deoxygenated air in the lungs moves into the anterior air sacs. During exhalation, the air in the posterior air sacs moves to the lungs, whereas the deoxygenated air in the lungs and the contracting anterior air sacs expels the air to the outside via the trachea [93]. In short, it is a predominantly one-way system in which gas exchange occurs only in the lungs. This constellation is considered particularly efficient for two reasons. First, it has minimal "dead space ventilation." In mammals, for example, the lungs are rather like a sack into which air flows tidally. This is just like air blowing into a paper bag via a tube, and once the bag fills up, one has to stop, with air still in the tube that has not reached the sack. This highlights the "mammalian problem" of having extensive "dead air" in the trachea, which never reaches the lungs to provide oxygen. This air sits there when the lungs are full until exhalation begins when it is expelled again. This is such a waste of space and oxygen! Second, because the bird respiratory system employs a countercurrent gas exchange system, with nearly all the inspired air passing through the lungs in the opposite direction to the blood flowing past them (producing a maximally advantageous diffusion gradient), the process effectively provides virtually continuous gas exchange to extract as much oxygen as possible. Again, this does not occur in mammals.

The respiration cycles of penguins on land, which generally reflect reduced activity, are normally manifested by low ventilation rates (i.e., breathing), varying around 8/min for Adélie and chinstrap penguins [55], 7/min for Magellanic [101] and gentoo penguins [55], and 6.9/min for little penguins [84]. The biggest penguins, emperors, breathe as little as four times per minute [52, 60]. As one might expect, low ventilation rates are principally associated with low amounts of air intake per breath and correspondingly low rates of air intake and exhalation within a single respiration cycle. We know this because, conveniently for penguin researchers, the clear inspiration and expiration phases of each of the breathing cycles are generally manifested in the angle between the two halves of the penguin beak (hereafter the "beak angle"). The beak opens and closes slightly with each respiration, and the opening extent is proportional to the flow rate of air [103]. However, penguins resting on land exhibiting the low breathing rates mentioned above appear to have their beaks closed (unless they are too hot, in which case they open their beaks much wider, and keep them open, presumably to lose heat via evaporation. That is, they pant, like a dog.). This effectively means that all the inspired and exhaled air passes through the tiny holes in the upper mandible- the nares, the bird equivalent to nostrils (Fig. 8.2a).

This indicates that the rates of inflowing and outflowing air are dramatically reduced because the rate of airflow is critically dependent on tube diameter (Box. 8.1) and the nares are tiny.

Fig. 8.2 (**a**) The head and bill of a Magellanic penguin brooding small chicks, showing the small nares/nostrils (located roughly at the center of its upper beak half) through which air passes during rest. At higher-intensity activities, the diameter of the nares is too small to allow for rapid passage of air. In that case, the beak is opened to allow air unhindered access to the trachea, as is apparent in panel (**b**), showing a porpoising macaroni penguin, which clearly has its beak open as it catches a breath. In panel (**c**), an erect-crested penguin replenishes its body oxygen levels on the surface between dives. Although during such surface pauses, birds are generally fairly quiet, their breathing patterns reflect the higher activities they undertake underwater, with the bill opening and closing with every breath (cf. also Fig. 8.3). (Photographs: Magellanic penguin – Rory Wilson; macaroni and erect-crested penguins courtesy of Peter Ryan and Eric Woehler, respectively)

Box 8.1

The flow rates of air critically depend on the diameter of any passage linking high to low pressure in fluids (and this includes air). Specifically, the Hagen–Poiseuille equation states that the flow rate of a fluid (d*V*/d*t*) along a tube is given by:

$$\mathrm{d}V/\mathrm{d}t = \left(\pi r^4/8\eta\right)\left(|\Delta P|/L\right)$$

where r is the internal radius of the tube, η is the dynamic fluid viscosity, ΔP is the pressure differential between one side and the other, and L is the length. The exponent of the tube radius to the power of four makes the dimensions of the nares critical when the flow rates of air are to increase, similar to what happens when penguins increase their metabolic rate and require more oxygen per unit time.

Where penguin activity necessitates higher airflow (when the birds use more oxygen per unit time), their closed beak with the passage of air through the nares option does not cut it. For this, penguins have to open their beaks to allow air to get into the respiratory system unhindered. An obvious case is during porpoising (see Chaps. 6 and 9 for fuller descriptions), a high-energy swimming behavior where penguins break the surface, leaping into the air to breathe. During the "air phase," which lasts for a fraction of a second, penguins have to inhale a substantial amount of air and open their beak to facilitate the high rates of airflow needed for this to happen (Fig. 8.2b). This effectively bypasses the nares and allows the air to enter the trachea unimpeded.

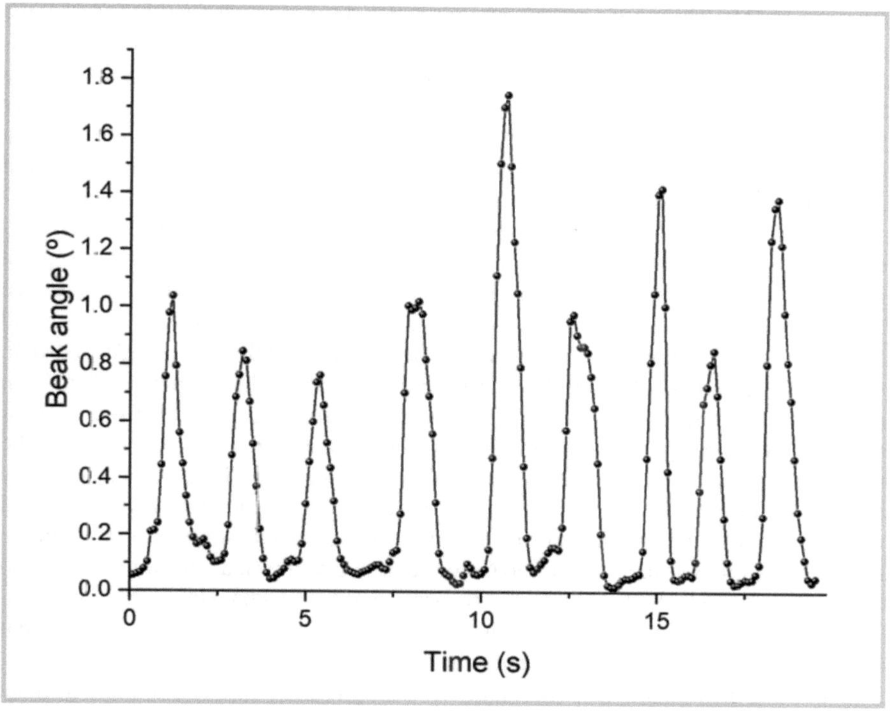

Fig. 8.3 The changing beak angle of a Magellanic penguin resting on the sea surface showing nine clear breaths over about 20 s. Note that the beak opens and closes by between 0.8 and 1.8° during each respiration

The same principle applies when penguins are resting on the surface between dives, during which they take a series of breaths to load up with oxygen (Fig. 8.2c), as they will not be breathing until they get back to the surface after a much longer period (Fig. 8.3). Examination of the beak angle over the full respiration cycle reveals that the beak is open the widest during the highest rates of inhalation and, in Humboldt penguins at least, the rate at which air is inhaled is proportional to the beak angle [103]. This exactly ties in with the simple physical principles relating the rates of airflow, which are greatly enhanced if the passages through which air is to flow are expanded (Box 8.1). What is particularly interesting, though, is that the beak angle rapidly closes down during exhalation (Fig. 8.3) [103].

The reason why this is the case is not necessarily obvious. Panting dogs, for example, do not open and close their jaws with each breathing cycle. The answer to this in penguins may relate to control of airflow. During inhalation, the bird should get air into its system as rapidly (and energetically efficiently) as possible, so low-resistance airways are important, a good reason to open the beak wide. The exhalation phase, however, is appreciably longer than the inhalation phase – in resting Magellanic penguins, for example, inhalation takes about 2 s, whereas exhalation takes 5 s [101]). Thus, because the amount exhaled roughly equals that inhaled, the exhalation phase must have lower rates of airflow, and we suggest that this is actually modulated by a reduced aperture for the exhaled air – the closed beak. What is important for this to work is that diving penguins inhale enough to have a slightly higher pressure within their respiratory system than the ambient air, a bit like a slightly inflated balloon. Indeed, Ponganis et al. [59] suggest that this may be about 3 kPa (Pascal (Pa) is a unit of pressure, for example, air pressure at sea level is typically around 101 kilopascals). In any case, there is a tendency for the air to be passively expelled from a penguin in a manner similar to mammals after great inspiration. Under these conditions, the closure of the beak will effectively reduce the maximum diameter of any airways that lead to the respiratory system and thereby act as a natural physical brake on the flow of air (Box 8.1) [103]. As such, this would modulate what would otherwise be an over-rapid expulsion of air from the pressurized anterior air sacs and perhaps premature passage of air from the posterior air sacs through the lungs, induced by the dramatic pressure drop between the two sites. We can demonstrate the principle on ourselves by inhaling substantially and then, during exhalation through our mouth, experiment with different sized mouth openings. In penguins, air from the posterior air sacs cannot pass through the lungs unless it has somewhere to go (into the partially emptied anterior air sacs), so the process of controlled exhalation, irrespective of the mechanism used, may effectively modulate airflow through the lungs. Clearly, this has to happen at a rate that allows gas exchange to be optimized [93, 103].

This optimization will depend on the necessity to acquire oxygen and eliminate carbon dioxide with respect to the current concentrations of both gases in the body tissues. This is the most critical in the periods that penguins spend on the surface between dives when the birds attempt to minimize time floating and maximize time underwater (which is where the magical fish-bird activities happen!). The first, and rather obvious point, is that penguins spend longer times on the surface recovering from dives if they have spent longer underwater (Chap. 9), and, unsurprisingly, they therefore take more breaths [103]. What is surprising though is the systematic change in beak angle that they incur over the surface pause – and since beak angle is linearly related to the tidal volume (air passing in or out) of single breaths [103], it shows us how much air is being pumped through the respiratory system over time. The pattern is consistent. On arrival to the surface, the emerging penguin takes a huge initial breath (it is what diving humans do, too), followed by other breaths of decreasing magnitude for a time (Fig. 8.4).

At some point, however, but generally about halfway through their time on the surface, penguins start to increase their tidal volume (breathing "deeper"), building it up with subsequent breaths until, just before diving, they take in a last huge breath (Fig. 8.4) (cf. Butler and Woakes [10]). One might expect that penguins, during the inter-dive pauses on the surface, would simply hyperventilate to the maximum because this should allow them to acquire most oxygen. However, our fixation with oxygen in diving animals tends to obscure the role of carbon dioxide in modulating breathing patterns [7]. Carbon dioxide has to be appropriately eliminated, as otherwise animals engaging in overly short surface pauses in repetitive dives will suffer

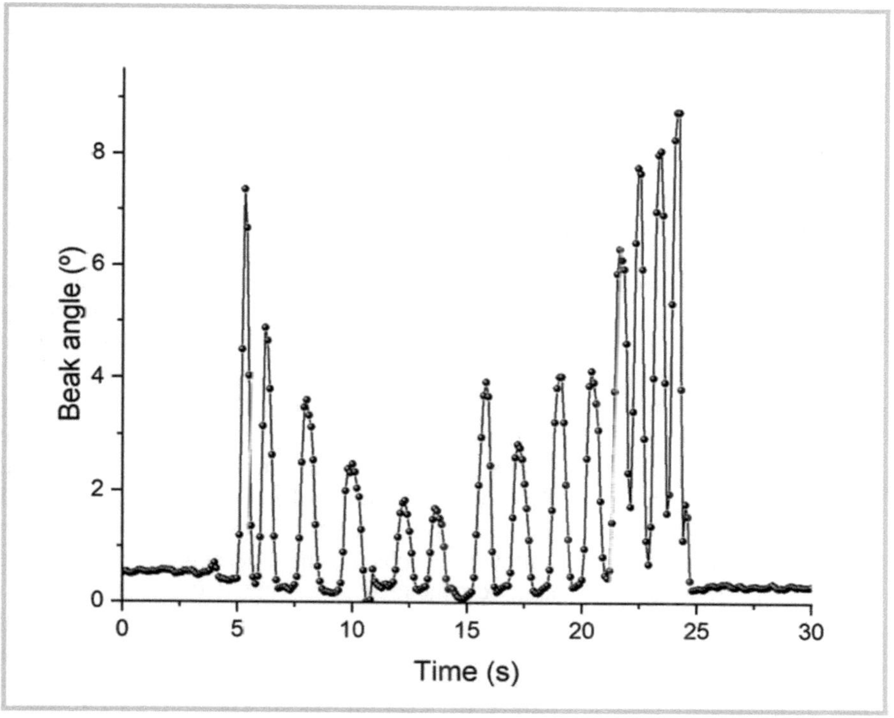

Fig. 8.4 The beak angle of a Magellanic penguin over time during a surface pause lasting about 20 s between dives (starting at 5 s and ending at 25 s). Since the maximum angle during each inspiration cycle is linearly related to tidal volume [103], the pattern shows how the first and last breaths on the surface are the largest, with tidal volumes tending to decrease to a minimum between them. Note also that the respiration rate is the highest closest to the dives, especially just prior to submergence, and that the maximum beak angle per breath close to dives is some four times higher than in a resting penguin (cf. Fig. 8.3)

from an overload of it (hypercapnia, which, in humans, occurs at carbon dioxide partial pressures greater than about 6 kPa – and leads to respiratory failure among other nasty things). The critical point for the between-dive surface pauses is that it takes more time to eliminate carbon dioxide from the body than to acquire oxygen. This is because the diffusion gradient for carbon dioxide is less than that for oxygen and the carbon dioxide has to diffuse from the tissues (where it is produced) into the blood to, then, be transported to the lungs before being exhaled. In addition, the release of carbon dioxide is enhanced when hemoglobin is saturated with oxygen [66, 77]. In other words, carbon dioxide movement into the lungs is promoted when the blood is oxygenated. So, surfacing penguins first take deep breaths to take in oxygen quickly and then settle down for a short period while they eliminate the carbon dioxide. After a period of this, they then hyperventilate again to load the blood with oxygen in preparation for the next dive.

Oxygen Management Underwater

Role of the Air Spaces

The penguin respiratory system is, like those of all birds, extraordinarily efficient at extracting oxygen quickly from the air, with this being a key element that has allowed birds to engage in not only high-cost aerial flight but also aerobic flight. In other words, taking flight does not leave the typical bird exhausted and thus vulnerable and needing to recover upon landing. This is, of course, different from being out of breath, which penguins experience after a deep dive (see above), an activity that many aerial bird species would not even be able to contemplate. Now, prepare your mind to learn about some astounding trickery on the part of penguins, beyond contemplation as to how evolution accomplished it (Chap. 1).

The prime party trick of the bird respiratory system is that air moves through the one-way lung system virtually continuously, as we have noted above. However, all that changes once birds dive. The wonderfully interlinked network of air sacs surrounding the lungs transforms into semi-"blind alleys" of air sacs filled with oxygen (or not – see below) that do not allow gas exchange but, at least theoretically in the case of the posterior air sacs, can access the lungs. However, to do so, the air has to have somewhere to go. If both the anterior and posterior air sacs are full, how might a pressure differential be built up between them that drives this movement, if indeed, it occurs? One possible explanation is that, during dives, birds exhale "used" air from their anterior air sacs to allow the air from the posterior air sacs to move through the lungs and forward [93]. This certainly seems like a possibility. As early as 1975, Gerald Kooyman [43] published a photograph of an Adélie penguin underwater with a large, long air bubble close to its head and neck with the comment "the large bubble…. is probably expirate" (p. 133). More recently, work using cameras on yellow-eyed penguins has noted air being exhaled by birds swimming along the seabed in the latter part of their dives [49]. If the oxygen in the air in any part of the respiratory system was so critical, why would they do that? Finally, our experience with African penguins also points to air being expelled in the latter part of an underwater submergence. We choose this term deliberately because our experience comes through serendipity (as does much science) while working with birds that were not actually diving in the normal sense. While we (Rory actually) were holding an African penguin (next to a bucket of water!), a kelp gull flew over and, by executing a dive bomb, managed to vent an impressive blob of feces onto the poor penguin's head, including in its eyes. Just wanting to sort that out, Rory dipped the penguin's head in the water. The idea was to just perform a brief dunk, but the bird immediately started looking round the inside of the bucket with interest (there was some writing at the bottom), so we left its head underwater while it did that. The bird seemed quite happy, so we decided to let it continue until it indicated otherwise – it could, in any event, take its head out of the water by bending its neck appropriately whenever it wanted. After some time, it began to dribble air from its bill before it lifted its head out to breathe. Fascinated by this, we tried a few more individuals, all of which displayed the same behavior, starting to expel air after about a minute for a duration of maybe 20 s before they raised their heads out of the water to breathe. Finally, some theoretical calculations were made about this, based on buoyancy, and the results indicate that emperor penguins may also exhale air during the latter part of their dives [74]. Again, as with yellow-eyed penguins, why would birds, which are constrained in actual breathing, exhale when seemingly intent on using the last molecules of oxygen during their dives?

Most light is shed on this, and other subsurface internal issues, by an extraordinary body of work conducted over a number of years, primarily by researchers from the West Coast of North America (including Gerry Kooyman, Paul Ponganis, Cassondra Williams, and Jessica Meir). They used emperor penguins as their model penguin because of their spectacular diving capacities as well as because they could keep them captive without restraint and feed them by themselves. They could do this by fencing in a 3-meter-wide hole drilled through fast sea ice too far from open water for the penguins to reach in one breath hold. Within the fence were several penguins, free to forage whenever they wanted to do so. The researchers could then conduct experiments with these captives– this was the so-called "Penguin Ranch," an incredibly creative idea! In one of their most recent offerings, Williams et al. [93] have measured the partial pressure of oxygen (PO_2) in the posterior and anterior air sacs of these birds as they dove. They point out that, far from being a one-way respiratory system, the air in the anterior sacs might oscillate with that in the posterior sacs, mixing their air in a way that does not normally occur in birds. They suggest two mechanisms for this. One is the movement of air between the anterior and posterior air sacs via the bronchial system, rather like roads bypassing a town, in this case, the "lung city." The other is movement of air between them via the city – the lungs. For either case to work, there has to be a difference in pressure between the two air sac sites to instigate a flow from higher to lower pressure. A potential mechanism for this was suggested to be the movement of the flippers during locomotion [5], presumably based on the fact that flighted birds have wingbeats that tie in with their breathing cycles [6, 9]. Certainly, in little penguins at least, there is a difference in pressure between the anterior and posterior sacs, and it fluctuates with flipper beats. However, this difference only amounts to 0.06 kPa [5]. Put in another way, this is the difference between something being in air or subjected to the pressure of just 6 mm underwater, hardly anything even to make ones' ears pop. Of course, this system would break down during a large part of the ascent to the surface, when responding to buoyancy, as flipper beats are reduced or absent (see Chap. 6). Given the dramatic effect of tube diameter on modulating flow rates (Box 8.1), the central question arising is whether the resistance to airflow of the lungs or the bronchial system, combined with the small and rapidly oscillating pressure differentials created by the flippers, is enough to obtain meaningful air mixing to allow penguins to use their respiratory oxygen levels fully.

In fact, aside from the minimally changing respiratory system pressures induced by flipper movement, there are much more radical, and differential, pressure changes happening in penguin air spaces during dives. Specifically, if there is differential pressure between different gas volumes within the respiratory system, there is a mechanism that could drive air movement within that system (Box 8.1). This is illustrated by the work conducted by Cassondra and colleagues showing that, as diving emperor penguins descend the water column, the partial pressure of oxygen (Box. 8.2) in both sets of air sacs increases, although the observed changes are not identical in both the air sacs (Fig. 8.5).

Box 8.2

In an ideal gas mixture, each gas in that mixture has a partial pressure. This is the pressure that it would have exerted if it occupied the full volume alone. This is given by the formula:

Partial pressure = total absolute pressure × volume of the gas component

This formula makes it clear why gas partial pressures depend on depth because, within this formula, the total absolute pressure is the pressure applied to the volume, which, in the case of the penguin, is the pressure applied by depth (see Chap. 6). As an example, air inspired on the surface that is 21% oxygen (therefore, with a partial pressure of 21 kPa at 1 atmospheric pressure (= 100 kPa)), taken down to 100 m depth (≈1000 kPa), has a partial pressure of 231 kPa.

Indeed, although Cassondra is rather critical of the ability of her sensor system to work effectively under high pressure (stemming primarily from depth), examination of her raw data shows that the partial pressure of oxygen in the posterior and anterior air sacs also varies fairly systematically with depth (Fig. 8.5).

This ties in with Boyle's Law (Chap. 6, Box 6.1), which details how air volumes should decrease with increasing pressure applied to the penguin by water (with pressure being a direct function of depth). So, the increasing oxygen partial pressure with depth shows that both the posterior and anterior air sacs change volumetrically with depth (although there seem to be interesting differences between the two air sacs and neither quite decreases as much as expected (Fig. 8.3), part of which may be explained by oxygen consumption during the descent as well as differences in gas solubility between oxygen, carbon dioxide, and nitrogen in the blood, plasma, and tissues [25]). It should be noted that it is relevant to assume that the increase in partial pressure with depth is translated to the lungs because it should help increase oxygen transfer to the blood, speeding up the process because the molecules move faster across higher concentration gradients (Box 8.3).

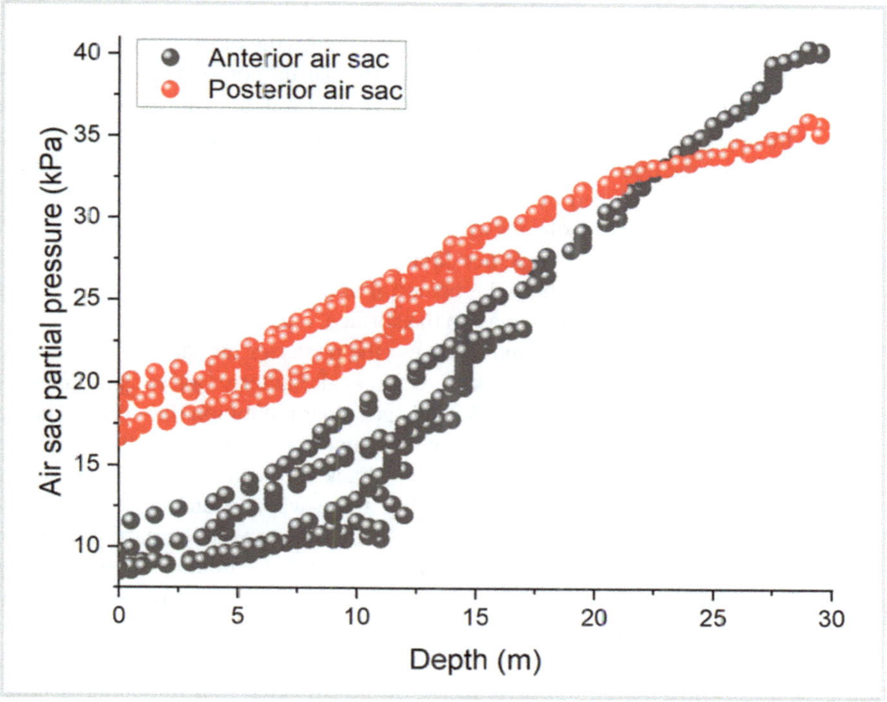

Fig. 8.5 Change in oxygen partial pressure with depth during the descent phase of four "shallow" dives executed by an emperor penguin (original data were obtained from Williams et al. [93]). The increase in partial pressure with depth demonstrates that volumetric changes occur within the air sacs (see Chap. 6, Box 6.1). In totally compliant air sacs, the partial pressure would be expected to approximately double in the first 10 m of descent, although the consumption of oxygen over time is likely to modify absolute values to some extent

Box 8.3
The rate of diffusion of a gas through an alveolar-capillary barrier (dV/dt) is well-described by Fick's Law of diffusion. It depends on the surface area (A) and thickness (T) of the barrier, the diffusion coefficient of the gas (D), and the difference in partial pressure of the gas across the barrier (P_1-P_2) and is described by the equation:

$$dV / dt = \left(AD\left[P_1 - P_2\right]\right) / T$$

Conversely, as penguins return to the surface from a dive, the process is reversed. Overlaid on this, of course, is the removal of oxygen by the birds, which affects the proportion of oxygen in the mix and therefore the partial pressure (Box. 8.2). This is further complicated by not knowing precisely how much air birds have inhaled prior to their dives (though there are some reasonable estimates – see below). The volume may change due to oxygen extraction [93], the fact that the volume of oxygen extracted is not replaced by the same volume of carbon dioxide [25], and the possibility that birds may exhale anyway (see above), thereby changing the volume.

Despite Cassondra's misgivings, her data show, in diving emperor penguins, that despite substantial variability, both the posterior and anterior air sac oxygen partial pressures measured when the bird was on the surface, regardless of its dive duration, decreased between the start and end of dives (the posterior air sac partial pressure decreased from about 9.6 kPa to 5.9 kPa, whereas the anterior air sac partial pressure decreased from 8.4 to 5.3 kPa) [93]. This is a clear indication that oxygen in the air sacs is being used, despite them having no structures for gas interchange within themselves. In other words, there must be some back and forth interchange between the air sacs and the lungs, rather than it being a one-way system, as is commonly cited for aerial birds. In fact, we can examine coarse changes in the partial pressure of oxygen in the posterior and anterior air sacs over the course of dives using Cassondra's data. Simplistically, if both the posterior and anterior air sacs were fully open to the lungs and each other (via, e.g., a low-resistance bronchial system), then we would expect oxygen partial pressures to be the same in both air sacs with, overall, both decreasing during the dive, as would be expected in a mammalian lung. Examination of four consecutive dives in an emperor penguin (to depths not exceeding a few tens of meters (see Fig. 8.5)) shows that, although the partial pressure in both sets of air sacs (depth effects taken into account; see Fig. 8.5) tends to decrease with time, there are substantial differences between the oxygen partial pressures between the two sets, with these varying semi-systematically as the dives progress (Fig. 8.6).

Specifically, the posterior air sac oxygen partial pressure is higher than that of the anterior air sac throughout much of the dive. However, as they both change with time and depth (Figs. 8.5 and 8.6a), the posterior air sac oxygen partial pressure decreases relatively faster [93] until it may at times undercut that of the anterior air sac (Fig. 8.6b) (cf. [93]). This is underpinned by the calculated depletion rate of oxygen in the posterior air sacs, which is calculated to be about eight times faster than that of the anterior air sacs [93]. So, what processes might account for these observations?

It seems that the compliancy of bird air sacs [35, 66], which contrasts with the rigidity of their lungs [46], coupled with the radical pressure changes that occur with depth, could be a major driver of air movement within the penguin respiratory system (cf. Ponganis et al. [59]). First, the increases in oxygen partial pressure with increasing depth demonstrate that the air in both sets of the air sacs is compressed (Fig. 8.5). However, the decrease in volume is not expected to occur uniformly throughout the system. Indeed, we suggest that the relative rigidity of the lungs leads to the compliant air sacs incurring virtually all the volumetric changes [59],

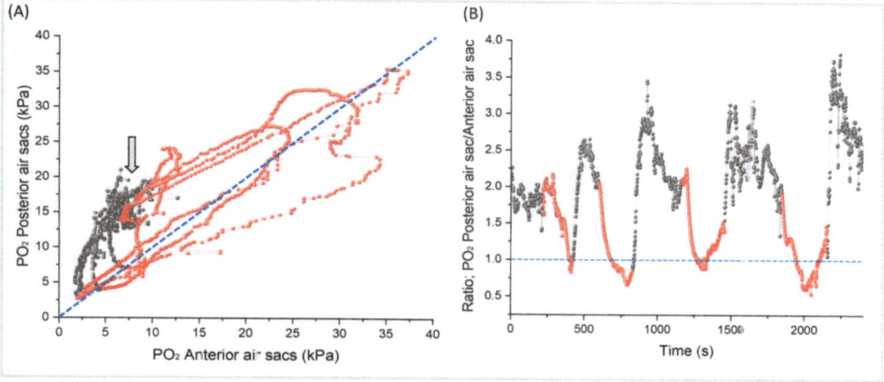

Fig. 8.6 The partial pressure of oxygen (PO_2) in the posterior air sacs compared to that in the anterior air sacs in an emperor penguin during four consecutive dives, depicted by the four red loops; the inter-dive intervals on the surface are shown in black. (**a**) The immediately pre-dive PO_2 (indicated by the gray arrow pointing to the start of the red-marked traces) is appreciably higher in the posterior than in the anterior air sacs (the blue dashed line shows equivalence). This condition is maintained for much of the dive (illustrated by the red-marked ovoids in the trace that progresses in time clockwise from the arrow, primarily increasing, and then decreasing with depth). The ratio of PO_2 posterior air sacs/PO_2 anterior air sacs in panel (**b**) makes it clear, though, that, as the dives progress, the posterior air sac oxygen partial pressure approaches, and possibly sometimes undercuts, the anterior air sac oxygen partial pressure. The black points show reoxygenation of the air sacs between dives. (Original data were obtained from Williams et al. [93] – subjected to a 50% median filter)

with the air from both sets of the air sacs – which during normal breathing flows in the one-way system – moving into the lungs (Fig. 8.7).

So, for example, an Adélie penguin diving to 10 m with surface lung and air sac volumes of around 25 and 296 mL/kg, respectively [59], would have an overall decrease in air volume of 160.5 mL/kg. If this were only enacted in the air sacs due to their flexibility, then they would incur a net movement of 135 mL/kg of air to the lungs, providing the oxygen in that volume for gas exchange (Fig. 8.8).

Importantly, the pressure difference between the air sacs and lungs during even such a shallow dive is very much higher than that between the posterior and anterior air sacs measured in penguins due to flipper beating (which, as noted above, amounts to just a 6-mm water pressure difference). The latter was proposed as a driver of air movement in the respiratory system [5]. Deeper dives would incur even greater amounts of air moving from the air sacs to the lungs using the same principle (Fig. 8.8).

This mechanism, driven by substantial forces, would allow a large proportion (perhaps all – depending on the depth) of the air in the air sacs to move to the lungs (see also discussion of barotrauma, below), allowing penguins to access the oxygen. The proximity of the collapsing air sacs to the lungs would also presumably tend to lead to the oxygen partial pressures of the lungs effectively mirroring those of the air sacs. Under these circumstances, given the higher oxygen partial pressures in the

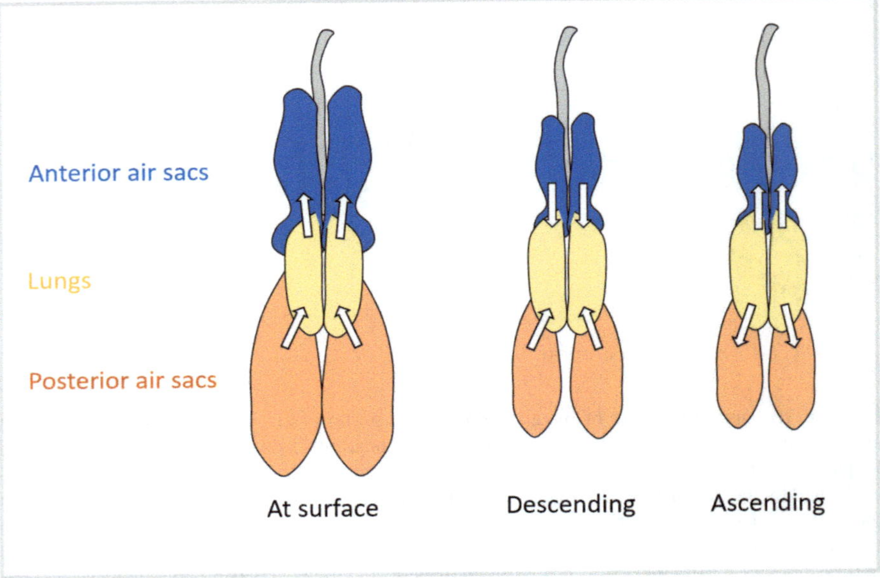

Fig. 8.7 A schematic diagram showing our interpretation of airflow (white arrows) within the lungs and air sacs of diving penguins. On the surface, the normal one-way system is used, with the air from the trachea entering the posterior air sacs before passing through the lungs and into the anterior air sacs before exiting via the trachea. Once a bird dives, however, the radically increasing pressure with depth will impose that pressure on the flexible air sacs, not only compressing their air volume but also forcing it from the air sacs into the noncompressible lungs. This movement of air presumably occurs from both the posterior and anterior air sacs, changing the original one-way system used during normal breathing. As a penguin returns to the surface, thus reducing depth and therefore pressure, the process is reversed

posterior air sacs than that in the anterior air sacs (because, initially at least, the air has not passed through the lungs), we might expect the oxygen at this site to be depleted faster (because it diffuses into the blood faster due to its higher partial pressure), which is what has been observed [93] (Fig. 8.6).

Upon the penguin's return to the surface, this situation is reversed. Air compressed within the lungs is expected to expand into the air sacs (Fig. 8.6). Here, too, water pressure may play a role in affecting posterior versus anterior air sac oxygen partial pressures. In emperor penguins, the difference in distance between the centers of the anterior and posterior air sacs is ~30 cm. In a bird traveling vertically upward, this translates to a difference in pressure of 3 kPa (again, some five times greater than the differential pressure induced by flipper movement – see above) – though an angled ascent would diminish that difference. This differential pressure should therefore affect the two sets of air sacs' volumes differentially, particularly nearer the surface (Chap. 6, Box 6.1). It may be, therefore, that in an ascending bird, the anterior air sacs tend to have lower pressure and thereby expand with lung air preferentially over the posterior air sacs, with the latter being filled with more oxygen-depleted air expanding out of the lungs later. Indeed, this process may

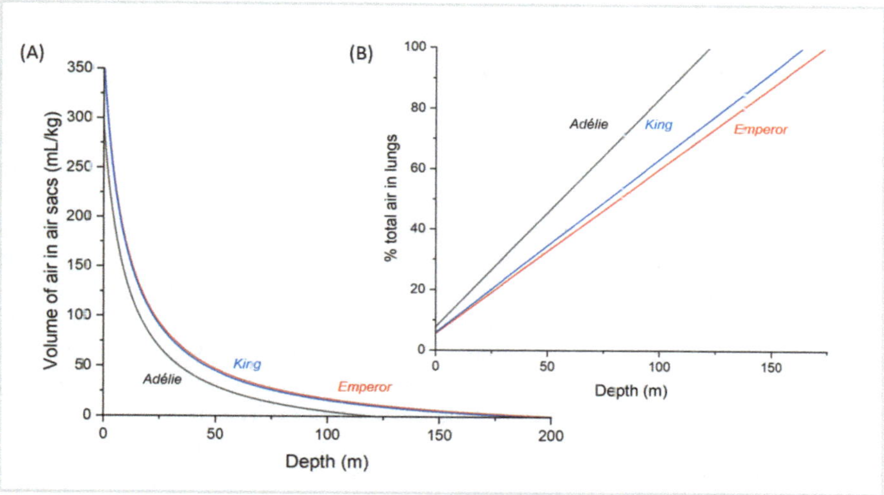

Fig. 8.8 (**a**) The volume of air (volume per unit body mass) that is predicted to occur in the air sacs of diving penguins as a function of depth and (**b**) the percentage of the total air inspired that is predicted to be in the lungs (and therefore available for gas exchange) as a percentage of depth. Data on lung and air sac volumes for penguins on the surface were taken from Ponganis et al. [59]. These graphs assume that the flexibility of the air sacs accounts for the total air volume decrease with depth, which is calculated using Boyle's Law (Chap. 6, Box 6.1). The volume of the trachea, which is relatively small, is not taken into account in this model

possibly account for the decreased oxygen partial pressure of the posterior compared to the anterior air sacs (Fig. 8.6b).

In a nutshell, the form and function of the avian one-way respiratory system, which presumably evolved to allow high rates of oxygen uptake to power energetically expensive flight, has particular meaning for a penguin that has to cope with no new oxygen intake or carbon dioxide outtake and dramatic changes in pressure the instant that it dives. The pre-dive scenario mirrors that expected of any birds, with the posterior air sacs with freshly inhaled air having the highest oxygen partial pressures and the anterior air sacs having lower oxygen partial pressures because that air has already passed through the lungs (Figs. 8.6 and 8.7). During diving, however, the assumed relative incompressibility of the lungs and the flexibility of the air sacs must mean that, as birds descend the water column, air would be compressed into the lungs and, presumably, from both the anterior and posterior air sacs. Thus, despite having an extremely reduced volume compared to the air sacs, the penguin lungs have access to most of the air sac oxygen, a process that is enhanced with depth (Fig. 8.8). On top of this, penguins appear to inhale greater volumes of air when they "know" that they will be diving to greater depths [74, 105, 106], thus providing them with greater oxygen stores for these longer dives. For both these reasons, greater depths should allow penguins to access more of their air sac oxygen stores so that, with all other things being equal, deep-diving penguins would remove proportionately more oxygen from their air sacs, a process that may also partly explain why dive duration tends to increase with depth (see below). We also note that we expect diffusion between the increasingly small (with depth) air sacs and the lungs to allow further oxygen transfer anyway. Overall, this would seem like a great way to get oxygen into their body efficiently and points to the importance of external pressure (depth) in being the driver of this.

These advantages just explained, however, are turned on their head when penguins return to the surface. As the pressure decreases, the air in the lungs must expand out into the air sacs again, effectively removing the oxygen from the site where it can move into the body, thereby precluding the bird from using it, aside perhaps from some limited diffusion. This is obviously not ideal but is a consequence of the way in which the respiratory system is constructed. It is therefore fortunate that penguins returning to the surface from depth have such low power and thus oxygen usage (see Chap. 6). It might also help explain why penguins returning to the surface after a dive tend to surface fairly promptly (Chap. 9) (but see [72]), rather than extending time near the surface. Superimposed on this, there may be an effect of penguin body angle, which effectively creates a differential partial pressure between the two air sac systems (see above), complicating the interpretation of oxygen partial pressures over time and depth (Fig. 8.6).

Gas Exchange to Body Tissues

The respiratory air spaces are not the only places that penguins store the oxygen that they can use during their dives. There are two other major places. First, the blood, where oxygen is bound to hemoglobin, and, second, the muscles, which store oxygen using myoglobin. These stores are, however, not substantial in the brain, which, together with the heart, needs special treatment during the dive, as we shall see later.

During the period on the surface between dives, penguins breathe deeply and hyperventilate (see above), thus exchanging the air in their air sacs and lungs so that the oxygen concentration increases [51, 60]. The increase in oxygen partial pressure, which contrasts the typically low partial pressure in the blood post-dive [41, 51, 52, 60], results in faster diffusion of oxygen in the lungs. This happens because the rate of diffusion is proportional to the difference in partial pressure. The process is enhanced by increased heart rates (tachycardia) [28, 51, 52], which sends the oxygen-poor blood rapidly past the lungs, whisking away the, thereby, oxygenated blood (to take it to the deoxygenated parts of the body). Thus, the oxygen gradient is maintained as high as possible. High heart rates also help "oxygenate" the venous side of the vascular system (arteralizing the venous system). Since this is usually the larger reservoir of blood, any increase in oxygen partial pressure on this side significantly increases the overall body oxygen store. In any event, once oxygen is in the blood, it can access all areas of the body, a process also enhanced by the higher heart-beating rates that provide the power to drive the blood throughout the full body blood network, providing the base for aerobic metabolism at any site that "needs it."

However, once penguins dive, all tissues that use oxygen are not considered equal in terms of how much oxygen they should receive. For penguins, and most vertebrates actually, there are two crucial places that must have oxygen all the time, and these are the central nervous system (mainly the brain) and the heart. Neither of these can be allowed to malfunction due to lack of oxygen because, otherwise, there will be severe consequences (like death)! In contrast, there are places within the body that can withstand low (even inadequate) oxygen levels for extended periods, and, therefore, the oxygen supply can be more relaxed. Early research pioneers, starting with Lawrence Irving and Per Scholander [78], identified that in diving animals, an important process for modulating the parts of the body that receive (how much) oxygen is "peripheral vasoconstriction"– selective constriction of the blood vessels due to contraction of the muscles in their walls [63]. Although most of their original work was conducted on seals, it seems that much (though not all) of what they reported may be applicable to penguins. The distribution of blood in a penguin body is exemplified by a huge interconnected network of canals (actually the blood vessels) of varying diameters. It therefore follows that the act of reducing the width of canals in certain areas will profoundly affect the amount of blood that flows through any affected canal and that part of the body (big rivers have plenty of water passing along– but, not so, little streams).

The work conducted by Handrich et al. [31] on variations in temperatures within penguin bodies (see Chap. 7) certainly points to changing vasoconstriction affecting blood flow by highlighting the great differences in temperatures between body parts, even some that are adjacent. The logic behind this is that if the blood flow is curtailed somewhere, then so is the heat that comes with it, hence causing the variation in temperature. In some senses, invoking vasoconstriction to explain how penguins might change the internal architecture of their blood flow is the easy part. What is horrendously complex is to figure out what is going on where and when. Not surprisingly, this is yet to be evaluated. Anyway, mirroring what is a generally accepted dogma for marine mammals, we would advocate that selective vasoconstriction of the blood vessels within penguin bodies has the potential to prioritize oxygen to the heart and the brain, the two critical components of the penguin physiology that cannot be allowed to be overly oxygen-deficient. This would happen to the detriment of other, less critical areas (Fig. 8.9).

If we simplistically represent this process as an actual break between the heart and the brain and the rest of the body, where does the oxygen in the heart–brain section come from, especially given that the brain has no oxygen-storing pigment within itself and so is a net oxygen "taker"? Well, in keeping with our discussions on barotrauma (below), we suggest that the lungs would be the oxygen providers and, therefore, that the critically sensitive section of the body to oxygen deficit consists of a "pool" involving the brain, the heart, and the lungs. Linked to this, as part of the general blood-redistribution tactic, diving penguins slow their heart rate (e.g., [12, 28]), also something that is considered to mitigate hypoxic damage [2]. In essence, assuming that blood redistribution, as suggested, takes place, the heart only has to ensure a fast track of oxygen between the lungs and the critical sites (primarily the brain and heart), and it does not have to beat hard and fast to push blood

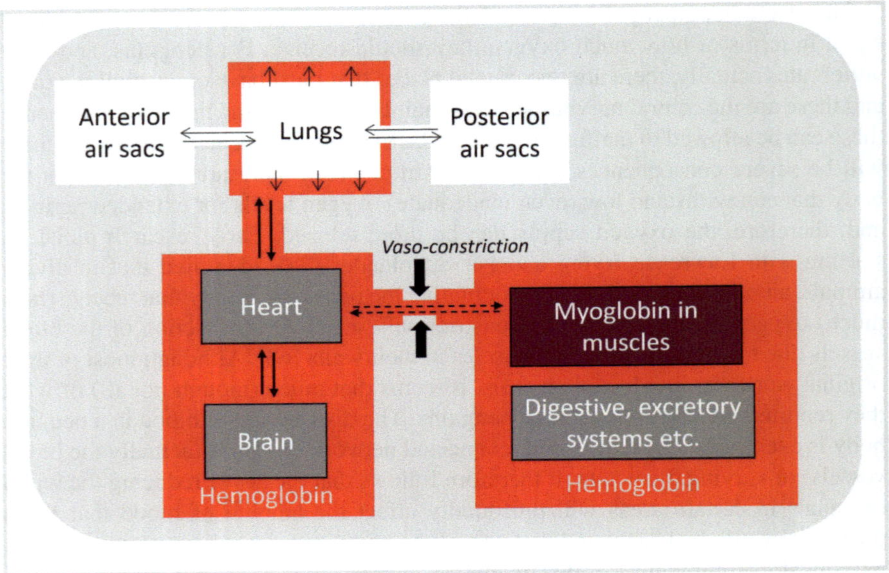

Fig. 8.9 A schematic diagram of theorized oxygen transfer pathways in a diving penguin. Oxygen coming from the lungs (open arrows – the two-way movement of air is discussed above) is contained in all parts of the system, but, in order to maximize the dive duration, birds may employ selective blood vessel vasoconstriction so that blood that has been oxygenated from air in the lungs – and is bound to hemoglobin proteins in the blood – is pumped around (closed arrows) so that it prioritizes the heart and brain. Oxygen deficits in other body parts can be overcome during post-dive surface intervals

throughout the entire blood network. Thus, slowing of the heart rate, which we shall call bradycardia during dives (although many authors do not regard it as true bradycardia because it is not below resting levels (see [10] for some discussion on this [52]), would be a manifestation of that.

Bradycardia is also important because it reduces the energy (oxygen) expenditure of the heart, which can be considerable and even greater the faster it beats [20]. In support of this, and how it relates to the general dive planning (which is a recurring theme throughout this book – see, e.g., Chaps. 9 and 10), bradycardia is more extreme in longer dives [52] when oxygen needs to be used particularly carefully. However, dive planning apart, things can change unexpectedly, and it would be nice to believe that penguins had a physiological plan for that too, which would accord with our suggestions. This does seem to be the case. In observations of captive Humboldt penguins that were allowed to dive as they pleased but then suddenly precluded from surfacing in their pool to breathe, the drop in heart rate form the surface levels (fell from around 180 beats per minute to about 130) became extreme (dropped to around 70 beats per minute) [10]. In this case, the birds were instantaneously faced with an unknown dive duration and had to throw out all the stops to plan for what might be a unpleasantly long dive ahead.

In tandem with all this, other parts of the body that are normally well-oxygenated may also shut down during dives. A great example of this is provided by Gerrit Peters' excellent PhD thesis [57], which demonstrates how both Adélie and gentoo penguins reduce their stomach churning during their dives, reinitiating it dramatically again the instant they return to the surface. Stomach churning not only needs oxygen for the muscles that elicit it but it also leads to higher oxygen demand to heat up the stomach contents (see Chap. 7). Thus, there are clear advantages to abstaining from digestive processes underwater (another advantage is that the food they capture is often intended for the chick waiting a long distance away, and, so, digesting it beforehand diverts the purpose of the trip). The value to the penguin in stopping digestion underwater is immediately obvious because the bird can digest at its leisure when the unlimited oxygen availability on the surface allows.

This situation is not so clear-cut for the swimming musculature though. The penguin flipper, that beats rapidly and efficiently, when swimming is powered by huge pectoral and supracoracoideus muscles (see Chap. 6), and the implication is that blood redistribution would at least partially preclude these muscles from acquiring the oxygen they need to function despite the oxygen bound within myoglobin and hemoglobin. However, they must function. The literature has been, perhaps purposefully, vague about this – at the very least, it is somewhat noncommittal [14]. However, researchers do all agree that in the absence of oxygen, muscles must work anaerobically, producing energy-supplying adenosine triphosphate (ATP) from carbohydrate, the end product of which is energy and lactic acid. Not only is this process unpleasant (lactic acid is what causes muscle stiffness and cramps in athletes) but it is also inefficient in terms of energy release (it provides just 1/18th as much energy as the aerobic pathway). Moreover, it takes an order of magnitude longer to reconvert the lactate back to a metabolic substrate that can be subsequently aerobically used when birds are on the surface. What this means is that, if the swimming muscles do sometimes function anaerobically, and this most likely happens in longer dives, the surface pauses following such dives should be greatly extended. Actually, there is little to no evidence that this occurs in Magellanic penguins (Fig. 8.10) nor in a suite of other species, including Adélies, gentoos, and chinstraps [96] (see also diving efficiency in Chap. 10). So, maybe the blood and muscle oxygen-binding pigments have enough oxygen in them to supply normal dives adequately anyway?

That surface periods immediately following dives should exactly mirror the length of the dives (or better still, the energy expended during the dive because that equates with the oxygen used) need not always be the case, though. There is nothing to stop penguins from diving again even with some lactic acid buildup, as long at the overall process of diving over time does not cause critical levels [13]. This is certainly not the case in king penguins, as existing data show, for example, birds diving to depths in excess of 140 m continuously over 5 h without any extended surface pauses (Fig. 8.11). This clearly allows no time for any lactate buildup to be eliminated.

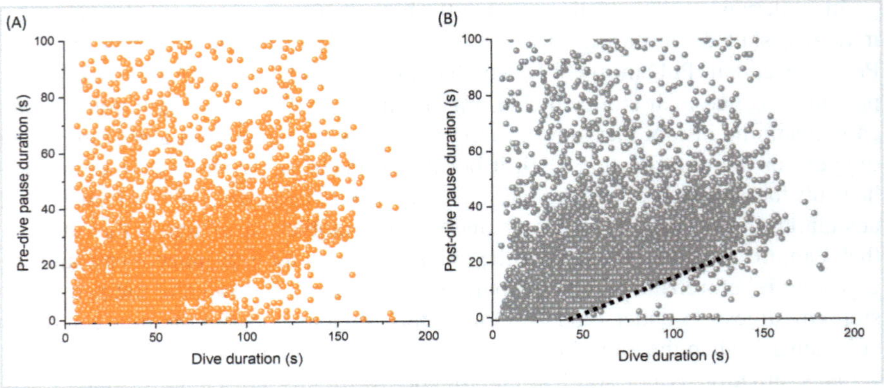

Fig. 8.10 Duration of surface pauses – during which time penguins take in oxygen and eliminate carbon dioxide – in relation to the duration of the (**A**) preceding dive duration and (**B**) following dive duration for Magellanic penguins engaged in foraging trips off Peninsula Valdes, Argentina. The surface pauses greater than 100 s are not shown since birds can spend extended periods on the surface for purposes other than gas exchange (preening, sleeping, etc.). The dark band of high point density extending linearly from zero, which probably represents pause durations uniquely associated with gas exchange, should be noted. As part of this, the lower limit of pause durations (delineated by the dotted line in panel (**B**)) is rarely undercut, with the extremely short durations presumably being followed by short dives and longer pauses. The generally linear relationship between pause duration and dive duration indicates that dives are probably aerobic. Were they to be anaerobic, particularly after long dives, the pause duration would be expected to become disproportionately long

One obvious case of lactic acid buildup was an incident reported by Sato and colleagues [74]. Here, an emperor penguin participating in "ice hole" experiments (Ponganis' "Penguin Ranch") dove for an astonishing 27.6 min. After this extraordinary feat, the penguin remained prone for 6 min, and took 20 min off before walking, and did not dive again for another 8.4 h. We were surprised it ever went back in the water at all after such an experience! Presumably, this bird simply had difficulty finding the exit hole and pushed its diving close to the limit. It is worth noting that free-living, foraging emperor penguins rarely dive for longer than 12 mins [92].

The Aerobic Dive Limit (ADL) and Beyond

The appearance of lactic acid in penguins, therefore, indicates anaerobic processes taking place, although, as has been pointed out, lactic acid is also produced in heavily exercising terrestrial animals that cannot provide oxygen fast enough to underpin the metabolic demands [14]- 100-m Olympic sprint athletes do not take a breath during their race, though those of longer races obviously have to do so. However, in general, an idea of the extent of the limits of aerobically powered diving in penguins, the aerobic dive limit (ADL), can be obtained either by measuring lactate

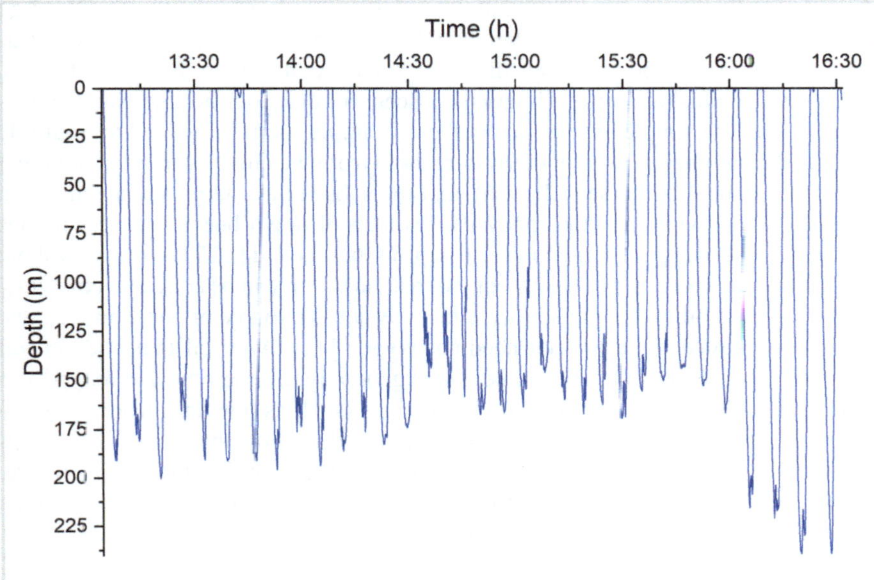

Fig. 8.11 The dive depth of a king penguin over a number of hours showing impressive, continuous diving with no extended surface periods within which lactate debt (or excessive nitrogen – see the bend below) might be dealt with. (Data courtesy of Charly Bost)

directly, a virtually impossible task in freely diving penguins, although Gerry Kooyman (who else?) did manage this in Weddell seals [39, 44]. On the other hand, this could be estimated simply by dividing the calculated oxygen stores by the rate at which they are consumed and calling it the cADL – the calculated aerobic dive limit [13]. This assumes that all the oxygen in the body has been used before lactate is produced, which is unlikely to be the case [13]. Ignoring this inconvenience, though, are the multiple studies that have attempted to determine the metabolic rate of penguins underwater (covered in some detail in Chap. 6) and a number that have attempted to quantify oxygen stores. The latter task is not trivial because it needs inter alia data on blood volume (and hemoglobin content) as well as the proportions that are venous and arterial, muscle masses (and myoglobin content), and the degrees of oxygen saturation of the various elements [63]. In a nutshell though, for, e.g., Adélie, emperor, and king penguins, respiratory air oxygen is calculated to be 200, 125, and 117 mL/kg, blood oxygen is 83, 87, and 100 mL/kg (with hemoglobin at 16, 18, and 18 g/dL), and muscles constitute 40, 33, and 12–25% of body mass (with myoglobin at 3, 4.3, and between 2.1 and 6.4 g/100 g), respectively [63]. All this translates to Adélie, emperor, and king penguins having 63, 55, and between 63 and 68 mL O_2 per kilogram body mass, respectively, with oxygen being fairly evenly distributed between the lungs, blood, and muscles.

Use of such information in calculations of how long penguins can stay underwater while swimming aerobically has proved to be somewhat of an embarrassment.

Multiple studies have concluded that penguins regularly exceed their ADL, e.g., gentoo and king penguins exceed their cADL in 40–55% of their dives [11], whereas Adélies do so in 14–50% of their dives [58]. They continue to dive at high rates, nonetheless, with short surface intervals, a scenario that does not indicate any lactate buildup. The simple answer is that there is ample room for (possible) multiple errors in the parameters used in the system, both from the energy use side, including the biomechanical costs (see Chap. 6), and the substantial energy savings that can be made by allowing selective body tissues to cool down because cooler tissues simply use less energy (detailed in see Chap. 7) (but see [61]). There is also wobble room within the oxygen storage aspect. To that, however, can be added "unknowns," such as the possibility that phosphocreatine may act as a source of phosphorus in the production of ATP, therefore not needing oxygen [11]. After all that, the general consensus now seems to be that all penguins generally dive aerobically, maybe as a result of a multitude of incremental clever tricks (such as low drag leading to low power use and tissue cooling etc.; Chaps. 6 and 7) that allow them to do this, keeping them for a maximum time underwater and minimum on the surface. Therefore, they can acquire their prey most efficiently, as fish-birds. Indeed, it is hard not be impressed when you learn about penguins' underwater capacities!

The Importance of Size in Dive Performance

In Chaps. 1 and 2, we reviewed the size range exhibited among penguins and some of its ecological ramifications (see Fig. 1.7). We warned the reader, too, about there being much more to the story. Here it is.

Duration

Ultimately, irrespective of whether aerobic or anaerobic, the exploitation of underwater prey might be reasonably assumed to be the best in species that can spend the longest amount of their time underwater. After all, if a penguin encounters a large, but rare, aggregation of prey, it can consume more if it can remain for extended periods exploiting it. As noted in Chap. 4, some prey are more accessible than are others, but they may differ in quality. Strategy is also used by penguins in a flock, as we expect all members of that flock to maximize their prey capture rate while the prey, of course, do their best not to be eaten! Some idea of projected maximum dive durations for such interactions can be obtained by dividing the total body oxygen stores by the rate at which they are used (the cADL discussed above). The other option is to cut to the chase and simply look at the maximum dive durations and the length of the period of repeated diving, in wild penguins (Fig. 8.12).

Dive durations show a pattern that is fairly consistent across air-breathing vertebrates – bigger animals can hold their breath for longer [79]. The details of how,

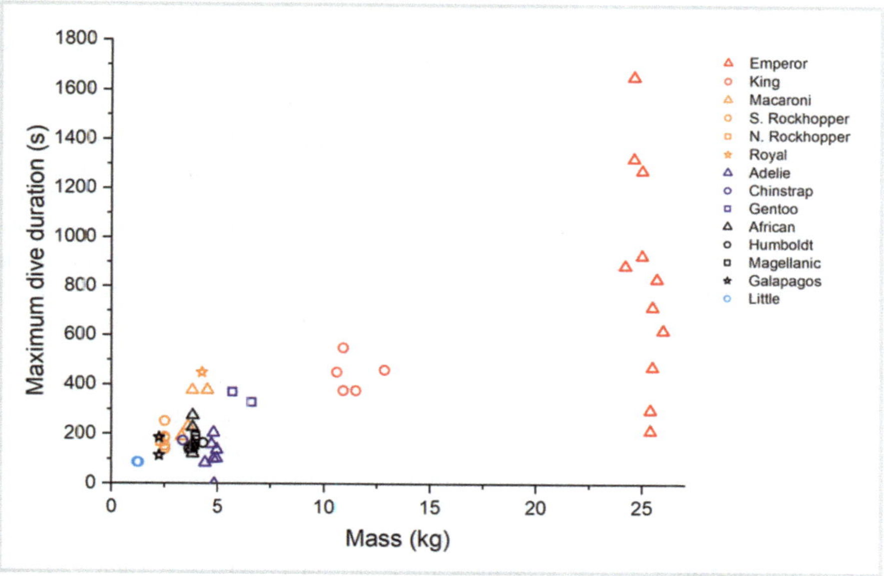

Fig. 8.12 The maximum dive duration reported for various penguin species (the genus is color coded) as a function of their mass for different studies (data were obtained from the Penguiness website (penguiness.net)). The variation in species-specific maxima, which is presumably due to many elements that are liable to affect dive duration, including water temperature (Chap. 7), swim speeds (Chap. 6), prey types (Chap. 10), prey depth, and the density between and within schools (Chap. 11) as well of, of course, sample size should all be noted

exactly, penguin species' maximum dive duration varies with their mass is an unfolding issue as researchers gather more data. This is because single maximum durations are improbable and therefore rare so that greater values are more likely to occur in larger datasets. However, researchers also like to factor in phylogenetic factors into their regressions of performance against mass to correct for biases introduced by species relatedness [30]. This apart, the general explanation as to why larger species of diving animals, including penguins, can dive for longer is grounded in the "oxygen store/usage hypothesis." This says that body oxygen stores increase linearly with mass so that each gram of animal tissue holds the same amount of oxygen, no matter how large the animal is (i.e., oxygen stores scale isometrically). In contrast, oxygen usage per gram of animal tissue (mass-specific oxygen use) decreases as animals get larger (the exponent of oxygen usage against mass is typically 0.67–0.75 [30]). So, weight for weight, small animals use their oxygen reserves faster, and this certainly seems to apply to penguins (Fig. 8.12).

It is important to stress, however, the role that ecological circumstance plays in the outcome of maximum dive duration versus mass representations. We have attempted to highlight this in our figure by producing study-specific values rather than simple absolute maxima as is done conventionally in the literature. The most extreme example is perhaps demonstrated by emperor penguins in which, for this one species, study-specific maxima vary between about 200 s and 1700 s (Fig. 8.12), almost an order of magnitude! Thus, if the circumstances of prey exploitation lead to birds being best served by relatively short, sharp dives (perhaps ice fish residing in pack ice, on the underside of floes), then this is what they will do, with an "unimpressive" maximum ([62]; Chap. 4). In contrast, deep-dwelling prey will require that emperor penguins make deep, and therefore long, dives. The take-home message for researchers seeking to determine absolute duration maxima for the different penguin species (to examine, for example, how the lines of best regression explain how oxygen usage varies with mass) is to ensure that the sample size is appropriately large and that, within the sample, there is opportunity for penguins to show the range of their capabilities. Accomplishing this more often than not requires multiple studies in multiple locations (Chaps. 3 and 4).

All that apart, penguins can stay underwater for an impressive length of time. Their phenomenal underwater excursions are possible because of their suite of anatomical adaptations (Chap. 6), coupled with some clever physiological tricks (Chap. 7, this chapter), not least an extraordinary ability to tolerate fantastically low oxygen partial pressures in the body [60]. For example, final venous oxygen partial pressures of emperor penguins at the end of longer dives can be as low as 200 Pa – noting that humans black out when the value is around 3300 Pa [45], some 25 times higher – although arterial minimum values would be more reliable as a measure of blood oxygen, being less liable to mixing.

Depth

The ability to hold their breath for extraordinary lengths of time means that penguins theoretically have time to dive deep, if they choose to do so. Whether they actually do this depends not just on their breath-holding capacity, though, because ambient pressure increases dramatically with depth– by about 1 atmosphere for every 10-m increase in depth. This equates to adding pressure of about 1.03 kg per square centimeter of body surface for every 10-m depth increment. Thus, the ability to remain underwater for long periods does not necessarily equate to correspondingly deep dives (dive durations of crocodiles, which may be in excess of 780 s but to depths of less than 2 m, spring to mind [15] as well as loggerhead turtles diving for up to 7 h at depths of just a few meters [32]). In fact, there is a reasonable relationship between the dive duration and the maximum dive depth shown by the different species with, however, extreme variation (Fig. 8.13).

This is particularly the case in emperor penguins, which have the greatest dive duration of any penguin (or bird) species (by some measure; cf. Figure 8.12) but which can clearly choose whether the extreme durations are invested in swimming horizontally or vertically (Fig. 8.13).

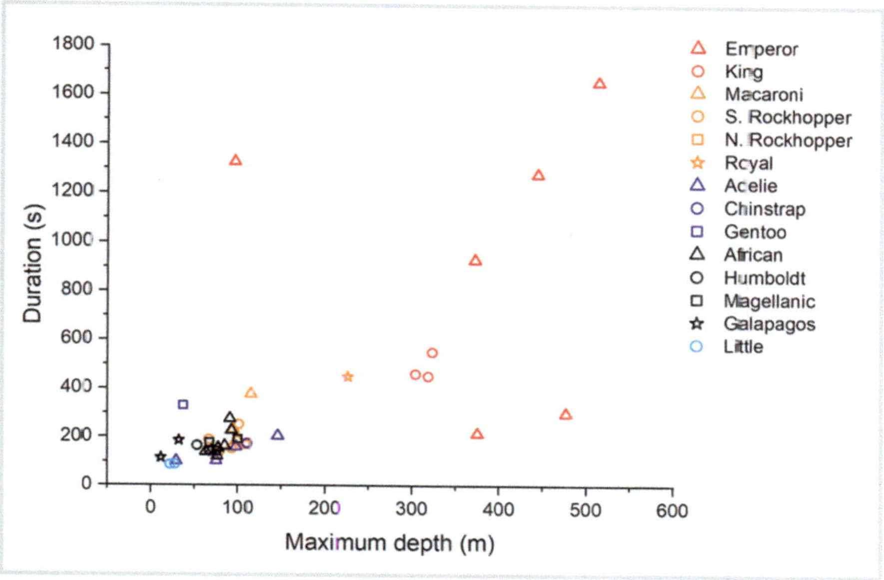

Fig. 8.13 The maximum depth recorded by various studies on different penguin species (the genus is color coded) in relation to the duration of that dive (data were obtained from the Penguiness website (penguiness.net)). Although there is a generally increasing dive duration with increasing maximum depth, the variation is considerable (as particularly manifested in the emperor penguin dataset)

A feature of the maximum depth versus duration graph, however, is that a certain minimum time is needed for transit between the surface and the point of greatest depth. Thus, greater maximum depths will also be accompanied by greater dive durations. The relationship between the maximum dive duration and body mass (Fig. 8.12), which is explained by the oxygen store/usage rule (see above), therefore allows bigger penguins the time to go deeper (see Fig. 8.14). Again, though, the variation between different studies (and sites) is appreciable, highlighting that the values expressed by the different publications are as much representative of penguin choice in how they decide to spend their time at depth. The latter will depend on a suite of factors, including light availability, prey density etc., as well as their physiological limitations. We saw in Chap. 5 that male penguins, which tend to be larger than females, in situations where food availability is less, choose to dive deeper rather than go farther from their colonies, which is the females' strategy.

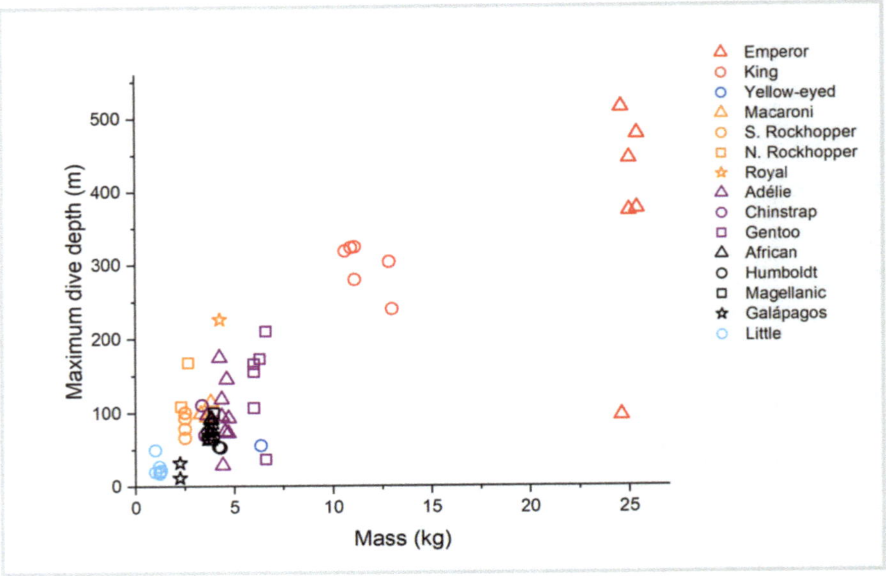

Fig. 8.14 The maximum dive depth reported for various penguin species (the genus is color coded) as a function of their mass for different studies (data were obtained from the Penguiness website (penguiness.net)). The variation in species-specific maxima, which is presumably primarily due to the prey depth distribution (Chap. 11) and sample size should be noted

Penguins under Pressure: Beating the Squeeze and the Bends

Barotrauma

Diving deep, i.e., more than, say, 5 meters, is not a trivial achievement for any animal that has air spaces, and this is because of the radical changes in pressure that vary dramatically in direct proportion to depth (Chap. 6, Box 6.1). Specifically, the change in pressure between the sea surface and just 10-m depth is 1 atmosphere, equivalent to that between the outer edge of Earth's atmosphere and sea level, encompassing a distance of some 10,000 km (6200 miles).

The proposed reduction in air sac volume with increasing depth (see above) should have benefits for enhancing oxygen uptake, but there is a limit to how far pressure can increase without causing harm to some parts of the body. One element of harm is barotrauma, known commonly, among human divers, as "the squeeze." This is where internal air spaces, e.g., the lungs in humans, are compressed by the pressure of the water so much that the natural elasticity of the thorax that surrounds them comes to an end. Further external pressure beyond this point then leads to extreme forces being exerted on the thorax – and that is not good because, unless the diving animal is built like a diving bell (impervious to pressure), it leads to tissue damage.

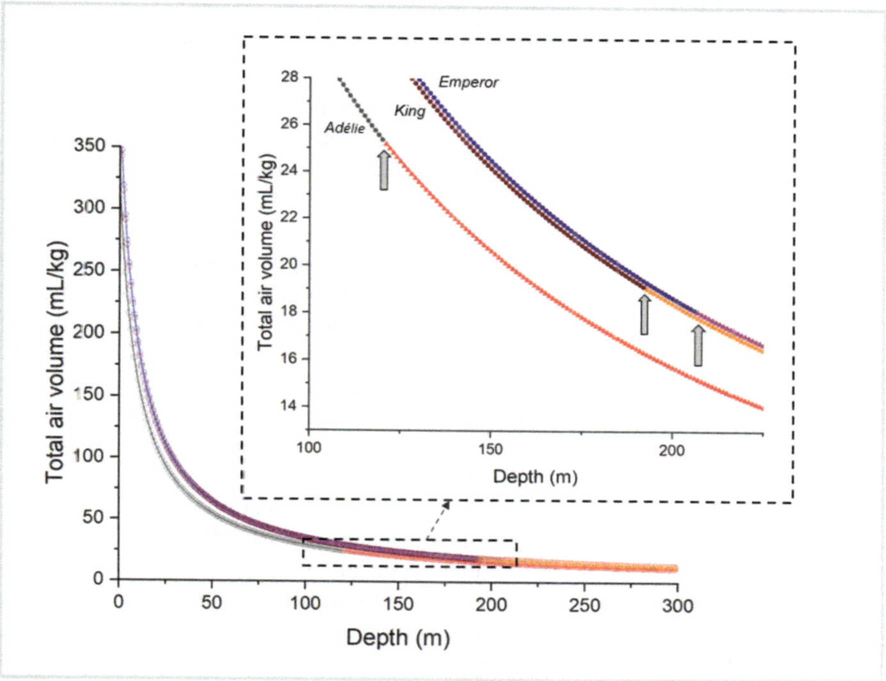

Fig. 8.15 The total air volume predicted to be in the respiratory system of three penguin species of differing sizes as a function of depth, highlighting the depth at which all the air in the air sacs has moved into the lungs (gray arrows). Assuming no elasticity in the lungs (the parabronchi), this is the point when pulmonary barotrauma (lung squeeze) should start to occur, predicted here to be 121 m, 193 m, and 207 m for Adélie, king, and emperor penguins, respectively. Lung volumes are taken to be 25, 18, and 19 mL/kg and air sac volumes are taken to be 296, 363, and 357 mL/kg for the same species, respectively. (Data were obtained from [59])

In the case of penguins, the whole issue revolves around the volumes of both the air sacs and the lungs and their respective elasticities. Actually, the trachea is important, too, but we will ignore that for convenience, although at extreme pressures, it starts to become problematic. Our point of departure, and we may be wrong, is that the air sacs are fully compliant and can collapse completely [35, 66]. Certainly, their extremely thin elastic walls make a good case for them doing that. In contrast, the bird lungs in general are quite rigid, with one study – albeit on ducks– indicating that they seem to contract minimally under pressure [46]. Using this, we can look at projected air sac versus lung volume changes with depth in three penguin species thanks to some marvelous work done by Paul Ponganis and colleagues to determine air sac and lung volumes using computerized tomography on live birds (Figs. 8.1 and 8.7). Specifically, we can look at the depth at which the air volume is so reduced that it is likely to start to cause tissue damage to the lungs. Figure 8.15 shows this, replicating Fig. 8.8 to some extent but highlighting the depth at which the air sacs are predicted to have collapsed completely, with all the air moving into the lungs.

This is, theoretically at least, the point when we might expect barotrauma to be problematic – the "lung squeeze danger point" – because further increases in depth will lead to compression of the lungs and, if they are not elastic, this is not good.

This exercise predicts that the air sacs would reduce to a volume of 0 mL/kg, with all air being compressed into the lungs at 121 m, 193 m, and 207 m for Adélie, king, and emperor penguins, respectively. These calculated lung squeeze danger points for these penguins accord basically with what we know about the diving habits of each. In general, and in keeping with body size (Chap. 1) and study site (i.e., the foraging situation), and as summarized in Appendix Table A3, depending on the situation, Adélies dive shallowly compared to the other two, typically on average 7–70 m compared to 50–250 m for kings and 50–300 m for emperors. Considering the means derived among individual studies, for example, it can be 7 m, 123 m, and 257 m, respectively [16, 36, 37, 90]. Thus, Adélie penguins have mean dive depths well within their predicted lung squeeze danger point, as do king penguins, but emperor penguins on average seem to dive deeper than their predicted lung squeeze danger point. Critically though, all three species, however, have maximum dive depths that exceed or greatly exceed these danger points (Appendix 3): Adélie, 180 m [90]; king, 343 m [67]; and emperor, 564 m [92].

In an attempt to look at options that might allow deeper dives without barotrauma, we played with the figures to some extent. Specifically, we increased the volume of the air sacs by one standard deviation of their mean measured value and reduced lung volume to one standard deviation of their measured values (both sets of data are provided in Ponganis et al. [59]), reasoning that this sort of change might be within measurement error or similar. Importantly, these maneuvers should help increase the critical depth at which there is danger from barotrauma. In fact, our renewed calculations led to lung squeeze danger points of 143 m, 222 m, and 215 m in Adélie, king, and emperor penguins, respectively. These are appreciable increases (up to almost 20% depending on the species) and show the potential importance of lung and air sac volumes in mitigating barotrauma. However, allowing the lungs to be compressed by 2% [46] or even 5% only increased this critical point marginally (2% increased it by between 1% and 2%, whereas 5% increased it by about 4.5%).

Seemingly, then, penguins can dive deeper than they theoretically should! Maybe this is yet another one of those animal mysteries? So, what needs to happen for the problem to be eliminated (as penguins appear to have accomplished)? In this, a central element must be the effective lung volume compared to the air sac volume because this makes a great difference to the predicted lung squeeze danger point (Fig. 8.16).

In particular, once the lung volume is appreciably reduced, even tiny changes in this volume make a disproportionate difference in the maximum depth that birds can attain without barotrauma (Fig. 8.17). Our calculations indicate that, to achieve their observed maximum depths, Adélie penguins would have to have their lung volume reduced to about 60%, compared to 58% for king penguins and about 38% for emperor penguins (Fig. 8.17). So, might penguins be fundamentally different from other birds in this respect?

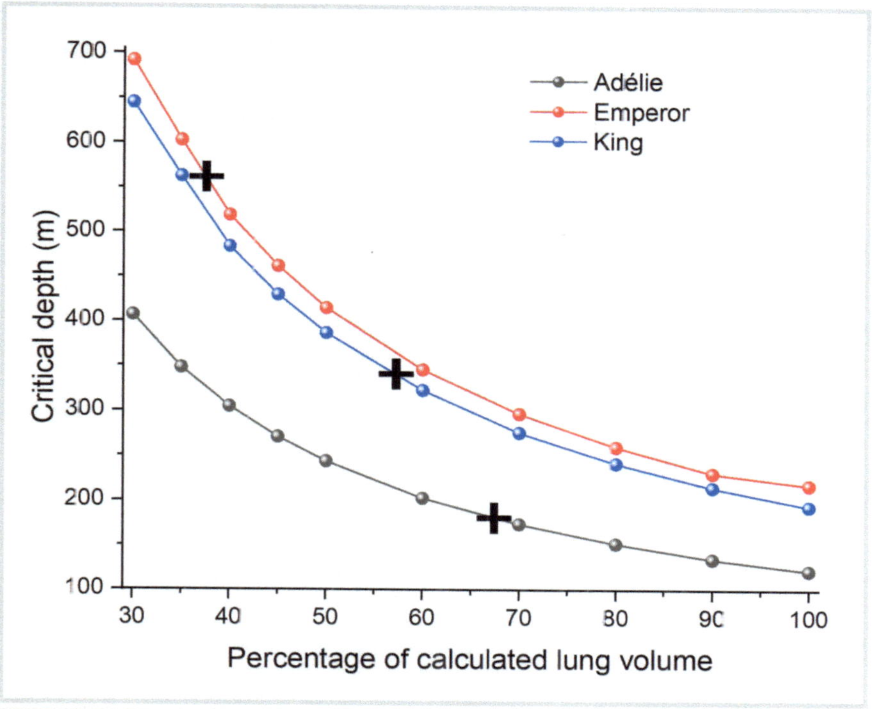

Fig. 8.16 The critical depth at which the pressure is predicted to have compressed all the respiratory air into the lungs of three penguin species, of varying size, thus exposing them to barotrauma (lung squeeze) as a function of lung volume (expressed as a percentage of its measured volume (data on the air sacs and lung volumes were obtained from Ponganis et al. [59]). Crosses show the maximum depths recorded for the three species and the necessary reduction in lung volume to mitigate barotrauma

Although Ponganis et al. [59], in their paper on barotrauma, write (p. 726) "The mechanisms contributing to the avoidance of barotrauma to the 'rigid' lungs of diving birds are unresolved," they do offer some tantalizing options that, particularly together, might effectively reduce lung volume. First, penguins have, overall, a larger ratio of air sac-to-lung volume than do non-diving birds (cf. [19]). As part of this, penguins may also have much more elastic lungs than other birds – Ponganis's work certainly suggests that this is the case, especially in the deeper-diving species – which would also help considerably. In addition, it has been suggested that the lung volume may be reduced by the surrounding smooth muscle [42], although the mechanism is not clear (to us at any rate). There may also be some thoracic pooling of blood, whereby the blood capillaries in the penguin lungs expand into the air capillary space [42], effectively reducing the lung air space still further.

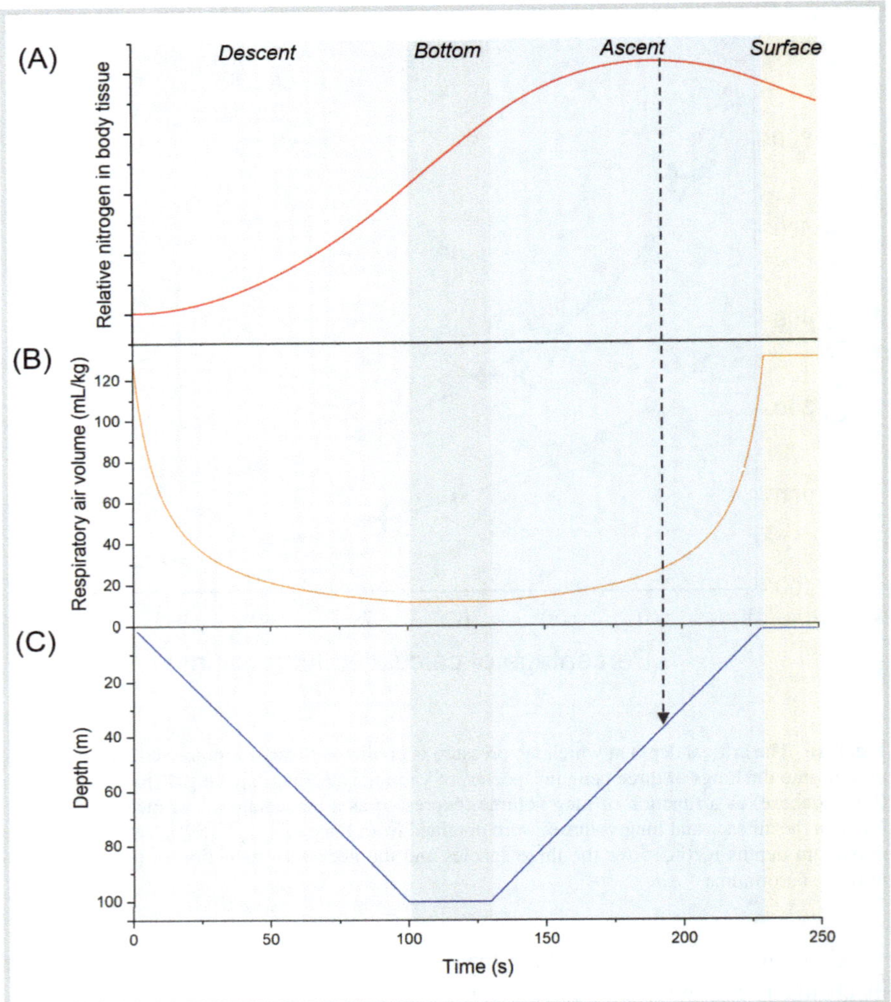

Fig. 8.17 (**a**) The generally expected pattern of body nitrogen accumulation over time in a penguin beginning a single U-dive after a long period on the surface and consequently (**b**) experiencing changes in respiratory air volume due to (**c**) depth affecting pressure (and thereby the partial pressure of the nitrogen in the respiratory air spaces). This simplistic modeled scenario shows how nitrogen levels in the body are expected to rise through much of the dive cycle, only beginning to drop when the bird is well into the ascent phase of the dive, when the partial pressure of the nitrogen in the respiratory air spaces is less than that of the body tissues (dashed black arrow). Data for the graph were generated from a simple model (based on Box 8.4) that considered the rate of transfer of nitrogen between the lungs and blood in either direction to be dependent on their relative partial pressures – these are dependent on depth and any nitrogen previously in the tissues

So, our current state of knowledge with respect to prevention of barotrauma in penguins is advancing, but we are still some way off from understanding how these birds achieve the currently theoretically impossible. The solution does seem to particularly relate to what happens to lung volumes in deep-diving instances, but, given the ease and speed with which penguins surge up and down the water column incurring huge pressure changes, this is likely to remain a particularly challenging research area for some time to come.

Beating the Bends

Decompression sickness, or "The Bends," is on the mind of any human diver who dives deep. The compression of air spaces with depth increases the partial pressure of gases within that air space (Box 8.2), which immediately leads to differences in pressure between the partial pressures of the gas constituents and those in the body tissues. This difference causes an increased transfer of gases from the higher gas concentration sites to the lower concentration sites (Box 8.3). Critically, this does not just occur with the gases that are important in metabolism, namely, oxygen and carbon dioxide, but it also occurs with nitrogen (N_2), and, since, nitrogen accounts for about 78% of the gases in air, this is not a trivial issue.

So, when a penguin dives, the higher pressure with depth causes the nitrogen to enter the body more rapidly, with the process being slowed as the nitrogen partial pressure in the tissues more closely approaches that of the lungs (Box 8.4).

Box 8.4
There are multiple unknowns in the rate at which nitrogen might be transferred from a diving penguin's respiratory system to the body tissues (and back) depending on both depth and time. However, the situation can be schematically considered to examine expected trends according to the way in which penguins allocate time to depth. To do this, we take Fick's Law (Box 8.3) and represent all elements within it that are likely to be constants (surface area and thickness of the gas exchange surface, and the diffusion coefficient of nitrogen) and represent them by a single constant (k1). We then apply the calculated rate of transfer to the amount of nitrogen in the penguin respiratory system at the instant the dive commences and remove that amount of nitrogen from the respiratory system, placing it in the body tissues (this process is reversed when the bird ascends the water column, though not immediately unless the nitrogen has already equilibrated). At a constant depth (= pressure), this also leads to a characteristic curve of increasing nitrogen in the tissues over time. There is thus an overall loss of nitrogen from the respiratory system during the descent phase of the dive due to the time-linked movement of nitrogen into the tissues, enhanced by increasing pressure. The body tissue nitrogen increases accordingly.

Indeed, when the nitrogen partial pressure of the lungs equals that of the body tissues, there is no net movement of nitrogen into, or out of, the body tissues. However, penguins that spend long periods at great depth have time for much of the nitrogen in their respiratory systems to move into the body. Of course, when these birds return to the surface, the process is reversed, and the nitrogen moves back out from the body into the respiratory air spaces. The problem, however, is that foraging penguins spend more than twice as long underwater (and sometimes at great depth) as they do between dives on the surface (see Chap. 10) and so should accumulate nitrogen in their tissues. This is expressed as a supersaturation (higher concentration than normal) of nitrogen when the penguins are back on the surface. If, under such circumstances, the nitrogen levels in the body are too high, then the gas does not diffuse out from the body tissues into the lungs in an orderly manner, thus to be expelled by exhalation. Rather, it bubbles out into the body tissues as a gas – one can think of opening a fizzy drink can! This, the nightmare of many a scuba diver, is known as decompression sickness or "the bends," and it can produce a variety of symptoms ranging from joint pain and dizziness through paralysis to death, certainly something to ruin your day.

All air-breathing diving animals are potentially prone to the bends, particularly those that move up and down the water column rapidly between the surface and great depths, spending extended time at depth and little time between dives on the surface. This is exactly what penguins do! It has been suggested that the problem may be solved by diving marine mammals by them generally exhaling before diving and having collapsible lungs (e.g., [24, 38, 50]). Both of these mean that there is no air–body tissue interface in the lungs so that nitrogen cannot move into the body. In addition, some species ascend the water column particularly slowly so that the nitrogen has time to move into the lungs from the tissues without bubbling out [23]. It should, however, be noted, that not all agree with this [24, 27]. However, in any case, penguins inhale before diving and have air sacs that seem to compress with depth in addition to having supposedly essentially incompressible lungs (see above). Thus, coupled with the penguin predilection for exploiting depths in multiple, serial high-speed dives, this seems to be a recipe for getting the bends. In contrast, there are no recorded cases of penguins suffering from the bends. So, what is going on?

To examine this, we need to think more specifically about the process of nitrogen accumulation by the body tissues. Before a penguin takes its first dive in a series, i.e., a diving bout, it has a finite, stable amount of nitrogen in its respiratory system and body (see time = 0 s in Fig. 8.17a), with the partial pressures of nitrogen in the lungs and body tissues being approximately equal. During the descent, the lung nitrogen will begin to make its way into the tissues, with the rate of tissue uptake increasing with both increasing depth and time (Box 8.4, Fig. 8.17).

This process will continue as long as the penguin keeps descending and for some time into the bottom phase of the dive, if there is one (see Chap. 9). Normally, even at the end of a bottom phase, the lung and body nitrogen will not have fully equilibrated, so there will be no net removal of nitrogen from the body to the lungs until the depth at which the partial pressure of the nitrogen in the lungs is less than that of the body (arrow in Fig. 8.17). Again, as the bird further ascends beyond this point, the rate of removal of the body tissue nitrogen will depend on the depth, with shallower depths removing nitrogen faster. Once on the surface, when the pressure difference between the depth of the bottom phase and the surface is the greatest, the rate of nitrogen removal will be the highest. In general, the conditions for nitrogen uptake depend on the relative lengths and depths of the descent, bottom, and ascent phases as well as the length of the surface period. However, serially deep-diving penguins spending little time on the surface (e.g., Fig. 8.11, also Chap. 10) are expected to accumulate nitrogen in their body tissues (compare body nitrogen levels pre- and post-dive in Fig. 8.17).

In order to try and explain why wild penguins can dive for prolonged periods without incurring any such problems, we are going to allow ourselves to get even more speculative than we were earlier (this is a book after all, though we do note that Andreas Fahlman, a diving physiology pundit who kindly reviewed this chapter for us, said, "I just don't believe what you are suggesting happens"!! We're feeling the fear but doing it anyway"! Earlier, we discussed about blood redistribution due to blood vessel vasoconstriction (see Fig. 8.9 cf. [27] for cetacean and turtle similarities), which could help separate the lung/heart/brain region (we will call this the LHB region for convenience) from the rest of the body as a mechanism to guarantee and regulate oxygen in the two most critical areas of the body. The removal of oxygen by the heart and brain in this LHB region will tend to keep partial pressure differences of this gas at a high level between the lungs and the associated blood, enhancing transfer (helped in no small measure by the water pressure). The pressure exerted by depth will similarly help push nitrogen from the lungs into this theorized region although, of course, it is not being metabolized so that the relative rate of transfer will be less than that of oxygen. This does not matter. It will still happen, and following the protocol outlined above, potential accumulation of extremely high levels of nitrogen could occur in that volume. What is critical, though, is that the extent to which the LHB region is discrete from the rest of the body will determine the total amount of nitrogen that goes into the whole of the body. Put bluntly, if the LHB region contains, for example, only 25% of the tissue of the total body, and that is saturated with nitrogen, once the vasoconstriction is relaxed and the whole body is reunited with the LHB region, the final nitrogen load per kilogram will only be a quarter of what it was in the upper region. Of course, it is unlikely that there is absolutely no flow between regions during diving, but even limiting the blood flow should help somewhat. So, one solution could conceivably be, for a penguin that has accumulated high levels of nitrogen into its LHB region during blood redistribution, the bird opening open the flood gates at an appropriate time – when the depth is much less and the partial pressure of nitrogen is less anyway– so that the high nitrogen blood of the LHB region mixes and dilutes with the total body blood to provide an overall much lower partial pressure of nitrogen across the penguin body (Fig. 8.18).

In other words, by constraining nitrogen accumulation primarily to just the small fraction of the blood volume allocated to the theorized LHB region, a penguin can avoid overloading its whole body. However, we can extend our speculations further still by asking when the penguin should end the vasoconstriction leading to blood

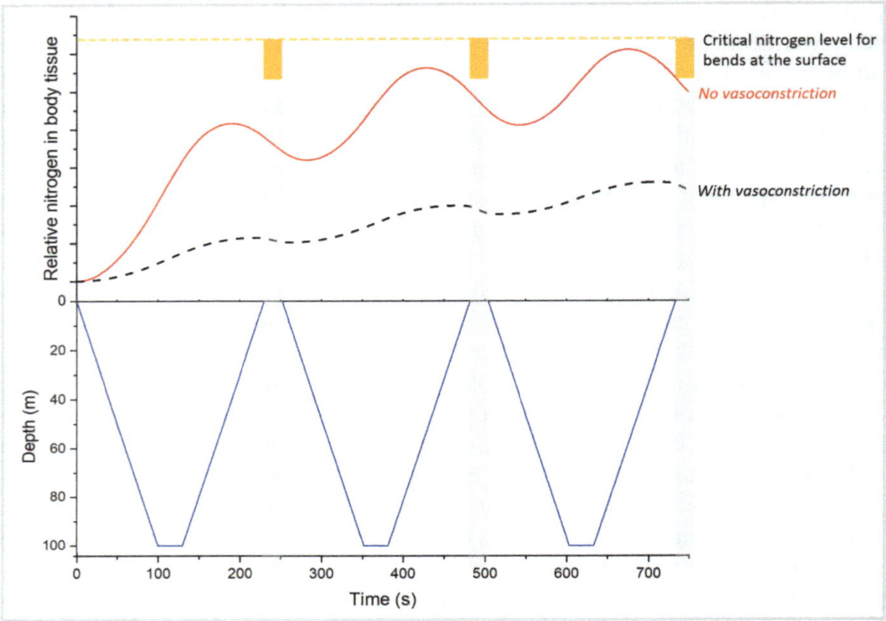

Fig. 8.18 Building on the example of a single dive shown in Fig. 8.17, illustrated here is the expected pattern of body nitrogen accumulation over time in a penguin executing three U-dives to the same depth (blue line) and interspaced by surface intervals (gray bars). The red line, with no vasoconstriction, illustrates how rapidly the nitrogen is taken up if all the body tissues are exposed to the high partial pressure nitrogen in the lungs. However, if blood in some portion of the body blood interacts little (or not at all) with the lungs due to circulatory changes following vasoconstriction, the lung-based nitrogen cannot supersaturate it at depth (in a manner analogous to the whole body of many marine mammals working with collapsible lungs) so that the total nitrogen flow into the body is reduced (dashed black line). Once on the surface during rest, the vasoconstriction is presumed to be over, which would mix the nitrogen-poor blood with the nitrogen-rich blood that had been in contact with the lungs. Data for the graph were generated from the simple model used for Fig. 8.17 but allocating different body tissue volumes for the tissue in vasoconstricted compared to the non-vasoconstricted states

redistribution that might help isolate the LHB region to maximize this effect. The first point is that the bird is obviously going to do this during all surface periods between dives anyway – it has to, in order to reoxygenate its whole body. However, it generally should not benefit from doing it during diving until the overall nitrogen levels in the body get critically high (which, given the above option may not happen much, if at all). The reason for this, as stated above, is that the rate of nitrogen transfer from the tissues back into the lungs is the highest when the partial pressure difference between the blood and lungs is the highest. Therefore, a penguin operating with high nitrogen levels within the LHB need not bother with reuniting the blood from this region with the rest of the body until the nitrogen levels reach such critical levels that they threaten to bubble out. In the end, if the LHB region is to be dismantled underwater, then we would expect it to occur only when the nitrogen levels are dangerously high.

In addition to the above explanation, two other behavioral options would also tend to mitigate the bends. One of these is for birds accumulating large amounts of nitrogen in the body to ascend slowly (as do scuba divers) to allow the body-associated nitrogen time to move into the lungs [26]. Indeed, this is the reason why Sato et al. suggested that king penguins sometimes ascended at relatively shallow angles, which, if the swim speed is constant, effectively reduces vertical velocity [73]. However, unless shallow return-to-surface angles have some other benefit (and they may well have – see Chap. 9 [72]), birds should not bother with them – because they are a waste of time– until nitrogen levels are high, which would be most likely during deep serial dives. Another option is that the short, shallow (<5-m deep) dives sometimes exhibited between serially deep dives by penguins such as kings may serve a similar purpose – effectively decompression "stops" in scuba diver lingo – whilst allowing them to snatch a breath or two for the reoxygenation in between.

A Gut Reaction in Fish-Birds

We saw earlier how penguins, through their evolution from aerial to aquatic creatures (Chap. 1), can do smart things to reduce their oxygen consumption during dives, which likely include redistribution of blood and cessation of stomach churning and these can change over short time frames – seconds to minutes. However, stomach issues are relevant to penguins over much longer time periods when adults are feeding chicks. All the food that they give to their brood is carried in the stomach, which, for this purpose, is used entirely as a sack for transportation. The time that the food has to be stored in this manner varies (depending on the species and colony size) between hours and days (Chap. 5, Appendix Table A3). This is longer than humans need, for example, to digest fairly substantial festive meals and certainly longer than penguins need when they do not have chicks [98]. At other times, though, the stomach has to function as part of the digestive system, with proteases ("digestive juices") to break down the dietary proteins, with the aid of stomach churning to mix the stew for better digestion, before the modified contents are passed out of the stomach or into their intestines (too bad for the chicks). These two conflicting demands have to be resolved and, like any sensible political situation, depend on a bit of give and take.

Gastric Emptying

The first part of this, or normal digestion, has been covered for penguins by a few authors, mainly by looking at stomach fullness over time since ingestion. As expected, this work has been carried out almost entirely in zoos, so we have had to assume that what is found in birds on land replicates what occurs at sea. However, one study conducted in 1957 shot some 250 African penguins far out at sea to determine their feeding habits. In this, Bob Rand [69] determined that birds shot in the early morning had empty stomachs, which became fuller as the day progressed. This tells us two things: that African penguins do not feed at night (more on this below; also Chap. 12) and that anything that they might eat during the day gets digested within the space of the night period. In other words, within 12 h, all stomach contents have been passed from the stomach. This ties in nicely with studies in which captive African penguins were noted to evacuate stomach contents consisting of 200 g of pilchard (ca. 7% of their body mass) completely within 4 h [21] and, in another study, where about 90% of stomach content mass (50–100 g) was evacuated within some 8 h [98]. Importantly, work on Humboldt penguin shows that the rate at which the digestive process takes place depends on the mass ingested, with larger meals being digested relatively faster [97]. All this makes sense, with penguins in this situation digesting all they have eaten, as would any normal animal.

Work in the 1980s, however, identified that African penguins that had returned to their nests with food in the stomach to give to their chicks, fed by regurgitation, had stopped digesting it [94]. This was ascertained because even small chicks continued to gain weight due to being fed by the adult for up to 24 h after the bird had returned from sea – a period very much longer than the normal few hours that it takes a penguin to evacuate its stomach contents in the normal process of digestion (see above). Indeed, emperor penguin females, returning to their newly hatched chick (thus relieving their mate of duties), feed their offspring small amounts of food over a few weeks; other penguin species exhibit a range of days in which they feed their small chick(s) during the brood or guard stage of chick raising (Chap. 2).

Gerrit Peters' PhD thesis [57], and later work by a French-German team [88], nicely explained how this happened. Studying gentoo, Magellanic, and king penguins using sensors that recorded stomach acidity (pH) and stomach churning, they noted that the stomach pH of these penguins foraging for chicks was low – at pH values between about 2 and 4, due to secretion of stomach acids – during the first part of the foraging trip. Values of pH then rose to become closer to neutral – pH 5–6 or so – for the later stages. In tandem, there was extensive stomach churning during the first part of the trip, which then reduced to almost zero for the latter part. In the king penguins, this situation was carried over onto land to last for days to weeks if the birds had to incubate the egg for some time before it hatched, although it was complicated by large variation [88]. The stomach proteases are only active when the pH is low [57], which is also when the stomach contents are being mixed. So, the first part of the foraging trip is dedicated to normal digestion. However, once the penguin has digested enough for its own energy needs, by increasing the pH and stopping churning, it stops digestion – effectively converting the stomach into a storage bag for the chicks at this time. These are two seemingly uncompromising tasks for one organ solved by being separated by time and a little chemistry!

Rotting Food

Although the variation in pH and stomach churning is an elegant way for penguins to modulate the function of the stomach, an obvious problem with keeping food in warm penguin stomachs for too long is that the high internal temperatures create a wonderful playground for bacteria, especially if the pH is equable. This might not be a problem for the species that spend a few hours between catching their prey and feeding their chicks, but there are some species, such as king and emperor penguins, that spend many days at sea and, as mentioned earlier, may return to store undigested food in their stomach for their chicks for up to 3 weeks. Although the French/German team mentioned above researching digestion in king penguins concentrated their work on incubating birds on land preserving food for their chicks for the moment they hatch, there is no reason to believe that what they found does not also hold for birds spending a long time foraging and having to return with food for their chicks over a number of days. In a nutshell, the team showed that king penguin stomach contents have strong antimicrobial peptides – which they called "spheniscin" – with a broad activity spectrum hindering the growth of pathogenic bacteria and fungi [86, 87]. This is a wonderful solution that must allow king penguins, at least, to be able to exploit sea regions hundreds of kilometers from their breeding sites (Appendix 3) and still return to feed their chicks food. To be sure though, anyone who has had the "fortune" to sift through king penguin stomach contents has to wonder whether there is not some rotting going on in there after all!

The Eyes Have It

Anyone who has opened their eyes underwater without a mask or goggles will appreciate what it is like to be radically long-sighted (hypermetropia – where, unhelpfully, the image is focused behind the retina instead of on it). If, somehow, our eyes were adapted to underwater vision, as are those of fish, for example (which have an almost spherical lens to bend the light properly [107]), we would have an equivalent problem in air. Welcome to the world of the diving seabird, which might have to have the visual acuity underwater to pick plankton up individually and where the idea of flying fast in visually challenging places (like landing on a thin rocky ledge on a cliff in high wind [80]) seems like a recipe for disaster. Perhaps penguins might relax on the flight score and maybe bumble about hopelessly short-sighted (myopic) on land, seeing little. Indeed, earlier accounts were convinced that this was the case [54], but, as discussed in Chap. 14, this is hardly the case. These ideas were perhaps generated by people who did not give much thought to other diving species with more challenging movement maneuvers that will have to have solved the problem.

So, before we review the literature on the "official capacities" of penguin eyes, what do our observations of penguins tell us about their visual abilities? We know from diet analyses that African penguins, for example, will catch individual prey down to 7 mm in size [95], and similar figures are available for other penguins and their prey (Appendix 4). So, there is no question that they see "well" underwater. However, we have also watched African penguins on land snapping at flies or near-to-fledging Adélie penguin chicks following the fall of individual snowflakes, snapping at them, during their first snow storms. Perhaps more importantly, we have also had the honor of raising abandoned African penguin chicks to independence during early work on Marcus Island, Saldanha Bay, South Africa, and therefore have been able to assess their "powers of observation." Specifically, these young birds like to chase flying insects and one in particular (Jean-Claude) took a real shine to a distinct species of fly, about as big as a mosquito, which used to fly erratically, somehow rapidly oscillating left and right by a few centimeters in the airspace. We used to watch how Jean-Claude's head motion would oscillate left and right in tandem as he tracked the passage of these flies, occasionally lunging at them – though we never saw him catch one. However, there is more. Come evening, lacking electricity, we had to switch to using a rather poor-quality gaslight, which did not allow us to be able to distinguish whether our strange flies were flying or not – except that Jean-Claude would periodically still display his fly-catching head movement and the odd lunge. So, penguin eyes are not only good on land but also at low light intensities. This capacity to deal with poor illumination makes perfect sense because they hunt in the attenuated light intensities underwater. In fact, if you hold a penguin's head and look closely at its pupils, and then move the head between direct sunlight and shadow, you can see that the pupil expands and contracts virtually instantaneously. This is in stark contrast to humans in whom, as we well know, we might have to wait minutes to adjust to darkness [4]. Again, rapid adaptation to low light by penguins is expected – it would be of little use to a bird that is going to spend just 2 min underwater if it had to hang around while its eyes adjusted to the gloom after the bright sunlight on the sea surface (which may be especially bright if there is highly reflective ice around).

There is appreciable literature on penguin eyes that points to some rather clever tricks for their fish-bird lives. The first major attempt to deal with the optical performance of penguin eyes in both air and water occurred in the 1970s [83]. It was pointed out that if the eye lens and the cornea in front of it did not have a markedly different refractive index from their environment (which is what forces the incoming light to bend so that it can focus on the retina), then there could be no focusing and the penguin would see the world blurred in that medium. Building on an earlier study by Sivak [81], one that proposed that a flat cornea would help (African) penguins see better underwater by minimizing the refractive effect of having to operate in both air and water (see also Martin [47] for king penguins), they used retinoscopy. This procedure involves projecting a beam of light into the eye to look at how various lenses placed in front of the eye change the direction and pattern of the reflection. In this way, they examined the refractive power of the eyes of rockhopper, king, and gentoo penguins. Cleverly though, they did this in both air and water – with the water bit being achieved by putting goggles filled with water on the birds. Their results suggested that these birds, and the African penguins used in the preliminary study by Sivak, should be slightly, though not consistently, short-sighted in air and appreciably long-sighted in water – actually about a quarter as bad as the human eye underwater. However, they found appreciable refractive power in the corneas of gentoo and king penguins (using a photokeratoscope, which measures the curvature of the cornea), a finding that led them to suggest that this might go some way to correcting this underwater problem. Finally, they suggested that eye accommodation, the muscle-based changing of the shape of the lens (which is so impressive in small children, demonstrated as they happily look at books from 5-cm distance), could actually sort out the rest. This would effectively have penguins seeing both nicely underwater and in air. Rather pleasingly for him, in his later work, Jacob Sivak, working with Howard Howland [33] confirmed this for rockhopper, gentoo, Magellanic, and king penguins and then later for Humboldt penguins [82]. For every black, however, there is a white in a "ying-yang"-structured world, so it is unsurprising that in a study on Humboldt penguins, Martin and Young found that the results of their schematic optics study indicated that this bird should see well underwater without accommodation, which should make them short-sighted in air [48]. To overcome this discrepancy, among various interesting options, they also suggested that Humboldt penguin eyes might be different from the other species examined. We doubt this, though, most particularly in the case of the, congeneric, African penguins. In terms of visual performance, it is hard to beat behavioral trials, and Jean-Claude is testament to that.

The value of vision for fish-birds is not just about clarity of image. Being pursuit predators, they stand to gain hugely by having effective binocular vision – anyone who does not believe this should try playing tennis or squash with one eye closed. In that respect, early reports, even by eye specialists, suggesting that penguins had monocular vision [89] (probably because many species will appear to look at human intruders by exposing repeatedly first one side of their head and then the other – the "alternate stare" (e.g., [22]), seem extraordinary. Detailed examination of Humboldt penguin visual fields by Graham Martin and S.R. Young indicates that, in air at least, this species has a binocular field (with a mean lateral max value of about 45° occurring above the bill but extending up to 125° vertically) with a total angular area that is somewhat larger than that of a tawny owl – and owls are considered to have the highest degree of binocularity among birds [53]. Interestingly, a penguin's bill intrudes into the binocular field, so, theoretically, it can virtually see their prey actually going in – at least that is what it would be in air. In fact, the optics indicate that the extent of the binocular field is dramatically decreased when the eyes are immersed in water, with the maximum width decreasing to about 17° and the vertical extent becoming about 80°, with the lion's share of that occurring above the bill [48]. This is perhaps another reason why penguins seem to prefer to capture their prey from underneath (see Chap. 11). Interestingly though, in his later work, Graham

Martin [47] calculated that king penguins should lose the binocular field entirely in water – more than inconvenient – but suggested that it is perhaps something that can be made good by convergent eye movements. In a previous publication, it had been suggested that penguins are like typical birds in having extremely limited eye movements [48]. This neither dovetails with our own observations – and we have been impressed by the mobility of penguin eyes compared to other birds – nor to measurements made by Graham himself, who reported eye movements of up to 15° in king penguins. In fact, there are compelling reasons for penguins to have more extensive eye movement than most birds because a simple observation of them swimming shows that their bills act as a highly effective frontal rudder (Chap. 6, Fig. 6.22). It would be awful if visual inspection of something, like a predator, resulted in the penguin directing its path toward it! As an aside, it would also be interesting to consider the shape of flying bird bills and eye movement capacity within this context, noting that in birds such as swifts, head turning presumably has a small effect compared to puffins or toucans, for example.

Penguin eyes also lose an appreciable amount of their uniocular field (the overall field of view seen by a single eye) when they are immersed in water (for example, reducing the width of the Humboldt penguin by 32° [48] and that of the king penguin by 28° [47]). This is presumably not without consequence since lateral vision must be important in searching for prey and, given the suite of tricks employed by prey to avoid being detected by penguins (see Chap. 10), penguins will benefit from every degree of uniocular field that they can muster up. The same goes for keeping an eye out for predators at sea.

Among the seemingly limitless issues that confront diving endotherms, of course, is reduction of available light with depth, although this varies in a horribly complex manner according to the sun angle, the scattering of the light in the atmosphere (including the effects of clouds), the structure of the sea surface, and the density and type of particulate matter within the water column, to name just a few [34]. This makes it difficult to know exactly what light conditions are experienced by penguins under given conditions. However, critics are quick to point out that there must be severe constraints for animals hunting at depth by vision, as was highlighted to us during our early days of using depth gauges on penguins: After discovering that African penguins routinely dive in excess of 30 m, we (Rory) were challenged by naysayers who said that there was not enough light at that depth for the birds to be able to forage. Subsequent scuba dives to 44 m in the areas used by these penguins for foraging showed, however, that there was plenty of light, even for us to see in the relatively turbid waters of the Cape, and we do not have the sophisticated low light-level adaptations shown by penguins. King penguins (and also emperors), for example, may dive to depths in excess of 300 m, and certainly regularly dive to 150 m (Appendix Table A3) so that, if the seawater in their foraging areas is classified as "type II and/or II clear oceanic," then this has them moving from bright sunlight on the surface to ambient light levels somewhere between sunset and moonlight at 100 m and at 300 m, encountering levels equivalent to minimum starlight [47]. This is perceived as absolute darkness by humans. Maybe it is relevant here to remember the penguin Jean-Claude, though! As one of the two supreme penguin divers, king penguins have a fascinating, and apparently unique, square pupil (our observations of a number of other species of penguin is that they are pentaradial), with maximal diameters that are similar to that of the tawny owl and at great variance to that of a domestic pigeon, suggesting facilitated low-light vision. Importantly, penguins can also reduce pupil diameter rapidly to a pinhole aperture during exposure to light on the surface. This undoubtedly helps them preadapt to the lower light levels at depth without having to incur the long acclimation times typical of most birds and mammals. Indeed, Graham Martin calculates that, even though penguin eyes in water probably experience about a 2.8-fold decrease in the illuminance of the retinal image compared to air [47, 48], the fully dilated pupil of the king penguin increases retinal brightness by about 316 times compared to the pinhole situation [47]. Although king penguins clearly capture individual fish during foraging (see Chap. 11), whether or how they can see them is a subject of debate. Certainly, many of their prime prey are lanternfish (see Chap. 4, Appendix Table A2), which have photophores that emit light, which the birds may hone in on [47]. In contrast, it is perhaps worth remembering that the supposed function of photophores in this taxon is to help conceal them from predators by counterillumination [56]. Ultimately, though, king penguins have relatively large eyes [47], which, as any owl will attest, helps catch photons when they become scarce. The big penguin eye challenge is, of course, the emperors, the deepest divers (exceeding the kings' maximum recorded depth by some 200 m (Appendix 2)) and presumably the species that consequently have to contend with the lowest light intensity. Quite what goes on at such extraordinary depths remains to be seen.

Just diving deep is only part of the issue for emperor penguins, as well as Adélie penguins, in that they forage beneath the shade of sea ice or even the shade of plankton blooms so thick that humans can hardly see a hand stretched out in front of their face [71]. While penguins avoid entering the water when it is dark, as they need to see whether predators are about (Chap. 12), they do forage into the darkness of the deep [1]. In a latter study, involving the measurement of light levels in >21,000 dives, Adélies experienced light levels averaging <11 lux and 97% averaged <400 lux (illuminance equivalent to civil twilight); 82% of dives were to depths at which light was 1 lux. They were, therefore, foraging in what we would call "the dark." The light level at depth is important to Adélies during foraging [3, 71], but whether it has to do with actually seeing prey or foraging where prey are most concentrated, i.e., at the edge of dense plankton blooms, remains to be determined.

With regard to exploiting the gloomy depths, another of water's tricks with light that penguins must deal with is that the depths attenuate wavelengths differently, removing the reds first and letting the greens and blues penetrate deeper [34] – which is why underwater photography is so blue unless a flash is used. In accordance, Humboldt penguin photoreceptors appear to have good wavelength discrimination in the blue–green region of the light spectrum, a capacity that undoubtedly helps them as they peruse the gloomy depths [8]. In this respect, it is interesting that Adélie penguins equipped with back-mounted tags are clearly more bothered by being equipped by blue and green units than they are by red ones, spending a significant amount of effort pecking at them while ostensibly leaving the

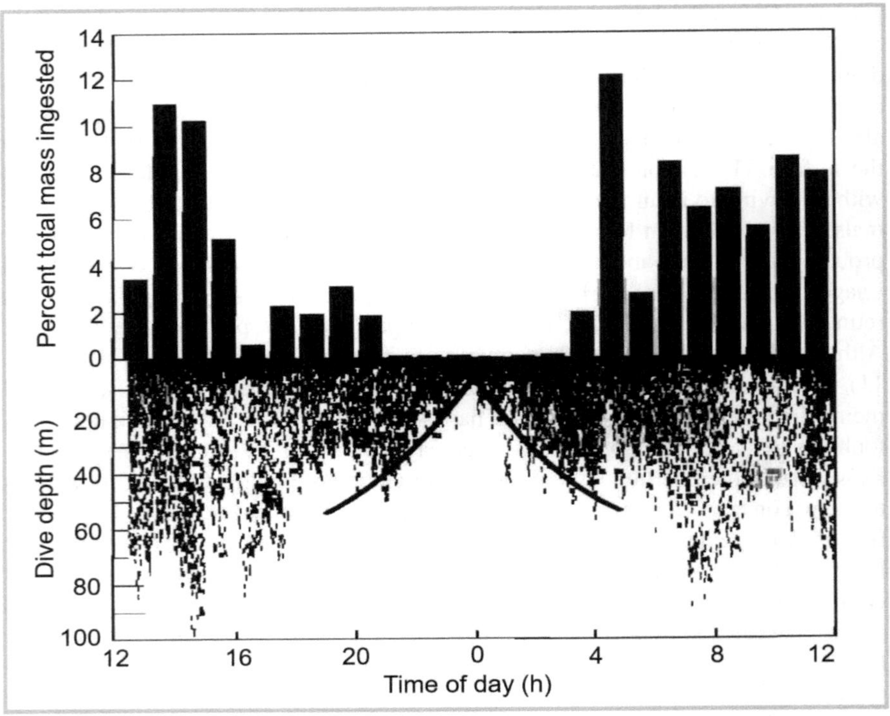

Fig. 8.19 The dive depths of Adélie penguins foraging on Ardley Island, Antarctica, in January as a function of local time of day. The continuous line indicates the depth at which the 1-lux light level was measured by the tags on the bird. Stomach temperature sensors also indicated when prey were caught (figure from [100] with permission from Inter-Research). It should be noted how there is minimal diving below the 1-lux threshold and how little, or no, prey were ingested for an extended period of Antarctic "night" even though, in absolute terms, there was still appreciable light

eye-catching reds alone [104]. This is presumably because they are seeing the reds as "gray," much closer to their perception of their dorsal plumage. It is a sober reminder that it is always wise to consider animal sensibilities when we are dealing with them and not to judge based on our own perceptions.

In the interests of completeness, it is only fair to mention the study by Poulter. He maintained that captive Humboldt penguins could find fish in tanks in complete darkness [64], and his suggestion was that they used sonar. This work has never been repeated or even given much consideration, although it must be said that penguins do not appear to have any specialized structure in their heads to allow this to happen. Rather, remembering Jean-Claude, it seems likely that some small amount of light must have been entering the experimental setup. Certainly, works on free-living penguins have demonstrated fairly convincingly that Adélie penguins do not generally dive deeper than a certain light threshold (about 1 lux) so that their depth use patterns tightly reflect the time of day (Fig. 8.19) [1, 100].

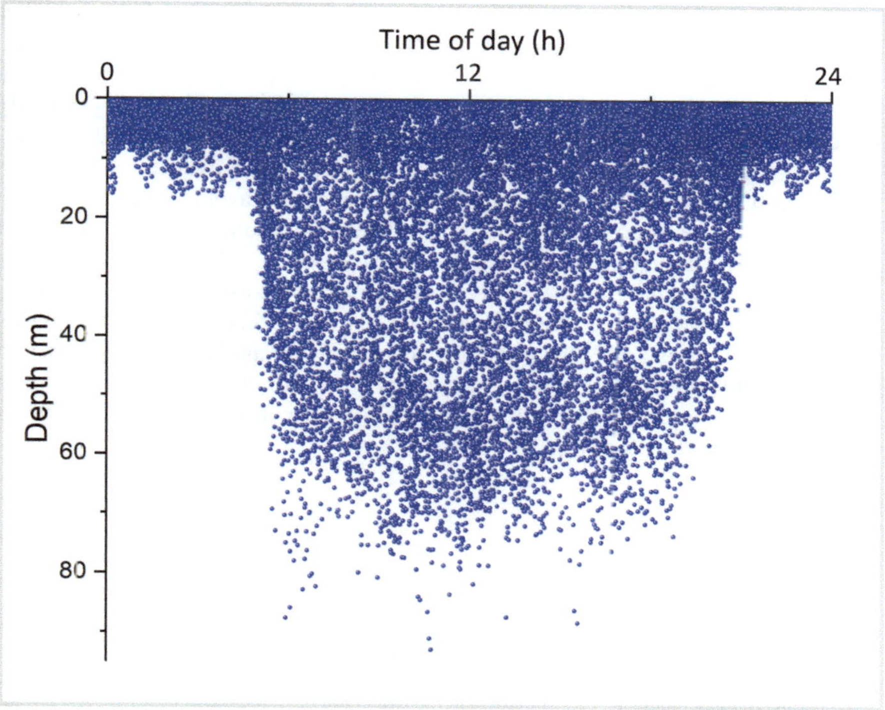

Fig. 8.20 Depth, displayed as one value per minute, versus time of day (local time) for 20 Magellanic penguins foraging for chicks from a colony in Peninsula Valdes, Argentina. The increase in depth associated with daylight is extremely obvious (though the local time does not equate with proper solar time). It should be noted that the vast majority of dives at night are to less than 10 m and none reach 20 m

Interestingly, at the same study site and time, chinstrap penguins dove to greater depths at night, stopping at a particular time of day-dependent depth maxima in a similar manner to the Adélies but at a light threshold of only 0.1 lux. This allowed them to exploit deeper depths at "night" and presumably stems from them having better low-light vision.

In fact, across penguin species, dive depth varies systematically with time of day in a similar manner, e.g., in Magellanic (Fig. 8.20) [102], emperor [36], king [40], macaroni [29], northern rockhopper [17], Fiordland [65], and little penguins [70]. Thus, the evidence that penguins are light-dependent for hunting is compelling.

That birds have time of day-dependent depth limits nicely underpins the theory of diel vertical migration by many penguin prey, whereby the prey seek to feed near the water surface but do so under cover of darkness when visually hunting predators cannot see them (see Chap. 10). Although many penguin species reduce or stop diving during the night, e.g., macaroni, eastern rockhopper [91], southern rockhopper [68], Galápagos [85], yellow-eyed [18], and emperor penguins [36], where continued diving does occur at night, it is more likely to represent a behavior other than foraging, such as traveling (see Chap. 9).

Finally, in a graphic demonstration of how sensory capacity affects ecology, the dependence of penguins on light to forage has knock-on effects regarding not only the depths that they can exploit, and therefore their access to prey, but also, consequently, on their ability to exploit particular latitudes as the available light varies across the seasons [36, 37]. For example, Adélie penguins avoid latitudes in which the sun does not rise in winter, but this is also likely the result of wanting to see that no predators are around before diving in (Chaps. 3 and 12). So, physiology affects depth exploitation (via, e.g., the ability to tolerate pressure as well as in the perception of prey in visually challenging scenarios), and this affects latitude – space use. It is worthwhile considering the truly tangled web of relevant factors that interact to produce the fish-birds we know, from the way they look, behave, and perform to their very distributions in three-dimensional (3D) space at the ends of the earth as they surf the world's oceans.

References

1. Ainley DG, Ballard G (2012) Non-consumptive factors affecting foraging patterns in Antarctic penguins: a review and synthesis. Polar Biol 35(1):1–13
2. Alboni P, Alboni M, Gianfranchi L (2011) Diving bradycardia: a mechanism of defence against hypoxic damage. J Cardiovasc Med 12(6):422–427
3. Ballard G, Schmidt AE, Toniolo V, Veloz S, Jongsomjit D, Arrigo KR, Ainley DG (2019) Fine-scale oceanographic features characterizing successful Adélie penguin foraging in the SW Ross Sea. Mar Ecol Prog Ser 608:263–277
4. Blough DS (1955) Method for tracing dark adaptation in the pigeon. Science 121(3150):703–704
5. Boggs D, Baudinette RV, Frappell PB, Butler P (2001) The influence of locomotion on air-sac pressures in little penguins. J Exp Biol 204(20):3581–3586
6. Boggs DF (1997) Coordinated control of respiratory pattern during locomotion in birds. Am Zool 37(1):41–53
7. Boutilier R, Reed J, Fedak M (2001) Unsteady-state gas exchange and storage in diving marine mammals: the harbor porpoise and gray seal. Am J Phys Regul Integr Comp Phys 281(2):R490–R494
8. Bowmaker J, Martin G (1985) Visual pigments and oil droplets in the penguin, *Spheniscus humboldti*. J Comp Physiol A 156(1):71–77
9. Butler P (1991) Exercise in birds. J Exp Biol 160(1):233–262
10. Butler P, Woakes A (1984) Heart rate and aerobic metabolism in Humboldt penguins, *Spheniscus humboldti*, during voluntary dives. J Exp Biol 108(1):419–428
11. Butler PJ (2001) Diving beyond the limits. Physiology 16(5):222–227
12. Butler PJ (2004) Metabolic regulation in diving birds and mammals. Respir Physiol Neurobiol 141(3):297–315
13. Butler PJ (2006) Aerobic dive limit. What is it and is it always used appropriately? Comp Biochem Physiol A Mol Integr Physiol 145(1):1–6
14. Butler PJ, Jones DR (1997) Physiology of diving of birds and mammals. Physiol Rev 77(3):837–899
15. Campbell HA, Sullivan S, Read MA, Gordos MA, Franklin CE (2010) Ecological and physiological determinants of dive duration in the freshwater crocodile. Funct Ecol 24(1):103–111

16. Charrassin J-B, Bost C-A (2001) Utilisation of the oceanic habitat by king penguins over the annual cycle. Mar Ecol Prog Ser 221:285–298
17. Cherel Y, Tremblay Y, Guinard E, Georges J-Y (1999) Diving behaviour of female northern rockhopper penguins, *Eudyptes chrysocome moseleyi*, during the brooding period at Amsterdam Island (southern Indian Ocean). Mar Biol 134(2):375–385
18. Chilvers B, Dobbins M, Edmonds H (2014) Diving behaviour of yellow-eyed penguins, port Pegasus/Pikihatiti, Stewart Island/Rakiura, New Zealand. N Z J Zool 41(3):161–170
19. Croll DA, Gaston AJ, Burger AE, Konnoff D (1992) Foraging behavior and physiological adaptation for diving in thick-billed murres. Ecology 73(1):344–356
20. Dasi LP, Pekkan K, Katajima HD, Yoganathan AP (2008) Functional analysis of Fontan energy dissipation. J Biomech 41(10):2246–2252
21. Davies DH (1955) The South African pilchard (*Sardinops ocellata*): bird predators, 1953–4. Division of Sea Fisheries
22. Eggleton P, Siegfried W (1979) Displays of the jackass penguin. Ostrich 50(3):139–167
23. Fahlman A (2009) How do marine mammals avoid DCS. In: The future of diving: 100 years of Haldane and beyond, pp 129–135
24. Fahlman A, Moore MJ, Wells RS (2021) How do marine mammals manage and usually avoid gas emboli formation and gas embolic pathology? Critical clues from studies of wild dolphins. Front Mar Sci 8:598633
25. Fahlman A, Sato K, Miller P (2020) Improving estimates of diving lung volume in air-breathing marine vertebrates. J Exp Biol 223(12):jeb216846
26. Fahlman A, Schmidt A, Jones D, Bostrom B, Handrich Y (2007) To what extent might N2 limit dive performance in king penguins? J Exp Biol 210(19):3344–3355
27. Garcia Parraga D, Moore M, Fahlman A (2018) Pulmonary ventilation–perfusion mismatch: a novel hypothesis for how diving vertebrates may avoid the bends. Proc R Soc B Biol Sci 285(1877):20180482
28. Green J, Butler P, Woakes A, Boyd I (2003) Energetics of diving in macaroni penguins. J Exp Biol 206(1):43–57
29. Green K, Williams R, Green M (1998) Foraging ecology and diving behaviour of macaroni penguins *Eudyptes chrysolophus* at Heard Island. Mar Ornithol 26:27–34
30. Halsey LG, Butler PJ, Blackburn TM (2006) A phylogenetic analysis of the allometry of diving. Am Nat 167(2):276–287
31. Handrich Y, Bevan R, Charrassin J-B, Butler P, Ptz K, Woakes A, Lage J, Maho YL (1997) Hypothermia in foraging king penguins. Nature 388(6637):64–67
32. Hochscheid S, Bentivegna F, Hays GC (2005) First records of dive durations for a hibernating sea turtle. Biol Lett 1(1):82–86
33. Howland HC, Sivak JG (1984) Penguin vision in air and water. Vis Res 24(12):1905–1909
34. Jerlov NG (2014) Optical oceanography. Elsevier, New York
35. Jones D, Furilla R (1987) The anatomical, physiological, behavioral, and metabolic consequences of voluntary and forced diving. Bird Respiration 2:76–119
36. Kirkwood R, Robertson G (1997) The foraging ecology of female emperor penguins in winter. Ecol Monogr 67(2):155–176
37. Kirkwood R, Robertson G (1997) Seasonal change in the foraging ecology of emperor penguins on the Mawson Coast, Antarctica. Mar Ecol Prog Ser 156:205–223
38. Kooyman G (1973) Respiratory adaptations in marine mammals. Am Zool 13(2):457–468
39. Kooyman G, Castellini M, Davis R, Maue R (1983) Aerobic diving limits of immature Weddell seals. J Comp Physiol 151(2):171–174
40. Kooyman G, Cherel Y, Maho YL, Croxall J, Thorson P, Ridoux V, Kooyman C (1992) Diving behavior and energetics during foraging cycles in king penguins. Ecol Monogr 62(1):143–163
41. Kooyman G, Drabek C, Elsner R, Campbell W (1971) Diving behavior of the emperor penguin, *Aptenodytes forsteri*. The Auk 88:775–795
42. Kooyman G, Ponganis P, Howard R (1999) Chapter 14: Diving animals. In: Lundgren C, Miller C (eds) The lung at depth. Marcel Dekker, New York, pp 587–620

43. Kooyman G (1975) Behavior and physiology of diving. In: Stonehouse B (ed) The biology of penguins. Macmillan Press, New York, pp 115–138
44. Kooyman GL (1967) An analysis of some behavioral and physiological characteristics related to diving in the Weddell seal. Antarctic Res Ser 11:227–261
45. Lindholm P, Lundgren CE (2009) The physiology and pathophysiology of human breath-hold diving. J Appl Physiol 106(1):284–292
46. Macklem PT, Bouverot P, Scheid P (1979) Measurement of the distensibility of the parabronchi in duck lungs. Respir Physiol 38(1):23–35
47. Martin GR (1999) Eye structure and foraging in King Penguins *Aptenodytes patagonicus*. Ibis 141(3):444–450
48. Martin GR, Young S (1984) The eye of the Humboldt penguin, *Spheniscus humboldti*: visual fields and schematic optics. Proc R Soc London Ser B Biol Sci 223(1231):197–222
49. Mattern T, Mcpherson MD, Ellenberg U, Van Heezik Y, Seddon PJ (2018) High definition video loggers provide new insights into behaviour, physiology, and the oceanic habitat of a marine predator, the yellow-eyed penguin. PeerJ 6:e5459
50. Mcdonald BI, Ponganis PJ (2012) Lung collapse in the diving sea lion: hold the nitrogen and save the oxygen. Biol Lett 8(6):1047–1049
51. Meir JU, Ponganis PJ (2009) High-affinity hemoglobin and blood oxygen saturation in diving emperor penguins. J Exp Biol 212(20):3330–3338
52. Meir JU, Stockard TK, Williams CL, Ponganis KV, Ponganis PJ (2008) Heart rate regulation and extreme bradycardia in diving emperor penguins. J Exp Biol 211(8):1169–1179
53. Meyer DB (1977) The avian eye and its adaptation. In: The visual system in vertebrates. Springer, New York, pp 549–611
54. Murray J (1910) Some notes by James Murray, biologist to the expedition. The heart of the Antarctic. In: British Antarctic Expedition, 1907-9, under the command of Sir E.H. Shackleton, c.v.o. Reports on the scientific investigations. W. Heinemann, London
55. Murrish DE (1982) Acid-base balance in three species of Antarctic penguins exposed to thermal stress. Physiol Zool 55(2):137–143
56. Paitio J, Yano D, Muneyama E, Takei S, Asada H, Iwasaka M, Oba Y (2020) Reflector of the body photophore in lanternfish is mechanistically tuned to project the biochemical emission in photocytes for counterillumination. Biochem Biophys Res Commun 521(4):821–826
57. Peters G (1997) Die Regulation der Verdauungsprozesse bei Pinguinen. Christian-Albrechts University, Kiel
58. Ponganis P, Meir J, Williams C (2010) Oxygen store depletion and the aerobic dive limit in emperor penguins. Aquat Biol 8(3):237–245
59. Ponganis P, St Leger J, Scadeng M (2015) Penguin lungs and air sacs: implications for baroprotection, oxygen stores and buoyancy. J Exp Biol 218(5):720–730
60. Ponganis P, Stockard T, Meir J, Williams C, Ponganis K, Van Dam R, Howard R (2007) Returning on empty: extreme blood O_2 depletion underlies dive capacity of emperor penguins. J Exp Biol 210(24):4279–4285
61. Ponganis P, Van Dam R, Knower T, Levenson D (2001) Temperature regulation in emperor penguins foraging under sea ice. Comp Biochem Physiol A Mol Integr Physiol 129(4):811–820
62. Ponganis P, Van Dam R, Marshall G, Knower T, Levenson D (2000) Sub-ice foraging behavior of emperor penguins. J Exp Biol 203(21):3275–3278
63. Ponganis PJ, Meir JU, Williams CL (2011) In pursuit of Irving and Scholander: a review of oxygen store management in seals and penguins. J Exp Biol 214(20):3325–3339
64. Poulter TC (1969) Sonar of penguins and fur seals. Proc Calif Acad Sci Fourth Ser 36(13):363–380
65. Poupart TA, Waugh SM, Bost CA, Kato A, Miskelly CM, Rogers KM, Arnould JP (2019) Foraging ecology of a winter breeder, the Fiordland penguin. Mar Ecol Prog Ser 614:183–197
66. Powell F (2015) Chapter 13: Respiration. In: Scanes CG (ed) Sturkie's avian physiology, 6th edn. Academic, New York

67. Pütz K, Cherel Y (2005) The diving behaviour of brooding king penguins (*Aptenodytes patagonicus*) from the Falkland Islands: variation in dive profiles and synchronous underwater swimming provide new insights into their foraging strategies. Mar Biol 147(2):281–290
68. Pütz K, Rey AR, Huin N, Schiavini A, Pütz A, Lüthi BH (2006) Diving characteristics of southern rockhopper penguins (*Eudyptes c. chrysocome*) in the Southwest Atlantic. Mar Biol 149(2):125–137
69. Rand RW (1960) The biology of guano-producing seabirds. The distribution, abundance and feeding habits of the Cape penguin, *Spheniscus demersus*, off the southwestern coast of the Cape Province. Investig Rep Div Fish S Afr 41:1–28
70. Ropert-Coudert Y, Kato A, Wilson RP, Cannell B (2006) Foraging strategies and prey encounter rate of free-ranging Little Penguins. Mar Biol 149(2):139–148
71. Saenz BT, Ainley DG, Daly KL, Ballard G, Conlisk E, Elrod ML, Kim SL (2020) Drivers of concentrated predation in an Antarctic marginal-ice-zone food web. Sci Rep 10(1):1–17
72. Sato K, Charrassin J-BT, Bost C-A, Naito Y (2004) Why do macaroni penguins choose shallow body angles that result in longer descent and ascent durations? J Exp Biol 207(23):4057–4065
73. Sato K, Naito Y, Kato A, Niizuma Y, Watanuki Y, Charrassin J, Bost C-A, Handrich Y, Le Maho Y (2002) Buoyancy and maximal diving depth in penguins: do they control inhaling air volume? J Exp Biol 205(9):1189–1197
74. Sato K, Shiomi K, Marshall G, Kooyman GL, Ponganis PJ (2011) Stroke rates and diving air volumes of emperor penguins: implications for dive performance. J Exp Biol 214(17):2854–2863
75. Scanes CG, Dridi S (2021) Sturkie's avian physiology. Academic, New York
76. Scheid P (1979) Mechanisms of gas exchange in bird lungs. Rev Physiol Biochem Pharmacol 86:137–186
77. Scheipers G, Kawashiro T, Scheid P (1975) Oxygen and carbon dioxide dissociation of duck blood. Respir Physiol 24(1):1–13
78. Scholander PF (1940) Experimental investigations on the respiratory function in diving mammals and birds. I kommisjon hos Jacob Dybwad
79. Schreer JF, Kovacs KM (1997) Allometry of diving capacity in air-breathing vertebrates. Can J Zool 75(3):339–358
80. Shepard E, Cole E-L, Neate A, Lempidakis E, Ross A (2019) Wind prevents cliff-breeding birds from accessing nests through loss of flight control. elife 8:e43842
81. Sivak J (1976) The role of a flat cornea in the amphibious behaviour of the blackfoot penguin (*Spheniscus demersus*). Can J Zool 54(8):1341–1345
82. Sivak J, Howland H, Mcgill-Harelstad P (1987) Vision of the Humboldt penguin (*Spheniscus humboldti*) in air and water. Proc R Soc Lond Ser B Biol Sci 229(1257):457–472
83. Sivak J, Millodot M (1977) Optical performance of the penguin eye in air and water. J Comp Physiol 119(3):241–247
84. Stahel C, Nicol SC (1988) Ventilation and oxygen extraction in the little penguin (*Eudyptula minor*), at different temperatures in air and water. Respir Physiol 71(3):387–398
85. Steinfurth A, Vargas FH, Wilson RP, Spindler M, Macdonald DW (2008) Space use by foraging Galápagos penguins during chick rearing. Endanger Species Res 4(1–2):105–112
86. Thouzeau C, Froget G, Monteil H, Le Maho Y, Harf-Monteil C (2003) Evidence of stress in bacteria associated with long-term preservation of food in the stomach of incubating king penguins (*Aptenodytes patagonicus*). Polar Biol 26(2):115–123
87. Thouzeau C, Le Maho Y, Froget G, Sabatier L, Le Bohec C, Hoffmann JA, Bulet P (2003) Spheniscins, avian β-defensins in preserved stomach contents of the king penguin, *Aptenodytes patagonicus*. J Biol Chem 278(51):51053–51058
88. Thouzeau C, Peters G, Le Bohec C, Le Maho Y (2004) Adjustments of gastric pH, motility and temperature during long-term preservation of stomach contents in free-ranging incubating king penguins. J Exp Biol 207(15):2715–2724
89. Walls GL (1942) The vertebrate eye and its adaptive radiation, Bulletin, 19. Cranbrook Institute of Science, Bloomfield Hills

90. Watanuki Y, Kato A, Naito Y, Robertson G, Robinson S (1997) Diving and foraging behaviour of Adélie penguins in areas with and without fast sea-ice. Polar Biol 17(4):296–304
91. Whitehead T, Kato A, Ropert-Coudert Y, Ryan PG (2016) Habitat use and diving behaviour of macaroni *Eudyptes chrysolophus* and eastern rockhopper *E. chrysocome filholi* penguins during the critical pre-moult period. Mar Biol 163(1):1–20
92. Wienecke B, Robertson G, Kirkwood R, Lawton K (2007) Extreme dives by free-ranging emperor penguins. Polar Biol 30(2):133–142
93. Williams CL, Czapanskiy MF, John JS, St Leger J, Scadeng M, Ponganis PJ (2021) Cervical air sac oxygen profiles in diving emperor penguins: parabronchial ventilation and the respiratory oxygen store. J Exp Biol 224(1):jeb230219
94. Wilson R, Ryan P, Wilson M (1989) Sharing food in the stomachs of seabirds between adults and chicks–a case for delayed gastric emptying. Comp Biochem Physiol A Comp Physiol 94(3):461–466
95. Wilson RP (1985) Seasonality in diet and breeding success of the jackass penguin Spheniscus demersus. J Ornithol 126(1):53–62
96. Wilson RP (2010) Resource partitioning and niche hyper-volume overlap in free-living Pygoscelid penguins. Funct Ecol 24(3):646–657
97. Wilson RP, Jackson S, Straten MT (2007) Rates of food consumption in free-living Magellanic Penguins Spheniscus magellanicus. Mar Ornithol 35:109–111
98. Wilson RP, La Cock G, Wilson M-P, Mollagee F (1985) Differential digestion of fish and squid in Jackass Penguins *Spheniscus demersus*. Ornis Scand 15:77–79
99. Wilson RP, Mcmahon CR (2006) Measuring devices on wild animals: what constitutes acceptable practice? Front Ecol Environ 4(3):147–154
100. Wilson RP, Puetz K, Bost CA, Culik BM, Bannasch R, Reins T, Adelung D (1993) Diel dive depth in penguins in relation to diel vertical migration of prey: whose dinner by candlelight? Mar Ecol Prog Ser 94:101–104
101. Wilson RP, Scolaro A, Quintana F, Siebert U, Mills K, Zimmer I, Liebsch N, Steinfurth A, Spindler G, Müller G (2004) To the bottom of the heart: cloacal movement as an index of cardiac frequency, respiration and digestive evacuation in penguins. Mar Biol 144(4):813–827
102. Wilson RP, Scolaro JA, Grémillet D, Kierspel MA, Laurenti S, Upton J, Gallelli H, Quintana F, Frere E, Müller G (2005) How do Magellanic Penguins cope with variability in their access to prey? Ecol Monogr 75(3):379–401
103. Wilson RP, Simeone A, Luna-Jorquera G, Steinfurth A, Jackson S, Fahlman A (2003) Patterns of respiration in diving penguins: is the last gasp an inspired tactic? J Exp Biol 206(10):1751–1763
104. Wilson RP, Spairani HJ, Coria NR, Culik BM, Adelung D (1990) Packages for attachment to seabirds: what color do Adelie penguins dislike least? J Wildl Manag 54:447–451
105. Wilson RP, Zimmer I (2004) Inspiration by Magellanic penguins: reduced swimming effort when under pressure. Mar Ecol Prog Ser 278:303–307
106. Wright AK, Ponganis KV, Mcdonald BI, Ponganis PJ (2014) Heart rates of emperor penguins diving at sea: implications for oxygen store management. Mar Ecol Prog Ser 496:85–98
107. Zigler MJ (1944) Review of the vertebrate eye and its adaptive radiation. Psychol Bull 41(6):403–405

Part IV
The Software of Fish-Birds

Chapter 9
Embracing the Depths: The Fish-Bird Dive

> "When you finally see what goes on underwater, you realize that you've been missing the whole point of the ocean. Staying on the surface all the time is like going to the circus and staring at the outside of the tent". — Dave Barry (Blub Story: A Very Deep Experience *(Miami Herald, 1989))*

The penguin "dive," a submergence underwater between two bouts of breathing on the oceanic surface, is a critical facet of their lives, allowing penguins to be penguins in a capacity unrivaled by any other bird, living or extinct. This capacity has influenced their morphology, physiology, and behavior certainly more than any other single activity has, although their attributes on land are shared with most other birds. Early reports of penguin diving described them as "feathered fish" (Chap. 1). Indeed, the early authors did not really make the connection between the birds they saw on land, which they assumed were some sort of goose, and the creatures they saw at sea.

Just a few decades ago, naturalists began to note how long penguins could remain underwater, eventually leading researchers, equipped with a stopwatch, to spend many land- and boat-based hours documenting reports of submergences varying between short dives of 10–20 s [4, 52] and longer dives of 70–150 s [9, 46]. However, the biases in collecting data on penguin dive duration were often surprisingly lost on these early workers due to their insufficient appreciation of penguins' aquatic prowess. Most notably, what should the researcher do if a penguin dived and was never seen to surface? We have been in this situation ourselves when a bird (or a group of birds) submerges and then seconds, sometimes minutes, tick by without seeing them returning to the surface. Appreciating the speed with which penguins travel underwater, there comes a point where the observer had to give up timing that particular bird (or group) because it had moved beyond the reasonable range of vision. The scientifically correct thing to do then should be to note the proportion of observed dives in which penguins were never seen to reappear. Few, if any, people did this, and these dives, with a start but no apparent end, generally never made it into the record. So for years, people believed that penguins dived for relatively short durations (typically <1 min) and consequently to shallow depths, similar to auks, grebes, and loons (typically imagined to be a few meters). It was only the advent of animal-attached technology that burst this bubble, with Gerald Kooyman first

D. G. Ainley, R. P. Wilson, *The Aquatic World of Penguins*, Fascinating Life Sciences, https://doi.org/10.1007/978-3-031-33990-5_9

publishing how emperor penguins could dive for as long as 18 min and to depths of 267 m [19]. We know today that emperors can dive for even longer periods of time (>20 min [18]) and to depths exceeding 500 m [33]! This information came as a huge shock to the scientific community and is a lesson to us all. It is summarily unwise and egoistical to presume that animal capacities should be measured in human terms.

As will be covered in other chapters, penguins use dives to travel, to escape and avoid predators, and to locate and capture food. A wild penguin will not catch prey on land. Indeed, Magellanic penguins on their way to sea to feed on Falkland/Malvinas sprats, at Cabo Virgenes, Argentina (see Chap. 4, Appendix Table A2), can occasionally be seen walking past thousands of flapping sprats washed onto the shore! However, walk past them they do, without fail. The Antarctic penguin species do a similar thing if krill are washed ashore! Penguins are the most specialized divers of the bird world, diving deeper and swimming faster than any other avian taxon, as we will show and explain below. Their sojourns underwater take them to a place without gaseous oxygen (that they breathe into their lungs), though some dives may enhance oxygen uptake (see later). In water, pressure also increases dramatically with depth, light intensity and colors are rapidly eroded by depth, and the temperature tends to drop. In short, the underwater world of penguins is profoundly different from their "air world" and it dramatically morphs with changing depth.

During foraging, and other activities, penguins may dive virtually continuously, with only short periods at the surface between dives – such periods are known as "bouts". A dive bout is considered to have terminated when a penguin spends a much longer interval at the surface to rest. These bouts, and the depths penguins reach during such periods have been repeatedly emphasised in the literature, most often by figures that plot depth against time of day (Fig. 9.1).

During individual dives, penguins may modulate their swim angle, their speed, and heading and, by so doing, they not only change the energy they expend on a second-by-second basis (see Chap. 6) but also change how they move through the three dimensions of space and how they operate in the fourth dimension that is time. They decide these factors between dives. This interaction with time is critically important because, for every second that penguins are underwater, they continue to consume their precious oxygen reserves, at a rate that is determined by their activity. Penguins have to ensure that they use their reserves judiciously – at the very least ensure that they make it back to the surface in time to renew the air in their lungs (Chap. 8). In all this, the specifics of penguin speed, dive angle, and heading determine the likelihood that they will find and actually capture prey, a prime motivation for diving at all. The fact that they apparently weigh the pros and cons of these various behavioral choices ahead of time is mind boggling (at least to us).

Thus, penguin dives are anything but simple submergences. Instead, they are hugely sophisticated strategies that must rapidly change often to allow them to find and capture prey with minimal energy expenditure and maximum energetic gain (maximizing net energy gain (cf. Watanabe et al. [42]). All these smart sojourns start with some inspection of the water column (Fig. 9.2) and the initial submergence.

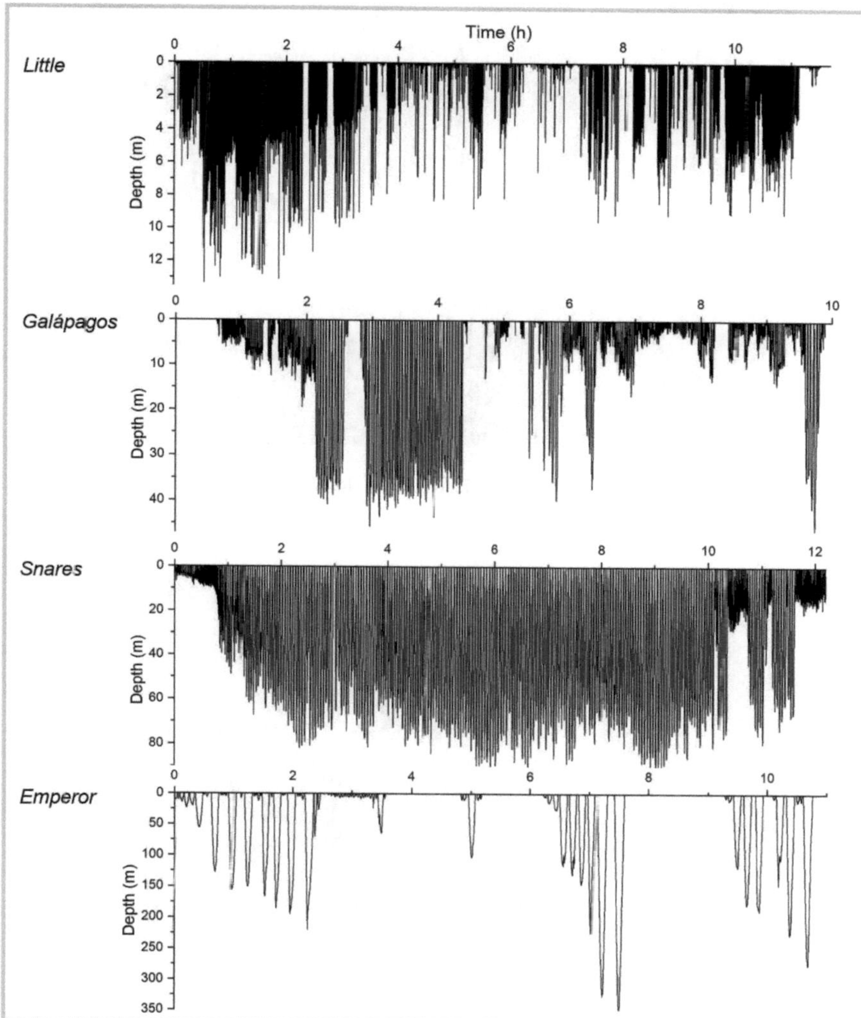

Fig. 9.1 Patterns of depth use during a typical day's foraging (10–12 h) by various penguin species of increasing mass (from bottom to top). Note the intra- and interspecific variability in maximum depth (or lack of it), how diving may sometimes be broken into dive bouts, with rests in between, and that there is substantial variation in time allocated to the specific depth strata over time. Finally, note how the depth and duration of dives tend to increase with penguin mass, to the point where even the descent and ascent phases of dives made by the emperor penguin are readily visible on this scale. (Data courtesy of Belinda Cannell (little penguins), Antje Steinfurth (Galápagos penguins), Thomas Mattern and Ursula Ellenberg (Snares penguins), and Ilka Zimmer (emperor penguins))

Fig. 9.2 A king penguin looking at its underwater world in the waters off Macquarie Island. (Photo courtesy of Louise Chilvers)

Fig. 9.3 A Magellanic penguin breathes on the water surface – as manifested by its open bill – before launching into its next dive with powerful strokes of its flippers

Submergence

Penguins on the water surface cut a low profile, with the head and neck protruding and the back virtually awash with water, so the submergence is easy (Fig. 9.3).

For this, when on the surface, they point their head down underwater, something they do so often, presumably scanning underwater [4]. Then, with an upward-directed flipper beat to bring the upper body underwater, they begin normal, regular flipper beating as they power down the water column. From inception to the point when the tail disappears takes about half a second (Fig. 9.4).

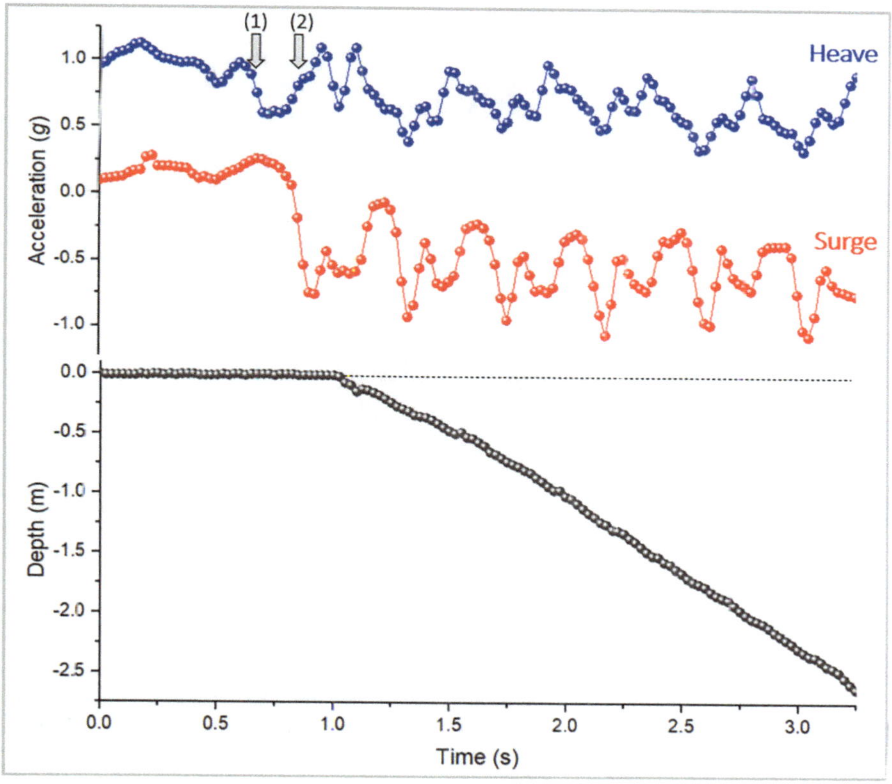

Fig. 9.4 Submergence of a Magellanic penguin as illustrated by, upper box, a lower back-mounted tag containing triaxial accelerometers (two sensors are shown here) and, lower box, a depth sensor, all recording at 40 Hz. Note how the heave (dorsoventral) acceleration (blue line) first decreases (gray arrow 1) just before the depth increases as the bird performs an upward stroke, rapidly followed by a downward stroke (gray arrow 2) during which surge (anterior–posterior) acceleration (red line) is generated. A change in depth (dark gray line, lower box) is first apparent about half a second after the initiation of the dive because the pressure sensor is close to the tail and is submerged last

The angle at which penguins descend the water column depends on a number of factors, most predominantly the final depth of the dive [28, 32, 47]. However, this also varies according to the penguin's perceived reason for diving, i.e., traveling, escaping predation, prospecting for food, or actually attempting to capture prey. Critically, penguin dives are rarely reactive, since they seem to quickly obtain enough information to plan a dive strategy [38]. That is, the birds prepare for the general expected purpose and performance at the outset rather than simply diving and changing strategy as the dive develops. The preparation for the dive (assessing information) actually happens on the surface between dives as penguins load oxygen into their systems (Chap. 8), but the extent of dive preparation is immediately apparent in the descent angle at the start of the dive, with birds taking a steeper

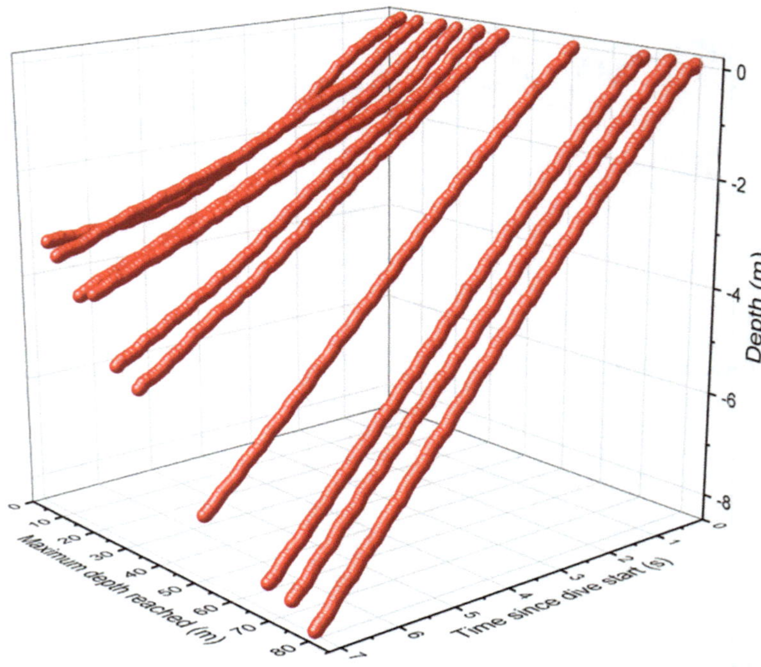

Fig. 9.5 Change in depth over time (for the first 7 s of a dive) for a Magellanic penguin as a function of the final maximum depth of the dive for "V-shaped prospecting" dives (see text). The more rapid increase in depth in deeper dives is instigated by steeper dive angles rather than faster swimming

descent trajectory if they are planning to dive deeper. Because swim speed is essentially invariant of maximum dive depth [28, 32, 44] (but see [17]) (Chap. 6), although there is plenty of variation, the descent angle determines the speed at which the birds descend the water column, which, therefore, is in accordance with the maximum depth to be reached (Fig. 9.5).

The Time Underwater: Basic Dive Descriptors

Although the concept of "a dive" may seem simple, penguin dives are highly complex and represent many different facets of diving behavior that need to be applied over time and space during the dive. These include efficient movement through water, including diving down at an angle that minimizes the vertical cost of transport according to the maximum depth to be reached (Chap. 6), gliding up the water column when allowed by the air-mediated upthrust [32] (Chap. 6), and chasing prey at speeds that allow their capture with minimal expenditure of oxygen [50]. All these are exemplified by a suite of measurable variables such as depth, speed (including

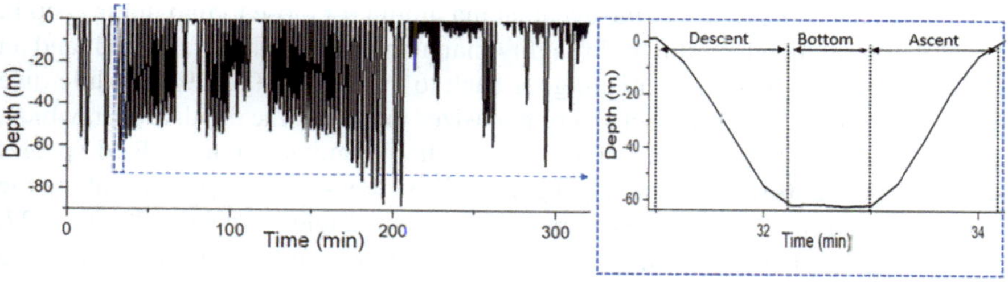

Fig. 9.6 Depth data from a foraging trip by a gentoo penguin in Antarctica. The left-hand panel shows 5 hours of virtually continuous diving, whereas the right-hand panel shows a single dive highlighting the stylized descent, bottom, and ascent phases of the dive, all critical elements of the dive profile

vertical and horizontal elements), swim angle, heading, flipper beat frequency and amplitude, and dynamic and static body acceleration. The key to understanding the putative function of dives lies in being able to represent them using various visualizations and metrics that can show how dives may be considered according to their functionality. Before describing this, however, we need to revert to a midpoint in our understanding of penguin dive behavior, where most of what was known for a long time depended on data derived from pressure sensors on tags attached to penguins. Equipped with only information on how dive depth varied over time, scientists, perhaps necessarily, grouped dive types according to "dive profiles" [12], which were graphs of inverted pressure against time (Fig. 9.6). They used these visualizations to describe many basic dive types, and we will adhere to these here while, at the same time, reporting how they seamlessly morph into one another and how underwater circumstances change the details of how penguins behave.

Again, because early work only used depth sensors to describe penguin diving behavior, dive types have been differentiated based on the characteristics of three main dive phases relating to depth, namely, the descent, bottom (i.e., the final depth), and ascent phases. These would seem to be fairly binary elements separated in time because penguins can only go down if they are on the surface (so it is the first thing that they will do) and can only come up if they are at depth (so this is the last thing that they will do). Actually, though, especially during prey capture, depths can change in either direction (up or down!) or not change at all. In other words, while submerged, especially near the bottom of the dive, there can be "wiggles" or "undulations" in the penguin's depth. Nonetheless, it is useful to consider the descent, bottom, and ascent phases as fundamental elements of penguin diving, and we adopt this approach here.

Dive Profiles

The standard visualization of the dive record has been pivotal in structuring how we think about penguins diving because, as mentioned above, researchers now describe "the dive" as being composed of a descent phase, during which time the penguin descends the water column to the eventual maximum (or close to maximum) depth; a bottom phase, in which the bird roughly maintains this depth for a period; and an ascent phase, in which the penguin goes back to the surface (Fig. 9.6). While dive profiles are useful, this approach overemphasizes the use of the depth aspect, which, though generally impressive, is only part of the penguins' strategy. Bird species such as cormorants (or even some penguins, like yellow-eyed [22]) forage along the seabed and thus have steep dive and return-to-surface angles (typically 70–90°) [27, 53]. They are well aware of the distance to the bottom, and this minimizes the time spent in transit between the surface and the bottom. Descent angles of penguins foraging in the water column are much lower (typically 10–50°) (cf. Watanuki et al. [44], reflecting the extent to which the horizontal component of the dive is important for them, for travel and/or for searching for prey in both vertical and horizontal components of underwater space. It is thus important that dive metrics should represent this too.

Dive Distance–Depth Profiles

To begin to build a better representation of space use during dives into our consideration of dive function, we suggest using the "distance–depth profile" to illustrate dive types [47]. In this, the dive is represented by the distance traveled against the depth of each dive. To do this properly requires information on the speed and swim angle of the birds so that the horizontal component can be extracted using trigonometry. However, a rough representation of it may be gained if birds are traveling at shallow descent and ascent angles by simply multiplying swim speed by duration.

Dive Aspect Ratios

To convert the dive–distance profile into a useful metric, we also consider a simple "dive aspect ratio." This is the total horizontal distance traveled by the penguin during the dive divided by the maximum depth reached. This effectively condenses the distance–depth profile into a single metric and provides an overall view of how penguins allocate horizontal and vertical distance components within their dives. Dive aspect ratios for Magellanic penguins can vary between 2 (where, for example, the penguin goes twice as far horizontally as it does vertically) and 300.

Horizontal Dive Directionality/Tortuosity

The cumulative horizontal distance covered during a dive used for the dive aspect ratio does not give an idea of dive path tortuosity, however, and this is important for indicating preferences for expending time (and energy) in particular areas (cf. Paiva et al. [25]). This tortuosity, the reverse of which is directionality, can be visualized at the level of the dive within circular rose plots, whereby a circular plot represents the 360° of the compass heading with segment radii indicating the fraction of time devoted to traveling in any given direction (Fig. 9.7). The "overall dive tortuosity" can be expressed as a metric, by dividing the straight-line distance between dive initiation and dive end by the total distance traveled during the dive. Thus, perfectly straight dives have a tortuosity of 1.0, whereas highly convoluted dives have trajectories tending toward 0.

Particular dive sections can also be treated using this metric to highlight directional changes, which are particularly prevalent during prey pursuit (see later). Unfortunately, tortuosity is only accessible from the few penguin studies that use dead reckoning [11] to determine the movements of birds underwater. Importantly, though, it is able to highlight the movement behaviors exhibited by birds in each of their three major dive phases as they adopt strategies to maximize their foraging and

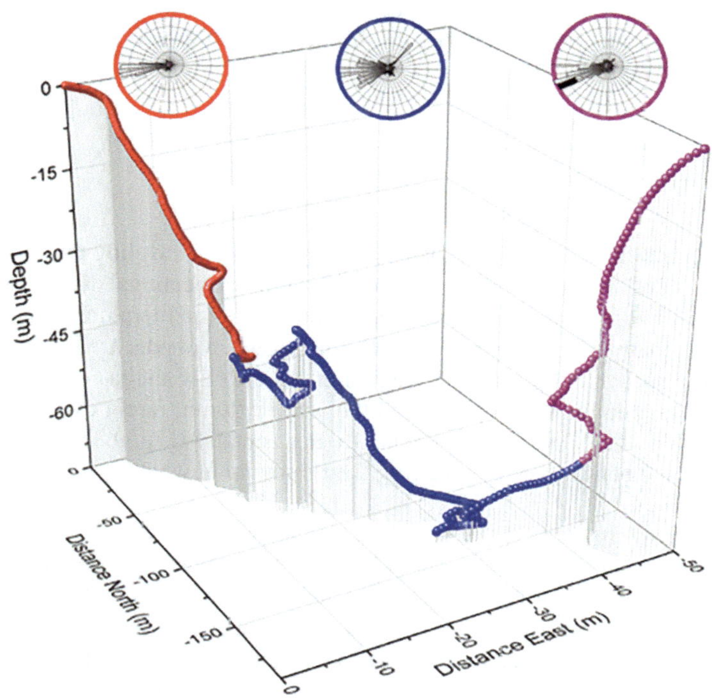

Fig. 9.7 The trajectory of an example dive by a Magellanic penguin (in waters off Argentina) beginning at coordinates 0,0,0 on the left-hand side and showing the descent (red), bottom (blue), and ascent (pink) phases. The directionality of the various phases is shown in the color-coded rose plots. This dive has an aspect ratio of 4.06 and an overall dive tortuosity of 0 69

traveling success. How dive aspect ratio and dive tortuosity relate to dive structure is illustrated in Fig. 9.7.

Basic Dive Types

A number of basic dive types – effectively combinations of dive behaviors – are recognized by biologists, although much of penguin dive behavior in the literature is simply lumped under "foraging." Where possible, we will be more specific about the putative function of dives. We also note that, although a penguin initially likely has a plan, one dive phase can morph readily into another if conditions change during the dive. For example, a "prey-seeking dive" may turn into a "prey-acquiring dive" if and when prey are encountered. Below, we document the main dive types considered as "standard" by researchers studying penguin foraging and discuss the extent to which dive descriptors change according to conditions. We primarily use data from Magellanic penguins to illustrate trends because we have detailed information on the underwater behavior of this species but allude to patterns in other species where appropriate.

T-Dives (for Traveling)

One important, but poorly appreciated, dive type is the "traveling dive," which we call the "T-dive" in an attempt to incorporate it into conventional dive classification (e.g., Schreer et al. [34]. Its minimal representation in the literature stems from the inability of many depth sensors to resolve diving when depths are less than 0.5 m. However, a substantial portion of penguin at-sea effort is allocated to this dive. In these T-dives, penguins barely descend the water column, rarely exceeding depths of 1.5 m (Fig. 9.8a), swimming essentially horizontally (Fig. 9.8b, c) with all the time spent just beneath the surface (Fig 9.8d) moving in what is virtually a straight line (Fig. 9.8e).

A surprising aspect of T-dives is the extreme proximity of penguins to the surface, especially in the shorter dives (Fig. 9.9). Animals swimming subsurface are reported to minimize surface drag [1, 2] by maintaining depths that are greater than three body diameters [10] and this accords with Yoda et al.'s [56] report that Adélies execute T-dives at mean depths of 0.9 m. However, this is clearly not adhered to all the time by penguins as Fig. 9.9 makes clear (most penguins have a body diameter, or girth, of 400–600 cm, larger for *Aptenodytes* and smaller for *Eudyptula*; Chap. 1, Table 1.2). Penguins at sea are not easily observed underwater, but we have witnessed African penguins so close to the surface that they cause a clear vertical bow wave and the tips of their flippers may occasionally break the surface. In Chap. 6, we discussed how penguins are positively buoyant, especially near the water

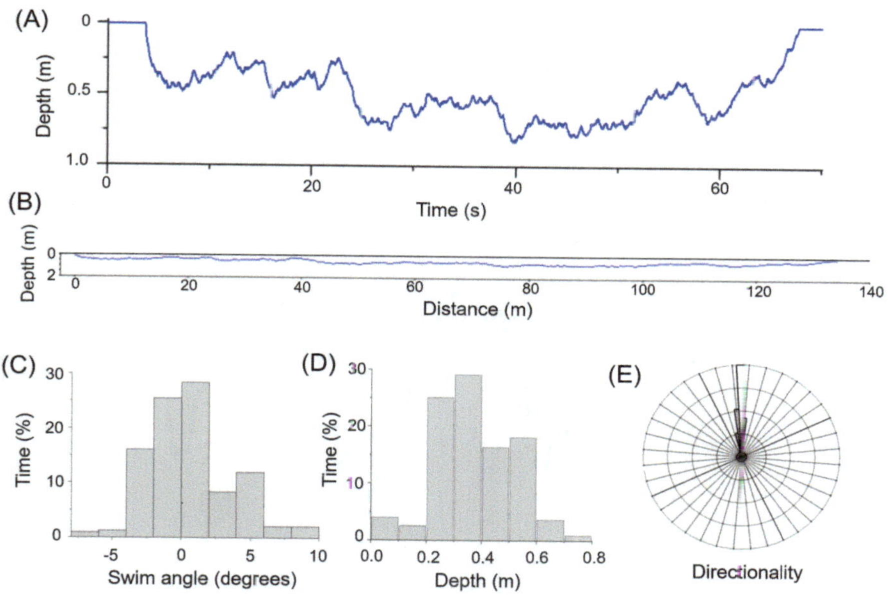

Fig. 9.8 An example T-dive ("traveling") of a Magellanic penguin (Peninsula Valdes, Argentina) showing (**a**) the dive profile, (**b**) the depth vs. distance plot (note that the dive–distance profile is even flatter than depicted since the depth scale is twice that of the distance scale), (**c**) the percentage time spent at different swim angles (0° is horizontal), (**d**) the percentage time spent at different depths, and (**e**) the directionality of the dive. Note the extreme time spent at body angles around the horizontal, the minimal depth (dive aspect ratios are usually >100), and highly directional swimming (dive tortuosities are close to 1). All this maximizes horizontal movement

surface, needing a downward-directed force to keep them stable and horizontal at their preferred depth. The presence of a visible wave above penguins swimming so close to the surface makes us wonder whether the birds benefit from the downward force of the increased volume of water directly above them that would serve this purpose [14] – a sort of reverse of the "ground effect" observed in some fish [45].

T-dives and their variants are likely to be extensively used by all penguin species and have been documented directly in emperor [18], chinstrap [40], Adélie [56], little [5], and yellow-eyed penguins [22]. These dives are most common in birds leaving their colonies to forage or returning to their colonies from foraging, especially during the sections close to the colony but traveling dives can also occur at almost any point within at-sea trips. The time spent on the surface between traveling dives is typically no more than a few seconds. This is generally not enough time for Global Positioning System (GPS) tags to acquire a fix, which explains the lack of detailed information for some periods of penguin-foraging trips documented by GPS systems alone (e.g., Cannell et al. [5]).

A particular variant of T-dives is "porpoising" [16, 56]. While engaged in such behavior, birds swim at depths of less than 1 m, but, instead of surfacing and

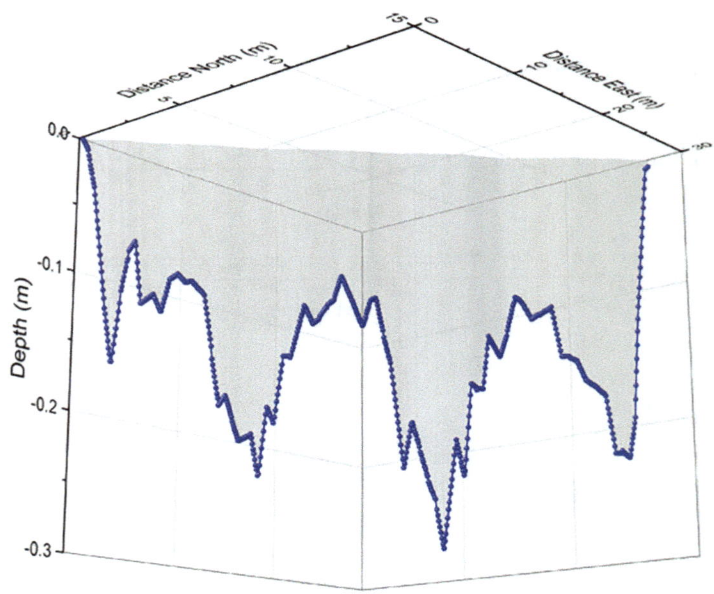

Fig. 9.9 A three-dimensional (3D) trajectory of a T-dive executed by a Magellanic penguin, illustrating the extreme directedness of the travel direction and minimal depth. With appropriate depth resolution, these dives appear to show some undulation in the depth (three undulations are apparent in this dive), which may indicate some form of burst and glide swimming

pausing to breathe, they snatch a breath by having their body briefly break the water surface [56], most spectacularly by having the body leave the water entirely (Fig. 9.10). Porpoising is commonly observed (or documented via accelerometry (cf. Yoda et al. [55, 56]) in all three *Pygoscelis* penguins, and we have seen it in macaroni and rockhopper penguins. However, it constitutes only a small fraction of the foraging trip's duration (typically <1%) [56]. It is infrequent in the four *Spheniscus* penguins and *Aptenodytes* (king and emperor) penguins, generally occurring only in the presence of predators. To our knowledge, and speaking to relevant researchers, it has never been observed in little penguins.

This mode of swimming is reported to be an energy-conserving behavior that allows penguins to breathe [16]. Thus, they come up for a breath at intervals, 15–20 m apart (for Adélies, this occurs every 6–12 s or roughly every 16 m traveled [56]). Porpoising speeds are higher than for normal swimming. For example, Yoda et al. [56] recorded a porpoising speed of 3 m/s for Adélie penguins, a figure echoed for the Humboldt penguins [16]. Critically, the travel must have enough momentum for the body to come out of the water high enough for the birds to take a breath (which is visible by the open beak in slow-motion films; Fig. 9.10). At ~3 m/s, porpoising is predicted to increase the costs of travel by a factor of about two (Chaps. 6 and 14). This, together with the fact that birds only have time for a single breath

Fig. 9.10 A chinstrap penguin catching a breath while porpoising. If the height reached above the water during the leap is relatively low, penguins angle their head upward slightly as they descend their arc, to allow them a fraction more time to inhale before their beak reenters the water. (Photo courtesy of Jean Pennycook)

during their ~0.5-s "air phase" [16, 56], may also explain why porpoising birds have to break the water surface so often.

Incomplete porpoising (where the whole body does not leave the water), such as occasionally observed in penguins of the genus *Spheniscus*, is less efficient because the penguins do not travel in a ballistic arc [16]. Instead, birds raise their head and neck out of the water to breathe, which degrades their fusiform shape (Chap. 6), increasing the drag for that period. Incomplete porpoising essentially occurs because swim speeds are too low to project the body fully out of the water, which, however, allows birds to swim underwater at speeds closer to their minimum costs of transport (see Chap. 6).

V-Dives (Prey Assessment)

Many penguin species are reported to execute what are called "V-dives." The name derives from a dive profile that shows a steep descent followed by an almost instantaneous reversal, the ascent (Fig. 9.11a).

The distance–depth profile shows a much greater allocation of space to the depth axis than traveling dives, although the horizontal distance covered can still be appreciable (Fig. 9.11b). Both descent and ascent angles are the same for both phases (Fig. 9.11c) so that vertical velocities are also fairly constant. This results in a roughly equal allocation of time to depth and distance right down the water column

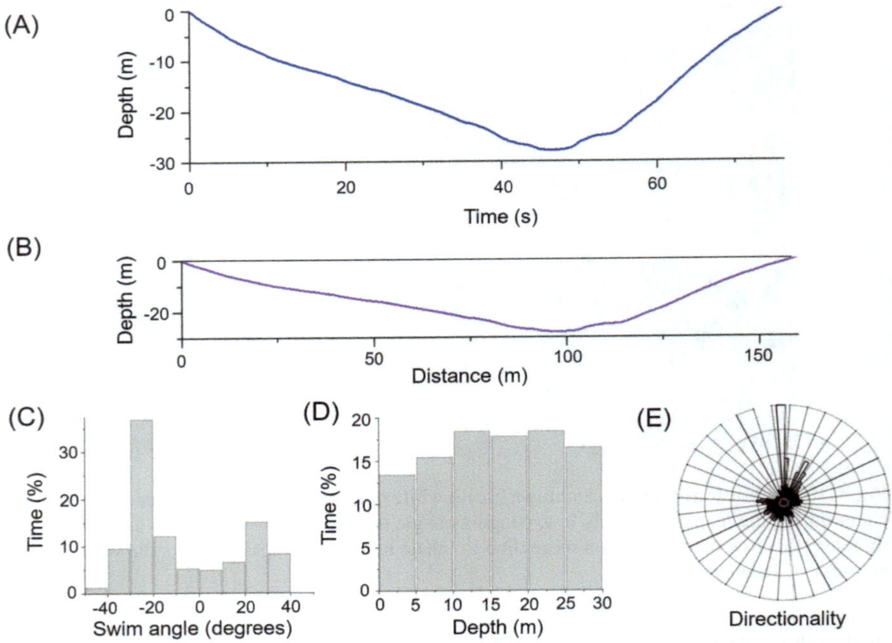

Fig. 9.11 An example V-dive executed by a Magellanic penguin showing (**a**) the dive profile, (**b**) the depth vs. distance plot, (**c**) the allocation of time to swim angle, (**d**) time allocated to depths down the water column, and (**e**) the directionality of the dive. Note how the depth–distance profile more closely resembles the dive profile and how, in contrast to traveling dives, the swim angle distribution is bimodal, with peaks corresponding to the (roughly constant) descent and ascent angles. A consequence of this is that birds executing V-dives spend a constant time at each depth down the water column. This coupled with the generally high directionality of the dive (though less than in T-dives) make its function likely to be that of prospecting the area, both vertically and horizontally

(Fig. 9.11d). These dives are also highly directional, with the overall dive tortuosities close to 1.0 (Figs. 9.11e and 9.12).

By definition, prey are never caught during V-dives (or only trivially) because if that happens, penguins immediately extend their time at depth, changing the form of the dive profile to that of a "U-dive" (see below). To do that, the bird then has oxygen issues to contend with, for which it may not have planned (see below; Chaps. 8 and 10). We surmise that V-dives allow birds to prospect both the water column and horizontal space efficiently. During such dives, although penguins may seek prey directly, we suggest that they have no prior information about the likely depth where prey might occur because no one depth is prioritized by this maneuver. They just go to the depth previously planned and return to the surface. Thus, if they do not actually spot prey, then V-dives help birds assess the water column for cues, such as water turbidity and thermoclines, which might enhance or diminish their hunting success (see Cimino et al. [7] and Saenz et al. [30]) (Fig. 9.13, Chap. 3).

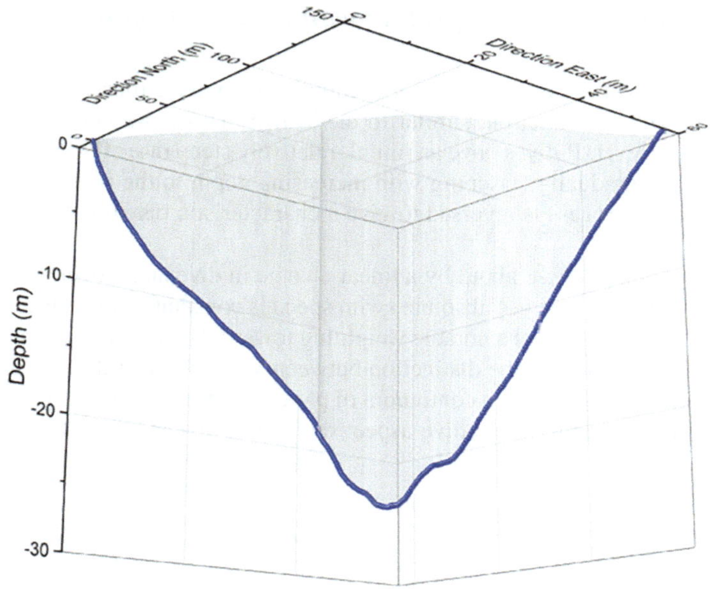

Fig. 9.12 The 3D trajectory of a Magellanic penguin V-dive showing high directionality (minimal tortuosity for the whole dive), with highly stylized descent and ascent patterns and minimal time used for the period between the descent and ascent

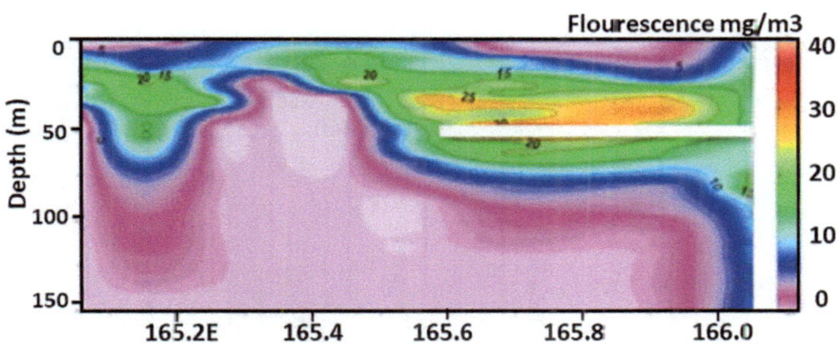

Fig. 9.13 The depth of the phytoplankton bloom along the McMurdo Sound fast ice edge, along which Adélie penguins were foraging in early January 2015 (as determined by a remotely operated vehicle). The horizontal white bar is the average depth at which the penguins were foraging, i.e., along the lower edge of the most turbid water. (From Saenz et al., unpublished)

V-dives have been documented for several penguin species, including Adélie, chinstrap [40], gentoo [21], macaroni [8, 31], emperor [18], and rockhopper penguins [39].

P (Parabolic)-Dives (Prospecting with No Prey Capture)

Parabolic dives ("P-dives"), with no prey capture, are named after a dive profile that shows no points of inflection, similar to the V-dive that has maximum depth at its midpoint. However, P-dives have an initial, relatively steep rate of descent down the water column, gradually lessening with increasing depth to the point of maximum depth when the process is reversed to give a characteristic bisymmetrical dive profile (Fig. 9.14a).

This pattern is brought about by a linear change in dive angle over the course of the dive (Fig. 9.14c) because absolute swim speed is constant. This results in a time-at-depth pattern that shows a notable weighting toward the point of maximum depth (Fig. 9.14d). Importantly, the distinction between V-dives and P-dives is not binary and many dive profiles show a continuum of patterns between the two types. P-dives for Magellanic penguins have dive aspect ratios of between about 15 and 35, are

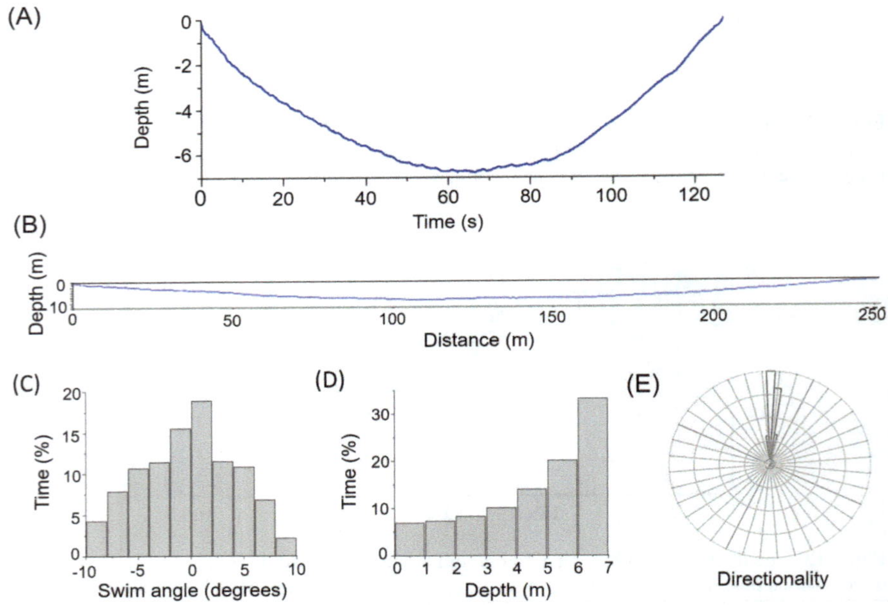

Fig. 9.14 An example P-dive (parabolic) executed by a Magellanic penguin showing (**a**) the dive profile, (**b**) the depth vs. distance plot, (**c**) the allocation of time to swim angle, (**d**) time allocated to depths down the water column, and (**e**) the directionality of the dive. In this case of a relatively shallow dive, note how, as with the traveling dives, the depth–distance profile departs considerably from the dive profile. Because, in contrast to V-dives, the swim angle changes linearly over the course of the dive, the time allocation to the swim angle has a defined distribution with a single maximum at the point of maximum depth. This, in turn, results in an increasing amount of time spent closer to the maximum depth, an indication that birds are prioritizing a particular depth while still assessing adjacent (shallower) depths. The high directionality is similar to that of T- and V-dives, indicating a priority for extensive horizontal searching

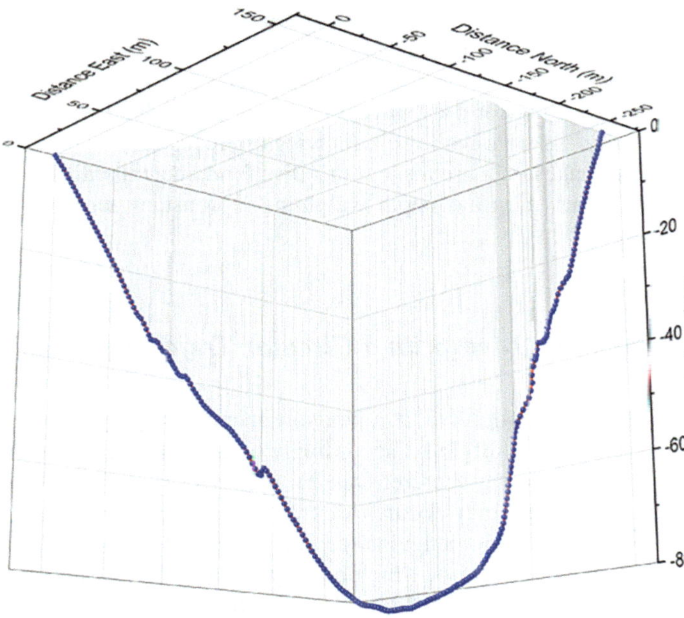

Fig. 9.15 The 3D trajectory of a Magellanic penguin executing a P-dive, showing the smooth transition from the descent to the ascent phase. Note that the amplified depth scale relative to the horizontal axes appears to show a tendency toward a V-dive, which disappears when the scales across dimensions are equalized (see Fig. 9.14b). Although there was nothing in this dive to indicate prey pursuit (see later), the minor irregularities in the trajectory may be due to the bird-inspecting objects/species of interest underwater, as we have seen them do on bird-borne video footage

highly directional (with the overall dive tortuosities close to 1.0) (Fig. 9.15), and occur most commonly in the outward leg of a foraging trip. Even so, they may occur at any time, either as singletons between other dive types or in a series.

P-dives have been documented for a number of penguin species, including emperor [18] and king penguins, in which they occur frequently [49], and gentoo penguins [12], in which they have also been described as “S-dives” [21].

One possible, and likely rather infrequent, function of the shallow (<5 m) P-dives has been suggested for cetaceans to help animals reoxygenate their body faster after taxing dives [15]. Penguins, specifically those that are at a great oxygen deficit but stand to benefit by recovering as rapidly as possible, may enhance their oxygen transfer rates by executing short, shallow P-dives. These increase the difference in oxygen partial pressure between the lungs and body tissues by compressing the lungs (the proportionate effects are greatest near the surface so even a dive to 5 m increases the oxygen partial pressure by about 50% – see Chap. 8 for more on this process). However, such dives have to be short because the oxygen in the lungs has to be continually replaced. As expected, such short, shallow dives do seem to occur between serial deep dives, notably in king penguins (unpublished data), but whether these are actually “intended,” as suggested, or are simply birds moving and searching efficiently (see below) is unclear.

Generally, though, with its form and directionality, the parabolic dive is likely to function as a “prospecting-for-prey” dive with a time-based depth weighting toward the deepest depth of the dive while still allowing appreciable horizontal movement (Fig. 9.14d). This makes the pure form of this dive fundamentally different from the pure V-dive, in which the time–depth weighting is constant across the full dive depth (Fig. 9.11d).

P_O-Dives (Parabolic Dives with a Circular Trajectory)

As with almost any foraging dive, if a penguin encounters prey, it will cease the prescribed dive pattern, modifying it accordingly or going to the surface to acquire the necessary oxygen for the venture. One particular form of the P-dive, though, which we term “P_O-dives,” may occur with prey circling from the outset. We have recorded these for Magellanic penguins feeding on anchovies. Such dives may be exclusive to *Spheniscus* penguins, which display a number of particular strategies for acquiring pelagic school fish [24, 51]. A major portion of the diet of all *Spheniscus* penguins includes anchovies or similar mid-water, schooling fish (Chap. 4, Appendix Table A2). Here, although the basic dive parameters remain essentially the same as in P-dives without prey encounter, the high directionality of the dive ceases. Instead, the three-dimensional (3D) dive trajectory profile shows birds descending in a circle, spiraling more tightly, to a slight degree, after the point of maximum depth before the path becomes much more tortuous on the ascent, finally giving way to a straight-line shallow ascent nearer the surface (Fig. 9.16).

Importantly, the highly stylized pattern of this dive makes its function to be unlike that of the standard P-dive. Instead, this P_O-dive has a functional difference from the outset because, we assume, the prey have already been found. This dive type seems to be associated with fish herding, in which the presence of the penguin at the periphery of the fish school during the descent may serve to drive the fish together, rather like a dog herding sheep. It, together with and other strategies for prey acquisition, is discussed in detail in Chap. 11.

U-Dives (Depth-Directed Prospecting)

“U-dives” (Fig. 9.17a) that do not involve prey pursuit, differ from both V- and P-dives in having constant descent and ascent angles for any one dive (Fig. 9.17c) coupled with constant vertical velocities. Therefore, the time spent at depth is constant down the water column until close to the point of maximum depth, although this tends to decrease as maximum dive depth increases (Fig. 9.5) [41]. At that

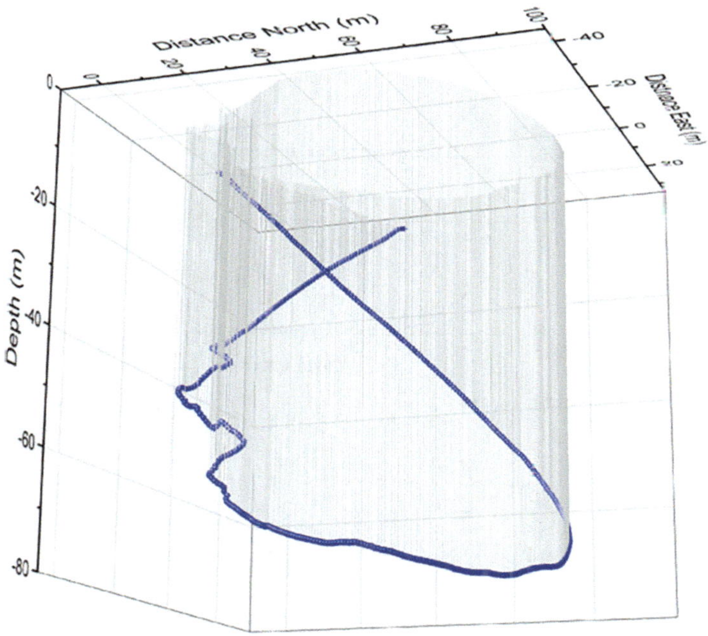

Fig. 9.16 A P_O-dive by a Magellanic penguin (the dive begins at point 0,0,0) showing how the dive profile is superficially similar to a P-dive but how the dive is clearly fundamentally different once directionality is considered. Tortuosities for such dives are around 0.2

point, time at depth increases disproportionately (Fig. 9.17d) and much more than in either V-dives or P-dives (cf. Figs. 9.11 and 9.14). This represents the U-dive format in its most definitive form. However, in reality, the difference between U- and P-dives may be less marked because, after a descent period at a constant angle, U-dives may sometimes show a gradual change in descent angles until the deepest point of the dives, as do P-dives, so intermediate forms certainly exist. The directionality of these dives tends to be less marked than both V- and P-dives (Figs. 9.17e and 9.18). The substantive allocation of time to a specific (maximum) depth with most horizontal travel occurring at this time indicates that the penguins concentrate their search effort to depth strata where the perceived probability of prey encounter is markedly higher than at other depths. This represents a fundamental difference to P-dives where the concentration of time occurs over a more expansive depth range.

An obvious example for such dive types is when penguins are searching close to the seabed such as been noted for rockhopper [39], chinstrap [37], yellow-eyed [23], and African penguins [36], where the bottom topography may be associated with higher densities of demersal (though not necessarily benthic) prey. However, these dives also occur frequently during mid-waters for probably all species [3, 6, 18, 26, 29, 48, 54].

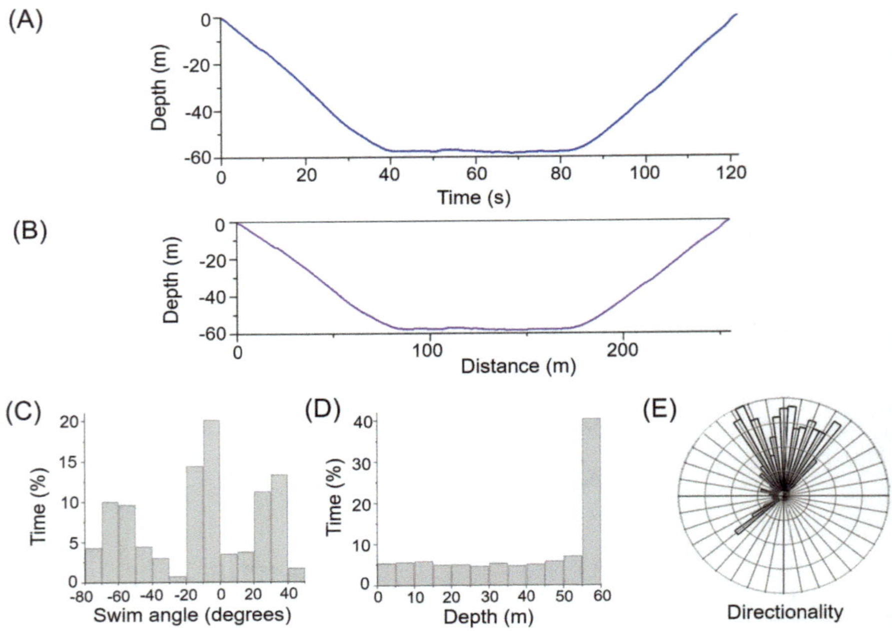

Fig. 9.17 An example U-dive executed by a Magellanic penguin during which no prey were encountered showing (**a**) the dive profile, (**b**) the depth vs. distance plot, (**c**) the allocation of time to swim angle, (**d**) time allocated to depths down the water column, and (**e**) the directionality of the dive. Note how, in contrast to T-, V-, and P-dives, the U-dive has a trimodal time allocated to body angle, representing the constant descent and ascent angles coupled with the horizontal body angle during the extended bottom phase of the dive. This pattern results in a swift descent and ascent of the water column but greatly extended time allocated to the point of maximum depth within a well-defined depth stratum. The directionality of this dive is also appreciably less than in T-, V-, and P-dives, although it is not tortuous either

W-Dives/U_p-Dives (U-Dives with Prey Pursuit)

Given that U-dives are believed to concentrate their foraging efforts at depths where prey are likely to be more readily encountered, it is little surprising that this dive type often morphs into the so-called "W-dive." This name is derived from the "wiggles" or "undulations" that many penguin species exhibit as they vary depth during the bottom phase of the dive (Fig. 9.19a) [13, 35]. A wiggle is usually equated to a prey capture event (e.g., Bost et al. [3], though the degree to which this is so is somewhat debatable, i.e., every wiggle does not necessarily represent a prey capture [43].

However, aside from the difficulties of deciding what depth change actually constitutes an "undulation" [12], this definition was brought about in the era when depth transducers were essentially the only sensors in penguin-borne tags [20]. Consequently, such undulations only represent a change in a single dimension when

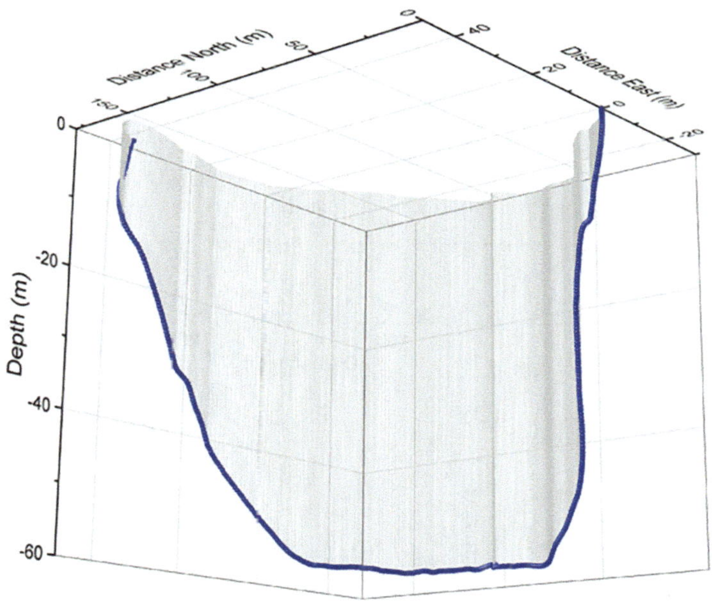

Fig. 9.18 The 3D trajectory of a Magellanic penguin executing a U-dive during which no prey were encountered

penguin use of space can change in three dimensions. Consideration of other metrics, such as speed and acceleration, may provide a more complete picture. Certainly, prey pursuit and capture occur frequently without changes in depth (see Chap. 11), and, so, we advocate calling U-dives in which prey are captured "U_{prey}-dives" or "U_p-dives" for short and not W-dives. Otherwise, the descent and ascent parameters of this dive are similar to those of the U-dive with fairly constant time allocated down the water column to the point of maximum depth whereupon the time component increases substantially (Fig. 9.19d). There is lower directionality in U_p-dives (Fig. 9.19e) than in U-dives due to the exploitation of prey, which leads to changes in the penguin heading during the bottom phase. Even so, in Magellanic penguins at least, the dives are surprisingly directional (Fig. 9.20), something that is discussed in relation to prey distribution in Chap. 10.

Depth Duration Effects Over Multiple Dives

The type of dive undertaken by penguins, specifically its form and function, determines how much time birds invest in the various phases of the dive. As noted, dives are typically divided into the descent, bottom, and ascent phases [12], with these phases being delineated by points of inflexion in the depth-versus-time profile.

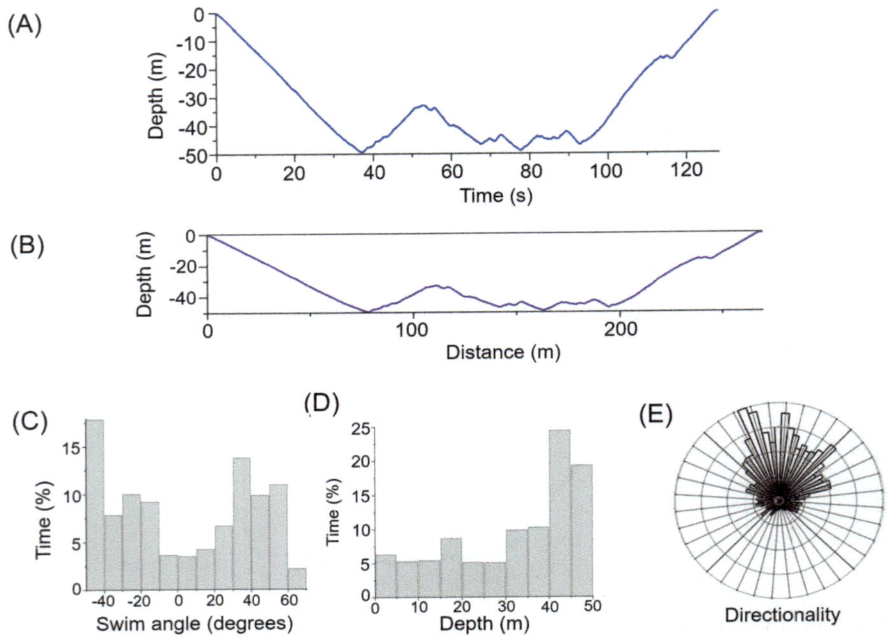

Fig. 9.19 An example W-dive (or U_p-dive) executed by a Magellanic penguin during which prey were encountered showing (**a**) the dive profile, (**b**) the depth vs. distance plot, (**c**) the allocation of time to swim angle, (**d**) time allocated to depths down the water column, and (**e**) the directionality of the dive. This dive shows the same basic features of the U-dive with some notable modifications. Where prey are pursued through the bottom phase of the dive, the trimodality in the frequency distribution of body angles in the non-prey U-dives (cf. Fig. 9.17c) becomes bimodal because the bird continually moves small distances up and down the water column, only being horizontal for short periods. The vertical movement during the bottom phase of the dive also results in an overall broader spread of time over extended depths during the bottom phase of the dive (**a**, **b**, and **d**). The pursuit of prey also results in more tortuosity within the dive (**e**)

Consideration of the "transit" phases, between the surface and the maximum depth reached during dives shows relatively little variation over multiple dives, with the duration increasing with maximum depth (Fig. 9.21a, c).

This pattern ensues because the function of the transit phase is to take the penguins from the surface down to the projected foraging depths and, because swim speed varies little, birds take longer to cover greater vertical distances (discussed in more detail in Chap. 10). Against this, there is appreciable variation in the length of the bottom phase. This tends to increase with increasing depth up to a point but then remains roughly constant, or even decreases, with increasing depth (Fig. 9.21b). The variation in the bottom duration versus maximum depth relationship is dependent on what happens during the bottom phase of dives, which is highly variable. In the two extremes, birds may execute a V-dive (a bounce dive) where there is no bottom duration (see the line of 0-s duration in Fig. 9.21b) or spend multiple minutes at the bottom depth (Fig. 9.21b). Factors that affect the amount of time spent in the

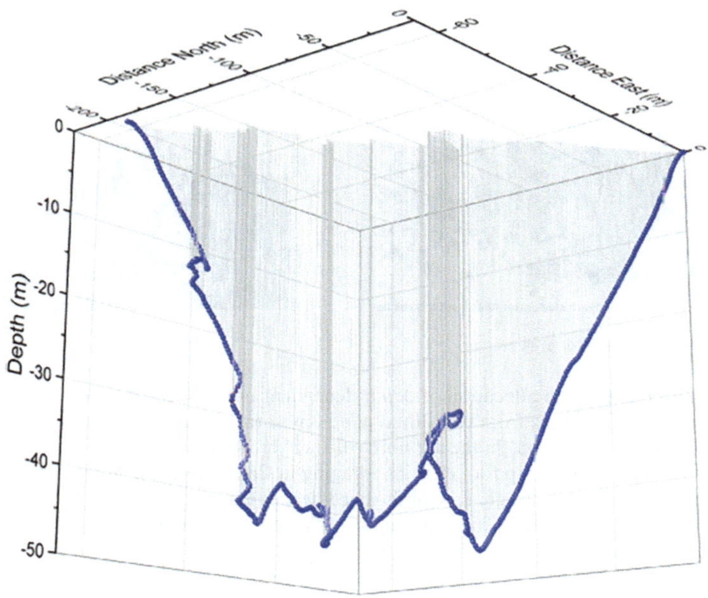

Fig. 9.20 The 3D trajectory of a Magellanic penguin executing a U-dive during which prey were encountered, made apparent here by the extensive tortuosity during the bottom phase of the dive. The start of the dive is at the position 0,0,0

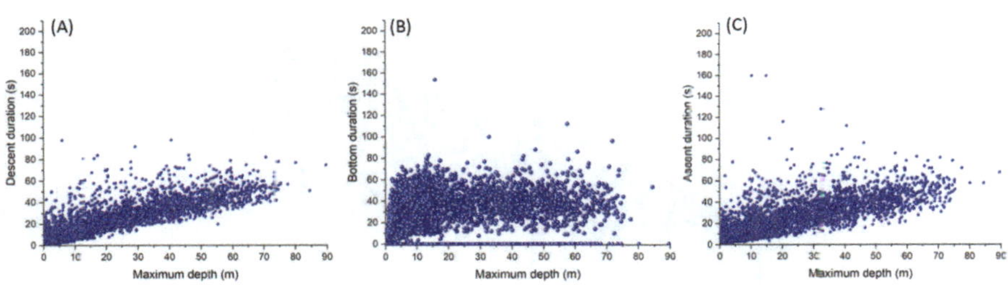

Fig. 9.21 An example of how the duration of the descent, bottom, and ascent phases vary with maximum depth reached by penguins, illustrated by 5000 dives made by 10 Magellanic penguins foraging from a colony at Peninsula Valdes, Argentina. Each point denotes a single dive. Note how there is relatively little variation in the (**a**) descent- and (**c**) ascent-versus-depth relationships but appreciably more in the (**b**) bottom duration versus maximum depth relationship

bottom phase include prey density, prey size (and therefore speed), penguin speed (because faster swimming uses disproportionately more energy and therefore oxygen; see Chaps. 6 and 14), and even water temperature (Chap. 7). It is thus hardly surprising that there are considerable inter-colonial differences in the distribution of bottom duration versus depth (Fig. 9.22).

Addition of the descent, bottom, and ascent durations together for each dive yields the much-vaunted dive duration versus depth plots, which have been adopted as a major descriptor of penguin diving habits (Fig. 9.23).

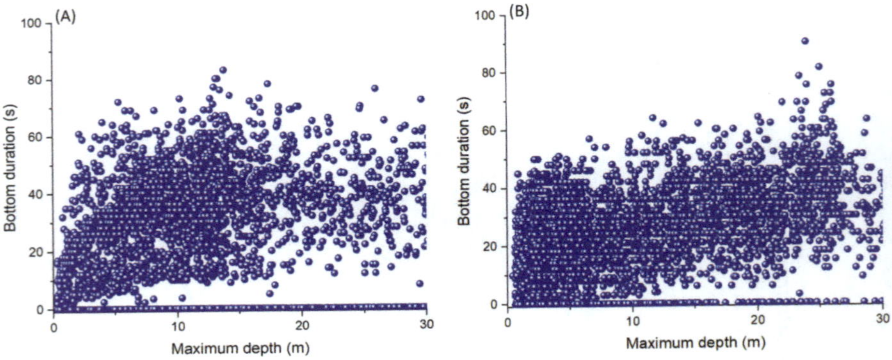

Fig. 9.22 Difference in the allocation of depth-dependent time allocated to the bottom phase of dives in Magellanic penguins from two different colonies. (**a**) The relationship for birds foraging in Peninsula Valdes, Argentina, in the far north of their breeding range (a zoomed-in section of Fig. 9.21b). (**b**) The relationship for penguins foraging in Cabo Virgenes, Argentina, in the southern portion of their breeding range. Few birds dived deeper than 30 m, so both datasets have been trimmed to facilitate comparison. The large number of dives with zero bottom durations are V-dives (bounce dives)

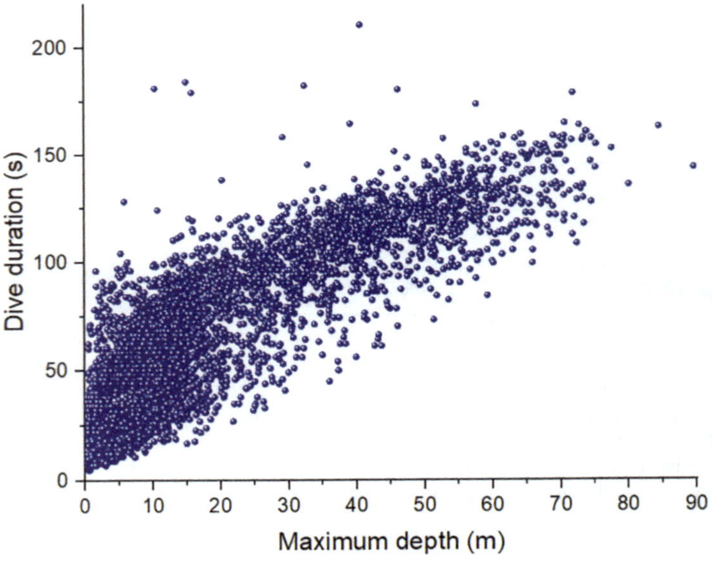

Fig. 9.23 A species-specific example of dive duration vs. depth provided by Magellanic penguins from Peninsula Valdes, Argentina. Note the generally increasing dive duration with depth and also the substantial variability in the relationship

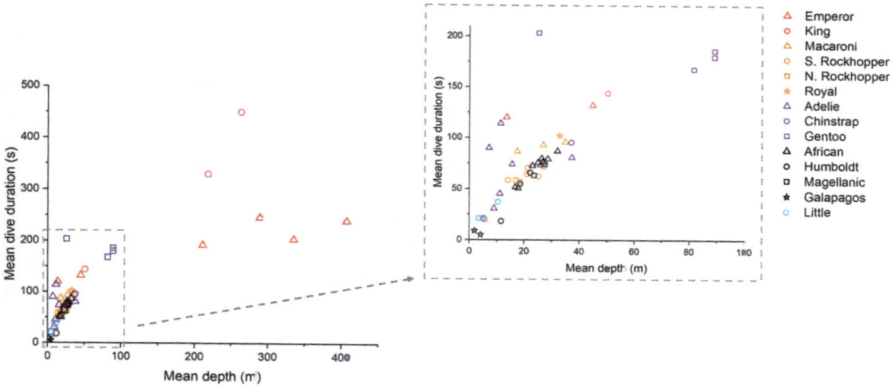

Fig. 9.24 The mean dive duration in relation to the mean dive depth for various penguins species (data from the Penguiness website (Penguiness.net)) showing that, generally, as dive depth increases, so too does dive duration. Note that there is a clear lower duration limit since it takes a finite amount of time to reach greater depths

Due to the way that both descent and ascent durations increase with maximum dive depth, superimposed on changes in bottom duration with dive depth, the standard pattern for this relationship is one of increasing duration with depth, although the relationship is not entirely linear (Fig. 9.23). This general pattern is also apparent when the mean dive durations are plotted against the mean maximum dive depth across species (Fig. 9.24).

Aside from the allometric implications of this relationship (Chap. 8), the overall decreasing proportion of time spent underwater with increasing depth demonstrates that the depth at which prey are encountered is a significant factor in determining how efficiently penguins can acquire them, and this is explicitly discussed in Chap. 10.

References

1. Au D, Weihs D (1980) At high speeds dolphins save energy by leaping. Nature 284(5756):548–550
2. Blake R, Smith M (1988) On penguin porpoising. Can J Zool 66(9):2093–2094
3. Bost C-A, Handrich Y, Butler P, Fahlman A, Halsey L, Woakes A, Ropert-Coudert Y (2007) Changes in dive profiles as an indicator of feeding success in king and Adélie penguins. Deep-Sea Res II Top Stud Oceanogr 54(3–4):248–255
4. Broni S (1985) Social and spatial foraging patterns of the jackass penguin *Spheniscus demersus*. Afr Zool 20(4):241–245
5. Cannell B, Ropert-Coudert Y, Radford B, Kato A (2020) The diving behaviour of little penguins in Western Australia predisposes them to risk of injury by watercraft. Aquat Conserv Mar Freshwat Ecosyst 30(3):461–474

6. Chiaradia A, Ropert-Coudert Y, Kato A, Mattern T, Yorke J (2007) Diving behaviour of little penguins from four colonies across their whole distribution range: bathymetry affecting diving effort and fledging success. Mar Biol 151(4):1535–1542
7. Cimino MA, Moline MA, Fraser WR, Patterson-Fraser DL, Oliver MJ (2016) Climate-driven sympatry may not lead to foraging competition between congeneric top-predators. Sci Rep 6(1):1–10
8. Croxall J, Briggs D, Kato A, Naito Y, Watanuki Y, Williams T (1993) Diving pattern and performance in the macaroni penguin *Eudptes chrysolophus*. J Zool 230(1):31–47
9. Duffy DC (1983) The foraging ecology of Peruvian seabirds. Auk 100(4):800–810
10. Gordon CN (1980) Leaping dolphins. Nature 287:759
11. Gunner RM, Holton MD, Scantlebury MD, Van Schalkwyk L, English HM, Williams HJ, Hopkins P, Quintana F, Gómez-Laich A, Börger L (2021) Dead-reckoning animal movements in R–A reappraisal using Gundog. Tracks
12. Halsey L, Bost C-A, Handrich Y (2007) A thorough and quantified method for classifying seabird diving behaviour. Polar Biol 30(8):991–1004
13. Hanuise N, Bost C-A, Huin W, Auber A, Halsey LG, Handrich Y (2010) Measuring foraging activity in a deep-diving bird: comparing wiggles, oesophageal temperatures and beak-opening angles as proxies of feeding. J Exp Biol 213(22):3874–3880
14. Hoerner SF (1965) Fluid-dynamic drag. S. F. Hoerner. Hoerner Fluid Dynamics, Brick Town, New Jersey
15. Holton MD, Wilson RP, Teilmann J, Siebert U (2021) Animal tag technology keeps coming of age: an engineering perspective. Philos Trans R Soc B 376(1831):20200229
16. Hui CA (1987) The porpoising of penguins: an energy-conserving behavior for respiratory ventilation? Can J Zool 65(1):209–211
17. Kato A, Ropert-Coudert Y, Grémillet D, Cannell B (2006) Locomotion and foraging strategy in foot-propelled and wing-propelled shallow-diving seabirds. Mar Ecol Prog Ser 308:293–301
18. Kirkwood R, Robertson G (1997) The foraging ecology of female emperor penguins in winter. Ecol Monogr 67(2):155–176
19. Kooyman G, Drabek C, Elsner R, Campbell W (1971) Diving behavior of the emperor penguin, *Aptenodytes forsteri*. Auk:775–795
20. Lescroël A, Ballard G, Toniolo V, Barton KJ, Wilson PR, Lyver POB, Ainley DG (2010) Working less to gain more: when breeding quality relates to foraging efficiency. Ecology 91(7):2044–2055
21. Lescroël A, Bost C-A (2005) Foraging under contrasting oceanographic conditions: the gentoo penguin at Kerguelen Archipelago. Mar Ecol Prog Ser 302:245–261
22. Mattern T, Ellenberg U, Houston DM, Davis LS (2007) Consistent foraging routes and benthic foraging behaviour in yellow-eyed penguins. Mar Ecol Prog Ser 343:295–306
23. Mattern T, Mcpherson MD, Ellenberg U, Van Heezik Y, Seddon PJ (2018) High definition video loggers provide new insights into behaviour, physiology, and the oceanic habitat of a marine predator, the yellow-eyed penguin. PeerJ 6:e5459
24. Mcinnes AM, Mcgeorge C, Ginsberg S, Pichegru L, Pistorius PA (2017) Group foraging increases foraging efficiency in a piscivorous diver, the African penguin. R Soc Open Sci 4(9):170918
25. Paiva VH, Geraldes P, Ramírez I, Garthe S, Ramos JA (2010) How area restricted search of a pelagic seabird changes while performing a dual foraging strategy. Oikos 119(9):1423–1434
26. Peters G, Wilson RP, Scolaro JA, Laurenti S, Upton J, Galleli H (1998) The diving behavior of Magellanic Penguins at Punta Norte, Península Valdés, Argentina. Colon Waterbirds:1–10
27. Ropert-Coudert Y, Grémillet D, Kato A (2005). Diving angle of great cormorants
28. Ropert-Coudert Y, Kato A, Baudat J, Bost C-A, Le Maho Y, Naito Y (2001) Time/depth usage of Adélie penguins: an approach based on dive angles. Polar Biol 24(6):467–470
29. Ropert-Coudert Y, Kato A, Wilson RP, Cannell B (2006) Foraging strategies and prey encounter rate of free-ranging Little Penguins. Mar Biol 149(2):139–148

30. Saenz BT, Ainley DG, Daly KL, Ballard G, Conlisk E, Elrod ML, Kim SL (2020) Drivers of concentrated predation in an Antarctic marginal-ice-zone food web. Sci Rep 10(1):1–17
31. Sato K, Charrassin J-BT, Bost C-A, Naito Y (2004) Why do macaroni penguins choose shallow body angles that result in longer descent and ascent durations? J Exp Biol 207(23):4057–4065
32. Sato K, Naito Y, Kato A, Niizuma Y, Watanuki Y, Charrassin J, Bost C-A, Handrich Y, Le Maho Y (2002) Buoyancy and maximal diving depth in penguins: do they control inhaling air volume? J Exp Biol 205(9):1189–1197
33. Sato K, Shiomi K, Marshall G, Kooyman GL, Ponganis PJ (2011) Stroke rates and diving air volumes of emperor penguins: implications for dive performance. J Exp Biol 214(17):2854–2863
34. Schreer J F, Kovacs K M, O'hara Hines R (2001). Comparative diving patterns of pinnipeds and seabirds. Ecol Monogr, 71(1): 137–162
35. Simeone A, Wilson RP (2003) In-depth studies of Magellanic penguin (*Spheniscus magellanicus*) foraging: can we estimate prey consumption by perturbations in the dive profile? Mar Biol 143(4):825–831
36. Sutton G, Pichegru L, Botha JA, Kouzani AZ, Adams S, Bost CA, Arnould JP (2020) Multi-predator assemblages, dive type, bathymetry and sex influence foraging success and efficiency in African penguins. PeerJ 8:e9380
37. Takahashi A, Dunn M, Trathan P, Sato K, Naito Y, Croxall J (2003) Foraging strategies of chinstrap penguins at Signy Island, Antarctica: importance of benthic feeding on Antarctic krill. Mar Ecol Prog Ser 250:279–289
38. Tessier E, Bost C-A (2020) Behavioural adjustments during foraging in two diving seabirds: king and macaroni penguins. Mar Biol 167(9):1–11
39. Tremblay Y, Cherel Y (2000) Benthic and pelagic dives: a new foraging behaviour in rockhopper penguins. Mar Ecol Prog Ser 204:257–267
40. Trivelpiece WZ, Bengtson JL, Trivelpiece SG, Volkman NJ (1986) Foraging behavior of gentoo and chinstrap penguins as determined by new radiotelemetry techniques. Auk 103(4):777–781
41. Watanabe H, Shiome K, Sato K, Takahashi A, Handrich Y, Bost C-A (In press) King penguins adjust their fine-scale travelling and foraging behaviours to spatial and diel changes in feeding opportunities. Marine Biology
42. Watanabe YY, Ito M, Takahashi A (2014) Testing optimal foraging theory in a penguin–krill system. Proc R Soc B Biol Sci 281(1779):20132376
43. Watanabe YY, Takahashi A (2013) Linking animal-borne video to accelerometers reveals prey capture variability. Proc Natl Acad Sci 110(6):2199–2204
44. Watanuki Y, Wanless S, Harris M, Lovvorn JR, Miyazaki M, Tanaka H, Sato K (2006) Swim speeds and stroke patterns in wing-propelled divers: a comparison among alcids and a penguin. J Exp Biol 209(7):1217–1230
45. Webb PW (2002) Kinematics of plaice, *Pleuronectes platessa*, and cod, *Gadus morhua*, swimming near the bottom. J Exp Biol 205(14):2125–2134
46. Wilson R (1985) The Jackass Penguin (*Spheniscus demersus*) as a pelagic predator. Mar Ecol Prog Ser:219–227
47. Wilson R, Culik B, Peters G, Bannasch R (1996) Diving behaviour of Gentoo penguins, *Pygoscelis papua*; factors keeping dive profiles in shape. Mar Biol 126(1):153–162
48. Wilson RP (2010) Resource partitioning and niche hyper-volume overlap in free-living Pygoscelid penguins. Funct Ecol 24(3):646–657
49. Wilson RP, Pütz K, Charrassin J-B, Lage J (1995) Artifacts arising from sampling interval in dive depth studies of marine endotherms. Polar Biol 15(8):575–581
50. Wilson RP, Ropert-Coudert Y, Kato A (2002) Rush and grab strategies in foraging marine endotherms: the case for haste in penguins. Anim Behav 63(1):85–95
51. Wilson RP, Ryan PG, James A, Wilson M-PT (1987) Conspicuous coloration may enhance prey capture in some piscivores. Anim Behav 35(5):1558–1560
52. Wilson RP, Wilson M-P, Duffy DC, Araya B, Klages N (1989) Diving behaviour and prey of the Humboldt penguin (*Spheniscus humboldti*). J Ornithol 130(1):75–79

53. Wilson RP, Wilson M-PT (1988) Foraging behaviour in four sympatric cormorants. J Anim Ecol:943–955
54. Wilson RP, Wilson M-P T (1990) Foraging ecology of breeding *Spheniscus* penguins. Penguin Biology:181–206
55. Yoda K, Naito Y, Sato K, Takahashi A, Nishikawa J, Ropert-Coudert Y, Kurita M, Le Maho Y (2001) A new technique for monitoring the behaviour of free-ranging Adelie penguins. J Exp Biol 204(4):685–690
56. Yoda K, Sato K, Niizuma Y, Kurita M, Bost C, Le Maho Y, Naito Y (1999) Precise monitoring of porpoising behaviour of Adélie penguins determined using acceleration data loggers. J Exp Biol 202(22):3121–3126

Chapter 10
Decisions, Decisions, and More Decisions: How Fish-Birds Search for Prey

The body of research conducted on penguins' activity in the ocean has progressed significantly toward understanding the hardware of these incredible birds and how, in the same way as vehicles, it determines the performance of the system and its energy expenditure. Compare, for example, a Ferrari to a lorry, which is like comparing a penguin to a steamer duck with respect to dealing with the sub-sea-surface world! In self-maintaining biotic systems, which means all living animals, energy expenditure has to be exceeded, or at least covered, by energy (food) acquisition – or the books do not balance, so to speak, and eventually the animal dies. Given the defined costs of fish-birds operating in water, as viscous, dense, and cold as it is (e.g., Chaps. 6 and 7), trying to find prey that would prefer not to be found, how do penguins do this by the way that they behave and perhaps more specifically by the way that they move? It should be obvious now that they do it like a Ferrari, but, for example, how do they decide the direction to take when they leave their breeding sites, ice floe, or sleeping spot at sea to find suitable prey? How do they decide on the best, sure-to-be energy-draining action even when they are in what seems to be a "good area" for prey? How do they decide the depths to visit during their dives? At what speeds should they travel? What descent and ascent angles are appropriate? What dive type should they execute (Chap. 9)? How long should they remain underwater anyway? Some potential answers are covered in the rather complex optimal foraging literature relating to diving animals (e.g., [38, 43, 107, 128]), but, as with all models (including any that we might conjure up), these are dependent on assumptions and, owing to their intricacy, we have not attempted to cover these here. Rather, we will examine some basic trends – such as how the various dive phases vary in duration and energy expenditure with depth and how this relates to their capacity to find prey. Doing this will help give a fundamental picture of penguin dive capacities and limitations and how they should relate to penguin movement choice. Importantly, optimal penguin movement, in terms of exploitation of a 3D world, is made more difficult because the very animals our fish-birds pursue do not "want" to be eaten. This inevitably makes such fish-bird movement choices more complex and actually requires "forethought" on the part of

D. G. Ainley, R. P. Wilson, *The Aquatic World of Penguins*, Fascinating Life Sciences, https://doi.org/10.1007/978-3-031-33990-5_10

the penguins or, at the least, "strategies." What faces the penguins is a fascinatingly fluid and multifaceted situation, and it all happens underwater, a place where most fish-eating birds would not dream of being. Indeed, they would have nightmares about staying away from air for longer than a few seconds.

As a convenient starting place, we can consider the current biological dogma that animals should attempt to try and maximize their net rate of energy gain [22, 74, 75] – that is, they should be maximizing the amount of energy taken in compared to that expended to acquire it and should always acquire more energy than they expend during foraging. Although a starting point, this concept is not necessarily true. To begin, a foraging animal has to deal with other factors that might confound its energy expenditure, such as attention to predators [28], especially true among penguins (Chap. 12). Moreover, to their advantage, to facilitate prey capture, penguins may operate in a flock, which involves some sort of "group thinking" (Chap. 14). Paying attention to such issues might upset calculations, but, also, the premise assumes that there are no time constraints on the foraging animal. For penguins, as with many birds, a key part of breeding is them supplying food to their brood at a rate that allows "appropriate" growth [77] as well as retaining enough of the catch to sustain themselves [4]. So, optimization of time, rather than energy, may become more important [89] – it is not good for adults to forage "slowly" to maximize their net rate of energy gain if their chicks are starving. Under such a scenario and to maximize their chances of producing fit chicks, there is selection pressure for breeding penguins to operate suboptimally in terms of energy optimization, although they may have to pay for it with their own energy reserves, losing weight over the course of the breeding season [4, 23].

In this chapter, we will look at a number of ways (though by no means all) in which penguins might make movement choices to enhance their foraging efficiency. In our considerations, though, we acknowledge that what we know about penguin movement choices during foraging comes from breeding birds that might well be subjected to time, as well as energy, constraints, plus any constraints coming from foraging in a flock (the "social pressure" that is so much a part of human existence). What goes on outside the breeding season – when penguins might have periods of less pressure to optimize time rather than energy, and may indeed accumulate an energetic cushion for future use – will remain a mystery, for now at least. Like migratory waterfowl and songbirds, might penguins exhibit some sort of pre-migratory or pre-breeding hyperphagia in order to fatten up for the breeding season? Would sedentary versus dispersing penguin species (Chap. 3) differ in that regard? Seems likely, but we just do not know yet.

Heading in the Right Direction

By basing our thinking on breeding penguins, with breeding in effect constituting an energy bottleneck, and as central place foragers (another bottleneck), we have a well-defined situation in terms of what direction the birds should take when they leave their colonies to find prey. As explained in Chap. 5, the concentrated numbers and activities of colonial seabirds result in a tendency, if enough of them are foraging simultaneously, to deplete prey in the vicinity of their breeding site – why travel far to a supermarket if there is food near at hand? Thus, theoretically, penguins should first head rapidly away from their colonies, moving across the prey-deficit area around the site quickly (the Storer–Ashmole halo [3, 109]). There are three elements to doing this: (i) moving at speeds that are close to the lowest costs of transport while (ii) traveling in a straight line through this area and (iii) minimizing any vertical displacement. Conversely, once an "appropriate" area for foraging has been reached, there may be a case for (i) still moving at speeds that are close to the lowest costs of transport but only until a high-quality prey patch is found, whereupon speeds may change (Chap. 11), while (ii) adopting a more meandering course (in other words, so to speak, if you have found a field with mushrooms, then do not walk straight across it [7]) and (iii) increasing the vertical dimension of movement to access prey that might be hiding at depth, though obviously trying to get away once pursued [21, 42]. Indeed, penguins restrict their horizontal but increase their vertical movement once they find an adequate prey patch [32, 57]. In some cases, they actually even have the wherewithal to know where they last found decent concentrations of prey (Chap. 14).

As pointed out above, factors other than energetically efficient foraging may tend to blur expected patterns, and this is certainly the case with speed. Although penguins do tend to swim at speeds that closely correspond to the minimum cost of transport (the maximum distance for minimum energy) for much of their foraging trips (Chap. 6), birds leaving their colonies across a zone where aquatic predators may lurk may initially increase their speed dramatically with many species porpoising extremely rapidly in the vicinity of their breeding sites (see Chap. 12). This makes perfect sense. If the probability of being caught by something nasty is a function of time spent in the danger zone, then the overall chances of being caught are diminished by reducing that time (Box 10.1). Fast-moving penguins, especially in a flock, may simply be harder to catch anyway.

Box 10.1
If the probability of a penguin being caught by a predator is a constant function of time (P_c) per second in a danger zone, then the overall probability that a penguin will be caught over a defined time that it spends in that danger zone is given by $(1\text{-}P_f)^n$, where P_f is the probability of not being caught and n is the number of seconds within the danger zone.

In the account of a leopard seal catching Adélie penguins at Esperanza Bay, as detailed in Chap. 12, it was extremely notable that the seal, which was simply lying in wait and ambushing the birds as they swam by (a typical penguin-catching strategy of pinnipeds trying to maintain their own foraging efficiency), was less successful when the penguin speed was high. If ever there was a case for penguins hurrying along, it is that! Fortunately for fish-birds, in the open ocean, these seals are not fast or maneuverable enough to catch them.

So, part of the trick for getting through the danger zone is also moving directly across it – if a lion is possibly in the bushes, you do not want to be going around in circles in that area. The reverse is true if, once out at sea, penguins move into an area where prey accessibility is high – they should stay in the "field of mushrooms." Quite what penguins do is indicated by their tracks, and dead reckoning – in which speed, heading, and depth data from penguin-attached tags are combined, constructing their 3D routes using vectors – is a great way to quantify this, especially underwater [37]. Global Positioning System (GPS) data on position in a horizontal plane (which only locates penguins on the surface) may also provide useful (but less resolved) information [78, 100]. In tandem with a simple examination of penguin routes through their seascape, it is also useful to consider the track tortuosity within their 3D space. Basically, tortuosity is a measure of how convoluted the tracks are (it is useful to think of tortuosity as a metric to indicate "dithering" in an area – one can think lions and mushrooms), and there are multiple ways to describe track tortuosity, most of which we shall ignore for convenience. Instead, we use a simple calculation to provide a metric that describes how convoluted (tortuous) animal tracks are (Box 10.2).

Box 10.2
A simple, direct method of calculating tortuosity is by dividing the straight-line distance between two points on a trajectory (separated by a defined time) by the total distance traveled between those two points. Values of 1 indicate straight-line travel. Increasing values above 1.0 indicate increasingly tortuous tracks.

Tortuosity is important to penguins for both avoiding predators and attempting to find prey and applies to the whole of the foraging trip. Accordingly, we expect low tortuosity – manifesting highly directional travel – to occur when birds leave their colonies and swim through the predator zone and prey-depleted areas out from their nesting site and when they are traveling through areas at sea where prey accessibility or availability is low. In contrast, we expect, once birds "discover" or re-gain an area where prey abundance is perceived, or is known to occur from previous experience, to be particularly high [104], that penguins will search that area more carefully. They do this by having more meandering tracks – again, do not walk straight through a field of mushrooms [100].

Inspection of the dead-reckoning tracks of *Pygoscelis* penguins (Fig. 10.1) shows variable tortuosities owing to changes in heading over the course of foraging trips, and this shows a clear movement strategy. Simplifying it, rather, the tracks may be broadly divided into straight-line sections, over which the penguins "know" or perceive that there is no prey, interspaced with "Gordian knots" of tortuosity in which the penguins spend time covering specific areas intensively because that is where

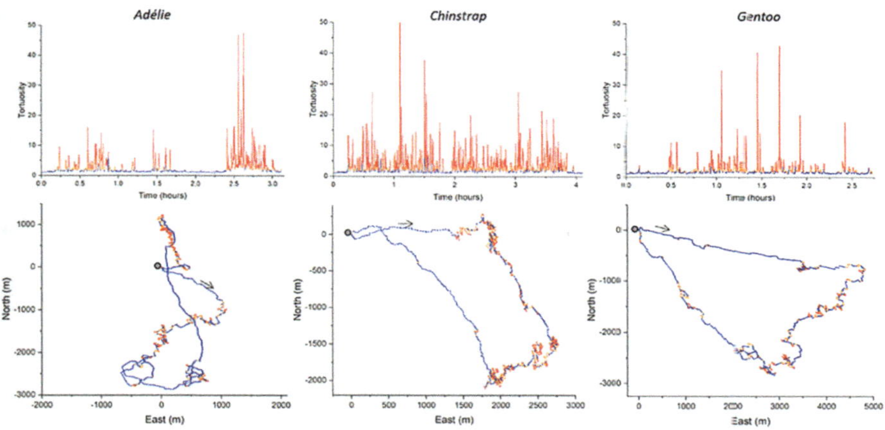

Fig. 10.1 Typical examples of foraging trajectories of the three *Pygoscelis* penguin species feeding on krill. All birds were breeding at the same colony on King George Island, Antarctica (gray circles), and their relative tortuosities ("*T*" – tortuosity metrics calculated over 1 min – see Box 10.1) are also shown for foraging trips of similar lengths (to facilitate comparison). The most convoluted movements are shown in red ($T > 5$), the most straight-line in blue ($T > 1 < 1.57$ (here, 1.57 is the ratio of the circumference of half a circle compared to it diameter)), and intermediate in orange ($T > 1.57 < 5$). The tracks (as if viewed from above) show clear periods of close to straight-line travel such as the initial movement away from the colony, as well as periods of straight-line travel between highly tortuous movement – the "knots in the string." These patterns presumably correspond to birds searching a particular water volume extensively, finding abundance to be higher, and/or the penguins were actively feeding, during which time the movements are dictated by the size of the krill swarm. They search again once the foraging flock has degraded the prey patch

the prey are available. In animal movement lingo, these knots are called area-restricted search (ARS).

The value of judicious choice in which penguins decide to spent time searching for prey can be easily demonstrated by running a simple model that ascribes a probability that penguins searching for a particular volume of water will find prey for every minute spent searching (assuming random encounter rates). Obviously, when prey are abundant in an area and penguin stay within that area (the Gordian knots in the track), the probability of finding something is high and birds acquire prey faster than when the situation is reversed (Fig. 10.2) [122].

There is, however, another particular and important aspect to this. As the prey accessibility decreases, and the probability of finding prey decreases with it, the variation in prey accumulation among individuals increases. In other words, by choosing to spend time searching in areas where prey are not abundant, individual penguins are running the risk of, by chance, just being unlucky and finding considerably less prey than the average penguin searching that area, or the other members of its flock (Fig. 10.2) [119, 122]. This is an excellent reason for penguins to travel efficiently through prey-poor areas by adopting straight-line travel so that they can allocate time to a more thorough search in better areas. The latter involves a

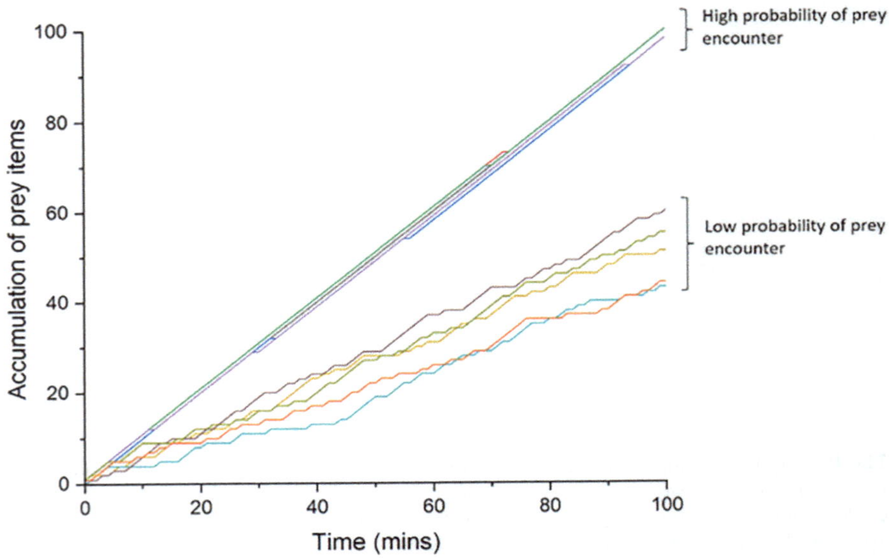

Fig. 10.2 Model-derived predicted accumulation of prey over time, based on the rate of accumulation of prey item encounters for five example penguins foraging in two different areas, one where there is a high (but random) probability of prey encounter (0.99 per minute searching – upper lines – each line represents an individual) and another where there is a lower probability (0.5 per minute searching – lower lines). Birds concentrating on the area with higher prey encounter probability obviously do better, but the variation in individual success also increases as the probability of finding prey decreases. So, one of the risks facing individual penguins allocating time to low-profitability areas is that they cannot guarantee that they will perform even as well as the "average penguin" (see, e.g., the cyan and orange lines). Instead, they might end up being substantially worse off than the average [122]. This is another compelling reason for penguins to be judicious in choosing to spend their searching time where the chances of finding prey are the highest. Group "thinking" and information transfer may also be involved in this process

meandering (ARS/Gordian knot) course, when they come across areas that appear profitable and which do not have this risk.

Although examination of time spent searching in particular areas (e.g., Fig. 10.1) points to penguins having mechanisms leading them to search particular areas for prey more thoroughly than others, it is not clear what cues birds use to inform them of prey densities, what thresholds incite a more tortuous searching pattern, and how birds actually decide on the specifics of the directionality of their tracks. We know that penguins are visual hunters [62, 63, 90], so they may assess prey densities using their eyes as well as by looking at what their flock members are doing, where appropriate. However, in a more broad sense, we also know that penguins, from their experience, favor areas where there are changes in water quality, i.e., fronts (Chap. 3). They also detect and react positively to smell, including to dimethyl sulfide (DMS), a breakdown product of algae being grazed by potential penguin prey (Chap. 14). Penguins may additionally use the presence of other penguin flocks as a cue to areas with greater prey abundance, although the success of this presumably depends on the number of penguins in the area (Chap. 14). We note, though, that in comparison with volant seabirds, which have a panoramic view of the ocean from up high, and can react to distant birds by seeing them easily and moving rapidly to reach them [39], penguins in the water, with their low profile and comparatively low traveling speeds, are distinctly less well placed to exploit patches in this manner, especially if they are ephemeral. However, then, too, some penguins species are known to "talk" underwater and on the surface [19, 96], and sound in, and just above, water travels a long way (Chap. 14).

The paths that penguins make through water, with their innumerable twists and turns, seem horribly complex, when seen in their entirety (Fig. 10.1). However, there is a start and end point of our resolved tracks. What happens in between Gordian knots is an unfolding pattern that depends on the dispersion of prey patches, and the way penguins decide to move at any moment in time, the pattern of which is superimposed on what happened before. It is reasonable to assume that turns in the pathway are a reaction to some sort of stimulus [69]. Beyond that, we feel that it is relevant that it costs animals more to turn than it does to continue in a straight line [115, 116], as has now been empirically demonstrated for terrestrial animals and the physical basis for it described in flying animals. Animals moving through water are no different because of turning stems from differential forces [111] and a selective breakdown in streamlining (i.e., an increase in drag). So, turns for penguins also require more energy. Based on this, we propose that the best way for a penguin to move when foraging is to make a decision about the best direction to head at the outset and swim, in a beeline, in that direction – after all, beelines are the shortest distance between two points. The decision need not be perfect though, just the most informed choice under the circumstances. Otherwise, the penguin should swim in the direction of choice, processing all sensory (and memory) information, until there is a compelling case for the swim direction to be changed. This occurs when the likelihood of encountering or capturing prey is perceived to be higher if the new heading is followed. The sensory information can be any one of the cues as to the whereabouts of prey discussed above, depending on the spatial scale (a DMS concentration gradient, calls from congeners, or simple observations of what other penguins in the flock are doing (Chap. 14)).

At the turn point, the increase in energy that the heading change necessitates is justified by the increased probability of encountering, or even catching, prey. So, we expect the movement paths of penguins (and actually the movement patterns of many other animals, including humans) to primarily consist of straight-line sections separated by discrete turns, with turn points being decisions based on information. To assess whether this might be true, we have used very highly resolved, and continuous, data of penguins in 3D space based on dead-reckoning systems that update penguin position, whether underwater or not, many times per second [37]! Using this, we note that, not only do Magellanic penguins, as an example, have tracks that appear to consist of straight-line sections broken by turns as predicted (Fig. 10.3) but that, when treated mathematically (using the definitions provided in Potts et al. [72]), more than 98% of their swim pathways consist of sections defined by straight lines.

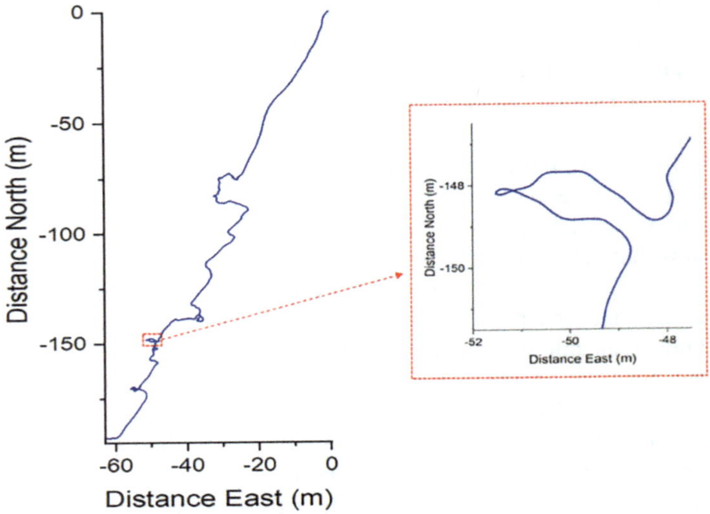

Fig. 10.3 A highly resolved movement path (using dead reckoning) of a Magellanic penguin during a U-dive (see Chap. 9). It should be noted how, on various scales, the track appears to consist of essentially straight-line sections separated by discrete turns

These straight lines are conventionally defined as "steps," with the rest being turns. Close examination of the lengths of these steps together with the turn angles adopted between the steps shows that both change over the course of foraging trips (Fig. 10.4).

Of course, we cannot yet know what most of the turns represent or what cue has resulted in any particular turn being elicited, except perhaps those executed during the active pursuit of prey (presumed to be the batches of acute turns). However, this will be a hugely exciting avenue to explore in the future, not least because it marries sensory and cognitive processes with actions. Such a procedure would help us understand the relative importance of different factors to penguins as they move through their ocean matrix, whether eating, avoiding being eaten, or some other factor.

Dealing with Prey Patchiness

The decision-based step length paradigm for foraging penguins described above has one sole purpose – to get the birds to where they can obtain prey as efficiently as possible. However, prey "accessibility"/"abundance"/"density" is, of course, relative across 3D space and time, and birds have to factor this into their decisions (Chaps. 4 and 5). Here, we use the term "accessibility" because it covers a multitude of sins. An absolute, more quantifiable measure of prey "density" is officially the

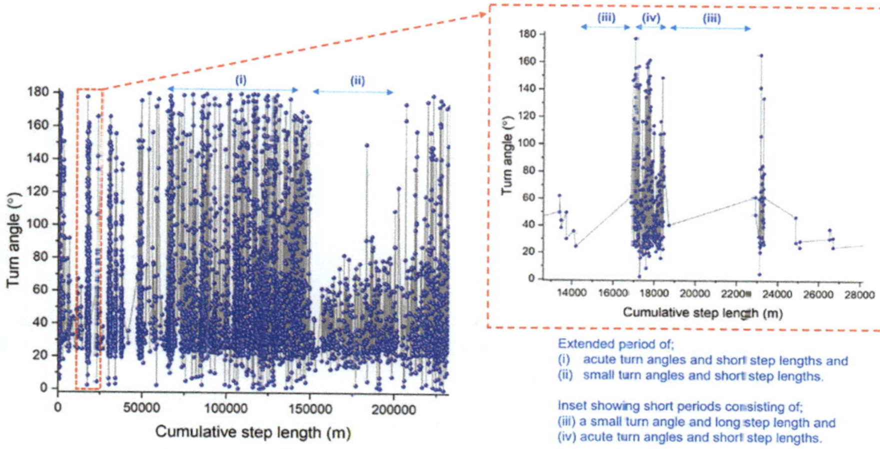

Fig. 10.4 Variation in the angle of identified turns in the horizontal movement of a Magellanic penguin during the course of a single foraging trip in relation to the (cumulative) length of the straight-line sections between them (indicating step lengths) (cf. Fig. 10.3). The inset (dashed red box) highlights the details over a shorter time interval. The variation in the turn angle and step length over the course of the trip over both timescales should be noted. We assume that this is due to the birds reacting to stimuli that occur at corresponding intervals. In this foraging trip, the longest period over which the bird maintained its heading was 5330 m and the shortest was less than 1 m. The paucity of points in the 0–20° turn angle bracket is due to the inability of the mathematical process used to identify turn points to resolve small changes in heading [72]

biomass of the prey within a certain volume of water (perhaps with units of gram per meter cubed), but even this is less helpful than it might seem. Prey can be distributed in a seemingly endless number of ways, even if the density is constant. For a start, the distance between prey items within "patches" and between patches can be highly variable, and anyway, things that we might describe as "patches," because they appear to us as such, might have prey between them, but at such low density that we might ignore them (though penguins might not). In short, prey distribution in all its various manifestations from "isolated individual prey" to varying patch size and within patch density as well as "patch shape" [8] will all have a profound effect on the probability that penguins will find them. Figure 10.5 takes a stab at showing how standardized prey density, but variable "conditions," within a modeled water volume can affect the likelihood of discovery by a penguin diving through the zone (Fig. 10.5).

In fact, many animals are considered to forage in a "patchy" environment [30, 35], where they are said to be moving (foraging) "in patches" or traveling "out of patches," and the literature on this is huge (and mainly does not involve penguins, although the concept is directly relevant, e.g., consider the black-legged kittiwake, a seabird that forages at the same spatial scale as do many penguins [31]). In fact, the original literature on this, which changed much about the way biologists thought about how hungry animals should move through the environment when resource (food) distribution was "patchy" [54], was vague. Because, food resources,

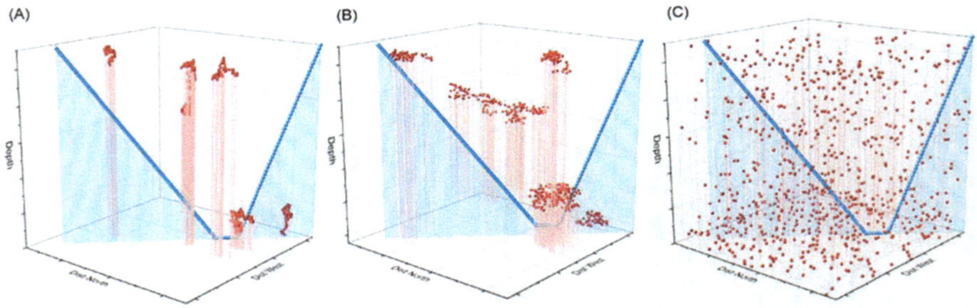

Fig. 10.5 Three identically modeled water volumes (the scales are not shown for convenience, though they are the same in all scenarios) within which the overall prey (each prey item is denoted by a single red sphere) density is the same (the total number of prey within the volume is constant). However, the model incorporates local variation in how prey density is distributed. (**a**) Prey highly concentrated in patches, (**b**) looser concentrations in patches, and (**c**) no patches but prey "everywhere" – and the effect that this has on the search radius of a penguin diving through the volume. This is shown by considering the proximity of the prey to the blue track depicting a U-dive (see Chap. 9 for information on this dive type). In both (**a**) and (**b**), there is a standardized start position for each prey aggregation while the nearest neighbor distances are held constant but are rarefied in (**b**) compared to (**a**) (they have relative distances of 1, 1, and 1 (vertically) in (**a**) and 5, 5, and 1 (vertically) in (**b**)). (**c**) A random distribution of the same abundance of prey within the water column but with a bias toward the bottom (which could, for example, be the seabed, an oxygen minimum zone, or, inverted, the underside of sea ice) (the bias has half the prey within 5 units of the bottom)

including penguin prey, can be distributed in space in a whole suite of ways, it would be nice to believe that our data on 3D movement of many penguin species would provide some information on just how prey are distributed in 3D space (Fig. 10.5). Certainly, meanderings deviating from the straight line are extremely obvious in some species at least (Fig. 10.1), which points to larger prey "patches" [11]. What is not always clear from these though, is how much the meanderings are due to penguins searching by moving within an area where they consider prey accessibility to be, overall, high – the field of mushrooms – and how much is due to penguin tracks that stem from birds actually chasing and consuming prey within an aggregation (see Chap. 11). Then, one has to also consider the "giving-up" time of a penguin or flock as it decimates a prey patch [32], causes it to fragment (see later), or incites it to move down the water column. To differentiate between chasing and consuming requires finely resolved movements and indications of when and where prey are caught, something that we do not have for our older *Pygoscelis* data (Fig. 10.1). Instead, we can examine the scales on which these penguins engage in their "Gordian knots" to get an idea about the prey accessibility associated with different areas (cf. Carroll et al. [11]). A glance at Fig. 10.1 shows that the diameter of the Gordian knots in the *Pygoscelis* data, with the penguins foraging on krill, varies between about 50 and 500 m. A study of Antarctic krill aggregation size noted that swarms could be grouped into four categories, ranging in size between (medians of) 30-m long by 4-m thick, up to 108-m long by 18-m thick (with densities of 11–226 individuals per m^3) [53], although on occasion, krill swarms can be much, much larger – super swarms [94]. However, the larger knots in the penguin data are unlikely to be due to birds actively feeding on a krill swarm, although the smaller knots could easily be so. On the other hand, we note that our tag data fail to reveal penguin flock size, the number of flocks, or whether competitors such as baleen whales are present. As noted, many penguin species do most things in flocks with, for example, all members of the flock diving simultaneously (Chap. 14, Fig. 14.1).

The way penguin prey are distributed in space, and specifically the size of the aggregations, is likely to be critical to penguin foraging success. Consider this: A penguin searching for food can be likened to a person searching for money in a large hall in the middle of the night with the light off – this represents the water turbidity limitations of penguin hunting using their underwater vision. The money, as coins, can either occur in one place, in one gloriously full bucket (a single large aggregation of all prey in the area – cf. Fig. 10.5a, b, which actually depicts five buckets within the defined volume) or it can be scattered, perhaps randomly, all over the hall (individual prey items or perhaps, more realistically, small aggregations spread randomly within the defined volume – cf. Fig. 10.5c). Our searcher crawls about on hands and knees, feeling with fingers if he/she can find the money (well, paying attention to what other flock members are doing, too – the more folk searching, the greater the chances of finding something useful owing to information transfer). The stretch of the arm is analogous to the visual field of the penguin. If the money is all in the one bucket and that person gets extremely lucky, then they might find the bucket promptly and collect all they want before moving back to the door. The chances of that happening are, however, low and get lower the more the money is aggregated. In contrast, if the money is scattered all over the hall, the searcher will find money promptly but will have to go on searching and collecting over an extended time to accumulate enough to satisfy needs (Fig. 10.6).

Box 10.3

A simple model to demonstrate the effect of prey patch size on encounter rates by a penguin. The probability that a single prey item (or more) will be discovered by the penguin depends on the "detection surface area" around the prey item (which becomes a volume as the penguin moves). This equates to the particular distance between the penguin and prey item at which the prey item will be detected and, assuming that penguins are primarily visual hunters, primarily relates to water turbidity. To represent a basic model, we assume that the overall biomass of prey is 1 prey item per 10,000 m^3 of water, that one prey item occupies a volume of 1 m^3 when in an aggregation, and, for simplicity, we will also consider this to represent the volume that gives it a distance at which it will be detected. Thus, assuming that the detection distance is represented by a sphere around the prey item (although data indicate that surface area–volume ratios of Antarctic krill and clupeid fish do not conform to

(continued)

this [8]), each individual prey item has a detection surface area of 4.83597 m^2. If we consider 10,000 prey as singletons swimming in 10^6 m^3 of water, then the probability of detection per unit time (assuming a random distribution) can be expressed by a probability of success, *Ps*. This is equivalent to the sum of the surface areas occupied by all 10,000 prey divided by the volume of water in which they are contained. Therefore, $Ps = 4.836 \times 10^4/10^6 = 0.04836$. If, however, the prey aggregate into a spherical "patch," then the surface area decreases relative to the volume so that the probability of encounter decreases accordingly. Here, the total surface area of all prey becomes 2244.661 m^2 so that the *Ps* is equivalent to $2244.661/10^6 = 0.0022447$. The detection probability of prey in this configuration is thus some 21.5 times lower than in the case above where the prey occur as singletons.

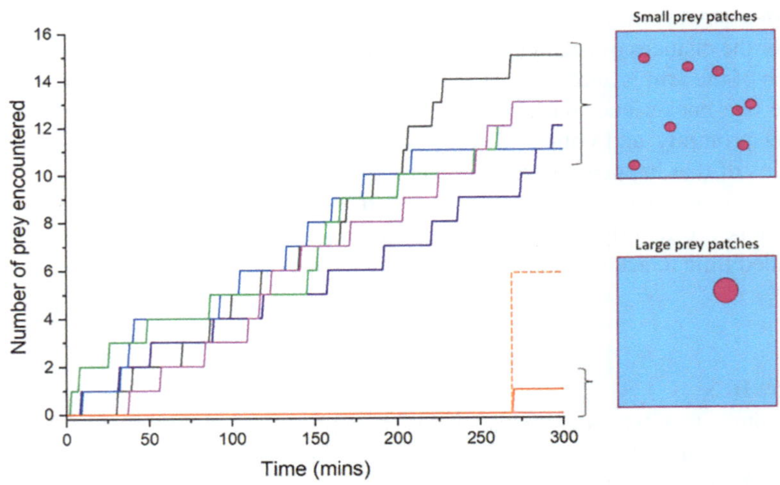

Fig. 10.6 Modeled prey accumulation rates (following [122]) (Box 10.3) of 10 penguins searching for prey of constant density per unit volume of water but according to two different aggregation sizes (details in Box 10.3). Five birds (upper lines of blue, black, green, pink, and purple) search the area where prey are randomly distributed but occur as singletons. These birds accumulate prey fairly constantly over the whole 300-min foraging period. Five other birds (warm colors at the bottom of the graph, although the lines are almost entirely overlaid) forage when the same prey are aggregated into one school or swarm. Although the overall prey density (and aggregation shape) is the same in both scenarios, in this case, only one individual finds the patch within 300 min. Although this bird may be able to capture many prey from that single patch (see dashed line), the conditions do not make for a guaranteed rate of food gain within the time frame considered

So, the specifics of how the biomass of penguin food is distributed in 3D space is hugely important with respect to how effectively penguins can accumulate food over time. Most importantly, once the aggregations become so large that the probability of them being discovered by penguins becomes critically small (the probability of discovery is a function of the surface area of the aggregation and larger aggregations have smaller surface area-to-volume ratios [8]), birds cannot assume that they will acquire the energy they need within a defined time. This is bad news, but it points to a pivotally important factor within the foraging ecology of penguins, and the decisions they make, that they must have a guaranteed rate of prey encounter over a defined period. This is particularly the case during the bottleneck of breeding when the brood has to be fed regularly. We know that penguin prey (perhaps with the exception of the yellow-eyed penguins, which, today, forage under changed circumstances – see Chaps. 5 and 13) form aggregations and that these aggregations vary in space and time (e.g., [40, 56]). Also of vital importance is the disturbance caused by boats [34], or whales, such as those that occurs during fishing pressure. Thus, as an aside, in our consideration of what penguins need for effective conservation at sea (Chap. 13), it is not enough to consider prey density vaguely because the specifics will make the difference! A primary point in those specifics is the scale at which prey patchiness is measured.

Our data on *Pygoscelis* penguins indicate that prey "patches" are (or were) relatively common and large at the time of the study (Fig. 10.1). Indeed, large aggregations are also fine as long as they are common. However, not all penguins show the "Gordian knot" pattern shown by the *Pygoscelis*. Magellanic penguins forging in northern Argentina have completely different foraging track characteristics (Fig. 10.7).

In stark contrast to *Pygoscelis* penguins feeding on krill, Magellanic penguins do not appear to show any "Gordian knots" in their movement patterns and a somewhat similar overall pattern has been noted for king penguins [100]. Both of the latter two species feed on more mobile fish, so why the difference? Fortunately, advances in technology allow us considerably more insight into prey distribution with Magellanic penguins today than we could glean 30 years ago when we collected the *Pygoscelis* data. Then, we had no precise indications of when the birds were feeding, so the dimensions of the aggregations were unclear (see above). Recently, though, highly resolved dead-reckoning technology has been able to show the details of penguin–prey interactions. These show that Magellanic penguins forging at Peninsula Valdes capture their anchovy prey from what are presumed to be small schools scattered widely over large areas rather than large schools (Fig. 10.8).

What is particularly notable is that Magellanic penguins travel in a fairly straight line, even at an individual dive level, catching prey as they move (Fig. 10.7a) – almost as if they were picking up sweets scattered all over the road! A recent work on king penguins has shown an extremely similar pattern, with dead-reckoning data showing how birds maintain a fairly constant heading, even when obviously in the foraging area [106]. This pattern closely accords with the "small aggregation" scenario modeled in Fig. 10.3, notably one that leads to a more "predictable"

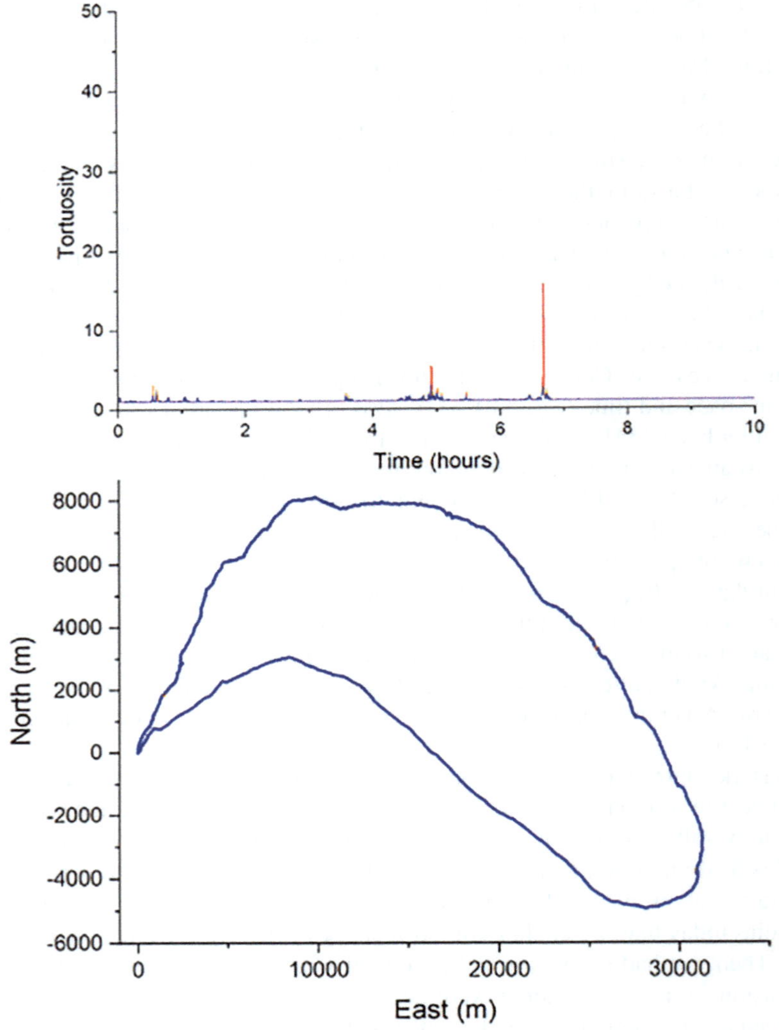

Fig. 10.7 A typical example of the foraging trajectories of a Magellanic penguin feeding on anchovies at Peninsula Valdes, Argentina. The metrics and color scheme are the same as in Fig. 10.1 to aid comparison between species ("*T*" – tortuosity metrics calculated over 1 min with the most convoluted movements shown in red ($T > 5$), the most straight line in blue ($T > 1 < 1.57$), and the intermediate in orange ($T > 1.57 < 5$)). The track (as if viewed from above) shows that, in stark contrast to the *Pygoscelis* pattern (Fig. 10.1), the bird engaged in periods of close to straight-line travel for almost all the trajectories

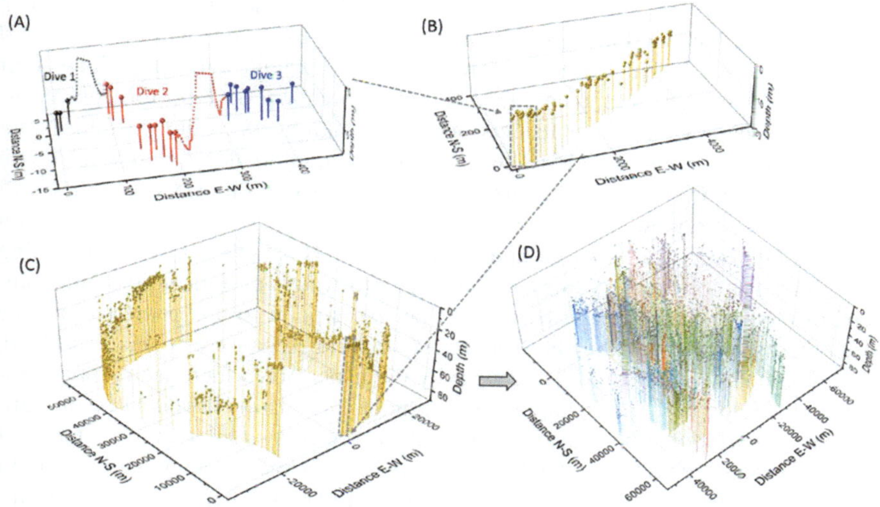

Fig. 10.8 3D tracks of Magellanic penguins foraging for anchovies at Peninsula Valdes, Argentina, resolved with 1/40th second resolution. (**a**) Prey capture (spheres) during three consecutive dives by one individual (the dotted lines show the period from the foraging depth to the surface and back), and prey capture was deduced from heightened VeSBA+ (vectorial sum of the static acceleration; see Chap. 11). (**b**) expands out from (**a**) and (**c**) expands out from (**b**). It should be noted how almost all prey are caught as the bird travels fairly linearly through the area. (**d**) Prey capture by 20 individuals foraging in the same area, showing that this general pattern is maintained across individuals. The fact that prey are caught in this manner makes a strong case for the prey being distributed in small, spaced aggregations rather than a few large shoals (cf. Fig. 10.1)

accumulation of prey, rather than one where all the prey may be contained in massive schools that are difficult to find.

The idea that penguins might benefit from, or even be dependent on, smaller schools of prey because it gives more consistent rates of prey accumulation, was aired over 30 years ago with respect to the African penguins feeding on anchovies [113]. It came about following discussions with local anchovy fishermen in South Africa. These people, who had to understand the anchovies in order to survive, were emphatic about two prime “states” adopted by the fish. They could either be “scattered” – where the fish were to be found in small schools throughout the water column (and such a configuration was useless to the fisherfolk) – or they could be in “big blocks” – huge aggregations of many tons of fish (which was exactly what the industry needed because they could encircle them with their nets). The fishermen pointed out that the fish transitioned seamlessly between these two states and, for instance, could change from “big blocks” to “scattered schools” if they became agitated by poor fishing practice, such as when they had to use multiple evasive maneuvers if the nets were cast badly. This rendered them uncatchable. The scattering into small schools could also be due to predation by creatures other than penguins.

Variable school sizes in pelagic shoaling fish is, in fact, well-documented, and various reasons have been suggested for the variation in state [68]. As pointed out above, large aggregations reduce the surface area-to-volume ratio of individual fish and thereby reduce the likelihood that fish in such aggregations will be discovered by predators [9, 25] (but see [87]). However, being in a huge aggregation is bad for fish wishing to feed. Anchovies, for example, a major food type for *Spheniscus* penguins (Chap. 4), are filter feeders of phytoplankton [48]. Filter feeding is not recommended if, as an individual fish, you are swimming along with your mouth open behind 10,000 other individuals all doing the same thing because they will have filtered everything out of the water before it gets to you [49]. So, one reason why large fish aggregations split up into small schools is believed to be for feeding [33] as well as for encountering more oxygenated water [8]. Indeed, the full stomachs of Cape anchovy caught by African penguins point to this process, with the birds therefore postulated to be feeding on small aggregations of this fish [113]. This was in stark contrast to fish caught by the fisherfolk, which had empty stomachs [113]. However, confusingly, small schools are also believed to be an anti-predator defense too. The value of decreased surface area–volume ratios in terms of encounter rates works best when the predators have to find the aggregation, exploit it, and then leave it (for example, seabirds feeding chicks) before returning to try again. Where it breaks down is when predators do not have to leave the aggregation. This is the case in water-breathing predators, for example, such as predatory fish like sharks or snoek. Once a large aggregation is encountered, the predators can remain with the large aggregation, feeding ad-lib. The upshot of this is that large aggregations of fish therefore tend to accumulate water-breathing predators over time, obviously not great news for the bait fish. The solution to this is for the large aggregation to disband when associated predator numbers get high because the predators cannot follow all of them. The regrouping can occur elsewhere, later.

So, seen from the perspective of the prey, fish can choose whether to adopt small or large schools (or something in between) [34] because each has its advantages with respect to avoiding predators and feeding. However, the precise benefits that come with school size also depend on whether their predators obtain their oxygen by water or air breathing. Maybe foraging in association with fish predators, to confuse the prey, could also, at times, be a penguin strategy? A similar scenario also applies to depth because water breathers do not have to worry about repeated trips to the surface unlike air breathers. Accordingly, prey fish can increase their chances of survival by judiciously exploiting depth, and how penguins, as air breathers, adopt a strategy to deal with that is complex.

In-Depth Considerations

Time-Based Efficiency

Much literature exists on the movements of the "prey species" within the water column, with one phenomenon, diel vertical migration (DVM), receiving particular attention [21, 42]. This is where the prey species spend time at greater depths during the day but migrate to the surface waters at night. The general belief is that these animals seek the protection of greater depths during the day when visual hunters do their foraging [36]. At night, when these visual hunters are compromised because they can see as well – penguins being a good example (see Chap. 8) – the prey species move into the surface photic zone to graze on phytoplankton and associated zooplankton because that is where all their food is. Sunlight penetrating the surface depths is utilized by the phytoplankton there, which, in turn, are food for the zooplankton and so are up the food chain. However, aggregating penguin prey do not all follow this pattern faithfully (e.g., [44]), and there are good reasons for this. For example, if water depth or deoxygenated water precludes the prey from going deep, then it is not necessarily advantageous for all individuals to go to the maximum depth possible for them. This is because this behavior would lead to high concentrations of prey at that depth, which would be reason enough for air-breathing predators to direct their dives there, rather than waste time and energy searching at shallower depths. Instead, there is a sort of cat and mouse behavior between air-breathing predator-like penguins and their prey, where the prey are best served by being where they are least likely to be discovered by the predators trying to acquire them – and that means that the prey do not always strictly adhere to diel vertical migration [123]. One thing that does emerge from studies of aggregating prey is that these marine animals are often distributed in quite precise layers within the water column [53], and this phenomenon is also apparent in data that show the depths at which penguins feed (Fig. 10.9).

Other than penguin gender and species size differences that relate to competition (Chap. 4), in order to forage optimally, how do fish-birds decide on the depth to which they will dive when trying to acquire such fickle prey? We have already noted the variation in the time allocated to depth among penguins executing their various dive types (see Chap. 9), but we have not considered what this means at the level of

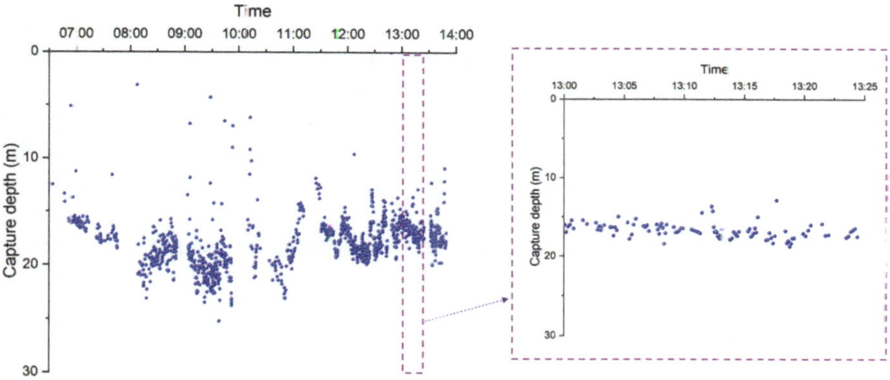

Fig. 10.9 The depths of prey capture (determined using a beak angle sensor) of a Magellanic penguin foraging for sprats at Cabo Virgenes, Argentina. It should be noted how the prey appear to be predominantly distributed in highly specific layers within the water column (the water depth was approximately 40 m for most of the period shown)

the full dive cycle, which must include the surface pauses between dives. Moreover, we have not attempted to consider the allocation of energy to depth for the different foraging dive types, either within individual dives or over complete dive cycles. Overall, the time and energy costs for penguins searching for prey vary with depth (Chap. 6), with the overall costs tending to increase with deeper dives as birds yo-yo between the surface and the foraging depths (cf. [107]). We can illustrate both time and energy costs of diving in penguins by using the Magellanic penguins as an example species.

In terms of time, deeper-diving Magellanic penguins increase the time they spend in transit between the surface and maximum depth and they increase this time linearly with depth during both the descent and the ascent (Fig. 9.21a, c in Chap. 9). This makes sense. The farther they have to go from the surface, the longer it takes to get there. This is true for both V- and U-dives (defined in Chap. 9). However, for U-dives, which constitute most foraging dives, the "bottom phase" of the dive increases with increasing depth (see also [60, 79]). It then either stabilizes or even decreases at the greatest depths (Fig. 9.21b), a pattern that has been noted for a number of penguin species, including all three of the genus *Pygoscelis* [118], northern rockhoppers [18], and emperors [129]. This is presumably because penguins cannot go on increasing their bottom time indefinitely with increasing depth because at some point they will have used all their body oxygen stores.

Any attempt to determine the time-based efficiency of dives also has to incorporate the surface pauses (dive recovery and preparation for the next dive) because a surface pause is an integral part of the dive. The length of a surface pause is somewhat of an anathema. Here, there are reasons to believe that penguins should have disproportionately longer surface intervals with longer dive durations (equivalent to depth) because the rate of oxygen transfer to the tissues is proportional to the partial pressure difference between the inhaled air and the body tissues (Chap. 8). So, penguins coming back to the surface after a dive with their depleted body oxygen reserves take up oxygen quickly. However, as they fill up their body with oxygen to higher levels, their "top-up" rates decrease. Accordingly, the surface pauses get longer with increasing dive depth and duration, as has been documented for many penguin species from the smallest, like the little penguins [6] to the largest, namely, the kings [24] and emperors [129]. Curiously though, in our Magellanic penguin data, although the recovery durations increase with depth (Fig. 10.10, see also Fig. 8.10 in Chap. 8), these seem to be proportionately less for deeper depths (which take longer).

However, it is hard to be definitive about this trend in Magellanic penguins because, within this apparent general pattern (Fig. 10.10), the broad picture is confused by post-dive surface pauses that are excessively short and excessively long. Penguins are at liberty to vary their surface pauses considerably because they do not have to dive with body oxygen stores replete – indeed, to do so would be inefficient because topping up the last little bit takes disproportionately long since the rate of oxygen uptake at such a time is particularly low (cf. [15]). So, for example, where prey are aggregated near the surface, penguins can emerge from a subsurface feeding bout and snatch a single breath (the oxygen within which will be incorporated

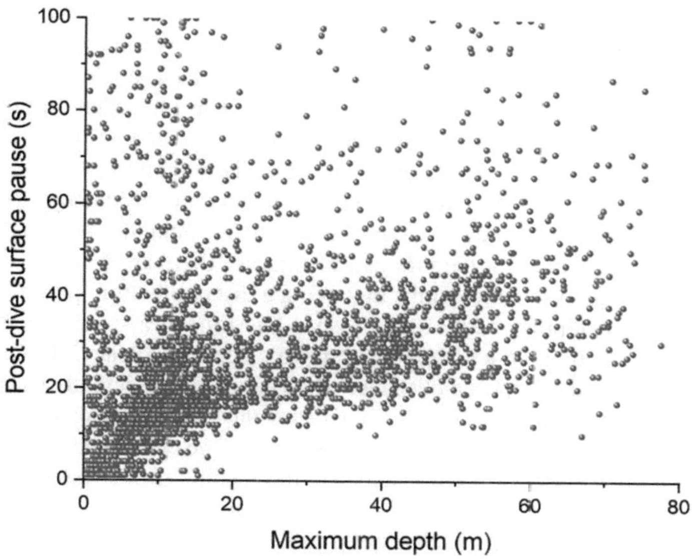

Fig. 10.10 Post-dive surface pauses (for those less than 100 s) in relation to the maximum depth of the preceding dive in Magellanic penguins foraging at San Julian, Argentina. Although there is a general increase in surface pauses with maximum depth, there is much variation because penguins can choose to dive with much reduced oxygen stores if conditions make that appropriate. Alternatively, they can spend excess time on the surface preening, interacting with other birds, etc. before undertaking the next dive

particularly rapidly into the oxygen-starved body) before re-diving into the fray, acquiring the prey before the aggregation moves to deeper waters. In fact, beak angle sensors and video footage show such behaviors clearly. Similarly, both king and macaroni penguins may curtail surface pauses when exploiting ephemeral prey patches in deeper waters [95]. Oxygen deficits can be overcome later. Among Adélie penguins, once they find a high-quality prey patch, they gain more mass by diving more frequently, i.e., shorter surface pauses [57]. In contrast, birds may also spend time between dives on the surface doing things other than preparing for the next dive such as preening or perhaps recovering from oxygen deficits incurred earlier (see above). Importantly, though, penguins have substantial flexibility in how long they choose to spend on the surface in between dives, which is why Fig. 10.10 is such a mess! Perhaps a more judicious way forward is to examine post-dive surface pauses in relation to dive depth or duration only during diving bouts – periods where penguins continually dive with high dive rates – something that all species seem to do [71, 107, 112] (see Chap. 9). This is especially true when they have found a high-quality patch [32, 57]. However, determination of when these bouts officially "start" and "stop" according to some "bout-starting" and "bout-ending" criteria is not a trivial matter [84].

To examine the time-based costs of dives to various depths in a quantitative manner, we can calculate time-based dive efficiency based on two possibilities. One is the efficiency of the total time spent underwater during a dive, which considers that prey may be encountered at any time during the dive. We know this to be true because various methods, such as video cameras and beak angle sensors show it clearly. If, therefore, the total time underwater is taken as important for prey exploitation, then the efficiency can be calculated by dividing the total time underwater during any dive by the dive cycle duration – the dive duration plus the following surface pause (Box 10.4). We have termed this "underwater efficiency." This definition of efficiency works for any dive type, be it U-dives (including W-dives), V-dives, or parabolic (see Chap. 9 for definitions). In contrast, there are dives in which penguins allocate time to specific depths by having a discrete bottom phase during the dive (e.g., U-dives), presumably because spending time at a specific depth is particularly profitable (e.g., [64, 101]) – the "specific depth efficiency." Extended time at specific depth is also important because when penguins are actually catching prey layered in the water column (e.g., Fig. 10.6), they must obviously extend the time at this depth. So, in these cases, dive efficiency can also be calculated by dividing the bottom duration by the dive duration added to the ensuing surface pause (Box 10.4). The latter assumes that the depth of the bottom duration is the only place where the penguin is actively foraging, which is not, strictly speaking, true since penguins will catch prey in transit and presumably curtail deeper dives if they encounter a prey aggregation while descending the water column [108].

Box 10.4

We consider two primary "time-based definitions of dive efficiency" (Et_{dive}) in the context of foraging. For this, we incorporate the descent duration (D_d), the bottom duration (B_d), the ascent duration (A_d), and the inter-dive surface pause duration (S_d).

1. The "underwater efficiency" is given by the duration of the total dive divided by the dive cycle duration:

$$Et_{\text{dive}} = (D_d + B_d + A_d)/(D_d + B_d + A_d + S_d) \tag{10.1}$$

This metric considers that prey may be found at any stage of the dive.

2. The "specific depth efficiency" is given by the duration of the bottom phase of the dive divided by the dive cycle duration [128].

$$ET_{\text{dive}} = (B_d)/(D_d + B_d + A_d + S_d) \tag{10.2}$$

This assumes that all, or most, of the prey is caught, or searched for, during the bottom phase of the dive.

Overall, though, using the definition of time-based underwater efficiency (Eq. 10.1 in Box 10.4), Magellanic penguins that dive within their normal depth ranges seem to maintain high levels of efficiency irrespective of depth (Fig. 10.11), although there is much variation in efficiency, particularly at the shallower depths.

Using relationships between the durations of various dive phases (e.g., descent, bottom, and ascent phases) and the maximum dive depth, we can construct best-fit lines of time-based underwater efficiency as a function of maximum depth for different dive types. This helps guide the eye through the variation (Fig. 10.11). The general pattern of having fairly constant time-based underwater efficiency that is apparent in Magellanic penguins is also mirrored in a number of other species and is likely to be a general penguin phenomenon within normal foraging depths (Fig. 10.12).

The notable consistency in efficiency irrespective of depth is relevant for authors who sometimes indicate "how hard" penguins are working, using "dive frequency" as an index, assuming that more dives per hour means that they are working harder. It is obviously impossible to dive deeply so quickly, with equivalent underwater efficiency; birds may be diving deep but working just as "hard" as shallow-diving individuals. A more realistic approach would be to consider that, only if, during specific bouts, birds actually increase their percentage underwater time and therefore might be considered to be working harder.

In a general sense though, that foraging penguins can spend between 65% and 80% of their time being underwater, irrespective of their dive depth (within reason), is testament to them being supremely adapted for high-speed movement underwater

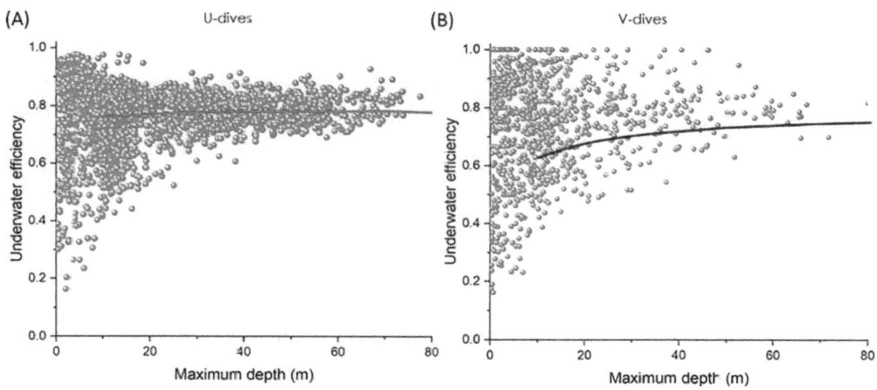

Fig. 10.11 The efficiency of the total time spent underwater (for definition, see Box 10.4) by Magellanic penguins (foraging at Peninsula Valdes, Argentina) showing (**a**) U-dives (extended bottom phase at depth) and (**b**) V-dives (no bottom phase) – the colored lines show calculated efficiencies based on best-fit lines of duration vs. depth for various dive phases. It should be noted that although U-dives appear to be slightly more efficient than V-dives in terms of time underwater, much of the time within the dive profile of U-dives is allocated to a specific depth range. As such, during these dives, penguins scan the complete water column down to the maximum depth less efficiently than V-dives. This is because they have decided to target a specific depth based on their previous searching (V-dives)

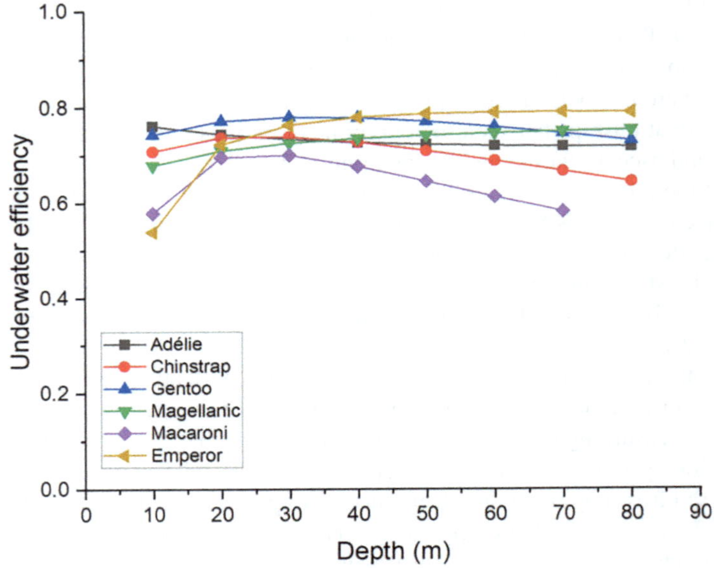

Fig. 10.12 Examples of the time-based "underwater efficiency" (for definition, see Box 10.4) of various foraging penguin species. It should be noted how all species generally spend more than 60% of their time underwater and that there are no particularly obvious overall trends with depth (although not shown, the emperor penguin efficiency remains above 0.7 up to depths exceeding 300 m). The *Pygoscelis* data are taken from [118], the emperor penguin data from [129], the Magellanic penguin data represent means of best fits for U- and V-dives in Fig. 10.10, and the macaroni penguin data are courtesy of Onno Huyser

and for dealing with depth. Indeed, this is part of the reason that they are such formidable predators in the top 100 m of the water column (and deeper for king and emperor penguins). This is what makes them penguins – glorious fish-birds (Chap. 15).

A comparable metric to underwater efficiency was extensively used a few decades ago to examine foraging in diving seabirds in general, with that being the "dive–pause ratio," first used in a charming account of bird diving by John Dewar in 1924 [26]. This began because John had no access to depth gauges but could determine how long birds spent underwater in relation to the surface pauses between dives by simply timing them. People also capitalized on the fact that radio signals do not pass through seawater, so birds equipped with radio transmitters (e.g., [105]) belayed whether they were underwater or on the surface by having their signals transmitted to the receiver. These sorts of documented dive–pause ratios result in underwater efficiencies in non-penguin diving birds of between about 40% and 60% [45, 105], which are generally notably less than those of penguins, again testimony to a high degree of specialization in fish-birds.

Considering that the dive–pause ratio indicates that for penguins, though, a difference does appear to exist in time-based "underwater efficiency" between U- and

V-dives, with U-dives being apparently slightly more efficient (Fig. 10.11), and we will look at the energetics of why later. However, it is important to remember that U-dives are distinguished by birds spending extended time at a specific depth (the bottom phase of the dive) and so, in the process, effectively scanning the complete water column down to the maximum depth less efficiently than V-dives. A penguin's choice of compromising its allocation of time equally down the water column by extending its bottom duration or giving all depths equal weighting depends presumably on its perception of prey densities with depth. Its sensing of the physical attributes of the water column, if, for instance, prey are associated with a strong thermocline (see Chap. 3), might also be involved. Consideration of the incidence of V-dives and U-dives in a general sense and specifically how they relate to each other in time as penguins begin searching will tell us much about what penguins "think" about depth-related prey densities. It does appear that penguins may well sense water column structure relative to prey, at least if king penguin behavior can be generalized to other species [16, 88]. Adélie penguins, and possibly gentoo penguins, also appear to make use of water column changes to decide where it would be most efficient to search for prey [5, 20]. A surface layer of dense phytoplankton may also be a cue [80]. To our knowledge, the distribution of the dive types has yet to be allocated to 3D space in any penguin species!

That penguins extend their time at specific depths via U-dives, though, is a reasonable strategy because, as we have seen, penguin prey aggregations are often located within defined water strata, e.g., related to the thermocline or phytoplankton concentration (see above). In addition, once penguins encounter aggregated prey within a specific water stratum, they tend to target and remain at this depth, which presumably also helps account for the preponderance of U-dives compared to V-dives. Both for reasons of prey location and particularly for prey exploitation, underwater time selectively allocated to specific strata can be examined using just the bottom phase of the dive to calculate the time-based efficiency of U-dives as a function of depth – the "specific depth efficiency" – (Eq. 10.2 in Box 10.4). This efficiency of exploiting specific depths shows a markedly different picture to the general "underwater efficiency" because it decreases with increasing depth (Fig. 10.13).

This pattern has been noted for many penguin species, including all three *Pygoscelis* penguins [14, 118], yellow-eyed penguins [29], and king [17, 51] and emperor penguins [129]. This occurs because the bottom duration is a tiny fraction of the total dive cycle, and as the maximum dive depth increases, there is an increasing amount of time dedicated to transit. Finally, as the depth increases, with increasing oxygen being used for surface-to-depth transit, penguins come to the end of their oxygen storage capacities to power their dives. As a result, the bottom phase duration levels off and eventually has to decrease (see Chap. 9, Fig. 9.22).

The take-home message from both V- and U-dives (as well as from parabolic dives, which are something in between; Chap. 9) is that, in terms of time, penguins are remarkably efficient at being underwater at most of the depths they habitually visit. However, despite their extraordinary diving abilities, it is still advantageous for penguins to acquire prey encountered near the surface (Fig. 10.13), and this is likely to be the reason why many species show a preference for searching for prey

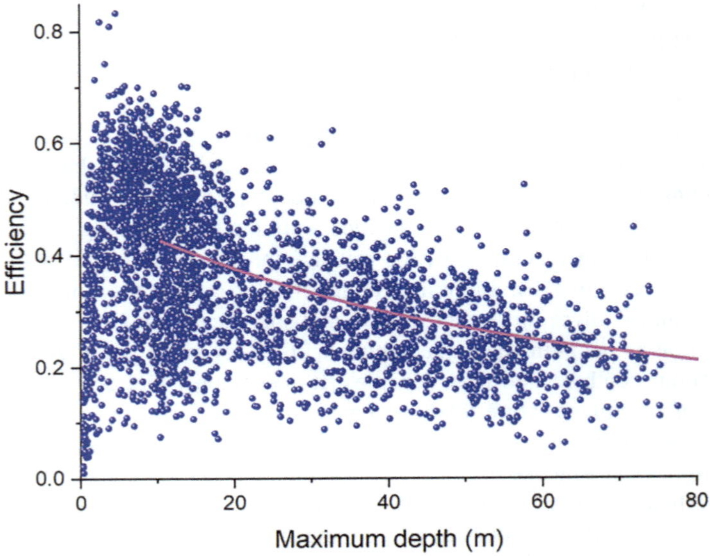

Fig. 10.13 The time-based efficiency of penguins foraging at specific depths – the "specific depth efficiency" – (during U- or W-dives) decreases with increasing depth for U-dives. Each point shows a single dive, whereas the pink line shows the overall trend based on the best-fit lines of durations of the various dive phases against depth. The example shows the efficiency of Magellanic penguins (foraging at Peninsula Valdes, Argentina). At shallow depths, birds may be able to allocate more than 50% of their dive duration to exploiting their preferred (bottom phase) depth, but this decreases to less than 30% for dives that exceed 60 m

at shallow depths, at least initially (Fig. 10.14). Owing to the competitive pressures of central place foraging, it is the shallow prey that are initially depleted, forcing Penguins to then forage deeper and farther, at least with regard to large colonies [2]. In the latter case, involving Adélie penguins at Cape Crozier, Ross Island, the shallower prey were less energy-dense (krill), yet the penguins chose the easier-to-obtain shallow ones, before diving to the more energy-dense prey (fish). Where there are not enough penguins to deplete the prey, they continue at shallow depths [80].

A preference for concentrating search time near the surface for a variety of penguins is illustrated well by the older literature when researchers were using penguin-mounted electronic depth gauges that recorded only the maximum depth per dive. This work clearly showed that penguins tend to dive more to shallower depths, for example, as seen in chinstrap [58] and king penguins [52], although the focus of attention then was on the maxima. However, even then, researchers could have deduced that lesser depths were important to penguins. Knowing that, for any one dive, the descent rate is constant down the water column (see Chap. 6) and that this rate is less for shallower dives (e.g., [41]), effectively produces an overall time-at-depth profile for penguins with more of a bias toward the shallower waters than consideration of depth maxima would imply (Fig. 10.15).

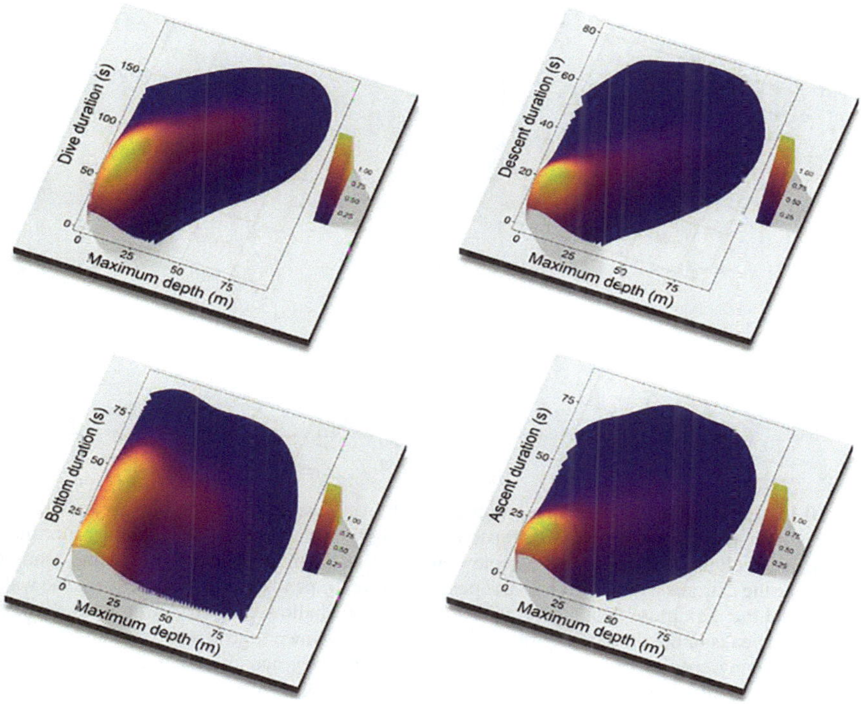

Fig. 10.14 Time versus depth allocation for Magellanic penguins (foraging at Peninsula Valdes, Argentina) – warmer colors indicate a greater proportion of dives. The decreasing time-based "specific depth efficiency" with increasing foraging depth (see Fig. 10.13) makes a compelling case for birds to forage shallowly, if they can. This, however, depends on the vertical distribution of the prey, which explains why penguins dive deep at all

This principle also works across individuals because, beyond individual penguin depth differences, compilation of depth use across multiple individuals very often results in a species-specific pattern of time against depth that illustrates how time at depth tends to decrease with increasing depth (Fig. 10.16), again emphasizing how much time penguins spend at shallower depths.

Penguin dive efficiency thus contrasts that of cormorants, for example, most of which forage primarily on the seabed and have descent angles of around 70–90° [76, 126]. These days, the latter may also apply to yellow-eyed penguins, which, thanks to humans, are mostly confined to searching for bottom prey (Chaps. 4, 5, and 13). From the prey perspective, the enhanced attention that penguins give to depths closer to the surface in their descent (and ascent) angles makes a generally compelling case for penguin food to make the decision to remain deep – if the bottom topography and water oxygen levels allow it [123] – when penguins are in the area (see also discussion of diel vertical migration earlier)!

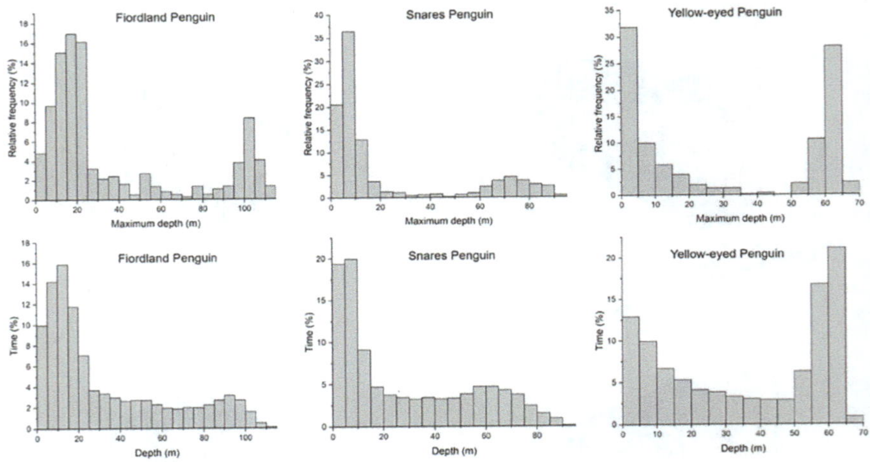

Fig. 10.15 Examples of how the maximum depth of dives (upper graphs) affects the overall time accumulation at depth (lower graphs) illustrated for single foraging trips, for example, Fiordland, Snares, and yellow-eyed penguins. Simple consideration of the maximum dive depth tends to emphasize the extremes. In reality, because penguins have to spend time moving between the water surface and the maximum depth, they accumulate time at shallower depths, "ironing out" maximum depth peaks in the time-based frequency distributions. Since time at any depth allows water-column-foraging species to perceive prey in those strata, they actually have a search bias that maximum depth distributions tend to underrepresent. We note that even species such as the yellow-eyed penguin, which forages predominantly on the seabed [64], spend appreciable time in the shallower depths, much of which is due to horizontal commuting and some of which is vertical commuting [29] (see Chap. 9). (Data courtesy of Thomas Mattern and Ursula Ellenberg)

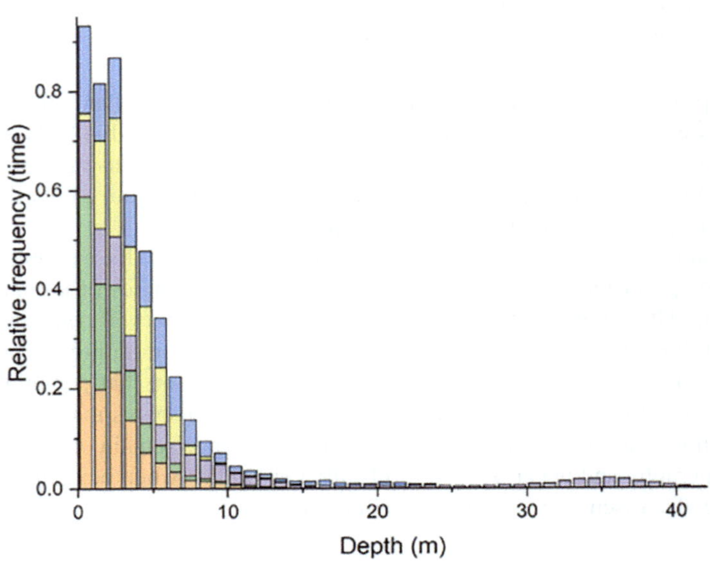

Fig. 10.16 Allocation of time to depth in five individual Galápagos penguins (indicated by the differently colored bars) each over a foraging day. It should be noted how clear depth preferences by individuals are largely "ironed out" when data are combined to produce a Galápagos penguin-specific picture. (Data courtesy of Antje Steinfurth)

Energy-Based Efficiency

The above considerations only apply to time-based efficiency, and although the passage of time and the energy expended are related – because animals, including penguins, cannot stop expending energy (Chap. 7) – the details of how much energy is expended according to activity are important. Options for calculation of energy-based efficiency basically mirror those of time-based efficiency, except that the efficiency of birds targeting particular depths (with U-dives) must be based on the duration of the bottom phase (Box 10.5) because increasing the duration of the bottom phase increases the probability of one or more prey encounters/captures (cf. Fig. 10.2).

To approximate energy expenditures, we can use the derivations for energy expended according to dive angle and instantaneous depth (see Chap. 6, Box 6.7) to graph trends in how penguin energy expenditure is expected to change during the course of a dive according to dive type (see Chap. 9) and dive depth, again using Magellanic penguins as an example species (Box 10.6).

This approach indicates how the details of the dive profile affect both the mean power usage over time, showing the huge variation depending on the phase of the dive, especially with respect to the descent and ascent phases (Fig. 10.18). It also points to the fundamental energetic differences between the different dive types. This is presumably relevant for penguins if they are attempting to optimize the net rate of energy gain, rather than basing foraging decisions on time alone. This is a

Box 10.5

We consider two primary "energy-based definitions of dive efficiency" (Ee_{dive}) in the context of foraging. For this, we incorporate the energy used in the descent phase of the dive (D_e), in the bottom phase (B_e), in the ascent phase (A_e), and in the inter-dive surface pause duration (S_e), as well as the duration of the bottom phase of the dive (B_d).

1. The "underwater efficiency" is given by the energy used during the dive divided by the energy used in the dive cycle duration:

$$Ee_{\text{dive}} = (D_e + B_e + A_e) / (D_e + B_e + A_e + S_e)$$

   This metric considers that prey may be found at any stage of the dive.

2. In an attempt to make the energy-based "specific depth efficiency" most similar to the definition of time-based "specific depth efficiency" (Box 10.4, Eq. 10.2), we divided the duration of the bottom phase of the dive by the energy used for the whole of the dive cycle (units s/J/kg).

$$Et_{\text{dive}} = (B_d) / (D_e + B_e + A_e + S_e)$$

This assumes that all, or most, of the prey is caught, or searched for, during the bottom phase of the dive.

long call from many energy-based optimal foraging models that effectively assume that energy expenditure is roughly constant for all aspects of foraging [75].

Defining the energetics of the different dive types allows us to derive the mean power (energy expended per unit time) for dive types as a function of dive depth (e.g., Fig. 10.19a). Indeed, these can be used to highlight "underwater efficiencies"

Box 10.6

To approximate the energetic efficiency of dives to show expected trends with depth, we present an example penguin, the Magellanic, and use the definitions of energetic efficiency provided in Box 10.5. To present a simple model that exemplifies trends, we used a bottom phase duration taken from Fig. 9.22a, given by:

$$\text{Duration} = 6.37\left(\ln\left(\text{max depth}\right)\right) + 17.8\left(r^2 = 0.20\right),$$

and ascent vertical velocity derived from the line of best fit for the plot of ascent duration against maximum depth (see Chap. 9, Fig. 9.21c) given by:

$$\text{Duration} = 0.68^* \,\text{max depth} + 5.9\left(r^2 = 0.76\right),$$

and used constant swim speeds of 1.6 m/s for the descent and bottom phases (see Chap. 6). The power used for the descent phase as a function of depth and dive angle is provided in Chap. 6 (Box 6.7), scaled up to be chemical power rather than biomechanical power (Box 6.8), whereas the chemical power to swim horizontally is taken to be 11.4 W/kg (Chap. 6). The power to ascend the water column depends on the depth at which penguins stop flipper beating (which is shown in Fig. 6.5) is best fitted by:

$$\text{Glide depth} = 0.58^* \,\text{max depth} + 0.35\left(r^2 = 0.84\right).$$

To provide a depth-dependent descent angle to inform the descent power, we use the equation based on the graph below derived from an example Magellanic penguin (Fig. 10.17) and assume that birds allocate decreasing power from swimming horizontally linearly with time between the bottom phase and the gliding depth.

For the gliding phase, we assume that the penguin's metabolic rates are equivalent to those of birds resting in water (of 6 W/kg – see Chap. 7, Table 7.1 using the value provided for Humboldt penguins). The same value is used for the surface dive recovery period, which was allocated a duration based on Fig. 10.7 after removal of all points > 50 s, with a line of best fit of:

$$\text{Pause duration} = 0.38^* \,\text{max Depth} + 14\left(r^2 = 0.33\right).$$

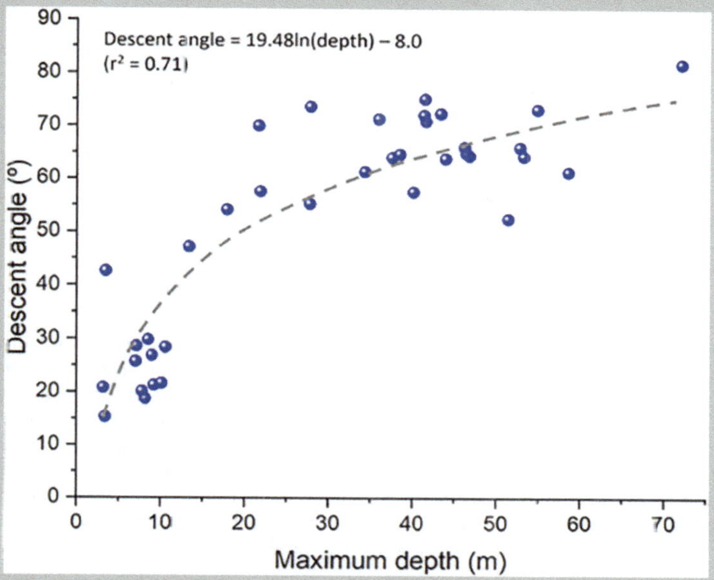

Fig. 10.17 The descent angle against depth for a Magellanic penguin foraging at Peninsula Valdes, Argentina, for U-dives before which no prey were captured and where the descent was entirely regular. This is a defined simplification since the descent angle varies according to the number of prey captured in dives (see later), but it serves to illustrate trends

(Fig. 10.19b) and specific "depth-based efficiencies" (in the bottom phase of their U-dives) (Fig. 10.19c).

The mean power calculations indicate that U-dives seem "cheaper" than V-dives (Fig. 10.19a), and this is primarily because the bottom phase has lower power requirements than any point on the descent. Otherwise, the descent and ascent phases of both U- and V-dives are roughly similar. Thus, extending the bottom phase, as happens with increasing depth, tends to result in decreasing the mean power use. So, penguins diving deeper use less oxygen per unit time. This would seem to make a case, all other things being equal, for penguins generally benefitting by diving deeper but we have glossed over an important element – that the bottom phase of U-dives is both used in looking for and in acquiring prey (Chap. 9). Incorporation of prey exploitation would change the energetics substantially. Lower mean power costs with increasing dive depth would generally hold for penguins that either maintain a roughly constant speed during their bottom phase or slow down to catch prey, such as the *Pygoscelis* species when foraging on krill [124] – remembering that lower speeds require less power. In contrast, any penguins that have to speed up to catch prey, such as the *Spheniscus* penguins foraging on fish [124], use

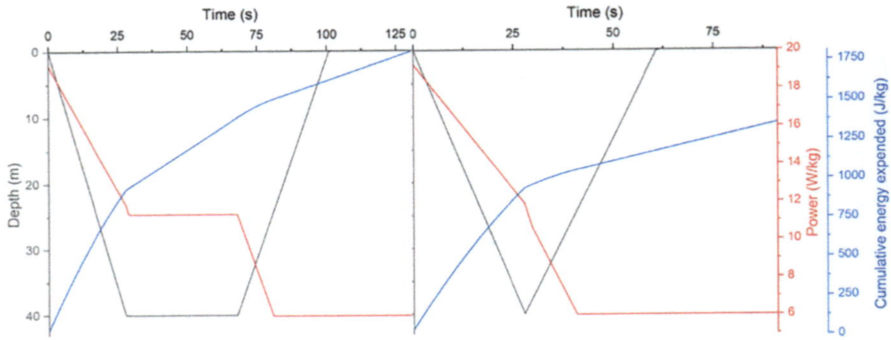

Fig. 10.18 Two representative dives made by a Magellanic penguin to 40 m over time to show how different dive types affect the energetics of the dives. These are constructed using standard (species-specific) relationships between durations and maximum depth and the approach provided in Box 10.6. The black line shows the depth over time of a U-dive followed by a V-dive, each with its respective post-dive rest periods on the surface (though the slightly changing *x*-axis should be noted to provide the best definition of events). The red line shows the instantaneous power, whereas the blue line shows the cumulative amount of energy used. The influence of the power use (red line) of the bottom phase in the U-dive is particularly obvious because it is missing from the V-dive and affects the overall energy expenditure (blue line) profoundly

much higher power during their bottom phase. This would tend to push the mean power costs up correspondingly, something our modeled energetic costs have (conveniently) ignored! It should be noted, however, that the fish usually being pursued by *Spheniscus* penguins, e.g., anchovies, are among the most energy-dense of any prey caught by penguins (Chap. 4, Appendix Table A4). Their increased effort, thus, is worth it.

As with time-based efficiency, overall energy-based "underwater efficiency" is fairly constant with maximum depth (cf. Figs. 10.11 and 10.19b), irrespective of whether dives are U- or V-shaped. Energy-based "underwater efficiency" is, however, calculated to be some 10% higher than time-based "underwater efficiency" (cf. Figs. 10.11 and 10.19b). As with time-based efficiencies underwater, the energetic efficiencies of V-dives are calculated to be less efficient than those of U-dives (Fig. 10.19b). This may indeed be the case when birds are searching for prey in V-dives but, as with time-based efficiency, the whole situation changes once prey are found at a certain depth or the penguin decides to concentrate time (its search) to specific depths. This is because, as noted, the energetic efficiency of dives decreases rapidly with increasing depth (Fig. 10.19c, cf. Fig. 10.13) owing to the commute between the sea surface and the foraging depth.

So, ignoring for now the differences in quality of the prey species (Chap. 4), what is the best dive strategy for penguins? It is clear that they generally favor shallow dives (Fig. 10.14) because that leads to the most efficient exploitation once a prey aggregation is found (Figs. 10.13 and 10.19c). However, they also have to deal with prey that do not "want" to be eaten and hence adopts vertical migration within the water column to avoid detection. From the prey perspective, if penguins in the area are diving deep, then the prey should "hide" at shallow depths, and vice versa

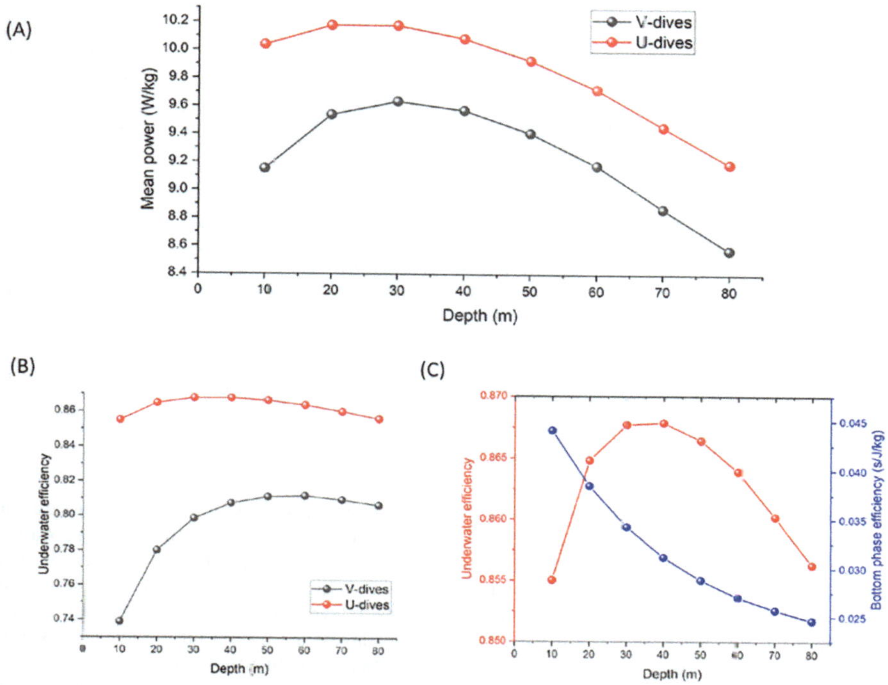

Fig. 10.19 (**a**) The calculated mean power per dive as a function of the maximum dive depth (for conditions, see Box 10.6) used by Magellanic penguins (foraging at Peninsula Valdes, Argentina) for two major dive types – U-dives, which have an extended bottom phase at depth (red line), and V-dives, which have no bottom phase (black line). (**b**) The "underwater energetic efficiency" (energy expended underwater/energy expended for the full dive cycle) as a function of the maximum dive depth for U- and V-dives. It should be noted that underwater efficiency is little affected by the maximum dive depth (as manifested by the reduced vertical scale). (**c**) The "energetic underwater efficiency" of U-dives (also shown in panel (**b**) but included for comparison) as well as the energy-based "specific depth efficiency" – the preferred foraging depth of U-dives – as a function of the maximum depth (blue line). The latter effectively shows how many seconds (including fractions of a second) during the bottom phase costs 1 J/kg. The descending value with depth shows that, in terms of energetics, Magellanic penguins are slightly less than twice as efficient during their bottom phases at 10 m as they are at 80 m

(see [20]). Penguins therefore have to decide how to allocate their time underwater to depth with the changing backdrop of prey depth selection [123]. Deeper-diving penguins, especially those adopting U-dives, travel more quickly through the upper water layers, and the reduced time that they spend at these depths decreases the overall probability that prey will be found at that depth (Fig. 10.20). In contrast, if the prey are deep and penguins only dive shallowly, they will never encounter any prey.

Presumably, penguins will also balance the inefficiencies in the bottom phase of deeper dives with prey density (and quality) in accord with the general optimal foraging principles [15]. Therefore, we expect the rate of prey ingestion (or "energy

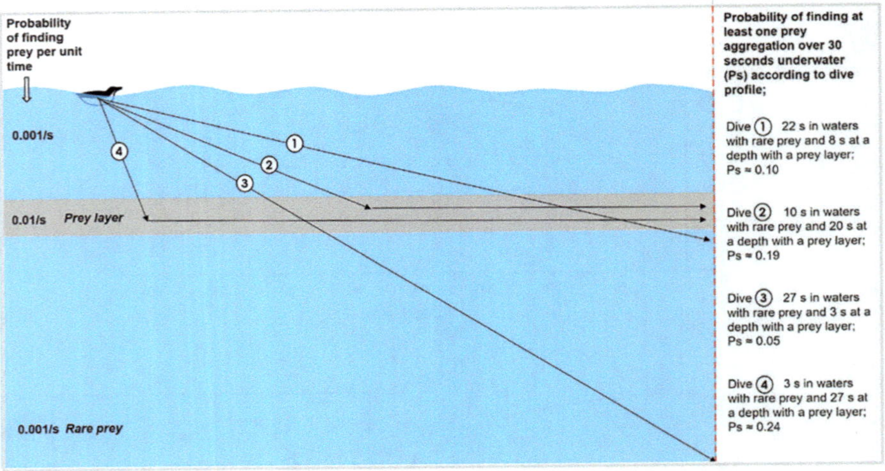

Fig. 10.20 A schematic showing various profiles of a penguin's V- and U-dives (only the descent and bottom phases are considered) and how adoption of any of these relates to the overall probability (based on simple probability theory) that at least one prey (or prey patch) will be discovered at any depth. In dive 1, which is a V-dive with no bottom phase, the dive angle is shallow, so the penguin takes a long time to traverse the high prey-density layer (shown in gray – probabilities of encountering prey per unit time for the different water layers are shown on the left-hand side). Thus, the overall probability of finding prey is higher than the V-dive shown in dive 3, where the bird descends the water column rapidly and thereby spends little time in the high prey-density layer. Dive 2 is a U-dive, where the penguin descends at a steeper angle than that in dive 1 but extends its bottom duration within the high prey-density layer that it has located and thereby has a higher probability of finding prey. Dive 4 is also a U-dive, but, in this case, the penguin has decided to dive steeply to the high prey-density layer (perhaps having caught prey there previously (see Sato et al. [83])). It thus spends longer in the profitable layer, with correspondingly higher chances of finding prey. This illustrates how penguin decisions regarding how to dive affect the chances of prey encounter and explains both variability in dive angle (descent rate) and the adoption of U- or V-dives

ingestion" if this relates to a change in prey type) to increase in deeper dives where penguins acquire their prey. There is certainly good evidence that this is the case for emperor penguins [129], but confirmation of such expected trends across species will have to depend on researchers determining prey capture events right down the water column. What already seems certain is that penguins really do direct their dives to depths where the likelihood of prey detection is high (perhaps because they already know that) and increase descent rate accordingly. Sato et al. indicated that this is the case in macaroni penguins, which descend at steeper angles to specific depths where they seem to have just acquired prey than they would otherwise [83]. A similar pattern has been noted for king penguins [73]. Although steep dive angles require higher power (see Chap. 6), this is more than compensated by the increased chances of eating – if you know the depth of the supermarket, it is silly to dally on the way down to it, especially if that supermarket is hell-bent on moving to another depth to avoid you finding it! Similarly, Adélie penguins will effectively reduce their depth-based underwater efficiency by reducing their surface interval once they have found and assessed a prey patch because the apparent loss in efficiency is more than compensated for by the higher rates of prey ingestion from a prey patch whose location in known [57].

Overall then, penguin decisions about the maximum depth of a dive together with variable descent rates and more or less extended bottom phases (specifically, the form of the dive profile), and maybe reduced surface times, seem to be responses to changes in the vertical distribution of prey. Furthermore, these behaviors indicate that penguins play a game of "cat and mouse" with their prey. Indeed, examination of maximum dive depth sequences coupled with time-at-depth trends over time will be a fascinating topic for future research [61, 65, 66], provided that we can quantify the preyscape and assess prey ingestion in all dives.

"Superficial" Considerations: Surface Pauses and Inspired Tactics

The general pattern of the dive trajectory – manifested by the descent angle applied at the moment penguins leave the sea surface (Chap. 9, Fig. 9.5) – makes it clear that the maximum depth of the dive is "decided" from the outset. However, it seems that the decision about depths to visit and time allocation to those depths is made even before that. For a start, penguins inhale more air if they are going to dive deeper, which allows them to modulate the upthrust during the bottom phase of dives [85, 127], saving energy and oxygen. However, research on the time penguins spend on the surface between dives indicates that the duration of this surface phase of the dive cycle is not only allocated to reoxygenation following depletion of oxygen stores due to the previous dive but also to loading of oxygen for the proximate dive [117]. Such "predictive oxygen loading" must take into account the time – or perhaps more properly the energy (and therefore oxygen) – used in the upcoming dive, and this varies with dive depth (Fig. 10.19a). In a nutshell, the length of the surface pause in a penguin actively foraging is related, in part, to the maximum depth (and duration) of the upcoming dive. However, penguins are seemingly even smarter than that. When attempting to predict the total energy (and oxygen) use of the upcoming dive, it would help if penguins could somehow incorporate the likelihood of finding prey, if such an extraordinary thing were possible. This could, theoretically, be used by them to guesstimate the oxygen needed to catch their prey as well as that used in the descent and ascent phases of the dive. Knowing this, birds could load their bodies with oxygen accordingly. In fact, this is exactly what they do. Research using beak angle sensors, which record when penguins catch each individual prey item [93] and when they take each breath on the surface [125], shows that penguins, as demonstrated by Magellanic penguins, have a well-defined pattern of inspiration of air that relates to both the maximum depth of the dive to come and the expectations with respect to prey capture (Fig. 10.21). During these surface periods, the maximum beak angle of each breath relates to the tidal volume (how much air is inhaled in that breath) (Chap. 8, Fig. 8.4), and examination of the details shows that when the penguins first break the surface, they have an initial huge inspiration (Fig. 10.21). Anybody who has snorkeled will appreciate that!

Following this, the tidal volume of each breath decreases to a minimum, at which point it is considered that penguins have repaid their oxygen debt from the preceding dive [125]. However, the trend then reverses, with the tidal volumes of serial breaths increasing until the last big breath before birds initiate their next dive (Fig. 10.21) [125]. The number of breaths in this second phase relates to the maximum depth of the following dive (Fig. 10.22a) [117]. This is obviously useful when planning for efficient diving! However, beyond this, for any given maximum depth with its allocated breath number, penguins put in extra breaths, with the number depending on the number of prey that they have caught in their previous dive (Fig. 10.22b) [117]. The critical point is that it is not efficient to spend long periods on the surface because the rate at which oxygen is accumulated decreases with increasing surface interval so penguins should only be diving with the

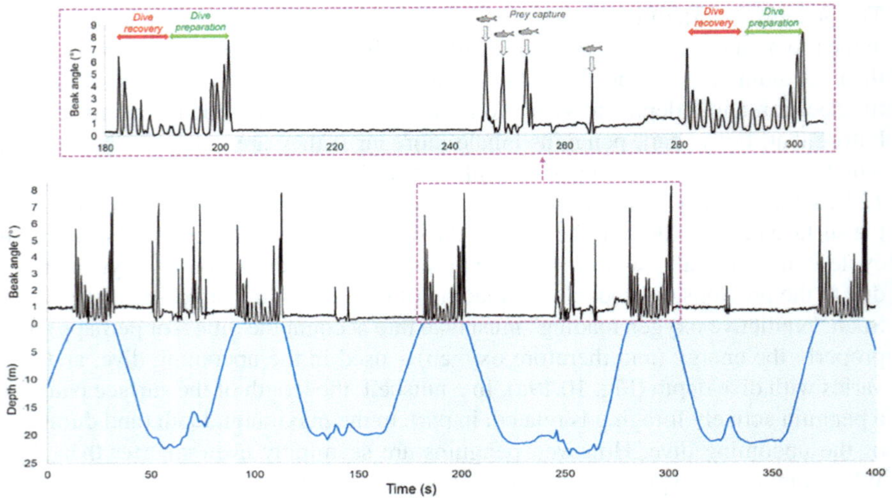

Fig. 10.21 The pattern of breathing – with each breath being manifested by a single waveform in beak angle – of a Magellanic penguin on the sea surface between dives (blue background) during which variable numbers of prey were caught (see inset for greater detail). It should be noted that the initially high beak angles (≡ tidal volume) decrease to a minimum – as the penguin recovers from the previous dive – before the trend is reversed, with the bird taking its largest breath just before diving again. The second phase of the surface pause, in which the tidal volume per breath increases, is due to the penguin preparing for the upcoming dive (cf. Fig. 8.4). Astonishingly, the number of breaths in this second phase is not only related to the maximum depth of that dive but also to the number of prey that the bird has caught in its previous dive (see Fig. 10.20) – as if penguins actually count! This indicates that penguins have highly sophisticated behavioral responses to prey distributions and densities as they vary in time and space up and down the water column – true fish-birds!

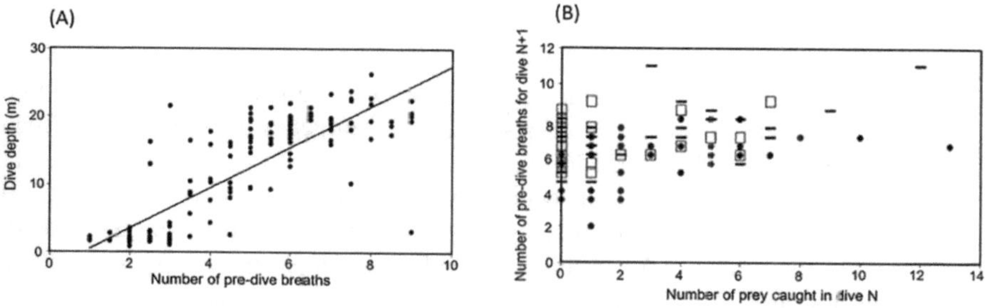

Fig. 10.22 (**a**) The number of breaths taken by a Magellanic penguin (foraging at Cabo Virgenes, Argentina) during surface pauses in preparation for its next dive as a function of the maximum dive depth showing an increasing number of preparative breaths with increasing depth. (**b**) Number of pre-dive breaths made by three different Magellanic penguins (shown by the different symbols), each of which was diving to a set depth range (and so would have had a reasonably constant number of pre-dive breaths (see panel **a**)) in relation to the number of fish caught in the previous dive. It should be noted that, as birds caught more fish in any one dive, they prepared for the next dive by taking more breaths. (Figures were obtained from [117] with permission from Inter-Research)

oxygen that they need for their projected performance and no more. However, having just caught prey from a patch at depth is a good reason to "believe" that the extra oxygen acquired during the otherwise inefficient extended surface pauses should allow the (aerobic) capture of more prey in the next dive – assuming, of course, that the patch can be relocated. No such strategy is necessary if the prey are shallow though because, as we mentioned above, penguins can exploit these patches easily, snatching breaths, dolphin-like, from time to time from the close water surface (cf. [57]).

However, that penguins might "plan" their breathing patterns depending on their proximate dive depth and the number of prey they are likely to encounter is extraordinary and shows the levels of behavioral sophistication that animals can achieve if there is appropriate selection pressure for a "clever strategy." It is even more extraordinary if all the members of the flock, having the same strategy, are doing the same thing (e.g., Fig. 14.1). It makes perfect sense because penguin prey usually occur in aggregations that the birds exploit during serial dives so the number of prey caught in a previous dive is likely to be a good measure of the number of prey that will be caught in the following dive – and this is exactly what has been found [117]. By doing this, when penguins have the option to catch many prey, they have prepared by loading extra oxygen. In contrast, when prey are few, penguins can recover fastest from dives by not spending excessive time on the surface topping up incremental levels of oxygen, a behavior that reduces their overall foraging efficiency [128]. So, it would seem that preparation is everything for animals that have to hold their breath to eat lunch! Who would have thought that these fish-birds would be effectively counting their mouthfuls as they flashed around in the twilight depths far beneath the glittering sea surface?

Being Picky About Food

Traveling down to Antarctica on one occasion, we were subjected to a lecture about predator–prey relationships by Professor Gotthilf Hempel, a truly impressive man who spent many years directing German polar research with verve. He simply, but emphatically, stated "the rule of the ocean is you eat what you meet." I had, and still have, the greatest respect for the man, but he was wrong. Certainly, penguins do not just eat what they meet as penguin-borne video cameras now neatly show [92]. Indeed, our footage taken from the backs of Magellanic penguins shows them happily swimming past aggregations of squid, crustaceans, and jellyfish (cf. [99]), with barely a glance. We should not be surprised. There is a wealth of literature that shows how animals concentrate their attention on food items that impart most energy (Chap. 4, Appendix Table A4). Examples range from sheep [91] to eider duck [103] and do not just deal with the general idea that animals should concentrate their attention on the best food but also actively avoid apparently unpalatable foodstuff [67].

To our mind, there are five major aspects that are important when animals make decisions about which food to eat to extract most energy, once food has been encountered (although sometimes studies need to consider critical nutritional elements rather than just energy [98]); (i) how much energy it takes to catch and ingest the foodstuff, (ii) the energy density of the food, (iii) the assimilation efficiency – i.e., how much of the energy contained within foodstuff can be extracted, (iv) the residence time of the food in the digestive system (because until it has been processed, nothing more can be ingested), and (v) the difference between the temperature of the food and the body temperature of the animal eating it. The latter is related to cold food needing to be heated up to body temperature for digestion, and this costs energy. All these are relevant to penguins, some of which are also discussed in Chap. 4, but we can even put figures on some aspects of this.

We shall first ignore the energetic costs of acquiring food because other parts of this book mention this in some detail, but it is important. This is particularly so if penguins have to travel substantial distances between their breeding colonies and the foraging site (cf. [10]), and at the depth of the dive, for example, affects how long penguin can stay underwater exploiting a prey patch (see above).

If we consider penguins' major prey types – fish, crustaceans, and squid (see Chap. 4), this gives wet mass energy densities that vary between about 2.9 and 9 kJ/g for fish, between 3.1 and 4.8 kJ/g for crustaceans, and between 3.2 and 5.6 kJ/g for squid (Appendix Table A4) (Fig. 10.23). To this, we will add jellyfish, with an energy density of about 0.14 kJ/g [27, 86] because recent work has suggested that under some circumstances, they may form an appreciable part of the diet [97, 99]. The highly variable absolute energy densities of these different foodstuffs get reduced when penguins try and digest them because the digestive process is not 100% efficient. Within birds in general, there is substantial variability in assimilation efficiency depending on the food type [12] and there is good evidence that penguins, as with other seabirds, have assimilation efficiencies that are higher if

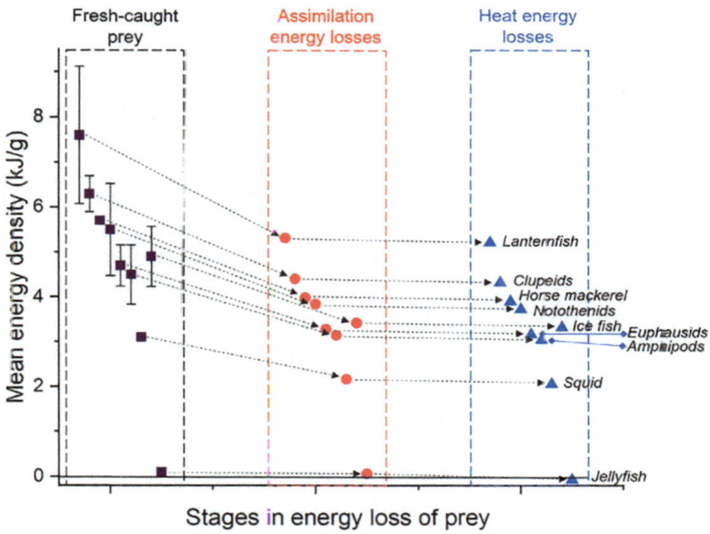

Fig. 10.23 The energy density of different prey taxa eaten by penguins (Data are from Appendix 4 – offset from each other in the *x*-axis to highlight standard deviation (SD) bars) and how these are eroded by assimilation efficiencies (standardized to be 70%) and the costs of heating cold food (sea temperature values approximated for the various taxa). Although it is clear that the assimilation efficiency results in more losses compared to the costs of heating ingesta, the thermal costs can result in some gelatinous prey taxa being energy-negative. The variability in energy density between the various taxa makes a compelling case for penguins to be discerning about what they should pursue and what they should ignore

they ingest their "preferred" prey (see [47] and the references therein). Sadly, there is not enough information to list penguin species and prey values within this context, but we can consider that energy densities of prey will be reduced to about 70% of their absolute value following digestion [1, 46, 50, 70] (Fig. 10.23).

As with assimilation efficiency, the time it takes for particular prey to be digested is likely to be important in modulating prey choice because, if the stomach is full, nothing more can be swallowed until it is empty, or at least has space for more [102]. In this regard, it has been noted that a fish-eating African penguin, for example, takes more than twice as long to empty the contents of its stomach if it has consumed squid than anchovy [121]. How this plays out for other penguin species and their normal, and not so normal, prey needs further work.

Finally, as discussed in Chap. 7, the cost of heating cold ingesta (which includes all penguin prey, i.e., the same temperature as the sea) also serves to diminish the energetic value of ingested food, and this may account for >10% of the daily energy expenditure of the polar penguin species [114]. These costs can be surprisingly high in penguins because the bulk of penguin prey is made up of water, and water, requiring 4.18 J to heat just a gram by 1 °C, is one of our planet's most expensive substances to heat (Box 10.7).

Box 10.7

Water temperature decreases the energetic value of penguin prey. The energy used to heat any material from a low temperature to a higher one can be calculated using:

$$\text{Energy}_{\text{heat}} = \text{SHC}_{\text{substance}} \times M \times \Delta T, \tag{10.3}$$

where $\text{SHC}_{\text{substance}}$ is the specific heat capacity of the substance to be heated (J/g/°C), M is its mass (grams), and ΔT is the required temperature increment (°C). In this case, $\text{SHC}_{\text{substance}}$ refers to the penguin prey, ΔT is the temperature difference between the water and the penguin core temperature, and M is the mass of food ingested.

The potential calorific value of the prey (see Appendix 4) ingested therefore becomes:

$$\text{Realized energy}_{\text{prey}} = \left(ED_{\text{prey}} \times M\right) - \left(\text{SHC}_{\text{substance}} \times M \times \Delta T\right), \tag{10.4}$$

where ED_{prey} is the available energy density of the prey to the penguin (J/g wet weight) after accounting for penguin assimilation efficiency.

The effect of this is particularly profound in penguins that forage in the colder water of penguindom and it makes some prey items, specifically jellyfish, energetically untenable. Such prey can require more energy just to heat than they provide (Fig. 10.23).

All these aspects together are good reasons for penguins to be judicious in what they decide to eat [92, 98]. Thus, we should expect to see favoritism in prey selection. However, we should also expect these preferences to change as the preyscape changes. For example, for penguins catching schooling fish in the water column, such as penguins of the genus *Spheniscus*, it is beneficial to initially "hold out" for high energy-dense fish when searching for prey, although we would also expect birds to lower their quality standards if their searches reveal little or no fish. Indeed, we have observed exactly this for Magellanic penguins foraging at San Julian in Argentina. Their breeding is normally timed so that it coincides with an influx of sprats as they migrate up from the south, but, if this does not occur, then the birds bring back squid to their chicks. This is catastrophic because the stomach contents are so poorly digested [121] by the time the adults reach the nest that small chicks cannot open their beaks wide enough to swallow the squid, which fall into the nest where they rot. Such happenings highlight the negative consequences of African penguins now having to consume low-quality "junk food," due to the depletion of high energy-dense anchovies and sardines ([59], Chap. 5). A similar scenario appears to be impacting another endangered species, the yellow-eyed penguin [110], in waters around New Zealand. On the west coast of the Antarctic Peninsula, apparently there is plenty of krill but not enough energy-dense Antarctic silverfish, with the fish disappearing with the sea ice, and so, in accord, Adélie penguin numbers have been decreasing (Chap. 13) [13, 55, 81].

Video cameras attached to penguins have been important in revealing what prey penguins catch and when and how they catch them, but future work should be able to examine how prey preferences change over time as penguins navigate through preyscapes, including prey quality. There is no doubt that penguins are fussy, and, so, they should be because there are substantial selective pressures that underpin this. Sadly, continued changes in marine ecosystems caused by poorly managed fishing may just be reducing the amount of choice that penguins normally have and that cannot be good. Indeed, the African and yellow-eyed penguins mentioned above are testament to what happens if birds have to compete with fisheries for their preferred prey (Chaps. 5 and 13).

Fish-Birds and Smart Foraging Strategies

Fish-birds walk, well actually swim, a difficult line, honing aspects of avian physiology so that they can operate in the domain of fish, where the physical properties of water are mercilessly punishing for animals of the higher airs. However, astonishingly, these fish-birds have adapted so far, including in their choices of behavior, that they can even outperform the animals they catch, all of which, in evolutionary time, have never left the sea. More than this, penguins have made the sea their first home, and their foraging success is such that they can accumulate energy at astonishing rates for themselves and their broods [82, 93, 120]. A substantial part of penguins' success in being able to effectively exploit the sea, from an evolutionary perspective (Chaps. 1 and 2), lies in their behavioral flexibility. As shown in this chapter, the flexibility ranges from choosing the number and depth of breaths they take on the surface to the rates at which they descend the water column, which depths they should prioritize, and what species they decide to swallow [67], thus to optimize efficiency in capturing the best prey. In fact, penguins manage a baffling gamut of options to make them such supremely efficient hunters in the fascinating, exigent, and mysterious sea, but there will be more, much more, "clever stuff" that they will be doing, and this should be cause for excitement and celebration.

References

1. Adams N (1984) Utilization efficiency of a squid diet by adult king penguins (*Aptenodytes patagonicus*). Auk 101(4):884–886
2. Ainley DG, Ballard G, Jones RM, Jongsomjit D, Pierce SD, Smith WO Jr, Veloz S (2015) Trophic cascades in the western Ross Sea, Antarctica: revisited. Mar Ecol Prog Ser 534:1–16
3. Ballance LT, Ainley DG, Ballard G, Barton K (2009) An energetic correlate between colony size and foraging effort in seabirds, an example of the Adélie penguin *Pygoscelis adeliae*. J Avian Biol 40(3):279–288

4. Ballard G, Dugger KM, Nur N, Ainley DG (2010) Foraging strategies of Adélie penguins: adjusting body condition to cope with environmental variability. Mar Ecol Prog Ser 405:287–302
5. Ballard G, Schmidt AE, Toniolo V, Veloz S, Jongsomjit D, Arrigo KR, Ainley DG (2019) Fine-scale oceanographic features characterizing successful Adélie penguin foraging in the SW Ross Sea. Mar Ecol Prog Ser 608:263–277
6. Bethge P, Nicol S, Culik B, Wilson R (1997) Diving behaviour and energetics in breeding little penguins (*Eudyptula minor*). J Zool 242(3):483–502
7. Bovet P, Benhamou S (1991) Optimal sinuosity in central place foraging movements. Anim Behav 42(1):57–62
8. Brierley AS, Cox MJ (2010) Shapes of krill swarms and fish schools emerge as aggregation members avoid predators and access oxygen. Curr Biol 20(19):1758–1762
9. Brock VE, Riffenburgh RH (1960) Fish schooling: a possible factor in reducing predation. ICES J Mar Sci 25(3):307–317
10. Burke C, Montevecchi W (2009) The foraging decisions of a central place foraging seabird in response to fluctuations in local prey conditions. J Zool 278(4):354–361
11. Carroll G, Cox M, Harcourt R, Pitcher BJ, Slip D, Jonsen I (2017) Hierarchical influences of prey distribution on patterns of prey capture by a marine predator. Funct Ecol 31(9):1750–1760
12. Castro G, Stoyan N, Myers J (1989) Assimilation efficiency in birds: a function of taxon or food type? Comp Biochem Physiol A Physiol 92(3):271–278
13. Chapman EW, Hofmann EE, Patterson DL, Ribic CA, Fraser WR (2011) Marine and terrestrial factors affecting Adélie penguin *Pygoscelis adeliae* chick growth and recruitment off the western Antarctic Peninsula. Mar Ecol Prog Ser 436:273–289
14. Chappell MA, Shoemaker VH, Janes DN, Bucher TL, Maloney SK (1993) Diving behavior during foraging in breeding Adélie penguins. Ecology 74(4):1204–1215
15. Charnov EL (1976) Optimal foraging, the marginal value theorem. Theor Popul Biol 9(2):129–136
16. Charrassin J-B, Bost C-A (2001) Utilisation of the oceanic habitat by king penguins over the annual cycle. Mar Ecol Prog Ser 221:285–298
17. Charrassin J-B, Le Maho Y, Bost C-A (2002) Seasonal changes in the diving parameters of king penguins (*Aptenodytes patagonicus*). Mar Biol 141(3):581–589
18. Cherel Y, Tremblay Y, Guinard E, Georges J-Y (1999) Diving behaviour of female northern rockhopper penguins, *Eudyptes chrysocome moseleyi*, during the brooding period at Amsterdam Island (Southern Indian Ocean). Mar Biol 134(2):375–385
19. Choi N, Kim J-H, Kokubun N, Park S, Chung H, Lee WY (2017) Group association and vocal behaviour during foraging trips in *Gentoo penguins*. Sci Rep 7(1):1–9
20. Cimino MA, Moline MA, Fraser WR, Patterson-Fraser DL, Oliver MJ (2016) Climate-driven sympatry may not lead to foraging competition between congeneric top-predators. Sci Rep 6(1):1–10
21. Cohen JH, Forward RB Jr (2016) Zooplankton diel vertical migration – a review of proximate control. In: Oceanography and marine biology. CRC Press, Boca Raton, pp 89–122
22. Cowie RJ (1977) Optimal foraging in great tits (Parus major). Nature 268(5616):137–139
23. Cresswell KA, Tarling GA, Trathan PN (2007) Weight loss during breeding is adaptive for female macaroni penguins, *Eudyptes chrysolophus*. Evol Ecol Res 9(7):1053–1076
24. Culik BM, Pütz K, Wilson R, Allers D, Lage J, Bost C, Le Maho Y (1996) Diving energetics in king penguins (*Aptenodytes patagonicus*). J Exp Biol 199(4):973–983
25. Cushing D, Jones FH (1968) Why do fish school? Nature 218(5145):918–920
26. Dewar JM (1924) The bird as a diver: a contribution to the natural history of diving birds. HF & G. Witherby, London
27. Doyle TK, Houghton JD, Mcdevitt R, Davenport J, Hays GC (2007) The energy density of jellyfish: estimates from bomb-calorimetry and proximate-composition. J Exp Mar Biol Ecol 343(2):239–252

28. Elgar MA (1989) Predator vigilance and group size in mammals and birds: a critical review of the empirical evidence. Biol Rev 64(1):13–33
29. Elley T, Mattern T, Ellenberg U, Young MJ, Hickcox RP, Van Heezik Y, Seddon PJ (2022) Consistent site-specific foraging behaviours of yellow-eyed penguins/hoiho breeding on Stewart Island. New Zealand Biol 11(6):844
30. Ford RG (1983) Home range in a patchy environment: optimal foraging predictions. Am Zool 23(2):315–326
31. Ford RG, Ainley DG, Brown ED, Suryan RM, Irons DB (2007) A spatially explicit optimal foraging model of Black-legged Kittiwake behavior based on prey density, travel distances, and colony size. Ecol Model 204(3–4):335–348
32. Ford RG, Ainley DG, Lescroël A, Lyver POB, Toniolo V, Ballard G (2015) Testing assumptions of central place foraging theory: a study of Adélie penguins *Pygoscelis adeliae* in the Ross Sea. J Avian Biol 46(2):193–205
33. Fréon P, Misund OA (1999) Dynamics of pelagic fish distribution and behaviour: effects on fisheries and stock assessment. Fishing News Books, Oxford
34. Gerlotto F, Castillo J, Saavedra A, Barbieri M, Espejo M, Cotel P (2004) Three-dimensional structure and avoidance behaviour of anchovy and common sardine schools in central southern Chile. ICES J Mar Sci 61(7):1120–1126
35. Gill FB, Wolf LL (1977) Nonrandom foraging by sunbirds in a patchy environment. Ecology 58(6):1284–1296
36. Gliwicz MZ (1986) Predation and the evolution of vertical migration in zooplankton. Nature 320(6064):746–748
37. Gunner RM, Holton MD, Scantlebury MD, Van Schalkwyk L, English HM, Williams HJ, Hopkins P, Quintana F, Gómez-Laich A, Börger L (2021) Dead-reckoning animal movements in R–A reappraisal using Gundog. Tracks Anim Biotelemet 9:23. https://doi.org/10.1186/s40317-021-00245-z
38. Halsey LG, Butler PJ (2006) Optimal diving behaviour and respiratory gas exchange in birds. Respir Physiol Neurobiol 154(1–2):268–283
39. Haney JC, Fristrup KM, Lee DS (1992) Geometry of visual recruitment by seabirds to ephemeral foraging flocks. Ornis Scand 23:249–262
40. Hansen J, Martos P, Madirolas A (2001) Relationship between spatial distribution of the Patagonian stock of Argentine anchovy, *Engraulis anchoita*, and sea temperatures during late spring to early summer. Fish Oceanogr 10(2):193–206
41. Hanuise N, Bost CA, Handrich Y (2013) Optimization of transit strategies while diving in foraging king penguins. J Zool 290(3):181–191
42. Hays GC (2003) A review of the adaptive significance and ecosystem consequences of zooplankton diel vertical migrations. Migrations and dispersal of marine organisms. Hydrobiologia 503(1):163–170
43. Houston AI, Carbone C (1992) The optimal allocation of time during the diving cycle. Behav Ecol 3(3):255–265
44. Ichii T (1988) Swarming ecology of Antarctic krill. Tokyo, Japan: Deep Sea Fisheries Research Institute News 67:1–4
45. Goudie RI (1999) Behaviour of harlequin ducks and three species of scoters wintering in the Queen Charlotte Islands, British Columbia: behaviour and ecology of sea ducks, Occasional paper No. 100. Canadian Wildlife Service, Ottawa, pp 6–13
46. Jackson S, Place AR (1990) Gastrointestinal transit and lipid assimilation efficiencies in three species of sub-antarctic seabird. J Exp Zool 255(2):141–154
47. Jackson S, Ryan PG (1986) Differential digestion rates of prey by white-chinned petrels (*Procellaria aequinoctialis*). Auk 103(3):617–619
48. James A (1987) Feeding ecology, diet and field-based studies on feeding selectivity of the Cape anchovy *Engraulis capensis* Gilchrist. S Afr J Mar Sci 5(1):673–692
49. Killen SS, Marras S, Steffensen JF, Mckenzie DJ (2012) Aerobic capacity influences the spatial position of individuals within fish schools. Proc R Soc B Biol Sci 279(1727):357–364

50. Kirkwood R, Robertson G (1997) The energy assimilation efficiency of emperor penguins, *Aptenodytes forsteri*, fed a diet of Antarctic krill, *Euphausia superba*. Physiol Zool 70(1):27–32
51. Kooyman G, Cherel Y, Maho YL, Croxall J, Thorson P, Ridoux V, Kooyman C (1992) Diving behavior and energetics during foraging cycles in king penguins. Ecol Monogr 62(1):143–163
52. Kooyman G, Davis R, Croxall J, Costa D (1982) Diving depths and energy requirements of king penguins. Science 217(4561):726–727
53. Krafft BA, Skaret G, Knutsen T, Melle W, Klevjer T, Søiland H (2012) Antarctic krill swarm characteristics in the Southeast Atlantic sector of the Southern Ocean. Mar Ecol Prog Ser 465:69–83
54. Krebs JR, Davies NB (2009) Behavioural ecology: an evolutionary approach. Wiley, New York
55. La Mesa M, Riginella E, Mazzoldi C, Ashford J (2015) Reproductive resilience of ice-dependent Antarctic silverfish in a rapidly changing system along the Western Antarctic Peninsula. Mar Ecol 36(2):235–245
56. Lascara CM, Hofmann EE, Ross RM, Quetin LB (1999) Seasonal variability in the distribution of Antarctic krill, *Euphausia superba*, west of the Antarctic Peninsula. Deep-Sea Res I Oceanogr Res Pap 46(6):951–984
57. Lescroël A, Schmidt A, Elrod M, Ainley DG, Ballard G (2021) Foraging dive frequency predicts body mass gain in the Adélie penguin. Sci Rep 11(1):1–10
58. Lishman G, Croxall J (1983) Diving depths of the Chinstrap Penguin *Pygoscelis Antarctica*. Br Antarct Surv Bull 61:21–25
59. Ludynia K, Roux J, Jones R, Kemper J, Underhill L (2010) Surviving off junk: low-energy prey dominates the diet of African penguins *Spheniscus demersus* at Mercury Island, Namibia, between 1996 and 2009. Afr J Mar Sci 32(3):563–572
60. Luna-Jorquera G, Culik BM (1999) Diving behaviour of Humboldt penguins *Spheniscus humboldti* in northern Chile. Mar Ornithol 27:67–76
61. Macintosh AJ, Pelletier L, Chiaradia A, Kato A, Ropert-Coudert Y (2013) Temporal fractals in seabird foraging behaviour: diving through the scales of time. Sci Rep 3(1):1–10
62. Martin GR (1999) Eye structure and foraging in King Penguins *Aptenodytes patagonicus*. Ibis 141(3):444–450
63. Martin GR, Young S (1984) The eye of the Humboldt penguin, *Spheniscus humboldti*: visual fields and schematic optics. Proc R Soc Lond B Biol Sci 223(1231):197–222
64. Mattern T, Ellenberg U, Houston DM, Davis LS (2007) Consistent foraging routes and benthic foraging behaviour in yellow-eyed penguins. Mar Ecol Prog Ser 343:295–306
65. Meyer X, Macintosh AJ, Chiaradia A, Kato A, Mattern T, Sueur C, Ropert-Coudert Y (2017) Shallow divers, deep waters and the rise of behavioural stochasticity. Mar Biol 164(6):1–16
66. Meyer X, Macintosh AJ, Chiaradia A, Kato A, Ramírez F, Sueur C, Ropert-Coudert Y (2020) Oceanic thermal structure mediates dive sequences in a foraging seabird. Ecol Evol 10(13):6610–6622
67. Miller AK, Trivelpiece WZ (2008) Chinstrap penguins alter foraging and diving behavior in response to the size of their principle prey, Antarctic krill. Marine Biol 154(2):201–208
68. Morrell LJ, James R (2008) Mechanisms for aggregation in animals: rule success depends on ecological variables. Behav Ecol 19(1):193–201
69. Munden R, Börger L, Wilson RP, Redcliffe J, Brown R, Garel M, Potts JR (2021) Why did the animal turn? Time-varying step selection analysis for inference between observed turning-points in high frequency data. Methods Ecol Evol 12(5):921–932
70. Nagy KA, Siegfried WR, Wilson RP (1984) Energy utilization by free-ranging jackass penguins, *Spheniscus demersus*. Ecology 65(5):1648–1655
71. Naito Y, Asaga T, Ohyama Y (1990) Diving behavior of Adélie penguins determined by time-depth recorder. Condor 92(3):582–586

72. Potts JR, Börger L, Scantlebury DM, Bennett NC, Alagaili A, Wilson RP (2018) Finding turning-points in ultra-high-resolution animal movement data. Methods Ecol Evol 9(10):2091–2101
73. Pütz K, Cherel Y (2005) The diving behaviour of brooding king penguins (*Aptenodytes patagonicus*) from The Falkland Islands: variation in dive profiles and synchronous underwater swimming provide new insights into their foraging strategies. Mar Biol 147(2):281–290
74. Pyke GH (1978) Optimal foraging: movement patterns of bumblebees between inflorescences. Theor Popul Biol 13(1):72–98
75. Pyke GH, Pulliam HR, Charnov EL (1977) Optimal foraging: a selective review of theory and tests. Q Rev Biol 52(2):137–154
76. Ropert-Coudert Y, Grémillet D, Kato A (2005) Diving angle of great cormorants. Polar Biosci 18:54–59
77. Ropert-Coudert Y, Wilson RP, Daunt F, Kato A (2004) Patterns of energy acquisition by a central place forager: benefits of alternating short and long foraging trips. Behav Ecol 15(5):824–830
78. Ryan P, Petersen S, Peters G, Grémillet D (2004) GPS tracking a marine predator: the effects of precision, resolution and sampling rate on foraging tracks of African Penguins. Mar Biol 145(2):215–223
79. Ryan P, Petersen S, Simeone A, Grémillet D (2007) Diving behaviour of African penguins: do they differ from other *Spheniscus* penguins? Afr J Mar Sci 29(2):153–160
80. Saenz BT, Ainley DG, Daly KL, Ballard G, Conlisk E, Elrod ML, Kim SL (2020) Drivers of concentrated predation in an Antarctic marginal-ice-zone food web. Sci Rep 10(1):1–17
81. Sailley SF, Ducklow HW, Moeller HV, Fraser WR, Schofield OM, Steinberg DK, Garzio LM, Doney SC (2013) Carbon fluxes and pelagic ecosystem dynamics near two western Antarctic Peninsula Adélie penguin colonies: an inverse model approach. Mar Ecol Prog Ser 492:253–272
82. Sala JE, Wilson RP, Quintana F (2012) How much is too much? Assessment of prey consumption by *Magellanic penguins* in Patagonian colonies. PLoS One 7(12):e51487
83. Sato K, Charrassin J-BT, Bost C-A, Naito Y (2004) Why do macaroni penguins choose shallow body angles that result in longer descent and ascent durations? J Exp Biol 207(23):4057–4065
84. Sato K, Mori Y, Yoda K (2001) Defining dive bouts using a sequential differences analysis. Behaviour 138(11–12):1451–1466
85. Sato K, Naito Y, Kato A, Niizuma Y, Watanuki Y, Charrassin J, Bost C-A, Handrich Y, Le Maho Y (2002) Buoyancy and maximal diving depth in penguins: do they control inhaling air volume? J Exp Biol 205(9):1189–1197
86. Schaafsma FL, Cherel Y, Flores H, Van Franeker JA, Lea M-A, Raymond B, Van De Putte AP (2018) The energetic value of zooplankton and nekton species of the Southern Ocean. Mar Biol 165(8):1–35
87. Scharf FS, Buckel JA, Juanes F (2002) Size-dependent vulnerability of juvenile bay anchovy *Anchoa mitchilli* to bluefish predation: does large body size always provide a refuge? Mar Ecol Prog Ser 233:241–252
88. Scheffer A, Trathan PN, Edmonston JG, Bost C-A (2016) Combined influence of meso-scale circulation and bathymetry on the foraging behaviour of a diving predator, the king penguin (*Aptenodytes patagonicus*). Prog Oceanogr 141:1–16
89. Shepard EL, Wilson RP, Quintana F, Gomez Laich A, Forman DW (2009) Pushed for time or saving on fuel: fine-scale energy budgets shed light on currencies in a diving bird. Proc R Soc B Biol Sci 276(1670):3149–3155
90. Sivak J, Howland H, Mcgill-Harelstad P (1987) Vision of the Humboldt penguin (*Spheniscus humboldti*) in air and water. Proc R Soc Lond B Biol Sci 229(1257):467–472
91. Stephens DW, Brown JS, Ydenberg RC (2008) Foraging: behavior and ecology. University of Chicago Press, Chicago
92. Sutton G, Arnould J (2022) Quantity over quality? Prey-field characteristics influence the foraging decisions of little penguins (*Eudyptula minor*). R Soc Open Sci 9(6):211171

93. Takahashi A, Dunn M, Trathan P, Croxall J, Wilson RP, Sato K, Naito Y (2004) Krill-feeding behaviour in a chinstrap penguin compared to fish-eating in *Magellanic penguins*: a pilot study. Mar Ornithol 32:47–54
94. Tarling GA, Fielding S (2016) Swarming and behaviour in Antarctic krill. In: Biology and ecology of Antarctic krill. Springer, London, pp 279–319
95. Tessier E, Bost C-A (2020) Behavioural adjustments during foraging in two diving seabirds: king and macaroni penguins. Mar Biol 167(9):1–11
96. Thiebault A, Charrier I, Aubin T, Green DB, Pistorius PA (2019) First evidence of underwater vocalisations in hunting penguins. PeerJ 7:e8240
97. Thiebot J-B, Ito K, Raclot T, Poupart T, Kato A, Ropert-Coudert Y, Takahashi A (2016) On the significance of Antarctic jellyfish as food for Adélie penguins, as revealed by video loggers. Mar Biol 163(5):1–8
98. Thiebot J-B, Mcinnes JC (2020) Why do marine endotherms eat gelatinous prey? ICES J Mar Sci 77(1):58–71
99. Thiebot JB, Arnould JP, Gómez-Laich A, Ito K, Kato A, Mattern T, Mitamura H, Noda T, Poupart T, Quintana F (2017) Jellyfish and other gelata as food for four penguin species–insights from predator-borne videos. Front Ecol Environ 15(8):437–441
100. Trathan PN, Bishop C, Maclean G, Brown P, Fleming A, Collins MA (2008) Linear tracks and restricted temperature ranges characterise penguin foraging pathways. Mar Ecol Prog Ser 370:285–294
101. Tremblay Y, Cherel Y (2000) Benthic and pelagic dives: a new foraging behaviour in rockhopper penguins. Mar Ecol Prog Ser 204:257–267
102. Van Gils JA, Schenk IW, Bos O, Piersma T (2003) Incompletely informed shorebirds that face a digestive constraint maximize net energy gain when exploiting patches. Am Nat 161(5):777–793
103. Varennes E, Hanssen SA, Bonardelli JC, Guillemette M (2015) A large molluscivore bird (Common Eider, *Somateria mollissima*) is able to discriminate quality of blue mussels (*Mytilus edulis*) based on size and provenance. Can J Zool 93(8):655–663
104. Wakefield ED, Phillips RA, Matthiopoulos J (2009) Quantifying habitat use and preferences of pelagic seabirds using individual movement data: a review. Mar Ecol Prog Ser 391:165–182
105. Wanless S, Corfield T, Harris M, Buckland S, Morris J (1993) Diving behaviour of the shag *Phalacrocorax aristotelis* (Aves: Pelecaniformes) in relation to water depth and prey size. J Zool 231(1):11–25
106. Watanabe H, Shiome K, Sato K, Takahashi A, Handrich Y, Bost C-A (in press) King penguins adjust their fine-scale travelling and foraging behaviours to spatial and diel changes in feeding opportunities. Mar Biol
107. Watanabe YY, Ito M, Takahashi A (2014) Testing optimal foraging theory in a penguin–krill system. Proc R Soc B Biol Sci 281(1779):20132376
108. Watanabe YY, Takahashi A (2013) Linking animal-borne video to accelerometers reveals prey capture variability. Proc Natl Acad Sci 110(6):2199–2204
109. Weber SB, Richardson AJ, Brown J, Bolton M, Clark BL, Godley BJ, Leat E, Oppel S, Shearer L, Soetaert KE (2021) Direct evidence of a prey depletion "halo" surrounding a pelagic predator colony. Proc Natl Acad Sci 118(28):e2101325118
110. Webster T (2018) The pathway ahead for hoiho Te ara whakamua. Impacts on hoiho: literature review and recommendations. Technical report, Yellow-Eyed Penguin Trust, Dunedin, New Zealand
111. Weihs D (1972) A hydrodynamical analysis of fish turning manoeuvres. Proc R Soc Lond B Biol Sci 182(1066):59–72
112. Williams T, Briggs DR, Croxall J, Naito Y, Kato A (1992) Diving pattern and performance in relation to foraging ecology in the gentoo penguin, *Pygoscelis papua*. J Zool 227(2):211–230
113. Wilson R (1985) The Jackass Penguin (*Spheniscus demersus*) as a pelagic predator. Mar Ecol Prog Ser 25:219–227
114. Wilson R, Culik B (1991) The cost of a hot meal: facultative specific dynamic action may ensure temperature homeostasis in post-ingestive endotherms. Comp Biochem Physiol A Comp Physiol 100(1):151–154

115. Wilson R, Griffiths I, Legg P, Friswell M, Bidder O, Halsey L, Lambertucci SA, Shepard E (2013) Turn costs change the value of animal search paths. Ecol Lett 16(9):1145–1150
116. Wilson R, Rose K, Metcalfe R, Holton M, Redcliffe J, Gunner R, Börger L, Loison A, Miloš J, Painter M (2021) Path tortuosity changes the transport cost paradigm in terrestrial animals. Ecography 44:1524–1532
117. Wilson RP (2003) Penguins predict their performance. Mar Ecol Prog Ser 249:305–310
118. Wilson RP (2010) Resource partitioning and niche hyper-volume overlap in free-living *Pygoscelid penguins*. Funct Ecol 24(3):646–657
119. Wilson RP, Holton MD, Neate A, Del Caño M, Quintana F, Yoda K, Gómez-Laich A (2022) Luck and tactics in foraging success: the case of the imperial shag. Mar Ecol Prog Ser 682:1–12
120. Wilson RP, Jackson S, Straten MT (2007) Rates of food consumption in free-living Magellanic Penguins *Spheniscus magellanicus*. Mar Ornithol 35:109–111
121. Wilson RP, La Cock G, Wilson M-P, Mollagee F (1985) Differential digestion of fish and squid in Jackass Penguins *Spheniscus demersus*. Ornis Scand 16:77–79
122. Wilson RP, Neate A, Holton MD, Shepard EL, Scantlebury DM, Lambertucci SA, Di Virgilio A, Crooks E, Mulvenna C, Marks N (2018) Luck in food finding affects individual performance and population trajectories. Curr Biol 28(23):3871–3877. e3875
123. Wilson RP, Puetz K, Bost CA, Culik BM, Bannasch R, Reins T, Adelung D (1993) Diel dive depth in penguins in relation to diel vertical migration of prey: whose dinner by candlelight? Mar Ecol Prog Ser 94:101–104
124. Wilson RP, Ropert-Coudert Y, Kato A (2002) Rush and grab strategies in foraging marine endotherms: the case for haste in penguins. Anim Behav 63(1):85–95
125. Wilson RP, Simeone A, Luna-Jorquera G, Steinfurth A, Jackson S, Fahlman A (2003) Patterns of respiration in diving penguins: is the last gasp an inspired tactic? J Exp Biol 206(10):1751–1763
126. Wilson RP, Wilson M-PT (1988) Foraging behaviour in four sympatric cormorants. J Anim Ecol 57:943–955
127. Wilson RP, Zimmer I (2004) Inspiration by *Magellanic penguins*: reduced swimming effort when under pressure. Mar Ecol Prog Ser 278:303–307
128. Ydenberg R, Clark CW (1989) Aerobiosis and anaerobiosis during diving by western grebes: an optimal foraging approach. J Theor Biol 139(4):437–447
129. Zimmer I, Wilson RP, Beaulieu M, Ropert-Coudert Y, Kato A, Ancel A, Plötz J (2010) Dive efficiency versus depth in foraging emperor penguins. Aquat Biol 8(3):269–277

Chapter 11
The Final Seconds: How Fish-Birds Catch Prey

Prey Acquisition: A Departure from the Norm

As prescribed as penguin dive types might seem from the initiation of the dive (Chap. 9), the development of events underwater can elicit a movement response from the penguins and this changes the characteristics of the dive. The most obvious event that can happen to a foraging penguin (a bird exhibiting V-, P-, or U-dives or some intermediate form) is prey encounter. However, even this does not necessarily invoke a clear-cut response because what we might consider to be potential prey can invoke highly varied responses on the part of the penguin, as has been revealed by penguin-attached video cameras. Specifically, the concept that animals should "eat what they meet" is clearly not true for penguins. We have video footage from back-mounted cameras on Magellanic penguins that show birds swimming past potential prey (crustaceans and squid in tight aggregations and jellyfish (see Chap. 4 for the calorific value of such prey for penguins)) (Fig. 11.1a) without any discernible change in any behavior that is associated with prey capture (see below). In short, anyone examining the dive profile of penguins that are ignoring potential prey would not know that anything had happened (because physically, from a penguin's perspective, it has not!). It is worth noting, though, that such potential prey can become actual prey if the preferred prey are not found (Chap. 10).

The next development in penguin behavior with respect to potential prey is the manifestation of underwater "pauses," in which birds may slow down, or even briefly stop, to examine possible prey (such as crustaceans in the case of Magellanic penguins) (Fig. 11.1). There is near-constant depth for a period in the dive profile if it occurs in the descent or ascent phase of the dive and/or if the rate of change of depth (the vertical speed) is reduced for that instant. However, prey encounter leading to pursuit, where birds find what they are looking for, usually results in substantial departure of the standard, measured constant descent-, ascent-, or bottom-phase swimming patterns (but see [60, 62, 63]). While the steady constant speed and directionality of normal searching allows penguins to scan the water column with

D. G. Ainley, R. P. Wilson, *The Aquatic World of Penguins*, Fascinating Life Sciences, https://doi.org/10.1007/978-3-031-33990-5_11

Fig. 11.1 The proximity of penguins to potential prey does not necessarily mean that foraging birds will engage. These images from back-mounted video footage of a Magellanic penguin shows the bird about to ignore two different groupings of crustaceans: a tight agglomeration of squat lobster (left) and loosely associated prawns (right). Whether penguins actually decide to pursue potential prey depends on many factors, including the calorific value and size of the prey, the residence time of the food in the stomach, the accessibility of other prey types, and the ease with which the prey can be caught (see Chaps. 8 and 10)

minimized energy expenditure (Chap. 6), energy-expensive, substantive departures from the norm make sense because they lead to acquisition of (energy-rich) prey. Depending on the penguin species and the prey type, this prey acquisition behavior ranges from minor deviations from the norm to what might appear to be seemingly frantic underwater chases, clearly involving a great deal of athleticism in terms of speed, acceleration, turn radius, angular velocity, and changes in depth. In fact, there is good evidence that, rather than being imprudent rushes at prey, penguins are highly strategic in how they pursue and secure prey, not least because indiscriminate high speed and high angular velocity turning maneuvers cost energy and oxygen. Those costs minimize duration underwater at a time in which an extended dive is highly beneficial. In order to examine penguin strategies for catching prey, though, we first need to consider how well we can identify prey pursuit and/or capture.

Direct visual observation is wonderfully informative, and the few lucky people who have watched penguins catching prey (and that includes us!) are in a strong position to describe the process, qualitatively at least. However, such observations are normally made by people looking down from above the surface of the water, so the information is restricted to penguins operating in the top couple of meters of the water column. As an extension of this, video cameras attached to penguins (e.g., [39, 82]) can provide much information about prey capture occurring at greater depths, but, although there is some measure of the context of prey pursuit (because, for example, some of the surrounding conditions can be seen [27, 73, 74]), these systems have their "visual field" restricted by the camera angle or the bird body. Thus, they do not work well, or at all, if the lighting conditions are poor (with increased depth, decreased water clarity, and/or at different light levels relative to day and night). Other systems that indicate prey capture are based on measuring beak angle [87], because penguins have to open their beak to swallow food, and/or esophageal [56] or stomach temperature [26], because ingestion of cold food produces a drop in temperature in the throat and stomach. Other than that, prey interactions have to be inferred from metrics measured by penguin-attached tags.

Performance Metrics for Prey Capture in Penguins

In the days when penguin-attached depth sensors, i.e., time–depth recorders, were the only remote-sensing tools available to help understand the underwater behavior of penguins, researchers naturally gravitated toward depth metrics in an attempt to understand prey capture. Fortunately, prey capture could be effectively validated by looking at more direct measures such as beak angle. A first cut into prey pursuit behavior was made by noting that, in Magellanic penguins at least, ingestion was often associated with sudden changes in the depth profile – an "undulation" [67] (subsequently also termed a "wiggle" [26]). More recent works have noted that such irregularities, in the otherwise smooth dive profile (Chap. 9), were either confirmed to be due to prey ingestion via beak angle sensors [76, 87]) or via well-founded assumptions [4, 6].

Working on the principle that animals expend no more energy than is "needed," with penguins especially being no exception, we can examine the movement metrics that describe the diving behavior of penguins motivated to catch prey by specifically looking for "exceptional" signals linked to high power use and heightened athleticism [58]. When, where, and how these signals occur can do much to help us understand the strategies penguins use to secure their prey. This approach was started by identification of "undulations" (see above), but it is now clear that it is better to use multiple metrics of performance [6]. Most notable of these are the absolute penguin swim speed, the vertical speed and acceleration, the vectorial sum of the dynamic body acceleration (VeDBA – a good proxy for energy expenditure [94]; which we will subsequently refer to as "effort") and the vectorial sum of the static acceleration (VeSBA+, a good metric to indicate when animals turn corners fast, which we will, for convenience, subsequently refer to as the "*g*-force") (Box 11.1; Fig. 11.2).

Box 11.1
VeSBA is the vectorial sum of the static body acceleration [85] and is determined by running a smoothing window through each of the three orthogonal acceleration axes (in an animal-mounted tag recording acceleration at high rates) and then adding the values at any moment in time vectorially. At constant velocity, the vectorial sum is 1.0 *g*, whereas during fast turning, this value may increase or decrease (depending on whether the animal is turning left or right and according to the way the sway acceleration is set to record) due to the centripetal acceleration. VeSBA+ corrects for decreases in VeSBA and makes them equivalent to increases in VeSBA by simply subtracting VeSBA values that are <1.0 *g* from 1.0 *g* and then adding the difference to 1.0 *g*. Thus, all VeSBA+ values are greater than 1.0 *g*. For the purposes of this book, we refer to VeSBA+ as "*g*-force."

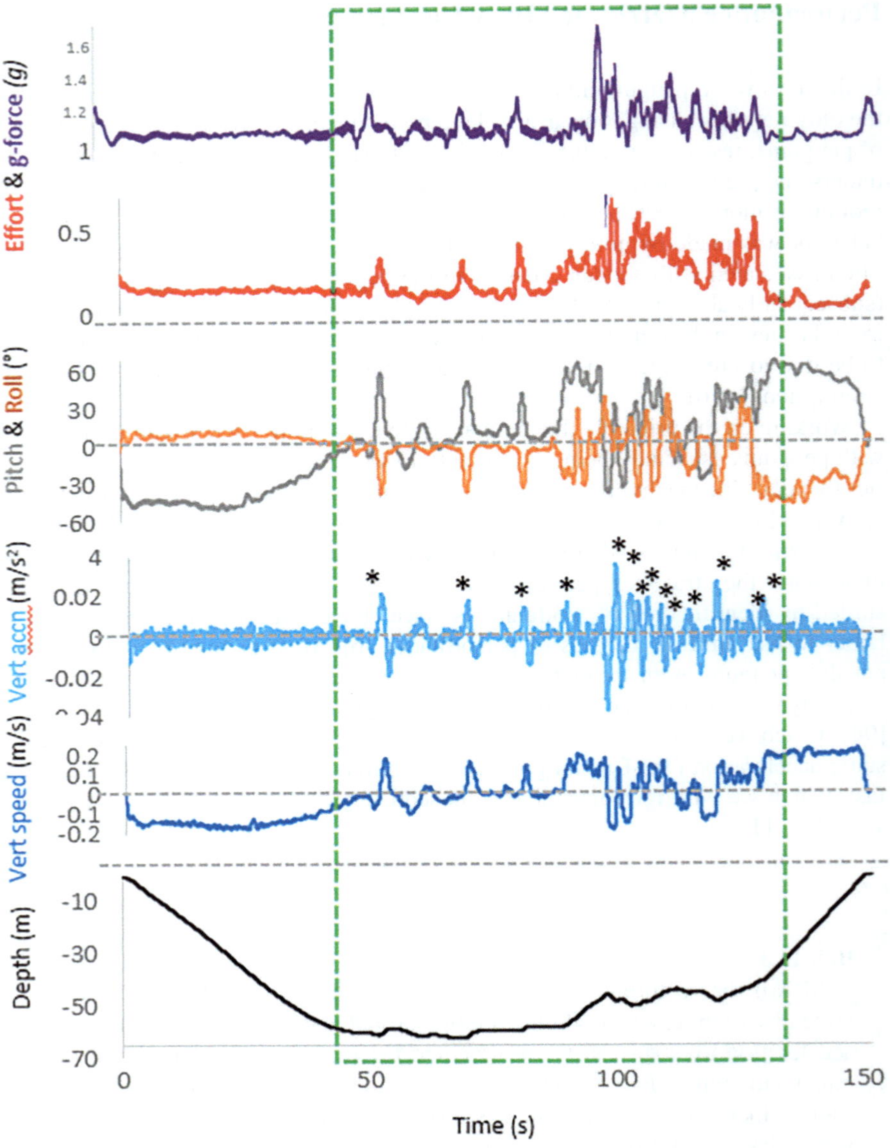

Fig. 11.2 In the absence of interactions with prey, most penguin dives are marked by smooth changes in the dive parameters (see Chap. 9). Once prey are encountered, however (see inside green box), all of this changes. This single dive by a Magellanic penguin shows possibly 14 prey interactions (asterixis) over 85 s in a dive lasting 152 s. Here, variations in the depth profile (black line), initiated by changes in bird swim (pitch) angle (gray line), produce highly variable vertical speeds (dark blue line), and associated vertical accelerations (turquoise line), driven, in part, by high flipper beat frequencies and amplitudes that are manifested in the increased effort (VeDBA) signal (red line). This, coupled with dramatic changes in the roll angle (orange line), causes radical changes in *g*-force (centripetal acceleration manifest by VeSBA+ (purple line; see Box 11.1,)

These can be calculated from either head- or body-mounted devices or both [14, 82], all being relevant because they highlight different aspects of penguin movement underwater. These metrics can be quantified thanks to the development of 3D accelerometers (which measure up-down, forward-backward, and sideways movements), rather than the more simplistic time–depth recorders.

Such metrics from the penguin body can be enhanced by similar metrics from tags attached to the penguin head [14, 82] because penguin head movement with respect to the body is also an important indicator of prey capture (see below). We will revisit these various metrics, where applicable, in our discussion of prey capture strategies.

Most penguin species hunt for prey in the water column (water lying between the surface and the bottom), rather than picking them from a substrate. The exceptions are yellow-eyed and gentoo penguins, which do both, as is also the case for emperor penguins (and also, though rarely, Adélie penguins) that can pick some of their prey from cavities on the underside of sea ice (see Chaps. 2 and 3). So, the concept that prey might hide in the way many terrestrial prey do generally does not apply for most penguins (but see [39]). Seemingly, once a member of one of these three penguin species has spotted their prey, there is some chance that the prey can bolt inside a hole. The pursuit by most penguins, and even most of the time for the three sometimes substrate foragers, generally happens above the seabed, where the best form of refuge is the turbidity of the medium, which might, for example, be offered by phytoplankton blooms or river outflow plumes (Chap. 3). Thus, because penguins are visual predators [9, 29, 69, 70], prey might conceivably disappear into the "fog" (Chap. 3). Other than that, there is the faint possibility that the prey might be able to outrun the penguin long enough for the bird's oxygen reserves to be used up (more on this below). Demersal prey often lie inconspicuously on the sea floor (something that is also replicated by prey that might be associated with the sea ice underside) to increase the probability that they will not be found – indeed, most such prey are colored cryptically (e.g., the notothenioid fish that polar penguins consume have no red blood and are thus "colorless"; Chap. 4). However, ultimately, whether the prey are solitary or aggregative, whether the penguins hunting them are doing so in groups or not, and the precise conditions under which prey are found all play a role in what might be considered the best strategy for a penguin to secure them.

Capture of Solitary Prey

Within the water column, penguin/solitary prey interactions have limited options for development. In the first instance, as with terrestrial cursorial predators that have been spotted by their prey, if the predator cannot travel faster than the prey, then there will be no capture [15]. Indeed, this has been suggested as one of the reasons why fast-swimming fish such as mackerel occur infrequently in penguin diet [88], although it may be that the power and, therefore, oxygen requirements for an extended high-speed chase are prohibitive [92], rather than the absolute maximum speed of the fish compared to that of the penguin (see Chap. 6). We also surmised earlier (Chaps. 1 and 2) that one reason that the penguin species are not found in warm waters (warmer than the Subtropical Front, 13–15 °C) is that the prey, being ectothermic, swim too fast for the penguins when the fish are warm [7]. Here, it is worth remembering that it is not just about the absolute maximum speeds of predators and prey but also the length of time that the chase might take at that speed because high-power chases deplete oxygen fast.

However, really, speed is (almost) everything, and solitary pelagic prey encountered by penguins in their cool-water environment, stand little chance of escape because most penguin prey species are small and the maximum swim speed in aquatic animals generally increases with body mass [12, 63, 81]. Crustacean-feeding penguins capture prey that each weigh less than a few grams (Chap. 4), which is about 1000 times smaller (or less) than the penguins pursuing them. Put bluntly, such animals will reach top speeds of generally <0.2 m/s, whereas their penguin predators will be cruising at ≥2 m/s (Fig. 11.3) (Chap. 6). Thus, the dice, so to speak, is loaded from the start. There is a marvelous video of an Adélie penguin surging upward to catch a single krill [82]. The krill swim pitifully away with

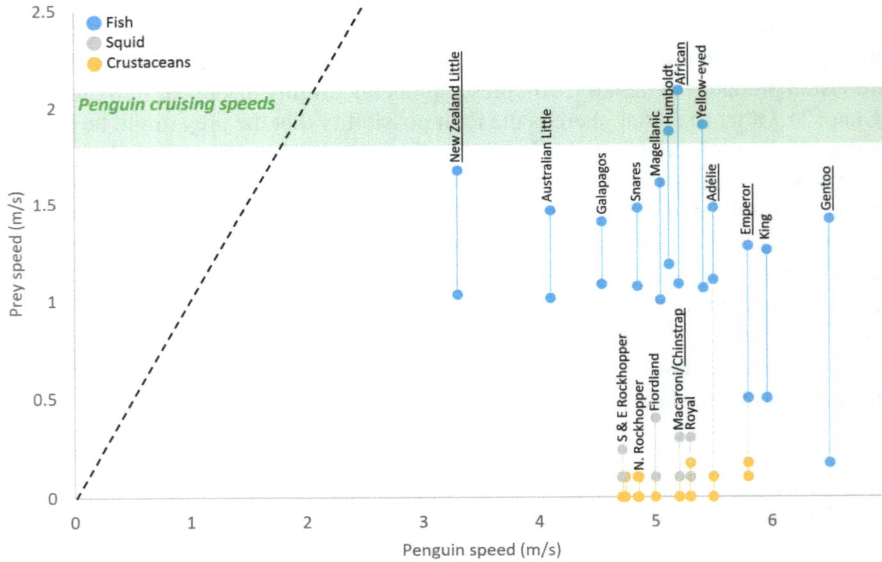

Fig. 11.3 The relationship between the approximate maximum speed of various penguin species (from published information (in which the species are underlined) and derived using a log relationship with mass based on the species for which data were available) and the approximate maximum speed of their prey (primarily derived for the literature). The dashed line shows equivalence between the penguins and prey and highlights what is normally an enormous difference in maximum speeds between the predator and prey, noting that even penguin cruising speeds generally exceed the maxima of their prey

multiple frantic flips of its tail only to be engulfed in a moment by its high-speed predator as it surges past, rather like a swallow or bat catching an insect.

However, as prey increase in size relative to the penguin, they become, theoretically at least, relatively faster compared to the penguin [15, 16]. This is particularly true for fish, in which even small subadults (though not juveniles or larvae) can reach comparatively high speeds for a short time (a second or two) at least [16, 83], though even these are generally appreciably slower over longer time intervals than penguins (Fig. 11.3).

Critically, when fish execute their absolutely maximum speed movements (e.g., "single-bend" and "double-bend" fast starts [16]), they are only maintained for about one tenth of a second. This allows the fish to move perpendicularly away from an approaching penguin's track by less than 20 cm (derived by multiplying maximum fish swimming speed by the time taken for a maneuver). The turn radius (of a circular path) of fish is independent of velocity [16] but increases with fish length [17, 18], something that generally holds true for aquatic vertebrates [15]. Thus, although larger fish have higher maximum speeds [12], they are predicted to be similar to penguins in their turn capacities. This, though, ignores the turn abilities of penguins, which are spectacular. For example, Clifford Hui analyzed footage of Humboldt penguins underwater to deduce that they can turn 180° (i.e., turn to go back from where they came) within a radius of just 0.14 m (about a quarter of their body length) irrespective of their speed (see Fig. 6.24)! This gives them an astonishing angular velocity of 576 degrees per second [30]! Given that 360° is a full circle, they can theoretically turn a full circle 1.6 times in a second. As such, although such perpendicular turns have real value in the predator–prey pursuits in terrestrial systems [89], they have limited value underwater for penguin prey.

Penguins, however, have one more trick that most other marine predators lack. This effectively renders escape impossible, at least for small prey, and has not been considered within the normal predator–prey literature. It simply involves rapid extension (including sideways) of their long flexible neck [8, 30] (Figs. 6.24 and 11.4). That penguins use this effectively was documented 100 years ago by Douglas Mawson, quoted by Falla [20], who observed Adélie penguins from a cliff vantage point above a shallow bay as follows (p. 76):

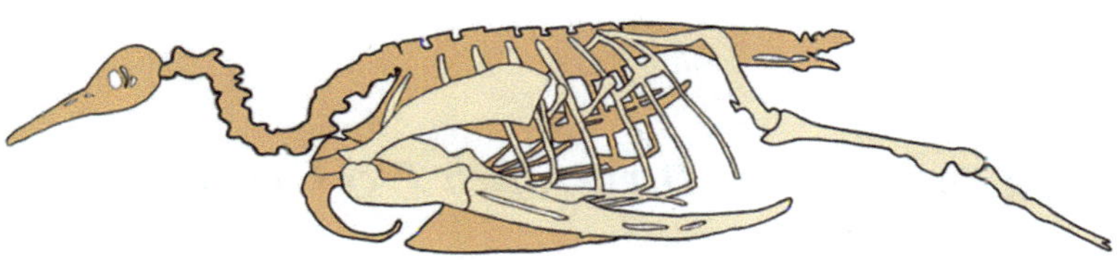

Fig. 11.4 The skeleton of a king penguin showing the extraordinary length of the neck. Although this is normally held bent during swimming, to form the typical penguin "torpedo shape" (Fig. 1.7), it can be substantially extended to capture prey items that attempt to escape by moving rapidly perpendicular to the penguin's track

"As they travelled in a zig-zag course backwards and forwards and round the little bay without slackening pace, one could observe their heads darting from time to time to the left and to the right, evidently picking up euphausians as they moved. In fact, their necks were jerking out and their beaks going just about as fast as a barnyard fowl feeds on grain thrown on the floor".

Such behavior is also apparent in bird-borne video cameras [82] as well as by substantial acceleration "strikes" in head-mounted accelerometers [34].

So, it is the difference in speed between penguin and prey that, first and foremost, facilitates capture. This is because the most sensible option for prey being approached by a penguin is to swim perpendicular to the penguin's track (to swim away or toward the penguin results in inevitable capture). However, the distance that the prey can swim away from the penguin depends on their relative swim speeds and the distance between the two animals. A penguin 4 m away from its prey and approaching at a cruising speed of 2 m/s will close the gap in approximately 2 s, during which time, the prey, depending on its size and taxon (Fig. 11.3), will have covered between 0.2 and 0.5 m to the side (because escape speeds cannot be maintained for more than a fraction of a second). The penguin, tracking this movement, will have altered its trajectory in accord (Fig. 11.5) and so approaches at speed. The best strategy for the prey then, is to wait until the bird is terrifyingly close before executing some lateral escape maneuver (perhaps in the direction it has come) in the hope that the penguin will sweep past and miss it (cf. [89]). However, the chances of this working are diminishingly small (Fig. 11.5) because penguins use their extensible neck to deal with last split-second maneuvers, as noted above. Importantly, the distance that the prey can move laterally between being sighted and caught can be reduced by the penguin swimming faster (Fig. 11.5). This reduces the distance that the penguin has to cover to reach its prey and also diminishes the distance that the prey can cover in its last drastic sideways maneuver, increasing the likelihood that it can be caught within the operational radius of the outstretched neck (Fig. 11.5). In other words, there are excellent reasons for penguins to swim extremely rapidly up to their prey. Against this, is the energetic cost of swimming fast, which rises as the cube of the swimming speed (Chap. 6). So, the strategic speed to approach and catch prey should depend on the energetic cost of swimming and the speed and energetic value of the prey. Perhaps unsurprisingly, there is good evidence that penguins modulate their swim speed according to the escape capacities of their prey [60]. We will revisit this when we examine hunting strategies for penguins catching aggregating prey.

Changing Buoyancy with Depth Affects Prey Capture Strategies

The scenario described above is somewhat simplistic because, although most penguin prey are close to neutral buoyancy [28], penguins are not neutrally buoyant at any depth [90] but most particularly not in the upper water column (Chap. 6). So, the substantial difference between pursuit by cursorial predators on land and

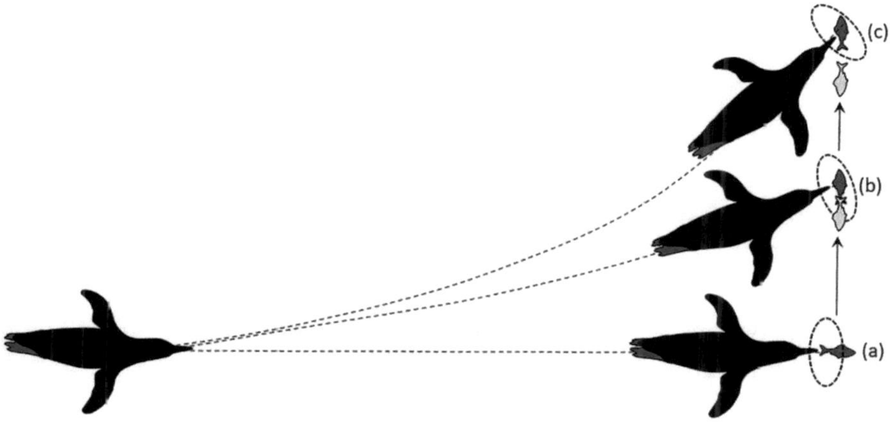

Fig. 11.5 A schematic diagram of solitary prey capture by a penguin. The speed differential between the penguin and its prey is important and is normally extremely substantial (cf. Fig. 11.3). A penguin approaching a prey that (**a**) simply swims away, will catch it as long as the bird swims appreciably faster and will do so with no advantage granted by extension of the neck laterally if necessary (represented by the dashed ovoid). If the prey swims perpendicular to the penguin's initial trajectory, then the bird must track it, continuously altering its course (cf. Tabachnikov [75]). If then, around the point of contact, the prey (**b**) doubles back (lightly shaded fish), then the penguin can still secure it as long as the swim speed of the prey relative to that of the penguin does not allow it to move beyond the action radius of the penguin's neck. If a perpendicularly moving prey swims (**c**) relatively rapidly compared to the penguin, and then doubles back just before contact, the action radius of the penguin neck may not be enough to secure the prey, in which case, the penguin would have to turn to realign with the prey and begin the process again. This last case is most likely for penguins pursuing large (and therefore fast) prey and is the standard situation for terrestrial cursorial predators (cf. [85, 89]), which do not have an option stemming from an extensible neck

penguins underwater is that underwater prey can attempt to evade capture by moving in three dimensions and these dimensions are not equal for the penguin in terms of effort used for pursuit. By opting to move sideways, maintaining constant depth, the prey do not change the physical properties of the pursuing penguin. However, if the prey move down the water column, they not only force the pursuing penguin to work harder against upthrust, because the air associated with their body gives penguins positive buoyancy (Chap. 6), but they also take the penguin farther from the surface. This necessitates time, oxygen, and energy investment by the penguin in order to retrace its steps and return to the surface. It is perhaps for this reason that encounters, in which solitary prey are first sighted by the penguin at greater depths, sometimes involve dramatic vertical descents by the bird (Fig. 11.6) and rapid prey capture. Doing so ultimately reduces the length and time of the return trajectory.

Such prey capture conditions will be particularly onerous for the penguins close to the surface where their upthrust is the highest. For example, at a depth of 3 m, the buoyancy force can be as high, or higher, than the force applied by the penguin to

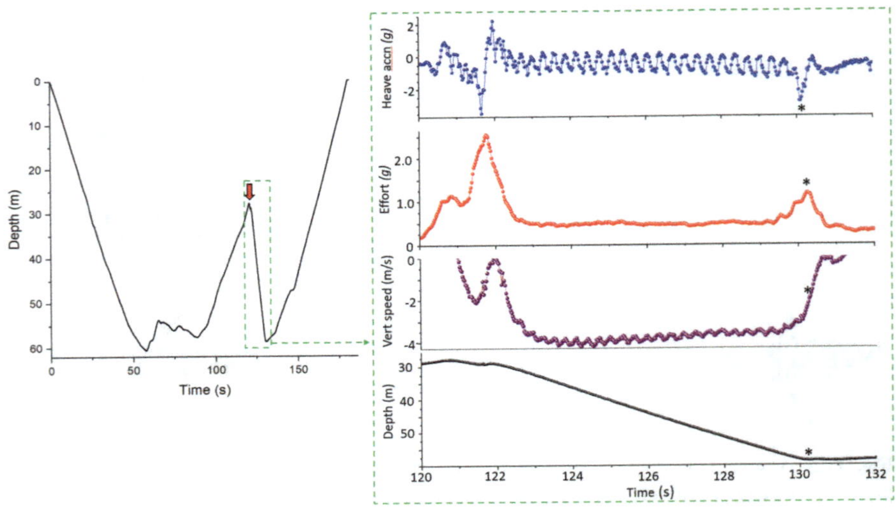

Fig. 11.6 Pursuit by a Magellanic penguin of what was assumed to be a solitary prey that reacted to being spotted by swimming rapidly down the water column. The left-hand panel shows that the penguin was already rising in the water column to terminate the dive when it saw this prey (red arrow in green box), at which point the bird rapidly changed its dive angle to descend sharply at about 55°, traveling at a vertical speed of around 4 m/s and at an absolute speed of 4.6 m/s. The penguin descended some 18 m before it (probably) caught its prey at a depth of 57.1 m (asterisks). The extraordinarily high effort (VeDBA) of the maneuver due to the high flipper beat amplitude and frequency (shown in the acceleration graph) as the bird increased its speed and worked against upthrust should be noted. This extreme reaction is unusual and makes it highly likely that the prey was large and fat (plenty of energy value), despite being fast

swim at normal cruising speeds (Chap. 6), and, therefore, rapid downward movements require penguins to expend dramatic amounts of energy to overcome this predicament.

Alternatively, prey may be encountered between the penguin and the water surface or at intense thermocline- or turbidity boundaries (see Chap. 3 and below), and/or the prey may elect to swim up to avoid capture. This is to the penguin's advantage for the reverse of the reasons stated above. Penguins near the surface can swim up at normal cruising speeds without any effort at all [61], actually increasing speed as they glide closer to the surface (Chap. 6). Certainly, penguins attempting to increase their speed to capture prey above them can do this easily. So, from a force and energy perspective, there are compelling reasons for penguins to catch prey by rising in the water column (the prey are also backlighted against the surface). This seems to be commonly the case for solitary prey items (Fig. 11.7) and may be the norm in aggregating prey (see below). This could be one reason for the "undulations" or "wiggles" in their dive profile, mentioned above.

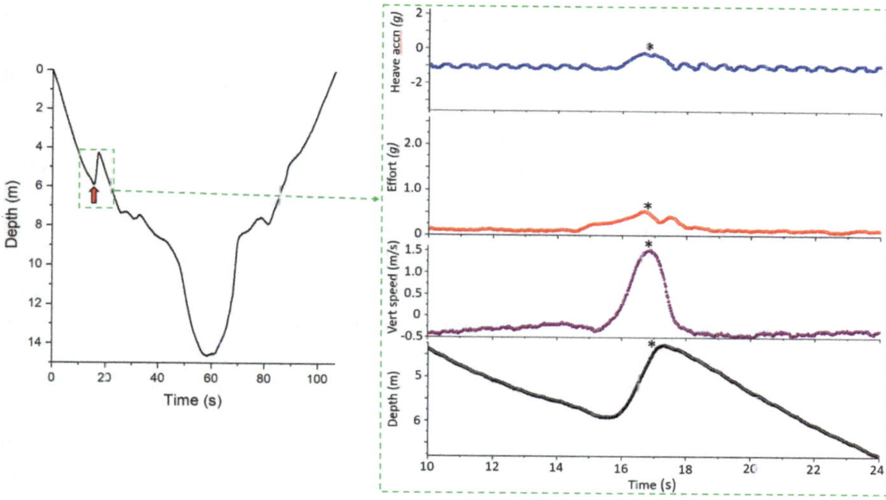

Fig. 11.7 The movement pattern of a Magellanic penguin reacting to (red arrow) and capturing what was assumed to be a solitary prey located above the bird (green dashed box). The inset of the green box shows the detail of the maneuver. It should be noted how (in stark contrast to the situation shown in Fig. 11.6) the capture strategy involved an appreciable increase in vertical speed to 1.5 m/s (and an absolute speed of 1.58 m/s) but involved little effort by the bird, which used upthrust to power its prey capture (indicated by the asterisks). The scaling of the effort and acceleration graphs are set the same as in Fig. 11.6 for comparison

Prey Pursuit Against Interfaces

The scenarios depicted above refer to mid-water pursuits. However, solitary prey associated with another medium may also be encountered by penguins. These media include hard surfaces, such as the seabed, against vertical rock faces, under sea ice, or at the water/air interface on the sea surface. Other such boundaries may be, for example, intense thermoclines, in which oceanic properties (such as temperature, salinity, and oxygen) change dramatically in just a short vertical distance, or turbidity and halocline boundaries related to river plumes or phytoplankton blooms (Chap. 3; see Fig. 9.13). The water properties above or below, or against, such boundaries may dissuade the prey from crossing due to less favorable conditions. Overall, interface hunting would seem to benefit the penguin because the escape movements of the prey are correspondingly constrained.

Where the interface is the seabed, prey might be bottom-living and specifically targeted by benthic-foraging penguins. For example, bottom-dwelling opalfish, flushed by yellow-eyed penguins swimming extremely close to the seabed, can only rise briefly in the water column, following which they appear to be caught extremely quickly by the bird located above and behind them [39]. In contrast, when flushed, much larger blue cod rise to just clear the benthos before swimming rapidly away from the penguin remaining notably parallel to the seabed and engaging in fast turns [39]. The latter is much in the manner of prey evading cheetahs [89]. Inspection of the video footage provided by Mattern et al. [39] indicates that the speed of the fish (by virtue of its size) and the diameter of these turns apparently lead to maneuvers that exceed the reach of the penguin's neck (cf. Fig. 11.5). Thus, chases may be protracted for periods even exceeding 20 s [39]. It should be noted, however, that the attached video recorder may diminish penguin performance [80], at a presumed enormous energetic (and oxygen) cost to the bird (see Chap. 14 for a discussion of how penguin-attached technology is particularly detrimental at high swim speeds). That these fish maintain proximity to the benthos is to their advantage since any upward motion benefits the penguin (see earlier), but it also means that the fish may be able to escape to bottom-associated shelter, if found [39]. As noted in Chaps. 5 and 13, due to competition with fisheries, yellow-eyed penguins have transitioned from a diet of smaller, water-column-inhabiting fish to larger, benthic blue cod. If not digested much (see Chap. 8 for details on how chick-provisioning penguins can delay digestion), some of these fish are too large for chicks to handle, when regurgitated to them by their parents, leading to the chick's undernourishment. They remain on the floor of the nest cavity. However, we might well speculate about the extent to which catching these fish has forced yellow-eyed penguins to operate at a corresponding energetic disadvantage during prey acquisition. Such would add to this endangered penguin species' plight. Seabed interface hunting has been reported for emperor, gentoo, rockhopper, and African penguins with birds executing what has been termed "benthic" dives [24, 37, 55, 73, 79]. The implication is that they capture prey associated with the benthos. However, whether these prey are taken from the seabed or whether they are simply demersal, being associated with the seabed although they may be some vertical distance from it (see [71] for an example), is often not clear. Ropert-Coudert et al. have also speculated that little penguins may use the seabed to assist them to trap prey [58].

Belinda Cannell's PhD work on little penguins in captivity [8] demonstrated, among other things, that they were adept at catching solitary prey against the walls of their tanks and that they took about 8 s to do so. This is longer than the prey pursuit durations of these birds in the wild, where they take a mean of 2.9 s (standard deviation (SD) 3.3) to secure prey [58] (but then data from the birds in the wild do not include any time for perception of prey). In the same way as penguins taking prey from the seabed, vertical walls presumably enhance prey capture because they constrain the movement of the prey. The ease with which the penguins did this in Belinda's study [8] raises the intriguing possibility that they might do this in the wild, perhaps using flotsam and jetsam, which are known to aggregate fish anyway, or rocky vertical surfaces. Similarly, some penguin species, such as the Galápagos penguin, regularly hunt in shallow waters adjacent to vertical surfaces, where similar advantages may apply (Fig. 11.8).

Emperor, and to a lesser extent Adélie penguins, have been documented to capture prey (especially bald rockcod) from the underside of floating ice. They do this by simple pursuit in a manner analogous to benthic-foraging penguins, such as the

Fig. 11.8 Left, Galápagos penguins routinely forage close to the shore [72] and by so doing may be able to use rocky outcrops to constrain the escape movements of their prey (photo courtesy of Tui De Roy). Right, although many studies assume that penguins only catch their prey in water deeper than 2 m, high-resolution data, including information captured by video recorders, show that this is not true. Here, a Magellanic penguin hunts anchovies close to the surface, which acts as an uncrossable interface for the fish. In addition though, if the penguin has provoked the fish to rise in the water column by herding, then the expansion of the swim bladder with decreasing depth will tend to make the fish positively buoyant. This, because they will have to use energy to counteract this force, will compromise their allocation of power to escape maneuvers and make them easier for the penguin to catch

yellow-eyed penguins, or by effectively cornering the fish in ice holes before simply plucking them out (Chaps. 3 and 4). Interestingly, the shallow depths at which this occurs means that the penguin upthrust is high so that the slow speeds used for capturing prey in crevices will be particularly onerous (see Chap. 6) because the body will tend to float up horizontally. Here, again, the extensible penguin neck is likely to be critical in providing the appropriate maneuverability. They make use of their neck action to take these fish from cavities as they pass on the underside of the sea ice and may capture several fish during one dive [52, 82]. On the other hand, these are notothenioid fish, which lack fast-contracting muscles (all they have is "white meat"), and thus cannot swim extremely fast [19].

Although many studies of penguins purport that no foraging occurs in the first couple of meters of the water column, video recorders and other "smart tags" have demonstrated that this is unequivocally not true. Indeed, *Spheniscus* penguins do this regularly (and we have seen it in vivo), with the birds undoubtedly benefitting from using the sea surface as an interface to constrain the escape movements of their prey, thus facilitating their capture (Fig. 11.8). This is particularly obvious when these birds are feeding on aggregating prey (see below) but also occurs for solitary prey items. Presumably, feeding this way has the added advantage that if the prey have been pushed toward the sea surface from depth, they will incur swim bladder expansion, which will require them to work against buoyancy rather than be able use all of their available power capacity to escape (Fig. 11.8).

Capture of Aggregating/Schooling Prey

The ease by which penguins can generally catch their prey when solitary (see above) is reason enough for their food to form aggregations. In fact, an aggregation is a common behavior observed across various animal taxa and is known to decrease the risk of predation. In part, this is due to what is known as the selfish herd effect [36] (as well as sometimes increasing the odds of locating food sources and potential mates [35, 50, 51]). However, not all aggregations function and react in the same way when it comes to minimizing predation, a particularly obvious dichotomy being between swarming crustaceans [44, 45] and pelagic school fish [47, 57] (see Chap. 3 for penguin diet). Although some similarities are apparent [44], we treat these two groups separately below.

Crustaceans

There is little information on the swarming behavior of many of the crustaceans eaten by penguins, with the notable exceptions of the two Antarctic krill species, Antarctic krill and crystal krill. Both have been intensively studied, particularly Antarctic krill owing to its commercial value (Chaps. 5 and 13). Data indicate that they both form swarms of highly variable density, ranging between 4 and 804 individuals per m^3 [5] although horizontal echo sounders indicate that the frequency distribution of krill density follows a power law [68]. This density variation in Antarctic krill may well be mirrored by other crustacea captured by, for example, *Eudyptes* penguins (e.g., *Euphausia vallentini*, *Thysanoessa gregaria*, no common names; Appendix Table A2). Such density variation poses an interesting problem for penguins because decisions have to be made about the density at which birds should stop searching for prey and forage. We have no immediate answer as to when this should occur. We speculate that, as with decisions on which prey types to pursue (e.g., fish vs. squid vs. crustaceans, see above), it is likely that penguins will base their decisions on their perceived abundance of krill in the foraging area, taking those from lower-density swarms when high-density aggregations are not, or rarely, encountered. Scientists using beak angle sensors [76] and bird-mounted videos [27] note capture rates that vary between 2 and 150 per dive for chinstrap penguins [76] and up to 2 per second for Adélie penguins [82]. Thus, it is clear that both these penguin species exploit high densities. A similar high ingestion rate was noted by Falla (see above). At a swim speed of around 2 m/s, this results in them catching a single prey item for every meter swum. However, it has also been noted that Adélie penguins at least, which swim at ca. 2.1 m/s during nonfeeding dives [56], reduce their speed to an average of 1.71 m/s when acquiring krill swarms [56], even dropping to modal swim speeds of 1.4 m/s in the case of some swarms [92]. It may be that variation in krill density is responsible for this. There is little point in penguins swimming fast through an aggregation of slowly moving crustaceans when the density is so high that the encounter rate with individual krill exceeds the processing rate of the penguin [92]. Indeed, the exponentially rising power costs that penguins have to pay with increasing speed (Chap. 6) are a compelling reason for birds to slow down under such conditions. Not only does the drop in speed not affect the rate at which krill are captured but also the energetic, and therefore oxygen, savings enable a slower-moving bird to stay longer underwater exploiting the aggregation.

Of course, krill that are exploited flee from their predators (cf. [44]), including research and fishing trawls, but do so in a simple manner that basically follows the "everyone-for-themselves" rule used by people fleeing from a burning cinema. So, individual penguins swimming through a krill aggregation create cylinders of reduced prey density around the penguin path. However, this also plays into the strategy of synchronously foraging penguins [27] because the krill fleeing from one bird may move into the path of another. Indeed, diving synchronicity has been observed in Adélie [1, 77], African [66], little [3], king [53], and rockhopper penguins [78], with this as a presumed advantage [40, 78, 95] (see penguin flocking in Chap. 14). Video data on two simultaneously foraging chinstrap penguins showed that the numbers of krill captured during dives were positively correlated between the two birds [27], indicating that the synchronicity enabled similar capture efficiencies within the krill swarms they encountered.

As part of the manner in which prey are caught, it has been noted that Adélie penguins exhibit characteristic undulations during the bottom phase of the dive associated with prey ingestion [56], including documenting rapid accelerations and decelerations: 43% of such accelerations were followed by one or more ingestion events with birds tending to swim upward (60% of the accelerations). Our data for Adélie penguins also show undulations (Fig. 11.9), but these are minimal compared to fish-eating dives (see below) and are unlikely to be linked to one undulation representing one prey item captured [82]. Such had been earlier suggested, albeit for fish-eating penguins [67]. At sites where emperor penguins capture some krill, they tend to dive relatively shallowly (for them), exploiting depths of less than 150 m. In these, they have been calculated to catch 115 individual krill per dive [32], though, as far as we know, nothing is known about their specific prey capture strategy. Moreover, nothing is known about how many more energy-dense fish they might catch on these dives (in terms of energy, one fish is the equivalent of many krill; Chap. 4). Most likely, when catching such krill, they act in much the same way as Falla [20] described for Adélie penguins (see above).

Fish

Conveniently, Antarctic silverfish, a herring-like, energy-rich, loosely schooling fish is also captured by Adélie and emperor penguins, which gives us an opportunity to see how penguins change their capture strategies according to prey type and, in this case, to highlight the substantial difference between densely packed crustaceans and loosely shoaled fish (2–4-m apart [22]). The most marked difference in the

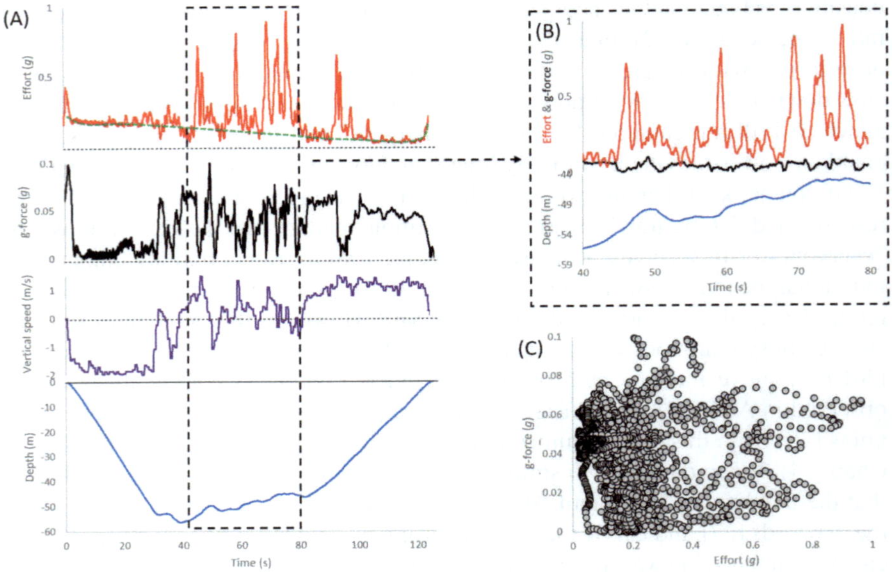

Fig. 11.9 (**a**) A prey capture dive assumed to represent exploitation of crystal krill by an Adélie penguin, showing the overall dive profile (blue), the vertical speed (in purple), the *g*-force (VeSBA+) (in black), and effort (VeDBA) (red). The green dashed line in the effort graph shows the approximate expected values in the absence of prey capture strategies, demonstrating that, in this case, the penguin was generally swimming faster than normal, although at times it swam slower. (**b**) Shown here is 40 s of an inset of the dive during which time the bird was catching prey. The weak undulations in the depth profile and the pulses of effort but only gentle turning maneuvers (extremely weak pulses of *g*-force – cf. Fig. 11.10), presumably due to birds turning in both the vertical and horizontal axes, as the bird moves through the krill swarm should be noted. (**c**) Bird effort and *g*-force are not linked across the whole dive, presumably because the turns are not sharp or fast

format of the prey exploitation dives by Adélie penguins is in the dynamism of the birds catching fish and the substantive changes in depth, producing the characteristic "undulations" in the dive profiles on which so many penguin researchers place importance (Fig. 11.10). Most interestingly in this case, the highest vertical speeds occur during the descent phases of the undulations, by which the bird would have had to work particularly hard against upthrust (although substantially reduced within the 50–70-m depth range used).

The strong implication is that (multiple) fish would have been caught as the penguin engaged in descents (otherwise it would have ascended faster and descended at a more measured pace – see below). This contradicts what has been suggested as the norm for many penguin species (e.g., [10, 67]). The fact that notothenioid fish, including silverfish, possess only white muscle and thus are slow swimmers, plus the fact that the cold water might slow them further, may allow the penguins to approach them from above without spooking them.

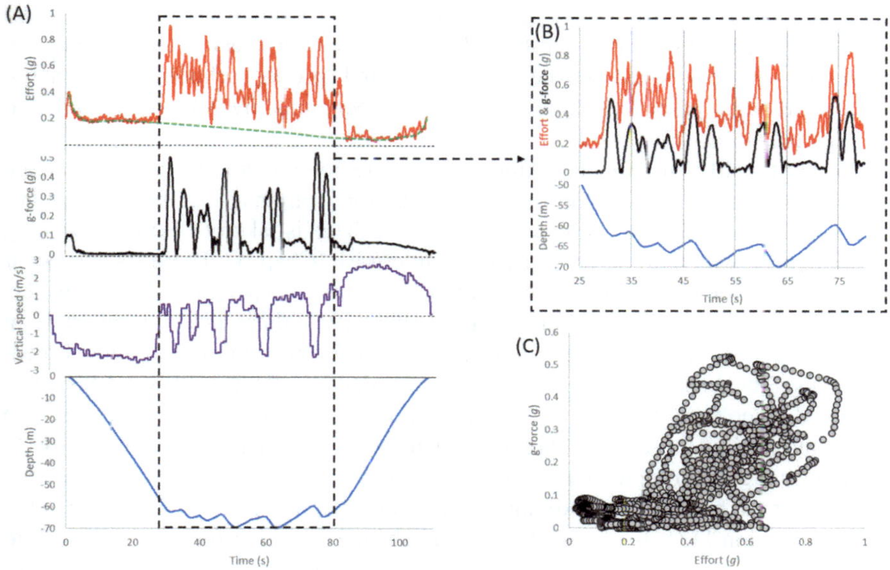

Fig. 11.10 (**a**) An example dive by an Adélie penguin presumed to be acquiring Antarctic silverfish showing the overall dive profile (blue), the vertical speed (in purple), the *g*-force (VeSBA+) (in black), and bird effort (VeDBA) (red). The green dashed line in the effort shows the approximate expected values in the absence of prey capture strategies and illustrates how, at all times during feeding, the bird was likely traveling much faster than its normal cruising speed. (**b**) Shown here is 55 s of an inset of the dive during which time the bird was catching prey. The clear undulations in the depth profile and the pulses of effort juxtaposed with radical turning maneuvers (pulses of *g*-force), primarily in the vertical axis, as the bird lunged up and down the water column at high speed (also apparent in the high and variable vertical speeds) should be noted. (**c**) The extent to which effort and *g*-force are linked across the whole dive because without the hard work involved by flipper beating, there is little speed. Thus, fast turns cannot take place. This is particularly the case at the appreciable foraging depth of this bird because the upthrust is greatly reduced

Emperor penguins feeding on Antarctic silverfish represent perhaps one of the most enigmatic cases of prey capture strategies. This is most particularly the case when they engage in their deep dives, which may be several hundred meters [32, 33, 84] and presumably conducted in an extremely light-limited environment [97]. What are assumed to be their feeding dives have clear undulations in the dive profiles [32], although these can be huge compared to Adélie penguins, with amplitudes of 25 m [98]. This makes a case for "rushing" at the prey, perhaps in a similar manner to Adélies (see above). However, whether they catch them on the descent or the ascent is unclear.

The bald notothen, another prey species captured by emperor penguins (Chap. 4), is similar to the Antarctic silverfish, in that it at times forms loose, ill-defined schools. They live underneath the sea ice but also within the water column down to several hundred meters [42]. Although these fish associate with each other, they are only loosely connected and do not possess the sophisticated antipredator communal response behaviors shown by the school fishes captured by *Spheniscus* and *Eudyptula* penguins. In this sense, and in their predator evasion tactics, they are likely to come somewhere between crustaceans and "proper" schooling fish (anchovies, sardines), probably being captured by penguins by simple pursuit (cf. Fig. 11.10).

Part of the difficulty in attempting to formulate notothenioid responses to attack by penguins lies in their "inconvenient" distribution, both in terms of depth and latitude, which makes them challenging to observe. This is taken to extremes in king penguin prey, which primarily consists of lanternfish (Chap. 4). These species occur in extremely loose, monospecific shoals [31, 96], in deeper mesopelagic waters [48]. During the day, when king penguins hunt [91], these fish are generally at depths in excess of 100 m, often hundreds of kilometers from the penguins' breeding sites. The lack of light in these gloomy depths effectively precludes bird-mounted video recorders (for the moment anyway), which is a pity since these fish adopt interesting loose aggregations, which may be a challenge for the birds to exploit (though then again, maybe not!) [64]. Such a state is possibly what is behind, at least in part, by their interindividual light-emitting photophores believed to aid in communication [13]. It seems that lanternfish may also use their photophores to render them inconspicuous to the predators beneath them because they emit light of variable intensities according to the intensity of downwelling light from the surface – a sophisticated form of counterillumination [11, 46]. How all this plays out in terms of king penguin fish capture strategy and fish escape responses is unclear, and it is a significant penguin challenge to find out. Work has shown, however, that king penguins acquiring these fish dive deep, employ undulations ("ragged bottoms") in the depth profile [4], have appreciable surges in acceleration during prey pursuit [6], and travel at both high and low swim speeds, with a mean of 1.91 m/s (SD 0.39) [92]) with variable vertical speeds and specific "rushes" at prey [59]. This, together with work using accelerometers, indicates that these penguins average 3.9 s (SD 3.5) per prey pursuit [6], with possible ingestion rates of between 0 and 12 per dive [4].

The textbook mid-water school fish such as anchovies, sardines, and sprats, the principal prey of the *Spheniscus* and *Eudyptula* penguins (Chap. 4), are fast swimmers (mostly "dark muscle"). Thus, they are able to exhibit some of the most sophisticated antipredator tactics on the planet, involving complex maneuvers such as "flash expansion," "fountain behavior," etc. [47]. These behaviors, with the fish also being "polarized" (all adopting the same orientation), necessitate that school members be appropriately spaced, move at similar speeds, and be aware of what their neighbors are doing [43]. Within large schools, appropriate spacing also prevents all the oxygen dissolved in the water within the school from becoming exhausted. Polarized fish are extremely hard for predators to catch [2]. This well-organized system breaks down, however, if the inter-fish distance reduces too much (cf. [21]), and this is what many predators, including penguins, seek to achieve in their prey capture strategy. They can do this by corralling the fish so tightly that the whole school volume decreases to a critical point, whereby the fish become depolarized and thereby easier to catch [93]. Does being oxygen-deprived, too, slow them down? This depolarization, however, can only happen if the school is relatively small because otherwise the predators cannot encircle them fast enough. Alternatively, or additionally, the predators can swim so rapidly in lunges straight through the school that the fish become locally depolarized and are also easier to catch. We have seen the corralling process firsthand, twice, after hearing about it extensively from local fisherfolk. On one occasion, a lone African penguin was swimming rapidly round a small (ca. 0.5-m diameter) school of mullet close to the sea surface. As it did this, the school compressed until the fish were milling frantically in the middle, whereupon the penguin surged through the middle of them, catching a fish as it did so, swallowing it immediately before continuing the process. It did this a few times while remaining underwater, before it eventually surfaced and the school dived and disappeared. We have observed an almost identical behavior in a Magellanic penguin, although it was not clear which fish species was being exploited. Similarly, little penguins have been observed to swim circles round small fish schools before driving through them to capture single individuals [65].

That fish schools become compressed because peripheral school members see and react to a circling penguin (or penguins, plus other predators) indicates a possible advantage if penguins were strikingly marked because being more visible might induce a more rapid inward movement of the outside fish, leading to depolarization [93]. Accordingly, we used captive Cape anchovy in a circular tank and whirled variously colored penguin models round the periphery of the tank. The models used the colors of krill-eating penguins (black dorsally and white ventrally (Fig. 11.11a)), *Spheniscus* penguins (colored the same as the krill eaters but with the characteristic lateral black stripe running longitudinally (Fig. 11.11b)) and using a transparent penguin model. The results were unequivocal. The transparent model produced no observable reaction from the fish, and the krill-eating penguin model

Fig. 11.11 A marked difference exists between the overall coloration of most penguin species and the school fish-specializing *Spheniscus* penguins (Fig. 1.7). While all other penguin species conform to the commonly seen pattern of dorsal dark and ventral light coloration such as (**a**) the macaroni penguin, *Spheniscus* penguins, such as (**b**) this African penguin, have an additional lateral stripe running the length of their right. There is evidence that this eye-catching feature may help them catch densely schooling fish by inducing depolarization, rendering the prey easier to catch. (Photographs: macaroni penguin courtesy of iStock; African penguin courtesy of Rory Wilson)

produced typical predator escape responses with the school remaining polarized. The *Spheniscus* penguin model, however, produced extensive depolarization [93].

Thus, the striking pattern on adult *Spheniscus* penguins would seem to help them catch prey. This does, though, lead to the intriguing question as to why neither juvenile *Spheniscus* penguins nor adult *Eudyptula* penguins have lateral stripes. In one of the earliest studies of African penguin diet in which the birds were simply shot at sea, juveniles were found to capture much slower-moving prey than the adults [54]. Thus, it is possible that they simply cannot swim fast enough to corral pelagic, high-speed fish schools. In the case of *Eudyptula* penguins, their markedly lower maximum swim speeds compared to that of the other penguin species (Fig. 11.3) might also have a similar effect. This subject is discussed a bit more in Chap. 14, which also considers inter-penguin communication at sea helped by bold coloration.

Dead-reckoning tags attached to Magellanic penguins show dive trajectories that provide compelling evidence of corralling behavior, with birds executing similarly shaped dives in 3D space repeatedly in one spot (Fig. 11.12). In this, the birds descend the water column describing a roughly even, constant diameter spiral before, at the maximum depth, turning and beginning an irregular ascent. The latter exhibits multiple undulations and rapidly changing vertical speeds, vertical accelerations with heightened effort, and *g*-force values.

In such cases, it is notable that the penguins always seem to approach the school from the same direction, moving toward the sun. Penguins often approach food aggregations from underneath, although some observations on little penguins report that they attack fish schools from the side [65]. Viewed from beneath, the prey silhouettes are seen against the surface [59]. Thus, sun orientation may enhance that process and/or may make the penguins more difficult for the fish to see. For the little penguin strategy to work, however, implies that the birds are cognizant of sun direction (which varies, of course, with the time of day). The process of using backlighting may be enhanced by the dark face mask of most penguins, something that is particularly apparent in *Spheniscus* penguins. Certainly, bite marks on African penguin prey indicate that they are invariably taken from underneath [86], something that is true for many other diving seabirds. The other notable thing about the bite marks on the African penguin prey was that all of the fish examined had the bite on, or directly behind, the opercula, concentrated on the soft body parts containing the heart and viscera (Fig. 11.13). Presumably, such a bite disables the fish most effectively. What this also means is that most prey capture is so assured that penguins can even choose where, precisely, to make their initial snap.

The details of movement involving the final approach used by *Spheniscus* penguins to secure prey have been particularly well-revealed using multichannel loggers that show how the behaviors adopted make energetic and strategic sense. As indicated above, it is in the penguin's interest to be moving at a much greater speed than the fish to maximize the chances of capture. So, typically, the penguin associated with a fish school may reside for a certain time at a fairly constant depth before lunging up through the school (Fig. 11.14), increasing its vertical acceleration and speed by both higher-frequency flipper beating and greater flipper beat amplitude. Both of these increase the effort signal (VeDBA), a measure of how hard the

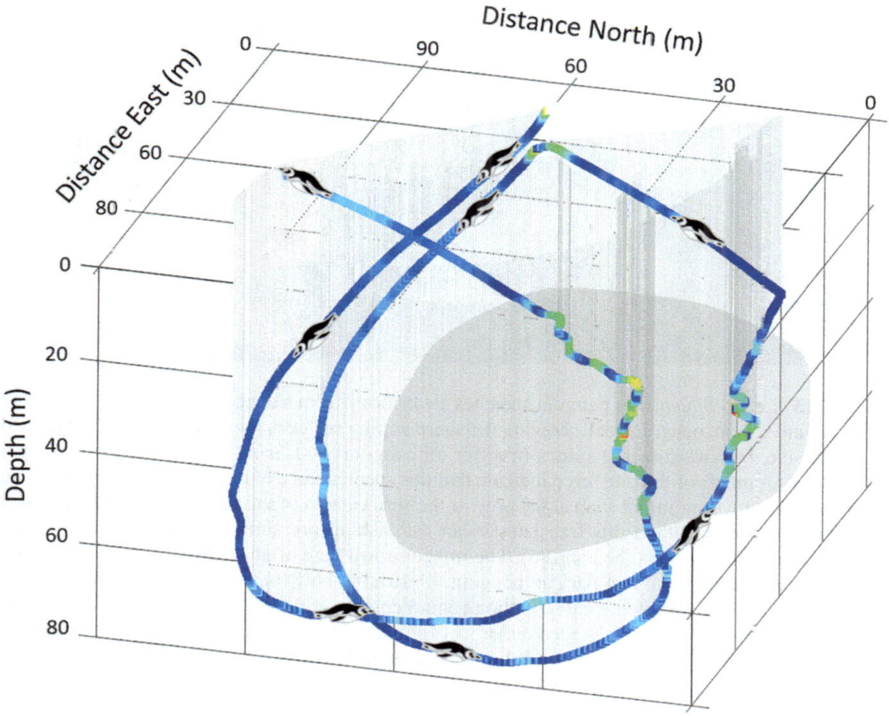

Fig. 11.12 The 3D structure of two successive dives (in a series of 16) during which a Magellanic penguin from a southern Argentina colony, apparently corralled fish (most likely anchovies; Appendix Table A2). The bird descended the water column before surging through the middle of the school, presumably catching prey on the way back to the surface (indicated by warmer colors, which show effort (VeDBA)). The fact that the circle diameter in this case was about 70 m indicated that the fish school (with a position indicated by the shaded gray area) was most likely being acquired by multiple penguins and/or possibly other species. It should be noted how the return to the surface direction was constant across both dives shown – and was across all dives in this herding series

penguin is working to increase its speed. However, by adopting a steep ascent angle, the bird also greatly benefits from the upthrust force (primarily generated by the air in its feathers and lungs; Chap. 6) that helps accelerate the bird even more at no extra energetic cost. The prey is snapped from below at the point of maximum vertical velocity, after which the bird, swallowing the prey, immediately changes direction and, using the momentum derived from its speed, generally heads rapidly down again. This turn is apparent in the centripetal acceleration, akin to driving round a corner at speed, and is manifest as a radical change in *g*-forces and some flipper beating (manifest by the effort metric; Fig. 11.14). Once the penguin has reached its approximate start depth, the process begins again, although in the end phase of dives, birds simply lunge upward (Fig. 11.15). The pattern is therefore fairly

Fig. 11.13 Left, a *Spheniscus* penguin snatches an anchovy from a depolarizing school. Right, the inside of an African penguin beak showing the sharp cutting surfaces (tomia) at the beak edges and the extensive, backward-facing spines (true for all penguins and many other diving seabird species). Measurements of the bite force indicate that this species can exert forces equivalent to 0.4 kg at the bill tip with their initial snap at the prey. If the surface area of the cutting edge at the bite site is taken into account, then this translates to an astonishing pressure of about 7 N/mm^2 (1015 pounds per square inch) on the fish [86]. The image shown here of an anchovy indicates the stylized bite marks inflicted by an African penguin, illustrating how the prey are usually taken from underneath (dorsally narrowing cuts) by the penguin coming at the fish perpendicularly. The consistency of bite position, being located on, or just behind the fish's gills is the body region containing vital organs and highlights an extremely effective immobilizing strategy as well as the precision with which these penguins can catch their prey

stylized and is what results in the characteristic undulations in the depth profile already noted by earlier studies [67].

Corralling of schooling prey by *Spheniscus* penguins may also benefit from similar behavior exhibited by other fish-eating species, i.e., cooperative foraging. Certainly, corralling has been demonstrated in many species such as dolphins feeding on pelagic school fish [23], and association with other species can enhance prey capture rates in African penguins [40]. Galápagos penguins feed regularly in mixed-species aggregations, which last longer if aerial predators, in their diving from above, do not participate because this may disrupt the corralled school [41]. These aggregations often also include predatory fish (Fig. 14.4).

Non-Corralling Feeding Behavior

In fact, with the exception perhaps of Galápagos penguins (above), it seems that quite a few prey caught by *Spheniscus* penguins, and probably *Eudyptula* penguins, are not secured by corralling, the extent to which is affected by the number of penguins (and other predators) relative to school size as well as the school size per se. Ideally, a lone penguin attempting to corral a school should be traveling at (cruise) speeds of around 2 m/s, which would allow it to encircle a school of diameter of ca.

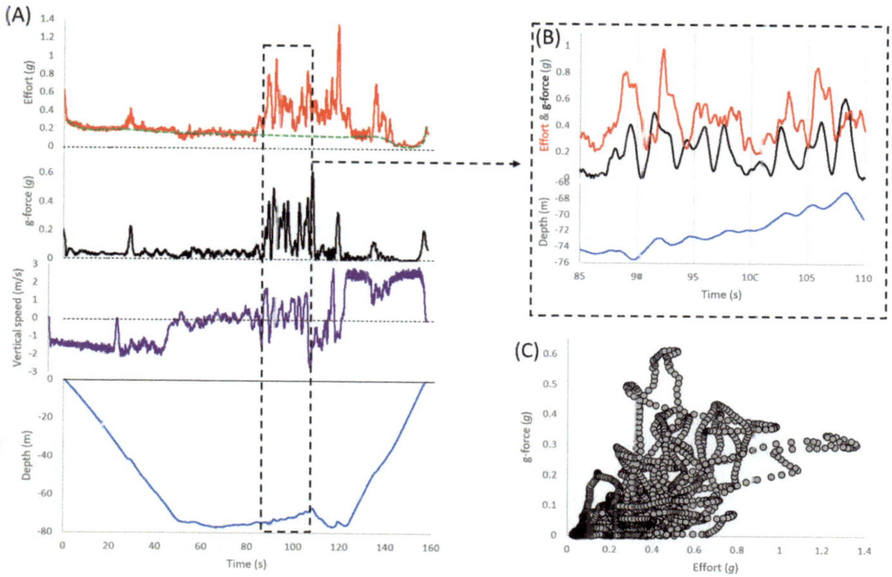

Fig. 11.14 (**a**) Prey capture dive by a Magellanic penguin showing the overall dive profile (blue), the vertical speed (in purple), the *g*-force (VeSBA+) (in black), and effort (VeDBA) (in red). The green dashed line in the effort graph shows the approximate expected values in the absence of prey capture strategies. (**b**) The 15 s during which the bird was catching prey. The undulations in the depth profile and the pulses of "effort" juxtaposed with radical turning maneuvers (shown by substantial pulses of *g*-force), primarily in the vertical axis, as the bird lunges up and down the water column at high speed should be noted. (**c**) The effort and *g*-force are linked across the whole dive because without the hard work involved by flipper beating (effort), there is little speed, and, thus, fast turns (with high *g*-forces) cannot take place. This is, particularly the case at the appreciable foraging depth of this bird because the upthrust is greatly reduced

0.5 m if it did so in a second, one of about 1 m diameter if it did it in 2 s. Many schools encountered by lone penguins are presumably appreciably larger than this. Definitely, corralling requires more than one penguin in most cases (or at least help from other predators such as sea lions and fish). In fact, most dead-reckoning trajectories of Magellanic penguins exploiting anchovies show that the birds are not traveling in circles but moving approximately linearly with small deviations (Chap. 10, Fig. 10.8) and otherwise the same features as noted in corralling prey bar the initial circling (Figs. 11.14 and 11.15). The average rates of prey capture range between 0.4 (SD 0.1) and 1.2 (SD 0.1) items per dive [73] in African penguins (equating to a mean of 10 prey/h, which seems pitifully little). By comparison, we have measured Magellanic penguins catching up to 22 sprats per dive during a 60-s bottom phase, amounting to about one fish every 3 s. Work on little penguins indicates that they exhibit an average 2.17 "pursuits" per dive (SD 1.53), with a maximum of 11 [58]. Importantly, the high speeds used by the penguin species that capture pelagic school fish necessitates high power and results in more rapid usage of oxygen stores than species able to travel slower to capture prey, such as the crustacean-feeding species.

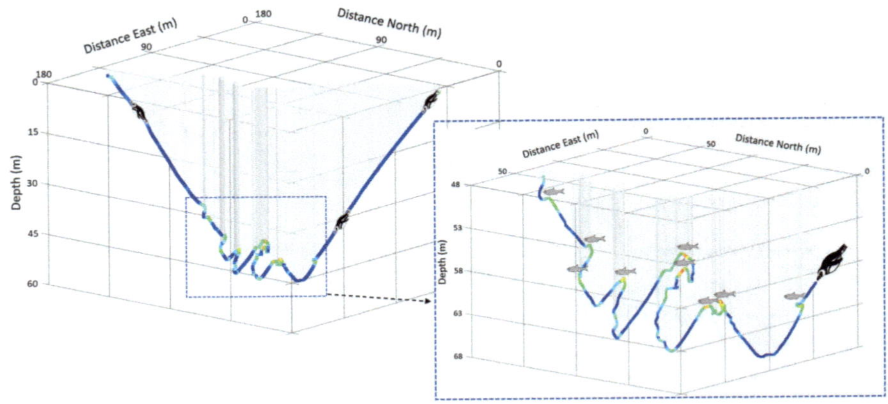

Fig. 11.15 The dead-reckoning track of a Magellanic penguin (colored according to *g*-force – a measure of how fast the bird changes direction – warm colors represent fastest cornering) presumed to be acquiring an anchovy school (or a series of small schools). During the 137-s U-dive, the penguins execute serial vertical lunges (cf. Fig. 11.14), redescending the water column after each lunge. The inset on the right shows 75 s during which the bird apparently caught multiple prey (with probable fish capture spots indicated)

Consequently, overall, "successful" dives for *Spheniscus* and *Eudyptula* penguins to any given depth (because the dive duration is dependent on the maximum dive depth; Chap. 9) are likely to be shorter than "unsuccessful" dives, with the reverse being true for *Eudyptes* and *Pygoscelis* penguins.

Clarity on Limitations of Penguin Underwater Vision

There is compelling evidence that penguins, even deep-diving species such as the emperor and king penguins, capture their prey using vision [9, 29, 38, 70], a subject covered in more detail in Chap. 8). Moreover, multiple lines of evidence indicate that diminished light precludes them from foraging, either at various times of the day/night or at greater depths [91, 97] (Chap. 8). That is not the same as penguins hesitating to dive into the ocean when they cannot see (an antipredator strategy; Chap. 12). Given that many penguin prey species engage in diel vertical migration, whereby they move from deeper waters during the day to near the surface at night to feed, which is supposedly a tactic to avoid predation [49], what is lacking in our understanding of penguin foraging strategies is precisely how operating at reduced light levels affects the efficiency of prey detection and capture by penguins and how it may affect their specific strategies. A tantalizing start in this direction has been made by identifying the number of prey captures of king penguins as a function of depth (and therefore light) [6] – where it was noted that shallower captures of prey took less time. Ultimately, we will need to use penguin-attached sensors recording light intensity, depth, effort, and prey ingestion as well as the necessary sensors for effective dead reckoning [25]. In doing so, we might better understand how water turbidity, time of day, depth, and prey density interact to compromise or enhance penguin foraging success.

References

1. Ainley D (1972) Flocking in Adélie penguins. Ibis 114(3):388–390
2. Becco C, Vandewalle N, Delcourt J, Poncin P (2006) Experimental evidences of a structural and dynamical transition in fish school. Phys A Stat Mech Appl 367:487–493
3. Berlincourt M, Arnould JP (2014) At-sea associations in foraging little penguins. PLoS One 9(8):e105065
4. Bost C-A, Handrich Y, Butler P, Fahlman A, Halsey L, Woakes A, Ropert-Coudert Y (2007) Changes in dive profiles as an indicator of feeding success in king and Adélie penguins. Deep-Sea Res II Top Stud Oceanogr 54(3–4):248–255
5. Brierley AS, Cox MJ (2010) Shapes of krill swarms and fish schools emerge as aggregation members avoid predators and access oxygen. Curr Biol 20(19):1758–1762
6. Brisson-Curadeau É, Handrich Y, Elliott KH, Bost C-A (2021) Accelerometry predicts prey-capture rates in the deep-diving king penguin *Aptenodytes patagonicus*. Mar Biol 168(10):1–10
7. Cairns DK, Gaston AJ, Huettmann F (2008) Endothermy, ectothermy and the global structure of marine vertebrate communities. Mar Ecol Prog Ser 356:239–250
8. Cannell B (1994) The Feeding Behaviour of Little Penguins, *Eudyptula minor* (Forster). Monash University
9. Cannell BL, Cullen J (1998) The foraging behaviour of little penguins *Eudyptula minor* at different light levels. Ibis 140(3):467–471
10. Carroll G, Slip D, Jonsen I, Harcourt R (2014) Supervised accelerometry analysis can identify prey capture by penguins at sea. J Exp Biol 217(24):4295–4302
11. Case JF, Warner J, Barnes AT, Lowenstine M (1977) Bioluminescence of lantern fish (Myctophidae) in response to changes in light intensity. Nature 265(5590):179–181
12. Cloyed CS, Grady JM, Savage VM, Uyeda JC, Dell AI (2021) The allometry of locomotion. Ecology:e03369
13. Davis MP, Holcroft NI, Wiley EO, Sparks JS, Smith WL (2014) Species-specific bioluminescence facilitates speciation in the deep sea. Mar Biol 161(5):1139–1148
14. Del Caño M, Quintana F, Yoda K, Dell'omo G, Blanco GS, Gómez-Laich A (2021) Fine-scale body and head movements allow to determine prey capture events in the Magellanic Penguin (*Spheniscus magellanicus*). Mar Biol 168(6):1–15
15. Domenici P (2001) The scaling of locomotor performance in predator–prey encounters: from fish to killer whales. Comp Biochem Physiol A Mol Integr Physiol 131(1):169–182
16. Domenici P, Blake R (1997) The kinematics and performance of fish fast-start swimming. J Exp Biol 200(8):1165–1178
17. Domenici P, Blake RW (1993) The effect of size on the kinematics and performance of angelfish (Pterophyllum eimekei) escape responses. Can J Zool 71(11):2319–2326
18. Domenici P, Blake RW (1993) Escape trajectories in angelfish (*Pterophyllum eimekei*). J Exp Biol 177(1):253–272
19. Eastman JT (1993) Antarctic fish biology: evolution in a unique environment. Academic, San Diego
20. Falla RA (1937) Birds. B.A.N.Z. Antarctic Research Expedition 1929–1931. Rep Ser B 2:1–288
21. Faucher K, Parmentier E, Becco C, Vandewalle N, Vandewalle P (2010) Fish lateral system is required for accurate control of shoaling behaviour. Anim Behav 79(3):679–687

22. Fuiman L, Davis R, Williams T (2002) Behavior of midwater fishes under the Antarctic ice: observations by a predator. Mar Biol 140(4):815–822
23. Gazda SK, Connor RC, Edgar RK, Cox F (2005) A division of labour with role specialization in group–hunting bottlenose dolphins (*Tursiops truncatus*) off Cedar Key, Florida. Proc R Soc B Biol Sci 272(1559):135–140
24. Goetz KT, Mcdonald BI, Kooyman GL (2018) Habitat preference and dive behavior of non-breeding emperor penguins in the eastern Ross Sea, Antarctica. Mar Ecol Prog Ser 593:155–171
25. Gunner RM, Holton MD, Scantlebury MD, Van Schalkwyk L, English HM, Williams HJ, Hopkins P, Quintana F, Gómez-Laich A, Börger L (2021) Dead-reckoning animal movements in R–A reappraisal using Gundog. Anim Biotelemetry 9:23. https://doi.org/10.1186/s40317-021-00245-z
26. Hanuise N, Bost C-A, Huin W, Auber A, Halsey LG, Handrich Y (2010) Measuring foraging activity in a deep-diving bird: comparing wiggles, oesophageal temperatures and beak-opening angles as proxies of feeding. J Exp Biol 213(22):3874–3880
27. Hinke JT, Russell TM, Hermanson VR, Brazier L, Walden SL (2021) Serendipitous observations from animal-borne video loggers reveal synchronous diving and equivalent simultaneous prey capture rates in chinstrap penguins. Mar Biol 168(8):1–12
28. Hoar J, Sim E, Webber D, O'dor R (1994) The role of fins in the competition between squid and fish. In: Mechanics and physiology of animal swimming. Cambridge University Press, Cambridge, pp 27–43
29. Howland HC, Sivak JG (1984) Penguin vision in air and water. Vis Res 24(12):1905–1909
30. Hui CA (1985) Maneuverability of the Humboldt penguin (*Spheniscus humboldti*) during swimming. Can J Zool 63(9):2165–2167
31. Hulley PA, Pa H (1981) Results of the research cruises of FRV "Walther Herwig" to South America. LVIII. Family Myctophidae (Osteichthyes, Myctophiformes). Arch FischWissenschaft 31:1–300
32. Kirkwood R, Robertson G (1997) The foraging ecology of female emperor penguins in winter. Ecol Monogr 67(2):155–176
33. Kirkwood R, Robertson G (1997) Seasonal change in the foraging ecology of emperor penguins on the Mawson Coast, Antarctica. Mar Ecol Prog Ser 156:205–223
34. Kokubun N, Kim J-H, Shin H-C, Naito Y, Takahashi A (2011) Penguin head movement detected using small accelerometers: a proxy of prey encounter rate. J Exp Biol 214(22):3760–3767
35. Krause J, Ruxton GD, Ruxton G, Ruxton IG (2002) Living in groups. Oxford University Press, Oxford
36. Lehtonen J, Jaatinen K (2016) Safety in numbers: the dilution effect and other drivers of group life in the face of danger. Behav Ecol Sociobiol 70(4):449–458
37. Lescroel A, Ridoux V, Bost CA (2004) Spatial and temporal variation in the diet of the gentoo penguin (Pygoscelis papua) at Kerguelen Islands. Polar Biol 27(4):206–216
38. Martin GR (1999) Eye structure and foraging in King Penguins *Aptenodytes patagonicus*. Ibis 141(3):444–450
39. Mattern T, Mcpherson MD, Ellenberg U, Van Heezik Y, Seddon PJ (2018) High definition video loggers provide new insights into behaviour, physiology, and the oceanic habitat of a marine predator, the yellow-eyed penguin. PeerJ 6:e5459
40. Mcinnes AM, Mcgeorge C, Ginsberg S, Pichegru L, Pistorius PA (2017) Group foraging increases foraging efficiency in a piscivorous diver, the African penguin. R Soc Open Sci 4(9):170918
41. Mills KL (1998) Multispecies seabird feeding flocks in the Galápagos Islands. Condor 100(2):277–285
42. Montgomery JC, Foster BA, Cargill JM (1989) Stomach evacuation rate in the planktivorous Antarctic fish *Pagothenia borchgrevinki*. Polar Biol 9(6):405–408
43. Norris KS, Schilt CR (1988) Cooperative societies in three-dimensional space: on the origins of aggregations, flocks, and schools, with special reference to dolphins and fish. Ethol Sociobiol 9(2-4):149–179

44. O'brien D, Ritz D (1988) Escape responses of gregarious mysids (Crustacea Mysidacea): towards a general classification of escape responses in aggregated crustaceans. J Exp Mar Biol Ecol 116(3):257–272
45. O'brien D (1987) Description of escape responses of krill (Crustacea: Euphausiacea), with particular reference to swarming behavior and the size and proximity of the predator. J Crustac Biol 7(3):449–457
46. Paitio J, Yano D, Muneyama E, Takei S, Asada H, Iwasaka M, Oba Y (2020) Reflector of the body photophore in lanternfish is mechanistically tuned to project the biochemical emission in photocytes for counterillumination. Biochem Biophys Res Commun 521(4):821–826
47. Parrish JK (1991) Do predators 'shape' fish schools: interactions between predators and their schooling prey. Neth J Zool 42(2-3):358–370
48. Perissinotto R, Mcquaid CD (1992) Land-based predator impact on vertically migrating zooplankton and micronekton advected to a Southern Ocean Archipelago. Mar Ecol Progr Ser Oldendorf 80(1):15–27
49. Pinti J, Visser AW (2019) Predator-prey games in multiple habitats reveal mixed strategies in diel vertical migration. Am Nat 193(3):E65–E77
50. Pitcher TJ (1986) Functions of shoaling behaviour in teleosts. In: The behaviour of teleost fishes. Springer, Boston, pp 294–337. https://doi.org/10.1007/978-1-4684-8261-4_12
51. Pitcher TJ (2012) The behaviour of teleost fishes. Springer Science & Business Media, Dordrecht
52. Ponganis P, Van Dam R, Marshall G, Knower T, Levenson D (2000) Sub-ice foraging behavior of emperor penguins. J Exp Biol 203(21):3275–3278
53. Pütz K, Cherel Y (2005) The diving behaviour of brooding king penguins (*Aptenodytes patagonicus*) from the Falkland Islands: variation in dive profiles and synchronous underwater swimming provide new insights into their foraging strategies. Mar Biol 147(2):281–290
54. Rand RW (1960) The distribution, abundance and feeding habits of the Cape Penguin (*Spheniscus demersus*) off the South-western coast of the Cape Province. Division of Fisheries, Cape Town
55. Rodary D, Bonneau W, Le Maho Y, Bost C (2000) Benthic diving in male emperor penguins *Aptenodytes forsteri* foraging in winter. Mar Ecol Prog Ser 207:171–181
56. Ropert-Coudert Y, Kato A, Baudat J, Bost C-A, Le Maho Y, Naito Y (2001) Feeding strategies of free-ranging Adélie penguins *Pygoscelis adeliae* analysed by multiple data recording. Polar Biol 24(6):460–466
57. Ropert-Coudert Y, Kato A, Bost C-A, Rodary D, Sato K, Le Maho Y, Naito Y (2002) Do Adélie penguins modify their foraging behaviour in pursuit of different prey? Mar Biol 140(3):647–652
58. Ropert-Coudert Y, Kato A, Wilson RP, Cannell B (2006) Foraging strategies and prey encounter rate of free-ranging Little Penguins. Mar Biol 149(2):139–148
59. Ropert-Coudert Y, Sato K, Kato A, Charrassin J-B, Bost C-A, Maho YL, Naito Y (2000) Preliminary investigations of prey pursuit and capture by king penguins at sea. Polar Biosci 101
60. Ropert-Coudert Y, Kato A, Sato K, Naito Y, Baudat J, Bost A, Le Maho Y (2002) Swim speed of free-ranging Adélie penguins *Pygoscelis adeliae* and its relation to the maximum depth of dives. J Avian Biol 33(1):94–99
61. Sato K, Naito Y, Kato A, Niizuma Y, Watanuki Y, Charrassin J, Bost C-A, Handrich Y, Le Maho Y (2002) Buoyancy and maximal diving depth in penguins: do they control inhaling air volume? J Exp Biol 205(9):1189–1197
62. Sato K, Shiomi K, Watanabe Y, Watanuki Y, Takahashi A, Ponganis PJ (2010) Scaling of swim speed and stroke frequency in geometrically similar penguins: they swim optimally to minimize cost of transport. Proc R Soc B Biol Sci 277(1682):707–714
63. Sato K, Watanuki Y, Takahashi A, Miller PJ, Tanaka H, Kawabe R, Ponganis PJ, Handrich Y, Akamatsu T, Watanabe Y (2007) Stroke frequency, but not swimming speed, is related to body size in free-ranging seabirds, pinnipeds and cetaceans. Proc R Soc B Biol Sci 274(1609):471–477

64. Saunders RA, Fielding S, Thorpe SE, Tarling GA (2013) School characteristics of mesopelagic fish at South Georgia. Deep-Sea Res I Oceanogr Res Pap 81:62–77
65. Schulz M (1987) Observations of feeding of a Little Penguin *Eudyptula minor*. Emu 87(3):186–187
66. Siegfried W, Frost P, Kinahan J, Cooper J (1975) Social behaviour of Jackass Penguins at sea. Afr Zool 10(1):86–100
67. Simeone A, Wilson RP (2003) In-depth studies of Magellanic penguin (*Spheniscus magellanicus*) foraging: can we estimate prey consumption by perturbations in the dive profile? Mar Biol 143(4):825–831
68. Sims DW, Southall EJ, Humphries NE, Hays GC, Bradshaw CJ, Pitchford JW, James A, Ahmed MZ, Brierley AS, Hindell MA (2008) Scaling laws of marine predator search behaviour. Nature 451(7182):1098–1102
69. Sivak J, Howland H, Mcgill-Harelstad P (1987) Vision of the Humboldt penguin (*Spheniscus humboldti*) in air and water. Proc R Soc Lond Ser B Biol Sci 229(1257):467–472
70. Sivak J, Millodot M (1977) Optical performance of the penguin eye in air and water. J Comp Physiol 119(3):241–247
71. Skaret G, Johansen GO, Johnsen E, Fall J, Fiksen Ø, Englund G, Fauchald P, Gjøsæter H, Macaulay GJ, Johannesen E (2020) Diel vertical movements determine spatial interactions between cod, pelagic fish and krill on an Arctic shelf bank. Mar Ecol Prog Ser 638:13–23
72. Steinfurth A, Vargas FH, Wilson RP, Spindler M, Macdonald DW (2008) Space use by foraging Galápagos penguins during chick rearing. Endanger Species Res 4(1–2):105–112
73. Sutton G, Pichegru L, Botha JA, Kouzani AZ, Adams S, Bost CA, Arnould JP (2020) Multipredator assemblages, dive type, bathymetry and sex influence foraging success and efficiency in African penguins. PeerJ 8:e9380
74. Sutton GJ, Hoskins AJ, Arnould JP (2015) Benefits of group foraging depend on prey type in a small marine predator, the little penguin. PLoS One 10(12):e0144297
75. Tabachnikov S (2009) Chases and escapes. The mathematics of pursuit and evasion by Paul J. Nahin. Math Intell 31(2):78–79
76. Takahashi A, Dunn M, Trathan P, Croxall J, Wilson RP, Sato K, Naito Y (2004) Krill-feeding behaviour in a chinstrap penguin compared to fish-eating in Magellanic penguins: a pilot study. Mar Ornithol 32:47–54
77. Takahashi A, Sato K, Nishikawa J, Watanuki Y, Naito Y (2004) Synchronous diving behavior of Adélie penguins. J Ethol 22(1):5–11
78. Tremblay Y, Cherel Y (1999) Synchronous underwater foraging behavior in penguins. Condor 101(1):179–185
79. Tremblay Y, Cherel Y (2000) Benthic and pelagic dives: a new foraging behaviour in rockhopper penguins. Mar Ecol Prog Ser 204:257–267
80. Vandenabeele S, Shepard E, Grémillet D, Butler P, Martin G, Wilson R (2015) Are biotelemetric devices a drag? Effects of external tags on the diving behaviour of great cormorants. Mar Ecol Prog Ser 519:239–249
81. Watanabe YY, Sato K, Watanuki Y, Takahashi A, Mitani Y, Amano M, Aoki K, Narazaki T, Iwata T, Minamikawa S (2011) Scaling of swim speed in breath-hold divers. J Anim Ecol 80(1):57–68
82. Watanabe YY, Takahashi A (2013) Linking animal-borne video to accelerometers reveals prey capture variability. Proc Natl Acad Sci 110(6):2199–2204
83. Webb P, Corolla R (1981) Burst swimming performance of northern anchovy, Engraulis mordax, larvae. Fish Bull 79:143–150
84. Wienecke B, Robertson G, Kirkwood R, Lawton K (2007) Extreme dives by free-ranging emperor penguins. Polar Biol 30(2):133–142
85. Wilson JW, Mills MG, Wilson RP, Peters G, Mills ME, Speakman JR, Durant SM, Bennett NC, Marks NJ, Scantlebury M (2013) Cheetahs, Acinonyx jubatus, balance turn capacity with pace when chasing prey. Biol Lett 9(5):20130620

86. Wilson R, Duffy D (1986) Prey seizing in African penguins Spheniscus-demersus: Nederlandse Ornithologische Unie C/O Paul Starmans. Oude Arnhemseweg 261:211–214
87. Wilson R, Steinfurth A, Ropert-Coudert Y, Kato A, Kurita M (2002) Lip-reading in remote subjects: an attempt to quantify and separate ingestion, breathing and vocalisation in free-living animals using penguins as a model. Mar Biol 140(1):17–27
88. Wilson RP (1985) Seasonality in diet and breeding success of the jackass penguin *Spheniscus demersus*. J Ornithol 126(1):53–62
89. Wilson RP, Griffiths IW, Mills MG, Carbone C, Wilson JW, Scantlebury DM (2015) Mass enhances speed but diminishes turn capacity in terrestrial pursuit predators. elife 4:e06487
90. Wilson RP, Hustler K, Ryan PG, Burger AE, Noldeke EC (1992) Diving birds in cold water: do Archimedes and Boyle determine energetic costs? Am Nat 140(2):179–200
91. Wilson RP, Puetz K, Bost CA, Culik BM, Bannasch R, Reins T, Adelung D (1993) Diel dive depth in penguins in relation to diel vertical migration of prey: whose dinner by candlelight? Mar Ecol Prog Ser 94:101–104
92. Wilson RP, Ropert-Coudert Y, Kato A (2002) Rush and grab strategies in foraging marine endotherms: the case for haste in penguins. Anim Behav 63(1):85–95
93. Wilson RP, Ryan PG, James A, Wilson M-PT (1987) Conspicuous coloration may enhance prey capture in some piscivores. Anim Behav 35:1558
94. Wilson RP, White CR, Quintana F, Halsey LG, Liebsch N, Martin GR, Butler PJ (2006) Moving towards acceleration for estimates of activity-specific metabolic rate in free-living animals: the case of the cormorant. J Anim Ecol 75(5):1081–1090
95. Wilson RP, Wilson M-P T (1990) Foraging ecology of breeding Spheniscus penguins. Penguin Biol:181–206
96. Zasel'sliy V, Kudrin B, Poletayev V, Chechenin SC (1985) Some features of the biology of Electrona carlsbergi (Taning)(Myctophidae) in the Atlantic sector of the Antarctic. J Ichthyol 25(2):163–166
97. Zimmer I, Wilson RP, Beaulieu M, Ancel A, Plötz J (2008) Seeing the light: depth and time restrictions in the foraging capacity of emperor penguins at Pointe Géologie, Antarctica. Aquat Biol 3(3):217–226
98. Zimmer I, Wilson RP, Beaulieu M, Ropert-Coudert Y, Kato A, Ancel A, Plötz J (2010) Dive efficiency versus depth in foraging emperor penguins. Aquat Biol 8(3):269–277

Part V
Penguins in a Fickle Environment

Chapter 12
Turning the Tables: Fish-Birds on the Menu

"It's much easier to be a predator than it is to be a prey." James Van Der Beek (Film, *Predator*, 1987).

"It is incredible that terrestrial predation [has been proposed] as an important factor in the penguins becoming aquatic [referring to the evolution of penguins]. The most effective enemies of adult penguins are aquatic (especially cetaceans and pinnipeds). That the land is for them safer than the water is deeply ingrained in penguin psychology… they do not flee to water when attacked or alarmed on land, and it is, indeed, almost impossible to drive them into the sea." George Gaylord Simpson [82], p. 90

Basic Law of the Sea: Big Fish Eat Little Fish

In 1556, Pieter Bruegel painted *Big Fish Eat Little Fish*, showing a large quantity of bait fish spilling out of the stomach of a huge predatory fish while a miniaturized person sliced it open, and in a discussion within Shakespeare's play *Pericles* (1608, Act 2), a fisherman expresses to his boss: "Master, I marvel how the fishes live in the sea," and the master replies, "Why, as men do a-land – the great ones eat up the little ones." With such ideas long in our collective wisdom, it thus is not surprising that penguins prey on great quantities of much smaller fish, squid, and pelagic crustaceans, which they are exquisitely adapted to do, as our chapters review (especially Chaps. 4 and 5). On the other hand, penguins are often referred to as "top predators," which clearly they are not. In fact, given that they are preyed upon readily by large fish, seals, fur seals, and whales (see above quote, and below), more correctly put, as they are not at the apex of marine food webs, rather penguins should be referred to as "mesopredators" [70]. While seabirds themselves have exhibited serious decreases in abundance especially owing to depletion of fish stocks, particularly forage fish [33, 63, 64] (Chap. 5), mesopredators have become more ecologically important as the truly top predators have increasingly become reduced in ocean food webs [31]. To some degree, positioning

D. G. Ainley, R. P. Wilson, *The Aquatic World of Penguins*, Fascinating Life Sciences, https://doi.org/10.1007/978-3-031-33990-5_12

penguins as mesopredators would be a bit tenuous since, on the one hand, penguins clearly are the largest, and heaviest of seabirds by a wide margin, but on the other, they are not the largest "fish" in the sea.

Classifying penguins as "top predators" is somewhat surprising since so much lore surrounds the predator–prey relationships between pinnipeds and penguins, especially the leopard seal and the Adélie penguin. Unlike predation at sea in general, the consumption of penguins by seals happens close to shore where ill-adapted (to the sea) humans can observe the process. Thus, it has been a topic of conversation by naturalists involved in the heroic days of Antarctic exploration, and a number of analyses since (see below). Of course, there is also the relationship between skuas and penguins, especially the South Polar Skua and Adélie penguin, in which the skuas take penguin eggs and small chicks. This interaction also has led to much interest and discussion. However, that relationship is not unique to penguins, nor just to penguins and skuas, but includes many seabird species. Since it occurs on land (where penguins feel somewhat safe; see above), however, we will not be covering it here (for the interested, the subject has been well covered elsewhere) [59, 94].

Seals as Predators

The seals that eat penguins are pretty much confined to the sea ice zone that surrounds Antarctica, in which there are four so-called "ice seals" (and we are talking here about true seals, i.e., Phocidae; see below for discussions of sea lions and fur seals): crabeater, leopard, Ross, and Weddell [30, 83, 84]. However, the crabeater seal lacks the dentition to tear flesh from a penguin, and little is known of the diet of the Ross seal. Several anecdotal observations indicate that Weddell seals might eat penguins (Adélie, gentoo, and chinstrap), including fledglings taking their first swim [15, 22, 50, 87]). However, Weddell seals would find eating a penguin problematic, as is indicated by how they daintily strip the flesh from large fish, being careful not to break any teeth on the fish bones – they need their teeth to gnaw breathing holes in sea ice during the winter [10, 85].

The leopard seal, on the other hand (Fig. 12.1), has long been known to hunt and consume polar penguins, and has the right set of teeth to do it (!). Indeed as Murray Levick [51], surgeon on Scott's second expedition, writes (pp. 87–88):

> "Some idea of the depredations committed by these animals [leopard seals] may be gathered from the fact that in the stomach of one which we shot I found the bodies of eighteen penguins, in various stages of digestion, the beast's intestines being literally stuffed with the feathers remaining from the disintegration of many more…. Evidence goes to show that the sea-leopard is the only living enemy, excepting man, that threatens the life of the adult Adélie penguin".

Edward Wilson [91], another medical officer on Scott's second expedition, and an accomplished artist, depicts a leopard seal chasing an emperor penguin in his collection of paintings. It would seem likely that he was depicting an actual event that

Fig. 12.1 A leopard seal. (Photo courtesy of Jean Pennycook)

he witnessed, since leopard seals also prey on other seal species of even larger size. Likely, however, at least chasing a penguin in the open ocean would not have been successful (see below).

Leopard seals have also been observed consuming other penguin species, such as king penguins at South Georgia [75]. The extent of that interaction is hard to ascertain since the leopard seal is closely associated with sea ice, but sea ice does not get very close to South Georgia, its maximum extent reaching waters well to the south of that island (Chap. 3). There are periodic spurts of leopard seals (mostly juveniles) that find their way to Macquarie, another sub-Antarctic island [86], and that phenomenon could be involved in leopard seal occurrence at South Georgia as well as Patagonia [1]. In the Macquarie case, the periodic invasions of leopard seals are related to the Southern Oscillation (see Chap. 13), with sea ice extent in that sector and leopard seal invasion being greater during El Niño (positive Southern Oscillation Index). What is the actual mechanism that leads to more seals visiting Macquarie (and perhaps South Georgia and Tierra del Fuego, too), in years of greater sea ice extent, is not known [86]. The fact that group predation characterizes the penguin–seal interaction at South Georgia, besides photos [75], is indicative that juveniles are involved. As noted below, adult leopard seals hold feeding territories thus spacing out along a penguin beach, and in no way cooperate in their predation.

Seals Hunting Penguins

Leopard seals catch penguins mostly by stealth. In this respect, these animals work like snakes. Most often, the seals wait at ice floe edges or in the water seaward of the ice ridge that piles up along beaches (Fig. 6.15). The ridge, constituting a seawall, is augmented by additional ice layers each time the wind blows the sea ice onshore during winter or spring. Otherwise, any floating sea ice abuts this shelf or foot and rises and falls with the tide. The seawall, resting on the beach, does not rise and fall with the tide. Having been heavily augmented with ice, this ice ridge wall remains along the length of beach long after most of the floating sea ice has disappeared. Leopard seals with just their nostrils above the water surface wait quietly for a penguin to pass close to, if not over, the seal, as the penguins accelerate to jump upward onto the ridge. Lying in shallow water, the seals might submerge for 2 min or so, then stick up their nostrils to breathe before submerging again. At low tide, a penguin might have to leap 2–3 m to get high enough to make it over the wall onto the ridge (Fig. 12.2). Or they may have to access an ice floe that has a high freeboard (a meter or so), which is characteristic of multiyear sea ice. They generally succeed (Fig. 12.3; also Fig. 6.15), but if they fail and fall back into the water, they are taken by leopard seals, if a seal has been lurking at that spot. This is why colony size matters [8]. At a very large colony (say, >300,000 breeding pairs, Cape Crozier), penguins stream to and from the sea by the thousands at any given time using places along the same beach, in what would be called "predator swamping" – too many and too fast for the predator to concentrate on any one individual. The very vast

Fig. 12.2 The low-tide conditions that discourage Adélie penguins from attempting to come ashore, since the height of the beach (in this case ice) is too high for their otherwise extraordinary leaping ability (see Fig. 6.15). During this time, any arriving penguins mill around away from this edge where leopard seals might be lurking and waiting for a penguin to fall back after attempting a leap. Actually in these conditions, the seals go to sleep on ice floes away from the beach because the penguin traffic is greatly reduced in this "cat and mouse" game. (Photo: David Ainley)

Fig. 12.3 The smaller penguin species, in order to come ashore or access an ice floe, at times need to overcome an appreciable height. They can propel themselves a meter or more from the water to achieve this task (see also Fig. 6.15). (Photo: David Ainley)

majority pass safely, with only one in thousands taken, especially given that the seals are spaced out by themselves, 50–100 m or more apart, defending their feeding territories. Capture incidence tends to increase in frequency as the number of passing penguins increases, at least to a point. At large colonies, once penguin passage rate exceeds 1200 penguins per hour, the capture rate no longer increases. The opposite occurs at small colonies, like Cape Royds (3000 penguin pairs), where a seal has very few chances to take a penguin and, being relatively inefficient at penguin capture, will not bother to hang out in the vicinity most of the time. Leopard seals are thus rare at colonies the size of Royds, and occasional visits last only for a few days. It is at medium-sized penguin colonies (e.g., Cape Bird, ~70,000 pairs) where seal predation has the most impact, because with lower penguin passage rate, "predator swamping" is less effective.

Adélie penguins (and presumably emperors) are most vulnerable to leopard seal predation in two situations. First, when the open water between the beach ice ridge/wall and offshore floating ice floes freezes, the new ice is little more than consolidated slush. The penguins become very hesitant to cross such ice, just thick enough to support their weight. So they wait for hours, as the wind may arise and break up the ice or the sun may simply melt it. If present, the seals lurk below the thin ice layer, waiting for a penguin to pass overhead, and then surge upward through the ice to snare it. The penguins are also especially vulnerable when the beach ice ridge is breaking up. At that time, large ice chunks (the size of cars) are washing about in the ocean swells (with their wash actually leading to their disintegration – kind of like stirring the ice in your drink, so it melts and cools). Penguins hesitate to barge through this because getting caught between ice blocks is very dangerous. If off the beach, coming ashore, they are again hesitant, looking for an ice-free passage, and in that situation they are easier prey to seals, which creep around offshore of the brash and cakes (like what is in a drink with crushed ice).

Using stealthy means and these sensitive situations, single seals have been observed making as many as 14 captures of Adélie penguins in 2 days (nine in 110 min, five in 134 min [77]). During one 70-min period at Cape Crozier, a single seal captured seven penguins, of which it ate six [45]. We have actually observed even greater hunting success than that though. One of us (Rory) watched a leopard seal at Esperanza Bay, catching Adélies from a large colony (>100,000 pair). Rory sat on the rocks overlooking a point that Adélie penguins had been using to exit the sea to access land, marveling at how they shoot out with such speed, gaining such height, taking the breath of the sea with them for an instant before becoming the waddling land birds of TV fame. He was right at the water edge and had noticed a number of rocks protruding above the calm sea surface close by. Suddenly, when it raised its head to protrude its nostrils above the water to breathe, he realized that one of these "rocks" was actually the back of a leopard seal, perhaps 3 m from him. The animal breathed slowly for a few breaths before resuming its immobile position, upper back protruding from the water and head hanging down underwater. Penguins swished past, directly in front of the seal, in their dozens, eliciting no movement in the predator until one came within maybe 30 cm of its head. The seal struck, snake-like, in a flash opening its huge mouth to catch the bird. As would be normal in such a situation, it bit the penguin squarely in the mid-part of the body whereupon the bird ceased all movement, as if stunned. Maybe the crushing bite precludes any movement or perhaps the playing-dead tactic is part of a hoped-for escape plan. It is certainly heart-stopping to see such vibrancy in the movement of penguins instantly frozen. After capture, the seal lazily swam out to about 10 m offshore where it proceeded to rip the bird open and, by biting on the breast and lashing its head violently from one side to the other, it skinned it and then proceeded to eat the contents of the penguin stomachs only – krill (Fig. 12.4). The rest of the carcass was left, with tiny bits curiously exploited by storm-petrels fluttering by in the area. Following this, the seal swam back to exactly its previous position and resumed the wait. Most penguins did not even see the seal, and probably most of those caught did not see it until caught. On one occasion, a bird swam the length of the seal body, from the tail to the head, passing by within centimeters of the jaws whereupon the seal, perhaps as surprised as the penguin, snapped at it. The bird passed between the huge jaws, which missed the penguin by millimeters, closing only on the trail of bubbles in the bird's wake. Rory watched how the leopard seal actually struck at 21 penguins, catching 16 of them, all within an hour and 20 min!

Under good hunting conditions, the seals can catch enough penguins that eventually they begin to play with the victims. In such circumstances, penguins play dead

Fig. 12.4 Left: a leopard seal tearing apart an Adélie penguin (Photo courtesy of Viola Toniolo). Right: Adélie penguin recovering from a near miss of a leopard seal. This bird stayed mostly motionless for a few days, allowing its wound to heal, before returning to nesting duties (Photo: David Ainley)

(see above) and then often escape but with very severe wounds (most of which heal; Fig. 12.4).

Leopard seals rarely chase penguins in open water, but if so perhaps out of frustration, at the penguins' speed and extraordinary underwater agility. Indeed, it seems mostly futile as the penguins essentially can swim circles around the seal. It might be comparable to bobbing for apples, an activity in which hands are not allowed, only the apples are moving, fast! Capture takes a huge amount of effort on the part of the seal, but occasionally a penguin gets worn out. However, by then, the seal lacks the energy to take two penguins successively in this manner without a long recovery rest. As testament to the superior acceleration and turning abilities of penguins compared to seals, we (Rory) have observed groups of gentoo penguins swimming, diving, and porpoising around leopard seals. It looked like the aqueous equivalent of land birds mobbing a hawk, and maybe it was.

As described by several authors (see above), when a seal captures a penguin, it grasps it by the head and flings it about to break its neck. It then grasps its skin and flings the penguin about successively to remove the skin (and feathers; Fig. 12.5). The skin strips off in pieces, which skuas fight over. As noted above, if far enough north, other avian species might partake of pieces, for example, giant petrels. Once it gets the skin off, the seal generally eats the very large breast or pectoral muscles, and leaves the rest, which the skuas also fight over. It is a pretty inefficient affair because only the seal's incisor and front teeth are designed for the task (the back teeth are more suited for straining krill). According to one study, it takes the seal 17–65 bites over 6–15 min to finish its meal [61]. During that time, many thousands of penguins have streamed by if the colony is large. If a penguin does escape the seal and can still swim away, it has usually sustained major wounds. However, penguins are incredibly tough, and most survive. These escapees are colored black and red (from blood), rather than black and white (Fig. 12.4). They sit still for days, moving only their eyelids. Eventually, they are as good as new except for the scar. Fairly often, one sees scarred penguins or penguins with a portion of a wing removed.

Fig. 12.5 Leopard seal flinging a penguin in order to skin it before dining. (Photo courtesy of Viola Toniolo)

Estimates of the number of Adélie penguins taken by the array of leopard seals that may frequent a beach have been made only at colonies large enough to keep the seals interested. Over the course of the penguin seasonal presence, the estimates range from 1.4% to 5% of the Adélie penguin breeding population at Cape Crozier [61, 65] to 2.7% at Magnetic Island [77]. However, it may be that it is the less-experienced nonbreeder penguins that tend to be taken. As mentioned earlier, only occasionally do leopard seals hunt penguins at small colonies, such as Cape Royds (3000 pairs). Predation by leopard seals is also rare in Arthur Harbor, Anvers Island (14,000 pairs), where the leopard seals eat a lot of krill, having their back teeth designed appropriately (straining-out the water, as in crabeater seals [38]).

The significance of predation away from colonies is more difficult to ascertain – refer to Oritsland's attempt [62] (below) – but it must be important given the relatively low survival rate apparent in Adélie penguins, compared to other seabirds, that is, 59–89% compared to the mid-90% [4, 9]. Along ice edges (and perhaps among the floes of pack ice), leopard seals are far denser where Adélie penguins are concentrated [77]. Predation is important enough that penguins refuse to dive into the water at night when they cannot see what dangers may lurk, although once in the water they are not afraid of the dark, diving into the very dark deeper depths [5] (see Chap. 8). The leopard seals also prey heavily on fledglings when the latter first venture into the water, helped by the fact that these birds exhibit no predator avoidance (see also below in regard to other predators). Against this, chick fledging in Adélie penguins is so synchronous that, unsurprisingly, relatively few are taken, another example of predator swamping. One study at Cape Crozier, where perhaps 100,000 or more Adélie penguin chicks fledge in a typical year, researchers estimated that just 0.63% of fledglings were taken by leopard seals [61]. How long it takes the young penguins to learn that leopard seals are dangerous is not known.

Penguins Avoiding Seals

Adélie penguins avoid being captured by leopard seals by adhering to certain rules [2, 3, 5, 61]. These general rules would apply to emperor penguins, and for those situations not involving ice would apply to other penguin species (as noted later). A pocket version is:

- Travel only in flocks, which enables penguins to detect predators (seals) more easily, and which of course is one of the reasons that birds in general form flocks. The confusion effect of congregated animals, where predators find it difficult to single out individual prey is likely also relevant.
- Spend as little time as possible in the water when within about 200 m of the beach, which is typically outside where predators (seals) patrol. In this zone, Adélie penguins (and other penguin species) are always moving, and if not for final approach or departure, parallel to the beach, waiting for the best conditions for approach. Only if conditions improve (e.g., high tide reduces the ice ridge leap), and they are certain that seals are not present will they approach. After departing the beach, when 200 m or more away, penguins will pause to scrape the mud from their feathers (which reduces their drag coefficient and therefore the energy they need to swim) and perhaps renew the air pockets between them (to increase their insulation in the cold Antarctic waters; discussed in Chap. 14).
- Where there are ice floes, walk on them, jumping from one to the next.
- Avoid ice thin enough that a seal can see a penguin walking overhead from below.
- When moving across thin ice, become motionless if a seal appears nearby or below. Eventually, the seal will go away (it has to find an opening from which to breathe). In one instance, observers noted a penguin on thin ice that did not move for at least 80 min.
- When a predator (seal) is seen in the water, the flock should scatter rapidly in all directions. This is so-called "flash escape behavior," which is also common among flying birds in the presence of a stooping falcon or of fish in the presence of a penguin (!).
- When at the water's edge, stay a few steps back until another flock of penguins swims safely by or comes ashore. Only upon observing that no predators are present, is it considered safe to dive into the water (Fig. 12.6).
- Do not enter the water if the presence of predator cannot be sensed, e.g., at night.

The second-listed rule has been well researched in waters off South Africa, though involving cape gannets and cape fur seals [25], but the penguin scenario would be the same. Although the seabirds and fur seals forage on the same schooling-fish prey (Chap. 5), competing for fish in the same schools during day, they separate during the night. Fur seal predation on gannets (and penguins) is mainly at night. In

Fig. 12.6 A flock of Adélie penguins that has accumulated at the edge of the beach, awaiting an incoming flock to be convinced that all is safe before diving in. (Photo: David Ainley)

their investigation of the "Seascape of Fear" hypothesis, these researchers, tracking both the birds and the seals, found that the gannets vacated near-shore water at night or when resting on the water at night did so well away from the beach. During the 1950–1960s, when there were very few fur seals in South African waters (population had yet to recover), African penguins violated this second rule by bathing immediately after entering the water, close to the beach [81]. Well, it was not really a violation if there are no pinniped predators about which the penguins should be worrying, and they would no doubt sense this (more below).

Paraphrasing, in part, a portion of the discussion of leopard seal predation on penguins contained in Ainley ([4] p. 203), the overall importance of seal predation to penguins can be summarized as follows. Through the 1970s, one study summarized several others that reported the remains of penguins in the stomachs of leopard seals [62]. One reported results of a sample of 121 seals collected 1901–1964, mostly near penguin colonies, and found 26% of stomachs containing penguin remains; another study of 84 stomachs, from leopard seals collected in the pack ice of the Scotia-Weddell Confluence, found an incidence of 16%. On the basis of these data, plus a global estimate of the number of leopard seals, T. Oritsland [62] estimated that they annually consume 0.38 tonnes of penguins, which of course translates to many thousands. Most would be Adélie and emperors penguins because these penguins, like the seal, are found only within the Antarctic sea ice zone, where other penguin species do not occur. Leopard seals are rare vagrants anywhere else (see above). Subsequently, analyzing leopard seal scats found on ice floes in Prydz Bay, East Antarctica, penguins were found to be the main prey item, though using scats mostly provides evidence of prevalence or frequency of occurrence rather than actual numbers or proportions of diet [37].

The behavior by which leopard seals hunt Adélie penguins and the latter's avoidance behavior have been studied along the landing beaches at Ross Island, Ross Sea, penguin colonies [8, 45, 46, 61, 65], and at Magnetic Island, Prydz Bay [77]. Having a full view of the length of the beaches, these researchers logged the coming and going of penguins, presence of seals, and the incidence/success of their predation attempts. At Magnetic Island, censuses of seals and penguins were also made by helicopter survey at the fast ice edge, some distance from the colony. The colonies at Cape Crozier and vicinity of Magnetic Island (Vestfold Hills) are large (>100,000 pairs [54]), and when the penguins are present, four to six leopard seals patrol the beaches at any one time. At a colony the size of Cape Bird North (25,000 pairs at the time), two to four seals are present, and at the tiny Cape Royds (<5000 pairs), single leopard seals are seen, but only on occasion [8]. Leopard seals are not all that adept at catching penguins, except by stealth as noted above, and so there must be enough penguin traffic to keep a seal interested and fed. Indeed, the presence of seals is persistent only at colonies in which more than 250 penguins per hour pass between ocean and beach [8]. The number of seals present at a penguin colony is further limited because they are territorial and space themselves along the beach a couple hundred meters apart. On the other hand, though important to penguins only by increasing the duration of predation, seals may take turns moving into waters near the colonies from offshore areas of sea ice [46]. At Magnetic Island, 19 seals were marked and 14 were resighted in the season of observation, but only 4 were present along the beach at any given time [77]. Individual seals each patrolled the penguin colony beach on average for 1.9 days, and then disappeared for an average 5.3 days.

If leopard seals come ashore, which is rarely, a territorial penguin might be easy prey, but lucky for penguins, leopard seals haul out only on ice floes or fast ice to rest or sleep. Why that is, is not known but after all they are "ice seals." If leopard seals did come ashore, then the penguins would be forced to nest much farther inland than they do (penguins do not mix with fur seals, either; see below). If penguins are present on an ice floe when a seal of any kind hauls out, they scatter quickly and maintain a space around the seal (Fig. 12.7). A common sight is a seal,

Fig. 12.7 Left: not much chance that this frustrated leopard seal will be catching this penguin (Photo: D. Ainley) Right: penguins maintaining the necessary distance from a leopard seal, who surprised them by hauling on onto their ice floe (Photo courtesy of Arvind Varsani)

of any species, sleeping on an ice floe surrounded by a halo of penguins 3–5 m away. For a seal to grab a penguin on the floe, or even by lunging from the water at those waiting at floe edges or on the beach, is a futile endeavor. The penguins are just too quick and wary.

Fur Seals as Predators

Fur seals of several species (Arctocephalinae) occur in waters of the sub-Antarctic and temperate zones. Fur seals exhibit extreme sexual size dimorphism, with males on the order of five times the size of females. The very large males hunt for and consume penguins, including capture far at sea, although direct observations, as available for leopard seals, are limited. On several research cruises across the Scotia Sea, one of us (David) was impressed by the high frequency of penguin skin pieces seen floating on the sea surface, and assumed these were the remains of predation by fur seals (see below for possibility of killer whale involvement). In the immediate region, large populations of Antarctic fur seals existed on South Georgia and other Scotia Sea islands [17], as well as large populations of penguins, especially macaroni penguins. The recent decrease in macaroni penguins at South Georgia (one-third decrease from ~5.4 million pairs) is most likely a case of them being outcompeted for food by the recovering fur seal population (going from near 0 to ~400,000 females during recent decades) – both species prey heavily on Antarctic krill [88], as detailed in Chap. 4. Predation of the penguins might well also be involved, but only a few direct data exist of predation rate [16]. Antarctic fur seals have been observed taking macaroni penguins, as well as king penguins, at a number of locations, e.g., Crozet Archipelago [19] and Marion Island [39].

The rallying populations of cape fur seals in South Africa and Namibia have been implicated in slowing the recovery of African penguins, due to trophic competition as well as predation [13, 26, 43]. Male cape fur seals were estimated recently to take 2–7% of the African penguins at Dyer Island, an activity by the fur seals in waters immediately off the beach as the penguins came and went to the colony (similar to the leopard seal–Adélie penguin interaction [29, 41, 55]). However, these days with a dramatic increase in the fur seal population, African penguins no longer hang out much in waters immediately off the beach (see above). In the Falkland Islands, Magellanic, rockhopper, and gentoo penguin colonies are absent in those parts of the coast where large colonies of sub-Antarctic fur seals are located [56]. Authors of the latter analysis thought that perhaps the penguins were avoiding predation, but certainly avoidance by the penguins of trophic competition could also be involved. Finally, Snares crested penguins are taken in low frequency by male New Zealand fur seals at The Snares, New Zealand, as judged from analysis of scats and spewings [48]. Both the penguin and the fur seals are protected species. Likewise, New Zealand fur seals have been observed taking southern rockhopper penguins at Macquarie Island as [28] (Fig. 12.8).

Fig. 12.8 Erect-crested penguins and New Zealand fur seal, Antipodes Islands, New Zealand. (Photo courtesy of Louise Chilvers)

One of us (Rory), as well as others [24, 29], has observed a few individuals of the Cape fur seal population repeatedly catching African penguins off Ichaboe Island, Namibia, over consecutive days. And it is known that it is particular fur seal individuals which seem to take penguins [43]. The island warden, Barny, in accord, reported that he had noted the same behavior from just one individual for months (being a penguin-fan, he was minded to shoot the animal but had no gun!). Capture took place 50–200 m offshore, the distance being relevant because if it is too far, then the density of birds moving between the island and the sea to forage is too low, as with Adélie penguins (see above), to make the encounter rate high enough to warrant the wait by the fur seal. Against this, because fur seals tend to capture penguins in the water during a chase, proximity to land increases the chances that the birds might make it to shore and escape.

During the observed Ichaboe penguin capture events, the fur seal simply tried to run the penguin down. On paper, African penguin maximum swim speeds are comparable to, and perhaps marginally less than, those of fur seals [18, 68, 92], so in a straight-line pursuit, the predator could well catch the prey. But, as has been documented for cursorial land predators such as cheetahs [93], there are physical reasons why smaller prey can turn more quickly and with a shorter turn radius than their predators, and this is certainly holds for penguins and those creatures that would eat them. This explains why the observed cape fur seal–African penguin pursuits consisted of straight-line sections of swimming, punctuated by rapid turns initiated by the penguin when the fur seal got too close. Importantly though, the penguin must turn with maximum angular velocity to ensure that its deviation minimizes the chance that the fur seal will catch it at that point. Such vigorous performances require extremely high energy, and therefore oxygen use. In contrast, although the fur seal must turn promptly, it can turn in a more relaxed manner just so long as the prey does not disappear beyond the range of vision – the worst thing that can happen is that its slow turns cause it to lose its supper, whereas such turns by a penguin may cause it to lose its life. So, repeated high-speed turns and straight-line sections with little chance to breathe (the Ichaboe penguins all breathed by porpoising – see Chaps. 6 and 9) gradually eroded both the energy and oxygen reserves of the penguins faster than they did the fur seals, which was what resulted in capture. On two occasions, out of perhaps a dozen, the penguins simply stopped at the water surface exhausted, to be taken by the fur seal at its leisure. But this game of cat and mouse presumably depends on water visibility, with more turbid conditions favoring the penguin, which, if turns are timed to perfection, should allow the bird to shoot off into the gloom and escape.

Sea Lions as Predators

Several species of sea lions (Otariidae) occur in sub-Antarctic and temperate/ Subarctic waters, breeding mostly on islands. In New Zealand, predation of yellow-eyed penguins by Hooker sea lions (also endangered) has been documented for the mainland (especially Otago Peninsula), as well as on islands of the region where both species occur [47, 58, 79]. The hunting behavior of the sea lions appears to be much the same as that displayed by leopard seals, i.e., lying in wait in the water just off the beach, intent on surprising a passing penguin [58]. The penguins' behavior is very much like that described for Adélie penguins dealing with leopard seals, above, swimming back and forth until the spotted sea lion goes away and/or not venturing into waters off the beach when it is dark. Because their behavior indicates that the penguins are very much aware of the danger of sea lions, it is further evidence that predation by sea lions could be fairly important. It has been suggested that a severe reduction in sea lion numbers with the arrival of humans in New Zealand, in part, facilitated the spread of yellow-eyed penguins from sub-Antarctic islands to the mainland [14]. The sea lion population is now recovering from this former decimation, and one hypothesis has it that, in doing so, they could be exerting a negative impact on local penguin numbers [47]. It appears that female Hooker sea lions may be mostly involved, with female sea lions being much larger than female fur seals (in fact sea lions prey on fur seals, too).

On The Snares Islands, the remains of crested penguins have been found in fur seal scats, although this was not the case for sea lion scats at this site [48]. In contrast, penguin feathers were found in a few Campbell Island sea lion scats, but which could not be identified to species [76]. Nevertheless, sea lion predation has been implicated in the population decrease of eastern rockhopper penguins on Campbell Island [60]. On Macquarie Island, male Hooker sea lions were observed to prey on gentoo penguins, but remains were not found in any scats [57]. On that basis, those authors surmised that, by relying on scats, the prevalence of penguins in pinniped diet may be underestimated.

Killer Whales as Predators

Much lore regarding killer whales and penguins remains from the heroic era of Antarctic exploration. In that body on information, the killer whale is considered to be a predator of Adélie penguins. The source of this idea may derive from observations such as those in the log of *HMS Terra Nova* – the ship used in Scott's second expedition to Antarctica (cited in Lowe and Kinnear [53], p. 111):

> "Adélie penguins do not seem to mind the killer whales unless very close, for though at times they will come shooting out of the water on to a floe in evident hurry and a killer appears soon afterwards, yet at other times they will remain in the water quite close to a whale. It seems as if the killers tried to get them under the ice, and not attack them in open water".

In that and other cases, there was no observation of the killer whale catching or eating an Adélie penguin, but the connection was inferred by the penguins' apparent fearful behavior. The fact that the penguins were observed "shooting out of the water" is no proof because they almost never leave the water in any other way (Fig. 12.3), and they are afraid of all big creatures, including minke whales (David, pers. obs). Another bit of lore from those early Antarctic expeditions confirms this conclusion, i.e., this somewhat common observation as reported by Levick ([51], p. 88):

> "One day, as I watched some hundreds of Adélies bathing in an open lead, suddenly the back of an enormous killer-whale … rose above the surface as it crossed the lead from side to side, appearing from beneath the ice on one side and disappearing beneath it on the other. To my surprise, not the slightest fear was shown by the birds in the water. Had this beast been a sea-leopard, there would have been a stampede, and every bird have leapt from the water on to the sea-ice. On this evidence I formed the opinion that in all probability killer-whales do no harm to Adélie penguins; later I saw it confirmed, when a school of killers shaved close past several floes that were crowded with Adélies, and made not the least attempt to get at them, as they might so easily have done by upsetting the floes. Very probably this is because the agile bird can escape with such ease from the ponderous whale, and fears it no more than a terrier fears a cow, though he thinks twice before coming within reach of its jaws".

There have been some recent reports of the behavior of killer whales that were seemingly hunting Adélie penguins, although the extent of actual predation was uncertain, with sightings in the western Ross Sea [49, 74] and along the Antarctic Peninsula [89]; Robert L. Pitman, pers. comm.). Type-C killer whales (see Pitman and Ensor [67] for definition of ecotypes; C's eat fish) have been observed playing with penguins, i.e., young animals on occasion herding a penguin flock or single individual, without capture, while adults foraged for fish nearby [12] (Fig. 12.9).

Reports indicate that killer whales on very rare occasions do kill and sometimes consume smaller penguins such as African, Magellanic, macaroni, and rockhopper, typically in the vicinity of pinniped rookeries (where the whales are actively hunting seals/sea lions/fur seals) [23, 40, 90]. African penguins appear to avoid close approach of killer whales, unlike other cetaceans, and thus there must be some merit to the idea that their behavior indicates some level of negative interaction [73]. In waters off the west coast of the northern Antarctic Peninsula, indeed, type-B killer

Fig. 12.9 Type-B killer whale chasing a gentoo penguin. (Photo courtesy of Robert Pitman)

whales (small form B, mostly eat seals) were observed on occasion chasing and about half the time capturing gentoo penguins (sometimes eating a small portion, only the large breast muscle), and in a couple of instances a chinstrap penguin [66]. According to Robert Pitman, the killer whales are mostly playing, as they rarely eat any that they capture. For all this, it does appear that killer whales do hunt and kill a small number of the small penguin species, but given that there is not much to eat, the food web or penguin demographic significance remains unclear. It happens often enough, however, that the penguins are somewhat wary of killer whales, but are pretty good at avoidance (Fig. 12.10).

Killer whales do prey extensively on the largest penguin species, king and emperor penguins. Recurring observations indicate that Type-D killer whales take king penguin fledglings as the latter first swim from the breeding island, at Crozet [34–36], Marion [23] and Macquarie islands (E. Woehler, pers. obs., in [7]). In the case of king and emperor penguins, the penguin prey is five to six times the size of the smaller penguin species, the victims are clustered together, they have high caloric value (being very fat), and in taking their first swims they are exceedingly slow and naïve. If they have a fair amount of down still adhering to their plumage, acting like a life preserver being so full of air, these fledglings can dive only with difficulty (David, pers. obs. of emperor penguins). Therefore, it makes sense that killer whales would show more than passing interest. Without question, king penguins are definitely worth the effort in caloric terms, being so fat that on many islands they were extirpated, along with southern elephant seals, by people in the business of acquiring animal oils during the eighteenth and nineteenth centuries [27, 78].

Emperor penguins have been found in the stomachs of killer whales, at least once [69]. Type-B killer whales (large form, also seal eaters [67]) have been observed chasing flocks of emperor penguins from the water onto fast ice [6]. That

Fig. 12.10 Type-B killer whales herding a gentoo penguin, perhaps just for play. (Photo courtesy of Holly Fearnbach)

Fig. 12.11 Type-B killer whales, both in the near and far background, patrolling the fast ice edge where emperor penguin chicks are awaiting arrival of a parent, with food, at Cape Washington. (Photo courtesy of Gerald Kooyman)

observation was near the Cape Royds Adélie penguin colony, at which there is no evidence of killer whale interest of Adélie penguins, though they have been seen taking Weddell seals there. Consistent with that observation, a type-B killer whale implanted with a satellite tag in the vicinity of that colony, within its pod, followed an apparent "trap-line" strategy visiting all emperor penguin colonies as it moved north along the Victoria Land coast over subsequent weeks [11]. Type-B killer whales have been observed patrolling one of these areas; the fast ice edge off the large emperor penguin colony at Cape Washington (Gerald Kooyman, pers. obs., discussed elsewhere [6]) (Fig. 12.11).

Large Fish as Predators

Sharks, especially white sharks, appear to be important predators of African penguins near colonies in Algoa Bay, south coast of South Africa, on the basis of the wounds inspected on penguins, both living and dead [72]. Authors of the latter analysis found that on the order of 30% of 70 penguin carcasses found on beaches over a 10-year period had clear signs of shark bites (they compared cut spacing in scars to actual shark jaw measurements). Seemingly, the sharks would take penguins that were resting on the surface, caught unaware. On the other hand, white sharks do not appear to prey on many African penguins at Dyer Island, a more westerly south coast colony [41]. We (Rory) have seen the sharks patrolling the channel between Dyer and Little Dyer islands, as have others [32]. It is thought that the presence of a large fur seal rookery at Dyer Island, with the sharks preying to a great extent on the fur seals, removes any interest among the sharks to be preying on the penguins, which are of much less nutritional value. Only a small fur seal rookery occurs in Algoa Bay, where shark predation of African penguins is more an issue.

It is interesting that Kearton [42], investigating African penguins in the 1920s, thought that white sharks were a much more important predator than appears to be the consensus opinion of researchers today. And, he did not include fur seals among what he thought were important predators (though he did include large, coastal octopus!). All that makes sense in that, during the early 1900s, there were very few cape fur seals, their population having been severely depleted by fur seal hunters (to the verge of near extinction [43]). Moreover, there probably were many more sharks than now, Kearton's experience being before the South African government began to greatly deplete the sharks to protect human swimmers. Indeed, once programs were begun in the 1950s to control sharks off South African beaches (using gill nets), reports and captures of white sharks decreased dramatically [20, 21, 52]. The white shark these days is classified as being "vulnerable," and decreasing, by IUCN, and the shark removal program in South Africa was ended in the 1990s (with shark sightings increasing since). On the other hand, the Cape fur seal population is now many times larger than it was not long ago, and exercises important predatory pressure on the penguins (see above).

Somewhat surprising is a lack of information on whether white sharks prey on penguins in Australia, New Zealand, or Patagonia, despite the sharks' presence. However, if the conclusions regarding African penguins not providing much interest from sharks where there are a lot of larger, more nutritious pinnipeds to be taken, then perhaps the large number of sea lions and fur seals that today inhabit the islands in Australia-Pacific region may divert sharks from penguins. This may be especially true given the large populations of elephant seals that inhabit many of those islands. Those seals are a specific target of white sharks owing to the seals' very high lipid content (Klimley [44], pp. 311–368). For sure, the tiny Little penguin of Australia and New Zealand does not offer much in the way of a meal, especially one that may take some effort to catch.

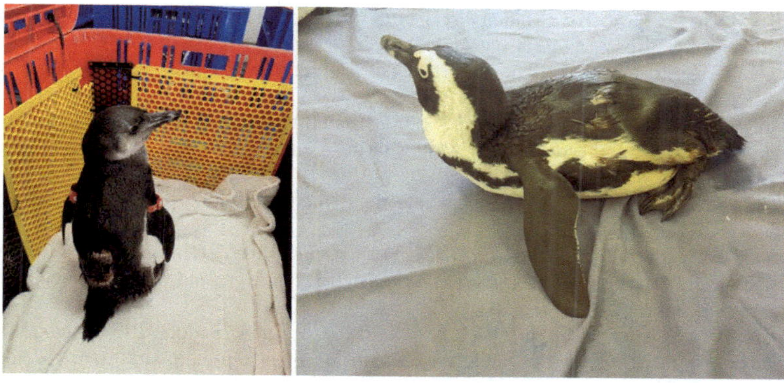

Fig. 12.12 African penguins showing nonfatal bites from sharks. (Photos courtesy of David Gordon Roberts, SANCCOB)

Sharks do not venture very far south of the Antarctic Polar Front. Penguin remains have been found in the stomachs of large Antarctic toothfish (~2 m in length), which in some circles is referred to as "the shark of the Antarctic," given its predatory nature at least with respect to fishes. Whether the penguins found in toothfish stomachs were scavenged from leopard seal kills, which often sink to the bottom, or are the result of outright predation, is not known [71, 80]. Large Antarctic toothfish do frequent the upper water column, where they compete with penguins for Antarctic silverfish (Chap. 4) (Fig. 12.12).

References

1. Aguayo-Lobo A, Acevedo RJ, Brito JL, Acuña GP, Bassoi M, Secchi ER, Dalla Rosa L (2011) Presence of the leopard seal, *Hydrurga leptonyx* (de Blainville, 1820), on the coast of Chile: an example of the Antarctica-South America connection in the marine environment. Oecol Aust 15:69–85
2. Ainley DG (1972) Flocking of Adélie penguins. Ibis 114:388–390
3. Ainley DG (1974) Comfort behavior of Adélie and other penguins. Behaviour 50:16–50
4. Ainley DG (2002) The Adélie penguin: bellwether of climate change. Columbia University Press, New York
5. Ainley DG, Ballard G (2011) Non-consumptive factors affecting foraging patterns in Antarctic penguins: a review and synthesis. Polar Biol 35:1–13
6. Ainley DG, Ballard G (2012) Trophic interactions and the decrease in killer whale (*Orcinus orca*) prevalence with reduced availability of large fish in the southern Ross Sea. Aquat Mamm 38:153–160
7. Ainley DG, Ballard G, Blight LK, Ackley S, Emslie SD, Lescroël A, Olmastroni S, Townsend SE, Tynan CT, Wilson P, Woehler E (2010) Impacts of cetaceans on the structure of southern ocean food webs. Mar Mamm Sci 26(2):482–489
8. Ainley DG, Ballard G, Karl BJ, Dugger KT (2005) Leopard seal predation rates at penguin colonies of different size. Antarct Sci 17:335–340

9. Ainley DG, Leresche RE, Sladen WJL (1983) Breeding biology of the Adélie penguin. University of California Press, Los Angles
10. Ainley DG, Siniff DB (2009) The importance of Antarctic toothfish as prey of Weddell seals in the Ross Sea: a review. Antarct Sci 21:317–327
11. Andrews RD, Pitman RL, Ballance LT (2008) Satellite tracking reveals distinct movement patterns for type B and type C killer whales in the southern Ross Sea, Antarctica. Polar Biol 31(12):1461–1468
12. Ballard G, Ainley DG (2005) Killer whale harassment of Adélie penguins at Ross Island. Antarct Sci 17:385–386
13. Barham PJ, Underhill LG, Crawford RJM, Leshoro TM (2007) Differences in breeding success between African penguins (*Spheniscus demersus*) that were and were not oiled in the MV Treasure oil-spill in 2000. Emu 107:7–13
14. Boessenkool S, Austin JJ, Worthy TH, Scofield P, Cooper A, Seddon PJ, Waters JM (2009) Relict or colonizer? Extinction and range expansion of penguins in southern New Zealand. Proc R Soc B Biol Sci 276:815–821
15. Bombosch A, Solovyev B (2017) Weddell seal (*Leptonychotes weddellii*) killing gentoo penguin (*Pygoscelis papua)* at Neko Harbour, Antarctic Peninsula. Polar Biol 40:1899–1902
16. Bonner WN, Hunter S (1982) Predatory interactions between Antarctic fur seals, macaroni penguins and giant petrels. Br Antarct Surv Bull 56:75–79
17. Boyd IL (1993) Pup production and distribution of breeding Antarctic fur seals (*Arctocephalus gazella*) at South Georgia. Antarct Sci 5(1):17–24
18. Boyd IL, Reid K, Bevan RM (1995) Swimming speed and allocation of time during the dive cycle in Antarctic fur seals. Anim Behav 50:769–784
19. Charbonnier Y, Delord K, Thiebot J-B (2010) King-size fast food for Antarctic fur seals. Polar Biol 33:721–724
20. Cliff G, Dudley SFJ, Jury MR (1996) Catches of White Sharks in KwaZulu-Natal, South Africa and environmental influences. In: Klimley AP, Ainley DG (eds) Great White Sharks: the biology of *Carcharodon carcharias*. Academic, San Diego, pp 351–362
21. Cliff G, Van Der Elst RP, Govender A, Withhuhn TK, Bullen EM (1996) First estimates of mortality and population size of White Sharks on the South African coast. In: Klimley AP, Ainley DG (eds) Great White Sharks: the biology of *Carcharodon carcharias*. Academic, San Diego, pp 393–400
22. Cobley ND, Bell G (1998) Weddell seal (*Leptonychotes weddellii)* feeding on gentoo penguins (*Pygoscelis papua*). Mar Mamm Sci 14:881–883
23. Condy PR, Van Aarde RJ, Bester MN (1978) The seasonal occurrence and behaviour of killer whales *Orcinus orca* at Marion Island. J Zool Lond 184:449–464
24. Cooper J (1974) The predators of the jackass penguin *Spheniscus demersus*. Bull Br Ornithol Club 94:21–24
25. Courbin N, Pichegru L, Seakamela M, Makhado A, Meÿer M, Kotze PGH, Mc Cue SA, Péron C, Grémillet D (2022) Seascapes of fear and competition shape regional seabird movement ecology. Commun Biol 5(1):1–7
26. Crawford RJM, David JHM, Shannon LJ, Kemper J, Klages NTW, Roux J-P, Underhill LG, Ward VL, Williams AJ, Wolfaardt AC (2001) African penguins as predators and prey – coping (or not) with change. S Afr J Mar Sci 23:435–447
27. Croxall JP, Prince PA, Hunter I, Mcinnes SJ, Copestake PG (1984) The seabirds of the Antarctic Peninsula, islands of the Scotia Sea, and Antarctic Continent between 80°W and 20°W: their status and conservation. In: Croxall JP, Evans PGH, Schreiber RW (eds) Status and conservation of the world's seabirds. International Council for Bird Preservation, Cambridge, pp 637–666
28. Csordas SE, Ingham SE (1965) The New Zealand fur seal, *Arctocephalus forsteri* (Lesson), at Macquarie Island, 1949–1964. CSIRO Wildl Res 10:83–99
29. Du Toit M, Bartlett PA, Bester MN, Roux J-P (2004) Seabird predation by individual seals at Ichaboe Island, Namibia. S Afr J Wildl Res 34:45–54

30. Erickson AW, Siniff DB, Cline DR, Hofman RJ (1971) Distributional ecology of Antarctic seals. In: Deacon G (ed) Symposium on antarctic ice and water masses. Scientific Committee for Antarctic Research, Cambridge, pp 55–95
31. Estes JA, Terborgh J, Brashares JS, Power ME, Berger J, Bond WJ, Carpenter SR, Essington TE, Holt RD, Jackson JBC, Marquis RJ, Oksanen L, Oksanen T, Paine RT, Pikitch EK, Ripple WJ, Sandin SA, Scheffer M, Schoener TW, Shurin JB, Sinclair ARE, Soulé ME, Virtanen R, Wardle DA (2011) Trophic downgrading of planet earth. Science 33:301–306
32. Ferreira CA, Ferreira TP (1996) Population dynamics of White Sharks in South Africa. In: Klimley AP, Ainley DG (eds) Great White Sharks: the biology of *Carcharodon carcharias*. Academic, San Diego, pp 381–391
33. Grémillet D, Ponchon A, Paleczny M, Palomares M-L D, Karpouzi V, Pauly D (2018) Persisting worldwide seabird-fishery competition despite seabird community decline. Curr Biol 28:4009–4013
34. Guinet C (1991) Intentional stranding apprenticeship and social play in killer whales (*Orcinus orca*). Can J Zool 69:2712–2716
35. Guinet C, Bouvier J (1995) Development of intentional stranding hunting techniques in killer whale (*Orcinus orca*) calves at Crozet Archipelago. Can J Zool 73:27–33
36. Guinet C, Tixier P (2011) Crozet Killer Whales: in a remote but changing environment. In: Pitman RL, Gorter U, Reznick K (eds) Killer Whale: the top, top predator. American Cetacean Society, pp 33–38
37. Hall-Aspland SA, Rogers TL (2004) Summer diet of leopard seals (*Hydrurga leptonyx*) in Prydz Bay, eastern Antarctica. Polar Biol 27:729–734
38. Hofmann RJ, Reichle RA, Siniff DB, Muller-Schwarze D (1977) The Leopard Seal (*Hydrurga leptonyx*) at Palmer Station, Antarctica. In: Llano GA (ed) Adaptations within Antarctic ecosystems. Smithsonian Institution, Washington, DC, pp 769–782
39. Hofmeyr GJG, Bester MN (1993) Predation on king penguins by Antarctic fur seals. S Afr J Antarct Res 23:71–74
40. Jehl JR Jr (1975) Mortality of Magellanic penguins in Argentina. Auk 92:596–598
41. Johnson RL, Ventger A, Bester MN, Oosthuizen WH (2006) Seabird predation by white shark and cape furs seal at Dyer Island. S Afr J Wildl Res 36:23–32
42. Kearton C (1931) The Island of penguins. National Travel Club, New York
43. Kirkman SP (2009) Evaluating seal–seabird interactions in southern Africa: a critical review. Afr J Mar Sci 31(1):1–18
44. Klimley AP (2013) The biology of sharks and rays. University of Chicago Press, Chicago
45. Kooyman GL (1965) Leopard seals of Cape Crozier. Animals 6:59–63
46. Kooyman GL (1981) Leopard Seal *Hydrurga leptonyx* Blainville, 1820. In: Ridgway SH, Harrison RJ (eds) Handbook of Marine Mammals. Academic, London, pp 261–274
47. Lalas C, Ratz H, Mcewan K, Mcconkey SD (2007) Predation by New Zealand sea lions (*Phocarctos hookeri*) as a threat to the viability of yellow-eyed penguins (*Megadyptes antipodes*) at Otago Peninsula, New Zealand. Biol Conserv 135:235–246
48. Lalas C, Webster T (2014) Contrast in the importance of arrow squid as prey of male New Zealand sea lions and New Zealand fur seals at The Snares, subantarctic New Zealand. Mar Biol 161:631–643
49. Lauriano G, Fortuna CM, Vacchi M (2007) Observation of killer whale (*Orcinus orca*) possibly eating penguins in Terra Nova Bay, Antarctica. Antarct Sci 19:95–96
50. Lee WY, Jung JW, Chung H, Kim JH (2019) Weddell seal feeds on Adélie penguins in the Ross Sea, Antarctica. Polar Biol 42:1621–1624
51. Levick GM (1914) Antarctic penguins: a study of their social habits. McBride Nast, New York
52. Levine M (1996) Unprovoked attacks by White Sharks off the South African Coast. In: Klimley AP, Ainley DG (eds) Great White Sharks: the biology of *Carcharodon carcharias*. Academic, San Diego, pp 435–448
53. Lowe PR, Kinnear NB (1930) Birds. British Museum, London, pp 103–193
54. Lynch HJ, Larue MA (2014) First global census of the Adélie penguin. Auk 131:457–466

55. Makhado AB, Crawford RJ, Waller LJ, Underhill LG (2013) An assessment of the impact of predation by Cape fur seals *Arctocephalus pusillus* at Dyer Island. Ostrich 84:191–198
56. Masello JF, Mundry R, Poisbleau M, Demongin L, Voigt CC, Wikelski M, Quillfeldt P (2010) Diving seabirds share foraging space and time within and among species. Ecosphere 1(6):1–28
57. Mcmahon CR, Holley D, Robinson S (1999) The diet of itinerant male Hooker's sea lions, *Phocarctos hookeri*, at sub-Antarctic Macquarie Island. Wildl Res 26:839–846
58. Moore PJ, Moffat RD (1992) Predation of yellow-eyed penguins by Hooker's sea lion. Notornis 39:68–69
59. Morandini V, Dugger KM, Ainley D, Ferrer M (2020) Rockhopper penguin–imperial cormorant mixed colonies in The Falkland Islands: a stroke of luck for late breeders. Ecosphere 11(11):e03272
60. Morrison KW, Armstrong DP, Battley PF, Jamieson SE, Thompson DR (2017) Predation by New Zealand sea lions and brown skuas is causing the continued decline of an eastern rockhopper penguin colony on Campbell Island. Polar Biol 40:735–751
61. Muller-Schwarze D, Muller-Schwarze C (1975) Relations between leopard seals and Adélie penguins. Rapport et Process-verbaux des Reunions International Pour LÕ Exploration de la Mer 169:394–404
62. Oritsland T (1977) Food consumption of seals in the Antarctic pack ice. In: Llano GA (ed) Adaptations within Antarctic ecosystems. Smithsonian Institution, Washington, DC, pp 749–768
63. Paleczny M (2008) An analysis of temporal and spatial patterns in global seabird abundance during the modern industrial era, 1950–2010, and the relationship between global seabird cecline and marine fisheries catch. MSc thesis. University of British Columbia
64. Paleczny M, Hammill E, Karpouzi V, Pauly D (2015) Population trend of the world's monitored seabirds, 1950–2010. PLoS One 10(6):e0129342
65. Penney RL, Lowry G (1967) Leopard seal predation on Adélie penguins. Ecology 48:878–882
66. Pitman RL, Durban JW (2010) Killer whale predation on penguins in Antarctica. Polar Biol 33(11):1589–1594
67. Pitman RL, Ensor P (2003) Three forms of killer whales (*Orcinus orca*) in Antarctic waters. J Cetacean Res Manag 5:1–9
68. Ponganis PJ, Ponganis EP, Ponganis KV, Kooyman GL, Gentry RL, Trillmich F (1990) Swimming velocities in otariids. Can J Zool 68:2105–2112
69. Prévost J (1961) Ecologie des manchots empereur. Actualities Scientifique et Industrielles, 1270
70. Prugh LA, Stoner CJ, Epps CW, Bean WT, Ripple WJ, Laliberte AS, Brashares JS (2009) The rise of the mesopredator. Bioscience 59:779–791
71. Prutko V (2004) Observer notes (Subarea 88.1). Commission for the Conservation of Antarctic Marine Living Resources, Hobart
72. Randall BM, Randall RM, Compagno LJV (1988) Injuries to jackass penguins (*Spheniscus demersus*): evidence for shark involvement. J Zool Lond 214:589–599
73. Randall RM, Randall BM (1990) Cetaceans as predators of jackass penguins *Spheniscus demersus*: deductions based on behaviour. Mar Ornithol 18:9–12
74. Richlen MF, Thomas JA (2008) Acoustic behavior of Antarctic killer whales (*Orcinus orca*) recorded near the ice edge of McMurdo Sound, Antarctica. Aquat Mamm 34:448–457
75. Robbins JR, Poncet D, Evans AR, Hocking DP (2019) A rare observation of group prey processing in wild leopard seals (*Hydrurga leptonyx*). Polar Biol 42:1625–1630
76. Roberts J, Lalas C (2015) Diet of New Zealand sea lions (*Phocarctos hookeri*) at their southern breeding limits. Polar Biol 38:1483–1491
77. Rogers T, Bryden MM (1995) Predation of Adélie penguins (*Pygoscelis adeliae*) by leopard seals (*Hydrurga leptonyx*) in Prydz Bay, Antarctica. Can J Zool 73:1001–1004
78. Rounsevell DE, Copson GR (1982) Growth rate and recovery of a king penguin, *Aptenodytes patagonicus*, population after exploitation. Aust Wildl Res 9:519–525
79. Schweigman P, Darby JT (1997) Predation of yellow-eyed penguins (*Megadyptes antipodes*) on mainland New Zealand by Hooker's sea lion (*Phocarctos hookeri*). Notornis 44:265–266

80. Shepherd LD, Miskelly CM, Cherel Y, Tennyson JD (2021) Genetic identification informs on the distributions of vagrant royal (*Eudyptes schlegeli)* and macaroni (*Eudyptes chrysolophus*) penguins. Polar Biol 44:2299–2306
81. Siegfried WR, Frost PGH, Kinahan JB, Cooper J (1975) Social behaviour of jackass penguins at sea. Zool Afr 10(1):86–100
82. Simpson GG (1946) Fossil penguins. American Museum of Natural History, New York
83. Siniff DB (1981) Seal population dynamics and ecology. J R Soc N Z 11:317–327
84. Southwell C, Bengtson J, Bester M, Blix AS, Bornemann H, Boveng P, Cameron M, Forcada J, Laake J, Nordøy E, Plötz J, Rogers T, Southwell D, Steinhage D, Stewart BS, Trathan P (2012) A review of data on abundance, trends in abundance, habitat use and diet of ice-breeding seals in the Southern Ocean. CCAMLR Sci 19:49–74
85. Stirling I (1969) Tooth wear as a mortality factor in the Weddell seal, *Leptonychotes weddelli*. J Mammal 50:559–565
86. Testa JW, Oehlert G, Ainley DG, Bengtson JL, Siniff DB, Laws RM, Rounsevell D (1991) Temporal variability in Antarctic marine ecosystems: periodic fluctuations in the phocid seals. Can J Fish Aquat Sci 48:631–639
87. Todd FS (1988) Weddell seal preys on chinstrap penguin. Condor 90:249–250
88. Trathan PN, Ratcliffe N, Masden EA (2012) Ecological drivers of change at South Georgia: the krill surplus, or climate variability. Ecography 35(11):983–993
89. Visser IN, Smith TG, Bullock ID, Green GD, Carlsson OGL, Imberti S (2008) Antarctic Peninsula killer whales (*Orcinus orca*) hunt seals and a penguin on floating ice. Mar Mamm Sci 24:225–234
90. Williams AJ, Dyer BM, Randall RM, Komen J (1990) Killer whales *Orcinus orca* and seabirds: 'play,' predation and association. Mar Ornithol 18:37–41
91. Wilson EA (1907) Aves. National Antarctic Expedition 1901–04, vol 2: Zoology, Part 2. London
92. Wilson RP (1985) The jackass penguin, *Spheniscus dernersus*, as a pelagic predator. Mar Ecol Prog Ser 25:219–227
93. Wilson RP, Griffiths IW, Mills MGL, Carbone C, Wilson JW, Scatlebury DM (2015) Mass enhances speed but diminishes turn capacity in terrestrial pursuit predators. eLife 4:e06487
94. Young EC (1994) Skua and penguin: predator and prey. Cambridge University Press, Cambridge

Chapter 13
Penguins Coping with a Changing Ocean

> "Just what is the penguin environment? I confess that I cannot define it precisely, nor can I suggest a common denominator….it is assumed that cold or cool water determines the distribution of penguins [though] there can be a difference of as much as 28 °C between the waters of the Antarctic…and the Galápagos. Yet there are penguins in both environments. An important factor seems to be the stability of the water temperature – sea currents where the annual fluctuation is not too great. Each species has evolved its own thermoregulation to cope with the environment in which it lives". R.T. Peterson [93] (pp. 97–98)

Earth's atmosphere, oceans, and weather are changing, no question about that. The full array of penguin species is distributed from polar to subtropical waters in the Southern Hemisphere, as Peterson noted, above (Chaps. 1 and 3). The degree to which a given species might cope with a changed ocean, other than indirectly through a change in prey availability (Chaps. 4 and 5), would have to do with its various tolerances, especially with regard to temperature range, as Peterson also noted. This tolerance in part would be related to the insulation versus conductance of a penguin's feather coat, and smaller penguins appear to have more issues with this (Chap. 7). There is a tendency for small-bodied penguins to frequent warmer waters, and for large-bodied penguins to frequent polar waters (Table 1.2). At least with respect to shifting ocean temperatures, which very much is happening and more so in the Southern Ocean [28, 29], despite penguins being homeotherms (warm blooded), it may well be that warmer-water penguins will be more sensitive to ocean temperature change than cold-water species. See below.

On the other hand, changing climate involves more than temperature. Incredibly important to penguins is the changing wind patterns of the Southern Hemisphere [115]. Southern Ocean winds have been increasing, and this is related to increasing temperature differential (and therefore atmospheric pressure) between the mid-latitude ocean and the Antarctic continent, but also due to the Antarctic ozone hole and the cooling of the stratosphere above Antarctica [118]. Wind is what generates surface ocean currents, sea ice extent, and other factors important to penguins [5, 79, 120].

Finally, a changing ocean involves pollution also, another topic that we will discuss in this chapter.

D. G. Ainley, R. P. Wilson, *The Aquatic World of Penguins*, Fascinating Life Sciences, https://doi.org/10.1007/978-3-031-33990-5_13

Penguins Have Always Been Challenged by a Changing Ocean

We briefly reviewed, in Chap. 1, the evolutionary radiation of the suite of extant penguin species, noting on a geological time scale that their speciation was a response to several factors: (1) plate tectonics and the opening of waterways (particularly in regard to the earliest, now long-extinct, penguin species); (2) the appearance of numerous oceanic islands in the climate transition of polar-to-temperate parts of the Southern Hemisphere (during the most recent 5M years); and (3) to changes of climate, with marked oscillations, in association with glacial and interglacial periods. Clearly, penguins have persisted through dramatic climate fluctuations. The Antarctic Ice Cap has been present for 30–35M years, forming after the Southern Ocean completely surrounded the continent, thus isolating it from the remainder of the world's oceans (finally via the opening of the Drake Passage, the current watery gap between Patagonia and the Antarctic Peninsula). This development of the Antarctic Circumpolar Current (ACC) thus predated all of the currently existing penguins [7] (Fig. 1.1), but overlapped their respective lineages (the direct ancestors of the current penguin species). The ACC's qualities as a faunal boundary of sorts also led to the unique fish, and to some degree zooplankton fauna, of the Southern Ocean [36], that is, the prey of existing polar and subpolar penguin species.

The Ice Cap's influence on the climate of the Southern Ocean – cooling – has been profound [7], as has been the response by penguins. However, more recently, during the Pleistocene (back to 2.5 Ma years), every 500,000 years or so "Ice Ages" came and went in the Northern Hemisphere, some, but not all, reflected by Southern Hemisphere changes, in reference to work of L. Thompson, W. Broeker and others investigating what glacial ice cores reveal about climate variation [6, 20, 82]. Clearly, the Antarctic ice sheet and its influence on Southern Ocean climate survived the repeated periods of cooling and warming as did the existing penguin species (other than those species exterminated by humans)! Sea ice, on the other hand, increased or decreased based on the climate regime. The currently existing penguin species not only survived, but in some cases also spread as climate and sea level fluctuated [32, 122]. We will get more into that somewhat, below, as we review penguins' responses to climate changes at the geological to decadal scale, and finally to interannual temporal scales.

Prehistoric Response to a Changing Ocean

While we do not know much about how most penguin species coped with climatic shifts at the long-to-medium temporal scale, we do know quite a bit about how Antarctic species, at least those of the genus *Pygoscelis,* have fared since the Last Glacial Maximum (LGM). That climate regime ended about 12,000 years ago (at the end of the Pleistocene and start of the Holocene; Fig. 1.4). However, most of what we know is related to variation in populations as a function of breeding habitat availability in turn the result of glacial movements and sea level, and not necessarily oceanographic/food web changes. Thus, we will touch on this only briefly. High latitude Pygoscelid penguins, but not emperor penguins, were likely forced away from nesting along the Antarctic coast during glacial maxima. The exception was the northwest Ross Sea and coast, which remained ice free with an adjacent recurring polynya (near-perpetual open water or loose ice cover [117]). The Transantarctic Mountains kept the East Antarctic Ice Sheet from expanding to cover that corner of the Ross Sea coast [67]. One result, as indicated by genetic analysis, are emperor and Adélie penguin variants associated with the Ross Sea [31, 100, 122, 129]. Also indicated by genetic analysis, Adélie and emperor penguin populations (as well as several other penguin species) actually increased during the LGM [122]. It would seem likely that this was a response to decreased sea level (much of the world's water became incorporated into glacial ice, and sea level dropped 200 m! [44]), which in turn provided much more nesting habitat, other than for the emperor (whose colonies are located on floating sea ice; Fig. 2.1). Populations returned to pre-LGM size after the LGM, possibly affected by rising sea level as Ice Caps transformed into liquid water.

Subsequently, these penguins persisted through the Holocene and its significant environmental fluctuations associated with the Medieval Warm Period (ca. 1000–1250 AD) and the cooler Little Ice Age (c. 1300-mid-1800s). Those climate fluctuations brought changes in sea ice extent as perhaps the most obvious factors affecting penguin life style. Penguins do not really like to walk, would rather swim, but extensive sea ice requires walking. From genetic analysis, as noted, we know that both Adélie and emperor penguins persisted through the glacial periods, with respective Ross Sea populations becoming genetically differentiated from the remainder. Based on biochemical analysis of sediment cores beneath the colony location, it appears that the emperor penguin colony in Amanda Bay, East Antarctica, increased in size slightly during the Little Ice Age [62], presumably because the presence of sea ice was more assured (resulting in fewer breeding failures?) [107].

We know quite a bit about how Pygoscelid penguins faired from their subfossil remains left in their colonies. Such deposits are not available for emperor penguins because they form colonies on fast ice, as noted, which fluctuated in location and local extent as ice sheets advanced or retreated in proximity to continental shelves, and which of course mostly disappear in the summer along with any penguin deposits [7]. The possible exception would be locations such as Amanda Bay, where the colony persisted to the degree that droppings reaching the ocean floor could accumulate to provide a record (see above). Due to the relatively dry conditions of the Antarctic, which is actually classified as a desert, there is an impressive record among subfossil Pygoscelid remains: bones, feathers, egg shells, prey parts, found beneath the surface in current and inactive colonies. While most deposits were derived during the Holocene, some deposits date back to the previous interglacial (i.e., back 45,000 years or even older (Steve Emslie, pers. comm.) [41, 44, 47]. Deposits of bones in the guano-laced soils of the Antarctic deteriorate very slowly, in stark contrast to the wet polar regions of the Northern Hemisphere. Yes, some remains of certain mammals, preserved by the cold and high acidity of tundra peat, have been recovered in the Arctic once the permafrost has thawed. But there is no peat in the Antarctic to protect nor, equally importantly, hide subfossils from researchers, which unfortunately seems to be the case for species nesting on subantarctic and even temperate islands (e.g., [57]). Helping in the Antarctic process to preserve bones has been the rapid accumulation of nest stones, as successive generations of Pygoscelid penguins build nests, carrying stones from nearby glacial moraines. The subcolonies become mounds of accumulated stones and (acidic) guano, with subfossil bones buried within, a tribute to times gone by, rather like Viking burial mounds (Fig. 12.1). The rocky surface layer shields what lies beneath from weather and ice. Overall, then, no other bird has a comparable subfossil record (S. Emslie, pers. comm.) (Fig. 13.1).

Owing to the availability of subfossils and interest by several researchers, Pygoscelid penguin response to a changing ocean during the Holocene is fairly well studied along the west coast of the Antarctic Peninsula and its northern offshore islands. The Southern Boundary of the Antarctic Circumpolar Current (SBACC) occurs close to that coast, as does the large scale marginal ice zone (MIZ; Chap. 3). Fluctuation in location and characteristics of the highly productive MIZ, so close to that coast, has had dramatic effects on penguins. The penguins have responded to fluctuation in sea ice cover (and accompanying biotic changes), reduced during the Medieval Warm Period and increased with the Little Ice Age, or even before [111]. The first investigations revealed that Adélie penguins nested in this region at least over the last several hundred years. However, chinstrap penguins nested most

Fig. 13.1 A mound of nest stones and guano in which Adélie penguin subfossils are found; Cape Royds, Antarctica, occupied by penguins for the last 1000 years [43]. (Photo: David Ainley)

abundantly among the northern South Shetland Islands only during short warm periods within the more recent Little Ice Age, and thus perhaps too much snow was present (particularly because it failed to dissipate in spring) when temperatures were colder [39, 45]. Snow covers nesting habitat [22, 92]. Subsequent work indicated that the youngest Adélie and chinstrap colonies occur in the very north of the Peninsula, compared to older ones in the south, further indicating that prevalence of sea ice was involved – the MIZ would shift farther north with the colder temperatures of the Little Ice Age and the penguins would also spread north [40]. Carbon dating of bones indicates that sea ice-avoiding gentoo penguins arrived to nest on islands in the very northern Antarctic Peninsula only recently, i.e., during the last thousand years [42]. Decrease in the prevalence of sea ice would likely also have been involved, as is the case in the present day (see below).

There is also a fairly impressive accumulation of knowledge about the prehistoric presence of Pygoscelid penguins along the coast of East Antarctica. Subfossil remains indicate that Adélie penguins have nested in coastal Wilkes Land (100–130 °E) for at least the last 9000 years [48], and even longer in other locations [129], once the Ice Cap began to retreat to begin the Holocene. Similarly, Adélies have been present in the southern Ross Sea back about the same number of years, in accord with a retreating West Antarctic Ice Sheet – youngest colonies toward the south [44]. Those southern, young colonies are <1000 years old, including what is now one of the largest for the species, Cape Crozier, directly adjacent to the outer cliffs of the West Antarctic Ice Sheet (called the Ross Ice Shelf, as it is floating on the sea). There are also Ross Sea deposits dating back 45,000 years indicating that the penguins came and went with retreat and then advance, followed by retreat again of the ice sheet/ice shelf [41]. Included as well, was loosening vs consolidation of sea ice, with concentrated sea ice cover inhibiting penguin presence [2, 43]. This was particularly evident 3000–4000 years ago, which was a warm period, and more colonies existed where there was less consolidated sea ice along the southern Ross Sea coasts [12, 43]. With return of colder temperatures, those colonies have disappeared with the re-appearance of extensive sea ice cover.

Lacking subfossil remains, all we are left with for conjecture on how other penguins responded to prehistoric climate change is through genetics. On the basis of recent analyses [32, 122], nine genetic lineages exhibited population increase and expansion coincident with the Last Glacial Period (king, chinstrap, yellow-eyed, gentoo [on Kerguelen Islands], Snares, Fiordland, Chatham Islands *Megadyptes* [since extirpated], African and Humboldt penguins). They then decreased as the LGP waned. In contrast, nine lineages (West Antarctic Peninsula and South Georgian gentoo, northern and eastern rockhopper, emperor, Magellanic, royal, erect-crested, and Adélie penguins) exhibited population decrease coincident with advance of the LGP, followed by population expansion toward its end. Otherwise, blue, little, and macaroni penguins exhibited population decreases spanning the entire LGP, while three taxa (Southern rockhoppers, Falkland/Malvinas gentoos, and Galápagos penguins) have been decreasing over a longer period. The authors of these studies were unable to provide an overall explanation, other than possible latitudinal shifts with climate that would bring different species into greater or lesser contact with each other, hence possible competition not just for nesting habitat but also for prey resources (see Chap. 5). Certainly though, as noted above, the dramatic decrease in sea level as glaciers and Ice Caps accumulated more of Earth's water during the LGP must have provided more nesting habitat, affected by local terrain, for many species and vice versa with disappearance of Ice Caps bringing return of high sea levels. For some, growing island Ice Caps and glaciers during the LGP could also have removed nesting habitat.

Possible Prehistoric Changes to Penguins' Food Web

The appearance and disappearance of penguins, "indicator species" that they are [17], is only an indirect measure of how the ocean might have changed, but we do have some direct clues about this. Fish otoliths and squid beaks are also preserved in the ornithogenic soils (soil mixed with penguin guano), along with nest pebbles that compose the mounds on which Pysgoscelid penguins build their nests (Fig. 12.1). In the Ross Sea, which is so far south that it experiences little in the way of between-year sea ice variation (actually the southernmost open ocean, seasonally; e.g., [5, 115]), penguin diet over the last several thousand years has exhibited little variation, unlike the northern edge of the penguin species' ranges in the Antarctic Peninsula. As is the case for the present day (Chap. 4), Antarctic silverfish has continued to dominate the diet among penguins foraging in waters overlying the continental shelf throughout the Holocene [75, 76]. Elsewhere in East Antarctica, along the Wilkes Land coast over the past 9000 years, Antarctic silverfish has also dominated the Adélie penguin diet; and owing to the narrow continental shelf, the lesser presence of glass squid indicates foraging, too, along the shelfbreak during summer [48]. Prevalence of that prey in the diet in that location is also comparable to the present day (Chap. 3).

The Antarctic Peninsula, with its greater climate and sea ice variability than the higher latitude Antarctica, shows a different pattern, its tip being 14 latitude degrees north of the southern Ross Sea (same, approximately, as distance from Anchorage AK to Portland OR, or London to Gibraltar). The Antarctic silverfish is associated with sea ice [72] (Chap. 4), which explains its slightly lesser representation in the diet at Anvers Island, mid-Peninsula, during warmer prehistorical periods; though certainly some, lesser amount, of sea ice was present [45]. There was probably a shorter sea ice season, as now [91, 115], and in accord fewer silverfish [73]. When there were fewer silverfish in the diet, there was a higher presence of glass squid. The shelf is narrow there, too, and thus glass squid as well as Antarctic lanternfish occurs more in the diet than in the Ross Sea (see Chap. 4). The idea is that, with less sea ice, there are fewer silverfish, forcing the penguins to forage farther away from colonies, out to the continental slope and beyond where they find different prey. Similar diet variation was found at sites a bit farther to the south in that region over the past several thousand years [46].

Overall then, there is no evidence that the Antarctic coastal food web has changed much during the Holocene, as Pygoscelid populations waxed and waned with the advance and retreat of ice sheets affecting availability of nesting and foraging habitat. When there was slightly less sea ice cover in the coastal ocean of the western Antarctic Peninsula, silverfish were less prevalent in the diet, consistent with the pattern today [104]. Unfortunately, given the relative fragility of fish bones and squid beaks there are no records to reveal how penguin prey may have changed their prevalence in waters north of the SBACC (Joe Eastman, pers. comm.).

The Anthropocene: How Will Penguins Cope, Now Also Dealing with Humans

Response to Long-Term Climate Change

While there are numerous reports of changes in penguin breeding patterns and populations as a response to short-term variation in ocean climate, such as to El Niño, we are not interested in that subject here (but see below). Rather, we are interested in multidecadal scale patterns. Earth's overall climate has been changing at an accelerated rate over the last ~170 years, but the over-exploitation of marine biological resources, i.e., penguin food, by humans has accelerated too, complicating penguin population climate responses and in some cases is more telling (Chap. 5, below). We, i.e., penguins and ourselves, are now living in what is referred to as the Anthropocene, having replaced the Holocene as a geological period, since about the mid-1800s [37, 123]. Human activity is now the major factor changing Earth's atmosphere and landscape. The greenhouse effect (warming) is progressing as humans release more carbon dioxide (CO_2) and methane (CH_4) into the atmosphere, and 90% of the excess heat generated has been, and is being, absorbed by the ocean. As a result the ocean has warmed on average >1 °C, more or less depending on region and depth, but which is huge considering the difficulty in heating water [28, 29]. In addition, the release into the atmosphere of chlorofluorocarbon refrigerants has led to a large, intense ozone hole over Antarctica (AOH). One result is that the stratosphere, which contains most of the ozone, now destroyed, over Antarctica has cooled by >9 °C! These factors, by increasing the atmospheric pressure differential between Antarctica and the mid-latitude ocean and between the stratosphere and surrounding troposphere, are changing the wind patterns over Antarctica and the Southern Ocean.

The winds have been increasing in strength. In turn, increased winds are affecting sea ice extent (greater, everywhere but the Antarctic Peninsula), the length of the sea ice season (less near to continent, more away from continent), and intrusion of warm water into continental shelf submarine canyons (increasing), though the degree of change depends on region within the Southern Ocean [23, 91, 114, 115, 118, 120, 130]. The response of penguins to these climate change patterns is easy to explain in a few examples. With the recent (last few decades) southern retreat of sea ice along the northwestern Antarctic Peninsula, as well as a shortening of the sea ice season there [115], in accord the ranges of the ice-avoiding gentoo and chinstrap penguins have been spreading south, while in contrast, at the same time populations of the ice-obligate Adélie penguin have also been shifting south [54, 79, 108, 111]. In other words, the Adélie penguin is being replaced by gentoo and chinstrap penguins. As sea ice extent expands elsewhere around Antarctica, also the effect of increasing wind but along with the wind-driven coastal polynyas with which they are associated, Adélie penguin colonies have been increasing in size [4, 78, 112]. Eventually, however, if atmospheric temperatures continue to rise (especially reaching an average 1.5 °C above preindustrial levels) with associated changes in other factors, sea ice will begin to retreat in a widespread fashion and models show that the sea ice obligate Adélie and emperor penguins will begin to decrease and become more confined to areas where sea ice still remains [5, 66] (see Chap. 2). 'If's' being what they are (guesses), recently an atmospheric generated increase in sea ice cover along the coast has led to decreased Adélie penguin populations along the coast of East Antarctica (see below).

Since sea ice inhibits industrial fishing vessels (most do not have ice-strengthened hulls), penguin species frequenting regions that have no sea ice have not been protected from human exploitation of penguin prey (see Chap. 5). Therefore in many cases, since increased resource exploitation and climate change have occurred simultaneously, it is difficult to separate the factors that have caused decadal changes in penguins' populations. Seabird populations, in general, have severely decreased over the past several decades as a result of over-fishing of their prey, especially forage fish, with help from climate change [55, 90]. Penguins are not exempt from this, which is an indirect effect of human activity (Chap. 5). The most extreme example of a recent human-caused penguin population shift, and this a direct effect, is the invasion of little penguins from Australia to occupy where New Zealand little penguins had been extirpated by humans during 1500–1900 [56].

Other penguin species, though less so, have also been affected by changes associated with direct human activity, as well as indirectly through climate change. Chinstrap penguins, after invading south in waters off the western Antarctic Peninsula and Scotia Sea in association with lessened prevalence of sea ice (noted above), more recently have been decreasing in those newly occupied waters [77, 116]. As discussed in Chap. 5, also involved in this trend is the recovery, return and reestablished trophic competition of baleen whales and Antarctic fur seals now in huge numbers, unlike back in the middle twentieth century. Chinstraps had become super abundant upon near extirpation of those competitors, leading to a hypothesized "krill surplus" [10, 60]. A similar scenario may be the same for macaroni penguins in the Scotia Sea, with recent decrease related more to fur seal recovery than to whale recovery (see Chap. 5). Although increasing sea temperatures have been implicated in the long-term decrease of eastern rockhopper penguins at Campbell Island, appreciable predation by recovering New Zealand sea lions likely is also involved [83, 84] (see Chap. 12).

On the other hand, the continued increase in the small colonies of gentoo penguins along the northern Antarctic Peninsula, after initially taking advantage of retreating sea ice, may be linked to recovery of stocks of its prey fishes following their commercial, near-extirpation in the 1960s. It was likely that populations of this penguin initially prospered by taking advantage of the temporary huge surplus of Antarctic Krill (due to the loss of whales and fur seals), with fish being more typical of this species' diet. Fish are far more energetically valuable (also discussed in Chaps. 4 and 5). In similar links, while the decrease of yellow-eyed penguins in New Zealand correlates to decades of rising sea surface temperatures and inferred reduced ocean productivity [81], the fact that its prey base has been severely depleted and altered by the fishing pressure that also grew during this time has to be included as a major causative factor [80, 124]. Ocean climate change in the Benguela Current appears to be altering the distribution of prey fish for African penguins, although the main culprit is likely over-fishing [110]. While aerial birds with their powers to search widely for food appear to be somewhat compensating, the changed prey distribution is having negative consequences for penguins, with their more limited prey searching capability [35] (discussed more fully in Chap. 5).

Trends related to climatic factors in other penguin populations are harder to explain, other than sea temperature (or the air above the sea) being a proxy for ocean climate and, it follows, food web change of unknown quality and degree. In the case of little penguins in Tasmania, the slightly rising sea temperatures over recent decades, as an index to food web processes (and behavior of the STF; see Chap. 3), appear to be beneficial with earlier laying, and heavier chicks, but with no population trend yet apparent [27]. Likewise, despite laying earlier over recent decades as sea temperatures warm, the royal penguin at Macquarie Island, a bit southwest of NZ, is also showing no population effect [58]. On the other hand, the northern rockhopper penguins of Amsterdam Island have been decreasing since the 1970s in accord with rising air temperatures (a proxy for sea temperature), in contrast to king penguins at Crozet and Kerguelen islands whose numbers have been increasing [125]. Cooling temperatures in the vicinity of Galápagos Islands are deemed beneficial to the penguins there, though trends are obscure, perhaps due to contrary influence of other anthropogenic factors, such as on-land predation [68]. Among other *Spheniscus* species, despite stable sea surface temperature around the Tierra del Fuegian islands during 1992–2012, populations of Magellanic (and gentoo) penguins have been increasing [98]. Finally, though climate-related storm frequency has been increasing, with negative effects when a storm occurs (e.g., chilling of wet chicks), the Magellanic penguin populations in central Argentina have not (yet) been affected [18].

That brings us back to penguins' tolerance, as homeotherms, to shifts in the temperature of their respective ocean (see start of this chapter). Irrespective of the degree to which temperature will be changing among Southern Ocean water masses and climatic zones in the future, smaller-bodied penguins, which live in warmer waters than large-bodied species, would appear to be more sensitive (Chap. 7). Having less tolerance to temperature range, they would have to shift their breeding locations in response. To date, there is no evidence that such a range shift has occurred, despite general warming of the world's oceans, with greater relative change in the cooler portions [28, 29].

Response to Short-Term Ocean Climate Variation

The phenomenon of El Niño-Southern Oscillation (ENSO) and the seabird response has been known for at least 100 years [64, 85, 86], and the ocean processes involved have been well known for the past 50 years [53]. However, it was not until the intense 1983–84 El Niño had dramatic negative effects on eastern and central Pacific Ocean seabird populations that marine ornithologists finally paid full attention [3, 109]. What followed was a couple of decades in which study after study attempted to relate interannual variability in seabird species' natural history patterns, particularly the negative excursions from normality in breeding success and related subjects, to subsequent ENSOs (particularly the El Niño portion, see below), or more generally variation in sea surface temperature (SST), as a proxy for food web structure and productivity. Indeed, it is many times easier to measure SST than prey availability, and then make the assumption that ENSO represents a model ("experiment") to explain most cases of interannual seabird natural history as a function of food web variability.

ENSO is generated by the periodic oscillation of wind strength along the Equatorial Pacific (responding to variation in atmospheric pressure between western Australia and the eastern Pacific), which in turn affects the depth of the mixed layer through propagation of long-period internal ocean waves, i.e., those that take weeks to travel along a thermocline (see Chap. 3). They then reflect/rebound off the America's coasts, spreading their influence to the north and south. A deep thermocline along the America's coasts, the location of the eastern boundary currents (Chap. 3), leads to the upwelling of warm, depleted water (El Niño), as opposed to a shallow thermocline bringing the upwelling of cold, highly productive water (La Niña). So, during El Niño, phytoplankton growth is diminished with the deeper mixed layer/warmer surface waters, and in turn, the abundance of grazers, e.g., anchovy, sardine, is reduced. The cycle between both extremes of El Niño and La Niña occurs over the course of 5–7 years. Seabird nesting success and survival vary accordingly, and basically this is why researchers identify warm water as poor, and cold water as beneficial for seabirds (including penguins). We saw above that this general rule of thumb has major exceptions. However, unfortunately, though the negative El Niño is given more attention than the positive La Niña, long-lived, late-maturing birds, like penguins, are geared to handle such variation (taking the good with the bad, as long as the bad does not occur very often). El Niño, the warm phase of ENSO, is not a proxy for warming of the ocean due to global atmospheric warming, owing to the difference in propagation and transmission. ENSO alters the ocean through internal waves and alteration of wind patterns as atmospheric temperature responds to regional changes in ocean temperature. Global warming alters the ocean through the absorption to depth of excess heat in the atmosphere, which contacts the ocean at its surface, and is not an episodic phenomenon.

While ENSO has long been recognized to affect the penguins of the Peru Current, including Galápagos penguins [15, 16, 86, 121], with changes in reproductive patterns and survival due to variation in prevalence of prey, it has since been found to correlate with variation in penguin natural history elsewhere – the so-called teleconnections among various portions of the globe. ENSO, though a tropical phenomenon, affects the wind patterns of the Southern Ocean, and in turn sea ice extent [71, 115, 119] and ultimately Adélie penguin natural history patterns [65, 126]. Influence through alteration of ocean climate, including mixed layer depth (Chap. 3), has also been detected among lower-latitude penguin species, e.g., among gentoo and macaroni penguins in the northern Scotia Sea [50, 61], macaroni, and southern rockhopper penguins at Marion Island, Indian Ocean [34], king penguins at Crozet Island [19], and African penguins in the Benguela Current [102]. During warm-water events, little penguins do not do well, although whether that has to do with temperature tolerance or changes in the food web remains to be seen. Pauline Reilly [99] documented their poor breeding success in response to two warm-water events during the 1970s thought to be an effect of reduced availability of appropriate prey. The results of subsequent research indicate that wind and wave characteristics, which are affected by ENSO, can have a direct effect on the foraging effort and efficiency of little penguin: in stronger winds and larger waves the penguins do more poorly [13, 106].

Another expression of climate variation that affects the ocean in which penguins find themselves is the Southern Annular Mode (SAM). It changes wind patterns owing to alteration of atmospheric pressure centers between the Amundsen Sea and Weddell Sea on the opposite side of Antarctica and, like ENSO, varies on a decadal time scale between two states, a positive SAM (stronger winds) and a negative SAM (weaker winds). However, due to the way the Antarctic ozone hole affects pressure centers (see above), SAM has become stuck in its positive mode for the last few decades [118, 120]. Like ENSO, it affects sea ice extent and sea ice season, as a function of wind strength. Because SAM has been stuck in its positive (windier) mode, as noted above, sea ice has mostly been increasing around most of Antarctica, with the exception of the southern Scotia Sea/northern Bellingshausen Sea (west coast of Antarctic Peninsula). There, sea ice cover and season has decreased [115]. The reason that sea ice is disappearing in northern Antarctic Peninsula/Scotia Sea region is that the increasing winds have forced a large, persistent wave in the South Polar Jet Stream, with a large southward bend sucking warm air from the South Atlantic into the Scotia Sea region [115]. Being stuck in its positive state, SAM has a major effect of global climate change (through ozone depletion), despite Southern Ocean sea ice mostly increasing in extent and season (which does not make the nightly news). Due to changed wind patterns and sea ice patterns in the Southern Ocean, its influence on Adélie and emperor penguins has been very much evident [4, 125]. Noted above, as a result of winds bringing more persistent open water near to colonies (polynyas), with which emperor and Adélie penguin colonies are associated [105], with large scale sea ice extent also increasing, Adélie penguin colonies have been increasing in much of Antarctica [78, 112]. On the other hand, most recently with a shift in SAM, winds apparently have slackened along East Antarctica, resulting in an increase in sea ice cover, further leading to a not surprising decrease in breeding productivity and decreasing Adélie penguin populations [38].

Finally, among the short-term climate factors that affect penguins are Marine Heat Waves (MHWs). These are defined as prolonged periods of anomalously warm water – SST exceeds the 90th percentile of the expected temperature for a period of at least 5 days [59, 88, 89], though obviously it's the MHWs that last for hundreds of days, covering hundreds of km^2 that matter the most. MHW generally occur due to atmospheric forcing, often extending from an El Niño, in which extreme waves in the subtropical and polar jet streams trap and thus perpetuate the conditions (a bit like the jet stream wave mentioned above). To be sure, the opposite/cold anomalies of extended duration also occur (www.marineheatwaves.org/tracker), but receive far less attention, as seabird breeding failures and high mortality, as in most cases of extreme El Niño, make better press, though for sure need to be documented. MHWs usually affect the upper water column, down to 50 m, although in some cases can affect the deeper waters. They change the depth and intensity of stratification through the surface warming and, thus, the surface mixed layer (see above), and in so doing make prey less available to seabirds (especially surface-foraging ones) for a given location. Perhaps the best documented MHW, as investigation included effects on the prey base as well as the seabird predators, was in the eastern North Pacific in 2015–16 [94]. The incidence of MHWs has been increasing over the past century, in part due to gradual overall warming of the oceans, but is largely confined to subtropical and temperate zones, and most of them thus far have occurred in Equatorial and Northern Hemisphere waters. Therefore, penguin populations have to date been little affected. Those that have been important to penguins, in this case little penguins, occurred in waters along the Western Australian coast in 2011 and off southeastern Australia/western Tasman Sea in both 2015–16 and 2017–18. A MHW occurred in the northern Benguela Current in 2001, but well north of African penguin range, and one occurred in the northern Peru Current during 2017, thus within the range of Humboldt and Galápagos penguins. However, no information is available on how it may have affected penguins.

The most thorough documentation of penguin response to MHW comes from waters off Western Australia in 2011 and Tasmania in 2015–16 [24, 49]. The changed structure of the water column affected phytoplankton and zooplankton. During the 2011 MHW, little penguins at a colony at the northern, most tenuous margin of the species' range, Penguin Island (near Perth), compared to the previous two decades, exhibited reduced breeding success, with lower fledgling success, fewer chicks per pair and generally a lower mean mass of fledglings. The usual fish fauna was replaced by a warmer-water community, with penguin diet changing accordingly [24, 25]. In the Tasman Sea events, more affected was the zooplankton which predominates more off the shelf, compared to fish, which are mostly on the shelf. Compared to the populations of petrel and albatross species in the area, which occurred mostly off the shelf, the prevalence and distribution of little penguins, here in the middle of the species' range, were little affected consistent with their diet, mostly fish (Chap. 4), and their close-to-coast occupation of the shelf (Chap. 5). The penguins sought the cooler patches of ocean, which remained available. It is interesting that the MHW off Tasmania did not affect the little penguins much, compared to isolated warm-water events related to ENSO [99] (see above).

Response to Marine Pollution

Humans have been disposing of their trash and wastes in the ocean for millennia, though any effects on the coastal biota for the longest time had been local, and for penguins likely not an issue. In the twentieth century, though, as the number of humans grew to reach unimaginable levels, with resources becoming much harder to come by, they have increasingly turned to synthetic materials, both to provide power and to make their day-to-day goods. In accord, dispersal and disposal of wastes and pollution have more and more affected the atmosphere and ocean, and in their turn its inhabitants. The synthetic materials tend not to be "eco-friendly," i.e., they persist for a very long time and don't do good things for the creatures contacting them. With the growing human population, these days as well there is more use of coastal waters, more ship traffic, and the increased noise appears to be having an impact.

Pointed out by Yan Ropert-Coudert and company [101], penguins increasingly have had to deal with plastic, mercury, persistent organic pollutants (POPs; and other chemicals) and oil pollution in the ocean. According to those authors' review, plastic has not been much of a problem because penguins only eat live prey (Chap. 4), and do not scavenge, as do many other seabirds. Penguins are consequently not interested in floating materials, unlike those other seabirds, which to an increasingly negative result, ingest plastic items thinking they are food [70, 113]. There are some isolated exceptions of penguins consuming plastic items, as judged by presence in their stomach contents, though whether it was them or their prey that consumed the plastic is not known (e.g., [97]). It remains to be seen whether microplastics will become an issue for penguins; those are microscopic particles derived from the breakup and deterioration of plastic items [9, 26]. Examination of penguin and other seabird scats can be used to monitor microplastic incidence in the food web, though microplastics mostly from atmospheric sources have been identified in the Antarctic [8]. So far, there does not appear to be an issue that penguins have been affected directly in the Antarctic or elsewhere by microplastics [14, 51, 74].

Penguins encounter POPs, mercury and other substances through the food web in a process called bio-magnification. With each stage in the food web, upward in the trophic pyramid, these compounds that do not break down are first consumed by the grazers, and then are passed on to each higher trophic level of predator as creatures of one level consume those of the next lower level. They accumulate, and then are passed on in an additive process, with each stage. Mercury, derived from methylmercury generated in the process of burning coal is everywhere, and owing to atmospheric processes involving the industrialized Northern Hemisphere (i.e., carried north by winds) is especially rampant in the Arctic, where species like polar bears (and humans) have become in a sense walking toxic waste dumps [33]. Concentrations of mercury in polar bears (which averages >20 ppm), can cause neurological and other problems. However, rates are more than an order of magnitude greater than what has been found in any penguin (e.g., 5.9 ppm in gentoo penguins at Crozet and Kerguelen islands [101]. Among penguins, owing to atmospheric and ocean processes that flow away from Antarctica (Chap. 3), and using the values summarized in the latter study, the average level of mercury found in penguin feathers at, or north of, the Antarctic Polar Front is 2.45 ± 1.6 SD ppm (25 studies), whereas the average to the south is 0.87 ± 0.5 ppm (17 studies). Also involved in the different levels between polar bears and penguins is their respective placements in the food web; bears are one trophic level higher – bears eat seals that eat fish but the penguins just eat fish [21]. There are no polar bears in the Southern Hemisphere, and that is a good thing for penguins.

The type of pollution that has had an incredibly large effect on penguins is that of oil floating on the sea surface, originating from oil platform blow outs but especially oil tanker wrecks. Oil spills in the penguin world therefore have been most pervasive where there is heavy, coastal tanker traffic, i.e., around Cape of Good Hope, South Africa, and Cape Horn, Patagonia [30, 69, 103]. Tens of thousands of African penguins have been killed, as well as significant numbers of other species. Given the importance of penguins' feathers to their diving and heat conservation capacities, as detailed in Chaps. 6 and 7, having their coats matted by oil is, to put it mildly, problematic. Indeed, losing their water repellency and insulation capacity is not survivable. Given the importance in several ways of their feathers (Chap. 7), it is not surprising that penguins normally spend a lot of time taking care of their feathers, washing and preening, and even applying a light, water-proofing oil from their uropygial gland (lower back, at base of tail). They do this every time they emerge from the sea in preparation for their next plunge [1] (Chap. 14). They might spend an hour or more preening upon arriving from the sea, before venturing to their nests to relieve mates or feed chicks. However, heavy petroleum oil coating their feathers is a totally different story, and cannot be removed by the oiled penguins.

Somewhat fortunate is that penguins, unlike other seabirds, do very well in captivity and thus are frequent inhabitants of aquaria and zoos. Therefore, they can be cleaned of oil, using mild dish washing detergent, but then often being held through a subsequent molt, before returning them to the sea; upon being returned they do fairly well [11, 63, 87] (Fig. 13.1). Moreover, they can be easily transported, especially by air, to a location away from the site of an oil spill, far enough that the site can being cleaned before penguins make it back [63, 103]. Aerial seabird species would make it back too soon! (Fig. 13.2).

To be sure, rehabituated penguins subsequently do not quite measure up to unoiled ones in regard to propensity to breed and breeding success, likely due to the ingestion of oil and its poisons while attempting to preen themselves clean before being cleaned by humans [11, 128]. Measuring that difference is not as easy as it seems, though, especially if penguins are migratory or if the oiling occurs on the wintering grounds. Thus, before and after comparisons cannot be made; it has best

Fig. 13.2 Top, Magellanic penguins heavily coated with oil, though even a small spot of oil can disrupt the powers of their feather coat to keep them warm; middle, a collection of penguins in various stages of being cleaned; bottom, penguins that have been cleaned, and upon release are headed for the ocean. (Photos courtesy of Valeria Ruoppolo)

been measured in the nonmigratory African and little penguins [11, 103, 128]. The issue of oil pollution has been well summarized elsewhere [52, 101].

Finally, there is still another kind of pollution. In coastal waters, there is increasing sound pollution where there is much boat and ship traffic. It has been known for some time that penguins have sensitive hearing [127] (Chap. 14), but only recently has it been discovered that they are averse to loud subsea noise [95]. Penguins do hear well beneath the waves, and actually elicit calls underwater, a subject to be discussed in the next chapter. The African penguin population nesting on islets In Algoa Bay, South Africa, has decreased noticeably in accord with a dramatic increase in ship traffic, and with it sound pollution, during the last two decades [96].

Increasingly, despite their love of penguins (Chap. 15) human activity, in several ways, is leading to the demise of a number of penguin species, and as detailed in that chapter, need our help.

References

1. Ainley DG (1974) Comfort behavior of Adélie and other penguins. Behaviour 50:16–50
2. Ainley DG (2002) The Adélie penguin: bellwether of climate change. Columbia University Press, New York
3. Ainley DG, Carter HR, Anderson DW, Briggs KT, Coulter MC, Cruz F, Cruz JB, Valle CA, Fefer SI, Hatch SA, Schreiber EA, Schreiber RW, Smith NG (1986) Effects of the 1982–1983 El Nino-Southern Oscillation on Pacific Ocean bird populations. National Museum of Canada, Ottawa, pp 1747–1758
4. Ainley DG, Clarke ED, Arrigo K, Fraser WR, Kato A, Barton KJ, Wilson PR (2005) Decadal-scale changes in the climate and biota of the Pacific sector of the Southern Ocean, 1950s to the 1990s. Antarct Sci 17:171–182
5. Ainley DG, Russell J, Jenouvrier S, Woehler E, Lyver POB, Fraser WR, Kooyman GL (2010) Antarctic penguin response to habitat change as earth's troposphere reaches 2°C above pre-industrial levels. Ecol Monogr 80:49–66
6. Alley RB (2000) The two mile time machine. Princeton University Press, Princeton
7. Anderson JB (1999) Antarctic marine geology. Cambridge University Press, Cambridge
8. Aves AR, Revell LE, Gaw S, Ruffell H, Schuddeboom A, Wotherspoon NE, Larue M, Mcdonald AJ (2022) First evidence of microplastics in Antarctic snow. Cryosphere 16:2127–2145
9. Avio CG, Gorbi S, Regoli F (2017) Plastics and microplastics in the oceans: from emerging pollutants to emerged threat. Mar Environ Res 128:2–11
10. Ballance L, Pitman RL, Hewitt RP, Siniff DB, Trivelpiece WZ, Clapham PJ, Brownell RL Jr (2006) The removal of large whales from the southern ocean: evidence for long-term ecosystem effects? In: Estes JA, Demaster DP, Doak DF, Williams TE, Brownell RL Jr (eds) Whales, whaling and ocean ecosystems. University of California Press, Berkeley, pp 215–230
11. Barham PJ, Underhill LG, Crawford RJM, Leshoro TM (2007) Differences in breeding success between African penguins (*Spheniscus demersus*) that were and were not oiled in the MV Treasure oil-spill in 2000. Emu 107:7–13
12. Baroni C, Orombelli G (1994) Abandoned penguin colonies as Holocene paleoclimatic indicators in Antarctica. Geology 22:23–26
13. Berlincourt M, Arnould JPY (2015) Influence of environmental conditions on foraging behaviour and its consequences on reproductive performance in little penguins. Mar Biol 162:1485–1501

14. Bessa F, Ratcliffe N, Otero V, Sobral P, Marques JC, Waluda CM, Trathan PN, Xavier JC (2019) Microplastics in gentoo penguins from the Antarctic region. Sci Rep 9:14191
15. Boersma PD (1977) An ecological and behavioral study of the Galapagos penguin. Living Bird 15:43–93
16. Boersma PD (1998) Population trends of Galapagos penguins: impacts of El Niño and La Niña. Condor 100:245–253
17. Boersma PD (2008) Penguins as marine sentinels. Bioscience 58:597–607
18. Boersma PD, Rebstock GA (2014) Climate change increases reproductive failure in Magellanic penguins. PLoS One 9:e85602
19. Bost CA, Cotté C, Terray P, Barbraud C, Bon C, Delord K, Gimenez O, Handrich Y, Naito Y, Guinet C, Weimerskirch H (2015) Large-scale climatic anomalies affect marine predator foraging behaviour and demography. Nat Commun 6:8220
20. Bowen M (2005) Thin ice: unlocking the secrets of climate in the world's highest mountains. Henry Holt & Co, New York
21. Brasso RL, Chiaradia A, Polito MJ, Raya Rey A, Emslie SD (2015) A comprehensive assessment of mercury exposure in penguin populations throughout the Southern Hemisphere: using trophic calculations to identify sources of population-level variation. Mar Pollut Bull 97:408–418
22. Bricher PK, Lucieer A, Woehler EJ (2008) Population trends of Adélie penguin (*Pygoscelis adeliae*) breeding colonies: A spatial analysis of the effects of snow accumulation and human activities. Polar Biol 31:1397–1407
23. Bronselaer B, Winton M, Griffies SM, Hurlin WJ, Rodgers KB, Sergienko OV, Stouffer RJ, Russell JL (2018) Change in future climate due to Antarctic meltwater. Nature 564(7734):53–58
24. Cannell B, Thomson P, Schoepf V, Pattiaratchi C, Fraser M (2019) Chapter 10: Impacts of marine heatwaves. In: Techera E, Winter G (eds) Marine extremes: ocean safety, marine health and the blue economy. Routledge, New York
25. Cannell BL, Chambers LE, Wooller RD, Bradley JS (2012) Poorer breeding by little penguins near Perth, Western Australia is correlated with above average sea surface temperatures and a stronger Leeuwin Current. Mar Freshw Res 63(10):914–925
26. Caruso G, Bergami E, Singh N, Corsi I (2022) Plastic occurrence, sources, and impacts in Antarctic environment and biota. Water Biol Secur 1(2):100034
27. Chambers LE, Devney CA, Congdon BC, Dunlop N, Woehler EJ, Dann P (2011) Observed and predicted effects of climate on Australian seabirds. Emu 111:235–251
28. Cheng L, Abraham J, Hausfather Z, Trenbeth KE (2019) How fast are the oceans warming? Science 363:128–129
29. Cheng L, Abraham J, Trenberth KE, Fasullo J, Boyer T, Mann ME, Zhu J, Wang R, Locarnini R, Li Y, Zhang B, Tan Z, Yu F, Wan L, Chen X, Song X, Liu Y, Reseghetti F, Simoncelli S, Gouretski V, Chen G, Mishonov A, Reagan J (2022) Another record: Ocean warming continues through 2021 despite La Niña conditions. Adv Atmos Sci 39:373–385
30. Clark RB (1986) Marine pollution. Clarenden, Oxford
31. Clucas GV, Younger JL, Kao D, Emmerson L, Southwell C, Wienecke B, Rogers AD, Bost CA, Miller GD, Polito MJ, Lelliott P, Handley J, Crofts S, Phillips RA, Dunn MJ, Miller KJ, Hart T (2018) Comparative population genomics reveals key barriers to dispersal in Southern Ocean penguins. Mol Ecol 27(23):4680–4697
32. Cole TL, Dutoit L, Dussex N, Hart T, Alexander A, Younger JL, Clucas GV, Frugone MJ, Cherel Y, Cuthbert R, Ellenberg U (2019) Receding ice drove parallel expansions in Southern Ocean penguins. Proc Natl Acad Sci 116(52):26690–26696
33. Cone M (2005) Silent snow: the slow poisoning of the arctic. Grove Press, New York
34. Crawford RJM, Dyer BM, Cooper J, Underhill LG (2006) Breeding numbers and success of *Eudyptes* penguins at Marion Island, and the influence of mass and time of arrival of adults. CCAMLR Sci 13:175–190

35. Crawford RJM, Whittington PA, Martin AP, Tree AJ, Makhado AB (2009) Population trends of seabirds breeding in South Africa's Eastern Cape and the possible influence of anthropogenic and environmental change. Mar Ornithol 37:159–174
36. Eastman JT (1993) Antarctic fish biology: evolution in a unique environment. Academic, New York
37. Edwards LE (2015) What is the anthropocene? Eos 96:6–7
38. Emmerson L, Southwell C (2022) Environment-triggered demographic changes cascade and compound to propel a dramatic decline of an Antarctic seabird metapopulation. Glob Change Biol:1–16. https://doi.org/10.1111/gcb.16437
39. Emslie SD (1995) Age and taphonomy of abandoned penguin colonies in the Antarctic Peninsula region. Polar Rec 32:408–418
40. Emslie SD (2001) Radiocarbon dates from abandoned penguin colonies in the Antarctic Peninsula region. Antarct Sci 13(3):289–295
41. Emslie SD (2020) Ancient Adélie penguin colony revealed by snowmelt at Cape Irizar, Ross Sea, Antarctica. Geology 49(2):145–149
42. Emslie SD, Baumann K, Van Tuinen M (2011) Late Holocene occupation of gentoo penguins (*Pygoscelis papua*) at Byers Peninsula, Livingston Island, Antarctica. Polar Biol 34(2):283–290
43. Emslie SD, Berkman PA, Ainley DG, Coats L, Polito M (2003) Late-Holocene initiation of ice-free ecosystems in the southern Ross Sea, Antarctica. Mar Ecol Prog Ser 262:19–25
44. Emslie SD, Coats L, Licht K (2007) A 45,000 yr record of Adélie penguins and climate change in the Ross Sea, Antarctica. Geology 35(1):61–64
45. Emslie SD, Fraser W, Smith RC, Walker W (1998) Abandoned penguin colonies and environmental change in the Palmer station area, Anvers Island, Antarctic Peninsula. Antarct Sci 10(3):257–268
46. Emslie SD, Mcdaniel J (2002) Adélie penguin diet and climate change during the middle to late Holocene in Marguerite Bay, Antarctic Peninsula. Polar Biol 25:222–229
47. Emslie SD, Polito MJ, Brasso R, Patterson WP, Sun L (2014) Ornithogenic soils and the paleoecology of pygoscelid penguins in Antarctica. Quat Int 352:4–15
48. Emslie SD, Woehler EJ (2005) A 9000-yr record of Adélie penguin occupation and diet at the Windmill Islands, East Antarctica. Antarct Sci 17:57–66
49. Evans R, Lea MA, Hindell MA (2021) Predicting the distribution of foraging seabirds during a period of heightened environmental variability. Ecol Appl 31(5):e02343
50. Forcada J, Trathan PN (2009) Penguin responses to climate change in the Southern Ocean. Glob Chang Biol 15:1618–1630
51. Fragão J, Bessa F, Otero V, Barbosa A, Sobral P, Waluda CM, Guímaro HR, Xavier JC (2021) Microplastics and other anthropogenic particles in Antarctica: Using penguins as biological samplers. Sci Total Environ 788:147698
52. García-Borboroglu P, Boersma PD, Reyes L, Skewgar E (2008) Petroleum pollution and penguins: marine conservation tools to reduce the problem. In: Hofer TN (ed) Marine pollution: new research. Nova Science Publishers, New York, pp 339–356
53. Glantz MH, Thompson JD (eds) (1981) Resource management and environmental uncertainty: lessons from coastal upwelling fisheries. Wiley, New York
54. Gorman KB (2015) Integrative studies of southern Ocean food-webs and *Pygoscelis* penguin demography: mechanisms of population response to environmental change. PhD dissertation, Bozeman, Montana: Montana State University
55. Grémillet D, Ponchon A, Paleczny M, Palomares M-L D, Karpouzi V, Pauly D (2018) Persisting worldwide seabird-fishery competition despite seabird community decline. Curr Biol 28:4009–4013
56. Grosser S, Rawlence NJ, Anderson CNK, Smith IWG, Scofield RP, Waters JM (2016) Invader or resident? Ancient-DNA reveals rapid species turnover in New Zealand little penguins. Proc R Soc B 283:20152879

57. Harsley-Smith G, Hamilton JE (1950) A recent deposit of bird bones in The Falkland Islands. Nature 166:198
58. Hindell MA, Bradshaw CJA, Brook BW, Fordham DA, Kerry K, Hull C, Mcmahon CR (2012) Long-term breeding phenology shift in royal penguins. Ecol Evol 2:1553–1571
59. Hobday AJ, Alexander LV, Perkins SE, Smale DA, Straub SC, Olive ECJ, Benthuysen J, Burrows MT, Donat MG, Feng M, Holbrook NJ, Moore PJ, Scannell HA, Gupta AS, Wernberg T (2016) A hierarchical approach to defining marine heatwaves. Prog Oceanogr 141:227–238
60. Hofman RJ (2017) Sealing, whaling and krill fishing in the Southern Ocean: past and possible future effects on catch regulations. Polar Rec 53(1):88–99
61. Horswill C, Matthiopoulos J, Ratcliffe N, Green JA, Trathan PN, Mcgill RAR, Phillips RA, O'onnell TC (2016) Drivers of intrapopulation variation in resource use in a generalist predator, the macaroni penguin. Mar Ecol Prog Ser 548:233–247
62. Huang T, Yang L, Chu Z, Sun L, Yin X (2016) Geochemical record of high emperor penguin populations during the Little Ice Age at Amanda Bay, Antarctica. Sci Total Environ 565:1185–1191
63. Hull CL, Hindell MA, Gales RP, Meggs RA, Moyle DI, Brothers NP (1998) The efficacy of translocating little penguins *Eudyptula minor* during an oil spill. Biol Conserv 86:393–400
64. Hutchinson GE (1950) The biogeochemistry of vertebrate excretion. American Museum of Natural History, New York
65. Jenouvrier S, Barbraud C, Weimerskirch H (2006) Sea ice affects the population dynamics of Adélie penguins in Terre Adélie. Polar Biol 29:413–423
66. Jenouvrier S, Holland M, Stroeve J, Serreze M, Barbraud C, Weimerskirch H, Caswell H (2014) Projected continent-wide declines of the emperor penguin under climate change. Nat Clim Chang 4(8):715–718
67. Johnson JS, Hillenbrand C-D, Smellie JL, Rocchi S (2008) The last deglaciation of Cape Adare, northern Victoria Land, Antarctica. Antarct Sci 20:581–587
68. Karnauskas KB, Jenouvrier S, Brown CW, Murtugudde R (2015) Strong sea surface cooling in the eastern equatorial Pacific and implications for penguin conservation. Geophys Res Lett 42:6432–6437
69. Kerley GIH, Erasmus T (1987) The management of oiled penguins. American Petroleum Institute, Washington, DC, pp 465–468
70. Kühn S, Van Franeker JA (2019) Quantitative overview of marine debris ingested by marine megafauna. Mar Pollut Bull 151:110858
71. Kwok R, Comiso JC (2002) Southern Ocean climate and sea ice anomalies associated with the Southern Oscillation. J Clim 15(5):487–501
72. La Mesa M, Eastman JT (2012) Antarctic silverfish: Life strategies of a key species in the high-Antarctic ecosystem. Fish Fish 13(3):241–266
73. La Mesa M, Riginella E, Mazzoldi C, Ashford J (2015) Reproductive resilience of ice-dependent Antarctic silverfish in a rapidly changing system along the Western Antarctic Peninsula. Mar Ecol Prog Ser 36:235–245
74. Le Guen C, Suaria G, Sherley RB, Ryan PG, Aliani S, Boehme L, Brierley AS (2020) Microplastic study reveals the presence of natural and synthetic fibres in the diet of King Penguins (*Aptenodytes patagonicus*) foraging from South Georgia. Environ Int 134:105303
75. Lorenzini S, Baroni C, Baneschi I, Salvatore MC, Fallick AE, Hall BL (2014) Adélie penguin dietary remains reveal Holocene environmental changes in the western Ross Sea (Antarctica). Palaeogeogr Palaeoclimatol Palaeoecol 395:21–28
76. Lorenzini S, Olmastroni S, Pezzo F, Salvatore MC, Baroni C (2009) Holocene Adélie penguin diet in Victoria Land, Antarctica. Polar Biol 32:1077–1086
77. Lynch HJ, Naveen R, Trathan PN, Fagan WF (2012) Spatially integrated assessment reveals widespread changes in penguin populations on the Antarctic Peninsula. Ecology 93(6):1367–1377

78. Lyver POB, Barron M, Barton KJ, Ainley DG, Pollard A, Gordon S, Mcneill S, Ballard G, Wilson PR (2014) Trends in the breeding population of Adélie penguins in the Ross Sea, 1981–2012: A coincidence of climate and resource extraction effects. PLoS One 93(3):e91188
79. Massom RA, Stammerjohn SE (2010) Antarctic sea ice change and variability – physical and ecological implications. Pol Sci 4:149–186
80. Mattern T, Ellenberg U (2018) Yellow-eyed penguin diet and indirect effects affecting prey composition – collation of biological information. Department of Conservation, Wellington
81. Mattern T, Meyer S, Ellenberg U, Houston DM, Darby JT, Young M, Van Heezik Y, Seddon PJ (2017) Quantifying climate change impacts emphasises the importance of managing regional threats in the endangered yellow-eyed penguin. PeerJ 5:e3272
82. Mayewski PA, White F (2002) The ice chronicles. University Press of New England, Hanover
83. Morrison KW, Armstrong DP, Battley PF, Jamieson SE, Thompson DR (2017) Predation by New Zealand sea lions and brown skuas is causing the continued decline of an eastern rockhopper penguin colony on Campbell Island. Polar Biol 40:735–751
84. Morrison KW, Battley PF, Sagar PM, Thompson DR (2015) Population dynamics of eastern rockhopper penguins on Campbell Island in relation to sea surface temperature 1942–2012: Current warming hiatus pauses a long-term decline. Polar Biol 38:163–177
85. Murphy RC (1925) The bird Islands of Peru. G.P. Putnam & Sons, New York
86. Murphy RC (1936) Oceanic birds of South America. Macmillan Company, New York
87. Nel DC, Crawford RJM, Parsons N (2003) The conservation status and impact of oiling on the African penguin. In: Nel DC, Whittington PA (eds) Rehabilitation of oiled African penguins: a conservation success story. BirdLife South Africa and the Avian Demography Unit, University of Cape Town, Cape Town, pp 1–7
88. Oliver E, Wernberg CJ, Benthuysen J, Chen K (2020) Advances in understanding marine heatwaves and their impacts. Frontiers Media SA, Lausanne
89. Oliver ECJ, Donat MG, Burrows MT, Moore PJ, Smale DA, Alexander LV, Benthuysen JA, Feng M, Gupta AS, Hobday AJ, Holbrook NJ, Perkins-Kirkpatrick SE, Scannell HA, Straub SC, Wernberg T (2018) Longer and more frequent marine heatwaves over the past century. Nat Commun 9(1):1–12
90. Paleczny M, Hammill E, Karpouzi V, Pauly D (2015) Population trend of the world's monitored seabirds, 1950-2010. PLoS One 10(6):e0129342
91. Parkinson CL (2002) Trends in the length of the Southern Ocean sea ice season, 1979–99. Ann Glaciol 34:435–440
92. Patterson DL, Easter-Pilcher AL, Fraser WR (2003) The effects of human activity and environmental variability on long-term changes in Adelie penguin populations at Palmer Station, Antarctica. In: Huiskes AHL, Gieskes WWC, Rozema J, Schorno RML, van der Vies SM, Wolf WJ (eds) Antarctic biology in a global context. Backhyus Publishers, Leiden, pp 301–307
93. Peterson RT (1979) Penguins. Houghton Mifflin, Boston
94. Piatt J, Parrish JK, Renner HM, Schoen SK, Jones TT, Arimitsu ML, Kuletz KJ, Bodenstein B, García-Reyes M, Duerr RS, Corcoran RM, Kaler RSA, Mcchesney GJ, Golightly RT, Coletti HA, Suryan RM, Burgess HK, Lindse J, Lindquist K, Warzybok PM, Jahncke J, Roletto J, Sydeman WJ (2020) Extreme mortality and reproductive failure of common murres resulting from the northeast Pacific marine heatwave of 2014-2016. PLoS One 15(1):e0226087
95. Pichegru L, Nyengera R, Mcinnes AM, Pistorius P (2017) Avoidance of seismic survey activities by penguins. Sci Rep 7:16305
96. Pichegru L, Vibert L, Thiebault A, Charrier I, Stander N, Ludynia K, Lewis M, Carpenter-Kling T, Mcinnes A (2022) Maritime traffic trends around the southern tip of Africa – Did marine noise pollution contribute to the local penguins' collapse? Sci Total Environ 849:157878
97. Pinto MBLC, Siciliano S, Di Beneditto APM (2007) Stomach contents of the Magellanic penguin *Spheniscus magellanicus* from the northern distribution limit on the Atlantic coast of Brazil. Mar Ornithol 35:77–78

98. Raya Rey A, Rosciano N, Liljesthröm M, Samaniego RS, Schiavini A (2014) Species-specific population trends detected for penguins, gulls and cormorants over 20 years in sub-Antarctic Fuegian Archipelago. Polar Biol 37:1343–1360
99. Reilly P (1983) Fairy penguins and earthy people. Lothian, Maryborough
100. Roeder AD, Marshall RK, Mitchelson AJ, Visagathilagar T, Ritchie PA, Love DR, Pakai TJ, Mcpartlan HC, Murray ND, Robinson NA, Kerry KR, Lambert DM (2001) Gene flow on the ice: Genetic differentiation among Adélie penguin colonies around Antarctica. Mol Ecol 10:1645–1656
101. Ropert-Coudert Y, Chiaradia A, Ainley D, Barbosa A, Boersma PD, Brasso R, Dewar M, Ellenberg U, García-Borboroglu P, Emmerson L, Hickcox R, Jenouvrier S, Kato A, Mcintosh RR, Lewis P, Ramírez F, Ruoppolo V, Ryan PG, Seddon PJ, Sherley RB, Vanstreels RET, Waller LJ, Woehler EJ, Trathan PN (2019) Happy feet in a hostile world? The future of penguins depends on proactive management of current and expected threats. Front Mar Sci 6:248
102. Roux J-P (2003) Risks. In: Molloy F, Reinikainen T (eds) Namibia's marine environment. Directorate of Environmental Affairs of the Ministry of Environment and Tourism, Namibia, pp 137–152
103. Ruoppolo V, Vanstreels RET, Woehler EJ, Heredia SAR, Adornes AC, Pinho Da Silva-Filho R, Matus R, Poleschi C, Griot K, Kolesnikovas CKM, Serafini P (2012) Survival and movements of Magellanic penguins rehabilitated from oil fouling along the coast of South America, 2000–2010. Mar Pollut Bull 64:1309–1317
104. Sailley SF, Ducklow HW, Moeller HV, Fraser WR, Schofield OM, Steinberg DK, Garzio LM, Doney SC (2013) Carbon fluxes and pelagic ecosystem dynamics near two western Antarctic Peninsula Adélie penguin colonies: an inverse model approach. Mar Ecol Prog Ser 492:253–272
105. Santora JA, Larue MA, Ainley DG (2020) Geographic structuring of Antarctic penguin populations. Glob Ecol Biogeogr 29(10):1716–1728
106. Saraux C, Chiaradia A, Salton M, Dann P, Viblanc VA (2016) Negative effects of wind speed on individual foraging performance and breeding success in little penguins. Ecol Monogr 86:61–77
107. Schmidt AS, Ballard G (2020) Significant chick loss after early fast ice breakup at a high-latitude emperor penguin colony. Antarct Sci 32:1–6
108. Schofield O, Ducklow HW, Martinson DG, Meredith MP, Moline MA, Fraser WR (2010) How do polar marine ecosystems respond to rapid climate change? Science 328:1520–1523
109. Schreiber RW, Schreiber EA (1984) Central Pacific seabirds and the El Nifio Southern Oscillation, 1982-1983 Perspective. Science 225:713–716
110. Sherley RB, Ludynia K, Dyer BM, Lamont T, Makhado AB, Roux JP, Scales KL, Underhill LG, Votier SC (2017) Metapopulation tracking juvenile penguins reveals an ecosystem-wide ecological trap. Curr Biol 27(4):563–568
111. Smith RC, Domack E, Emslie S, Fraser WR, Ainley DG, Baker K, Kennett J, Leventer A, Mosley-Thompson E, Stammerjohn S, Vernet M (1999) Marine ecosystem sensitivity to historical climate change: Antarctic Peninsula. Bioscience 49:393–404
112. Southwell C, Emmerson L, Mckinlay J, Newbery K, Takahashi A, Kato A, Barbraud C, Delord K, Weimerskirch H (2015) Spatially extensive standardized surveys reveal widespread, multi-decadal increase in East Antarctic Adélie penguin populations. PLoS One 10(10):e0139877
113. Spear LB, Ainley DG, Ribic CA (1995) Incidence of plastic in seabirds from the tropical Pacific, 1984–91: relation with distribution of species, sex, age, season, year and body weight. Mar Environ Res 40:123–146
114. Stammerjohn S, Massom R, Rind D, Martinson D (2012) Regions of rapid sea ice change: An inter-hemispheric seasonal comparison. Geophys Res Lett 39:L06501
115. Stammerjohn SE, Martinson DG, Smith RC, Yuan X, Rind D (2008) Trends in Antarctic annual sea ice retreat and advance and their relation to El Niño–Southern Oscillation and Southern Annular Mode variability. J Geophys Res 113:C03S90

116. Strycker N, Wethington M, Borowicz A, Forrest S, Witharana C, Hart T, Lynch HJ (2020) A global population assessment of the Chinstrap penguin (*Pygoscelis antarctica*). Sci Rep 10:19474
117. Thatje S, Hillenbrand CD, Mackensen A, Larter R (2008) Life hung by a thread: endurance of Antarctic fauna in glacial periods. Ecology 89(3):682–692
118. Thompson DWJ, Solomon S (2002) Interpretation of recent Southern Hemisphere climate change. Science 296:895–899
119. Turner J (2004) The El Niño–Southern Oscillation and Antarctica. Int J Climatol 24:1–31
120. Turner J, Comiso JC, Marshall GJ, Lachlan-Cope TA, Bracegirdle T, Maksym T, Meredith MP, Wang Z, Orr A (2009) Non-annular atmospheric circulation change induced by stratospheric ozone depletion and its role in the recent increase of Antarctic sea ice extent. Geophys Res Lett 36:L08502
121. Vargas FH, Lacy RC, Johnson PJ, Steinfurth A, Crawford RJM, Boersma PD, Macdonald DW (2007) Modelling the effect of El Niño on the persistence of small populations: the Galpágos penguin as a case study. Biol Conserv 137:138–148
122. Vianna JA, Fernandes FA, Frugone MJ, Figueiró HV, Pertierra LR, Noll D, Bi K, Wang-Claypool CY, Lowther A, Parker P, Le Bohec C (2020) Genome-wide analysis reveals drivers of penguin diversification. Proc Natl Acad Sci 117(36):22303–22310
123. Waters CN, Zalasiewicz J, Summerhayes C, Barnosky AD, Poirier C, Gałuszka A, Cearreta A, Edgeworth M, Ellis EC, Ellis M, Jeandel C, Leinfelder R, Mcneill JR, Richter D Deb, Steffen W, Syvitski J, Vidas D, Wagreich M, Williams M, Zhisheng A, Grinevald J, Odada E, Oreskes N, Wolfe AP (2016) The Anthropocene is functionally and stratigraphically distinct from the Holocene. Science 351(6269):2622
124. Webster T (2018) The pathway ahead for hoiho. Yellow-Eyed Penguin Trust, Dunedin
125. Weimerskirch H, Inchausti P, Guinet C, Barbraud C (2003) Trends in bird and seal populations as indicators of a system shift in the Southern Ocean. Antarct Sci 15:249–256
126. Wilson PR, Ainley DG, Nur N, Jacobs SS, Barton KJ, Ballard G, Comiso JC (2001) Adélie penguin population change in the Pacific sector of Antarctica: Relation to sea-ice extent and the Antarctic Circumpolar Current. Mar Ecol Prog Ser 213:301–309
127. Woehler EJ (2004) Hearing abilities in Antarctic penguins. Polarforschung 72:95–98
128. Wolfaardt AJ, Underhill LG, Altwegg R, Visagie J (2008) Restoration of oiled African penguins *Spheniscus demersus* a decade after the Apollo Sea spill. Afr J Mar Sci 30:421–436
129. Younger J, Emmerson L, Southwell C, Lelliott P, Miller K (2015) Proliferation of east Antarctic Adélie penguins in response to historical deglaciation. BMC Evol Biol 15:236
130. Zwally HJ, Comiso JC, Parkinson CL, Cavalieri DJ, Gloersen P (2002) Variability of Antarctic sea ice 1979–1998. J Geophys Res 107(C5):9-1–9-19

Chapter 14
Not Forgetting…

The Social Side – Behavior and Communication at Sea

Many animals form groups ranging vastly in sizes depending on species and context. A huge amount of literature covers the subject, dealing from issues such as aggregations in general (in 2D or 3D), to costs and benefits of varying group sizes, how decisions are made with respect to movement and behavior [15, 54], and who makes them [16, 33]. Penguins are no exception to the phenomenon. It is well documented that they are generally social birds on land [2, 59], traveling in flocks, as well as occurring in nesting groups that range from diffuse clusters of singletons, such as in the yellow-eyed penguin, to colonies that may number hundreds of thousands, as can be the case for, e.g., king, Adélie, and macaroni penguins (Chap. 2). Accordingly, it's not surprising that sociality among penguins occurs also at sea, and indeed has been repeatedly confirmed by vessel-based surveys (Tables 14.1 and 14.2).

The popular view, found often in cartoons and jokes, is of a line or flock of penguins, supposedly following the leader. Really, though, it appears to be a group effort in "deciding" where they are going or what they are doing [14] (and see below) (Fig. 14.1).

The critical difference between penguin flocks at sea and most other aggregating animals is that fish-birds have to operate in two media, and in one of these they have to breath-hold! Thus, penguins spending extended periods at the surface are analogous to many terrestrial animals operating in 2D space, such as might occur in volant birds associating together to sleep for the night. In such "roosts" they organize according to their respective "personal distance." That is, even penguins operate as if they are like poles of magnets, repelling one another to space in a very regular fashion [2, 72]. On the other hand, diving penguins are only underwater for as long as they can hold breath and so have little time for the niceties of sorting themselves by personal space. The problem becomes acute when penguin flocks

D. G. Ainley, R. P. Wilson, *The Aquatic World of Penguins*, Fascinating Life Sciences, https://doi.org/10.1007/978-3-031-33990-5_14

Table 14.1 Characteristics of penguin flocks at sea, away from colonies: maximum number of penguins seen in a flock, mean ± SD flock size, number of flocks quantified, and where and during what time period at-sea surveys were conducted

	Max size	Mean size ± SD, N flocks[a]	Location, decade (19xx)	Source
Emperor	23	2.7 ± 4.7, 22	Weddell Sea, 80s	D. Ainley, unpubl.
	6	2.0 ± 1.4, 29	Ross Sea, 70s	D. Ainley, unpubl.
King	4	1.8 ± 1.2, 8	APF, Pac & Atl, 80s	D. Ainley, unpubl.
	116	6.9 ± 16.7, 129	Indian Ocean, 80s	J.-C. Stahl, unpubl.
Adélie	25	4.0 ± 7.1, 11	Weddell Sea, 80s	D. Ainley, unpubl
	163	12.2 ± 23.2, 152	Ross Sea, 70s	D. Ainley, unpubl
Chinstrap	19	3.0 ± 2.6, 216	Weddell Sea, 80s	D. Ainley, unpubl
Gentoo	25	3.8, 33	Scotia Sea, 90s	[31]
	4	2.5, 2	Scotia Sea, 80s	D. Ainley, unpubl.
	45	3.3 ± 7.6, 36	Indian Ocean, 80s	J.-C. Stahl, unpubl.
African	19		Benguela Current, 80s	[84]
	15	2.1 ± 2.1	Benguela Current, 80s	[83]
	26	4.8 ± 4.5, 260[b]	Benguela Current, 50s	[72]
Galápagos	10+	5.5 ± 4.0, 316	Cromwell Current, 90s	[49]
Humboldt	3	1.1 ± 1.5, 14	Peru Current, 80s	[20]
	3	1.6 ± 1.7, 11	Peru Current, 80s	L. Spear, unpubl.
Magellanic	6	2.6 ± 1.6, 14	Patagonia, 80s	D. Ainley, unpubl.
	8	2.0 ± 1.6, 78	Peru Current, 80s	L. Spear, unpubl.
		1.6 ± 1.0, 82[b]	Peru Current, 80s	[80]
Macaroni	64	20.5, 33	Scotia Sea, nr SG, 90s	[31]
	8	1.9 ± 2.1, 11	Scotia Sea, 80s	D. Ainley, unpubl.
	35	6.4 ± 16.0, 114	Indian Ocean, 80s	J.-C. Stahl, unpubl.
Rockhopper	21	3.4 ± 4.3, 72	Indian Ocean, 80s	J.-C. Stahl, unpubl.
	12	1.6 ± 1.5, 78	Patagonia, 80s	D. Ainley, unpubl.

In the case of Antarctic species, flocks seen on ice floes or icebergs have not been included

[a]Note, that upon an observer encountering a flock of porpoising penguins (Figs. 1.1 and 9.3), it is very difficult to determine just how many penguins there are since only a few at a time surface for air (Chap. 9). With experience, an observer can get closer to the actual number, as they have seen how many penguins there actually are upon the penguins getting out of the water. Regardless, the information in these tables likely represents a slight underestimate of true flock size

[b]Size of flocks arriving on the beach from foraging at sea

have to somehow coordinate surface pauses with periods underwater, thus to operate efficiently as a group beneath the surface. Otherwise, much has been written about fission-fusion dynamics in birds in general [73], and from that it does appear that understanding penguin flocks is likely to remain challenging for years to come.

Table 14.2 Frequency distribution of the proportion of penguin flocks of various size categories as seen at sea

Proportions of flocks of various size categories								
No. penguins/flock	1	2–5	6–10	11–20	21–50	51+	Flocks	Source
Emperor	55	39	4	2			51	D. Ainley, unpubl.
King	36	43	8	4	5	1	129	J.-C. Stahl, unpubl.
Adélie	21	38	15	8	11	5	152	D. Ainley, unpubl.
Chinstrap	34	51	13	2			216	D. Ainley, unpubl.
Gentoo	67	22	3	6	3		36	J.-C. Stahl, unpubl.
African	29	40	16	7	5	3	282	[72]
	35	38	5	1			149	[83]
Magellanic	52	41	7				78	L. Spear, unpubl.
Macaroni	46	32	10	5	4	3	114	J.-C. Stahl, unpubl.
Rockhopper	43	38	8	4	3		72	J.-C. Stahl, unpubl.
	67	31	1	1			78	D. Ainley, unpubl.

Fig. 14.1 Penguin flocks at sea do not obviously adhere to a particular formation although they generally have appreciable distance between birds, as shown by these royal penguins. (Photo courtesy of Peter Ryan)

Penguin Flock Fusion/Cohesion

A convenient place to start considering the nuances of penguins in flocks at sea is to examine the advantages for them doing so. As with all birds, flocking is a predator avoidance behavior (more individuals staying aware of surroundings; Chap. 12), but also a prey-capture strategy. Kearton [37] describes African penguins at sea tightening their flocks once a prey school is encountered (typically clupeid, schooling fish). Then when that school is fully acquired or it breaks up and escapes, the flock disperses a bit until another school is found. When that occurs, the frequency of uttering contact calls increases [48] (see also below).

Group foraging can be the most effective strategy, as several penguins working together are able to “herd” prey leading to better foraging success [47] (Chap. 11). Hints of penguins’ proclivity of doing this have been seen on occasions of encountering a flock of Adélie penguins at sea, especially when confined to a small open area in the sea ice (the water between ice floes, a “lead”). The penguins dive in unison and then surface in unison a few moments later, continually repeating the sequence (Fig. 14.2). Individuals walking up to encounter the scene dive in to become part of the cohesive activity. This activity has been timed in African penguins showing that all members of foraging flocks dive within a few seconds of one another and, likewise, all surface within a few seconds of one another sometime later [72].

These sorts of observations were seen as well by the attachment at the same time of data loggers to a number of chinstrap penguins on the beach. Subsequently, the data showed that they foraged together, diving synchronously, as they fed on a school of krill [34]. Thus, apparently, the flock on the beach stayed together once they went to sea to forage (perhaps all were foraging to feed chicks, and thus of the same “state of mind”; see below). In many situations, especially when the prey of choice is a fast-swimming, schooling fish in a 3D environment (mid-water column), such as anchovy, sardine or other clupeids, a lone penguin has greater difficulty in

Fig. 14.2 A flock of Adélie penguins surfacing in unison, then diving in unison to pursue prey. Note the birds on the ice, diving in to join the fray. (Photo courtesy of John Weller)

Fig. 14.3 Here, a solitary Galápagos penguin demonstrates the issues of trying to catch fast-swimming fish if they are allowed to maintain their highly organized, "polarized," state – the fish work as one cohesive body that attempts to keep a "safe distance" from the penguin as it approaches, with the school splitting and swimming to the left and right and around the bird. The large numbers of organized fish create a confused matrix which makes it difficult for the penguin to pick out individuals. A similar phenomenon is observed in flocking birds reacting to falcons. (Photo courtesy of Tui De Roy)

catching these fish, especially if the water is somewhat clear [47]. In that situation, the fish easily alter the shape of their school to avoid the single penguin (Fig. 14.3). A flock of penguins – sometimes aided by other predators – can better herd the fish, rendering individual fish on the periphery easier to catch (Fig. 14.4). On the other hand, and this may be another reason for flock fragmentation when foraging (see below), too many penguins may just confuse the situation, whereas just a few penguins foraging together are much better at operating like a team.

Indeed, it has been noted that African penguins have apparently fully coordinated diving and surfacing behaviors if groups are less than about 15 birds but that this breaks down when the groups are larger [84]. One idea is that "head dipping" – penguins at the sea surface are perpetually dipping their heads underwater, presumably to look down and around – may serve as a cue to initiate diving by the group [72]. Another idea is that "head dipping" is displacement behavior that has acquired signal quality, with more rapid dipping showing birds ready to dive. There is also the "ark" (flock contact) call that appears to maintain flock cohesion or formation. First reported for penguins by E.A. Wilson [79], it's essentially a short, loud chirp or short "wail," the sound of which carries over long distances (100s of meters).

Fig. 14.4 Galápagos penguins, with side-stripe in full display, herding a school of fish. Notice that the prey fish, the long, thin silver ones, are all heading in same direction. Some larger predatory fish are also involved in this foraging event. (Photo courtesy of Tui De Roy)

Seemingly, in the case of lone penguins, if another penguin answers, they do their best to find one another. Penguins utter these calls not only when at the water surface, but when underwater as well [77]. More generally, the cohesiveness of flocks is maintained by these "contact calls," a behavior commonly exhibited by all bird species that form in flocks – the more excited they are, the more they call [4, 48, 72]. In this regard, it is highly relevant that penguins have hearing that is particularly sensitive to underwater sounds [74, 86], in accord with aquatic birds in general having inner ear structures suitable to the task [88]. In noisy modern times, this turns out not to be the advantage it might be (see Chap. 13).

Besides behaviors such as head dipping and contact calling, flock cohesion is also facilitated by penguin coloration. For example, it has been suggested that the often dramatic coloration of penguin heads exemplified by the yellows in the *Aptenodytes'* "commas" and *Eudyptes'* "tassles," as well as the departure from the simple dorsally dark and ventrally pale coloration in gentoo and chinstrap penguins, might help them to keep track of each other at the sea surface. Otherwise, the general color pattern of penguins, black dorsally and white ventrally is very much a part of their at-sea strategy. The "counter shading," as it is called, hides them against the dark sea when viewed from above, and hides them against the bright light from the surface when viewed from below [40].

An interesting further development of penguin color patterning that helps in their at-sea communication is shown by the genus *Spheniscus* – the African, Galápagos,

Humboldt, and Magellanic penguins. These penguins dwell in waters in which the forage species are fast-swimming, schooling, mid-water fish, such as anchovy and sardine (Chaps. 3 and 4). In addition to the dorsal dark and ventral pale plumage, they have a bold lateral dark stripe that runs down the length, both sides of their body (Fig. 14.3). During SCUBA diving with these species, we have noticed how eye-catching this pattern is, which would seem lunacy as all these species also share their waters with predatory sharks and fur seals (Chap. 12). Despite the predatory aspect (being eaten), which may be overcome by simple speed and tactics, the likely function of the stripe is to facilitate the "herding" of small schools of the clupeid fish, a subject treated in detail in Chap. 11.

It is notable, and addressed to some degree in Chap. 11, that the juveniles of these penguin species do not have the lateral stripe, which has been suggested to be due to their slow swimming speed and general inefficiency at catching mid-water, schooling fish [66]. It's not that they wouldn't benefit from some aspects of the bold, striped coloration in prey capture, but there seems little point increasing the chances of being predated if the benefits are not substantial. And, if speed is lacking anyway, not only are many of the pelagic fish out of reach, but predators have a better chance of catching you. Siegfried and colleagues [72] noted that flocks containing juveniles were about double the typical size of adult flocks. It's not clear what that indicates, other than perhaps staying together as an antipredator strategy until they figure out more of the sea's mysteries? In a further twist to this story, some juveniles undergo partial head molt at sea, acquiring head coloration that may even mirror the head plumage of adults (Fig. 14.5) although they never have the lateral body stripe. In a series of elegant experiments backed up by observations of African penguins at sea, it was shown that juveniles with the adult head plumage tend to be less the subjects of aggression by the adults than normal juveniles, with aggression occurring more often in adult groups than expected by chance [66]. It seems, therefore, that the partially head-molted juveniles trick the adults into believing that they too are adults. These juveniles benefit from the fish-herding expertise of the adults without having to sport the predator-attracting coloration of their elders. Thus, it's a temporal cushion in which foraging techniques are learned.

Penguin Flock Fission/Fragmentation

In Chap. 10, we reveal that foraging penguins spend about 70% of their time underwater when on a trip to sea, with a critical amount of time needed in surface pauses for birds to replenish oxygen stores before undertaking the next dive. Obviously, if they were totally fish, they'd be underwater 100% of the time (as are penguin prey)! So, individuals of a flock returning "late" to the surface after a dive may not have enough time to re-stock with oxygen before the group dives again. If they ran out of oxygen too soon, they'd no longer be of use to the flock (or themselves). All this makes effective group cohesion paramount, with such cohesion clearly evident (see

Fig. 14.5 The plumage of African penguin juveniles (the three at the back of the photo) are normally distinctly different from that of adults, being more uniformly brown/grey and having no clear face mask. However, some individuals undergo a partial head molt at sea that results in head plumage that begins to resemble that of the adults. Thus, the bird in the foreground, which is a juvenile, has a reasonably striking mask like the adult just behind it on the right-hand side, and even has the beginnings of the lateral line across the upper chest (another feature normally absent in juveniles). It is thought that this "deception" allows juveniles, which are likely to be non-expert fishers, to mix with adult flocks to benefit from their expertise while the adults are "unaware" that they have a learner in their midst. (Photo courtesy of Valeria Ruoppolo, with that of head-molted juvenile, from Peter Ryan, inserted)

above). That being said, it makes a case for distinguishing between penguins foraging (or traveling) close to the surface, where it is easy to snatch a breath without deviating greatly from the group – porpoising penguins, during which a few at a time take a breath, are a prime example of this (Chap. 9). However, the situation is different for birds that are diving to depths especially greater ones, because any individual returning early to the surface to breathe puts extended distance between the "breather" and the group. Given that penguins diving to depth generally travel tens or hundreds of meters horizontally (see Chaps. 9 and 10), is it even realistic for a bird that breaks ranks to breathe to relocate the flock, especially since, by hurrying to rejoin, energy – and therefore oxygen – costs skyrocket? This does not put the miscreant in a good position with regard to preparation for the next dive.

All this, and the fact that simple observations of penguin flocks at sea often do not reveal the specific conditions leading to the flocking and how they might relate to flock size, make any interpretation of penguin flocking challenging. Flock size is relevant because it is harder to co-ordinate movement in our breath-holders if flock size is large, and this may be a major factor in the rather small flocks that are found at sea, away from colonies (Tables 14.1 and 14.2). In particular, individuals in a larger 2D flock may find it more difficult to assess all individuals for movement co-ordination because nearer birds block line of sight. Sophisticated electronic tags, that have revealed so much about penguins, unfortunately, are of little help in providing information on flocking in penguins because, although video cameras may sometimes show other associated penguins [76], it is neither clear how transient these associations are, nor how many individuals occur together. What these tags have very occasionally shown, though, because multiple tagged individuals have serendipitously foraged together, is that associations may last hours [30, 60]. It does seem, though, that one individual makes most of the decisions about when to dive and when to return to the surface, with depth differences between individuals generally being less than 1 m and event timings occurring within 4 s of each other [72, 78].

One confounding factor about penguins staying together at sea is that individuals within the flock may not all be in the same "state of mind" or condition. Penguins foraging just for themselves likely have different motivation, and level of satiation, than those foraging to also feed chicks. Related to this, inappropriate direction by the group, taking some individuals farther than they wish from their nest/colony/island, also may be reason enough to split from a group. We saw in Chap. 5, for instance, that females often tend to forage farther from the colony than males, and it is unlikely that when massing at the beach for departure there is any sorting according to gender. That would happen, for whatever reason, once the penguins got to sea.

In further regard to flock fragmentation, in waters near large penguin colonies, flocks are much larger than those seen at sea. In African penguins, it has been observed that flock size of departing birds is larger, about three times larger, than that of returning birds, and that flock size decreases with distance from shore [72]. Numbers of penguins accumulate on beaches prior to them diving in, but not until they are convinced no predators are around (Chap. 12). Then a horde of them debarks, at least among the more populous species (Figs. 2.1, 2.2, 2.3, 2.4, 2.5 and 2.6). Similarly, if penguins headed to the shore sense a predator, such as a pinniped that is working the beach, then numbers accumulate, too, farther off the beach, as the penguins swim back and forth until assessing that it is safe to proceed ashore. These sorts of flocks can reach dozens if not low hundreds of birds. Again though, flock size is also likely to be a function of bird density in the area, with groups coalescing more where the density is highest close to colonies. Clearly, these large flocks do fragment, as the information in Tables 14.1 and 14.2 illustrate. Farther at sea, the average size of flocks of most species generally numbers 3–5 penguins (Table 14.1), although in general a third to half of any penguins encountered at sea are of lone birds (Table 14.2). It's not clear whether these lone birds may simply be doing better for their particular situation (see discussion of "state of mind" above as well as solitary species such as the yellow-eyed penguin) or whether these single birds might not do better in company, hence the "arks" calling out to other penguins that could be in the area.

Navigation

Long-Range

In Chap. 2, we discussed the issues of penguins coming to land which, unlike most real fish, they need to do in order to raise young and molt. Otherwise, they are sea creatures far more than they are land creatures. Being fish-birds, they do seem to know well their way around the ocean, as reviewed in Chaps. 3 and 5, and rarely stray from that part of the ocean that is theirs, for that particular penguin species. As a "vagrant," occasionally one does show up outside its normal range (e.g [70]), that being the ocean adjacent to those points of land that they do use for breeding (see Chap. 3: penguins are "dispersive" in their seasonal movements, i.e., not true migrants). However, when they are way out to sea for months or years, how is it that they so successfully find their way back? It seems that they do so in a number of ways, which differ depending on the spatial scale involved.

Other marine creatures have the same "problem" of finding their way back and forth in areas that interest them, usually in different seasons. Sharks, for instance, know their way around and can find their way back to where they want to be by orienting to Earth's magnetic field [38, 39], just like sea captains in the sailing days who used a compass, along with their sextant and a clock to do the same. Every scout, who has a map orientation merit badge, knows how compasses can be used to find their way. Birds can do the same despite their maps, and perhaps clocks, being mental [53, 85], though no one has investigated the phenomenon in penguins. It would be surprising if penguins do not use Earth's magnetic field for orientation.

Birds are well known to use celestial cues, such as the stars and sun, not just for orientation but for navigation as well, a subject that has gotten much investigation [12, 69]. They have been placed in cages in planetariums to indicate which way they would like to go as humans change the night sky projected above them. In similar ways, it has been shown that honey bees use the sun for orientation, thus to be able to find their way to and from nectar sources as well as to communicate the information to other occupants of their hives [25]. It is the polarized direction of the sun's rays that are important, and trustworthy in their lack of variation.

The first inklings that penguins use the sun for navigation came from personnel on the ship *SS Aurora*, which in 1916 was supposed to offload supplies at Cape Evans, Ross Island, to support Ernest Shackleton's ill-fated trans-Antarctic crossing. However, in May just after tying up and before off-loading, a fierce gale broke the lines and blew the ship north in the pack ice along the Victoria Land coast and eventually out of the Ross Sea. It remained stuck in the pack ice for the entire winter (same for Shackleton's ship on the other side of Antarctica!) well north of Ross Island and the Ross Sea, ~150 km south of the Balleny Islands. There had been no penguins around during the winter, given that it was dark 24 h and penguins need to be able to see before they will dive into the ocean [5] (see Chap. 12). However, on 5 August, the first day that the sun rose where the ship was located (~69°S), Lt. A.H. Ninnis noted that Adélie penguins appeared with their tracks coming across the ice from the north headed toward the southern horizon [26]. The ship *Belgica* (R. Amundsen, First Mate) also spent a winter (1898) locked in the Antarctic ice, and no penguins were seen during the 60 days of darkness (~70°S). They came with the dawn, however, July 23, and thereafter Amundsen warded off scurvy by including their fresh meat in his meals (unlike other crew members [67]).

Penguins' use of the sun for orientation was next demonstrated experimentally. In the 1960s, one of the foremost researchers on bird migration, John Emlen and his student Richard Penney, captured several Adélie penguins at Cape Crozier, Ross Island, and flew with them to the South Pole [24]. You can imagine that the polar plateau was not actually novel to the penguins, as they are used to wide expanses of ice, though it is sea ice not glacial ice. But the reason for taking them to the polar plateau was that it had no topographic features that penguins could use for navigation. It was a featureless plane in all directions, quite like the sea ice surrounding the *Aurora* or *Belgica*. The researchers, however, didn't have to travel to sea for their experiment. Upon release, as long as the sun was visible, the penguins headed northeast in a straight line. That is, until clouds obscured the sun, upon which they meandered, directionless. Emlen and Penney concluded that the Adélie penguins could use the sun for navigation as long as the sun's rays were not interrupted.

One of us actually experienced this phenomenon of penguin directionality when there was sun during a cruise along the Marie Byrd Land and Ellsworth Land coasts [6, 7]. It was March, and penguins were all around until one day, air temperatures plunged and the sea surface became a continuous sheet of ice. Penguins were nowhere to be seen except for their footprints, in the snow and across the ice heading northeast. It then became evident why the northeast direction taken by the Adélie penguins in Emlen and Penney's experiment. Since the flow of coastal ice is east to west, by heading northeast the penguins were compensating the ice flow, "theoretically" (penguins likely don't think about such stuff) in order to head directly north, which in general would be the closest distance to the outer pack and its marginal ice zone (Chap. 3). This was observed of emperor penguins, too, which actually swam between breathing holes punched through the thin ice by minke whales (whose scientific name refers to their long, hard beak – *Balaenoptera acutorostrata*). Thus, it may well be that the whales were using the sun for orientation, too, to follow shortest route to less ice-covered seas!

Finally, this northeastern direction that Antarctic, pack ice penguins take when they are not sure what direction they should be heading, presented itself in a strange interaction that one of us (David) had with the film-maker Werner Hertzog, while he was making his film, *Encounters at the End of the World* (2007, Discovery Films). A penguin had wandered through Herzog's dive camp (underwater filming) that was on the sea ice along the coast, far south in McMurdo Sound, Victoria Land. The penguin kept walking to shore and then southwest into the interior of Antarctica. Herzog was curious about the apparent suicidal behavior of the penguin. From time to time, penguins do this as indicated by a number of mummies that have been found far up the coastal Dry Valleys in that part of Antarctica. The occasional seal does this, too [75]. Herzog wondered if penguins were plagued by strange behavior, and came to Cape Royds, on the other side of McMurdo Sound, to ask David Ainley about it. Hertzog would not reveal why his interest in the question, and a possible explanation occurred to David only after seeing the film. Had David known the reason for the questioning, he would have offered an explanation that involved what is known as "mirror-image misorientation," which explains the occurrence of a number of vagrant land birds (birds found a long way from where they "should be"). David Ainley's friend, David DeSante, had "flown" vagrant birds in flight cages in planetariums, the birds having been captured in coastal California (by mist-net). DeSante found that with normal star patterns, these birds that originated in north-eastern North America, and should have been migrating southeast to reach Brazil, rather took the mirror-image of the direction they should have taken, southwest. Thus, they ended up where they had not intended (central California), as vagrants. Changing the star patterns still led to them taking the mirror image of where they should be going. These vagrant songbirds were, in a way, dyslexic. The penguins wandering up the McMurdo Dry Valleys, seen by Hertzog, were headed southwest, which is the mirror image of the northeast direction that they "should" have been taking, as deduced by Emlen and Penney [24]. David Ainley explained all this to the BBC in their *Winter's Weirdest Events* program, aired in 2017, the BBC producers having become intrigued by Hertzog's presentation.

As far as we know, there has not been any analogous research done on the long-range navigation of other penguin species, and its connection to the polarized rays of the sun. But we have no reason to think that it is just the two high-latitude penguin species that possess the skill.

Medium- to Short-Range

The foraging of penguins, when acting like central-place foragers, can extend a few hundred kilometers from large colonies and shorter distance at small colonies (Chap. 5) but it also depends on the species. In the vast majority of cases, especially if their colony is in a mountainous location, e.g., created by a volcano (Chap. 1), distinctive landmarks can be seen from a long way away. Indeed, J.C. Ross's book [65] that summarized his voyages to the Southern Ocean in the 1840s is replete with sketches prepared by the expedition artist of islands and coasts where we know now are penguin colonies, though rarely did persons from that expedition actually land on the shores. These vistas were viewed from a long way off – a wise strategy in those days of sailing ships when no one was around to help should a rescue be needed. Yet Ross named many of those mountains and geographic spots. In the cruises that one of us (David) has made to the Ross Sea, and elsewhere in Antarctica, we've seen the same vistas, from far at sea, depicted in the Ross expedition drawings.

With little doubt, penguins can see these land features, too, and use vision to identify distinct aspects of their neighborhood. In fact, David worked at the Cape Crozier Adélie penguin colony at the end of a 10-year demographic study in which chicks were banded annually. A few thousand were banded every year in four localities, one at the east end, another at the west end, and two areas in the middle. The colony extended about 5 km along a beach and 2.5 km inland, backed by impressive, mountainous terrain. The population exceeded 100,000 breeding pairs. In subsequent years, efforts were made to find all the banded penguins when they came back as subadults and adults [8]. Repeatedly, when walking the beach, one would encounter a banded penguin where it had landed. It was always adjacent to where its nest was located up slope (all nests found were marked). In other words, the penguin swam along the shore (a goodly distance out to avoid predatory seals; Chap. 12) and ventured to the beach only when it found the point that would require the minimum of walking to reach the nest. Clearly, they knew the terrain from a long distance away. On a few occasions, a banded penguin was caught, put in a duffle bag, and carried to the opposite end of the colony. Upon release, it would make a bee-line to its nest cutting diagonally across the colony and not walking first along the beach to become opposite to the nest location. Somewhat of a similar study has been conducted on Magellanic penguins to determine how they find their nest so readily among thousands spread along the coast in Argentina [61]. Moreover, an analogous pattern of using visual cues to find their way across sometimes immense colonies has been detected in king penguins by comparing Global Positioning System (GPS) tracks repeated over time [52]. Lance Richdale [62], who conducted a long-term study of yellow-eyed penguins, concluded that when they came ashore, which they do every evening, the vast majority would do so close to, if not where their nest was located, regardless of season. An extreme example of close range at-sea navigation by a penguin based on familiarity with the local coastline was demonstrated by a Magellanic penguin, which was found oiled on a beach near Rio de Janeiro, Brazil (https://www.youtube.com/watch?v=oks2R4LqWtE). Rio de Janeiro is well within the area where Magellanic penguins occur during the non-breeding season (Chap. 3). A fisherman captured and cleaned the bird, and then fed it daily through its molt, before talking it by boat to an offshore island, releasing it there a few months after capture. Within a day, the penguin showed up at the fisherman's beach house, but that was its first trip to sea in months! The location of this penguin's nesting area is not known, though certainly a good distance south, with closest nesting colony in northern Argentina being about 1400 km away. However, this particular bird also showed up at the fisherman's house for the next few years.

Even more impressive, perhaps, is the ability of penguins to leave the colony and return to the same at-sea location where they had foraged most successfully a few days previously, as deduced by satellite tags. This ability was exhibited by Adélie penguins foraging within the water column in waters far deeper than they can dive, which rules out benthic topography as a cue [28] (also yellow-eyed penguins [44]). Such a location can be many dozens of kilometers out to sea. That Adélie study, too, was done at Cape Crozier and so the penguins, still able to see distant mountains, could locate the general vicinity of the recent successful foraging area rather than spend time searching more widely. If they went to sea again more than a few days later, they did not go back to the previous foraging spot, either not being able to remember that long or "knowing" that over several days the mid-water preyscape with little doubt would have changed. The yellow-eyed penguin, which forages along the bottom in shallow water (Chap. 5) and appears to use benthic geographic cues for orientation as well as above-water geographic information, has favored foraging locations, even from one year to the next as observed on NZ mainland locations [50, 51]. In fact, as observed at Stewart Island to the south, individuals have preferred foraging locations to which they return repeatedly [22]. Using satellite imagery, Mattern et al. [45] were impressed by these penguins' ability to follow straight-line courses but then found that what the satellite positions represented was that penguins were following scars left on the bottom by fishing trawls. The trawls are pulled along a linear path. Along those scars the penguins found that scavenger fish tend to concentrate. Thus, in one respect the penguins benefitted from the trawlers to a small degree but in another respect the total availability of fish has been reduced by the industrial fishing (Chap. 5).

In Chap. 3, we described the general trend of penguins foraging downstream of their colonies, and ascribed that to the wakes, which increase prey availability, generated by respective islands or capes. It is possible, too, that penguins can use current flow to facilitate meso-scale navigation, i.e., find their way back to the colony. On the other hand, since penguins have a good sense of smell, it's possible that odor plumes from guano washed from colonies might help them find their way back, though that is speculation. What is not speculation is that some penguins at least, such as African penguins, can find schools of diatom-chomping fish, such as anchovy, by honing in on the odor of DMS (dimethyl sulfide). That compound is released when the cell walls of diatoms are broken as the fish forage on them. Today, however, these penguins' olfactory ability is leading them astray, because outside of marine protected areas, the fishing industry is removing the fish from the phytoplankton patches [71]. Thus the penguins, following the odor corridor to phytoplankton blooms, find nothing to eat.

Understanding the Daily "Wash"

It is typical for penguins, upon departure from the beach and before going very far out to sea, to "bathe" [3, 72]. How far off they go before they involve themselves in this activity depends on whether predators are present, doing so just beyond the breakers if none are around but 200–400 m out if predators are present (Chap. 12). Among Adélie penguins, which don't begin a bathing bout until ~200 m from the beach, they may subsequently stop to bathe as many as seven times (average 3.6 times, $n = 32$ flocks) within the first 400 m of travel; bouts of bathing last 5–35 s (mean 13.7 s, $n = 107$) [3]. With increasing distance from the beach (when predators are present), duration of bouts increases and time between bouts decreases. While individuals in a flock are slow to begin, they all stop instantaneously and swim farther off, as if someone spotted some danger.

At least initially they bathe to clean their plumage, which can become very dirty, having accumulated squirts of guano from their neighbors while stationed at their nests or soiled from digging. Antarctic species sometimes go through the bathing

Fig. 14.6 Various behaviors associated with "washing," exemplified by African penguins. The top left photo shows simple preening in a bird that is almost on its back (note the position of the feet) while the other three show various stages of "body roll" associated with what we term "body-roll-swiping" (variously "thrashing," "splashing," etc. by other authors) where the body is rotated from the norm so that the sides and/or belly is uppermost while the flippers are swiped across the flanks. We believe that this process, together with careful preening (see top left photo) introduces air into the plumage to help insulation. (Photos courtesy of Peter Ryan)

motions in snow banks, but only rarely do so in pools of fresh water (unless it is clean). Included are five behaviors: "body-roll-swiping" (termed "thrashing" in Ainley [3]), head-dips, head-shakes, tail-wags, preening and washing [3] (Fig. 14.6).

Vigorous head-shakes occur after each head-dip underwater, and tail-wags sometimes occur as the penguin changes from one side to the other during "body-roll-swiping." Body-roll-swiping, is a major part of washing behavior out at sea after the initial wash and, as the name suggests, consists of the bird at the sea surface rolling the body to one side by about 90° and repeatedly swiping the upper flipper across the air-exposed surface. Typically, the flipper is swiped 5–10 times before the bird rolls back to horizontal before immediately rolling to the other side and repeating the process (Fig. 14.7).

Out at sea, in Magellanic penguins at least, each body side will be treated in this manner between 5 and 12 times before the behavior finishes, often to be followed by simple preening, most particularly of the lower abdomen (Fig. 14.6). Such preening often involves more extensive body rolls (up to about 120°) while the penguin preens the exposed feathers on the side and belly (Fig. 14.6). Part of the function of preening is to introduce oil from the preen gland at the base of the tail, a behavior easily seen in penguins on land. To our knowledge, penguin feathers are never preened when they are underwater. We suggest that washing behavior in general,

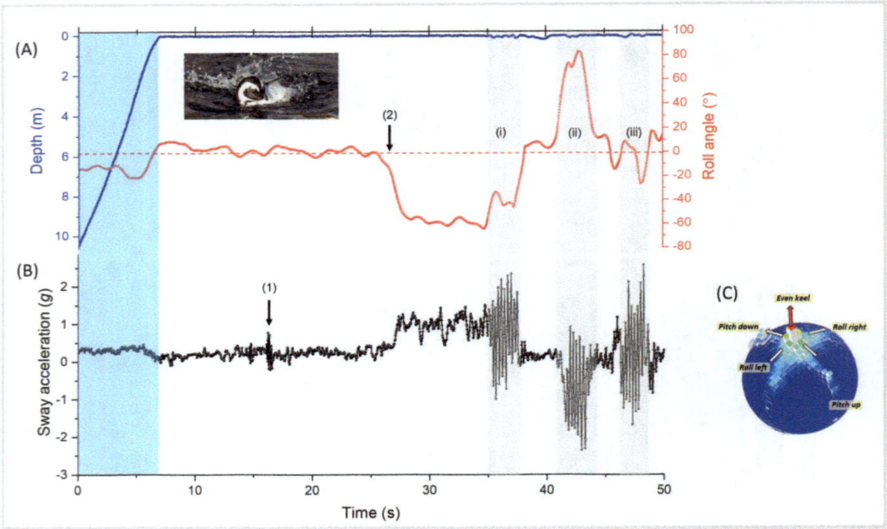

Fig. 14.7 Graphic depiction of the onset of the "body-roll-swiping" behavior with data from a Magellanic penguin (though the photo is of an African penguin). (**a**) Shows the depth in blue [with the blue overlay highlighting the terminal phase of a dive] and the body roll angle in red, while (**b**) shows the raw sway acceleration in black. After reaching the surface, the penguin shown here rested prone for about 20 s – although it performed a head shake about half way through (arrow (1)) – before rolling its body to one side by some 60° (arrow (2)), and remaining in this position for about 8 s before beginning "body-roll-swiping" (gray bar (i)). This is manifest by the large peaks and troughs in the sway acceleration, with one waveform representing one full "swipe" and recovery. Two other body-roll-swipes occur subsequently (gray bars (ii) and (iii)). The radical departure from normal body angle changes during such rolling behavior, as is made apparent in a 3D histogram showing pitch and roll on a sphere (**c**). During normal diving, penguins exhibit primarily pitch-down and pitch-up (descent and ascent, respectively), but once washing starts, addition of the roll onto the sphere creates a characteristic cross shape. (Photo courtesy of Peter Ryan)

and body-roll-swiping in particular, serves to replenish the air layer in their feathers, not least because in body-roll-swiping, the flippers are swiped across the plumage only when it is exposed to the air (see below). We also note that penguins on land often perform what are called "ruffle shakes," in which they puff out their feathers and then shake vigorously. Air in their feathers helps in thermoregulation on land, just like a down jacket does for a person.

The critical role of air in helping minimize heat loss in penguins at sea is discussed in Chap. 7, with the air layer thickness generally assumed to be constant (though it compresses with depth). This is an over-simplification because there is good evidence that there is appreciable flux of air in and out of the feathers during penguin diving.

Air Flux in Diving Penguins, an Aspect of "Washing"

Air trapped by penguins' feathers is hugely important to their relative buoyancy (Chap. 6). Loss of air has been documented for diving pinnipeds as they return to the surface after dives, with the belief that this air comes from the lungs and serves to prevent "shallow water blackout." That occurs when there is a reversal of the oxygen gradient between the body tissues and lungs so that oxygen is actually drawn out of the body and returned to the lungs [35, 36]. Although penguins also show air being lost in a similar manner as they surface, we do not believe that this air is respiratory. Rather, we suggest that this air comes from the plumage and is a consequence of changing air volumes with depth, that interact with physical forces acting across the penguin body length and the site-specific ability of the penguin feathers to hold air volumes (Fig. 6.4). Specifically, videos taken of diving penguins with the field of the view directed toward the head show that, at a certain point during the ascent of dives, air begins to be lost from a single spot in the plumage in the head area (Fig. 14.8a).

Fig. 14.8 Stages of air loss from a Magellanic penguin's head in a bird surfacing toward the end of a dive showing: (**a**), the first bubbles appearing, popping out of the feather matrix that breaks its contiguity at that point, followed by (**b**), where increasing numbers of bubbles are released from multiple breaks in the feather matrix as the bird continues upward. This air loss is followed by (**c**) a short period at the surface (of about 1 s) where the feather matrix is fully coherent again, manifest by the sleek head plumage, still covered by a thin layer of water, before (**d**) the coherence of the plumage surface breaks down as large cracks appear. This is often initiated by a head shake

As the penguin continues its ascent, increasing amounts of air are lost through multiple holes that appear in the plumage matrix (Fig. 14.8b). At the instant the penguin surfaces, the plumage is once again coherent, with no cracks apparent (Fig. 14.8c). That is, until, a second or so later, the plumage surface separates, with individual feathers discernable and large horizontal cracks breaking the surface (Fig. 14.8d).

Despite the seemingly large quantities of air being lost during the surfacing process, it is hard to image how a penguin exhaling through its highly stylized, anatomically static beak, could somehow get the air to leave the beak and move into the plumage (which is thinnest around the beak anyway; see Chap. 7, Fig. 7.11) before, later, being lost from that plumage. Indeed, footage of penguins underwater does sometimes show clear exhalations, with the air leaving the beak directly as a relatively large volume (Chap. 8) and immediately moving rapidly up the water column [46].

In accordance with the visually apparent increase in air lost as penguins approach the surface, more formal analysis of the rough quantities of air being lost from Magellanic penguin plumage during ascents shows more clearly how air loss from the plumage during any given dive increases with decreasing depth (Fig. 14.9).

However, the point at which air starts to be lost by the plumage also depends on the maximum depth achieved during the dive, with air being lost at greater depths at the end of deeper dives (Fig. 14.10a).

This relationship has appreciable variation and it is useful to speculate about what might account for this by examining air loss during dives to relatively constant depth (which takes the depth-dependency out of the relationship). This is easier said than done because using video systems on penguins does not allow much of a time window to assess this. Typically, we can record for 1–2 h during which the equipped birds have to be at sea and where we might hope that they would be diving serially to a constant depth. However, we did manage to find a few instances where Magellanic penguins dived serially to a relatively constant depth. On these occasions, there seemed to be a fairly systematic decrease over time in the depth at which bubbles first appeared (Fig. 14.11b, c). In other words, the amount of air lost from the plumage appeared to decrease as the birds continued with their uninterrupted dive sequence. This implies that air is lost by diving.

This contrasts the situation following washing, where the depth at which bubbles are first released increases substantially postwash (Fig. 14.11).

A final indication that lost air is reacquired by "washing" is provided by examination of penguin body pitch angles immediately preceding and following a washing bout. Accelerometers deployed on penguins can give us precise information on the pitch angle of the birds, in much the same way as they could with a boat. In particular, we can examine the pitch angle of penguins resting at the sea surface immediately prior to a wash, and immediately after the wash. As with boats, this pitch angle will depend on the distribution of buoyancy forces (which don't change per se in a boat) and the downward forces due to the load on the vessel. Normally, boat pitch angle will only change if any added load changes from more stern-laden or more bow-laden, for example – and if the boat is loaded equally, there will

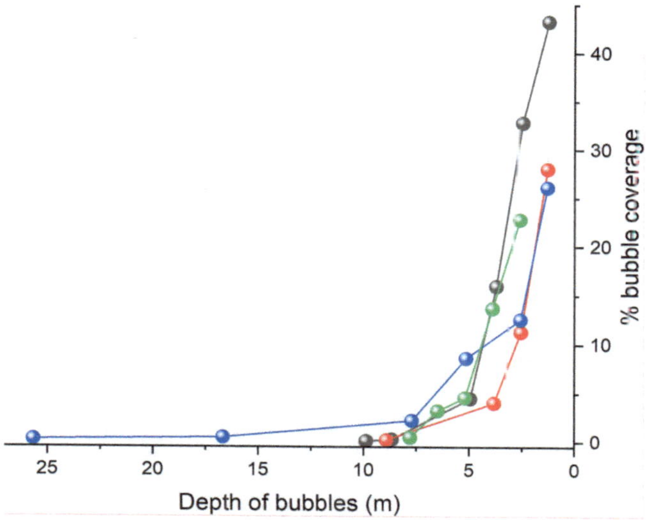

Fig. 14.9 A rough measure of the amount of air lost from the plumage of a surfacing penguin can be obtained by using specialist software to identify the bubbles in video footage (see Fig. 14.8a, b) and express them as a percentage of the total image area. Although not measuring volumes of air directly, the expectation is that greater volumes equate to higher area coverage with bubbles. This graph shows how this measure of lost air increases at shallower depths during the ascent of four Magellanic penguins (denoted by different colors), each over a single dive (dives executed to maximum depths between 45 and 65 m). Note how, in accordance with relative changes in air volume being greatest near the surface, the amount of air being released increases radically as the penguins approach the surface

actually be no change in pitch. For a penguin resting at the sea surface, there is no change in the downward acting forces because the birds are not taking on any load (such as food). However, there can easily be changes in pitch because the distribution of upward-acting forces (buoyancy) changes if air is selectively introduced to certain parts of the body. Imagine the effect of adding small balloons to a penguin's lower abdomen. In this respect, we think that it is relevant that, during 76 washing bouts undertaken by 10 Magellanic penguin, resting penguin body pitch (measured by accelerometers mounted on the lower back) changed during the wash – it increased in 60 cases and decreased by just 16, resulting in an overall mean change in pitch of 3.2° (SD 4.9). The relationship between body pitch angle before and after washes is clarified in Fig. 14.12.

So how does this all fit together? We have spoken about how air volume decreases with increasing depth according to Boyle's Law (Chap. 6, Box 6.1), but our consideration has for convenience, up until now, primarily considered the body-associated air in penguins as a single discrete unit. We noted in Chap. 8, though, how anterior and posterior air sacs will have volumes that depend on their relative depths. That is, even though the depth difference between them may seem small, being anything between maybe 0 and 50 cm (depending on the size of the penguin species and how

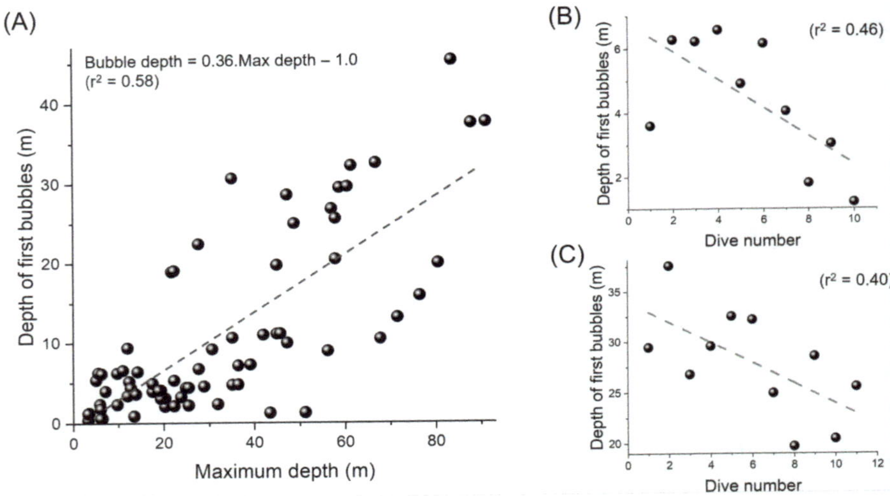

Fig. 14.10 (**a**) Depth at which the first bubbles are observed to be lost from Magellanic penguin plumage (identified using bird-attached video) as a function of the maximum depth reached during the dive (data from seven individuals). (**b**) and (**c**) show this same first bubble release depth for sequential dives recorded by two individuals diving serially to 10–20 m and 50–90 m, respectively. These latter indicate that some of the variability in (**a**) can be explained by the passage of time

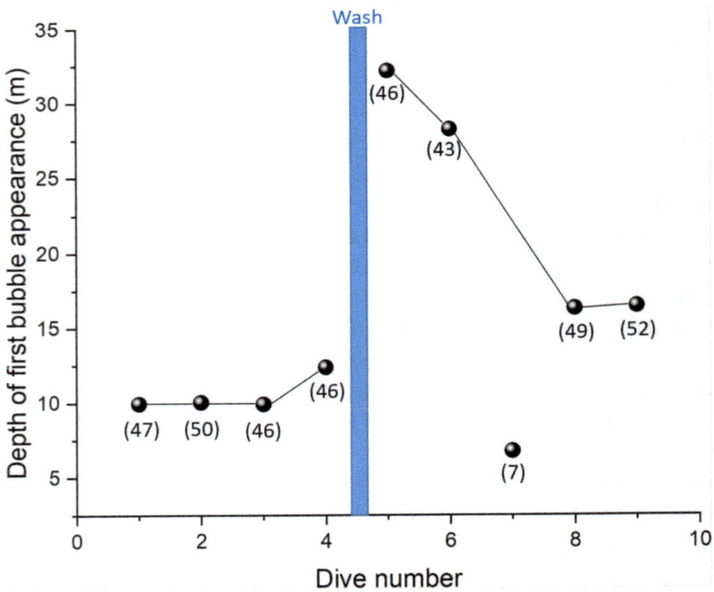

Fig. 14.11 Depth of first bubble release in a Magellanic penguin diving to fairly constant depths (indicated by the figures in brackets where depth is shown to the nearest meter), for periods pre- and postwashing (8 min). The points of comparable depths are joined by lines, which excludes a single dive to 7 m postwashing – expected to have a much lower depth of first bubble release anyway (see Fig. 14.10a). Note how the depth of first bubble release increases dramatically following the wash, implying that the plumage has acquired air during the wash

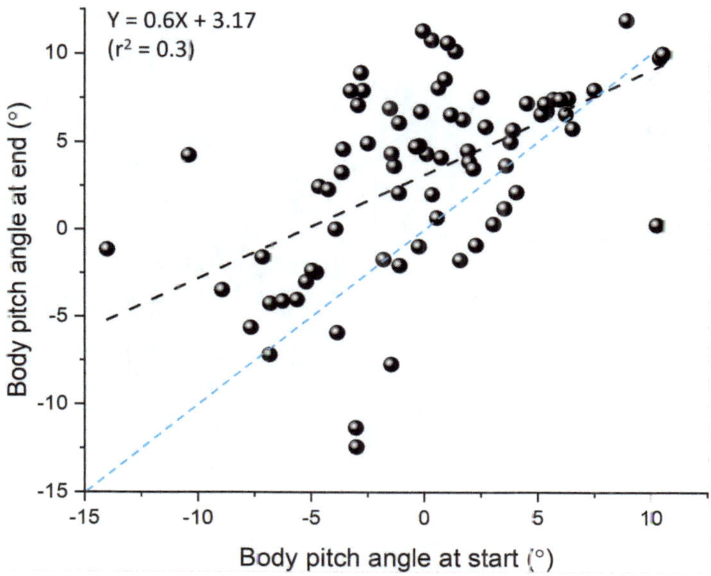

Fig. 14.12 Pitch angle of lower back-mounted accelerometers – indicative of penguin body pitch angle (the variation is due to the precise position of the tag on the bird) – before washes compared to the pitch angle following washes for 76 washes executed by 10 Magellanic penguins. The line of equivalence – where the prewash pitch angle would equal the postwash pitch angle (i.e. no change in body angle) – is shown by the blue dashed line. Points are above this line indicate that, overall, the birds had their tail ends floating higher in the water at the end of the wash than at the beginning. This accords with the idea that the main loss of air during diving would be toward the rear of the penguin (see text) but that this air can be replaced by "washing"

vertically it is orientated in the water column), the pressure difference can be substantial because water is so dense that the air volume is similarly affected. Extending this to the air contained in penguin feathers, we know that water does not penetrate to the skin so that the air effectively forms a (variable thickness with depth) layer that encompasses the body, rather like a sock. The thickness of this "air sock" will, however, nominally be greatest in the most dorsal part of the body. If the penguin is diving down vertically, the thickest layer will be in the tail region, the reverse being true if the bird is returning to the surface. But there is another process going on with the penguin air sock that is relevant to our findings. We all know, if we hang a wet T-shirt out to dry (sorry about another clothing analogy), that after a time water starts to drip off the bottom. It has effectively run down due to gravity. Air in penguin feathers is no different in principle, except that the process happens in reverse. The predominant force on the air underwater is one of upthrust (see Chap. 6), which will tend to make the air move up the body toward the water surface within the plumage, with the effect being more pronounced with time at any particular body angle. The bird body angle itself will also play a role, with steeper angles producing a more marked effect. In this context, and in the context of air within insulating

Fig. 14.13 Schematic diagram to illustrate the theorized distribution of air in a penguin descending and ascending the water column. The descending bird has its anterior end under higher pressure, which differentially compresses the air in that region. At the same time, the air tends to percolate toward the tail (red arrows), leading to relatively greater air layer thicknesses (shown in yellow) in the posterior regions. It can be held in that region by the longer feathers. The reverse is true of the ascending bird: In particular, the expanding and moving anteriorly air occupies the head region, which, with its shorter feathers cannot contain all the air which bubbles out at increasing rates as the bird approaches the surface (cf. Fig. 14.9)

layers in general, it is worth mentioning that dry suits worn by divers (rather like penguins) often have a valve at the top of the hood – to let the air that accumulates there out!

Chapter 7 waxes lyrical about the magic of penguin feathers and how, when the birds are underwater, they interlock together, providing a wonderful continuous matrix, through which water does not penetrate, even under high water pressures. But as part of this, the ingenuity of penguin feathers is that they can also adapt to changing pressures, even though they are effectively having to deal with variable feather air volumes as the birds zip up and down the water column. Perhaps sadly for us, because we think most of what penguins "do" is incredible, our data indicate that their feather layer does not function perfectly – if it did, no air would be lost at any time!

The process of air loss can be described as follows: When a penguin dives, the volume of air within the feathers is increasingly compressed, even as it is tending to be redistributed somewhat toward the tail (Fig. 14.13).

However, the feather air layer thickness in the lower body region is one of the highest of the body (see Chap. 7, Fig. 7.11). Therefore, there is little danger of air percolating from the anterior parts of the penguin not being accommodated by the feather matrix, especially since the volume is being reduced all the time due to the increasing water pressure anyway. When the bird returns to the surface, all this is reversed (Fig. 14.11), with the added effect of relative volume changes being highest nearer the surface (see Chap. 6, Fig. 6.2). Critically, though, the feather air layer thickness, which is determined by the length of the individual feathers, is smallest in the head area (Chap. 7 – to avoid the “highland cow” effect, penguins obviously cannot have long feathers near their eyes, for example). So, all the air expands as the bird rises and tends to percolate slowly toward the head where the feathers are least able to contain it. The upshot is that the outward-acting force from the expanding air “blows” holes in the wonderful penguin feather matrix and the air pearls out (Fig. 14.8a, b). In fact, it is nothing short of incredible, that the penguin feather matrix is so constructed that it can resist both inward-acting forces (the water pressure as the bird descends the water column) and outward-acting forces as the air expands on the way back to the surface although, as we have said, it’s not quite perfect!

Less obvious, is why the initiation depth for plumage air loss occurs at a deeper depth for dives that have deeper maximum depths (Fig. 14.10). However, deeper dives have longer transit times, and therefore there is more time for the air to percolate anteriorly. In addition, deeper dives occur with steeper descent and ascent angles (Chap. 9), which should help speed up the rate of air migration anteriorly. The also explains why little or no air is lost during shallow dives. Theoretically, penguins could dive all day without losing plumage air if they restricted their maximum depths to just a few meters.

But under normal dive depths, penguins clearly cannot go on losing air from their plumage as they dive indefinitely. What we observe, though, is that on contact with the air, penguins can reinflate the feathers on their head and neck, leaving gaping gashes through which air can enter (Fig. 14.8d) to reinsulate the region. This process may be helped by head shakes and it is interesting to speculate that tail shakes may do the same thing (though with obviously less potential because there is less of that region that clears the water – although who knows whether the long tails of *Pygoscelis* penguins, which are shaken often during surface periods, might not just enhance that process?). Interestingly, reinflation of the head and neck region can presumably also help put air in the lower body regions because, when the birds dive, there will be some percolation of the anterior air to posterior regions. This does not seem to be enough to maintain air layers appropriate over longer time periods though, especially when dives are to greater depths. Indeed, this process will tend to leave posterior parts of the body with less air, which is perhaps why body temperatures in this region are so low (see Chap. 7 for details). As a measure against this, we suggest that penguins at sea “wash” to introduce air into the posterior regions, noting too, that penguins washing at sea concentrate their efforts on their lower halves. And in support of that, the effect of greater air volumes in the tail end will also lead to a predicted, and observed, change in pitch acquired during washing.

So consideration of penguin washing has led to a picture that points to appreciable air flux in serially diving birds, with air entering and leaving the plumage under defined conditions, but also percolating up and down the body, like a reverse version of the dripping T-shirt on the washing line. All this has consequences for heat loss because air is such an important part of their insulation (see Chap. 7). Little wonder that penguins are so obsessed with washing and we would expect judicious management of when and how long to wash to depend on inter alia on dive depth and water temperature as well as perhaps on plumage condition. This is yet another important optimality aspect of penguins' multifaceted lives that could do with further study.

Are Auks Really "Northern Penguins"?

It would be remiss of us, in a book about penguins, not to provide a nod in the direction of auks (bird family, Alcidae), often referred to as "the penguins of the North." Why do people do this? Because some auks, which live in the Northern Hemisphere, can have similar basic coloration to penguins, sporting dark dorsal and pale ventral plumage (18 of 25 species), some stand erect like penguins, and all move underwater using their wings for propulsion (though some, such as the pigeon guillemot, also use their feet at times, something penguins never do). It is also germane to note that the morphologically similar diving petrels (bird family, Procellariidae), which live in the Southern Hemisphere like penguins, are never referred to as "also southern penguins" or even, perhaps more appropriately, "southern auks." And of course, referring to auks as "northern penguins" is also somewhat catalyzed by biologists who are often happy to find ecological equivalents – animals that perform the same function in different ecosystems, the best showing convergent evolution of their various traits.

The extreme specialization in anatomy, physiology and behavior that allows some birds to forage underwater in a cold, dense, viscous medium – one of earth's hardest taskmasters for warm-blooded creatures – should lead to extreme convergent evolution if solutions are to be optimal. In fact, comparisons have shown that auklets and diving petrels are prime examples of convergent evolution, having similar color pattern, shape, flight, molt and other aspects.

What's the case for auks and penguins then? To compare, and cut to the chase, we will examine the accelerations to which these birds subject their bodies as they swim underwater. Acceleration is linked to force via mass (Force = Mass × Acceleration – Thank you Isaac), so if we create plots that display the forward and vertical accelerations of swimming birds as determined by bird-attached accelerometers, we are effectively also showing force plots (see Chap. 6). Furthermore, to do the job properly, it is useful to compare the mass-corrected force (no point comparing movement forces of an aircraft carrier with those of a pedalo), which can be done by dividing the forces by the bird mass. Doing this actually leads us back to where we were when we just showed acceleration, and so we'll stick with that. In a nutshell, simple acceleration bi-plots give us a good idea of how much force auks and penguins need to swim. Importantly, the greater the forces required to swim, the less efficient the birds are at swimming. Creating these plots for various auks and penguins reveals astonishing differences between the two groups, with auks exerting much greater forces to swim than penguins do so to swim (at lower speeds too!) (Fig. 14.14).

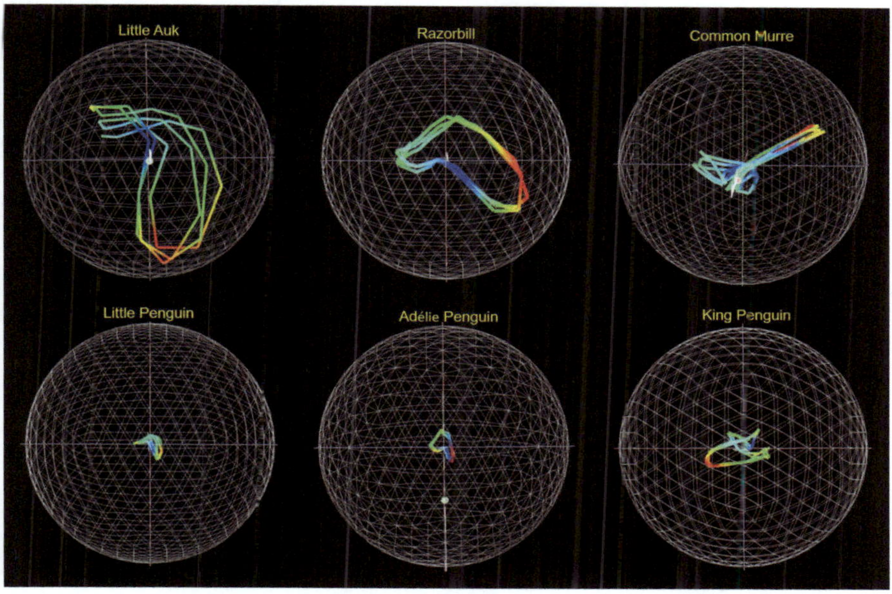

Fig. 14.14 Acceleration bi-plots (static acceleration corrected) of 3 wingbeats for three species of auk and three species of penguin of varying sizes (little auks ≈ 0.15 kg, razorbill ≈ 0.75 kg, common murre ≈ 0.99 kg, little penguin ≈ 1.0 kg, Adélie penguin ≈ 4.7 kg, king penguin ≈ 12 kg). All individuals were diving down to start dives and measurements were extracted 5 wingbeats after leaving the water surface. All data are shown on the same scale (acceleration limits across axes are −1.5 to 1.5 *g*) to facilitate comparison. Although the precise swimming conditions could not be standardized, all birds will have been working against buoyancy and swimming at roughly similar speeds (although the penguins will have been traveling faster, which will tend to increase the size of their acceleration footprint). Note how, despite faster speeds, the penguin acceleration (which equates to force) footprint is just a tiny fraction of that of the auks. (Data courtesy of David Grémillet (little auk), Emma Cole (razorbill and murre), Belinda Cannell (little penguin), and Charles-Andre Bost (king penguin))

There are multiple reasons for this, including the higher buoyancy of auks due to greater amounts of body-associated air [82], but it is primarily the auk wing that gets them into trouble. In having to fly, auks cannot allow their wing to completely adapt to efficient locomotion underwater (nor indeed for efficient locomotion in the air [23]). Well, unless they evolved to become great auks (Chap. 1). Compared to penguins, auks have a much greater wing surface area and they have to drag this up and down through viscous water with every wing beat as they swim underwater. That's hard work, and it makes auks energetically inefficient, particularly when compared to penguins (Fig. 14.14). So it is flying that stymies auks with swimming efficiency compared to penguins. But in this respect auks are little different to many other diving, flighted seabirds that swim with their wings. In these birds, on the one hand, the bigger the wing for flight, the less costly flight can be [55]. On the other hand however, the greater is the penalty for swimming underwater, and so flying birds that "wing-swim" all lie on a continuum for swimming and flying costs. Penguins have stepped (swum?) off this continuum and, as such, are in a league of their own, fantastically efficient underwater.

The truth about auks being "northern penguins," then is that they are not, or at least no more than many other seabirds that fly and swim with their wings. It does raise the question about how great auks would have fitted into this picture, though. Sadly, we'll never know – another secret buried on the back of our outrageous behavior.

Research Tags – The Flip Side for Evolutionarily Honed Fish-Birds

A large part of what we understand about the behavioral ecology and energetics of fish-birds in their true element comes from tags attached to them that record, sometimes with extraordinary detail, what these birds do when they are far from land or at depth underwater. Indeed, penguins are impossible to track at sea without them (though not impossible to investigate, as noted in Chaps. 3 and 5). But, as we have seen in Chaps. 2 and 6 (and above), penguins have been honed by draconian forces to become sublime beings that slip through the ocean waters at high speed with minimal effort, and yo-yo between their land and marine habitats. As such, we have to think seriously about messing with their exquisite design by putting instruments and tags on them without expecting consequences – in fact, it would be naïve to consider otherwise.

We can say, however, that a primary tag characteristic that is commonly applied to aerial birds, that of the weight of the tag, is likely to be trivial for penguins underwater. This is because the actual weight of the tag in air is largely compensated by the upthrust of water so that the overall downward-acting force of a tag on a penguin underwater is far less than in air. Against this, any tag that is out of the water when the penguin is at the surface – a tag that is on the penguin's back, for example – will exert the full force of that weight, acting downward. If the tag is mounted in the mid-line, and the penguin is perfectly level, the effect of this will be to push the penguin down and its water-line up, to an extent dependent on the weight of the tag and the buoyancy of the penguin. This, in itself, is trivial given the size of penguins and the tags put on them, though it is perhaps worth remembering that penguin feather thickness is geared to deal with the amount of time that various parts of the body are underwater to minimize heat loss (see Chap. 7). For this reason, the situation of increased weight due to a back-mounted tag is not comparable to an increased weight of food in the stomach, which is underwater, and thus is essentially compensated by the upthrust from the seawater (see above).

Perhaps the most important concern of a back-mounted tag for a penguin at the surface is its propensity to cause the penguin to roll (Fig. 14.15). Penguins do not, of course, live and operate at the surface of a mirror-calm sea, so their bodies are

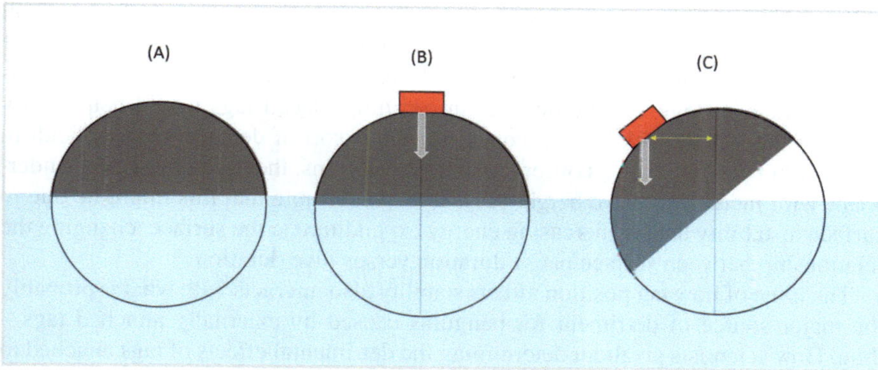

Fig. 14.15 Schematic of the cross-section of a penguin at sea showing how the weight of a back-mounted tag (red rectangle) affects the way a penguin floats. (**a**) The norm, with a penguin wearing no tag. The water-line reaches around the intersection between the dark back and the white belly. (**b**) Shows the situation when the tag is placed directly above the bird's center line, where the force stemming from the weight of the tag simply pushes the penguin (marginally) deeper in the water so that it sits lower. The extent to which the penguin is pushed down depends on the weight of the tag and the buoyancy of the penguin. (**c**) Shows the situation when the bird begins to roll, due to irregularities in the water surface such as waves and chop. Here, the force exerted by the tag applies a "force moment," tending to make the bird roll more. The force promoting the roll is given by the distance between the tag and the center line (shown by the yellow arrow in (**c**) multiplied by the mass of the tag). Thus, heavier tags and more acute roll angles make penguins more unstable at the water surface

subject to the vagaries of waves, chop and weather, all of which conspire to move the body off the horizontal. Once this happens, the downward-acting force of any back-mounted tag, instead of simply acting to push the whole bird lower in the water (compare Fig. 14.15a, b), will tend to pull the bird over to one side – to roll (Fig. 14.15c). The extent of the rolling force is dependent on the mass of the tag and the distance it has rolled from the center point (yellow arrow in Fig. 14.15c). This has implications because it will tend to cause parts of the feather layer that are normally exposed to the comparatively mild heat loss conditions of being in air at the surface to move underwater (conditions with which the feathers are less well able to deal). But also, within the inherent instability that comes with a high-mounted tag, if birds wish to maintain their position close to horizontal, they will have to be continually correcting for the roll with their flippers, and this costs energy.

This is presumably what researchers had in mind when they investigated the roll stability of little penguins wearing back-mounted tags in captivity [32]. They noted that birds had trouble balancing and were disinclined to dive, especially when the tags were mounted on the lower back. The problem, apparently, was lessened (or disappeared) when the tags were mounted in the mid-back. However, the water is completely calm under such conditions and does not really replicate what happens at sea. Theoretically, a tag mounted on the high middle back will exert a greater destabilizing force to the penguin (at the surface) than one lower down because it can move farther from the center line (yellow arrow in Fig. 14.15c). However, this will only be the case when sea surface conditions cause the penguin body to roll anyway. So, in a nutshell, we expect balancing problems for penguins at the surface to be greatest in rough weather, greatest in birds with high-mounted tags and greatest with heavier tags. Accelerometers in carefully placed tags would help clarify much of this. The knock-on consequences of this sort of detriment are difficult to quantify but we note that, as observed in little penguins, they spend less time underwater with increasing wave height [11]. It is conceivable that this might be due to surface instability and/or increasing energy expenditure at the surface, changing the relationship between surface pause duration versus dive duration.

The issue of how tag position affects stability also interacts with what is probably the major source of detriment for penguins caused by externally attached tags – drag. How scientists go about determining the detrimental effects of tags attached to penguins, and drag in particular, depends fundamentally on their discipline. Biologists look for changes in penguin behavior, performance, provisioning, survival etc., perhaps as a function of tag size, mass etc. [63], while physicists work from "first principles" (the basic laws of nature), such as the Newtonian framework. The physicist's approach has much to recommend because it deals with certainties, deriving other important certainties from these. Our calculations indicate a drag coefficient of a Magellanic penguin of 0.028 (Chap. 6), close to that projected by Bannasch [10] of 0.03. This is even lower than the most efficient body constructed by man for movement through water, and about three times lower than other diving birds [42]. As such, it is unthinkable that an attached "foreign body" could do anything other than increase the drag substantially. Importantly, the extraordinarily low drag coefficient of penguins (see Chap. 6) means that even small increases in this coefficient stemming from an attached tag can have a disproportionate effect on the energy required to swim (Chap. 6, Box 6.8) – and efficient swimming is pivotal to penguin existence.

It is the extent to which an attached foreign body (a tag or a flipper band) disrupts laminar (smooth and regular) water flow over the penguin body (or wing) that is responsible for the increased drag. This was neatly shown by Rudi Bannasch's work using penguin models in an air stream in a wind tunnel [10] (air can be used to simulate water flow under the right conditions since both air and water are "fluid media"). Rudi's work also showed that the position of a back-mounted tag was critical in the degree to which water flow was disrupted, with tags placed further back – toward the tail – causing less disruption than those mounted more anteriorly. Perhaps the single most important, and generally unappreciated, thing about tag-induced drag is that even very small projections from the body can cause massive increases in the drag. For example, the late, great Collin Pennycuick, a biologist who advocated a physical framework to understand the intricacies and energetics of animal movement, including bird flight (see Chap. 3), determined that a tiny protrusion (a baseplate) from the back of a rose-colored starling flying in a wind tunnel, increased the drag by a horrifying 50% [56]. No less is expected for a penguin carrying a tag. Indeed, this is exactly the reason why the addition of a tiny band on Adélie penguin flippers was found to increase the energy expenditure to swim at normal speeds in a swim channel by 24% [17]. Perhaps the most important thing to note, here, is that there is no way for a penguin to swim that will reduce this speed-dependent drag. It is not something that can easily be avoided by canny swimming tricks, any more than it is possible to reduce the drag on a car carrying a roof rack. It is a purely physical constraint. That is not to say, however, that at least some penguin species cannot learn or develop to compensate in some way, with regard to small wing bands [21].

For example, penguins can modulate the energy they use to swim while wearing a tag by simply swimming slower because drag increases with speed (Chap. 6, Box 6.4). In fact, faced with the increased drag imposed by having to swim with an external tag, penguins can choose some options; swim at their usual speeds but undergo increased energy expenditure (Chap. 6, Box 6.8), maintain their usual energy expenditure by slowing down (Fig. 14.16a) or something that lies midway between the two.

There is some evidence that penguins actually reduce their normal cruising speed when their drag is artificially increased by the addition of a tag and, by so doing, they maintain their energy use as normal. For example, African penguins swim more slowly as attached tags get bigger [81], which is expected if birds were keeping energy expenditure constant. This raises the intriguing question about what penguin swim speeds would be without tags – by measuring it, we affect the bird performance – so we really don't know! Where we would expect penguins not to compromise their speed would be in scenarios that require high speed, such as catching highly mobile prey, e.g., schooling fish such as anchovy, or avoiding predators. Here, the effect of increased drag results in a decrease in the maximum speed

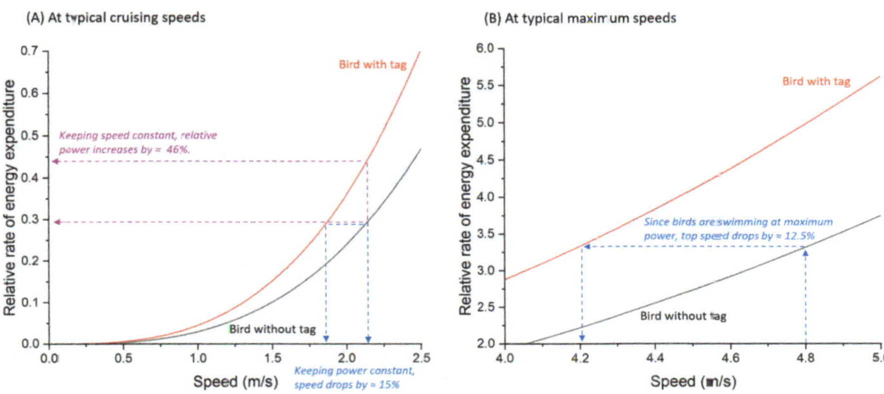

Fig. 14.16 The effect of increased drag due to an externally attached tag to a penguin manifest in changes in speed or energy (or both). In both graphs, the red line shows the relative rate of energy expenditure (relative power) against speed of the bird swimming with a tag that causes the drag coefficient to rise from the norm at 0.03 to 0.045 – the case of the penguin without a tag is shown in black. In (**a**), the penguin can choose to maintain the same power to swim, in which case the speed must drop by some 15% (blue arrows), or maintain speed, in which case the power must increase by 46% (pink arrows). In (**b**), where the penguin is operating at maximum power, the addition of a tag can only reduce maximum speed, in this case by about 12.5%. How often this sort of issue leads to reduced capture of prey or the bird being caught by predators is not known

(Fig. 14.15b) because at their top speeds penguins are operating at maximum power. Whether this means that they miss potential prey or fall to the jaws of predators, we can only guess.

We can, however, see how increases in drag, quantification of which is the province of the physicist, translate into changes in behavior, possibly affecting fitness, these being areas covered by biologists. Importantly, there is a line of reasoning from drag to behavioral ecology that proceeds seamlessly from purely physical change to changes in behavior that depend on how the birds react to the physical changes. An example of this is shown in Fig. 14.17 whereby tag-derived increases in drag are either met by penguins that drop speed to maintain "normal" energy expenditure or keep speed constant by increasing energy expenditure.

In the first case, lower speed results in a lower vertical velocity during dives (birds could also increase descent angle to counteract this but, as Chap. 6 shows, this requires higher power). Lower velocity increases the time taken to reach the target depth. Since the transit time must be subtracted from the total time underwater (a function of oxygen store), this increase in transit time reduces the duration of the bottom, prey-catching, phase. As with almost any animal foraging, if the foraging duration is reduced, less food can be ingested (Fig. 14.17). Otherwise,

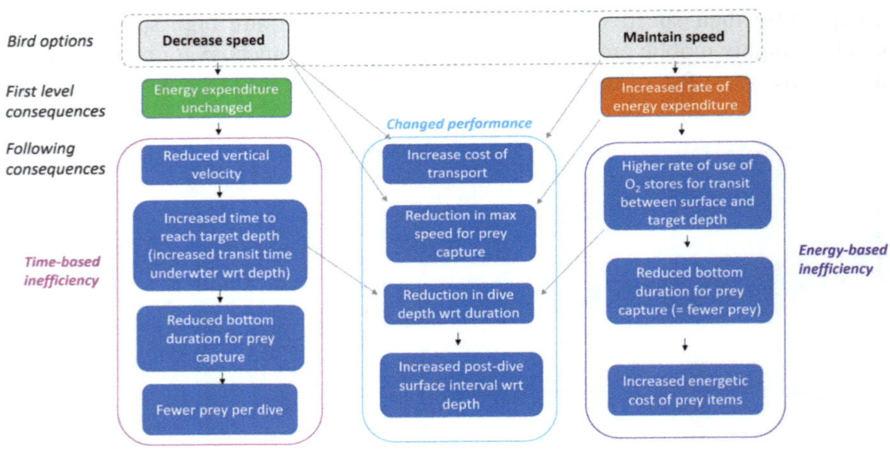

Fig. 14.17 Schematic diagram to show how penguins may deal with increased drag resulting from the attachment of an external tag. On the left-hand side, they may decrease speed to conserve their normal rates of energy expenditure during swimming. However, this leads to increased transit times between the surface and the foraging depth, which reduces the time available for prey capture, ultimately leading to birds catching fewer prey per dive. On the right-hand side, penguins can maintain speed but use more energy, and therefore oxygen, per unit time to do so. This reduces the overall duration of the dive, increasing the energetic cost of acquiring prey so that the net return is reduced. Note that both these cases are not mutually exclusive. In both cases, penguins incur increased costs of transport because they are not operating at the minimum point on the costs of transport versus velocity curve (see Chap. 6). Rather, they have a reduction in maximum speed and it changes the relationship between dive duration and depth as well as necessitates a relative increase in the postdive recovery phase

maintenance of speed with higher energy expenditure results in oxygen stores being used during the dive faster than usual, which also reduces the duration of the bottom phase, correspondingly increasing the energetic cost of prey items. In both cases, penguins have an increased cost of transport (the energetic cost of travel per unit distance – see Chap. 6), have a reduced maximum speed attainable (Fig. 14.16b), a reduction in the time they can spend underwater as a function of depth (Chap. 11) and a correspondingly increased postdive surface interval (to recover oxygen), reducing their overall foraging efficiency (Chap. 11).

So, incrementally, it has become clear how something as simple as an increase in drag can have huge ramifications, affected by, it would seem, prey availability (how difficult it is to obtain prey; Chaps. 4 and 5). It is therefore entirely understandable that external tags change penguin dive behavior [27, 43, 64], and tag-wearing, breeding birds may extend foraging periods at sea in order to acquire the amount of food they "intended" to capture, either for their own energy budget or that of their chicks [43]. It is worth noting, though, in the interests of appropriate brood provisioning, that flipper-banded penguins may not always extend foraging duration [13], presumably shifting the energetic inefficiencies to maintaining their own body condition. Noted by Ballard et al. [9], some penguin individuals begin the breeding season with condition to spare, i.e. an energetic "cushion" (to then invest the surplus in chick provisioning), while others show up in lower condition with the result that less food goes to the chick as the parent uses more of the food it captures to maintain its own condition. In all this, prey availability is critical [19] because, when food is plentiful, even increased foraging costs and decreased efficiencies may be tenable, as assessed in little penguins [1]. Thus, it would be naïve to expect penguins to always manifest changes in all their foraging outcomes when they are dealing with additional drag. Indeed, in a comparison of foraging trip duration and food loads brought to Adélie penguin chicks, banded versus unbanded (but RFID-tagged), in a population growing significantly (food plentiful), food loads did not differ although in a few, but not all, years foraging trip durations were slightly longer (~8%) in banded penguins [21].

Without consequence, decreasing body condition, however, can only occur within a limited time frame, i.e., a single foraging trip, and in this respect, it is relevant that penguins tagged for long periods, e.g., a measure of annual survival, may incur increased mortality. This has been shown in the banding of king penguins [29, 68], Adélie penguins [8, 21], and little penguins [18]. In the latter two species, the effect was most apparent during the first year, thus applicable to the query expressed above about whether at least some penguins may learn to compensate for the increased drag. Another indication of that is the observation in the Adélie penguin study that despite banded penguins exhibiting longer foraging trips, they actually brought back more prey than the unbanded cohort, perhaps compensating for the longer time that they were away. These studies also showed that the effect is most apparent during trying foraging conditions, also observed in other species, such as African penguins [87]. This issue of the pros and cons of penguin banding have been thrashed out in a number of reviews and we will not here consider it any further [57, 58] (also see the more recent papers cited above). Possibly having more effect than bands are the much larger instruments glued or taped to penguins' backs or heads, but much less study has been addressed at them, possibly because they are not worn for a penguin's lifetime. Some back-mounted tags have been attached for months, however, with results (though not any negative effects if the occurred) presented in Chap. 3.

The take-home message from this in the context of penguins' capabilities of exploiting their respective preyscapes is twofold. First, th ere are aspects of what penguins do at sea, particularly underwater, that we are only approximating [64, 81], though we hope to do better by making tags smaller and more streamlined [17]. Second, the longer we equip penguins with external tags, the more detriment we will be causing [68]. If penguins are equipped with back-pack type tags for just one or a few foraging trips (to feed chicks, for example), there may be no effects that can be detected [41]. Against that, without tags, we would know virtually nothing about the incredible underwater prowess of these magical beings, nor much about their demography and how sea conditions affect that. This is not just knowledge for knowledge sake. It translates to a real understanding of what penguins need in terms of prey fields, densities, and distributions to survive and breed, and such information is needed for conservation initiatives. Whether governments take heed of recommendations of, for example, the effectiveness of aspired fishing policies or proposed marine protected areas, is another question entirely (see Chaps. 5 and 13)!

References

1. Agnew P, Lalas C, Wright J, Dawson S (2013) Effects of attached data-logging devices on little penguins (*Eudyptula minor*). Mar Biol 160:2375–2382
2. Ainley DG (1972) Flocking of Adélie penguins. Ibis 114:388–390
3. Ainley DG (1974) Comfort behavior of Adélie and other penguins. Behaviour 50:16–50
4. Ainley DG (1975) Displays of Adélie penguins: a re-interpretation. In: Stonehouse B (ed) The biology of penguins. Macmillan, London, pp 503–534
5. Ainley DG, Ballard G (2011) Non-consumptive factors affecting foraging patterns in Antarctic penguins: a review and synthesis. Polar Biol 35:1–13
6. Ainley DG, Ballard G, Ackley S, Blight LK, Eastman JT, Emslie SD, Lescroël A, Olmastroni S, Townsend SE, Tynan CT, Wilson P, Woehler E (2007) Paradigm lost, or, is top-down forcing no longer significant in the Antarctic marine ecosystem? Antarct Sci 19(3):283–290
7. Ainley DG, Jacobs SS, Ribic CA, Gaffney I (1998) Seabird distribution and oceanic features of the Amundsen and southern Bellingshausen seas. Antarct Sci 10:111–123
8. Ainley DG, Leresche RE, Sladen WJL (1983) Breeding biology of the Adélie penguin. University of California Press, Los Angles
9. Ballard G, Dugger KM, Nur N, Ainley DG (2010) Foraging strategies of Adélie penguins: adjusting body condition to cope with environmental variability. Mar Ecol Prog Ser 405:287–302
10. Bannasch R (1995) Hydrodynamics of penguins – an experimental approach. In: Dann P, Norman I, Reilly P (eds) The penguins: ecology and management. Surrey Beatty, Sydney, pp 141–176
11. Berlincourt M, Arnould JPY (2015) Influence of environmental conditions on foraging behaviour and its consequences on reproductive performance in little penguins. Mar Biol 162:1485–1501

12. Berthold P, Bauer H-G, Westhead V (2001) Bird migration: a general survey. Oxford University Press, Oxford
13. Boersma PD, Rebstock GA (2009) Foraging distance affects reproductive success in Magellanic penguins. Mar Ecol Prog Ser 375:263–275
14. Conradt L, Roper TJ (2005) Consensus decision making in animals. Trends Ecol Evol 20:449–456
15. Couzin ID (2009) Collective cognition in animal groups. Trends Cogn Sci 13:36–43
16. Couzin ID, Krause J, Franks NR, Levin SA (2005) Effective leadership and decision-making in animal groups on the move. Nature 433:513–516
17. Culik BM, Wilson RP, Bannasch R (1993) Flipper-bands on penguins: what is the cost of a life-long commitment? Mar Ecol Prog Ser 98:209–214
18. Dann P, Sidhu LA, Jessop R, Renwick L, Healy M, Dettmann B, Vaker B, Catchpol EA (2014) Effects of flipper bands and injected transponders on the survival of adult Little Penguins *Eudyptula minor*. Ibis 156:73–83
19. Dehnhard N, Ludynia K, Poisbleau M, Demongin L, Quillfeldt P (2013) Good days, bad days: wind as a driver of foraging success in a flightless seabird, the southern rockhopper penguin. PLoS One 8(11):e79487
20. Duffy DC (1983) The foraging ecology of Peruvian seabirds. Auk 100(4):800–810
21. Dugger K, Ainley DG, Ballard G, Barton KJ (2006) Effect of flipper-bands on the foraging behavior and survival of Adélie penguins on Ross Island. Auk 23:858–869
22. Elley T, Mattern T, Ellenberg U, Young MJ, Hickcox RP, Van Heezik Y, Seddon PJ (2022) Consistent site-specific foraging behaviours of yellow-eyed penguins/hoiho breeding on Stewart Island. New Zealand Biol 11(6):844
23. Elliott KH, Ricklefs RE, Gaston AJ, Hatch SA, Speakman JR, Davoren GK (2013) High flight costs, but low dive costs, in auks support the biomechanical hypothesis for flightlessness in penguins. Proc Natl Acad Sci 110(23):9380–9384
24. Emlen JT, Penney RL (1964) The navigation of penguins. Nat Hist 73:105–113
25. Evangelista C, Kraft P, Dacke M, Labhart T, Srinivasan MV (2014) Honeybee navigation: critically examining the role of the polarization compass. Philos Trans R Soc Lond Ser B Biol Sci 369(1636):20130037
26. Falla RA (1937) Birds. In: B.A.N.Z. Antarctic research expedition, 1929-1931. Hassell Press, Reports-Series B, Adelaide
27. Fallow PM, Chiaradia AF, Ropert-Coudert Y, Kato A, Reina RD (2009) Flipper bands modify the short-term diving behaviour of little penguins. J Wildl Manag 73:1348–1354
28. Ford RG, Ainley DG, Lescroël A, Lyver POB, Toniolo V, Ballard G (2015) Testing assumptions of central place foraging theory: a study of Adélie penguins *Pygoscelis adeliae* in the Ross Sea. J Avian Biol 46(2):193–205
29. Gauthier-Clerc M, Gendner JP, Ribic CA, Fraser WA, Woehler EJ, Descamps S, Gilly C, Le Bohec C, Le Maho Y (2004) Long-term effects of flipper-bands on penguins. Proc R Soc B 271:S423–S426
30. Handley JM, Pistorius P (2016) Kleptoparasitism in foraging gentoo penguins *Pygoscelis papua*. Polar Biol 39:391–395
31. Harrison NM, Whitehouse MJ, Heinemann D, Prince PA, Hunt GLJ, Veit RR (1991) Observations of multispecies seabird flocks around South Georgia. Auk 108(4):801–810
32. Healy M, Chiaradia A, Kirkwood R, Dann P (2004) Balance: a neglected factor when attaching external devices to penguins. Mem Nat Inst Polar Res 58(Special Issue):179–182
33. Herbert-Read JE (2016) Understanding how animal groups achieve coordinated movement. J Exp Biol 219:2971–2983
34. Hinke JT, Russell TM, Hermanson VR, Brazier L, Walden SL (2021) Serendipitous observations from animal-borne video loggers reveal synchronous diving and equivalent simultaneous prey capture rates in chinstrap penguins. Mar Biol 168:135
35. Hooker SK, Andrews RD, Arnould JP, Bester MN, Davis RW, Insley SJ, Gales NJ, Goldsworthy SD, Mcknight JC (2021) Fur seals do, but sea lions don't – cross taxa insights into exhalation during ascent from dives. Philos Trans R Soc B 376(1830):20200219

36. Hooker SK, Miller PJ, Johnson MP, Cox OP, Boyd IL (2005) Ascent exhalations of Antarctic fur seals: a behavioural adaptation for breath–hold diving? Proc R Soc B Biol Sci 272(1561):355–363
37. Kearton C (1931) The Island of penguins. National Travel Club, New York
38. Keller BA, Putman NG, Grubbs RD, Portnoy DS, Murphy TP (2021) Map-like use of earth's magnetic field in sharks. Curr Biol 31:2881–2886
39. Klimley AP (1993) Highly directional swimming by scalloped hammerhead sharks, *Sphyrna lewini*, and subsurface irradiance, temperature, bathymetry, and geomagnetic field. Mar Biol 117:1–22
40. Ksepka DT (2016) The penguin's palette: more than black and white. Am Sci 104(1):36–44
41. Lescroël A, Ballard G, Toniolo V, Barton KJ, Wilson PR, Lyver POB, Ainley DG (2010) Working less to gain more: when breeding quality relates to foraging efficiency. Ecology 91(7):2044–2055
42. Lovvorn JR, Liggins GA, Borstad MH, Calisal SM, Mikkelsen J (2001) Hydrodynamic drag of diving birds: effects of body size, bodyshape and feathers at steady speeds. J Exp Biol 204:1547–1557
43. Ludynia K, Dehnhard N, Poisbleau M, Demongin L, Masello JF, Quillfeldt P (2012) Evaluating the impact of handling and logger attachment on foraging parameters and physiology in southern rockhopper penguins. PLoS One 7(11):e50429
44. Mattern T, Ellenberg U, Houston DM, Davis LS (2007) Consistent foraging routes and benthic foraging behaviour in yellow-eyed penguins. Mar Ecol Prog Ser 343:295–306
45. Mattern T, Ellenberg U, Houston DM, Lamare M, Davis LS, Van Heezik Y, Seddon PJ (2013) Straight line foraging in yellow-eyed penguins: new insights into cascading fisheries effects and orientation capabilities of marine predators. PLoS One 8(12):e84381
46. Mattern T, Mcpherson MD, Ellenberg U, Van Heezik Y, Seddon PJ (2018) High definition video loggers provide new insights into behaviour, physiology, and the oceanic habitat of a marine predator, the yellow-eyed penguin. PeerJ 6:e5459
47. Mcinnes AM, Mcgeorge C, Ginsberg S, Pichegru L, Pistorius PA (2017) Group foraging increases foraging efficiency in a piscivorous diver, the African penguin. R Soc Open Sci 4:170918
48. Mcinnes AM, Thiebault A, Cloete T, Pichergru L, Aubin T, Mcgeorge C, Pistorius PA (2020) Social context and prey composition are associated with calling behaviour in a diving seabird. Ibis 162(3):1047–1059
49. Mills KL (1998) Multispecies seabird feeding flocks in the Galápagos Islands. Condor 100(2):277–285
50. Moore PJ (1999) Foraging range of the yellow-eyed penguin *Megadyptes antipodes*. Mar Ornithol 27:49–58
51. Moore PJ, Wakelin MD, Douglas ME, Mckinlay B, Nelson D, Murphy B (1995) Yellow-eyed penguin foraging study, South-Eastern New Zealand, 1991–1993. Department of Conservation, Wellington
52. Nesterova AP, Le Bohec C, Beaune D, Pettex E, Le Maho Y, Bonadonna F (2010) Do penguins dare to walk at night? Visual cues influence king penguin colony arrivals and departures. Behav Ecol Sociobiol 64(7):1145–1156
53. Ossenkopp K-P, Barbeito R (1978) Bird orientation and the geomagnetic field: a review. Neurosci Biobehav Rev 2:255–270
54. Parrish JK, Hamner WM (1997) Animal groups in three dimensions: how species aggregate. Cambridge University Press, Cambrige
55. Pennycuick CJ (2008) Modelling the flying bird. Elsevier
56. Pennycuick CJ, Fast PLF, Ballerstädt N (2012) The effect of an external transmitter on the drag coefficient of a bird's body, and hence on migration range, and energy reserves after migration. J Ornithol 153:633–644
57. Petersen SL, Branch GM, Ainley D, Boersma D, Cooper J, Woehler EJ (2005) Is flipper banding on penguins a problem? Mar Ornithol 33:75–79

58. Petersen SL, Branch GM, Crawford RJM, Cooper J, Underhill LG (2005) The future for flipper banding African penguins: discussion, recommendations and guidelines. Mar Ornithol 33:E1–E4
59. Pettingill S Jr (1964) Penguins ashore at the Falkland Islands. Living Bird 3:45–64
60. Putz K, Cherel Y (2005) The diving behaviour of brooding King Penguins (*Aptenodytes patagonicus*) from the Falkland Islands: variation in dive profiles and synchronous underwater swimming provide new insights into their foraging strategies. Mar Biol 147:281–290
61. Quintana F, Gómez-Laich A, Gunner RM, Gabelli F, Omo GD, Duarte C, Brogger M, Wilson RP (2022) Long walk home; Magellanic penguins have strategies that lead them to areas where they can navigate most efficiently. Proc R Soc B 289(1977):20220535
62. Richdale LE (1951) Sexual behavior in penguins. University of Kansas Press, Lawrence
63. Ropert-Coudert Y, Knott N, Chiaradia A, Kato A (2007) How do different data logger sizes and attachment positions affect the diving behaviour of little penguins? Deep-Sea Res II 54:415–423
64. Ropert-Coudert Y, Wilson RP, Yoda K, Kato A (2007) Assessing performance constraints in penguins with externally-attached devices. Mar Ecol Prog Ser 333:281–289
65. Ross JC (1847) A voyage of discovery and research in the Southern and Antarctic regions during the years 1939–43. John Murray, London
66. Ryan PG, Wilson RP, Cooper J (1987) Intraspecific mimicry and status signals in juvenile African penguins. Behav Ecol Sociobiol 20:69–76
67. Sancton J (2021) Madhouse at the end of the earth. Diversified Publishing, Portland
68. Saraux C, Le Bohec C, Durant J, Viblanc VA, Gauthier-Clerc M, Beaune D, Park Y-H, Yoccoz NG, Stenseth NC, Le Maho Y (2011) Reliability of flipper-banded penguins as indicators of climate change. Nature 469:203–206
69. Sauer EGF (1958) Celestial navigation by birds. Sci Am 199:42–47
70. Shepherd LD, Miskelly CM, Cherel Y, Tennyson AJD (2021) Genetic identification informs on the distributions of vagrant royal (*Eudyptes schlegeli*) and macaroni (*Eudyptes chrysolophus*) penguins. Polar Biol 44:2299–2306
71. Sherley RB, Ludynia K, Dyer BM, Lamont T, Makhado AB, Roux JP, Scales KL, Underhill LG, Votier SC (2017) Metapopulation tracking juvenile penguins reveals an ecosystem-wide ecological trap. Curr Biol 27(4):563–568
72. Siegfried WR, Frost PGH, Kinahan JB, Cooper J (1975) Social behaviour of jackass penguins at sea. Zool Afr 10(1):86–100
73. Silk MJ, Croft DP, Tregenza T, Bearhop S (2014) The importance of fission–fusion social group dynamics in birds. Ibis 156:701–715
74. Sørensen K, Neumann C, Dähne M, Hansen KA, Wahlberg M (2020) Gentoo penguins (*Pygoscelis papua*) react to underwater sounds. R Soc Open Sci 7:191988
75. Stirling I, Kooyman GL (1971) The crabeater seal (*Lobodon carcinophagus*) in McMurdo Sound, Antarctica, and the origin of mummified seals. J Mammal 52:175–180
76. Takahashi A, Sato K, Naito Y, Dunn MJ, Trathan PN, Croxall JP (2004) Penguin-mounted cameras glimpse underwater group behaviour. Proc R Soc Lond B 271:S281–S282
77. Thiebault A, Charrier I, Aubin T, Greenm DB, Pistorius PA (2019) First evidence of underwater vocalisations in hunting penguins. PeerJ 7:e8240
78. Tremblay Y, Cherel Y (1999) Synchronous underwater foraging behaviour in penguins. Condor 101:179–185
79. Wilson EA (1907) Aves. National Antarctic expedition 1901–04, vol 2: zoology, Part 2. London
80. Wilson RP, Duffy DC, Wilson MP, Araya B (1995) Aspects of the ecology of species replacement in Humboldt and Magellanic penguins in Chile. Le Gerfaut 85:49–61
81. Wilson RP, Grant WS, Duffy DC (1986) Recording devices on free-ranging marine animals: does measurement affect foraging performance? Ecology 67:1091–1093
82. Wilson RP, Hustler K, Ryan PG, Burger AE, Noldeke EC (1992) Diving birds in cold water: do Archimedes and Boyle determine energetic costs? Am Nat 140(2):179–200

83. Wilson RP, Wilson M-PT, Duffy DC (1988) Contemporary and historical patterns of African penguin *Spheniscus demersus*: distribution at sea. Estuar Coast Shelf Sci 26(4):447–458
84. Wilson RP, Wilson M-PT, Mcquaid L (1986) Group size in foraging African penguins (*Spheniscus demersus*). Ethology 72(4):338–341
85. Wiltschko R, Wiltschko W (2019) Magnetoreception in birds. J R Soc Interface 6:20190295
86. Woehler EJ (2004) Hearing abilities in Antarctic penguins. Polarforschung 72:95–98
87. Wolfaardt AC, Nel DC (2003) Breeding productivity and annual cycle of rehabilitated African penguins following chronic oiling. In: Nel DC, Whittington PA (eds) The rehabilitation of oiled African penguins: a conservation success story. BirdLife South Africa and Avian Demography Unit, University of Cape Town, Cape Town, pp 18–24
88. Zeyl JN, Snelling EP, Connan M, Basille M, Clay TA, Joo R, Patrick SC, Phillips RA, Pistorius PA, Ryan PG, Snyman A, Clusella-Trullas S (2022) Aquatic birds have middle ears adapted to amphibious lifestyles. Sci Rep 12:5251. https://doi.org/10.1038/s41598-41022-09090-41593

Part VI
Our Final Thoughts

Chapter 15
Penguins: Why the Hype?

Sources of Hype

Us

Is there anybody out there who has had the luck, or honor, to see wild penguins but who does not remember the first moments? David saw his first wild penguin in Antarctica upon arriving by helicopter at Cape Crozier, Ross Island, in December 1968. The bird was introduced by pioneering penguinologist, Richard Penney. Rory came across his first wild penguin in a storm while in a fishing vessel, modified (rather badly) for doing research just off Ichaboe Island, Namibia. The wind was doing the thing it does sometimes, of whipping the surface layers off the sea, especially at the wave crests, pulling a reluctant heaving ocean into the air to create a lacework of flying white. Seemingly everything was moving – a picture of dynamism and drama. And in the midst of the surging, frothing chaos, there was a juvenile African penguin, gently idling at the surface but rolled on one side, one flipper aloft, carefully and meticulously preening its lower breast feathers, a nucleus of profound calm in the maelstrom. It epitomized the bond between penguins and the ocean – a marriage so profound that we, as land creatures, could not hope to grasp how such calm and understanding could prevail in the storm. It was like seeing someone knitting in a supernova! We've also been on ocean research vessels during which the waves were so large that the penguins could be well above our head, i.e., we looked up at them! The penguin looked down seemingly bemused of us! For both of us, this sort of experience has driven the desire to contemplate the depth of the union between fish-bird and sea, and it has catalyzed much of our research over the years.

D. G. Ainley, R. P. Wilson, *The Aquatic World of Penguins*, Fascinating Life Sciences, https://doi.org/10.1007/978-3-031-33990-5_15

Many People

Trends by inspecting the web (as of this writing) tell us that we are not alone in our profound fascination and respect for penguins, and in our desire for information about them. Punching in "penguin bird" indicates that from 2004 to the present, the number of people looking at penguins on the web is slowly increasing, with the world regions showing most interest, understandably, being Australia, New Zealand, and South Africa but, less obviously, there are also plenty of "clicks" coming from Canada, India, Malaysia, Pakistan, the United Kingdom, and the United States. On our own website, penguinscience.com, developed as an outreach portion of our research project, we get 1 to 2 million hits per month during the midst of the Austral summer, which is also the midst of the school year in the United States. The web also tells us that penguins are about as popular as elephants or lions or tigers, but 10 times more popular than sloths (and sloths are cool, though in an understated way), only 10% as popular as sharks, and less than 3% as popular as whales! Maybe it's a biomass thing with whales?

So what is it that drives people to penguins, birds that, for the most part, live far from most of us? The most obvious reason is that penguins, like other favored animals, somehow incite considerable emotion, and we like that – it's the reason we watch rom-coms, horror films and read books.

So what emotions do penguins incite and why? Again, web search trends have something to say. Looking at interest in penguins over time, it is clear that there is a substantial ground swell with small pulses (sometimes with figures that denote a doubling in public attention, lasting about a year, with the highest points in the end-of-year season of religious holidays!) and occasional irregular large peaks, typically lasting 2–3 months. The timing of these larger peaks generally seems to be associated with ocean catastrophes, such as oil slicks or bits of Antarctica drifting off and possibly upsetting penguin lives in some way. So penguins can act as a vehicle for environmental awareness, which is a great thing because we badly need to pay attention to our beleaguered oceans. The religious holiday thing does, after all, present an opportunity for people to spread penguin paraphernalia, but this is just a specific point in general interest. What of the ground swell? Maybe the penguin aficionados can tell us?

Here are quotes from two penguin enthusiasts:

> "Penguins are habit forming. How I became addicted is related [later]. I hope [my writing] will lead others into this pleasant and generally harmless addiction.....Most penguin addicts have to obtain their fixes at zoos and in books." (G.G. Simpson [10], p. ix)

> "I've been extraordinarily lucky – one of the very, very few people who's logged the equivalent of three to four entire years working in this dream location [Antarctica], firsthand, with marquee animals. Field researchers only reluctantly reveal the intense attachment they have for the subjects of their investigations. It feels almost petty, perhaps even childish, but I'm not ashamed to rave on about penguins. I can't help it. They've grabbed me and simply won't let go." (R. Naveen [9], pp. 7–8)

Actually, these people are remarkably nonspecific. They don't tell us *why* they became attached or addicted. Perhaps the film *Happy Feet* gives a window to that attachment though. We note that the film is called *Happy Feet* and not *Happy Flippers* – it's about a penguin who "tap dances" – but why shouldn't it dance at sea, where it spends most of its time? The film did end with a penguin crying out, "Stop eating our fish!" It apparently grossed $384 million against a production budget of $100 million and undoubtedly started a penguin trend. Then there's the penguin team from Madagascar (dare we say Skipper, Kowalski, Rico, and Private), Mr. Popper's Penguins and innumerable nature movies, like *March of the Penguins*, all of which concentrate on nesting season events. That is all, bar the rather lovely, beautifully honest but sadly somewhat obscure film, entitled *Meet the Real Penguin* involving Lloyd Spencer Davis (a penguin researcher). Then there is the film *The Last Ocean* in which the thought is expressed that the next generation of penguins may not find what their parents experienced owing to what people are doing to the ocean. Clearly, people seem to love penguins for the way they do things on land, which is fine – why not indeed? That's where most people dwell, and where most people can most easily see them.

If you punch "why do people like/love penguins?" into the web, it comes back with things like: "they have the cutest walk," "they are loyal and hopeless romantics," "they aren't the least bit scared of humans," and "like us, they love snow." So people pick out media-spawned attributes, many of which are not true, or only partially true, and use them to paint their own personal image of what penguins are – and they do sound lovely. The fact that wild penguins are keen proponents of self-defense if you mess with them, and will slap you hard and fast with rigid flippers while also savaging you with their bills, seems to have eluded the press. We won't go into details of penguin divorce rates either. No point bursting the media bubbles.

There is certainly a case for airing the truth about penguins, though – because the media shouldn't thrive on untruths and the public have a right to have the true facts (crazy dream, we know). In that vein, maybe penguin researchers have a more balanced view on why they suffer from the penguin hype.

Researchers

One of the first to be explicit was Bernard Stonehouse, a pilot who crashed in Antarctica and was forced to contemplate the crowds of penguins contemplating him (penguins are insatiably curious) and became spellbound by his onlookers, taking to science and publishing many of the first accounts of these birds. He was also the first to really try to bring disparate, ragged penguin science together. In fact, in the 1975 compilation that he edited, he wrote

> "We know very little about penguins in the medium for which they are best adapted. They are too small and too difficult to watch closely from the deck of a ship, and nobody has yet had the determination to spend much time in the water with them. This is a pity, for the

behaviour of birds so well adapted to water could not fail to be interesting. The average zoological garden pool cannot do justice to their swimming ability, or tell us much about how they dive and catch their prey. They take all their food at sea, and most adult penguins die by violence at sea….." (B. Stonehouse [11], p. 74)

So Bernard really touched on the truth about penguins being birds of water rather than land. This is the single most incredible thing about penguins for us.

We are not the only people who find penguins simply incredible. As George Gaylord Simpson says, watching penguins is addictive, which also explains why they are among the most favorite attraction at zoos and aquaria. Watching them can be life-affirming, in part because they appear to have no self-doubts and can inspire spine tingling by demonstrating continually that they are "workmen" (or women) of no parallel who know fully what they are doing both on land and at sea. Their various purposes are clearly obvious, and as we have attempted to lay out, they adjust their aquatic behavior to fit the scene following their quick and spot-on assessment of what needs to be done and how. While some people may laugh at penguins– the way they do some things like people, such as tripping over a rock– they are no laughing matter, at all, particularly at sea. Yes, they are totally serious as they cope with their very dangerous world, inherent within the ecological niche they have succeeded in occupying. What they do in the ocean is essentially beyond words, though we have tried our best to tell their extraordinary fish-bird story in the previous 14 chapters.

The Transition

There are many incredible creatures that inhabit Earth, doing amazing things. Near the top of the list are penguins, which do what no other bird can do– make like a fish. It's remarkable that it is a bird that does this, rather than a reptile or mammal. Way back in time, reptiles did do this, but those creatures are long gone (except iguanas). It was actually fish that emerged from the sea that eventually became lizards, mammals, and birds. Certain mammals then reverted back to the sea becoming pinnipeds and cetaceans, and there are a few mammals that frequent the ocean margin (otters). However, it is penguins that occupy the ecological niche in which a highly mobile, acrobatic, small predator occurs in regions that range from near the coast to far out at sea, and can dwell there, well, almost indefinitely. Like all birds (and pinnipeds), they do need to change their coats each year, and penguins need to do this where, for a few weeks, they are high and dry.

But they are birds, damn it, though everything about them shows they are not traditionalists! They evolved from ancestors that could fly!! But when they ditched that option, evolution morphed their bodies into that characteristic, incredible (almost) torpedo shape with tiny, powerful flippers and a suite of other anatomical and physiological changes that allowed them to coalesce with the sea. The transition, when they lost (renounced?) the power of aerial flight, that most extraordinarily empowering faculty given to birds, must have been incredible. After all, flighted birds- and some 99.5% of the ca. 10,000 species can fly – may cover crazy distances across the globe in a ridiculously short time and can almost "warp" themselves from one place to another in a manner akin to time travel, certainly compared to even the fastest terrestrial mammals. This means that birds can rapidly migrate to where the weather is most "lovely" as the Earth turns parts of herself toward, and away from the sun and summer twice a year. And on a daily basis, some birds can contemplate provisioning their young with food from realms that might be 500 km away. Who in their right mind would do that for lunch? Penguins gave all of that up! Any reason for refusing that jackpot must have been compelling. Casting back our mind's eye to the time when the penguin ancestor got a better deal by renouncing flight, conjures up images of proto-penguins that could just about fly because their tiny wings, perhaps similar to those of auks today, only allowed them to flutter hopelessly over the sea surface, maybe at times bouncing unwanted into big waves as diving petrels sometimes seem to do. The reduced size of these wings, however, made proto-penguins quite good at swimming underwater with their small flippers. If only the wings were a little smaller, though, they would have been even better at underwater travel! But it was on this knife edge of evolutionary indecision, once all aspiration to grapple inadequately with air had gone, that penguins came into being, and the evolution of the true penguin as we know it must have been rapid. Once there was no compunction to try and use flippers for flight, there was no selective pressure for proto-penguins to try and sit uncomfortably between two vastly different worlds – the rarified airs and gusts of the atmosphere and the dense, viscous, cold of the oceans. These birds could finally commit to a proper union with the sea, and their gains by doing this outweighed the losses they incurred by dropping the power of aerial flight. The result of this is, of course, what we see in penguins today. We would argue that they are the most extremely divergent avian taxon from the "classic bird" body plan on the planet. Their extreme specialization – many of the things that we have talked about in this book – makes them truly unique. There is nothing like a penguin, with its torpedo shape, weird "futuristic" flippers, and exquisite feathers, and that's why we, and seemingly much of the rest of the human planet, love them.

The Fascination of Species

We wonder, though, how much the human planet distinguishes between penguin species. We saw within the earlier pages of this book, by our comparative approach, that there are major similarities among penguin species, as well as major differences, in regard to their marine ecology and behavior or otherwise how they use the ocean to their clever advantage, their true "penguin-ness." A number of penguin attributes became apparent to us only as a result of our approach, and we have recognized how these attributes vary between species. Humans mostly "cope" generically with the ocean, which is a concept in the background of most penguin summaries written to date (as touched on at end of Chap. 1). That is, the opinion is derived by viewing penguins from a human perspective. However, as we have detailed, penguins do more than make use of the ocean (that's the extent of what humans do, ultimately to their downfall), but penguins thrive in it, each species in its own way. Were there no ocean, there would be no penguins.

Recently, making use of the recent revolution in genetics, some very interesting presentations of penguin lineages have been made, detailed somewhat in Chap. 1 (see Fig. 1.4) [2, 3, 12]. This research, as well as that following the more classical mode of investigation, i.e., paleontology [7], has pretty much stuck to explaining penguin species evolution and radiation in terms of physical factors, i.e., changing climate, the opening of water ways and oceans, and provision of nesting habitat. They have given very little attention to interspecific interaction and how penguins relate to the ocean, i.e., the ecological niches of the different penguins, except perhaps for the idea that the earlier, large penguin species had to make way for the appearance of dolphins and pinnipeds, somewhat the same size as them. The entire niche of a species, however, includes not just the physical but also its social/behavioral aspects [5, 8].

OK, we admit that the above thoughts require a bit of elaboration, in particular our comparative approach! Bigger animals, like elephants, do not have the available power to do onerous things that smaller animals can [13, 14]. For the large *Aptenodytes* penguins, their most strenuous tasks are those required on land, where they are a bit slow and ponderous, and where even a gentle slope can be taxing. But these are the most specialized penguins for deep diving, for which large size is an advantage (Chap. 8). The truth is that the more adapted a species is for swimming/diving, the less proficient it is at doing stuff in a completely different environment – land. Thus, indeed, with the presence of large penguins, as well as large(r) marine mammals efficiently exploiting the world's oceans at depth, there was selection pressure for new penguin arrivals to be small, able to exploit the empty niches. This involved using the difficult to access land areas (especially those penguins of the genus *Eudyptes*), but which, because they had to be small to do this, had repercussions for their diving capacities. So the bitter irony was that, in their evolution, penguins had given up flight for the oceans but then the smaller species had to make concessions to the sea to exploit the land for breeding. The pay-back, however, was that although the existing, larger penguin species had taken the more productive stretches and depths of the ocean, the smaller size, later-appearing penguins could at least exhibit quicker breeding. This was particularly the case if they reverted back to the one-chick brood strategy that requires less prey capture. So, perhaps in a re-enactment of what happened in times gone by, *Eudyptes* penguins always lay two eggs but then never rear more than one chick to adulthood; how weird is that!

But size has more consequences for penguins. Smaller penguins have higher mass-specific metabolic rates, making them less able to fast during long trips to distant foraging grounds than the larger species. This also means that the chicks of smaller species have to be fed more often (Chaps. 2 and 5). On top of that, penguins that have two-chicks broods must divide the prey they catch between their offspring, so resources – which can only fit in one stomach– are, in any event, effectively halved, and so used up twice as fast. Such species, therefore, have to forage closer to their breeding spots, shuttling more often between their ocean element and land.

Curiously, though, ostensibly to deal with the above problems, some penguin species show marked differentiation in roles among members of pairs, in a sense a further evolutionary radiation – with one member guarding the chick(s), and the other one foraging, as is especially evident among *Eudyptes* penguins. For example, in the macaroni penguin, the male guards the newly hatched chick(s) for about 25 days, while the female shuttles back and forth between the breeding site and the sea, bringing food back to provision the chick. During this time, the male gets progressively thinner and thinner – which cannot be good. Why on earth wouldn't they switch roles, like most sensible middle mass penguins, not least because if the females actually brooded the chick, they could serially feed it, especially when it was small, over time while the males capitalized on that time by foraging? Weird! The *Eudyptes* penguins have much to explain!

At the other end of the size spectrum, also feeding just one chick, as well as having temporally separate ocean vs. land (for emperors this is ice) lives among pairs, the largest penguins can hold an impressive amount of food in their stomach, especially compared to the size of their chicks. So, using delayed gastric emptying, they can come back to a small chick possessing lots of food that has come from afar but been "put on ice" for days. In that way, the chick can then be continuously provisioned, a little at a time, when it is most susceptible to starvation [4]. The extended time that it takes for brooding parents with small chicks to empty their extensive stomach reserves is capitalized on by the partners, which go to sea to fill up for the next take-over (*Eudyptes* penguins take note!).

Beyond the Transition

One of the things that theoretical ecologists repeatedly tell us is that highly specialized animals, such as penguins, are playing a dangerous game. Such specializations make such animals supreme at "doing what they do." They do it better than other would-be competitors, reason enough for evolution to keep pushing that design. But what happens if conditions change, if the Formula 1 car suddenly has to operate on a muddy track? Humans are doing their best to create that muddy track for penguins (and a whole suite of highly specialized animals), and we're doing it fast, which doesn't give evolution time to push for change that might be a solution. In the last hundred years, our track record of "sustaining" the oceans has been abysmal. We have overfished, polluted, filled with plastic, and generally messed up the world's seas to an extraordinary degree (Chap. 13). Although we have to "feed the world," let's be honest, the people on the cutting edge of this exploitation often do it to make money, and the most radical of these have not seriously considered what actually happens if we aspire to "wanting ever more" in a finite system (and rumor has it that the Earth is finite). On top of all that, we have managed to change the temperature of the whole planet! That's some achievement. In Victorian times, they wouldn't have considered that even remotely credible, and some people even aspire to that belief today, though maybe these are those that stand to gain the most – dare we say it – money. We used to do a party piece in a penguin colony to demonstrate what penguins think of man-conceived "values" to students. It consisted of placing a $100 note in a penguin nest. The penguins on such nests always looked at the note being placed in position but then, within a few seconds, looked elsewhere. Typically, the note would blow away in the wind without the penguins giving it a second thought – an animal statement about the value of money in natural systems! This dismissive attitude that penguins have with regard to cash, which would presumably be replicated by animals everywhere, sadly doesn't recognize the destructive power of high finance. If penguins only knew that it was as dangerous to them and their bloodlines as the most awful of predators. Although some penguin species and populations are doing fine, some are not. And, in an effort to help, many penguin researchers repeatedly visit penguin trouble spots, highlighting recommendations to human practice, such as reducing fishing pressures, to change that. Might we do that and turn some of our "hard-earned cash" into rebuilding? Certainly, it is reason enough for a global penguin hype. Penguins, with their enormous public appeal, are a bellwether of human-induced change (*The Adélie Penguin: Bellwether of Climate Change*, is one book [1]). They speak to us with changes in their populations. We need to listen.

Role in Ecosystems

We were once asked, "What use are penguins to us?" Us, meaning humans. This is an extraordinary question, as if the justification for everything on Earth, living and nonliving, from penguins to volcanoes, should relate to humankind. We are going to ignore that question here. But, as Chap. 1 notes, penguins are (sometimes) numerous and, as Chap. 5 notes, may eat substantial quantities of seafood. A follow-on question to that posed above is, "If penguins consume so much, wouldn't we benefit by having more seafood if we got rid of them?" In fact, this question has also been posed to us, more than once. The interesting thing about animals in ecosystems is that, left to their own devices, they do not ruin their environment (excepting us, of course). Populations, even those of species that eat large amounts, ebb and flow, but balanced ecosystems remain. Before exploitation of the Peru/Humboldt and Benguela upwelling areas, this was wonderfully obvious, with, as documented by Alexander von Humboldt, Robert Cushman Murphy and the like, unthinkable numbers of marine birds and mammals exploiting a seeming never-ending supply of fish that just went on and on and….

Ecosystems are complex things, a wonderfully tangled, interconnected matrix of living organisms, each doing something that affects system function to create a living mosaic. So, for their part, penguins, feeding on their prey out at sea may take these nutrients to other places on the planet, for example bringing them back to their colonies to feed their chicks. This leads to local guano build-up and, inevitably, guano run-off, whereby the phosphorus and nitrogen and countless other chemicals that underpin life, slip into the sea to enhance local productivity, and make their way into the oceanic current systems where they affect living systems across the planet. A very large, entire book was written about that in regard to the Peru/Humboldt Current [6]. What happens, then, if you take out the middle person (penguin)? That is a procedure, in fact, that has been underway for the last few decades, as responding to advice of the World Bank, hundreds of years of mining the guano deposits of penguins and other seabirds, and conserving the bird populations as a resource gave way to mining the sea directly for the forage fish, forgetting about the birds. Bird populations and guano deposits dwindled. Stay tuned for the eventual full answer to the question. Humboldt and African penguin populations, once so dominant in the very rich Peru/Humboldt and Benguela currents (Chap. 3), are on the way out (no prizes for guessing who's done that; Chaps. 5 and 13). Our penguin hype should extend to understanding this, not just for the people that ask those questions, and not just for our understanding of penguins and of ecosystems, but because penguins and upwelling ecosystems are as much a fundamental part of life on Earth as we are.

Biomimicry

To humanity's credit, we have paid some attention to how we might learn to "do things better" by looking at features of penguin design. For example, we spoke of the absurdly low drag coefficient of penguins underwater, and how the pundits in a giant Berlin water flume could not believe the values that their system threw back at them when Rudi Bannasch tested his gentoo penguin model (Chap. 6). How could a bird possibly outperform the finest human brains when it comes to designing ocean-going ships? The answer is simple. Evolution is one of the hardest taskmasters. It is meticulous, pushing the tiniest incremental advantages in system efficiencies. It doesn't stop on Friday at five O'clock. It is relentless and perpetual, honing its products day in and day out, aspiring to the perfection of animals in their environments. We have much to learn, then, particularly from looking at the most specialized animals – a science called biomimicry. Encouragingly, penguins have been the inspiration behind simulated "feathered" water vessel hulls, but this is the tip of the iceberg. We need to keep looking to appreciate the fineries of penguin design, and maybe to build it into the way we construct and operate our boats so that they can move with ease and grace (and critically using less energy) through the world's seas. Penguins have been doing this for millennia. We need to learn from them – a better reason for penguin hype we could hardly imagine.

Our Last Word

Finally, for our part, we have been honored to be able to "open the book" on penguins at sea, both within our own researches, and in considering what has been done by the scientific community. It's a long way from our own "early days" when we both stood firmly on land at penguin colonies and wondered what went on when penguins disappeared into the waves – knowing that nobody knew. One of us took to research vessels, investigating penguins' habitat, where they went, and their place in it; the other of us took to playing with, and trying to perfect, electronic devices that quantified how the penguins exploited the sea. And it may seem as if we (collectively as a penguin community), using both methods, have made some progress in demystifying the enigmatic lives of penguins at sea.

We used to wonder whether the scientific process of slowly unpicking the penguin's water world would rob them of their mystery and fascination, and turn them into the mundane – another victim of man's supposed domination, intellectual, or otherwise, of the world. Not at all. We have just scratched the surface, lifted the lid on a huge box of treasure, and peeked inside. But even when, after many more years of exhaustive research, we have examined each and every single piece of this treasure in our hands, and weighed it, quantified the glitter, and catalogued it together within the collective knowledge on animals in general, nothing will take away the giddying rush that we feel when nature allows us to watch, perceive, or even imagine these exquisite beings of the far oceans as they skid through the thermocline eddies, plucking secrets from the deep.

We could go on and on here detailing the remarkable and varied fish-bird attributes and capabilities that they do not hide– unlike a lot of creatures, penguins do things in full view or at least lend themselves to us measuring what they do when out of our sight (at sea). These ocean spirits have truly captured our imagination and energy, obvious, yes, but if you're still not sure why, just read (or re-read) the previous chapters of this book? There we justify why we are not bashful to confess to be penguin aficionados.

References

1. Ainley DG (2002) The Adélie penguin: bellwether of climate change. Columbia University Press, New York
2. Cole TL, Dutoit L, Dussex N, Hart T, Alexander A, Younger JL, Clucas GV, Frugone MJ, Cherel Y, Cuthbert R, Ellenberg U (2019) Receding ice drove parallel expansions in Southern Ocean penguins. Proc Natl Acad Sci 116(52):26690–26696
3. Cole TL, Zhou C, Fang M, Pan H, Ksepka DT, Fiddaman SR, Emerling CA, Thomas DB, Bi X, Fang Q, Ellegaard MR, Feng S, Smith AL, Heath TA, Tennyson AJD, Borboroglu PG, Wood JR, Hadden P, Grosser S, Bost C-A, Cherel Y, Mattern T, Hart T, Sinding M-HS, Shepherd LD, Phillips RA, Quillfeldt P, Masello JF, Bouzat JL, Ryan PG, Thompson D, Ellenberg U, Dann P, Miller G, Boersma PD, Zhao R, Thomas M, Gilbert P, Yang H, Zhang D-X, Zhang G (2022) Genomic insights into the secondary aquatic transition of penguins. Nat Commun 13(1):1–13
4. Gauthier-Clerc M, Le Maho Y, Clerquin Y, Drault S, Handrich Y (2000) Penguin fathers store food for their chicks. Nature 408:928–929
5. Grant PR, Grant BR (2008) How and why species multiply: the radiation of Darwin's finches. Princeton University Press, Princeton
6. Hutchinson GE (1950) The biogeochemistry of vertebrate excretion. American Museum of Natural History, New York

7. Ksepka D, Ando T (2011) Penguins past, present, and future: trends in the evolution of the Sphenisciformes. In: Dyke G, Kaiser G (eds) Living dinosaurs: the evolutionary history of modern birds. Wiley, New York, pp 155–186
8. Lack DL (1947) Darwin's finches. Cambridge University Press, Cambridge
9. Naveen R (1999) Waiting to Fly: my escapades with the penguins of Antarctica. William Morrow and Company, New York
10. Simpson GG (1976) Penguins: past and present, here and there. Yale University Press, New Haven
11. Stonehouse B (1968) Penguins. Golden Press, New York
12. Vianna JA, Fernandes FA, Frugone MJ, Figueiró HV, Pertierra LR, Noll D, Bi K, Wang-Claypool CY, Lowther A, Parker P, Le Bohec C (2020) Genome-wide analysis reveals drivers of penguin diversification. Proc Natl Acad Sci 117(36):22303–22310
13. Wall J, Douglas-Hamilton I, Vollrath F (2006) Elephants avoid costly mountaineering. Curr Biol 16:R527–R529
14. Wheatley R, Buettel JC, Brook BW, Johnson CN, Wilson RP (2021) Accidents alter animal fitness landscapes. Ecol Lett 24:920–934

Appendices

Appendix Table A1

The common and Latin names of vertebrate and invertebrate species mentioned in the text

Common name	Latin name
PENGUINS	
Emperor Penguin	*Aptenodytes forsteri*
King Penguin	*Aptenodytes patagonica*
Adélie Penguin	*Pygoscelis adeliae*
Chinstrap Penguin	*Pygoscelis antarctica*
Gentoo Penguin	*Pygoscelis papua*
Yellow-eyed Penguin	*Megadyptes antipodes*
NZ Little Penguin	*Eudyptula novaehollandiae*
Australian Little Penguin	*Eudyptula minor*
African Penguin	*Spheniscus demersus*
Magellanic Penguin	*Spheniscus magellanicus*
Humboldt Penguin	*Speniscus humboldti*
Galápagos Penguin	*Spheniscus mendiculus*
Royal Penguin	*Eudyptes schlegeli*
Macaroni Penguin	*Eudyptes chrysolophus*
Southern Rockhopper Penguin	*Eudyptes chrysocome*
Northern Rockhopper Penguin	*Eudyptes moseleyi*
Eastern Rockhopper Penguin	*Eudyptes filholi*
Erect-crested Penguin	*Eudyptes sclateri*
Fiordland Penguin	*Eudyptes pachyrhynchus*
Snares Penguin	*Eudyptes robustus*

D. G. Ainley, R. P. Wilson, *The Aquatic World of Penguins*, Fascinating Life Sciences, https://doi.org/10.1007/978-3-031-33990-5

Common name	Latin name
OTHER SEABIRDS	
Albatross species	Diomedeidae
Shearwater species	*Ardenna* spp.
Petrel species	*Pterodroma* spp.
Northern Gannet	*Morus bassana*
Cape Gannet	*Morus capensis*
Imperial Shag	*Phalacrocorax atriceps*
Gull species	*Larus* spp.
Black-legged Kittiwake	*Rissa tridactyla*
Sooty Tern	*Onychoprion fuscatus*
Great Auk	*Pinguinus impennis*
Common Murre	*Uria aalge*
Thick-billed Murre	*Uria lomvia*
Black Guillemot	*Cepphus grylle*
Pigeon Guillemot	*Cepphus columba*
Puffin	*Fratercula* spp.
Razorbill	*Alca torda*
Little Auk	*Alle alle*
Common Eider	*Somataria mollisima*
LANDBIRDS	
Galapagos Finches	*Geospiza* spp.
Hawaiian Honeycreeper species	Drepanidae
Hoatzin	*Opisthocomus hoazin*
Flightless Steamer Duck	*Tachyeres* spp.
Andean Condor	*Vultur gryphus*
Tawny Owl	*Strix aluco*
Toucan	*Ramphastos* spp.
Swifts	Apodinae
Domestic Pigeon	*Columba livia*
Rose-colored Starling	*Pastor roseus*
PINNIPEDS	
Antarctic Fur Seal	*Arctocehalus gazella*
South American Fur Seal	*Arctocehalus australis*
South African or Cape Fur Seal	*Arctocephalus pusillus pusillus*
Australian Fur Seal	*Arctocephalus pusillus doriferus*
New Zealand or Southern Fur Seal	*Arctocephalus forsteri*
Subantarctic Fur Seal	*Arctocephalus tropicalis*
New Zealand or Hooker's Sea Lion	*Phocarctos hookeri*
South American Sea Lion	*Otaria flavescens*
Leopard Seal	*Hydruga leptonyx*
Crabeater Seal	*Lobodon carcinophaga*
Weddell Seal	*Leptonychotes weddellii*
Ross Seal	*Ommnaphoca rossii*
Polar Bear	*Ursus maritimus*
Sea Otter	*Enhydra lutris*
Cheetah	*Acinonyx jubatus*

Common name	Latin name
CETACEANS	
Killer Whale	*Orcinus orca*
Humback Whale	*Megaptera novaeangliae*
Antarctic Minke Whale	*Balaenoptera bonaerensis*
SHARKS	
White Shark	*Charcharodon carcharias*
BONY FISH	
BOVICHTIDAE	
Thornfish	*Bovichtus psychorolutes*
CLUPEOIDS	
Anchoveta	*Cetengraulis mysticetus*
Australian Anchovy	*Engraulis australis*
South African Anchovy	*Engraulis capensis*
European Anchovy	*Engraulis encrasicolus*
Pilchard/Sardine	*Sardinops sagax*
Australian Pilchard or Sardine	*Sardinops neopilchardus*
European Sardine	*Sardinia pilchardus*
Common Sardine	*Strangomera bentinki*
Mullet species	*Mugil* spp.
Pacific Thread Herring	*Opisthonema libertate*
Falkland Sprat	*Sprattus fuegensis*
NZ Blueback Sprat	*Sprattus antipodum*
Sandy Sprat	*Hyperlophus vittata*
Blue Sprat	*Spratelloides robustus*
Peruvian Silverside	*Odontehthese regia*
Silverside	*Argentina elongata*
Australian Silverside	*Atherinason dannevigi*
Silverside species	Atherinid
Chilean Silverside	*Odontesthes regia*
Smelt	*Retropinna retropinna*
Smelt species	*Austroatherina* spp.
Red-Eye Round Herring	*Etrumeus teres*
Whitehead's Round Herring	*Etrumeus whiteheadi*
Araucanian Herring	*Strangomera bentincki*
Pacific Saury	*Scombersox saurus*
California Grunion	*Leuresthes tenuis*
MYCTOPHIDAE	
Rhombic Lanternfish	*Kreffichthys anderssoni*
Electron Subantarctic Lanternfish	*Electrona carlsbergi*
Antarctic Lanternfish	*Electrona antarctica*
Tenison's Lanternfish	*Protomyctophum tenisoni*
Norman's Lanternfish	*Protomyctophum normnni*
Bolin's Lanternfish	*Protomyctophum bolini*
Gaptooth Lanternfish	*Protomyctophum choriodon*

Common name	Latin name
Hector's Lanternfish	*Lampanyctodes hectoris*
Flaccid Lanternfish	*Metelectrona ventralis*
Fraser's Lanternfish	*Gymnoscopelus fraseri*
Southern Blacktip Lanternfish	*Gymnoscopelus piabilis*
Nichol's Lanternfish	*Gymnoscopelus nicholsi*
Brauer's Lanterfish	*Gymnoscoelus braueri*
Southern Barracudina	*Magnisudis prionosa*
Silvery Lightfish	*Maurolicus muelleri*
Slender Lightfish	*Vinciguerria attenuata*
NOTOTHENIIDAE	
Marbeled Rockcod	*Notothenia rossi*
Triangular Rockcod	*Notothenia acuta*
Magellanic Rockcod	*Notothenia magellanica*
Narrowheaded Rockcod	*Notothenia angustifrons*
Yellowbellied Rockcod	*Notothenia neglecta*
Larsen's Rockcod	*Nototheniops larseni*
Grey Rockcod	*Notothenia squamifrons*
Cod Icefish	*Nototheniops nybelini*
Yellowfin Rockcod	*Nototheniops nudifrons*
Bald Rockcod	*Pagothenia borchgrevinki*
Striped Rockcod	*Pagothenia hansoni*
Emerald Rockcod	*Pagothenia bernachii*
Dusky Rockcod	*Trematomus newnesi*
Slender Scalyhead	*Trematomus lepidorhinus*
Blunt Scalyhead	Trematomus eulepidotus
Humped Rockcod	*Gobionotothen gibberifrons*
Painted Notie	*Lepidonotothen larseni*
Antarctic Bullhead	*Lepidonotothen nudifons*
Smalleye Moray Cod	*Muraenolepis microps*
Eel Cod species	*Muraenolepis sp.*
Patagonian Notothen	*Patagonotothen sp.*
Antarctic Silverfish	*Pleuragramma antarctica*
Patagonia Toothfish	*Dissostichus eleginoides*
Longtail Southern Cod	*Patagontothen ramsayi*
CHANNICHTYIDAE	
South Georgia Icefish	*Pseudochaenichthys georgianus*
Mackerel Icefish	*Champsocephalus gunnari*
Unicorn Icefish	*Channichthys rhinoceratus*
Spiny Icefish	*Chaenodraco wilsoni*
Blackfin Icefish	*Chaenocephalus aceratus*
Crocodile Fish	*Chionodraco hamatus*
Icefish Sp.	*Pageotopsis sp.*
Jonah's Icefish	*Neopagetopsis ionah*
Antarctic Jonasfish	*Notolepis coatsi*

Common name	Latin name
Slender Escolar	*Paradiplospinus gracilis*
South Georgia Spiny Plunderfish	*Harpagifer georgianus*
BATHYDRACONIDAE	
Dragonfish Sp.	*Gerlachea australis*
Dragonfish Sp.	*Parachaenychthis georgianus*
Klipfish	*Bovichthys gouldi*
HARPAGIFERIDAE	
Antarctic Spiny Plunderfish	*Harpagifer georgianus*
MISCELLANEOUS	
Snake Mackerel	*Thyrsites atun*
Horse Mackerel	*Trachurus trachurus*
Peruvian Horse Mackerel	*Trachurus murphyi*
South African Horse Mackerel	*Trachurus capensis*
Mackerel	*Scomber japonicus*
Sharpfin Barracuda	*Sphyraena acutipinnis*
Peruvian Hake	*Merluccius gayi*
Argentine Hake	*Merluccius hubbsi*
South African Hake	*Merluccius capensis*
Southern Blue Whiting	*Micromesistius australis*
Southern Garfish	*Hyporhamphus melanochir*
Beaked Salmon	*Gonorhynchus gonorhynchus*
Trumpeter Fish	*Mendosoma lineaturn*
Long-Snouted Pipefish	*Leptonotus norae*
Redbait	*Emmelichthys nitidus*
Red Cod	*Pseudophycis bacus*
Blue Cod	*Parapercis colias*
Unicorn Cod	*Bregmaceros macclellandi*
Rock Cod	*Lotella sp.*
Hoki	*Macruronus novaezelandiae*
Ahuru	*Aucheroceros punctatus*
Rattail	*Coelorhincus* sp.
Ling	*Genypturus blacodes*
Sea Perch	*Helicolenus percoides*
Tarakihi	*Nemadactylus macropterus*
Stargazer	*Kathetostoma giganteum*
Opalfish	*Hemerocoetes artus*
Triplefin Sp.	*Tripterygiidae*
Gudgeon	*Grahamichthys radiata*
Barracouta	*Thrysites atun*
Blue Warehou	*Seriotella brama*
Silver Warehou	*Seriotella puntata*
Sole	*Peltorhamphus novaezeelandiae*
Leatherjacket	*Parika scaber*

Common name	Latin name
Filefish Sp.	*Meuschenia sp.*
Barred Fingerfish	*Chirodactylus pixi*
Red Coronetfish	*Fistularia petimba*
Greater Pipefish	*Syngnathus acus*
Bluefish	*Pomatomus saltatrix*
Combtooth Blenny Sp.	*Parablennius sp.*
Channel Bull Blenny	*Cottoperca gobio*
Cape Rockling	*Gaidropsarus capensis*
Whitelegi Sp.	*Psenes whitelegii*
Moontail Bullseye	*Priacantus hamrur*
Honeycomb Filefish	*Cantherines pardalis*
Flying Gurnard	*Catyloptena orientalis*
Pelagic Goby	*Sufflogobius bibarbatus*
Antarctic Deepsea Smelt	*Bathylagus antarcticus*
Pouched Lamprey	*Geotria australis*
Antarctic Horsefish	*Zanchlorhynchus spinifer*
Blue Grenadier	*Macruronus novaezelandiae*
Pigfish	*Congiopodus* spp.
CEPHALOPDS	
Longfin Squid	*Doryteuthis gahi*
Arrow Squid	*Nototodarus sloani*
Gould's Arrow Squid	*Nototodarus gouldi*
Giant Warty Squid	*Kondakovia longimana*
Violet Squid	*Histioteuthis atlantica*
Glacial Squid	*Pyschroteuthis glacialis*
Glass Squid	*Galiteuthis glacialis*
Antarctic Neosquid	*Alluroteuthis antarcticus*
Greater Hooked Squid	*Onykia ingens*
Antarctic Arm Squid	*Slosarczykovia circumantarctica*
Luminous Bay Squid	*Loliolus noctiluca*
Sevenstar Flying Squid	*Martialia hyadesi*
Cape Hope Squid	*Loligo reynaudi*
Argentine Shortfin Squid	*Illex argentinus*
Antarctic Flying Squid	*Todarodes fillippovae*
Armhook Squid	*Gonatus antarcticus*
Pelagic Octopus	*Ocythoe tuberculata*
Common Octopus	*Octopus vulgaris*
CRUSTACEA	
Note: Many Prey Are Amphipods That Don't Have Common Names, Same For Most Krill Species	
Squat Lobster	*Munida gregaria*
Crystal Krill	*Euphausia crystallorophias*
Antarctic Krill	*Euphausia superba*
No common name	*Thysanoessa gregaria*
Caridean shrimp	*Nauticaris marionis*

Appendix Table A2

Summary of literature on the diet of penguins (Tables A2.1, A2.2, A2.3, A2.4, A2.5, A2.6, A2.7, and A2.8)

Table A2.1 The diet of emperor and king penguins, mainly during spring chick feeding unless noted; where information available, the main (though not necessarily exclusive) prey species are listed; percent by mass or volume, by species if data available

Location	Habitat	Main prey species	References
EMPEROR PENGUIN			
Weddell Sea			
Scotia-Weddell Confluence	Pelagic	Fish, 5% Myctophidae	Autumn, Winter [1]
		Crustacean, <1%	
		Cephalopod, 94% Glacial Squid, 48% Giant Warty Squid, 32% Glass Squid, 14%	
Dresher Inlet (72°52′S, 19°25′W)	Shelf/slope	Fish, 48% Antarctic Silverfish Bald Rockcod	Autumn [2, 3]
		Crustacean, 52% Antarctic Krill	
EAST ANTARCTICA			
Auster Glacier (67°12′S, 50°45′E	Shelf/slope	Fish, 55% Slender Scalyhead, 31% Antarctic Silverfish, 5%	[4]
		Cephalopod, 45% Glacial Squid, 13% Antarctic Neosquid, 20%	
		Fish, 27% Antarctic Silverfish, 13% Bald Rockcod, 3%	Winter [5]
		Crustacean, 70% Antarctic Krill	
		Cephalopod, 2%	
		Glacial Squid	
Taylor Glacier (67°27′S, 60°50′E)	Shelf/slope	Fish, 31% Spiny Icefish, 5%	[4]
		Cephalopod, 69% Glacial Squid, 22% Antarctic Neosquid, 44%	
Amanda Bay (69°17′S, 76°46′E)	Shelf/slope	Fish, 100% Antarctic Silverfish, 78% Bald Rockcod, 9%	[6]
		Cephalopod, 3% Glacial Squid	

(continued)

Table A2.1 (continued)

Location	Habitat	Main prey species	References
Prydz Bay (68°30′S, 78°00′E)	Shelf/slope	Fish, 100% Antarctic Silverfish Bald Rockcod	[7]
Point Géologie (66°39′S, 139°55′E)	Shelf/slope	Fish, 95% Nototheniidae Bald Rockcod	[8, 9]
		Crustacean, 2% Antarctic Krill	
		Cephalopod, 3% Glacial Squid	
		Fish, 95%	[3]
ROSS SEA			
Cape Roget (71°59′S, 170°37′E)	Shelf	Fish, 93% Antarctic Silverfish	[10]
		Crustacean, <1%	
		Cephalopod, 7% Glacial Squid	
Coulman Island (73°28′S, 169°45′E)	Shelf	Fish, 88% Antarctic Silverfish	
		Crustacean, <1%	
		Cephalopod, 12% Glacial Squid	
Cape Washington (74°39′S,165°22′E)	Shelf	Fish, 89–95% Antarctic Silverfish	
		Crustacean, 5–11% Amphipoda	

Location	Habitat	Main prey species	Source
KING PENGUIN			
Marion Island (46°52′S, 37°51′E)	Pelagic	Fish, 87% Rhombic Lanternfish Tenison's Lanternfish	[11]
		Crustacean, <1%	
		Cephalopod, 12% Giant Warty Squid	
Crozet Islands (46°25′S, 51°59′E)	Pelagic	Fish, 92% Rhombic Lanternfish, 29% Tenison's Lanternfish, 24% Electron Subantarctic Lanternfish, 36%	Winter [12]
		Cephalopod, 7%	
Possession Island	Pelagic	Fish Autumn 99%, winter 43%, spring 97% Rhombic Lanternfish Tenison's Lanternfish Electron Subantarctic Lanternfish	[13, 14]

(continued)

Table A2.1 (continued)

Location	Habitat	Main prey species	Source
		Cephalopod Autumn 1%, winter 57%, spring 3%	
		Fish, 99% Rhombic Lanternfish, 16–37% Tenison's Lanternfish, 14–22% Electron Subantarctic Lanternfish, 15–70%	[15]
Kerguelen Islands (48°45–50′S, 68°45–70′E	Pelagic	Fish, 100% Opalfish, 30% Tenison's Lanternfish, 32% Bolin's Lanternfish, 14%	[16]
		Fish, Pelagic, 97%	[17]
		Benthic, <1%	
		Cephalopod, 3%	
Heard Island (53°05′S; 73°30′E)	Pelagic	Fish, 99% Electron Subantarctic Lanternfish Rhombic Lanternfish	[18]
		Crustacean, <1%	
		Cephalopod, >1%	
		Fish, 72% Rhombic Lanternfish, 48% Tenison's Lanternfish, 2%	[19]
		Crustacean, <1%	
		Cephalopod, 28% Sevenstar Flying Squid, 25% Giant Warty Squid, 2%	
Macquarie Island (54°30′S, 158°57′E)	Pelagic	Fish, 97% Rhombic Lanternfish, 38% Electron Subantarctic Lanternfish, 53%	[20]
		Cephalopod, 2% Hooked Squid spp.	
Falkland Islands (51°29′S, 57°50′W)	Pelagic	Fish Autumn 99%, Winter 43%, Spring 97% Tenison's Lanternfish Rhombic Lanternfish, 38% Electron Subantarctic Lanternfish, 53%	[21]

(continued)

Table A2.1 (continued)

Location	Habitat	Main prey species	Source
		Cephalopod Autumn 1%, Winter 57%, Spring 3% Greater Hooked Squid	
South Georgia (54°09′S, 36°48′W)	Pelagic	Fish, 97% Rhombic Lanternfish Electron Subantarctic Lanternfish Antarctic Jonasfish	[22]
		Cephalopod, 2%	

Table A2.2 The diet of yellow-eyed penguins during chick-feeding season unless noted; where information available, the main (though not necessarily exclusive) prey species are listed; percent by mass or volume, by species if data available

Location	Habitat	Main prey species	References
YELLOW-EYED PENGUIN			
Otago Coast (45°06′S, 170°58′E)	Shelf	Fish Red Cod, 36% Opalfish spp., 22% NZ Blueback Sprat, 12% Ahuru, 7% Blue Cod, 2% Silverside, 6%	[23, 24]
		Cephalopod Arrow Squid, 9%	
		Fish Opalfish, 28% Blue Cod, 18% Silverside, 10% Red Cod, 7% Pouched Lamprey, 6% Ahuru, 5% Warehou spp., 4% NZ Blueback Sprat, 2%	[25]
		Cephalopod Arrow Squid, 14%	
Codfish Island (46°45′S, 167°40′E)	Shelf	Fish Blue Cod, 77% Opalfish spp., 20%	[23, 26]
		Fish Blue Cod, 70% Opalfish, 28% NZ Blueback Sprat, 2%	[27]
Stewart Island (46°53′S, 168°10′E)	Shelf	Fish Blue Cod, 99%	[27]

Table A2.3 The diet of Gentoo penguins during summer chick feeding unless otherwise noted; where composition information is generalized, if available the main (though not exclusive) prey species are listed; percent by mass or volume, by species if data available

Location	Habitat	Main prey species	References
GENTOO PENGUIN			
Marion Island (46°52′S, 37°51′E)	Shelf	Fish, 70% SG Spiny Plunderfish Rockcod spp.	Winter [28]
		Crustacean Caridean Shrimp, 30%	
		Fish, 53% Grey Rockcod Unicorn Icefish	[29, 30]
		Crustacean, 44–46% *Euphausia vallentini* Caridean Shrimp	
		Cephalopod, 1–2% *Octopus* sp.	
		Fish Juv. Rockcod spp. Painted Notie	[31]
		Crustacean *E. vallentini* Caridean Shrimp	
		Cephalopod *Octopus* sp.	
Crozet Islands (46 25′S, 51 59′E)	Shelf/slope	Fish, 44% Nichol's Lanternfish Rhombic Lanternfish Tenison's Lanternfish juv. Patagonian Toothfish Painted Notie	[32]
		Crustacean, 54% *E. vallentini* Caridean Shrimp	
		Cephalopod, 2%	
Kerguelen Archipelago (48°45–50′S, 68°45–70′E)	Shelf	Fish, 2%	Winter [32]
		Crustacean *E. vallentini*, 70% *Themisto gaudichaudii*, 26%	
	Coastal Bay	Fish, 39–95% Mackerel Icefish Antarctic Horsefish	[33]
		Crustacean, 1–28% *Th. gaudichaudii* *E. vallentini*	

(continued)

Table A2.3 (continued)

Location	Habitat	Main prey species	References
		Cephalopod, 2% *Octopus* sp.	
	Open sea	Fish, 13%	
		Crustacean, 84%	
Heard Island (53°05′S; 73°30′E)	Shelf/slope	Fish, 90% Rhombic Lanternfish, 57% Rockcod spp., 24% Mackerel Icefish,	[18]
		Crustacean, 8% *E. vallentini*	
		Cephalopod, 2%	
Macquarie Island (54°30′S, 158°57′E)	Slope	Fish, 80% Magellanic Rockcod, 18% Electron Subantarctic Lanternfish, 28% Rhombic Lanterfish, 15% Lanterfish sp., 16%	Winter [34]
		Crustacean, <1%	
		Cephalopod *Octopus* sp., 19%	
		Fish, 86% Antarctic Horsefish, 17% Electron Subantarctic Lanternfish, 24% Lanternfish spp., 14% Magellanic Rockcod, 9%	[35]
		Cephalopod, 15% Sevenstar Flying Squid, 9%	
	Shelf/slope	Fish, 88% Antarctic Horsefish, 17% Lanternfish spp., 14% Magellanic Rockcod, 28% Grey Rockcod, 23%	[36]
		Crustacean, <1%	
		Cephalopod, 11%	
Falkland/Malvinas Islands (51°26′S, 57°52′W)	Shelf	Fish, 50–55% Inshore species	[37, 38]
		Crustacean, 18–50% Squat Lobster	
		Cephalopod Longfin Squid Armhook Squid Greater Hooked Squid	

(continued)

Table A2.3 (continued)

Location	Habitat	Main prey species	References
		Fish, 86% Patagonian Notothen Channel Bull Blenny Falkland Sprat	[39]
		Crustacean, 1% *Th. gaudichaudii*	
		Cephalopod, 13% Longfin Squid	
Beauchene Island		Fish	[40]
		Patagonian Notothen, 9% Falkland Sprat, 84%	
		Crustacean Squat Lobster, 3%	
Hope Harbor		Fish Channel Bull Blenny, 2% Falkland Sprat, 27%	
		Crustacean Squat Lobster, 4%	
		Cephalopod Greater Hooked Squid, 58%	
New Island		Fish Patagonian Notothen, 1% Falkland Sprat, 99%	
		Fish, 42%	[38]
		Crustacean, 38%	
		Cephalopod, 20%	
Saunders Island		Fish Patagonian Notothen, 7%	[40]
		Crustacean Squat Lobster, 52%	
		Cephalopod Longfin Squid, 22% Greater Hooked Squid, 17%	
Weddell Island		Fish Patagonian Notothen, 37%	
		Crustacean Squat Lobster, 46%	
		Cephalopod Longfin Squid, 13% Greater Hooked Squid, 3%	
Whaler Bay		Crustacean Squat Lobster, 99%	
		Cephalopod Longfin Squid, 1%	
Kidney Cover		Fish Patagonian Notothen, 34%	Winter [41]

(continued)

Table A2.3 (continued)

Location	Habitat	Main prey species	References
		Crustacean Squat Lobster, 12%	
		Cephalopod Longfin Squid, 53%	
Volunteer Beach		Fish, 90%	[38]
		Crustacean, 2%	
		Cephalopod, 7%	
Seal Bay		Fish, 99%	
		Crustacean, 1%	
Cow Bay		Fish, 99%	
		Crustacean, 1%	
Bird Island, South Georgia (54°S, 38.3°W	Shelf	Fish, 33% Marbled Rockcod Larsen's Rockcod Mackerel Icefish	[42]
		Crustacean Antarctic Krill, 77%	
		Fish, 1–85%	Winter [43]
		Crustacean, 24–100% Antarctic Krill	
		Fish, 13%	Winter [44]
		Larsen's Rockcod	
		Crustacean, 87% Antarctic Krill *Thysanoessa macrura* *Th. gaudichaudii*	
		Fish, 5–19%	[45]
		Crustacean Antarctic Krill, 81–95%	
		Fish, 9–85% Mackerel Icefish Painted Notie	[46]
		Crustacean, 13–81% Antarctic Krill *Th. gaudichaudii*	
		Cephalopod, 0–1%	
		Fish, 25% SG Dragonfish Mackerel Icefish Painted Notie	Winter [47]
		Crustacean *Th. gaudichaudii* Antarctic Krill	

(continued)

Table A2.3 (continued)

Location	Habitat	Main prey species	References
South Orkney Islands Laurie Island (60°46′S, 44°42′W)	Shelf	Fish, 25–81% Humped Rockcod, 39–74% Antarctic Bullhead, 2–18%	Autumn [48]
		Crustacean Antarctic Krill, 17–70%	
		Cephalopod, 1–3%	
Antarctic Peninsula King George Island (62°10′S, 58°30′W)	Shelf	Fish, 24%	[49]
		Crustacean Antarctic Krill, 76%	
		Crustacean Antarctic Krill, 100%	[50]
		Fish, 53% Antarctic Bullhead Antarctic Silverfish	[51]
		Crustacean Krill spp., 47%	
Livingston Island (62°28′S, 60°47′W)	Shelf	Fish, 21% Rockcod spp. Icefish spp.	[52]
		Crustacean Antarctic Krill, 79%	
		Fish, 16–31% (31–47% SIA) Mackerel Icefish	[53, 54]
		Crustacean Antarctic Krill, 68–84%	
Anvers Island (64°W, 64°S)	Shelf	Fish, 0–4%	[55]
		Crustacean Antarctic Krill, 95–100% *Thy. macrura*, 0–5%	

Table A2.4 The diet of the four penguin species of the genus *Spheniscus* during chick-feeding season unless noted; where information available, the main (though not necessarily exclusive) prey species are listed; percent by mass or volume, by species if data available

Location	Habitat	Main prey species	References
GALÁPAGOS PENGUIN			
ECUADOR			
Galápagos Islands	Shelf	Small fish	[56]
		Small fish	[57]
		Fish Anchovetta Pacific Sardine Mullet spp.	[58]
HUMBOLDT PENGUIN			
PERU			
Chincha Islands (13°38′S, 76°24′W)	Shelf	Fish Anchovetta, 100%	[56]
Punta San Juan (15°22′S, 75°10′W)	Shelf	Fish Anchovetta Peruvian Silverside	[59]
CHILE			
Pan de Azucar Island (26°09′S, 70°41′W)	Shelf	Fish Pacific Saury, 66% Horse Mackerel, 20% Anchovetta, 12% Sardine, <1%	[60]
Isla Chañiaral (29°01′S, 71°50′W)	Shelf	Fish Anchovetta, 72% Sardine, 4%	[61, 62]
Isla Algarrobo (33°30′S, 71°45′W)	Shelf	Fish Pacific Saury, 94% Anchovetta, 3% Sardine, <1%	
		Cephalopod Antarctic Flying Squid, 14%	
Punihuil Islands (41°55′S, 74°02′W)	Shelf	Fish Anchovetta, 80% Sardine, 20%	[61]
		Fish Peruvian Silverside, 32% Anchovetta, 30% Sardine, 30% Peruvian Hake, 7% Horse Mackerel, <1%	[60]
MAGELLANIC PENGUIN			
CHILE			
Punihuil Islands (41°55′S, 74°02′W)	Shelf	Fish Anchovetta, 61% Sardine, 6%	

(continued)

Table A2.4 (continued)

Location	Habitat	Main prey species	References
		Cephalopod Antarctic Flying Squid, 19%	[61]
Magdalena Island (52°55′S, 70°34′W)	Shelf	Fish Falkland Sprat, 100%	[63]
Otway Sound	Shelf	Fish Falkland Sprat, 90% Other species, 5%	
		Cephalopod Longfin Squid, 5%	
FALKLAND/MALVINAS ISLANDS (51°26′S, 57°52′W)			
New Island	Shelf	Fish, 24–55% Southern Blue Whiting Falkland Sprat	[64]
		Crustacean, 4–15% Squat Lobster	
		Cephalopod Armhook Squid	
		Fish, 40%	[37, 38]
		Crustacean, 19% Squat Lobster	
		Cephalopod Longfin Squid	
Seal Bay	Shelf	Fish, 74%	
		Crustacean, 20%	
		Cephalopod, 5%	
Volunteer Beach	Shelf	Fish, 40%	
		Crustacean, 15%	
		Cephalopod, 44%	
Bull Point	Shelf	Fish, 67%	
		Crustacean, 20%	
		Cephalopod, 13%	
Keppel Island	Shelf	Fish, 98%	
		Crustacean, 1%	
		Cephalopod, 1%	
West Point Island	Shelf	Fish, 7%	
		Crustacean, 85%	
		Cephalopod, 8%	
ARGENTINA			
Martillo Islands (54°53′S, 67°34′W)	Shelf	Fish, 77% Falkland Sprat, 75%	[65]
		Crustacean Squat Lobster, 15%	

(continued)

Table A2.4 (continued)

Location	Habitat	Main prey species	References
		Cephalopod Longfin Squid, 10% *Enteroctopus* sp., <1%	
Cabo Virgenes (52°20′S, 68°21′W)	Shelf	Fish Falkland Sprat, 85% Smelt sp., 9%	[66]
		Cephalopod Longfin Squid, 6%	
		Fish Falkland Sprat, 100%	[67]
Monte Leon (50°17′S, 68°51′W)	Shelf	Fish Peruvian Silverside, 25%	[67]
		Cephalopod, 75%	
San Julian (49°16′S, 67°42′W)	Shelf	Fish Silverside spp., 50%	[67]
		Cephalopod Longfin Squid, 50%	
Isla Chaffers to Bahía Laura (47°46′S, 65°52′W to 48°21′S, 66°21′W)	Shelf	Fish, 70% Falkland Sprat, 15% Argentine Hake, 1% Other fish, 54%	[66]
		Cephalopod Longfin Squid, 30%	
Golfo San Jorge (46°55, 67°W)	Shelf	Fish Argentine Anchovy, 55% Argentine Hake, 32%	[68]
		Crustacean Squat Lobster, 2%	
		Cephalopod Longfin Squid, 8%	
Cabo Dos Bahias (44°54′S, 65°32′W)	Shelf	Fish Argentine Anchovy, 54% Argentine Hake, 45%	[66]
		Cephalopod *Illex* sp., <1% Octopus sp., <1%	
Punta Loberia (44°35′S, 65°22′W)	Shelf	Fish Argentine Anchovy, 90% Silverside, 5%	[69]
		Cephalopod, 5%	
San Lorenzo and Punta Clara (42°04′S, 63°′21′W) (43°58′S, 65°16′W)	Shelf	Fish Argentine Anchovy, 52% Silverside, 25% Argentine Hake, 3%	[70]
		Cephalopod Longfin Squid, 16%	

(continued)

Table A2.4 (continued)

Location	Habitat	Main prey species	References
		Fish Argentine Anchovy, 95%	[69]
		Cephalopod, 5%	
		Fish Argentine Anchovy, 100%	[67]
AFRICAN PENGUIN			
NAMIBIA			
Walvis Bay (22°57′S, 14°30′E)	Shelf	Fish Sardine, 30–99% Anchovy, 19–31% Horse Mackerel, 0–16% Mackerel, 0–3% Redeye Round Herring, 0–6% Other fish, 0–26%	[71]
		Cephalopod, 1–4%	
Mercury Island (25°38'S, 14°50′E)	Shelf	Fish Pelagic Goby, 17–83% South African Anchovy, 1–35% Sardine, 0–19% South African Hake, 0–11% Horse Mackerel, 0–21% Whitehead's Round Herring, 0–6%	[72]
		Cephalopod *Onchoteuthid* sp., <1–5%	
SOUTH AFRICA			
Cape Province (32.0–34.5°S, 17.5–20.0°E)	Shelf	Fish Sardine, 32% South African Anchovy, 21% Horse Mackerel, 15% Whitehead's Round Herring, 5% Other fish, 21%	[73]
		Crustacean, 1%	
		Cephalopod, 5%	
Robben Island (33°47′S, 18°22′E)	Shelf	Fish South African Anchovy, 48–100% Sardine, 0–70% Horse Mackerel, 0–24% South African Hake, 0–50%	[74]
		Fish South African Anchovy, 98% Sardine, 2% Beaked Salmon, <1%	[75]

(continued)

Table A2.4 (continued)

Location	Habitat	Main prey species	References
Boulders (34°12′S, 18°27′E)	Shelf	Fish South African Anchovy, 73% Sardine, 2%	
		Cephalopod Onychoteuthid sp.	
St. Croix Island (33°48′S 25°46′E)	Shelf	Fish South African Anchovy, 56% Sardine, 23% Red-eyed Round Herring, 5% Beaked Salmon, 2%	[76]
		Cephalopod *Loligo reynaudi*, 13%	
Marcus Island (33°03′S, 17°58′E)	Shelf	Fish South African Anchovy, 80% Horse Mackerel, 5% Whitehead's Round Herring, 3% Smelt sp., 2% Saury, 2% Snake Mackerel, 5%	[77, 78]
		Cephalopod *Lolliguncula* spp./*Loligo* spp., 1%	

Table A2.5 The diet of Little Penguins during chick-feeding season unless noted; where information available, the main (though not necessarily exclusive) prey species listed; percent by mass or volume (*WFO* weighted frequency of occurrence), by species if data available

Location	Habitat	Main prey species	References
AUSTRALIAN LITTLE PENGUIN			
WESTERN AUSTRALIA			
Penguin Island (32°18′S., 115°42′E)	Shelf	Fish Blue Sprat Sandy Sprat Australian Sardine Southern Garfish	[79]
VICTORIA			
Port Campbell (38°7′S, 143°04′E)	Shelf	Fish, 71% WFO Australian Sardine, 20% Australian Anchovy, 18% Sandy Sprat, 12%	[80]
		Crustacean, 6%	
		Cephalopod, 22% Gould's Arrow Squid, 6%	
Port Phillip Bay (37°51′S, 144°57′E)	Shelf 2007–2008	Fish, 91% Australian Anchovy, 73% Southern Garfish, 12% Other fish species, 7%	[81]
		Cephalopod Bay Squid, 9%	
	Shelf 2011–2012	Fish Australian Anchovy, 15% Australian Sardine, 33% Sandy Sprat, 13% Blue Sprat, 22% Southern Garfish, 12%	
Phillip Island (38°15′S, 145°30′E)	Shelf	Fish, 90% Australian Anchovy, 37% Australian Sardine, 34%	[82]
		Cephalopod, 10%	
		Fish, 80% Australian Anchovy, 40% Australian Sardine, 33%	[83]
		Crustacean, 1%	
		Cephalopod, 19% Gould's Arrow Squid,	
		Fish, 73% WFO Australian Anchovy, 17% Australian Sardine, 30% Sandy Sprat, 12%	[80]
		Crustacean, 5%	
		Cephalopod, 22% Gould's Arrow Squid	

(continued)

Table A2.5 (continued)

Location	Habitat	Main prey species	References
		Fish, 82% WFO Australian Anchovy Red Cod Snake Mackerel Blue Warehou	[84, 85]
		Crustacean, 3% *Nyctiphanes australis*	
		Cephalopod, 12% Gould's Arrow Squid, 9%	
Rabbit Island (38°55′S, 146°31′E)	Shelf	Fish, 96% WFO Australian Anchovy, 28% Australian Sardine, 30% Sandy Sprat, 20%	[80]
		Crustacean, <1%	
		Cephalopod, 4% Gould's Arrow Squid, 3%	
TASMANIA			
Fort Direction (43°05′S, 147°25′E)	Shelf	Fish, 99%	[6]
		Crustacean, <1%	
		Cephalopod, <1%	
Albatross Island (40°24′S, 144°32′E)	Shelf	Fish, 68% Blue Grenadier Barracouta	
		Crustacean, 6% *Nyctiphanes australis*	
		Cephalopod, 26% Gould's Arrow Squid	
Marion Bay (42°50′S, 147°52′E)	Shelf	Fish, 71%	
		Crustacean, 9%	
		Cephalopod, 20%	
NZ LITTLE PENGUIN			
South Island, Omaru (45°06′S, 170°58′E)	Shelf	Fish NZ Blueback Sprat, 73% Gudgeon, 7% Pigfish sp., 8% Smelt sp., 2%	[86]
		Cephalopod Arrow Squid, 5%	
		Fish	[87]
		Gudgeon, 92% NZ Blueback Sprat, 3%	
		Cephalopod Arrow Squid, 5%	

(continued)

Table A2.5 (continued)

Location	Habitat	Main prey species	References
South Island, Otanerito Bay (43°50′S, 173°4′E)	Shelf	Fish NZ Blueback Sprat, 34% Ahuru, 37% Red Cod, 14%	
		Cephalopod Arrow Squid, 15%	
Codfish Island (46°45′S, 167°40′E)	Shelf	Fish Ahuro, 37%	[26]
		Cephalopod Arrows Squid, 58% *Octopus maorum*, 3%	
Stewart Island, Ackers Point (46°53′S, 168°10′E)	Shelf	Fish NZ Blueback Sprat, 10% Red Cod, 9%	[87]
		Crustacean Mantis Shrimp larvae, 6%	
		Cephalopod Arrow Squid, 73%	

Table A2.6 The diet of Adélie penguins during chick-feeding season unless noted; where information available, the main (though not necessarily exclusive) prey species are listed; percent by mass or volume, by species if data available

Location	Habitat	Main prey species	References
ADÉLIE PENGUIN			
South Orkney Islands (60°S, 44–46°W)	Shelf/slope	Crustacean Euphausiids	[88]
		Antarctic Krill, 100%	[89]
		Antarctic Krill, some fish	[90]
		Antarctic Krill, 99%	[91]
		Antarctic Krill, 99%	[92]
Antarctic Peninsula			
East Coast (50°W)	Shelf/slope	Crustacean Euphausiids	[88, 93]
West Coast			
King George Island (62°S, 58°15′W)	Shelf/slope	Fish, 2%	[49, 94]
		Crustacean Antarctic Krill, 95–99%	
		Fish, 7%	[51]
		Crustacean Euphausiids spp.	
		Fish, <1%	[95]
		Crustacean	
		Antarctic Krill, 99%	
		Crustacean Antarctic Krill, 100%	[50]
Esperanza Bay (63°23′S, 57°00′W)	Slope	Fish, 3% Antarctic Silverfish Antarctic Lanternfish	[96]
		Crustacean Antarctic Krill, 96%	
Anvers Island (64°33′S, 63°35′W)	Shelf/slope	Nototheniid fish & Antarctic Krill	[93]
		Fish, Antarctic Krill, *Thy. macrura*	[97]
		Fish Antarctic Lanternfish, 3–12%	[98]
		Crustacean Antarctic Krill, 68–75% *Thy. macrura*, 20–42%	
		Crustacean Antarctic Krill, 63–100% *Thy. macrura*, 0–35%	[55]
Avian Island (67°46′S 68°54′W)	Shelf/slope	Fish Antarctic Lanternfish, 5–8% Antarctic Silverfish, 4%	[98]

(continued)

Table A2.6 (continued)

Location	Habitat	Main prey species	References
		Crustacean Antarctic Krill, 56–62% Crystal Krill, 12–15% *Thy. macrura*, 15–20%	
Charcot Island (69°45′S, 75°15′W)	Shelf/slope	Fish Antarctic Silverfish, 4%	[98]
		Crustacean Antarctic Krill, 44% Crystal Krill, 52%	
East Antarctica			
Hukuro Cove (69°10′S, 39°30′E)	Shelf/slope	Fish, 26–34% Antarctic Silverfish	[99]
		Crustacean Antarctic Krill, 58–64% Crystal Krill, 2–4%	
Cape Hinode (68°07′S, 42°38′E)	Shelf/slope	Fish, 4% Antarctic Silverfish	[100]
		Crustacean Antarctic Krill, 79% Crystal Krill, 10%	
MacRobertson Land (65–67°S, 48–61°E)	Shelf/slope	Fish Antarctic Silverfish, 100% (shelf)	[101]
		Crustacean Antarctic Krill, 100% (slope)	
	Shelf/slope	Fish, 42% Antarctic Silverfish (shelf)	[102]
		Crustacean Antarctic Krill, 40% Crystal Krill, 8% Amphipoda, 3%	
Prydz Bay (69°00′S, 75°00′E)	Shelf/slope	Fish, 32% Antarctic Silverfish	[93]
		Crustacean Antarctic Krill, 24% Crystal Krill, 42%	
		Fish, 8% Antarctic Silverfish	[99]
		Crustacea Antarctic Krill, 85% Crystal Krill, 7%	
		Fish, 28% Antarctic Silverfish	[103]
		Crustacean Antarctic Krill, 66% Crystal Krill, 3%	

(continued)

Table A2.6 (continued)

Location	Habitat	Main prey species	References
		Fish, 31% Antarctic Silverfish	[104]
		Crustacean Antarctic Krill, 59% Crystal Krill, 3%	
Wilkes Land (Shirley Is) (66°17′S, 110°30′E)	Shelf/slope	Crustacean Euphausiids	[105]
		Fish & Krill	[106]
		Fish, 49% Antarctic Silverfish, 42%	[102]
		Crustacean, 48% Antarctic Krill, 40% Crystal Krill, 8%	
		Fish, 26–30% Antarctic Silverfish	[107]
		Crustacean, 51–53% Crystal Krill	
Adélie Land (Petrel Is) (66°40′S, 140°01′E)	Shelf/slope	Fish Antarctic Silverfish, 26–30%	[107]
		Crustacean Antarctic Krill, 27–38% Crystal Krill, 22–39%	
		Fish, 1% Antarctic Silverfish	[108]
		Crustacean Antarctic Krill, 4% Crystal Krill, 94%	
Victoria Land			
Cape Adare (71°18′S, 170°09′E)	Shelf/slope	Mostly Antarctic Krill	[109, 110]
Cape Hallett (72°19′S, 170°12′E)	Shelf	Fish, 2% Crustacean (krill), 97%	[93]
		Fish, 17% Antarctic Silverfish	[111]
		Crustacean Antarctic Krill, 80% Crystal Krill, 3%	
Franklin Island (76°05′S, 168°19′E)	Shelf	Crustacean Crystal Krill, 99%	[112]
Beaufort Island (76°56′S, 166°56′E)	Shelf	Fish Antarctic Silverfish, 39%	
		Crustacean Crystal Krill, 60%	

(continued)

Table A2.6 (continued)

Location	Habitat	Main prey species	References
Ross Island			
Cape Crozier (77°31′S, 169°24′E)	Shelf	Fish Silverfish, 39%	[112]
		Crustacean Crystal Krill, 60%	
		Fish Antarctic Silverfish, 30–80%	[113–115]
Cape Bird (77°10′S, 166°41′E)	Shelf	Fish Antarctic Silverfish, 44%	[116]
		Crustacean Crystal Krill, 46% Amphipoda, 10%	
		Crustacean Crystal Krill, 99%	[117]
		Fish Antarctic Silverfish, 10–75%	[113–115]
		Crustacean Crystal Krill, 25–70%	
		Other, 1–3%	
Cape Royds (77°33′S, 166°09′E)	Shelf	Fish and Euphausiids	[110]
		Fish Antarctic Silverfish, 40–75%	[113–115]
		Crustacean Crystal Krill, 25–60%	
OPEN OCEAN			
Ross Sea	Slope	Crustacean, 98–99% Antarctic Krill Crystal Krill	Spring [118]
Weddell Sea	Pelagic	Fish *Myctophid* spp., 14%	Winter [1]
		Crustacean Antarctic Krill, 28%	
		Cephalopod, Squid, 54%	
Indian Ocean	Pelagic	Euphausiids, some fish	Summer [119]

Table A2.7 The diet of six crested penguin species (see Table A2.8 for macaroni penguin) during chick-feeding season unless noted; where information available, the main (though not necessarily exclusive) prey species listed; percent by mass or volume, by species if data available

Location	Habitat	Main prey species	References
NORTHERN, EASTERN ROCKHOPPER PENGUIN			
Gough Island (40°19′S, 9°56′W)	Pelagic	Fish, 6% Silvery Lightfish Trumpeter Fish	[131]
		Crustacea *Thy. gregaria* *Nematoscelis megalops* *Euphausia similis* *E. lucens*	
		Cephalopod, 2%	
Marion Island (46°52′S, 37°51′E)	Slope/pelagic	Fish Rhombic Lanternfish Tenison's Lanternfish Norman's Lanternfish	[125]
		Crustacea Caridean shrimp, 30% *E. vallentini*	
Kerguelen Islands (48°45–50′S, 68°45–70′E)	Shelf	Fish, benthic, 20%	[17]
		Crustacea Euphausiids, 57% Amphipoda, 2%	
		Cephalopod, 19%	
Amsterdam Island (37°50′S 77°33′E)	Shelf/slope	Fish, 28% Slender Lightfish	[132]
		Crustacea, 58% *Thy. gregaria*	
		Cephalopod, 14% Antarctic Flying Squid	
Heard Island 53°05′S, 73°30′E)	Slope/pelagic	Fish, 8% Rhombic Lanternfish Bolin's Lanternfish Humped Rockcod	[127]
		Crustacean, 91% *E. vallentini* *Thy. macrura* *Th. gaudichaudii*	
		Cephalopod, 1%	
Macquarie Island (54°30′S, 158°′57′E)	Slope/pelagic	Fish, 18% Myctophid spp.	[133]
		Crustacea *E. vallentini* *Thy. gregaria*	
		Fish, 29%	[134]

(continued)

Table A2.7 (continued)

Location	Habitat	Main prey species	References
		Rhombic Lanternfish, 16% Lanternfish sp., 5% Electron Subantarctic Lanternfish, 2% Magellanic Rockcod, 4% Antarctic Horsefish, 3%	
		Crustacea, 69% *E. vallentini*, 62% *Thy. gregaria*, 5%	
		Cephalopod Sevenstar Flying Squid, 2%	
SOUTHERN ROCKHOPPER PENGUIN			
Campbell Island, NZ (52°32′S, 169°9′E)	Shelf/slope	Fish	[135]
		Crustacean	
		Cephalopod Greater Hooked Squid Antarctic Neosquid Giant Warty Squid Sevenstar Flying Squid *Octopus campbelli*	
Falkland/Malvinas Islands (52°50′S, 59°10′W)			
Beauchene Island	Slope/pelagic	Fish, 2%	[136]
		Crustacean, 45% *E. vallentini* *E. lucens* *Thy. gregaria*	
		Cephalopod, 53%	
New Island	Slope/pelagic	Fish, 1%	[137]
		Crustacea, 45% Euphausia sp. Cephalopod, 46% Armhook Squid	
		Fish, 41%	[38]
		Crustacea, 39%	
		Cephalopod, 29%	
Seal Bay	Slope/pelagic	Fish, 47%	
		Crustacea, 52%	
Staten Island, Patagonia (54°50′S, 64°40′W)	Slope/pelagic	Fish, 3%	[138, 139]
		Crustacea *E. vallentini*, 8% *Thy. gregaria*, 10% *Th. gaudichaudii*, 4%	

(continued)

Table A2.7 (continued)

Location	Habitat	Main prey species	References
		Cephalopod Armhook Squid, 17% Greater Hooked Squid, 10% Longfin Squid, 27% *Enteroctopus megalocyathus*, 21%	
South Georgia, Bird Island (54°00′S, 38°03′W)	Slope/pelagic	Crustacea Antarctic Krill, 100%	[42]
		Fish, 9% Mackerel Icefish Rhombic Lanternfish	[130]
		Crustacea Antarctic Krill, 64% *Thysanoessa* sp. 7% *Th. gaudichaudii*, 20%	
		Cephalopod, <1%	
ROYAL PENGUIN			
Macquarie Island (54°30′S, 158°57′E)	Slope/pelagic	Fish, 54%	[133]
		Crustacean, 17% *E. vallentini* *Thy. gregaria* *Primno macropa* *Th. gaudichaudii*	
		Cephalopod, 27%	
		Fish, 46%	[140]
		Rhombic Lanternfish, 24% Lanternfish spp., 8% Electron Subantarctic Lanternfish, 10% *Magnisudis prionosa*, 4%	
		Crustacean, 51% *E. vallentini*, 32% *Thy. gregaria*, 10%	
		Cephalopod, 3% Hooked Squid sp., 2%	
FIORDLAND PENGUIN			
Fiordland, NZ (43°58′S 168°37′E)	Shelf	Fish, 2%	[141, 142]
		Crustacean, 13% *Nyctiphanes australis* Squillid shrimp	
		Cephalopod Hooked Squid, 71% *Octopus maorum*, 14%	

(continued)

Table A2.7 (continued)

Location	Habitat	Main prey species	References
SNARES PENGUIN			
Snares Islands, NZ (48°09′S, 166°36′E	Slope/pelagic	Fish, 30% Long-snouted Pipefish, 12% Redbait, 7% Hector's Lanternfish, 2% Thornfish, 3%	[143]
		Crustacean, 60% Caridean shrimp, 54%	
		Cephalopod Arrow Squid, 2%	

Table A2.8 The diet of chinstrap and macaroni penguins, mainly during summer chick feeding unless otherwise noted; where information available, the main (though not necessarily exclusive) prey species are listed; percent by mass or volume, by species if data available

Location	Habitat	Main prey species	References
CHINSTRAP PENGUIN			
South Orkney Islands			
Signy Island (60°43′S 45°36′W)	Slope/pelagic	Crustacean Antarctic Krill, 100%	[89]
		Fish, 3%	[90]
		Crustacean Antarctic Krill, 97–100%	
Laurie Island (60°46′S, 44°42′W	Slope/pelagic	Crustacean Antarctic Krill, 100%	[120]
Antarctic Peninsula			
Elephant Island (61°08′S 55°07′W)	Slope/pelagic	Crustacean Antarctic Krill, 96%	[121]
Seal Island (60°58′S 55°24′W)	Slope/pelagic	Crustacean Antarctic Krill, day 98% Antarctic Krill, night 42%	[122]
		Fish, night 43%	
		Cephalopod, night <1%	
		Fish, myctophids, night 25%	[123]
		Crustacean Antarctic Krill, day 99% Antarctic Krill, night 74%	
King George Island Admiralty Bay (62°10′S, 58°25′W)	Shelf/slope	Fish, 11%	[49, 94]
		Crustacean Antarctic Krill, 84%	
		Fish, 7%	[51]
		Crustacean Antarctic Krill, 93%	
		Crustacean Antarctic Krill, 100%	[50]
Harmony Point (62°18′S, 59°15′W)		Crustacean Antarctic Krill, 100%	[124]
Nelson Island (62°18′S 59°03′W)	Slope	Crustacean Antarctic Krill, 100%	
Livingston Island (62°36′S 60°30′W)	Slope	Fish, 1% Antarctic Jonasfish	[53, 54]
		Crustacean Antarctic Krill, 99%	
		Fish (SIA), 10–16% Crustacean (SIA) Antarctic Krill, 84–89%	

(continued)

Table A2.8 (continued)

Location	Habitat	Main prey species	References
Anvers Island (64°33′S, 63°35′W)	Slope	Fish, 4–12% Antarctic Lanternfish	[98]
		Crustacean Antarctic Krill, 50–74% *Thy. macrura*, 22–46%	
MACARONI PENGUIN			
Marion Island (46°52′S, 37°51′E)	Slope/pelagic	Fish, 5% Rhombic Lanternfish Tenison's Lanternfish	[125]
		Crustacean, 93% Caridean shrimp, 21% *E. vallentini*, 23% *Thy. vicina*, 21% *Th. gaudichaudii*, 21%	
		Cephalopod, 2%	
		Fish, 18%	[126]
		Crustacean, 79% *Nauticaris marionis* *E. vallentini* *Thy. vicina* *Th. gaudichaudii*	
		Cephalopod, 3%	
Kerguelen Islands (48°45–50′S, 68°45–70′E	Slope/pelagic	Fish (pelagic), 76%	[17]
		Crustacean Euphausiid, 22% Amphipoda, 1%	
		Cephalopod, <1%	
Heard Island (53°05′S, 73°30′E)	Slope	Fish, 23% Electrona Subantarctic Lanternfish Rhombic Lanternfish Bolin's Lanternfish Mackerel Icefish	[127]
		Crustacean, 77% *E. vallentini* *Thy. macrura* *Th. gaudichaudii*	
		Cephalopod, <1%	
		Fish, 22% Rhombic Lanternfish	[128, 129]
		Crustacean, 69% *E. vallentini* *Thy. macrura* *Th. gaudichaudii*	
		Cephalopod, <1%	

(continued)

Table A2.8 (continued)

Location	Habitat	Main prey species	References
South Georgia, Bird Island (54°00′S, 38°03′W)	Slope/pelagic	Fish, 2%	[42]
		Crustacean Antarctic Krill, 98%	
		Fish, 9%	[130]
		Crustacean, 91% Antarctic Krill, 64% Euphausiids spp., 7% *Th. gaudichaudii*, 20%	
Antarctic Peninsula Elephant Island (61°15′S, 55°30′W	Slope/pelagic	Fish, 25%	[121]
		Crustacean Antarctic Krill, 37% *Thysanoessa* spp., 38%	

Appendix Table A3

Aspects of foraging behavior of penguins during the chick period unless otherwise indicated, as follows: E, incubation; G, guard; C, crèche; W, winter or nonbreeding. Foraging range, i.e., distance from colony of foraging tracks, among penguin species fitted with GPS, satellite, or other transponders. In most cases, shown in italics, average maximum distance and in parenthesis range. If just a single number, then distance of outer-most foraging was interpolated from a map included in publication; or author stated that individuals of colony foraged no more than or within a specified distance. For dive depths, in italics are average and range or a range of means; otherwise, a single, hyphenated number gives range (average not available). In some cases, values were averaged if from multiple years or genders. Breeding sites are listed in order of most to least abundant

Colony/Breeding site	Year		Colony size (pairs)	Foraging range (km)	Depth (m)	Source
EMPEROR PENGUIN						
Taylor Glacier	1993		2460	75	20–360 (470)	[144]
Auster	1994	G	13,400	*133 (91–565)* G	~50 (444)	[145]
		E		*96 (62–125)*		
	1993	E	13,300	110	20–150 (300)	[144]
	1988–1994	E	13,500		<200 (564)	[146]
Pt Géologie	2005	E	2500	*94 (62–116)*	56 (339)	[147]
		E		*106 (78–133)*	70 (438)	
		C		*647 (425–838)*	99 (328)	
	1990	G	3000	*92 (50–157)*		[145]
Cape Roget	1993		6000	180		Kooyman, pers comm
Coulman Island	1993	C	20,000	130		Kooyman, pers comm
Cape Washington	1990	C	23,500	*232 (160–450)*	~150 (400)	[145]
	1992	G	26,500	200	75–300 (>500)	[148]
	2011	G	20,400	160	(470)	[148]
KING PENGUIN						
Kerguelen Island	2006–2011	E	70,000	683	130 (260)	[149]
		G		278	140 (260)	
	1995	G	50,000	*202 (104–292)*		[150]

Colony/Breeding site	Year		Colony size (pairs)	Foraging range (km)	Depth (m)	Source
	1995	C	50,000	*267 (129–409)*	150 (20–250)	[151]
	1998–2001	C	40,000	304 (max 603)	148 (272)	Charrassin
Possession Island	1992–1994	C	40,000	*377 (264–558)*	135 (261)	[152–154]
Crozet Island	1994–1997	E	40,000	*448 (304–644)*	50–200 (300)	[155]
		G		*338 (236–428)*	100–250 (300)	
		C		*900 (605–1532)*	150–200 (275)	
		C		*1609 (1475–1856*	175–250 (325)	
	1994–1995	C	40,000	*406 (260–796)*		[150, 156]
		C		*1707 (725–2239)*		
Heard Island	1992	C	8400	370	70–160 (10–235)	[157]
Falkland Islands	1996	G	360	*412 (207–580)*		[150]
		C		*727 (224–1547)*		
South Georgia	1996	G	80	*463 (301–626)*		[150]
		C		661 C		
	2005–2006	E	450,000	~24,480	106–109	[158]
ADÉLIE PENGUIN						
Anvers Island	1990–1992	E-C	8000	*9 (1–36)*	*26* (3–98)	[159]
	2002–2011	G-C	3000	12		[160]
	2011	G-C			5–60	[161]
	2012–2015	G-C	2000	*7 (1–34)*	~ 40 (5–95)	[162]
Admiralty Bay	1981	G-C	18,046	50		[163]
		G-C	2900	*29 (9–48)*	*31 (15–35)*	[164]
Signy Island	2000–2001	G-C	37,200	*100 (56–177)*		[165]
Hukuro Cove	1996–1997	G-C	250	*4 (3–6)*	*7–11 (5–105)*	[166, 167]
Shirley Island	1995	G-C	93,000	*113 (94–144)*	~50 (5–125)	[168]

Colony/Breeding site	Year		Colony size (pairs)	Foraging range (km)	Depth (m)	Source
Béchervaise Island		G-C	80,000	200		[169]
		E		*250 (164–351)*		[170]
		G		12–110		
Petrel Island	2012	G-C	54,000	102		[171]
	1995	G-C	41,000	*36 (6–58)*	~50 (5–125)	[168]
Cape Hallett	2004	G-C	25,000	*95 (49–183)*	50–67	[172]
	2005	G-C	43,000	*111 (50–247)*	50–67	[172]
Cape Crozier	2005–2013	C	272,000	240	40–60 (134)	[173]
	2012–2013	G-C	272,000	70	~30 (10–80)	[174]
		C		170	~70 (10–95)	
	2005–2007	G-C	141,000	180		[175]
	2003	G-C	135,000	*142 (5–250)*		[176]
Cape Bird	2005–2007	G-C	29,500	97		[175]
	2009–2010	G-C		*26–34*	*45*	[177]
Cape Royds	2003–2004	G-C	3900	14		[176]
	2014–2015	G-C	2000	*11 (1–15)*	*47–54*	[178]
GENTOO PENGUIN						
Near Is, Falklands Islands						
North End	2008	G	5500	*15 (4–23)*	5–120	[179]
South End		G	?	*21 (11–30)*		
Bird Island, So Georgia	2009	W	4500	*<30 (1–30)*		[180]
Admiralty Bay, King George Island	1981	G-C	2602	24		[163]
	2009–2014	G-C	250	*15 (4–26)*	*52 (10–55)*	[164]
Anvers Island	2011	G-C	3000	31	5–140	[161]
	2012–2015	G-C	3600	*7 (1–34)*	~70 (5–145)	[162]
Cape Shirreff, Livingston Island	2002–2003		300	20	*41 (5–150)*	[181]
	2009–2014	G-C	850	*16 (10–22)*	*52 (10–58+)*	[164]
Kildalkey, Marion Island	1984	E-C	?	14		[182]

Colony/Breeding site	Year		Colony size (pairs)	Foraging range (km)	Depth (m)	Source
Hasselborough Bay, Macquarie Island	1993	G-C	110	*5 (1–26)*		[183]
Kerguelen Island	2002–2003	G-C	110 200	*31 (23–39)*		[184]
Trypot Beach, Marion Island	2014	E-G	99	*15 (4–26)*		[185]
Kerguelen Island, Pt Suzanne/Estacade	2014–2015	C	50	*22 (3–78)*		[186]
CHINSTRAP PENGUIN						
Signy Island	2000–2001	G-C	13,000	*58 (18–112)*		[165]
		G-C		60		[187]
	2016	G	19,500	*52 (5–160)*	*61 (10–135)*	[188]
Laurie Island	2012	G	143,800	18	*70 (20–130)*	
Seal Island	1988	G-C	20,000	3–20	*31 (10–115)*	[189]
Admiralty Bay, King George Island	1981	G-C	7662	33		[163]
		G-C	320	*17 (6–28)*		[164]
Cape Shirreff, Livingston Island	2009–2014	G-C	4100	*24 (12–36)*	*31 (15–35+)*	[164]
Barton Pen, King George Island	2007–2015	G	2961	*18 (8–28)*	*33*	[190]
Ardley Island King George Island	1991, 95	G	238	*12 (3–36)*	~20 (10–100)	[191]
Bouvetoya Island	2007–2008	G-C	<100	*10 (1–57)*	*20 (4–99)*	[192]
YELLOW-EYED PENGUIN						
South Island, Oamaru	2003–2005	G	7	*18 (6–20)*	25 (42)	[193]
		G-C		20–30		[194]
South Island, Otago						
Boulder Beach	1990–1993	G-C	40	*15 (2–57)*	10–65	[195]
Long Point	1991–1993	G-C	10	*10 (1–25)*	10–125	
	2003–2018	G-C		10–35		[194]
South Island, Aramoana	2003–2018	G-C		5–35		[194]
Stewart Island, Fouveaux Strait	2003–2018	G-C		4		[194]
Stewart Island, Port Pegasus	2011	E	7	10	*61 (25–116)*	[196]

Colony/Breeding site	Year		Colony size (pairs)	Foraging range (km)	Depth (m)	Source
Port Pegasus	2020	G	8	5	60–120	[197]
Paterson Inlet	2020	G	8	10	20–45	[197]
Codfish Island	2003–2018	G-C	46	50		[194]
	2020	C	24	23	20–80	[197]
Enderby Island	2016–2017	G	200–300	*11–28 (10–47)*	*67–93*	[198, 199]
ERECT-CRESTED PENGUIN						
SNARES PENGUIN						
FIORDLAND PENGUIN						
Jackson's Bay	1984	G	160	10		[200]
NORTHERN ROCKHOPPER PENGUIN						
Amsterdam Island	1995	G	25,000	6 (1–22)	18 (109)	[201]
SOUTHERN ROCKHOPPER PENGUIN						
Islas de los Estados	2011–2013	G	13,0000	*30 (15–45)*	*16* (69+)	[202]
Marion Island	1985	G		*36 (4–157)*		[203]
Falkland Island, New Island	2008–2010	G	5000	*16 (7–55)*	*26* (86)	[204]
	2008	G	5500	*16 (4–28)*	5–60	[179]
MACARONI PENGUIN						
Marion Island	1985	G-C	405,000	*398 (59–1180)*		[203]
Bird Island, South Georgia	2001–2002	E	50,000	*570 (274–891)*		[205]
Capsize Beach, Heard Island	2003–2004	G	50,000	*47 (1–150)G*	*26 (12–42)*	[206]
		C		*169 (53–347)*	*29 (19–54)*	
Bouvetoya Island	2007–2008		1100	*48 (2–284)*	*32 (4–133)*	[192]
ROYAL PENGUIN						
Macquarie Island	1994–1995	G	500,000	~700		[207]
		C		~650		
GALÁPAGOS PENGUIN						
Fernandina/Bartolomé	1994	G-C	110 360	1		[208]
HUMBOLDT PENGUIN						
Pan de Azucar Island, Chile	1994–1995	G-C	2900	35	3–53	[209, 210]
Punta San Juan, Peru	2000	G-C	1000	*15 (8–42)*		[211]
MAGELLANIC PENGUIN						
Punta Tombo	1990–2000	G-C	225,000	*77 (30–186)*		[211]

Colony/Breeding site	Year		Colony size (pairs)	Foraging range (km)	Depth (m)	Source
	1996–2006	E	225,000	*411 (605)*		[212]
	1996–2006	C	225,000	*111 (330)*		[212]
Cabo Virgenes	1994–1997	E	89,200	90	~100	[213]
Cape Lorenzo	1994–1997	E	56,740	100		[213]
San Julian	1994–1997	E	37,150	150		[213]
Punta Loberia	1994–1997	E	6745	190		[213]
New Island, Falklands	2008	G	5500	*25 (18–37)*	5–80	[179]
	1994–1997	E	3725	45		[213]
Isla Martillo, Beagle Channel	2006	G	2000	*24 (15–45)*	*17*	[214]
Islas de los Estados	2011–2013	G	1600	*30 (19–41)*	*11* (76+)	[202, 215]
AFRICAN PENGUIN						
St Croix, Algoa Bay	1984–1985	G-C	19,450	*40 (10–75)*		[216]
	2008–2009	G-C	7800	*20–30 (5–44)*	*26 (85)*	[217]
Bird Island, Algoa Bay	2008–2009	G-C	2700	*14 (4–30)*	*27 (93)*	[217]
Robben & Dassen Islands	2003	G-C	7000–21,000	*9 (6–17)*	*52 (29–69)*	[218]
Boulders	2003	G-C	1000	*19 (6–28)*	*51 (29–63)*	[218]
LITTLE PENGUIN, AUSTRALIA						
Albatross Island, Tasmania	1987	G-C	500	13		[219]
Phillip Island, Victoria	1986–1988	G-C	550	*8 (1–20)*		[220]
	2005	G-C	"small"	*20*	*9–13*	[221]
	2008	W	26,000	*9–13*		[222]
	2010	G-C		*20 (5–30)*		[223]
		G			*10 (22)*	[224]
		G			*11–15*	[225]
Kanowna Island		G-C	"small"	*17*	*4–7*	[221]
Rabbit Island		G-C	"small"	*18*	*8–9*	[221]
LITTLE PENGUIN, NZ	2010–2012	G-C	220	*12–23 (5–35)*	*5–13*	[226]
Motuara Island	2000–2001	G	300	*9*	*10*	[226, 227]
Omaru		G	300	*20–30 (6–35)*	*6–13*	

Appendix Table A4

Average prey size and the energy density (ED) among various fish, squid, and crustacean species found in penguin diets, and in which ED has been measured[a]

	Length (cm)	Mass (g)	ED (kJ g^{-1} WM)	Mean ED ± SD	Source
FISH					
MYCTOPHIDAE				7.6 ± 1.52	
Antarctic Lanternfish	8.1	7.4	9.0		[228]
"			7.9		[229]
"			8.6		[230]
Electron Subantarctic Lanternfish	7.2	5.5	6.1		[228]
"	7.9		7.0		[231]
"			6.6		[229]
"			6.9		[230]
"	2.4–5.3				[232]
Brauer's Lanternfish	10.1	9	9.3		[228]
Nichol's Lanternfish	14.9	37	10.3		[233]
"			5.8		[230]
Tenison's Lanternfish			5.5		[233]
"			6.1		[229]
"	1.2–3.2				[232]
Bolini's Lanternfish			5.8		[233]
"	5.6–6.0				[232]
Rhombic Lanternfish			8.8		[231]
"	4.8		8.1		[228]
"			8.2		[229]
"	2.8–5.5				[232]
Clupeoids				6.3 ± 0.40	
Sardine spp.			6.0		[234]
Sardine spp.			5.0–10.5		[235]
Anchovy spp.			6.6		[234]
European Anchovy			4.5–7.0		[235]
Southern Anchovy		5.5	5.2		[236]
Sandy Spratt		2.6	5.0		[236]
CARANGIDAE					
Horse Mackerel spp.			5.7	5.7	[234]
NOTOTHENIIDAE				5.5 ± 1 03	
Antarctic Silverfish	13.4	23.7	5.0		[228]
"	7.0, 12.0		5.2		[237]
Marbled Rockcod			6.0		[233]
Yellowbellied Rockcod			5.4		[233]
Grey Rockcod	22.4	291	4.0		[228]

	Length (cm)	Mass (g)	ED (kJ g^{-1} WM)	Mean ED ± SD	Source
Bald Rockcod	18.2	88.7	5.6		[228]
Emerald Rockcod	14.6	48.7	4.7		[228]
Striped Rockcod	21.1	148	5.4		[228]
Painted Notie			4.9		[229]
Blunt Scalyhead	19.6	121.8	5.7		[228]
Slender Scalyhead	27.4	217	7.7		[228]
CHANNICHTHYIDAE				4.7 ± 0.46	
Mackerel Icefish			4.9		[233]
"			5.4		[229]
"	12.1				[232]
Unicorn Icefish			4.0		[233]
South Georgia Icefish			4.6		[229]
Blackfin Icefish			4.6		[233]
Slender Escolar	16.9		4.6		[231]
MISCELLANEOUS					
Pelagic Goby			3.7	3.6 ± 0.11	[234]
Hake spp.			3.5		[234]
BATHYLAGIDAE					
Antarctic Deepsea Smelt	15.1	38.6	2.9	2.9	[233]
CRUSTACEANS					
EUPHAUSIACEA				4.5 ± 0.66	
Antarctic Krill			4.8		[233]
"			3.8		[238]
"			4.6		[229]
"			4.2		[230]
Crystal Krill			5.1		[233]
"	1.7–2.2, 2.5–3.5		4.6		[237]
Euphausia frigida			4.6		[233]
Thysanoessa macrura			5.1		[233]
"			5.0		[229]
Nyctiphanes australis		0.02	3.0		[236]
AMPHIPODS					
Themisto gaudichaudii			3.1		[229]
SQUID				4.9 ± 0.67	
Longfin Squid			3.1–5.0		[233]
Greater Hooked Squid			4.3–5.6		[233]
Argentine Shortfin Squid			5.0		[233]
Squid			3.2		[234]
Sevenstar Flying Squid	5.2–6.1				[232]
Giant Warty Squid	5.8–6.1				[232]
JELLYFISH				0.04	
Cyanea spp.		2.9	0.04		[236]

[a]Standard length in fish, mantle length in squid, total length in crustaceans

References

References for Appendix Table A1

1. Ainley DG, Ribic CA, Fraser WR (1992) Does prey preference affect habitat choice in Antarctic seabirds? Mar Ecol Prog Ser 90:207–221
2. Pütz K (1995) The post-moult diet of emperor penguins (*Aptenodytes forsteri*) in the eastern Weddell Sea, Antarctica. Polar Biol 15(7):457–463
3. Zimmer I, Piatkowski U, Brey T (2007) The trophic link between squid and the emperor penguin *Aptenodytes forsteri* at Pointe Géologie, Antarctica. Mar Biol 152(5) 1187–1195
4. Robertson G, Williams R, Green K, Robertson L (1994) Diet composition of emperor penguin chicks *Aptenodytes forsteri* at two Mawson Coast colonies, Antarctica. Ibis 136(1):19–31
5. Kirkwood R, Robertson G (1997) The foraging ecology of female emperor penguins in winter. Ecol Monogr 67(2):155–176
6. Gales RP, Pemberton D (1990) Seasonal and local variation in the diet of the little penguin, *Eudyptula minor*, in Tasmania. Wildl Res 17(3):231–259
7. Green K (1986) Food of the emperor penguin *Aptenodytes forsteri* on the Antarctic fast ice edge in late winter and early spring. Polar Biol 6(3):187–188
8. Offredo C, Ridoux V (1986) The diet of emperor penguins *Aptenodytes forsteri* in Adélie Land, Antarctica. Ibis 128(3):409–413
9. Offredo C, Ridoux V, Clarke MR (1985) Cephalopods in the diets of emperor and Adélie penguins in Adélie Land, Antarctica. Mar Biol 86(2):199–202
10. Cherel Y, Kooyman GL (1998) Food of emperor penguins (*Aptenodytes forsteri*) in the western Ross Sea, Antarctica. Mar Biol 130(3):335–344
11. Adams NJ, Klages NT (1987) Seasonal variation in the diet of the king penguin (*Aptenodytes patagonicus*) at sub-Antarctic Marion Island. J Zool 212(2):303–324
12. Ridoux V (1994) Diets and dietary segregation at subantarctic Crozet Islands. Mar Ornithol 22:1–64
13. Cherel Y, Ridoux V (1992) Prey species and nutritive value of food fed during summer to king penguin *Aptenodytes patagonica* chicks at Possession Island, Crozet Archipelago. Ibis 134(2):118–127
14. Cherel Y, Verdon C, Ridoux V (1993) Seasonal importance of oceanic myctophids in king penguin diet at Crozet Islands. Polar Biol 13(5):355–357
15. Bost CA, Georges JY, Guinet C, Cherel Y, Pütz K, Charrassin JB, Handrich Y, Zorn T, Lage J, Le Maho Y (1997) Foraging habitat and food intake of satellite-tracked king penguins during the austral summer at Crozet Archipelago. Mar Ecol Prog Ser 150:21–33
16. Bost CA, Zorn T, Le Maho Y, Duhamel G (2002) Feeding of diving predators and diel vertical migration of prey: King penguins' diet versus trawl sampling at Kerguelen Islands. Mar Ecol Prog Ser 227:51–61
17. Cherel Y, Bost CA, Guinet C, Weimerskirch H (2005) Feeding habits of seabirds and marine mammals of the Kerguelen Archipelago. Model Antarctic Mar Ecosyst Fish Centre Res Rep 13(7):31–36
18. Klages NTW, Pemberton D, Gales RP (1990) The diets of king and gentoo penguins at Heard Island. Wildl Res 17(1):53–60
19. Moore GJ, Robertson G, Wienecke B (1998) Food requirements of breeding king penguins at Heard Island and potential overlap with commercial fisheries. Polar Biol 20(5):293–302
20. Hindell MA (1988) The diet of the king penguin *Aptenodytes patagonicus* at Macquarie Island. Ibis 130(2):193–203
21. Cherel Y, Pütz A, Hobson KA (2002) Summer diet of king penguins (*Aptenodytes patagonicus*) at The Falkland Islands, southern Atlantic Ocean. Polar Biol 25(12):898–906
22. Olsson O, North AW (1997) Diet of the king penguin *Aptenodytes patagonicus* during three summers at South Georgia. Ibis 139(3):504–512

23. Van Heezik Y (1988) The growth and diet of the yellow-eyed penguin, *Megadyptes antipodes*. PhD dissertation. Dunedin, New Zealand: University of Otago
24. Van Heezik Y (1990) Seasonal, geographical, and age-related variations in the diet of the yellow-eyed penguin (*Megadyptes antipodes*). N Z J Ecol 17(2):201–212
25. Moore PJ, Wakelin MD (1997) Diet of the yellow-eyed penguin *Megadyptes antipodes*, South Island, New Zealand, 1991-1993. Mar Ornithol 25:17–29
26. Van Heezik Y (1990) Diets of yellow-eyed, Fiordland crested, and little blue penguins breeding sympatrically on Codfish Island, New Zealand. N Z J Ecol 17(4):543–548
27. Browne T, Lalas C, Mattern T, Van Heezik Y (2011) Chick starvation in yellow-eyed penguins: evidence for poor diet quality and selective provisioning of chicks from conventional diet analysis and stable isotopes. Austral Ecol 36(1):99–108
28. Lacock GD, Hecht T, Klages N (1984) The winter diet of gentoo penguins at Marion Island. Ostrich 55(4):188–191
29. Adams NJ, Klages NT (1989) Temporal variation in the diet of the gentoo penguin *Pygoscelis papua* at sub-Antarctic Marion Island. Colon Waterbirds 12:30–36
30. Adams NJ, Wilson MP (1987) Foraging parameters of gentoo penguins *Pygoscelis papua* at Marion Island. Polar Biol 7(1):51–56
31. Carpenter-Kling T, Handley JM, Connan M, Crawford RJM, Makhado AB, Dyer BM, Froneman W, Lamont T, Wolfaardt AC, Landman M, Sigqala M, Pistorius PA (2019) Gentoo penguins as sentinels of climate change at the sub-Antarctic Prince Edward Archipelago, Southern Ocean. Ecol Indic 101:163–172
32. Bost CA, Koubbi P, Genevois F, Ruchon L, Ridoux V (1994) Gentoo penguin *Pygoscelis papua* diet as an indicator of planktonic availability in the Kerguelen Islands. Polar Biol 14(3):147–153
33. Lescroël A, Ridoux V, Bost CA (2004) Spatial and temporal variation in the diet of the gentoo penguin (*Pygoscelis papua*) at Kerguelen Islands. Polar Biol 27(4):206–216
34. Hindell M (1989) The diet of gentoo penguin *Pygoscelis papua* at Macquarie Island: winter and early breeding season. Emu 89(2):71–78
35. Hindell M, Robertson GG, Williams R (1995) Resource partitioning in four species of Sympatrically breeding penguins. In: Dann P, Norman I, Reilly P (eds) The penguins: ecology and management. Surrey Beatty, Chipping Norton, pp 196–215
36. Robinson SA, Hindell MA (1996) Foraging ecology of gentoo penguins *Pygoscelis pupua* at Macquarie Island during the period of chick care. Ibis 138(4):722–731
37. Clausen A, Pütz K (2002) Recent trends in diet composition and productivity of gentoo, magellanic and rockhopper penguins in The Falkland Islands. Aquat Conserv Mar Freshwat Ecosyst 12(1):51–61
38. Pütz K, Ingham RJ, Smith JG, Croxall JP (2001) Population trends, breeding success and diet composition of gentoo *Pygoscelis papua*, Magellanic *Spheniscus magellanicus* and rockhopper *Eudyptes chrysocome* penguins in The Falkland Islands. Polar Biol 24(11):793–807
39. Handley JM, Baylis AMM, Brickle P, Pistorius P (2016) Temporal variation in the diet of gentoo penguins at The Falkland Islands. Polar Biol 39(2):283–296
40. Clausen AP, Arkhipkin AI, Laptikhovsky VV, Huin N (2005) What is out there: diversity in feeding of gentoo penguins (*Pygoscelis papua*) around The Falkland Islands (Southwest Atlantic). Polar Biol 28(9):653–662
41. Clausen A, Pütz K (2003) Winter diet and foraging range of gentoo penguins (*Pygoscelis papua*) from Kidney Cove, Falkland Islands. Polar Biol 26(1):32–40
42. Croxall JP, Prince PA (1980) The food of gentoo penguins *Pygoscelis papua* and macaroni penguins *Eudyptes chrysolophus* at South Georgia. Ibis 122(2):245–253
43. Kato A, Williams TD, Barton TR, Rodwell S (1991) Short-term variation in the winter diet of gentoo penguins *Pygoscelis papua* at South Georgia in July 1989. Mar Ornithol 19:31–38
44. Williams TD (1991) Foraging ecology and diet of gentoo penguins *Pygoscelis papua* at South Georgia during winter and an assessment of their winter prey consumption. Ibis 133(1):3–13

45. Berrow SD, Taylor RI, Murray AW (1999) Influence of sampling protocol on diet determination of gentoo penguins *Pygoscelis papua* and Antarctic fur seals *Arctocephalus gazella*. Polar Biol 22(3):156–163
46. Waluda CM, Hill SL, Peat HJ, Trathan PN (2017) Long-term variability in the diet and reproductive performance of penguins at Bird Island, South Georgia. Mar Biol 164:39
47. Xavier JC, Trathan PN, Ceia FR, Tarling GA, Adlard S, Fox D, Edwards EWJ, Vieira RP, Medeiros R, De Broyer C, Cherel Y (2017) Sexual and individual foraging segregation in gentoo penguins *Pygoscelis papua* from the Southern Ocean during an abnormal winter. PLoS One 12(3):e0174850
48. Coria N, Libertelli M, Casaux R, Darrieu C (2000) Inter-annual variation in the autumn diet of the gentoo penguin at Laurie Island, Antarctica. Waterbirds 23:511–517
49. Trivelpiece WZ, Trivelpiece SG, Volkman NJ (1987) Ecological segregation of Adélie, gentoo, and chinstrap penguins at King George Island, Antarctica. Ecology 68(2):351–361
50. Wawrzynek-Borejko J, Panasiuk A, Hinke JT, Korczak-Abshire M (2022) Are the diets of sympatric Pygoscelid penguins more similar than previously thought? Polar Biol. https://doi.org/10.1007/s00300-00022-03090-00309
51. Karnovsky NJ (1997) The fish component of *Pygoscelis* penguin diets. Msc thesis. Bozeman, Montana: Montana State University
52. Miller AK, Karnovsky NJ, Trivelpiece WZ (2009) Flexible foraging strategies of gentoo penguins *Pygoscelis papua* over 5 years in the South Shetland Islands, Antarctica. Mar Biol 156(12):2527–2537
53. Polito MJ, Trivelpiece WZ, Karnovsky NJ, Ng E, Patterson WP, Emslie SD (2011) Integrating stomach content and stable isotope analyses to quantify the diets of pygoscelid penguins. PLoS One 6(10):e26642
54. Polito MJ, Trivelpiece WZ, Patterson WP, Karnovsky NJ, Reiss CS, Emslie SD (2015) Contrasting specialist and generalist patterns facilitate foraging niche partitioning in sympatric populations of *Pygoscelis* penguins. Mar Ecol Prog Ser 519:221–237
55. Pickett E (2016) Foraging niche separation of Adélie (*Pygoscelis adeliae*) and gentoo (*P. papua*) penguins during the breeding season at Palmer Station, Antarctica. MSc thesis, Corvallis, Oregon: Oregon State University
56. Murphy RC (1936) Oceanic birds of South America. Macmillan Company, New York
57. Boersma PD (1977) An ecological and behavioral study of the Galapagos penguin. Living Bird 15:43–93
58. Boersma PD, Frere E, Kane O, Pozzi LM, Pütz K, Rey AA, Rebstock GA, Simeone A, Smith J, Van Buren A, Yorio P, Borboroglu PG (2013) Magellanic penguin (*Spheniscus magellanicus*). In: Borboroglu PG, Boersma PD (eds) Penguins: natural history and conservation. University of Washington Press, Seattle, pp 233–263
59. Zavalaga C, Paredes R (1997) Humboldt Penguins in Punta San Juan, Peru. Penguin Conserv 10:6–8
60. Herling C, Culik BM, Hennicke JC (2005) Diet of the Humboldt penguin (*Spheniscus humboldti*) in northern and southern Chile. Mar Biol 147(1):13–25
61. Wilson RP, Duffy DC, Wilson MP, Araya B (1995) Aspects of the ecology of species replacement in Humboldt and Magellanic penguins in Chile. Le Gerfaut 85:49–61
62. Wilson RP, Wilson MP, Duffy DC, Araya B, Klages N (1989) Diving behaviour and prey of the Humboldt penguin (*Spheniscus humboldti*). J Ornithol 130(1):75–79
63. Radl A, Culik BM (1999) Foraging behaviour and reproductive success in Magellanic penguins (*Spheniscus magellanicus*): a comparative study of two colonies in southern Chile. Mar Biol 133(3):381–393
64. Thompson KR (1993) Variation in Magellanic penguin *Spheniscus magellanicus* diet in The Falkland Islands. Mar Ornithol 21:57–67
65. Scioscia G, Rey AR, Saenz Samaniego RA, Florentín O, Schiavini A (2014) Intra- and interannual variation in the diet of the Magellanic penguin (*Spheniscus magellanicus*) at Martillo Island, Beagle Channel. Polar Biol 37(10):1421–1433

66. Frere E, Gandini P, Lichtschein V (1996) Variación latitudinal en la dieta del Pingüino de Magallanes (*Spheniscus magellanicus*) en la costa Patagónica, Argentina. Ornitología Neotropical 7:35–41
67. Wilson RP, Scolaro JA, Grémillet D, Kierspel MA, Laurenti S, Upton J, Gallelli H, Quintana F, Frere E, Müller G, Straten MT, Zimmer I (2005) How do Magellanic penguins cope with variability in their access to prey? Ecol Monogr 75(3):379–401
68. Gandini PA, Frere E, Pettovello AD, Cedrola PV (1999) Interaction between Magellanic penguins and shrimp fisheries in Patagonia, Argentina. Condor 101(4):783–789
69. Scolaro JA, Wilson RP, Laurenti S, Kierspel M, Gallelli H, Upton JA (1999) Feeding preferences of the Magellanic penguin over its breeding range in Argentina. Waterbirds 22:104–110
70. Scolaro JA, Badano LA (1986) Diet of the Magellanic penguin (*Spheniscus magellanicus*) during the chick rearing period at Punta Clara, Argentina. Cormorant 13:91–97
71. Crawford RJM, Shelton PA (1978) Pelagic fish and seabird interrelationships off the coasts of South West and South Africa. Biol Conserv 14(2):85–109
72. Ludynia K, Roux JP, Jones R, Kemper J, Underhill LG (2010) Surviving off junk: low-energy prey dominates the diet of African penguins *Spheniscus demersus* at Mercury Island, Namibia, between 1996 and 2009. Afr J Mar Sci 32(3):563–572
73. Rand RW (1960) The biology of guano-producing seabirds. The distribution, abundance and feeding habits of the Cape penguin, *Spheniscus demersus*, off the southwestern coast of the Cape Province. Investig Rep Div Fish S Afr 41:1–28
74. Crawford RJM, Dyer BM (1995) Responses by four seabird species to a fluctuating availability of Cape anchovy *Engraulis capensis* off South Africa. Ibis 137(3):329–339
75. Petersen SL, Ryan PG, Grémillet D (2006) Is food availability limiting African penguins *Spheniscus demersus* at boulders? A comparison of foraging effort at mainland and Island colonies. Ibis 148(1):14–26
76. Randall RM, Randall BM (1986) The diet of jackass penguins *Spheniscus demersus* in Algoa Bay, South Africa, and its bearing on population declines elsewhere. Biol Conserv 37(2):119–134
77. Duffy DC, Wilson RP, Ricklefs RE, Broni SC, Veldhuis H (1987) Penguins and purse seiners: competition or coexistence? Natl Geogr Res 3(4):480–488
78. Wilson RP (1985) Seasonality in diet and breeding success of the jackass penguin *Spheniscus demersus*. J Ornithol 126(1):53–62
79. Klomp NI, Wooller RD (1988) Diet of little penguins, *Eudyptula minor*, from Penguin Island, Western Australia. Mar Freshw Res 39(5):633–639
80. Cullen JM, Montague TL, Hull C (1991) Food of little penguins *Eudyptula minor* in Victoria: comparison of three localities between 1985 and 1988. Emu 91(5):318–341
81. Kowalczyk ND, Chiaradia A, Preston TJ, Reina RD (2015) Fine-scale dietary changes between the breeding and non-breeding diet of a resident seabird. R Soc Open Sci 2(4):140291
82. Montague TL (1982) The food and feeding ecology of the little penguin *Eudyptula minor* at Phillip Island, Victoria, Australia. MSc thesis. Melbourne, Australia: Monash University
83. Montague TL, Cullen JM (1988) The diet of the little penguin *Eudyptula minor* at Phillip Island, Victoria. Emu 88(3):138–149
84. Chiaradia A (1999) Breeding biology and feeding ecology of little penguins *Eudyptula minor* at Phillip Island – A basis for a monitoring program. PhD dissertation. Hobart, Australia: University of Tasmania
85. Chiaradia A, Costalunga A, Kerry K (2003) The diet of little penguins (*Eudyptula minor*) at Phillip Island, Victoria, in the absence of a major prey – pilchard (*Sardinops sagax*). Emu 103(1):43–48
86. Fraser MM, Lalas C (2004) Seasonal variation in the diet of blue penguins (*Eudyptula minor*) at Oamaru, New Zealand. Notornis 51(1):7–15
87. Flemming SA, Lalas C, Van Heezik Y (2013) Little penguin (*Eudyptula minor*) diet at three breeding colonies in New Zealand. N Z J Ecol 37:199–205

88. Sladen WJL (1958) The Pygoscelid penguins. I. Methods of study. II. The Adélie penguin *Pygoscelis adeliae* (Hombron and Jacquinot). London
89. White MG, Conroy JWH (1975) Aspects of competition between pygsocelid penguins at Signy Island, South Orkney Islands. Ibis 117(3):371–373
90. Lishman GS (1985) The food and feeding ecology of Adélie penguins (*Pygoscelis adeliae*) and chinstrap penguins (*P. Antarctica*) at Signy Island, South Orkney Islands. J Zool 205(2):245–263
91. Libertelli MM, Coria N, Marateo G (2003) Diet of the Adélie penguin during three consecutive chick rearing periods at Laurie Island. Polish Polar Res 24:133–142
92. Lynnes AS, Reid K, Croxall JP (2004) Diet and reproductive success of Adélie and chinstrap penguins: linking response of predators to prey population dynamics. Polar Biol 27(9):544–554
93. Puddicome RA, Johnstone GW (1988) The breeding season diet of Adélie penguins at the Vestfold Hills, East Antarctica. Hydrobiologia 165(1):239–253
94. Volkman NJ, Presler P, Trivelpiece WZ (1980) Diets of pygoscelid penguins at King George Island, Antarctica. Condor 82(4):373–378
95. Juáres MA, Casaux R, Corbalán A, Blanco G, Pereira GA, Perchivale PJ, Coria NR, Santos MM (2018) Diet of Adélie penguins (*Pygoscelis adeliae*) at stranger point (25 de Mayo/King George Island, Antarctica) over a 13-year period (2003–2015). Polar Biol 41(2):303–311
96. Coria NR, Spairani H, Vivequin S, Fontana R (1995) Diet of Adélie penguins *Pygoscelis adeliae* during the post-hatching period at Esperanza Bay, Antarctica, 1987/88. Polar Biol 15(6):415–418
97. Fraser WR, Ainley DG (1992) Monitoring activities at Palmer Station, Antarctica. National Marine Fisheries Service, Southwest Fisheries Science Center, La Jolla
98. Gorman KB (2015) Integrative studies of southern Ocean food-webs and *Pygoscelis* penguin demography: mechanisms of population response to environmental change. PhD dissertation. Simon Fraser University, Burnaby, British Columbia
99. Watanuki Y, Kato A, Naito Y, Robertson G, Robinson S (1997) Diving and foraging behaviour of Adélie penguins in areas with and without fast sea-ice. Polar Biol 17:296–304
100. Watanuki Y, Kato A, Mori Y, Naito Y (1993) Diving performance of Adélie penguins in relation to food availability in fast sea-ice areas: comparison between years. J Anim Ecol 62:634–646
101. Kerry KR, Clarke JR, Else GD (1995) The foraging range of Adélie penguins at Béchervaise Island, MacRobertson land, Antarctica, as determined by satellite telemetry. In: Dann P, Norman I, Reilly P (eds) The penguins: ecology and management. Surrey Beatty, Chipping Norton, pp 216–243
102. Kent S, Seddon J, Robertson G, Wienecke BC (1998) Diet of Adélie penguins at Shirley Island, East Antarctica, January 1992. Mar Ornithol 26:7–10
103. Clarke J, Kerry K, Irvine L, Phillips B (2002) Chick provisioning and breeding success of Adélie penguins at Béchervaise Island over eight successive seasons. Polar Biol 25(1):21–30
104. Tierney M, Emmerson L, Hindell M (2009) Temporal variation in Adélie penguin diet at Béchervaise Island, East Antarctica and its relationship to reproductive performance. Mar Biol 156(8):1633–1645
105. Penney RL (1968) Territorial and social behavior in the Adélie penguin. In: Austin OL (ed) Antarctic bird studies. American Geophysics Union, Washington, DC, pp 83–131
106. Cowan AN (1983) A modified penguin stomach tube. Corelia 7:59–61
107. Wienecke BC, Lawless R, Rodary D, Bost C-A, Thomson R, Pauly T, Robertson G, Kerry KR, Lemaho Y (2000) Adélie penguin foraging behaviour and krill abundance along the Wilkes and Adélie land coasts, Antarctica. Deep Sea Res II 47:2573–2587
108. Cherel Y (2008) Isotopic niches of emperor and Adélie penguins in Adélie Land, Antarctica. Mar Biol 154(5):813–821
109. Levick GM (1915) Natural History of the Adélie Penguin. In: British Antarctic ("Terra Nova") Expedition, 1910–1913. British Museum of Natural History, pp 55–84

110. Wilson EA (1907) Aves. John Murray, London
111. Lyver POB, Macleod CJ, Ballard G, Karl BJ, Barton KJ, Adams J, Ainley DG, Wilson PR (2011) Intra-seasonal variation in foraging behavior among Adélie penguins (*Pygoscelis adeliae*) breeding at Cape Hallett, Ross Sea, Antarctica. Polar Biol 34(1):49–67
112. Emison WB (1968) Feeding preferences of the Adélie penguin at cape crozier, Ross Island. In: Austin OL (ed) Antarctic bird studies. American Geophysics Union, Washington, DC, pp 191–212
113. Ainley DG, Ballard G, Barton KJ, Karl BJ, Rau GH, Ribic CA, Wilson PR (2003) Spatial and temporal variation of diet composition and quality within a presumed metapopulation of Adélie penguins. Condor 105(1):95–106
114. Ainley DG, Dugger KM, La Mesa M, Ballard G, Barton KJ, Jennings S, Karl BJ, Lescroël A, Lyver POB, Schmidt A, Wilson P (2018) Post-fledging survival of Adélie penguins at multiple colonies: chicks raised on fish do well. Mar Ecol Prog Ser 601:239–251
115. Ainley DG, Wilson PR, Barton KJ, Ballard G, Nur N, Karl B (1998) Diet and foraging effort of Adélie penguins in relation to pack-ice conditions in the southern Ross Sea. Polar Biol 20(5):311–319
116. Paulin CD (1975) Feeding off the Adélie penguin *Pygoscelis adeliae*. Mäuri Ora 3:27–30
117. Van Heezik Y (1988) Diet of Adélie penguins during the incubation period at cape bird, Ross Island, Antarctica. Notornis 35(1):23–26
118. Ainley DG, O'connor EF, Boekelheide RJ (1984) Ecology of seabirds in the Ross Sea, Antarctica. Ornithol Monogr 32:iii–97
119. Falla RA (1937) Birds. In: B.A.N.Z. Antarctic research expedition, 1929–1931, Reports-series B. Hassell Press, Adelaide
120. Rombolá E, Marschoff E, Coria N (2006) Interannual study of chinstrap penguin's diet and reproductive success at Laurie Island, South Orkney Islands, Antarctica. Polar Biol 29(6):502–509
121. Croxall JP, Furse JR (1990) Food of chinstrap penguins *Pygoscelis antarctica* and macaroni penguins *Eudyptes chrysolophus* at Elephant Island Group, South Shetland Islands. Ibis 122(2):237–245
122. Jansen JK, Boveng PL, Bengston JL (1998) Foraging modes of chinstrap penguins: contrasts between day and night. Mar Ecol Prog Ser 165:161–172
123. Ichii T, Bengston JL, Boveng PL, Takao Y, Jansen JK, Hiruki-Raring LM, Cameron MF, Okamura H, Hayashi T, Naganobu M (2007) Provisioning strategies of Antarctic fur seals and chinstrap penguins produce different responses to distribution of common prey and habitat. Mar Ecol Prog Ser 344:277–297
124. Rombolá E, Marschoff E, Coria N (2010) Inter-annual variability in chinstrap penguin diet at south Shetland and south Orkneys Islands. Polar Biol 33(6):799–806
125. Brown CR, Klages NT (1987) Seasonal and annual variation in diets of macaroni (*Eudyptes chrysolophus*) and southern rockhopper (*Eudyptes crestatus*) penguins at sub-Antarctic Marion Island. J Zool 212(1):7–28
126. Crawford RJM, Cooper J, Dyer BM (2003) Population of the macaroni penguin *Eudyptes chrysolophus* at Marion Island, 1994/95–2002/03, with Information on breeding and diet. Afr J Mar Sci 25(1):475–486
127. Klages NTW, Gales RP, Pemberton D (1989) Dietary segregation of macaroni and rockhopper penguins at Heard Island. Wildl Res 16(6):599–604
128. Deagle BE, Gales NJ, Evans K, Jarman SN, Robinson S, Trebilco R, Hindell MA (2007) Studying seabird diet through genetic analysis of faeces: a case study on macaroni penguins (*Eudyptes chrysolophus*). PLoS One 2(9):e831
129. Deagle BE, Gales NJ, Hindell MA (2008) Variability in foraging behaviour of chick-rearing macaroni penguins *Eudyptes chrysolophus* and its relation to diet. Mar Ecol Prog Ser 359:295–309
130. Waluda CM, Hill SL, Peat HJ, Trathan PN (2012) Diet variability and reproductive performance of macaroni penguins *Eudyptes chrysolophus* at Bird Island, South Georgia. Mar Ecol Prog Ser 466:261–274

131. Klages NT, Brooke MDL, Watkins BP (1988) Prey of northern rockhopper penguins at Gough Island, South Atlantic Ocean. Ostrich 59(4):162–165
132. Cherel Y, Tremblay Y, Guinard E, Georges JY (1999) Diving behavior of female northern rockhopper penguins, *Eudyptes chrysocome moseleyi*, during the brooding period at Amsterdam Island (Southern Indian Ocean). Mar Biol 134(2):375–385
133. Horne RSC (1985) Diet of royal and rockhopper penguins at Macquarie Island. Emu 85(3):150–156
134. Hindell MA (1988) The diet of the rockhopper penguin *Eudyptes chrysocome* at Macquarie Island. Emu 88(4):227–233
135. Xavier JC, Cherel Y, Ceia FR, Queirós JP, Guimarães B, Rosa R, Cunningham DM, Moors PJ, Thompson DR (2018) Eastern rockhopper penguins *Eudyptes filholi* as biological samplers of juvenile and sub-adult cephalopods around Campbell Island, New Zealand. Polar Biol 41(10):1937–1949
136. Croxall JP, Prince PA, Baird A, Ward P (1985) The diet of the southern rockhopper penguin *Eudyptes chysocome chysocome* at Beauchene Island, Falkland Islands. J Zool 206(4):485–496
137. Thompson KR (1989) An assessment of the potential for competition between seabirds and fisheries in The Falkland Islands. Falkland Islands Foundation, Brighton
138. Rey AR, Schiavini A (2005) Inter-annual variation in the diet of female southern rockhopper penguin (*Eudyptes chrysocome chrysocome*) at Tierra del Fuego. Polar Biol 28(2):132–141
139. Schiavini A, Rey AR (2004) Long days, long trips: foraging ecology of female rockhopper penguins *Eudyptes chrysocome chrysocome* at Tierra del Fuego. Mar Ecol Prog Ser 275:251–262
140. Hindell MA (1988) The diet of the royal penguin *Eudyptes schlegeli* at Macquarie Island. Emu 88(4):219–226
141. Van Heezik Y (1989) Diet of the Fiordland crested penguin during the post-guard phase of chick growth. Notornis 36:151–156
142. Warham J (1974) The Fiordland crested penguin *Eudyptes pachyhynchus*. Ibis 116(1):1–27
143. Mattern T, Houston DM, Lalas C, Setiwan AN, Davis LS (2009) Diet composition, continuity in prey availability and marine habitat – keystones to population stability in the snares penguin (*Eudyptes robustus*). Emu 109(3):204–213

References for Appendix Table A3

144. Kirkwood R, Robertson G (1997) The foraging ecology of female emperor penguins in winter. Ecol Monogr 67(2):155–176
145. Ancel A, Kooymant GL, Ponganis PJ, Gendner J-P, Lignon J, Mestre X, Huin N, Thorsont PH, Robisson P, Le Maho Y (1992) Foraging behaviour of emperor penguins as a resource detector in winter and summer. Nature 360:336–339
146. Wienecke B, Robertson G, Kirkwood R, Lawton K (2007) Extreme dives by free-ranging emperor penguins. Polar Biol 30(2):133–142
147. Zimmer I, Wilson RP, Gilbert C, Beaulieu M, Ancel A, Plötz J (2008) Foraging movements of emperor penguins at Pointe Géologie, Antarctica. Polar Biol 31(2):229–243
148. Kooyman GL, Goetz K, Williams CL, Ponganis PJ, Sato K, Eckert S, Horning M, Thorson PT, Van Dam RP (2020) Crary bank: a deep foraging habitat for emperor penguins in the western Ross Sea. Polar Biol 43(7):801–811
149. Scheffer A, Trathan PN, Edmonston JG, Bost CA (2016) Combined influence of meso-scale circulation and bathymetry on the foraging behaviour of a diving predator, the king penguin (*Aptenodytes patagonicus*). Prog Oceanogr 141:1–16
150. Pütz K (2002) Spatial and temporal variability in the foraging areas of breeding king penguins. Condor 104(3):528–538

151. Bost CA, Zorn T, Le Maho Y, Duhamel G (2002) Feeding of diving predators and diel vertical migration of prey: King penguins' diet versus trawl sampling at Kerguelen Islands. Mar Ecol Prog Ser 227:51–61
152. Bost CA, Georges JY, Guinet C, Cherel Y, Pütz K, Charrassin JB, Handrich Y, Zorn T, Lage J, Le Maho Y (1997) Foraging habitat and food intake of satellite-tracked king penguins during the austral summer at Crozet Archipelago. Mar Ecol Prog Ser 150:21–33
153. Bosta C-A, Handrichb Y, Butler PJ, Fahlman A, Halsey LG, Woakes AJ, Ropert-Coudert Y (2007) Changes in dive profiles as an indicator of feeding success in king and Adelie penguins. Deep-Sea Res II 54:248–255
154. Charrassin JB, Park Y-H, Le Maho YL, Bost CA (2002) Penguins as oceanographers unravel hidden mechanisms of marine productivity. Ecol Lett 5(3):317–319
155. Charrassin JB, Bost CA (2001) Utilisation of the oceanic habitat by king penguins over the annual cycle. Mar Ecol Prog Ser 221:285–297
156. Pütz K, Ropert-Coudert Y, Charrassin JB, Wilson RP (1999) Foraging areas of king penguins *Aptenodytes patagonicus* breeding at Possession Island, southern Indian Ocean. Mar Ornithol 27:77–84
157. Moore GJ, Wienecke B, Robertson G (1999) Seasonal change in foraging areas and dive depths of breeding king penguins at Heard Island. Polar Biol 21(6):376–384
158. Trathan PN, Bishop C, Maclean G, Brown P, Fleming A, Collins MA (2008) Linear tracks and restricted temperature ranges characterise penguin foraging pathways. Mar Ecol Prog Ser 370:285–294
159. Chappell MA, Shoemaker VH, Janes DN, Bucher TL, Maloney SK (1993) Diving behavior during foraging in breeding Adélie Penguins. Ecology 74:1204–1215
160. Oliver MJ, Irwin A, Moline MA, Fraser W, Patterson D, Schofield O, Kohut J (2013) Adelie penguin foraging location predicted by tidal regime switching. PLoS One 8(1):e55163
161. Cimino MA, Moline MA, Fraser WR, Patterson-Fraser DL, Oliver MJ (2016) Climate-driven sympatry may not lead to foraging competition between congeneric top-predators. Sci Rep 6:18820
162. Pickett E (2016) Foraging niche separation of Adélie (*Pygoscelis adeliae*) and gentoo (*P. papua*) penguins during the breeding season at Palmer Station, Antarctica. MSc thesis, Corvallis, Oregon: Oregon State University
163. Trivelpiece WZ, Trivelpiece SG, Volkman NJ (1987) Ecological segregation of Adélie, gentoo, and chinstrap penguins at King George Island, Antarctica. Ecology 68(2):351–361
164. Hinke JT, Cossio AM, Goebel ME, Reiss CS, Trivelpiece WZ, Watters GM (2017) Identifying risk: concurrent overlap of the Antarctic Krill fishery with krill-dependent predators in the Scotia Sea. PLoS One 12(1):e0170132
165. Lynnes AS, Reid K, Croxall JP, Trathan PN (2002) Conflict or co-existence? Foraging distribution and competition for prey between Adélie and chinstrap penguins. Mar Biol 141(6):1165–1174
166. Kato A, Watanuki Y, Naito Y (2003) Annual and seasonal changes in foraging site and diving behavior in Adelie penguins. Polar Biol 26:389–395
167. Watanuki Y, Kato A, Naito Y, Robertson G, Robinson S (1997) Diving and foraging behaviour of Adélie penguins in areas with and without fast sea-ice. Polar Biol 17:296–304
168. Wienecke BC, Lawless R, Rodary D, Bost C-A, Thomson R, Pauly T, Robertson G, Kerry KR, Le Maho Y (2000) Adélie penguin foraging behaviour and krill abundance along the Wilkes and Adélie land coasts, Antarctica. Deep Sea Res II 47:2573–2587
169. Clarke J, Manly B, Kerry K, Gardner H, Franchi E, Corsolini S, Focardi S (1998) Sex differences in Adélie penguin foraging strategies. Polar Biol 20(4):248–258
170. Clarke J, Kerry K (1994) The effect of monitoring procedures on Adelie penguins. CCAMLR Sci 1:155–164
171. Widmann M, Kato A, Raymond B, Angelier F, Arthur B, Chastel O, Pellé M, Raclot T, Ropert-Coudert Y (2015) Habitat use and sex-specific foraging behaviour of Adélie penguins throughout the breeding season in Adélie Land, East Antarctica. Mov Ecol 3:30

172. Lyver POB, Macleod CJ, Ballard G, Karl BJ, Barton KJ, Adams J, Ainley DG, Wilson PR (2011) Intra-seasonal variation in foraging behavior among Adélie penguins (*Pygoscelis adeliae*) breeding at Cape Hallett, Ross Sea, Antarctica. Polar Biol 34(1):49–67
173. Ballard G, Schmidt AE, Toniolo V, Veloz S, Jongsomjit D, Arrigo KR, Ainley DG (2019) Fine-scale oceanographic features characterizing successful Adélie penguin foraging in the SW Ross Sea. Mar Ecol Prog Ser 608:263–277
174. Ainley DG, Ballard G, Jones RM, Jongsomjit D, Pierce SD, Smith WO Jr, Veloz S (2015) Trophic cascades in the western Ross Sea, Antarctica: revisited. Mar Ecol Prog Ser 534:1–16
175. Ford RG, Ainley DG, Lescroël A, Lyver POB, Toniolo V, Ballard G (2015) Testing assumptions of central place foraging theory: a study of Adélie penguins *Pygoscelis adeliae* in the Ross Sea. J Avian Biol 46(2):193–205
176. Ainley DG, Ballard G, Dugger KM (2006) Competition among penguins and cetaceans reveals trophic cascades in the Ross Sea, Antarctica. Ecology 87(8):2080–2093
177. Lescroël A, Lyver POB, Jongsomjit D, Veloz S, Dugger KM, Kappes P, Karl BJ, Whitehead AL, Pech R, Cole TL, Ballard G (2020) Inter-individual differences in the foraging behavior of breeding Adélie penguins are driven by individual quality and sex. Mar Ecol Prog Ser 636:189–205
178. Saenz BT, Ainley DG, Daly KL, Ballard G, Conlisk E, Elrod ML, Kim SL (2020) Drivers of concentrated predation in an Antarctic marginal-ice-zone food web. Sci Rep 10:7282
179. Masello JF, Mundry R, Poisbleau M, Demongin L, Voigt CC, Wikelski M, Quillfeldt P (2010) Diving seabirds share foraging space and time within and among species. Ecosphere 1(6):1–28
180. Tanton JL, Reid K, Croxall JP, Trathan PN (2004) Winter distribution and behaviour of gentoo penguins *Pygoscelis papua* at South Georgia. Polar Biol 27(5):299–303
181. Miller AK, Karnovsky NJ, Trivelpiece WZ (2009) Flexible foraging strategies of gentoo penguins *Pygoscelis papua* over 5 years in the South Shetland Islands, Antarctica. Mar Biol 156(12):2527–2537
182. Adams NJ, Wilson MP (1987) Foraging parameters of gentoo penguins *Pygoscelis papua* at Marion Island. Polar Biol 7(1):51–56
183. Robinson SA, Hindell MA (1996) Foraging ecology of gentoo penguins *Pygoscelis pupua* at Macquarie Island during the period of chick care. Ibis 138(4):722–731
184. Lescroël A, Bost CA (2005) Foraging under contrasting oceanographic conditions: the gentoo penguin at Kerguelen Archipelago. Mar Ecol Prog Ser 302:245–261
185. Carpenter-Kling T, Handley JM, Green DB, Reisinger RR, Makhado AB, Crawford RJM, Pistorius PA (2017) A novel foraging strategy in gentoo penguins breeding at sub-Antarctic Marion Island. Mar Biol 164:33
186. Camprasse E, Cherel Y, Bustamante P, Arnould J, Bost CA (2017) Intra and inter-individual variation in the foraging ecology of a generalist subantarctic seabird, the gentoo penguin. Mar Ecol Prog Ser 578:227–242
187. Takahashi A, Dunn MJ, Trathan PN, Sato K, Naito Y, Croxall JP (2003) Foraging strategies of chinstrap penguins at Signy Island, Antarctica: importance of benthic feeding on Antarctic krill. Mar Ecol Prog Ser 250:278–289
188. Phillips JA, Fayet AL, Guilford T, Manco F, Warwick-Evans V, Trathan P (2021) Foraging conditions for breeding penguins improve with distance from colony and progression of the breeding season at the South Orkney Islands. Mov Ecol 9:22
189. Bengtson JL, Croll DA, Goebel ME (1993) Diving behaviour of chinstrap penguins at Seal Island. Antarct Sci 5:9–15
190. Kokubon N, Lee WY, Kim J-H, Takahashi A (2015) Chinstrap penguin foraging area associated with a seamount in Bransfield Strait, Antarctica. Pol Sci 9:393–400
191. Wilson RP, Peters G (1999) Foraging behaviour of the chinstrap penguin *Pygoscelis Antarctica* at Ardley Island, Antarctica. Mar Ornithol 27:85–95
192. Blanchet MA, Biuw M, Hofmeyr GG, De Bruyn PN, Lydersen C, Kovacs KM (2013) At-sea behaviour of three krill predators breeding at Bouvetøya—Antarctic fur seals, macaroni penguins and chinstrap penguins. Mar Ecol Prog Ser 477:285–302

193. Mattern T, Ellenberg U, Houston DM, Davis LS (2007) Consistent foraging routes and benthic foraging behaviour in yellow-eyed penguins. Mar Ecol Prog Ser 343:295–306
194. Mattern T (2020) Modelling marine habitat utilisation by yellow-eyed penguins along their mainland distribution: baseline information. Fisheries New Zealand, Wellington
195. Moore PJ (1999) Foraging range of the yellow-eyed penguin *Megadyptes antipodes*. Mar Ornithol 27:49–58
196. Chilvers BL, Dobbins ML, Edmonds HK (2014) Diving behaviour of yelloweyed penguins, Port Pegasus/Pikihatiti, Stewart Island/Rakiura, New Zealand. N Z J Zool 41:161–170
197. Elley T, Mattern T, Ellenberg U, Young MJ, Hickcox RP, Van Heezik Y, Seddon PJ (2022) Consistent site-specific foraging behaviours of yellow-eyed penguins/hoiho breeding on Stewart Island, New Zealand. Biology 11(6):844
198. Muller CG, Chilvers BL, Chiaradia A, French RK, Kato A, Ropert-Coudert Y, Battley PF (2021) Foraging areas and plasticity of yellow-eyed penguins *Megadyptes antipodes* in their subantarctic range. Mar Ecol Prog Ser 679:149–162
199. Muller CG, Chilvers BL, French RK, Battley PF (2020) Diving plasticity in the ancestral range of the yellow-eyed penguin *Megadyptes antipodes*, an endangered marine predator. Mar Ecol Prog Ser 648:191–205
200. Van Heezik Y (1989) Diet of the Fiordland crested penguin during the post-guard phase of chick growth. Notornis 36:151–156
201. Cherel Y, Tremblay Y, Guinard E, Georges JY (1999) Diving behavior of female northern rockhopper penguins, *Eudyptes chrysocome moseleyi*, during the brooding period at Amsterdam Island (Southern Indian Ocean). Mar Biol 134(2):375–385
202. Rosciano NG, Polito MJ, Rey AR (2016) Do penguins share? Evidence of foraging niche segregation between but not within two sympatric, central-place foragers. Mar Ecol Prog Ser 548:249–262
203. Brown CR (1987) Traveling speed and foraging range of macaroni and rockhopper penguins at Marion Island. J Field Ornithol 58:118–125
204. Ludynia K, Dehnhard N, Poisbleau M, Demongin L, Masello JF, Quillfeldt P (2012) Evaluating the impact of handling and logger attachment on foraging parameters and physiology in southern rockhopper penguins. PLoS One 7(11):e50429
205. Green JA, Wilson RP, Boyd IL, Woakes AJ, Green CJ, Butler PJ (2009) Tracking macaroni penguins during long foraging trips using behavioural geolocation. Polar Biol 32(4):645–653
206. Deagle BE, Gales NJ, Hindell MA (2008) Variability in foraging behaviour of chick-rearing macaroni penguins *Eudyptes chrysolophus* and its relation to diet. Mar Ecol Prog Ser 359:295–309
207. Hull CL, Hindell MA, Michael K (1997) Foraging zones of royal penguins during the breeding season, and their association with oceanographic features. Mar Ecol Prog Ser 153:217–228
208. Mills KL (1998) Multispecies seabird feeding flocks in the Galápagos Islands. Condor 100(2):277–285
209. Culik BM, Luna-Jorquera G (1997) The Humboldt penguin *Spheniscus humboldti*: a migratory bird? J Ornithol 138:325–329
210. Luna-Jorquera G, Culik BM (1997) Diving behaviour of Humboldt penguins *Spheniscus humboldti* in northern Chile. Mar Ornithol 27:67–76
211. Boersma D, Rebstock GA, Stokes DL, Majluf P (2007) Oceans apart: conservation models for two temperate penguin species shaped by the marine environment. Mar Ecol Prog Ser 335:217–225
212. Boersma PD, Rebstock GA (2009) Foraging distance affects reproductive success in Magellanic penguins. Mar Ecol Prog Ser 375:263–275
213. Wilson RP, Scolaro JA, Grémillet D, Kierspel MA, Laurenti S, Upton J, Gallelli H, Quintana F, Frere E, Müller G, Straten MT, Zimmer I (2005) How do Magellanic penguins cope with variability in their access to prey? Ecol Monogr 75(3):379–401
214. Raya Rey A, Bost C-A, Schiavini A, Pütz K (2010) Foraging movements of Magellanic penguins *Spheniscus magellanicus* in the Beagle Channel, Argentina, related to tide and tidal currents. J Ornithol. https://doi.org/10.1007/s10336-010-0531-y

215. Rosciano NG, Pütz K, Polito MJ, Rey AR (2018) Foraging behaviour of Magellanic penguins during the early chick-rearing period at Isla de los Estados, Argentina. Ibis 160(2):327–341
216. Heath RGM, Randall RM (1989) Foraging ranges and movements of jackass penguins (*Spheniscus dernersus*) established through radio telemetry. J Zool (Lond) 217:367–379
217. Pichegru L, Grémillet D, Crawford RJM, Ryan PG (2010) Marine no-take zone rapidly benefit threatened penguin. Biol Lett 6(4):498–501
218. Petersen SL, Ryan PG, Grémillet D (2006) Is food availability limiting African penguins *Spheniscus demersus* at Boulders? A comparison of foraging effort at mainland and Island colonies. Ibis 148(1):14–26
219. Gales RP (1989) Feeding ecology and free-living energetics of the little penguin, *Eudyptula minor*, in Tasmania. University of Tasmania, Hobart
220. Weavers BW, Dowling B, Smout AD (1991) Radio-tracking the little penguin (*Eudyptula minor*) off Phillip Island, South-Eastern Australia: techniques and field data. Arthur Rylah Institute for Environmental Research
221. Hoskins AJ, Dann P, Ropert-Coudert Y, Kato A, Chiaradia A, Costa DP, Arnould JPY (2008) Foraging behaviour and habitat selection of the little penguin *Eudyptula minor* during early chick rearing in Bass Strait, Australia. Mar Ecol Prog Ser 366:293–303
222. Mccutcheon C, Dann P, Salton M, Renwick L, Hoskins AJ, Gormley AM, Arnould JPY (2011) The foraging range of little penguins (*Eudyptula minor*) during winter. Emu 111(4):321–329
223. Pelletier L, Chiaradia A, Kato A, Ropert-Coudert Y (2014) Fine-scale spatial age segregation in the limited foraging area of an inshore seabird species, the little penguin. Oecologia 176(2):399–408
224. Kato A, Ropert-Coudert Y, Grémillet D, Cannell B (2006) Locomotion and foraging strategy in foot-propelled and wing-propelled shallow-diving seabirds. Mar Ecol Prog Ser 308:293–301
225. Ropert-Coudert Y, Knott N, Chiaradia A, Kato A (2007) How do different data logger sizes and attachment positions affect the diving behaviour of little penguins? Deep Sea Res II 54:415–423
226. Agnew P (2014) Demographic parameters, foraging and responses to environmental variation of little penguins (*Eudyptula minor*). PhD dissertation. Dunedin, New Zealand: University of Otago
227. Mattern T (2001) Foraging strategies and breeding success in the little penguin, *Eudyptula minor*: a comparative study between different habitats. MSc thesis, Dunedin, New Zealand: University of Otago

References for Appendix Table A4

228. Lenky C, Eisert R, Oftedal OT, Metcalf V (2012) Proximate composition and energy density of nototheniid and myctophid fish in McMurdo Sound and the Ross Sea, Antarctica. Polar Biol 35(5):717–724
229. Waluda CM, Hill SL, Peat HJ, Trathan PN (2012) Diet variability and reproductive performance of macaroni penguins *Eudyptes chrysolophus* at Bird Island, South Georgia. Mar Ecol Prog Ser 466:261–274
230. Ichii T, Bengston JL, Boveng PL, Takao Y, Jansen JK, Hiruki-Raring LM, Cameron MF, Okamura H, Hayashi T, Naganobu M (2007) Provisioning strategies of Antarctic fur seals and chinstrap penguins produce different responses to distribution of common prey and habitat. Mar Ecol Prog Ser 344:277–297
231. Cherel Y, Ridoux V (1992) Prey species and nutritive value of food fed during summer to king penguin *Aptenodytes patagonica* chicks at Possession Island, Crozet Archipelago. Ibis 134(2):118–127

232. Cooper J, Brown CR, Gales RP, Hindell MA, Klages NT, Moors PJ, Van Heezik Y (1990) Diets and dietary segregation of crested penguins (*Eudyptes*). In: Davis LS, Darby JT (eds) Penguin Biology. Academic, San Diego, pp 131–156
233. Schaafsma FL, Cherel Y, Flores H, Van Franeker JA, Lea M-A, Raymond B, Van De Putte AP (2018) Review: the energetic value of zooplankton and nekton species of the Southern Ocean. Mar Biol 165(129). https://doi.org/10.1007/s00227-00018-03386-z
234. Ludynia K, Roux JP, Jones R, Kemper J, Underhill LG (2010) Surviving off junk: low-energy prey dominates the diet of African penguins *Spheniscus demersus* at Mercury Island, Namibia, between 1996 and 2009. Afr J Mar Sci 32(3):563–572
235. Campanini C, Albo-Puigserver M, Gérez S, Lloret-Lloret E, Giménez J, Pennino MG, Bellido JM, Colmenero AI, Coll M (2021) Energy content of anchovy and sardine using surrogate calorimetry methods. bioRxiv. https://doi.org/10.1101/2021.02.01.429042
236. Sutton GJ, Hoskins AJ, Arnould JPY (2015) Benefits of group foraging depend on prey type in a small marine predator, the Little Penguin. PLoS One 10:e0144297
237. Ainley DG, Ballard G, Barton KJ, Karl BJ, Rau GH, Ribic CA, Wilson PR (2003) Spatial and temporal variation of diet composition and quality within a presumed metapopulation of Adélie penguins. Condor 105(1):95–106
238. Bedford M, Melbourne-Thomas J, Corney S, Jarvis T, Kelly N, Constable A (2015) Prey-field use by a Southern Ocean top predator: Enhanced understanding using integrated datasets. Marine Ecology Progress Series, 526: 169–181

Glossary

Aerobic dive limit approximately the length of time that a penguin can remain underwater using only its oxygen reserves, not powering its metabolic rate anaerobically (Chap. 8).

Antarctic ozone hole the destruction of ozone in the atmosphere above the Antarctic, owing to accumulation of aerial pollutants, which changed the wind regimes in the Southern Ocean; in turn ocean currents and sea ice were affected (Chap. 13).

Barotrauma; also known as "the squeeze" radical compression air-filled body cavities due to increasing pressure with depth to the point that internal organs can be compromised (Chap. 8).

Bends (the); also called decompression sickness the loading of nitrogen in the blood and body tissues during repeated dives to depth, with pauses at the surface not leaving enough time for the nitrogen to be eliminated. The condition can be fatal in humans (Chap. 8).

Biologging the fixing of instruments having microelectronic sensors to animals – here penguins – in order to obtain information on their movements (both 2D and 3D), energy expenditure, behavior, and physiology (most chapters have results; Chap. 14, effects).

Body size and physical capacity Bergman's Rule; physical attributes that facilitate arrival from or departure to the sea, and capability of foraging relative to depth, heat budget, prey capture (Chaps. 1, 2, 7, 8, and 11).

Bottom phase of dives The period during a dive when penguins remain at roughly constant depth and are generally considered to be feeding or concentrating their effort at particularly profitable depths (Chap. 10).

Buoyancy the tendency of a penguin to be pushed upwards to the water's surface, primarily due to the air contained in its feathers and respiratory spaces (Chap. 6) [see also upthrust].

D. G. Ainley, R. P. Wilson, *The Aquatic World of Penguins*, Fascinating Life Sciences, https://doi.org/10.1007/978-3-031-33990-5

Colony an assembly of breeding penguins that is spatially and demographically distinct from neighboring assemblies. Therefore, there could be uninhabited space between segments, like being on separate islets in a cluster, but there is movement between and together they constitute a colony (Chaps. 1 and 2, and 5; Appendix 3).

Competition, exploitative the process by which a penguin or its colony (population) removes enough prey to deplete actual abundance and thus affect the foraging efficiency of co-occurring (meso-)predators; can also involve human fisheries exploiting more than their fair share (Chaps. 5 and 13).

Competition, interference the process by which a penguin or its colony (population) by its foraging alters the availability, though not the overall abundance, for co-occurring (meso-) predators (Chap. 5).

Competition, interspecific the process of different species, such among penguins or penguins vs whales, or penguins vs humans (fisheries), by the intensity of their foraging/exploitation affecting the availability of prey to one another (Chaps. 5 and 13).

Competition, intraspecific the process of depleting or altering the availability of prey among member of the same penguin colony, facilitated by colony size (Chap. 5).

Central place foraging with the center being a persistently used location, such as a colony, the area used by penguins to acquire food and then return to where they started. Also known as Storer-Ashmole foraging halo (Chaps. 5, 9, and 10; Appendix 3).

Climate oscillation periodic swings in climate that vary in temporal scale, from maximal to minimal ice ages, to Little Ice Age vs Medieval Warm Period, to decadal swings such as Southern Annular Mode and El Nino-Southern Oscillation (ENSO); penguin populations responded, sometimes positively and sometimes negatively, depending on their range and ecological niches (Chap. 13).

Conductance the opposite of insulation, it is the extent to which the outer body layer allows heat to pass from a hot penguin to the cold sea (Chap. 7).

Cost of transport the energetic cost of traveling of penguins swimming underwater, expressed in joules per meter swum (with an equivalence to miles per gallon in cars) (Chap. 6).

Diet the prey that a penguin consumes, with composition varying by species, season, location, and year, owing to degree of specialization of the penguin as well as prey availability (Chaps. 2 and 4; Appendix 1, 2).

Dispersion the behavior of spreading the foraging area once central place foraging, necessary for breeding, is relaxed, and at which time the adult breeders join the portion of the population (juveniles, pre-breeders, floaters) that has not been visiting land in waters that still provide abundant food (Chaps. 1, 2, 3, 14, and 15).

Dive angle the angle between the trajectory of a penguin swimming underwater and the sea surface (Chaps. 6, 9, and 10).

Dive efficiency split into "time-based efficiency" and "energy-based efficiency," these metrics indicate the proportion of time or energy allocated to the bottom phase of dives in relation to the total time/energy invested in the full dive cycle (Chap. 10).

Dive profile a graphical representation of depth on the y-axis against time on the x-axis of penguin dives (Chaps. 6 and 9).

Dive type description of various dives with different dive profiles (see above) and other dive metrics that indicate dive function (Chap. 9).

Drag the force experienced by a swimming penguin that opposes motion and increases with the square of the penguin velocity (Chaps. 6 and 14).

Drag coefficient a unitless figure attributed to bodies moving through water (or air) that indicates how easily a particular body shape slips through the water (Chaps. 6 and 14).

Ecological niche the assembly of a penguins' physical, physiological, and behavioral attributes that allow usage of resources more efficiently than penguins not having those attributes (Chaps. 1, 2, 5, 8, 9, 10, and 11).

"Featherometer" an instrument that applies a constant known pressure to penguin outside body surfaces to measure the air layer thickness at different sites (Chap. 7).

Flock fusion/cohesion the coalescing in space of individual penguins for purposes of increasing foraging efficiency, mediated by voice and coloration (Chap. 14).

Flock fission/fragmentation the separation in space of individual penguins, usually to foil predators (Chaps. 12 and 14).

Foraging range the area covered away from a site through central place foraging by individual penguins, as well as the sum total of all the penguins in a colony; can vary by sex, colony size, and species; see also Central place foraging (Chaps. 5, 9, 10, and 14; Appendix 3).

Foraging volume the combination of foraging range and the depth of foraging, which can vary by sex, colony size, body size, and species (Chaps. 5, 9, 10, and 11; Appendix 3).

Gastric emptying the passage of food from the stomach to the lower parts of the digestive system (Chap. 8).

Heat loss the rate at which energy is lost by a penguin at sea due to heat moving from the penguin into the water. This depends on water temperature, penguin insulation, swimming activity and dive depth (Chap. 7).

Hyperphagia intensive feeding to prepare for and recover from activities that deplete a penguins' body condition, i.e., fat stores, lost during periods of intensive activity or when foraging is not possible (as in Molt) (Chaps. 2, 3, 4, 5, 7, 10, and 14).

Insulation the extent to which the outer body layer that separates the warm penguin body interior from the cold sea exterior acts to minimize heat loss. It is composed primarily of air, skin and some sub-dermal fat. The better the insulation, the less heat is lost. Insulation is the opposite of conductance (Chaps. 7 and 14).

Marginal ice zone the transition from an ocean entirely covered by sea ice to adjacent open water, with ocean productivity enhanced within the zone owing to changes in phytoplankton abundance and species composition (upon which penguin prey forage) (Chap. 3). Scale can range from the edge of polynyas (open water areas in the pack ice) to the edge of a regional extent of sea ice (Chap. 3).

Mesopredator a marine creature, like penguins, that at the same time are predators and prey, i.e., they are not "top" or "apex" predators within a food web (Chaps. 4, 5, and 12).

Migration the process of seasonally changing locality going from one species place to another and then back. Owing to their high energy lifestyle penguins, at most, exhibit partial migration, which is the movement of adults from the habitat where non-breeders occupy in order to nest in a location somewhat separate (Chap. 3).

Navigation finding a penguins' way in the ocean, and/or knowing where it is, through local knowledge (vision, landmarks) at regional/local scale or using celestial cues at large scale (see Dispersion) (Chaps. 3 and 14).

Ocean currents the large scale, directional flow of waters that are distinct from adjacent waters; the most important for penguins are the Antarctic Circumpolar Current, the Falklands, Leeuwin and Creswell currents, and the eastern boundary currents: Humboldt (Peru) and Benguela (Chap. 3).

Ocean fronts the sharp boundary between waters of different temperature and salinity that tend to separate different marine faunas, including penguins and their prey; most important for penguins are the Antarctic Polar Front (APF), Southern Boundary of Antarctic Circumpolar Current (SBACC), Subantarctic Front (SAF), and Subtropical Front (STF); see Shelfbreak Front (Chap. 3).

Penguin populations the sum total of penguins at various spatial scales, from colony (see Colony), to region, to species (Chap. 1).

Penguins as fish the view of penguins by the European mariners who first encountered them upon venturing into waters of the far Southern Hemisphere, and before connecting the lumbering birds found on land and the exquisitely functioning creatures they encountered at sea (Chaps. 1 and 2).

Pinniped aquatic, air breathing mammals, seals and fur seals, some of which compete with penguins for prey, or are predators of penguins (Chaps. 5 and 12); their evolution may have contributed to the disappearance of earlier penguin lineages (Chaps. 1 and 2).

Plumage maintenance taking care to not destroy feathers or leave them soiled thus to breakdown their ability to further penguin swimming and diving, included is bathing or washing; also molting to renew featheration (Chaps. 2 and 14).

Pollution novel to penguins' existence is the appearance in the ocean of the wastes of human existence, such as petroleum, other chemicals, or even noise, that break down a penguins' feather properties, heat balance, or physiological wellbeing (Chap. 14).

Power to swim a measure of the rate at which energy is used by penguins to swim underwater, this varying with speed, depth, and descent/ascent angle (Chaps. 6 and 7).

Predator avoidance the strategies of penguins to avoid being eaten by sharks, pinnipeds, and killer whales, including vigilance and spatial separation; one measure is predator swamping, which is confusing the predators by sheer size of penguin flocks (Chaps. 5 and 12).

Prey capture metrics derived from biologgers/tags these indicate unusual movement metrics of penguins underwater that are considered to be indicative of prey pursuit or capture. They include "wiggles" (undulations) in the depth profile, sharp turns and increases in speed (Chap. 11).

Prey energy density measured in kJ/g, the value of a prey to satisfy the energetic needs of a penguin in its high-energy life style (Chaps. 5, 6, 7, and 10; Appendix 4).

Prey size relative to a penguins' bill size, and the amount of time taken to manipulate the prey in order to swallow it, the length of the prey which, in the vast majority of cases for penguins, is fusiform in shape (Chaps. 4, 5, 10, and 11; Appendix 4).

Provisioning efficiency the energy required to obtain prey relative to the energetic value of the prey consumed (Chaps. 2, 4, 6, 10, and 11; Appendix 4).

Resting metabolic rate the rate at which energy is used by a penguin resting at sea. This varies with water temperature (Chap. 7).

Sea ice frozen ocean, either attached to land (fast ice) or floating as an array of floes; separates species (consistent with their niches) or either inhibits or facilitates foraging efficiency depending on a penguins' capabilities (Chaps. 3 and 13).

Shelfbreak Front the oceanographic transition from waters overlying continental shelves to the oceanic waters seaward; prey tend to differ between the two water types (Chap. 3).

Species radiation in the process of evolution, i.e., a penguin form changing enough to be recognized as being different (with restricted genetic interchange, i.e. breeding), the diversion to occupy separate ecological niches (see Ecological niche) (Chaps. 1 and 2).

Surface duration the length of time that a penguin remains at the surface between dives replenishing oxygen and eliminating carbon dioxide (Chap. 8).

Tag (see Biologging).

Upthrust the force acting so as to push a penguin towards the water surface [see also buoyancy] (Chap. 6).

Whales aquatic, air breathing mammals, some of which prey on penguins (killer whales) and some of which compete with penguins for prey (baleen whales) (Chaps. 5 and 12); the appearance of dolphins, through competition, may have led to the disappearance of early penguin lineages (Chaps. 1 and 2).

MIX
Papier aus verantwortungsvollen Quellen
Paper from responsible sources
FSC® C105338

If you have any concerns about our products,
you can contact us on
ProductSafety@springernature.com

In case Publisher is established outside the EU,
the EU authorized representative is:
Springer Nature Customer Service Center GmbH
Europaplatz 3, 69115 Heidelberg, Germany

Printed by Libri Plureos GmbH
in Hamburg, Germany